Salinity and Drought Tolerance in Plants

Ashwani Kumar • Pooja Dhansu •
Anita Mann
Editors

Salinity and Drought Tolerance in Plants

Physiological Perspectives

Springer

Editors
Ashwani Kumar
ICAR – Central Soil Salinity Research Institute
Karnal, Haryana, India

Pooja Dhansu
ICAR – Sugarcane Breeding Institute, Regional Centre
Karnal, Haryana, India

Anita Mann
ICAR – Central Soil Salinity Research Institute
Karnal, Haryana, India

ISBN 978-981-99-4668-6   ISBN 978-981-99-4669-3 (eBook)
https://doi.org/10.1007/978-981-99-4669-3

This Springer imprint is published by the registered company Springer Nature Singapore Pte Ltd.
The registered company address is: 152 Beach Road, #21-01/04 Gateway East, Singapore 189721, Singapore

Preface

This book aims to bring out a comprehensive collection of scientific research on different plants under drought and salt stress conditions. The main focus of this book is to elaborate the mechanisms operative in plants under stress and how various biological factors mitigate the adverse effects for better plant productivity. This book covers all physiological, biochemical, and molecular mechanisms operating under drought and saline stresses. The current status and impact of drought and salinity on various crop plants have been elaborated in different chapters. Agricultural lands are either turning barren or becoming more saline and drought prone with increasing temperatures, decreasing water tables, untimely rainfall, and other environmental factors. In India, salt-affected soils occupy an area of about 6.73 million ha of which saline and sodic soils constitute roughly 40% and 60%, respectively. All these factors, individually or cumulatively, ultimately affect plant growth and development, thus impacting crop productivity and causing monetary loss. The inbuilt plant's ability with modified/acclimatized mechanisms has been described in various chapters with stepwise descriptions. The role of various plant growth-promoting agents including nano-particles, micro-organisms, metabolites, or phytohormones in mitigating the adverse effects of drought and salinity has been explained precisely. The authors have briefly compiled updated information on the use of speed breeding, proteomics, epigenetics, and transcriptomics in different crops along with high-throughput technologies with the cross talk of various network mechanisms.

This book will be helpful for readers in knowing salinity and drought at physiological, biochemical and genetic molecular levels to understand plant behavior under stress conditions. The authors, being plant physiologists, have wonderfully explored the plant's mechanisms operative under stress conditions from which students may cultivate new information on new ideas by relating the current requirement of plant research. National and international agricultural scientists, as well as policy makers, will also find this to be a useful read for understanding the plant system under stressful environments.

All the authors have made efforts to present the information in a well-structured way to elaborate the concepts in an easy way. This book focuses on the area of future research, driving us all towards new developments in the changing climate scenario.

Karnal, Haryana, India Ashwani Kumar
Karnal, Haryana, India Pooja Dhansu
Karnal, Haryana, India Anita Mann

Contents

Editors and Contributors

About the Editors

Ashwani Kumar Senior Scientist (Plant Physiology), has been working on various aspects of crop improvement for salt-affected areas through institutional and externally funded research projects at ICAR-CSSRI, Karnal. He has made significant contributions for studying the physiological and biochemical mechanisms of different crop plants in varying environments of salinity and drought. His scientific efficiency can be evaluated based on his association in inter-divisional/inter-institutional research programs with funding of more than 3 crores from different agencies. The Google Scholar Citation of 1354 with h-index of 20 depicts scientific output in the form of more than 50 peer-reviewed research papers, 15 book chapters, 15 popular articles, 4 technical bulletins and nine training manuals, success stories, etc. along with teaching experience of more than 10 years.

Pooja Dhansu Senior Scientist (Plant Physiology), has been working on various aspects of sugarcane improvement through institutional and externally funded research projects at ICAR-SBI, Regional Center, Karnal. She has made significant contributions for preservation of ready to serve sugarcane juice that had the shelf life of 30 days without preservatives and 90 days with addition of preservatives. In addition, she has been working on physiological and biochemical mechanisms of sugarcane clones in varying environments of salinity and drought. She has also received funding of more than 3.5 crores from various agencies. She has published more than 25 peer-reviewed research papers, 10 book chapters, 10 popular articles, 6 technical bulletins and eight training manuals, success stories, etc. along with research guiding experience of more than 10 years to M.Sc. and Ph.D. students.

Anita Mann Principal Scientist (Plant Physiology), is working on molecular plant physiology under abiotic stress in different crop plants. She has been working on gene identification from salt-tolerant halophytes through transcriptomics and transformation of important gene(s) for salt tolerance in rice for enhanced salt tolerance. Simultaneously, she has been associated in identifying the salt tolerance in chickpea, pomegranate, okra, tomato, etc. through physiological, biochemical, molecular, and breeding approaches. She has also been awarded research fellowship from CSIR and Department of Science and Technology, New Delhi. She has published more than 60 research papers in national and international peer-reviewed journals including book chapters, general articles, bulletins, etc. She has been guiding M.Sc. and Ph.D. students of plant physiology, plant biotechnology, soil science, agroforestry, etc.

Contributors

Hitashi Aggarwal School of Agricultural Biotechnology, Punjab Agricultural University, Ludhiana, Punjab, India

Pooja Ahlawat CCS Haryana Agricultural University, Hisar, Haryana, India

Akshay ICAR – Central Soil Salinity Research Institute, Karnal, Haryana, India

Sunder Singh Arya Department of Botany, Maharshi Dayanand University, Rohtak, India

Ashok Department of Genetics and Plant Breeding, CCS Haryana Agricultural University, Hisar, Haryana, India

Gaurav Augustine School of Agricultural Biotechnology, Punjab Agricultural University, Ludhiana, Punjab, India

S. Avinash ICAR – Indian Institute of Millets Research, Hyderabad, India

Avni ICAR – Central Soil Salinity Research Institute, Karnal, Haryana, India

Ram Avtar Oilseeds Section, Department of Genetics and Plant Breeding, CCS Haryana Agricultural University, Hisar, India

Parvin Kumar Bagaria ICAR – Indian Institute of Maize Research, Ludhiana, Punjab, India

Aradhana Bali Krishi Vighyan Kendra, CCS Haryana Agricultural University, Dambla, Yamunanagar, India

Rakesh Banyal ICAR – Central Soil Salinity Research Institute, Karnal, Haryana, India

Nirmalendu Basak ICAR – Central Soil Salinity Research Institute, Karnal, Haryana, India

Ajay Kumar Bhardwaj ICAR – Central Soil Salinity Research Institute, Karnal, Haryana, India

Arshvir Kaur Boparai Department of Plant Breeding and Genetics, Punjab Agricultural University, Ludhiana, Punjab, India

Priyanka Chandra ICAR-Central Soil Salinity Research Institute, Karnal, Haryana, India

Rukoo Chawla Department of Genetics and Plant Breeding, Maharana Partap University of Agricultural & Technology, Udaipur, India

Shiksha Chourasia ICAR – Central Soil Salinity Research Institute, Karnal, Haryana, India

Avni Dahiya ICAR – Central Soil Salinity Research Institute, Karnal, Haryana, India

Hemant Dasila Eternal University, Baru Sahib, Himachal Pradesh, India

Geeta Devi Rashtriya Kisan Post Graduate College, Shamli, Uttar Pradesh, India

Sarita Devi Department of Botany & Plant Physiology, CCS Haryana Agricultural University, Hisar, India

Sunita Devi ICAR – Central Soil Salinity Research Institute, Karnal, Haryana, India

Pooja Dhansu ICAR – Sugarcane Breeding Institute, Regional Centre, Karnal, Haryana, India

Seva Nayak Dheeravathu ICAR – Indian Institute of Millets Research, Hyderabad, India
ICAR – Indian Grassland and Fodder Research Institute, Jhansi, Uttar Pradesh, India

Nilamani Dikshit ICAR – Indian Grassland and Fodder Research Institute, Jhansi, Uttar Pradesh, India

Jyothi Duhan Department of Biochemistry, CCS Haryana Agricultural University, Hisar, Haryana, India

Vikash Dutt ICAR – Central Soil Salinity Research Institute, Karnal, Haryana, India

Alisha Goyal CCS Haryana Agricultural University, Hisar, Haryana, India
ICAR – Central Soil Salinity Research Institute, Karnal, Haryana, India

Prinka Goyal Department of Botany, Punjab Agricultural University, Ludhiana, Punjab, India

Heena Department of Botany, Maharshi Dayanand University, Rohtak, India

M. L. Jakhad Sri Karan Narendra Agriculture University, Jobner, Rajasthan, India

Minakshi Jattan Department of Genetics and Plant Breeding, CCS Haryana Agricultural University, Hisar, Haryana, India

Norah Johal Department of Agronomy, Punjab Agricultural University, Ludhiana, Punjab, India

Karuna Department of Genetics and Plant Breeding, CCS Haryana Agricultural University, Hisar, Haryana, India

Gurpreet Kaur ICAR – Central Soil Salinity Research Institute, Karnal, Haryana, India

Deepak Kaushik Department of Genetics and Plant Breeding, CCS Haryana Agricultural University, Hisar, Haryana, India

Y. KeerthiSree College of Biotechnology, CCS Haryana Agricultural University, Hisar, Haryana, India

Kiran CCS Haryana Agricultural University, Hisar, Haryana, India

Kuldeep College of Agriculture, CCS Haryana Agricultural University, Hisar, Haryana, India

Neeraj Kulshreshtha ICAR – Central Soil Salinity Research Institute, Karnal, Haryana, India

Ajay Kumar Department of Botany, Maharshi Dayanand University, Rohtak, India

Aman Kumar School of Agricultural Biotechnology, Punjab Agricultural University, Ludhiana, Punjab, India

Arvind Kumar ICAR – Central Soil Salinity Research Institute, Karnal, Haryana, India

Ashwani Kumar ICAR – Central Soil Salinity Research Institute, Karnal, Haryana, India

Manish Kumar ICAR – Central Soil Salinity Research Institute, Karnal, Haryana, India

Naresh Kumar Eternal University, Baru Sahib, Himachal Pradesh, India

Naveen Kumar Department of Botany, Maharshi Dayanand University, Rohtak, India

Pradeep Kumar Department of Agricultural Biotechnology, SVP University of Agriculture & Technology, Meerut, Uttar Pradesh, India

Pushpendra Kumar Department of Agricultural Biotechnology, SVP University of Agriculture & Technology, Meerut, Uttar Pradesh, India

Raj Kumar ICAR – Central Soil Salinity Research Institute, Karnal, Haryana, India

Sandeep Kumar Department of Genetics and Plant Breeding, CCS Haryana Agricultural University, Hisar, Haryana, India
Department of Botany, Baba Mastnath University, Rohtak, India

Sunil Kumar Department of Botany, CRM Jat College, Hisar, India

Upendra Kumar Department of Molecular Biology and Biotechnology, CCS Haryana Agricultural University, Hisar, Haryana, India

Vikram Kumar Department of Botany, Maharshi Dayanand University, Rohtak, India

Anita Kumari Department of Botany & Plant Physiology, CCS Haryana Agricultural University, Hisar, Haryana, India

Archana Kumari Department of Plant Breeding and Genetics, Punjab Agricultural University, Ludhiana, Punjab, India

Nisha Kumari Department of Biochemistry, CCS Haryana Agricultural University, Hisar, Haryana, India

Sushil Kumari Department of Forestry, CCS Haryana Agricultural University, Hisar, Haryana, India

Chayanika Lakhar Department of Plant Breeding and Genetics, Punjab Agricultural University, Ludhiana, Punjab, India

Nita Lakra College of Biotechnology, CCS Haryana Agricultural University, Hisar, Haryana, India

Charu Lata ICAR – Indian Institute of Wheat and Barley Research, RRS, Shimla, Himachal Pradesh, India

Sukham Madaan CCS Haryana Agricultural University, Hisar, Haryana, India

Mitali Mahajan CSIR – Institute of Himalayan Bioresource Technology, Palampur, Himachal Pradesh, India

Mamta Department of Botany, Maharshi Dayanand University, Rohtak, India

Manisha ICAR – Central Soil Salinity Research Institute, Karnal, Haryana, India

Anita Mann ICAR – Central Soil Salinity Research Institute, Karnal, Haryana, India

Hemanthkumar Manne Department of Biochemistry, CCS Haryana Agricultural University, Hisar, Haryana, India

K. Manorama College of Biotechnology, CCS Haryana Agricultural University, Hisar, Haryana, India

Mukesh Kumar Mehla ICAR – Central Soil Salinity Research Institute, Karnal, Haryana, India

J. L. Michael College of Biotechnology, CCS Haryana Agricultural University, Hisar, Haryana, India

Monika Department of Environmental Science, Maharshi Dayanand University, Rohtak, India

Naresh Department of Genetics and Plant Breeding, CCS Haryana Agricultural University, Hisar, Haryana, India

Neha ICAR – Central Soil Salinity Research Institute, Karnal, Haryana, India

Junya Onishi Japan International Research Center for Agricultural Sciences (JIRCAS), Tsukuba, Ibaraki, Japan

Parul ICAR – Central Soil Salinity Research Institute, Karnal, Haryana, India

Adhini S. Pazhany ICAR – Sugarcane Breeding Institute, Coimbatore, India

D. S. Phogat Department of Genetics and Plant Breeding, CCS Haryana Agricultural University, Hisar, Haryana, India

Pooja ICAR – Sugarcane Breeding Institute, Regional Centre, Karnal, Haryana, India

Govind Prasad ICAR – Central Soil Salinity Research Institute, Karnal, Haryana, India

Arvind Kumar Rai ICAR – Central Soil Salinity Research Institute, Karnal, Haryana, India

Navreet Kaur Rai Department of Genetics and Plant Breeding, CCS Haryana Agricultural University, Hisar, Haryana, India

Prashant Singh Rai Department of Biochemistry, Central University of Rajasthan, Ajmer, India

Om Prakash Raigar Department of Plant Breeding and Genetics, Punjab Agricultural University, Ludhiana, Punjab, India

Yadav Rajesh Department of Genetics and Plant Breeding, CCS Haryana Agricultural University, Hisar, Haryana, India

G. Rajesha ICAR – Indian Institute of Millets Research, Hyderabad, India

Rajkumar ICAR – Central Soil Salinity Research Institute, Karnal, Haryana, India

Anita Rana ICAR – IARI, Regional Station, Katrain, Himachal Pradesh, India

Babita Rani CCS Haryana Agricultural University, Hisar, India

Ravika Oilseeds Section, Department of Genetics and Plant Breeding, CCS Haryana Agricultural University, Hisar, India

Sandeep Rawal Krishi Vighyan Kendra, CCS Haryana Agricultural University, Dambla, Yamunanagar, India

Reetu ICAR – Indian Grassland and Fodder Research Institute, Jhansi, Uttar Pradesh, India

Sachin Department of Genetics and Plant Breeding, CCS Haryana Agricultural University, Hisar, India

Neha Saini College of Agriculture, CCS Haryana Agricultural University, Hisar, Haryana, India

Sakshi Saini Department of Botany, Maharshi Dayanand University, Rohtak, India

Varun Saini College of Agriculture, CCS Haryana Agricultural University, Hisar, Haryana, India

Omender Sangwan Department of Genetics and Plant Breeding, CCS Haryana Agricultural University, Hisar, Haryana, India

Satish Kumar Sanwal ICAR – Central Soil Salinity Research Institute, Karnal, Haryana, India

Sapna CCS Haryana Agricultural University, Hisar, Haryana, India

D. Seva Nayak ICAR – Indian Grassland and Fodder Research Institute, Jhansi, Uttar Pradesh, India

Aarju Sharma ICAR – Central Soil Salinity Research Institute, Karnal, Haryana, India

Amit Sharma Department of Genetics and Plant Breeding, CCS Haryana Agricultural University, Hisar, India

Anubhuti Sharma ICAR – Directorate of Rapeseed-Mustard Research, Bharatpur, India

Asha Sharma Department of Botany, Maharshi Dayanand University, Rohtak, India

Himanshu Sharma Department of Plant Breeding and Genetics, Punjab Agricultural University, Ludhiana, Punjab, India

Jyoti Sharma Department of Botany, Maharshi Dayanand University, Rohtak, India

K. D. Sharma CCS Haryana Agricultural University, Hisar, Haryana, India

Prashant Sharma YSP University of Horticulture and Forestry, Solan, India

Priyanka Sharma Department of Botany, Maharshi Dayanand University, Rohtak, India

Pooja K. Shekhawat ICAR – Central Soil Salinity Research Institute, Karnal, Haryana, India
Sri Karan Narendra Agriculture University, Jobner, Rajasthan, India

Parvender Sheoran ICAR – Central Soil Salinity Research Institute, Karnal, Haryana, India

ICAR – Agricultural Technology Application Research Institute, Ludhiana, Punjab, India

Ravika Sheoran Department of Genetics and Plant Breeding, CCS Haryana Agricultural University, Hisar, Haryana, India

Sanjoy Shil Bidhan Chandra Krishi Viswavidyalaya, Bankura, West Bengal, India

Shweta Department of Genetics and Plant Breeding, CCS Haryana Agricultural University, Hisar, Haryana, India

Monika Sihmar Maharishi Dayanand University, Rohtak, India

Jogendra Singh ICAR – Central Soil Salinity Research Institute, Karnal, Haryana, India

Pooja Singh Department of Botany, Maharshi Dayanand University, Rohtak, India

Tejveer Singh ICAR – Indian Grassland and Fodder Research Institute, Jhansi, Uttar Pradesh, India

Vijayata Singh ICAR – Central Soil Salinity Research Institute, Karnal, Haryana, India

Kapil Singla Krishi Vighyan Kendra, CCS Haryana Agricultural University, Dambla, Yamunanagar, India

Sinky CCS Haryana Agricultural University, Hisar, Haryana, India

Pooja Gupta Soni Department of Genetic and Plant Breeding, CCS Haryana Agricultural University, Hisar, Haryana, India
ICAR – Central Soil Salinity Research Institute, Karnal, Haryana, India

Sonia Department of Biochemistry, CCS Haryana Agricultural University, Hisar, Haryana, India

Ashutosh K. Srivastava Department of Basic Science, Rani Lakshmi Bai Central Agricultural University, Jhansi, Uttar Pradesh, India

S. Srividhya ICAR – Indian Institute of Millets Research, Hyderabad, India

Parul Sundha ICAR-Central Soil Salinity Research Institute, Karnal, Haryana, India

Suniti Department of Agricultural Biotechnology, SVP University of Agriculture & Technology, Meerut, Uttar Pradesh, India

Sushil Department of Entomology, CCS Haryana Agricultural University, Hisar, Haryana, India

Manish Kumar Suthar ICAR – Directorate of Medicinal & Aromatic Plant Research, Anand, Gujarat, India

R. Swarna ICAR – Indian Institute of Millets Research, Hyderabad, India

Maharishi Tomar ICAR – Indian Grassland and Fodder Research Institute, Jhansi, Uttar Pradesh, India

T. N. Usha Zonal Agricultural and Horticultural Research Station, Keladi Shivappa Nayaka University of Agricultural and Horticultural Sciences, Shivamogga, Karnataka, India

Thulasi Bai Vadithe Department of Microbiology, Acharya Nagarjuna University, Guntur, Andhra Pradesh, India

K. Venkatesh ICAR – Indian Institute of Millets Research, Hyderabad, India

D. Venkateshwarlu CSIR – Central Food Technological Research Institute, Mysore, Karnataka, India

Kamlesh Verma ICAR – Central Soil Salinity Research Institute, Karnal, Haryana, India

Reetu Verma ICAR – Indian Grassland and Fodder Research Institute, Jhansi, Uttar Pradesh, India

Vikas Verma School of Agricultural Biotechnology, Punjab Agricultural University, Ludhiana, Punjab, India

Gajender Yadav ICAR – Central Soil Salinity Research Institute, Karnal, Haryana, India

Neha Yadav Department of Botany, Maharshi Dayanand University, Rohtak, India

Praveen Yadav ICAR – Indian Grassland and Fodder Research Institute, Jhansi, Uttar Pradesh, India

Rajender Kumar Yadav ICAR – Central Soil Salinity Research Institute, Karnal, Haryana, India

Rajesh Yadav Department of Genetics and Plant Breeding, CCS Haryana Agricultural University, Hisar, Haryana, India

Renu Yadav Central University of Haryana, Mahendragarh, India

Sujata Yadav ICAR – Central Soil Salinity Research Institute, Karnal, Haryana, India

Sunaina Yadav Department of Genetics and Plant Breeding, IARI, New Delhi, India

Occurrence of Salinity and Drought Stresses: Status, Impact, and Management

1

Pooja Gupta Soni, Nirmalendu Basak, Arvind Kumar Rai, Parul Sundha, Priyanka Chandra, and Rajender Kumar Yadav

The world's population is expected to reach nine billion by 2050 (FAO 2009), and the climate is also changing. Abiotic and biotic stresses are the major challenges in crop production worldwide, and climate change will likely lead to more severe abiotic and biotic stress conditions (Cobb et al. 2013). By the year 2050, 50% of all arable lands will be consequently threatened by global climate changes, low water availability, and salinization (Wang et al. 2003). Salinity and drought are the most severe abiotic stresses that threaten crop productivity worldwide (Guo et al. 2014). Drought is expected to increase in frequency and severity in the future due to climate change, mainly due to decreases in regional precipitation but also because of increasing evaporation driven by global warming (Lobell et al. 2008). Drought affects more than 10% of arable land, causing desertification, especially in arid and semiarid areas, while salinization is rapidly increasing on a global scale, declining average yields for most major crops (Bray et al. 2000). Soil salinization is one of the severe forms of soil degradation, which can arise from natural causes and human-mediated activity, such as irrigation in arid and semiarid regions (Rengasamy et al. 2010).

Water stress is found to be the most important limiting factor controlling primary production in terrestrial environments. Water stress is a limited water supply to plant roots, which reduces plants' transpiration rate. It is mainly caused by water deficit as a result of drought conditions or soil salinity. The effects of drought in agriculture are aggravated due to the depletion of water resources and the increased food demand

P. G. Soni
Department of Genetic and Plant Breeding, CCS Haryana Agricultural University, Hisar, Haryana, India

ICAR – Central Soil Salinity Research Institute, Karnal, Haryana, India

N. Basak (✉) · A. K. Rai (✉) · P. Sundha · P. Chandra · R. K. Yadav
ICAR – Central Soil Salinity Research Institute, Karnal, Haryana, India
e-mail: nirmalendu.basak@icar.gov.in; ak.rai@icar.gov.in

A. Kumar et al. (eds.), *Salinity and Drought Tolerance in Plants*,
https://doi.org/10.1007/978-981-99-4669-3_1

from an alarming world population growth (O'Connell 2017). Drought is one of the most common environmental stresses affecting plant growth and development and is a challenge to agricultural researchers. Water comprises about 80–95% of the fresh biomass of the plant body and plays a vital role in various physiological processes, including many aspects of plant growth, development, and metabolism (Brodersen et al. 2019; Abbasi and Abbasi 2010). The drought's unpredictable nature depends upon various factors, such as uneven and erratic distribution of rainfall, high evapotranspiration demand, and low water-holding capacity around the rhizosphere (Passioura and Angus 2010; Devincentis 2020). Water stress numerously affects plants' growth and development. These changes depend on the severity, duration, and time course of water stress (Bradford and Hsiao 1982). Drought stress often induces stomatal closure that restricts the diffusion of CO_2 into the leaf or is due to non-stomatal limitations, which leads to a decrease of carbon assimilation and other processes of photosynthesis (Ashraf and Harris 2013; Guan et al. 2015; Paul et al. 2016).

More than 800 million ha of land worldwide are salt-affected, which has important consequences for the productivity of crops. Salinity is major stress limiting the increase in the demand for food crops. More than 20% of cultivated land worldwide (~about 45 ha) is affected by salt stress, and the amount is increasing day by day. Increased soil salt concentrations decrease the ability of plants to take up water leading to apparent water limitation or can lead to the accumulation of salt in the shoots, which negatively affects growth by impairing metabolic processes and decreasing photosynthetic efficiency, partly through stomatal closure (Flowers and Yeo 1995; Maser et al. 2002; Munns 2002; Roy et al. 2014). Salinity affects almost all aspects of plant development, including germination, vegetative growth and reproductive development, and decreasing water and nutrient uptake (Akbarimoghaddam et al. 2011; Singh and Chatrath 2001). Salinity can affect plant functions via two main mechanisms (Munns and James 2003; Munns and Tester 2008; Arzani and Ashraf 2016): (1) via inducing external osmotic pressure around the roots in the soil, which decreases the uptake of water leading to symptoms similar to those caused by drought, and (2) via toxic effect of salt ions, mostly Na^+ and Cl^-, which accumulate in the plant tissues, mostly in the leaves. Soil salinity is known to repress plant growth through osmotic stress, which is then followed by ion toxicity (James et al. 2011; Rahnama et al. 2010). During the initial phases of salinity stress, the water absorption capacity of root systems decreases, and water loss from leaves is accelerated due to the osmotic stress of high salt accumulation in soil and plants, and therefore salinity stress is also considered hyperosmotic stress.

Within an agricultural context, drought is a prolonged period of deficient precipitation that negatively impacts crop growth or yield. An increasingly warming climate is expected to intensify the frequency and severity of drought in the near future. As a consequence of the ongoing global climate changes, low water availability and salinization are expected to affect up to 50% of all arable lands by the year 2050 (Wang et al. 2003), which will hamper efforts to meet the dramatically increasing demand for food predicted by the same year (Cobb et al. 2013). Salinity affects almost all aspects of plant development, including germination, vegetative growth, and reproductive development. Soil salinity imposes ion toxicity, osmotic stress, nutrient (N, Ca, K, P, Fe, Zn) deficiency, and oxidative stress on plants, thus limiting water uptake from soil. The adverse effects of salinity on plant development are more profound during the reproductive phase. Therefore, it is necessary to improve management techniques to reduce the damage caused by drought and salinity. These management strategies are being useful for stress management in arid and semiarid climates.

1.1 Impact of Drought and Salinity

1.1.1 Drought

Drought, or lack of or insufficient rain for an extended period, causes a considerable hydrologic (water) imbalance and, consequently, water shortages, stream flow reduction, groundwater depletion, soil moisture, and crop damage. It occurs when evaporation and transpiration (water movement in the soil through plants into the air) exceed precipitation for a considerable period. Drought is the most severe physical hazard to agriculture in nearly every part of the world. Drought is expected to increase in frequency and severity in the future due to climate change, mainly due to decreases in regional precipitation but also because of increasing evaporation driven by global warming (Lobell et al. 2008). Several factors can cause plant water deficit, including inadequate precipitation, high evaporative demand, decreased groundwater level, and water retention by soil particles (Gimenez et al. 2005; Salehi-Lisar and Bakhshayeshan-Agdam 2016). Three main mechanisms that reduce crop yield by soil water deficit include (1) reduced canopy absorption of photosynthetically active radiation, (2) decrease in radiation use efficiency, and (3) reduced harvest index (Earl and Davis 2003).

1.1.1.1 Plant Growth and Development

Drought stress is a limiting factor that alters plant growth and development aspects. The qualitative and quantitative attributes of plant growth result from interactive phenomena among genetic, physiological, ecological, and morphological characteristics under drought conditions (Wang et al. 2003; Farooq et al. 2009). Drought events limit plant performances in different developmental stages. Seed germination is the primary aspect of growth sensitive to drought stress. Germination is one of the most sensitive plant growth stages to water deficit (Farooq et al. 2009). Drought delays germination onset (El-Midaoui et al. 2001) with a substantial decrease in sunflower germination (Sajjan et al. 1999). Ample soil moisture is required to initiate the germination process. Imbibition by seeds is the first step, which depends on the water potential gradient between seed and soil. Any decrease in soil water potential causes a linear decline in seed germination (Wen 2015).

Water scarcity outbreaks are due to the occurrence of less or the absence of rainfall resulting in low soil moisture content and low water potential in aerial parts of the plant, such as leaves and stems (Ristvey et al. 2019). A decrease in growth rate under drought could be attributed to the inhibition of cell elongation because the water flow is interrupted from the xylem to the surrounding cells (Nonami 1998). Furthermore, shoot growth seemed to be more adversely affected than root growth (Bassiri et al. 1977). The soil moisture stress causes a decrease in seed germination, shoot length, fresh and dry weights of seedlings, shoot and root dry matter, and relative growth rate in safflower (Hojati et al. 2011). The roots strive to uptake more water through their expansion, which ultimately adapts plants to minimize stomatal water loss when there is a water deficit (Martínez-Vilalta and Garcia-Forner 2017). Typical drought stress symptoms in plants include leaf rolling, stunning plants, yellowing leaves, leaf scorching, and permanent wilting (Corso et al. 2020). Moreover, plant response to a given water deficit is strongly dependent on the previous occurrence and intensity of other drought stress events (Adnan et al. 2020; Battaglia et al. 2020; Hafez et al. 2015) and the presence of other stresses (Thomason and Battaglia 2020).

Water stress results in a decrease in the photosynthetic assimilation of carbon dioxide. This decrease is due to two reasons. First, the restricted diffusion of carbon dioxide in the leaf is due to the stomatal closure, and second, carbon dioxide metabolism is inhibited due to water stress. Experiments conducted on cowpea show that the decrease in carbon dioxide assimilation due to water stress is mainly due to stomatal closure, which reduces internal carbon dioxide and restricts water loss through transpiration (Souza et al. 2004). Drought stress after germination has systematic effect on growth with a reduction of water potential, relative water contents (Ünyayar et al. 2004), and turgor of plant cells (Benlloch-González et al. 2015), which elevates the concentration of solutes in the cytosol. These changes decrease cell elongation, thus leading to growth inhibition (Lisar et al. 2012). Growth inhibition is followed by less carbon assimilation, imbalanced mineral nutrition, and accumulation of abscisic acid (ABA), which causes wilting of plants (Farooq et al. 2012; Lisar et al. 2012). The adverse effects of drought stress on mineral nutrition and metabolism result in reduced leaf area and disruption of assimilate partitioning.

1.1.1.2 Plant Water Relations

Certain factors influence water relations, including the leaf water potential, leaf and canopy temperature, transpiration rate, and stomatal conductance. Exposure to drought stress disturbs all these factors in plants. However, stomatal conductance is most affected (Farooq et al. 2009). The increase in stomatal resistance reduces the rate of transpiration and, therefore, leads to an increase in leaf temperature because the process of transpiration is the crucial mechanism that controls leaf temperature (Arbona et al. 2013; Sapeta et al. 2013). Water potential is considered a reliable indicator of plants' response to water stress and decreases the water loss through stomata, which maintain the turgor pressure (Siddique et al. 2000; Terzi and Kadioglu 2006; Bayoumi et al. 2008). Plant water potential (PWP) and turgor are decreased under water-limited conditions; thus, plants cannot function correctly under these conditions (Zlatev and Lidon 2012; Osakabe et al. 2014; Zare et al.

2011). Drought stress decreases the water supply to the xylem, which reduces an adequate nutrient supply to the phloem, thus resulting in lower water potential. The relative water content (RWC) is decreased in the initial consequence of drought stress (Farooq et al. 2009). The lower RWC reflects substantial reductions of leaf water potential, which lead to the closing of stomata. A significant reduction in the leaf water potential and transpiration rate was observed under drought conditions, ultimately increasing the leaf and canopy temperature (Turner et al. 2001). The higher leaf temperature denatures various proteins and enzymes and reduces membrane permeability, and resultantly, multiple facets of plant metabolism are affected. These changes are the main reasons for the disturbance in photosynthesis, respiration, mineral nutrition, and synthesis of proteins and amino acids (Sapeta et al. 2013; Tiwari and Yadav 2020).

In summary, drought stress decreases the water supply to the xylem which decreases an adequate nutrient supply to the phloem, thus resulting in lower water potential. However, huge variability exists among different genotypes for maintaining water potential mainly due to the ability of water absorption and root system. A decrease in leaf water potential may provoke osmotic adjustment, which helps maintain leaf hydration at low leaf water potential. Leaf relative water content and leaf water potential in plants were affected by water deficit.

1.1.1.3 Mineral Uptake and Assimilation

Drought stress decreases transpiration rates and impairs active transport and membrane permeability, reducing crop plants' absorbing power (Kramer and Boyer 1995). Thus, the nutrient transport from root to shoot is restricted due to the weak absorbing power of the crop plants under drought stress. Drought stress also reduces the nutrient uptake by the roots and their translocation in the plant due to low transpiration rates, diminished active transport, and impaired membrane permeability (Hu and Schmidhalter 1998). Drought stresses significantly impacts the nutrient relations of plants. Many essential nutrients, including nitrogen, silicon, magnesium, and calcium, are uptaken by roots along with water; the drought conditions limit the movement of these nutrients via diffusion and mass, which leads to retarded plant growth (Barber 1995). Plants increase roots' length and surface area and change their architecture to capture less mobile nutrients (Lynch and Brown 2001). The soil moisture deficit at times reduces the growth of the roots and, hence, reduces the uptake of less mobile nutrients such as phosphorus (Garg 2003). Drought stress reduces soil N mineralization, which ultimately lowers N availability. A decreased transpiration due to drought stress is the other factor that lowers N transport from roots to the shoots (Tanguilig et al. 1987). The P uptake is hampered under moisture deficit conditions (Pinkerton and Simpson 1986). The primary reason for reduced P uptake is the restricted translocation of P to shoots, even under mild drought stress (Resnik 1970). The N and K utilization under drought stress at different growth stages is influenced by several factors, including physiochemical characteristics of the soil, duration, and intensity of drought relative to phenology, and the organism's evolutionary history (Killingbeck 2004; Silla and Escudero 2006).

The stomatal closure is hampered by lower K supply because of loss of epidermal cell turgidity (Rahbarian et al. 2011) as stomatal closure needs back pressure exerted by fully turgid epidermal pavement cells, whereas the K accumulation is responsible for the required pressure (Roelfsema and Hedrich 2002; Habibi 2013). Root-microbe interactions also play an essential role in the nutrient relations of a plant. The impaired carbon and oxygen flux to the nodules and N accumulation under drought stress inhibited the N-fixing ability of certain legumes (Ladrera et al. 2007). The composition and activity of the soil microbial colonies are negatively affected by the soil water deficit, which eventually disturbs the plant nutrient relations (Schimel et al. 2007). In conclusion, drought significantly reduces mineral uptake and disturbs nutrient balances (Gunes et al. 2008). Nutrient imbalances ultimately seriously affect various growth and developmental processes. However, plant species and genotypes within species vary in their response to water deficit stress in this regard (Garg 2003).

1.1.2 Salinity

Salinity is the primary environmental stress source that restricts agricultural productivity and sustainability in arid and semiarid regions by reducing the germination rate and delaying germination and subsequent seedling establishment. Salt negatively affects crop production worldwide. Crops exhibit a spectrum of responses under salt stress. Salinity not only decreases the agricultural production of most crops but also affects soil physicochemical properties and the ecological balance of the area. The saline condition causes many adverse effects on plant growth due to the low osmotic potential of the soil solution (osmotic stress), specific ion effects (salt stress), nutritional imbalances, or a combination of these factors (Ashraf 2004). All these factors cause adverse effects on plant growth and development. Salinity affects almost all aspects of plant development, including germination, vegetative growth, and reproductive development. Soil salinity imposes ion toxicity, osmotic stress, nutrient (N, Ca, K, P, Fe, Zn) deficiency, and oxidative stress on plants, thus limiting water uptake from the soil. This number could be increased in the future due to increased land salinization due to contaminated artificial irrigation, climate change, and unsuitable land management. Salinity is a significant stress responsible for the inhibition of seed germination or reduction in germination percentage and a delay in crop germination time.

1.1.2.1 Plant Growth and Development

Salinity is the most important abiotic stress that inhibits growth and productivity of crops, and it is one of the world's primitive and most widely distributed environmental challenges. Salinity is defined as the presence of an excessive concentration of soluble salts in the soil which suppresses plant growth (Zaki 2011). The drastic effect of salt stress can be seen in terms of yield loss. The primary effects related to crop yield can be germination, which either decreases or sometimes ceases under extreme saline conditions. Soil salinity has an overall detrimental impact on plants' health. High levels of soil salinity can significantly inhibit seed germination and seedling growth, due to high osmotic potential outside the seed inhibiting the absorption of water or the toxic effect of Na^+ and Cl^- (Khajeh-Hosseini et al. 2003). Salt stress had adverse effects on the functioning and metabolism of plants considerably hindering the productivity (Khan and Srivastava 1998). Salinity has diverse outcome on plants; for example, salt in the soil solution diminishes the accessibility of water to the roots, and the salt reserved in the plant will rise to toxic effect in several tissues of plants (Munns et al. 1995). Plant growth depends on photosynthesis; therefore, environmental stresses affecting growth also affect photosynthesis (Taiz and Zeiger 1998). Iyengar and Reddy (1996) attributed decreases in photosynthetic rate as a result of salinity to a number of factors: dehydration of cell membranes which reduces their permeability to CO_2. High salt concentration in soil and water creates high osmotic potential which reduces the availability of water to plants. A decrease in water potential causes osmotic stress, which reversibly inactivates photosynthetic electron transport via the shrinkage of intercellular space. Salt toxicity is caused particularly by Na^+ and Cl^-. According to Banuls et al. (1990), Cl inhibits photosynthetic rate through its inhibition of NO_3-N uptake by the roots. Stomata closing reduces the CO_2 supply. The reduction in stomatal conductance results in restricted availability of CO_2 for carboxylation reactions (Brugnoli and Bjorkman 1992). Iyengar and Reddy (1996) reported that stomatal closure minimizes loss of water by transpiration, and this affects chloroplast light harvesting and energy conversion systems, thus leading to alteration in chloroplast activity.

1.1.2.2 Effects on Plant Water Uptake

Salinity is an important environmental factor that can severely inhibit plant growth and agricultural productivity. In addition to the toxic effects of the sodium and chloride ions, salinity disturbs the plant's water relations due to decreased availability of water from soil solution due to lowered osmotic potential (Munns 2005). Root is the primary site for plants to uptake water. Root hydraulic conductance represents water uptake capacity and mainly depends on the driving force, root anatomy, and root water permeability (Steudle 2000; Sutka et al. 2011). Osmotic stress is the first stress experienced when a plant is exposed to saline soil; it immediately influences plant growth (Horie et al. 2011). One of the primary responses of plants to osmotic stress is a decrease in root hydraulic conductance (Lp) (Boursiac et al. 2005). High concentrations of salt in solution result in increased osmotic stress, which limits water absorption by the plant and in turn affects leaf water content, stomatal conductance (gs), leaf growth, and photosynthesis (Boursiac et al. 2005; Munns and Tester 2008). High concentrations of salts outside the roots result in increased osmotic stress, which induces root water uptake difficulties, causing leaf water imbalance and ultimately reducing plant growth (Boursiac et al. 2005; Munns and Tester 2008).

Salinity has a dual effect on plant growth via an osmotic effect on plant water uptake and specific ion toxicities. Osmotic stress and ionic toxicity both affect all major plant processes (Yadav et al. 2011). Plants are able to take up water and essential minerals because they have a higher water pressure than the soil under normal conditions. When salt stress occurs, the osmotic pressure of the soil solution is greater than that in plant cells. Thus, the plant cannot get enough water (Kader and Lindberg 2010). By decreasing the osmotic potential of the soil solution, plant access to soil water is decreased, because of the decrease in total soil water potential. As the soil dries, the concentration of salt in the soil solution increases, further decreasing the osmotic potential. In order to maintain water uptake from a saline soil, plants must osmotically adjust. When plants are exposed to osmotic stress, their immediate response is to close the stomata to decrease the transpiration rate and thereby to reduce water loss (Cornic 2000). The closure of the stomata also reduces CO_2 fixation and decreases the photosynthetic rate. In order for plant growth to continue, however, the plant must maintain an optimal stomata aperture. Due to salt stress, cells will have decreased turgor and its stomata will close to conserve water. Stomatal closing can lead to less carbon fixation and the production of reactive oxygen species (ROS) such as superoxide and singlet oxygen. Reactive Oxygen Species disrupt cell processes through damage to lipids, proteins, and nucleic acids (Parida and Das 2005).

The osmotic and ionic stress induced by salinity can halt plant growth as the plant focuses its energy on conserving water and improving ionic balance. With the reduction in water potential gradient between soil and plant under soil moisture deficit conditions, the uptake capacity of roots becomes limiting. Root hydraulic conductivity decreases under drought, limiting roots' water uptake capacity from the soil (Zhu et al. 2021). Aquaporin controls the root hydraulic conductivity, and generally under moisture deficit condition, the expression of genes coding for aquaporin is downregulated, which ultimately lowers down the root water uptake capacity (Mukarram et al. 2021). Under stress, due to continuous transpiration loss and lower soil moisture, the xylem cavitation gradually increases which ultimately reduces the hydraulic conductance in plants and blocks the water movements in the plant (Mukarram et al. 2021).

1.1.2.3 Effect on Mineral Uptake, Nutrient Imbalance, and Specific Ion Toxicity

Plants absorb nutrients from the soil through water, which provides a medium for nutrients to move within the soil matrix and from soil to plants. Soil moisture deficit condition decreases the nutrient uptake from soil. The reduction in nutrient uptake may attribute to a decrease in the nutrient supply through mineralization, reduction in diffusion, and mass flow of nutrients in the soils (Bista et al. 2018). The kinetics of nutrient uptake by the roots also reduced the rate of nutrient uptake under drought stress (Luo et al. 2018). The reduced translocation of nutrients from root to shoot also contributes to a reduction in the nutrient status of different plant parts (Luo et al. 2018). The microbial growth in the rhizosphere is also affected under deficit soil moisture, which ultimately affects the nutrient uptake by roots (Karlowsky et al. 2018).

Salt accumulation in the root zone causes the development of osmotic stress and disrupts cell ion homeostasis by inducing both the inhibition in uptake of essential elements such as K^+, Ca^{2+}, and NO_3^- and the accumulation of Na^+ and Cl^-.

Accumulation of injurious ions may inhibit photosynthesis and protein synthesis, inactivate enzymes, and damage chloroplasts and other organelles (Paranychianakis and Chartzoulakis 2005). These effects are more critical in older leaves, as they have been transpiring the longest so they accumulate more ions (Munns et al. 2002). Plant deficiencies of several nutrients and nutritional imbalances may be caused by the higher concentration of Na^+ and Cl^- in the soil solution derived from ion competition (i.e., Na^+/Ca^{2+}, Na^+/K^+, Ca^{2+}/Mg^{2+}, and Cl^-/NO_3^- in plant tissues) (Grattan and Grieve 1992). Ca deficiency symptoms are common when the Na^+/Ca^{2+} ratio is high in soil water. Soil salinity significantly reduces plant phosphorus (P) uptake because phosphate ions precipitate with Ca ions (Bano and Fatima 2009). Some elements, such as sodium, chlorine, and boron, have specific toxic effects on plants. Excessive sodium accumulation in cell walls can rapidly lead to osmotic stress and cell death (Munns 2002). Plants sensitive to these elements may be affected at relatively low salt concentrations if the soil contains enough of the toxic element. Because many salts are also plant nutrients, high salt levels in the soil can upset the nutrient balance in the plant or interfere with the uptake of some nutrients (Blaylock 1994).

Salinity has direct effects on the nutrient imbalance between soil and plant. The most important harmful effect of salinity is the sodium and chloride ion accumulation in plant tissues and soil (Nishimura et al. 2011). High sodium ion (Na^+) concentration has an antagonistic effect on potassium (K^+) ions (Jung et al. 2009). Moreover, N uptake reduction by the plant has also been observed under high salt conditions (Abdelgadir et al. 2005). Similarly, salinity has an antagonistic effect on P, K^+, Zn, Fe, Ca^{2+}, and Mn, while it has a synergistic effect on N and Mg in field crops such as rice (Jung et al. 2009; Garcia et al. 2010). The entrance of sodium and chloride ions into the plant cell from the soil causes ion imbalance in plant and soil, and excessive uptake of these ions by plants causes many problems related to the physiology of plant tissues such as root, leaf, grain, fruit, or fiber (James et al. 2011). Similarly, the reduction of plant osmotic potential, excessive uptake of Na^+ and Cl^- in the cell, and disruption of cell metabolic functions are due to ion toxicity (James et al. 2011). Excessive sodium ion in plant tissues harms the cell membrane and plant organelles, resulting in cell death of plant (Siringam et al. 2011).

Ionic toxicity occurs when concentrations of salts are imbalanced inside cells and inhibit cellular metabolism and processes. Sodium ions at the root surface disrupt plant nutrition of the similar cation potassium by inhibiting both potassium uptake and enzymatic activities within the cell. Ion toxicity is the result of replacement of K^+ by Na^+ in biochemical reactions, and Na^+ and Cl^- induced conformational changes in proteins. For several enzymes, K^+ acts as a cofactor and cannot be substituted by Na^+. High K^+ concentration is also required for binding tRNA to ribosomes and thus protein synthesis (Zhu 2002). Ion toxicity and osmotic stress cause metabolic imbalance, which in turn leads to oxidative stress (Chinnusamy et al. 2006). Specific ion toxicity, which results from the excessive uptake of certain ions, is the primary cause of growth reduction under salt stress (Chinnusamy et al. 2005). Toxic ions in salt-affected soils are usually sodium, chloride, and sulfate (Ghassemi et al. 1995; Munns and Tester 2008). Excessive sodium ion (Na^+)

accumulation causes ion toxicity and interferes with plant metabolism, while accumulation of potassium ion (K^+) can alleviate Na^+ toxicity by adjusting osmotic potential through ion balance. Many physiological studies have demonstrated that Na^+ toxicity is not only due to toxic effects of Na^+ in the cytosol but also because K^+ homeostasis is disrupted possibly due to the ability of Na^+ to compete for K^+ binding sites. Similarly, the reduction of plant osmotic potential, excessive uptake of Na^+ and Cl^- in the cell, and disruption of cell metabolic functions are due to ion toxicity (James et al. 2011). Excessive sodium ion in plant tissues harms the cell membrane and plant organelles, resulting in cell death of plant (Siringam et al. 2011).

1.2 Management Strategies

1.2.1 Drought

In the theory of evolution by Darwin, he describes "survival of the fittest" which means that a fit individual survives and others diminish. Only those crops which are successful under a harsh environment can withstand stress. It is crucial for crop scientists to develop strategies to make the crop plants fit for harsh environmental conditions under climate change. Various management strategies have been opted by different scientists in different agroclimatic regions to improve crop performance under drought stress. These strategies are discussed critically in the following section. These approaches have significant potential to overcome the drastic effects of drought, but their application is mostly limited to lab conditions.

1.2.1.1 Use of Mineral Nutrients and Organic Manures

Water stress induces a reduction in plant tissue water levels and subsequently affects leaf water potential, leaf elongation, leaf photosynthesis, protein synthesis, N metabolism, and cell membrane properties which leads to a reduction in plant productivity (Shangguan et al. 2000). Mineral nutrients are usually taken from soils as inorganic ions required for plant growth and development. Under drought conditions, nutrient uptake is impaired due to reduced soil moisture, leading to the slow diffusion of mineral nutrients from the soil to the root surface. Hence, the translocation speed to the leaves is also reduced. Drought induces early closure of stomata, thus reducing the transpiration rate, and also limits the transport of nutrients from the root to the shoot. Thus, drought stress reduces the availability and transport of nutrients in the soil matrix and plant tissues (Silva et al. 2011). However, adequate nutrition of plants under water deficit may improve the performance of crops. N and K are primary macronutrients required by plants in large amounts and govern several developmental processes such as photosynthesis, translocation of photosynthates from roots to shoots, protein synthesis, stomatal closure, water use efficiency, and regulation of enzymes (Salami and Saadat 2013). The application of nitrogen significantly improved crop performance under drought stress. N also plays a significant role in preventing plasma membrane damage and osmotic adjustment. N application under drought stress increases N, K, Ca, and glycine betaine concentrations in leaf tissues. Drought stress enhances malondialdehyde (MDA) concentration in leaves, while nitrogen supplementation reduces MDA in both control and water-stressed plants (Saneoka et al. 2004). Under drought stress, nitrogen supply improves photosynthetic pigment contents and photosynthetic capacity by increasing leaf area (LA), enhancing photosynthetic efficiency, and alleviating photo-damage under water stress (Wu et al. 2008). Crops significantly enhance their water usage ability and help in drought resistance with phosphorus application (Hansel et al. 2017). Under drought stress P nutrition increases the root growth (Singh and Sale 1998), increases stomatal conductance and faster the nitrate reductase activity (Oliveira et al. 2014), leaf area and photosynthesis (Singh et al. 2006), higher cell-membrane stability and water relations (Kang et al. 2014). Potassium is well-known for its osmoregulatory functions in crops. It regulates stomatal conductance and water uptake; the optimum K application increases WUE (Jatav et al. 2014). Potassium fertilization facilitates plant tolerance via different mechanisms such as osmotic adjustment, maintaining the activity of aquaporins and hence water uptake, cell elongation, promotion of root growth and cell membrane stability, stomatal regulation, as well as detoxification of reactive oxygen species resulting in improved drought stress tolerance (Wang et al. 2013). These nutrients enhance the tolerance against drought stress by improving protein synthesis, stomatal regulation, homeostasis, and osmoregulation through quenching the ROS (Cakmak 2005). Organic manures are another viable option that improves drought tolerance when applied alone or in combination with synthetic fertilizers (Esmaeilian et al. 2012). These manures are a beneficial source of significant nutrients and affect the temporal dynamics of nutrient availability through improving soil physicochemical properties (Paul and Beauchamp 1993). Vermicomposts have consistently improved seed germination, growth, and development more than converting mineral nutrients into more plant-available forms. Taleshi et al. (2012) detected that seed yield and yield components increased with the application of vermicomposts under water stress.

1.2.1.2 Seed Priming

Drought stress exposure adversely affects plant growth and productivity via non-normal physiological processes. Various seed priming techniques have been experimented to mitigate the adverse effect of drought stress on plant performance. Priming is an alternative technique to overcome these limitations and serves as a means to boost the stress tolerance potential of plants (Sen and Puthur 2020a; Thomas and Puthur 2020). Seed priming is a pre-sowing seed treatment that allows the controlled hydration of seeds to imbibe water and go through the first stage of germination but does not allow radical protrusion through the seed coat (McDonald 2000). Pre-sowing seed priming consists of priming the seeds in water with or without organic and inorganic salts in a controlled environment, followed by shade drying before sowing. The hydration process is performed using different techniques, e.g., immersion of seeds in water (hydropriming), osmotic solution (osmotic priming), chemicals (chemical priming), or hormones (hormonal priming) (Nawaz et al. 2013).

Priming can improve germination by enhancing the physiological metabolism like the activity of alfa amylase and increase in soluble and proline contents during germination without involving any lag phase or activation period, which increases seedling vigor in normal and stress conditions (Singhal and Bose 2020). Seed priming helps to ameliorate drought stress by adopting several strategies such as early mobilization of seed food reserves, elongation of embryo cells, endosperm weakening, etc., which enhances the pre-germination metabolic activities resulting in uniform and high germination percentages (Chen and Arora 2013). Priming has a critical function in improving germination and growth in a variety of crops under various abiotic stressors (Paparella et al. 2015; Zheng et al. 2016; Hussain et al. 2017). It activates numerous stress-responsive genes, enabling earlier germination and greater abiotic stress tolerance (Manonmani et al. 2014; Paparella et al. 2015; Wojtyla et al. 2016). Seed priming induces mild plant stress and activates stress-responsive genes and proteins, like late embryogenesis abundant (LEAs), that potentially cause drought stress tolerance (Chen and Arora 2013; Sen and Puthur 2020b; Thomas and Puthur 2020). Seed treatments prior to germination induce a particular physiological state called primed state, which augments several cellular responses (Wojtyla et al. 2016). As a result, plants are equipped to respond quickly to further stress exposure (Farooq et al. 2020). The seedlings emerging from primed seeds are characterized by early and uniform germination, and an overall enhancement in various growth features can be noted in its life span (Jisha et al. 2013; Huang et al. 2020; Khalaki et al. 2021).

Seed priming is an easy and non-monitoring technique that shows noticeable results in improving plant establishment and growth development, particularly under abiotic stress conditions. Therefore, seed priming can be used to ameliorate the negative effects of drought stress on germination, growth, and yield of crops to a significant extent.

1.2.1.3 Application of Plant Antitranspirants

Water deficit is one of the main abiotic stresses, restricting the growth and productivity of plants and causing alternations in plant physiology and biochemistry (Bakry et al. 2016). Nearly 95–98% of the water absorbed by the plant is lost in transpiration (Prakash and Ramachandran 2000; Gaballah et al. 2014). Antitranspirants are chemical compounds that favor reduced transpiration rates from plant leaves by reducing the size and number of stomata and gradually hardening them to stress (Ahmed and Ahmed 2014; El-Khawaga 2013). Applying plant antitranspirants is one of the main tools to balance leaf transpiration and water loss prevention (Goreta et al. 2007). It is a substance involved in increasing drought stress resistance. Foliar sprays markedly increase all growth parameters and relative water content and may reduce transpiration. Based on the mode of action, antitranspirants are classified into three types. The metabolic or stomata closing type such as exogenous abscisic acid (ABA) reduces transpiration by physiologically inducing stomatal closure (AbdAllah et al. 2018). The reflective class (e.g., kaolin) enhances leaf surfaces' light reflectance properties to minimize leaf temperature and consequently the transpiration rate (Glenn 2012). The third group, called the film-forming antitranspirants (e.g. di-1-*p*-menthene), as used here, reduce transpiration by physically blocking stomatal pores when a spray application has dried on the leaf surface (Palliotti et al. 2010). Kaolin spray was found to decrease leaf temperature by increasing leaf reflectance and to reduce transpiration rate more than photosynthesis in many plant species grown at high solar radiation levels (Nakano and Uehara 1996). Film-forming and reflecting antitranspirant were found to be nontoxic and have a longer period of effectiveness than metabolic types. Kaolin spray reduces leaf temperature through rising leaf reflectance which decreases the transpiration rate more than the photosynthesis of plants grown at high solar radiation levels (Nakano and Uehara 1996). Studies by Cantore et al. (2009) reported that on tomato and potato, the foliar application of kaolin suspension reduces plant stress which is essential for the best plant growth, yield, and quality. Water stress substantially impacts yield. Hence, the application of antitranspirant immediately prior to this stage may conserve water and improve grain set which could outweigh the photosynthetic limitations (Kettlewell et al. 2010).

1.2.1.4 Planting Density and Planting Date

Beyond the use of water deficit, another option to increase yield is the planting density technique. Optimization of plant density is the main strategy for increasing yield. The increase in planting density should be carefully chosen so that intraspecific competition does not happen and it results in the best use of available resources for grain growth and yield. Higher planting densities have enabled earlier canopy closure, which increases the season total light interception per ground area (Thornley 1983) and also reduces soil evaporation (Richards 1991) which otherwise helps limit water available for transpiration that supports photosynthesis. Nielsen et al. (2002) and Parker et al. (2016) pointed out that too early maize planting was associated with potentially under-optimal soil and weather planting conditions, while too late planting exposes plants to a reduced growing season length, low temperatures, and low-income solar radiation. Crop dry biomass and kernel weight decrease with delayed planting date due to low and decreasing temperature and radiation during the grain filling stage (Andrade et al. 1993).

1.2.2 Salinity

Soil salinization is a serious land degradation problem in most coastal regions worldwide due to the use of saline water for irrigation (Chi et al. 2021; Huang et al. 2022; Sun et al. 2020). Appropriate land management is an important prerequisite for improving the soil quality and productivity of saline soil (Basak et al. 2022). There are two groups of management strategies against salinity: first one is natural adaptation responses toward salinity, and second are human-made management strategies to handle the salinity stress in field crops or plants. Salinity occurs because of excessive accumulation of soluble salts via soil chemical properties and irrigated water. As a result of salinity stress and ion (Na^+ and Cl) toxicity, the disturbance of ion imbalance occurs. The natural management strategies by the plants to salinity stress are based on three strategies: (1) exclusion of Na^+ from the cytoplasm due to low uptake or pumping out of the ion from the cell by active mechanisms, (2) requisitioning of Na^+ into the vacuole, and (3) preferential accumulation in the leaf tissues. Among the human-made management strategies, the salinity problems can manage plant growth by adopting agronomic strategies such as water and nutrient management to improve soil health, plant growth, and input use efficiency (IUE) under salinity (USSLS 1954).

1.2.2.1 Physical Management of Saline Soil

1. Scraping: When soluble salt accumulates on the soil surface, scraping helps to remove salts. The salts accumulated on the surface can be removed by mechanical means. This is the simplest and most economical way to reclaim saline soils if the area is very small, e.g., small garden lawn or a patch in a field. This improves plant growth only temporarily as the salts accumulate again and again.
2. Subsoiling: Soil in deep layers has less salt content compared to above layers. Subsoiling breaks the top custard soil and makes more permeable. Subsoiling is "deep ripping" to improve soil properties at deeper layers where a dense soil layer (or hard pan) exists, thereby limiting the penetration of roots and water infiltration.
3. Deep plowing: Chisel plow is needed for deep plowing in order to increase the permeability for better leaching.
4. Leveling: Surface leveling is to get uniform leaching entire land should be levelled avoiding unnecessary wastage of water.

   Flushing: Washing of surface salts by flushing water. This is especially practicable for soils having a crust and low permeability. However, this is not a sound method of practice.
5. Sand mixing: Permeability was very low in heavy clay soil. Applying and mixing sand in soil with 30–40% clay content increases the permeability and gets higher leaching efficiency.
6. Leaching: When soil elements, after dissolving with water, go down from the upper parts to the lower level, it is called leaching. Dissolve and translocate the soluble salts downward below 45–60 cm. This water, along with the water supply to crops, takes away salts after dissolving them. That is why more water is required in this method. This process is mostly adopted in the dry season. The fields are divided into small fragments through bunding on the boundary so that water is conserved there, reducing the effect of salts. This salt accumulation can be controlled by applying water in addition to the ET water requirement of the crop. This extra water will usually push the salts below the root zone. The amounts of water required for leaching (leaching requirement—LR) can be calculated by standard procedures (Ayers and Westcot 1985).

   Leaching requirement: The amount of water needed to remove the excess soluble salts from the saline soils is called the leaching requirement or the fraction of the irrigation water that must be leached through the root zone or soil profile to control soil salinity at any specific level (salt balance):

$$\text{Leaching Requirement (LR)} = \frac{EC_{iw}}{EC_{dw}} = \frac{D_{dw}}{D_{iw}}$$

where
EC = electrical conductivity in dS m^{-1}
iw = EC of irrigation water in dS m^{-1}
dw = EC of drainage water in dS m^{-1}
D_{dw} = depth of drainage water in in.
D_{iw} = depth of irrigation water in in.

7. Drainage: An effective and good drainage system is required to flush excess salts from the root zone. It requires lowering the groundwater table and effectively leaching salts from the root zone. "Biodrainage" is also practiced as it involves growing certain trees along the canal or field boundaries, of which water demand is very high. It controls salinity and rising water tables.
8. Subsurface drainage: It is used to remove excess water present in the root zone. It brings down the water table and provides air circulation in the root zone. Tiles improve subsurface drainage. Besides this, tube wells are also installed in the area of the shallow water table to improve vertical drainage which helps in lowering water table.

1.2.2.2 Chemical Method

Gypsum Application

Sodium content and presence of carbonate and bicarbonate of the soil increased ESP more than 15 or pH more than 8 (saline-sodic soil) for replacing the Na^+ by Ca^{2+} and subsequent leaching of Na^+. Gypsum is primarily used on Na-affected soils as a source of Ca^{2+} ions to displace Na^+ ions, which tend to disperse soil particles and restrict water infiltration. The resulting displaced Na^+ ions are leached readily from the soil profile. Gypsum is a neutral salt that does not directly reduce pH. However, it can indirectly lower the pH of sodic soils by reducing the hydrolysis reactions associated with Na^+ ions on the exchange complex.

Nutrient Addition

Salt-affected soils suffer from many troubles around the world, such as limited crop production due to their abiotic stresses, particularly in arid and semiarid regions (Nan et al. 2016; Zhang et al. 2017). Proper plant nutrition is one of the most important strategies to alleviate this salt stress in crop production. In general, plants uptake their nutrients from the soil solution and/or by foliar application for plant growth, development, and other processes. The bioavailability of these soil nutrients is totally controlled by many factors, including soil characterization (e.g., soil pH, salinity, nutrient biogeochemical cycles, and physicochemical processes) and environmental and climatic changes. Concerning the effects of soil salinity on the nutrition of plants, nutrient plant disturbances reduce plant growth by affecting the transport and partitioning of different nutrients. Soil salinity also may cause deficiencies or imbalances in plant nutrients, due to the competition of Na^+ and

Cl^- with many plant nutrients such as Ca^{2+}, K^+, and NO_3^--N. Plant nutrients give the plants full power during their entire life and help plants ameliorate different stresses including abiotic and biotic. The use of fertilizers in somewhat greater quantities than normal in saline soils is beneficial. Application of nutrients like NPK and magnesium reduces the toxicity effects of saline soil and raise optimum crop growth and yield. Nitrate reduces chloride uptake, while potassium reduces Na uptake (Martinez and Cerda 1989). K^+ foliar and soil application significantly reduces the toxic effect of saline soil by maintaining the water balance and ion ratio (Golezani and Abriz 2018). Normally, deficit of zinc, iron, manganese, and nitrogen elements is found in saline soils. Hence, productivity can be increased by the use of these elements. Foliar selenium and silicon in combination or alone improved transpiration rate, water relations, photosynthetic attributes, chlorophyll contents, and the growth of wheat seedlings under stressed conditions. This increase is due to the accumulation of osmoprotectants (e.g., proline, soluble protein, and soluble sugar) and the increase in antioxidant enzyme activity (Sattar et al. 2017).

1.2.2.3 Organic Manure

Soil salinization is a serious land degradation problem in most coastal regions worldwide due to the use of saline water for irrigation (Chi et al. 2021; Huang et al. 2022; Sun et al. 2020). Appropriate soil management is an essential prerequisite for improving the soil quality and productivity of saline soil. Soil management practices include tillage, mulching, and crop residue management. In general, a large amount of organic residue and waste are produced in the world every year; more importantly, the improper disposal of these organic residues has led to severe environmental pollution and nutrient losses (Bai et al. 2016; Hazrati et al. 2020; Jia et al. 2018). Recycling these organic residues to farmland is a common agricultural practice for increasing soil fertility and agricultural productivity during decades (Meena et al. 2016; Sun et al. 2020; Wu et al. 2021). Organic amendments, such as livestock manure, plant residue and waste, and bioorganic fertilizer, are better practices to reclaim saline soil by alleviating soil salinity, improving soil fertility, and promoting crop growth (Chen et al. 2021; Cui et al. 2021; Huang et al. 2019; Leogrande and Vitti 2018; Wu et al. 2019). Straw mulching is very promising option for farmers to control soil salinity as it reduces soil water evaporation and regulates soil water and soil movement. Straw mulching seems to decrease the salt content on the surface of the soil by regulating the salt vertical distribution, which could reduce the salt damage to the crops, enhance their yields, and reduce the risk of soil salinization and erosion. Mulching with crop residues or live crops reduces the evaporation of moisture from the soil surface compared to barren soil. It reduces salts' upward pull from groundwater table (Siyal et al. 2001). Organic manure has a high water-holding capacity. When a sufficient amount of organic manure is added, the water holding capacity of soil increases. As a result, the conductivity of the soil solution decreases.

1.2.2.4 Seed Priming

Salt stress substantially reduces and delays germination in crops due to salinity-induced osmotic stress and toxic effects of Na^+ and Cl^- ions on germinating seeds. Different strategies are used to improve the salinity tolerance of other cultivated crops; one of the best approaches is seed priming, which has been evaluated as an active method to alleviate salinity stress (Munns and Gilliham 2015; Banerjee and Roychoudhury 2018; Farooq et al. 2019). Seed priming is a short-term and pragmatic approach to coping with salt stress. In seed priming, seeds can imbibe in low water potential, permitting partial imbibition without radicle protrusion. Seed priming enables faster and better germination in plants under stressful conditions. Primed seeds can activate the signal pathways during the early growth stage and trigger a faster stress response. This condition stimulates many of the metabolic processes involved with the early phases of germination, and it has been noted that seedlings from primed seeds emerge faster, grow more vigorously, and perform better in adverse conditions (Cramer 2002).

1.2.2.5 Selection of Crops and Crop Rotations

Using salt-tolerant crops is one of the most important strategies to solve the problem of salinity. Salt-tolerant agricultural crops are recommended to grow in salt-affected soils to reduce crop yield losses under saline conditions. Salt-tolerant varieties have been developed which grow very well in these soils. Salt-tolerant crops and cultivars capable of increasing in unreclaimed or partially reclaimed soils represent such a strategy. Salt-tolerant cultivars give stable yields and significantly reduce the need to apply amendments to enhance soil productivity (ICAR–CSSRI 2015).

On the basis of crop tolerance to the quality of irrigation water or soil salinity, the crops can be classified in four groups.

1	Highly tolerant crops	Barley, sugar beet, date palm, cotton
2	Tolerant	Wheat, rice, sorghum, maize, tapioca, mustard, coconut, spinach, *Amaranthus*, pomegranate, guava, ber
3	Semi-tolerant	Ash gourd, bitter gourd, brinjal, cabbage, cluster bean, pea, lady's finger, muskmelon, onion, potato, dolichos, sweet potato, tomato, turnip, water melon
4	Sensitive	Radish, carrot, tomato, potato, onion, coriander, cumin, mint, grape, sweet orange

S. No.	Crops	Varieties
1	Rice	CSR10, CSR13, CSR23, CSR27, Basmati CSR30, CSR36, CSR43, CSR46, CSR49, CSR52, CSR56, CSR60, and CSR76
2	Wheat	KRL283, KRL213, KRL210, KRL19, KRL 1-4
3	Mustard	CS 60, CS 58, CS 56, CS 54, CS 52
4	Chickpea	Karnal Chana-1
5	Lentil	PDL-1, PSL-9

(continued)

S. No.	Crops	Varieties
6	Dhaincha	CSD 137, CSD 123

These soils should not be kept fallow, but cultivation should be continuously done according to crop rotation. Crops should be sown which can bear salts, mainly beet, potato, barley, wheat, cotton, etc.

1.2.2.6 Irrigation Practices (Method of Water Application and Frequency of Irrigation)

Water management is very crucial to manage salt-affected soils. Efficient irrigation through modification in irrigation scheduling and decreasing water consumption may reduce salinization process. The irrigation method and volume of water applied have a pronounced influence on salt accumulation and distribution. Flood irrigation and an appropriate leaching fraction generally move salts below the root zone. Different irrigation methods like drip irrigation, sprinkler irrigation, and furrow-irrigated plot are helpful in reducing the saline condition. Drip irrigation is considered the most efficient because it applies water precisely at the root zone. The water application is more or less uniform and can be operated frequently. It maintains high soil matric potential (SMP) in the root zone and thus compensates for the decreased osmotic potential caused by irrigation with saline water (Goldberg et al. 1976). Drip irrigation has the potential of increased yield under saline soil conditions. In drip-irrigated plots, water moves away from the emitter and salts concentrate where the water evaporates. Unlike flood irrigation, drip irrigation might increase the risk of salinization of upper soil horizons but prevents salt from leaching to groundwater (Marchand and Abd El Hadi 2002). Similar results can be obtained with a properly managed sprinkler irrigation system. In furrow-irrigated plots, water moves from the furrow into the bed via capillary flow. When adjacent furrows are irrigated, salts concentrate in the center of the intervening bed. Manipulating bed shape and planting arrangement are often used to avoid salt damage in furrow-irrigated row crops.

When adopted, an irrigation system should permit frequent, uniform, and efficient water application with as minimum percolation loss as possible, but without curtailing essential leaching requirements. In addition, a good irrigation system should also avoid using saline water at the seed germination stage (a very sensitive stage). Where appropriate and good quality water is also available, farmers should practice using recycled water for irrigation. Salinity affects the sprouting of seeds. Hence, during the first irrigation, more quantity of water should be supplied to neutralize the effect of salts. Thereafter, even during subsequent irrigation, along with reducing the quantity of water, even the duration between two irrigations should be reduced so that soil does not get dry and the dissolution of salts continues in the soil. Deficit irrigation involves the application of water below full crop water requirements, so that a mild crop water stress is allowed with bearable effects on yield. Deficit irrigation strategies save water but also have the potential to improve the management of soil salinity by better control of rising water tables and by reducing the ingress of salts by irrigation water.

1.3 Conclusions

Salinity and drought are two of the most serious abiotic stresses that threaten crop productivity worldwide. Drought affects more than 10% of arable land, causing desertification, especially in arid and semiarid areas, while salinization is rapidly increasing on a global scale, declining average yields for most major crops. Drought is the most severe hazard to agriculture in nearly every part of the world. Drought stress is well recognized as a limiting factor that alters multiple aspects of plant growth and development. Salinity not only decreases the agricultural production of most crops but also affects soil physicochemical properties and ecological balance of the area. Therefore, it is necessary to improve management techniques to reduce the damage caused by drought and salinity. Therefore, a holistic approach considering the different management options to deal with drought and salinity stress may be a win-win approach in the future. Therefore use of mineral nutrients and organic manures, seed priming, application of plant antitranspirants, planting density, planting date, selection of tolerant crops, varieties, crop rotation, method of water application, and frequency of irrigation play critical roles in plant adaptation to salinity and drought stress. These management strategies are useful for stress management in arid and semiarid climates.

References

Abbasi T, Abbasi SA (2010) Biomass energy and the environmental impacts associated with its production and utilization. Renew Sust Energ Rev 14:919–937

AbdAllah AM, Burkey KO, Mashaheet AM (2018) Reduction of plant water consumption through anti-transpirants foliar application in tomato plants (Solanum lycopersicum L). Sci Hortic Amsterdam 235:373–381. https://doi.org/10.1016/j.scienta.2018.03.005

Abdelgadir EM, Oka M, Fujiyama H (2005) Nitrogen nutrition of rice plants under salinity. Biol Plant 49(1):99–104

Adnan M, Fahad S, Zamin M, Shah S, Mian IA, Danish S, Zafar-ul-Hye M, Battaglia ML, Naz RMM, Saeed B et al (2020) Coupling phosphate-solubilizing bacteria with phosphorus supplements improve maize phosphorus acquisition and growth under lime induced salinity stress. Plants 9:900

Ahmed Y, Ahmed M (2014) Impact of spraying some antitranspirants on fruiting of Williams bananas grown under Aswan Region conditions. Stem Cell 5(4):34–39

Akbarimoghaddam H, Galavi M, Ghanbari A, Panjehkeh N (2011) Salinity effects on seed germination and seedling growth of bread wheat cultivars. Trakia J Sci 9(1):43–50

Andrade FH, Uhart SA, Cirilo A (1993) Temperature affects radiation use efficiency in maize. Field Crop Res 32:17–25

Arbona V, Manzi M, de Ollas C, Gómez-Cadenas A (2013) Metabolomics as a tool to investigate abiotic stress tolerance in plants. Int J Mol Sci 14:4885–4911

Arzani A, Ashraf M (2016) Smart engineering of genetic resources for enhanced salinity tolerance in crop plants. Crit Rev Plant Sci 35:146–189. https://doi.org/10.1080/07352689.2016.1245056

Ashraf M (2004) Some important physiological selection criteria for salt tolerance in plants. Flora 199:361–376
Ashraf M, Harris PJC (2013) Photosynthesis under stressful environments: an overview. Photosynthetica 51:163–190. https://doi.org/10.1111/plb.12014
Ayers RS, Westcot DW (1985) Water quality for agriculture. FAO irrigation and drainage paper 29 Rev. 1. FAO, United Nations, Rome, p 174
Bai Z, Ma L, Jin S, Ma W, Velthof GL, Oenema O, Liu L, Chadwick D, Zhang F (2016) Nitrogen, phosphorus, and potassium flows through the manure management chain in China. Environ Sci Technol 50:13409–13418
Bakry AB, Ibrahim FM, Abdallah MMS, El-Bassiouny HMS (2016) Effect of banana peel extract or tryptophan on growth, yield and some biochemical aspects of quinoa plants under water deficit. Int J Pharm Tech Res 9(8):276–287
Banerjee A, Roychoudhury A (2018) Seed priming technology in the amelioration of salinity stress in plants. In: Rakshit A, Singh H (eds) Advances in seed priming. Springer, Singapore, pp 81–93. https://doi.org/10.1007/978-981-13-0032-5_5
Bano A, Fatima M (2009) Salt tolerance inZea mays (L.) following inoculation with RhizobiumandPseudomonas. Biol Fertil Soils 45:405–413
Banuls J, Legaz F, Primo-Millo E (1990) Effect of salinity on ion content, water relations and gas exchange parameters in some scion-rootstock combinations. J Hort Sci 65:714–724
Barber SA (1995) Soil nutrient bioavailability: a mechanistic approach, 2nd edn. Wiley, New York, NY
Basak N, Rai AK, Sundha P, Meena RL, Bedwal S, Yadav RK, Sharma PC (2022) Assessing soil quality for rehabilitation of salt-affected agroecosystem: a comprehensive review. Front Environ Sci 10. https://doi.org/10.3389/fenvs.2022.935785
Bassiri A, Khosh-Khui M, Rouhani I (1977) The influences of simulated moisture stress conditions and osmotic substrates on germination and growth of cultivated and wild safflowers. J Agric Sci 88(1):95–100. https://doi.org/10.1017/S0021859600033815
Battaglia ML, Lee C, Thomason W, Van Mullekom J (2020) Effects of corn row width and defoliation timing and intensity on canopy light interception. Crop Sci 59:1718–1731
Bayoumi TY, Eid M, Metwali EM (2008) Application of physiological and biochemical indices as a screening technique for drought tolerance in wheat genotypes. Afr J Biotechnol 7:2341–2352
Benlloch-González M, Quintero JM, García-Mateo MJ, Fournier JM, Benlloch M (2015) Effect of water stress and subsequent re-watering on K+ and waterflows in sunflower roots: a possible mechanism to tolerate water stress. Environ Exp Bot 118:78–84
Bista DR, Heckathorn SADM, Jayawardena S, Mishra JK (2018) Boldt effects of drought on nutrient uptake and the levels of nutrient-uptake proteins in roots of drought-sensitive and-tolerant grasses. Plants 7:28
Blaylock AD (1994) Soil salinity, salt tolerance and growth potential of horticultural and landscape plants. Co-operative Extension Service, University of Wyoming, Department of Plant, Soil and Insect Sciences, College of Agriculture, Laramie, WY
Boursiac Y, Chen S, Luu DT, Sorieul M, van den Dries N, Maurel C (2005) Early effects of salinity on water transport in Arabidopsis roots. Plant Physiol 139:790–805. https://doi.org/10.1104/pp.105.065029
Bradford KJ, Hsiao TC (1982) Physiological responses to moderate water stress. In: Physiological plant ecology. II. Springer, Berlin, pp 263–324
Bray EA, Bailey-Serres J, Weretilnyk E (2000) Responses to abiotic stresses. Biochemistry and molecular biology of plants. American Society of Plant Physiologists, Rockville
Brodersen CR, Roddy AB, Wason JW, McElrone AJ (2019) Functional status of xylem through time. Annu Rev Plant Biol 70:407–433
Brugnoli E, Bjorkman O (1992) Growth of cotton under continuous salinity stress: influence on allocation pattern, stomatal and non-stomatal components of photosynthesis and dissipation of excess light energy. Planta 128:335–337

Cakmak I (2005) The role of potassium in alleviating detrimental effects of abiotic stress in plants. J Plant Nutr Soil Sci 168:521–530

Cantore V, Pace B, Albrizio R (2009) Kaolin-based particle film technology affects tomato physiology, yield and quality. Environ Exp Bot 66:279–288

Chen K, Arora R (2013) Priming memory invokes seed stress-tolerance. Environ Exp Bot 94:33–45

Chen M, Zhang S, Liu L, Wu L, Ding X (2021) Combined organic amendments and mineral fertilizer application increase rice yield by improving soil structure, P availability and root growth in saline-alkaline soil. Soil Tillage Res 212:105060

Chi Z, Wang W, Li H, Wu H, Yan B (2021) Soil organic matter and salinity as critical factors affecting the bacterial community and function of Phragmites australis dominated riparian and coastal wetlands. Sci Total Environ 762:143156

Chinnusamy V, Xiong L, Zhu J (2005) Use of genetic engineering and molecular biology approaches for crop improvement for stress environments. Abiotic stresses: plant resistance through breeding and molecular approaches. Food Product Press, New York, NY

Chinnusamy V, Zhu J, Zhu J-K (2006) Gene regulation during cold acclimation in plants. Physiol Plant 126(1):52–61

Cobb JN, De Clerck G, Greenberg A, Clark R, McCouch S (2013) Next-generation phenotyping: requirements and strategies for enhancing our understanding of genotype-phenotype relationships and its relevance to crop improvement. Theor Appl Genet 126:867–887. https://doi.org/10.1007/s00122-013-2066-

Cornic G (2000) Drought stress inhibits photosynthesis by decreasing stomatal aperture–not by affecting ATP synthesis. Trends Plant Sci 5:187–188. https://doi.org/10.1016/S1360-1385(00)01625-3

Corso D, Delzon S, Lamarque LJ, Cochard H, Torres-Ruiz JM, King A, Brodribb T (2020) Neither xylem collapse, cavitation, or changing leaf conductance drive stomatal closure in wheat. Plant Cell Environ 43:854–865

Cramer GR (2002) Sodium-calcium interactions under salinity stress in Läuchli A, Lüttge salinity. Environ Plan 4:205–227

Cui Q, Xia J, Yang H, Liu J, Shao P (2021) Biochar and effective microorganisms promote Sesbania cannabina growth and soil quality in the coastal saline-alkali soil of the Yellow River Delta, China. Sci Total Environ 756:143801

Devincentis AJ (2020) Scales of sustainable agricultural water management. Ph.D. Thesis, University of California, Davis, CA, USA

Earl HJ, Davis RF (2003) Effect of drought stress on leaf and whole canopy radiation use efficiency and yield of maize. Agron J 95(3):688–696

El-Khawaga AS (2013) Response of Grand Naine banana plants grown under different soil moisture levels to antitranspirants application. Asian J Crop Sci 5:238–250

El-Midaoui M, Talouizte A, Benbella M, Serieys H, Griveau Y, Berville A (2001) Effect of osmotic pressure on germination of sunflower seeds (Helianthus annuus L.). Helia 24:129–134

Esmaeilian Y, Sirousmehr AR, Asghripour MR, Amiri E (2012) Comparison of sole and combined nutrient application on yield and biochemical composition of sunflower under water stress. Int J App 2(3):214–220

FAO (2009) How to feed the world in 2050. FAO, Rome

Farooq M, Wahid A, Kobayashi N, Fujita D, Basra SMA (2009) Plant drought stress: effects, mechanisms and management. Agron Sustain Dev 29:185–212. https://doi.org/10.1051/agro:2008021

Farooq M, Hussain M, Wahid A, Siddique KHM (2012) Drought stress in plants: an overview. In: Aroca R (ed) Plant responses to drought stress: from morphological to molecular features. Springer, Berlin, pp 1–36

Farooq M, Usman M, Nadeem F, Rehman HU, Wahid A, Basra SMA et al (2019) Seed priming in field crops: potential benefits, adoption and challenges. Crop Past Sci 70(9):731–771. https://doi.org/10.1071/CP18604

Farooq M, Romdhane L, Al Sulti MK, Rehman A, Al-Busaidi WM, Lee DJ (2020) Morphological, physiological and biochemical aspects of osmopriming-induced drought tolerance in lentil. J Agron Crop Sci 206(2):176–186. https://doi.org/10.1111/jac.12384

Flowers TJ, Yeo AR (1995) Breeding for salinity resistance in crop plants: where next? Aust J Plant Physiol 22:875–884. https://doi.org/10.1071/PP9950875

Gaballah MS, Shaaban SM, Abdallah EF (2014) The use of anti-transpirants and organic compost in sunflower grown under water stress and sandy soil. Int J Acad Res 6:211

Garcia MJ, Lucena C, Romera FJ, Alcántara E, Perez-Vicente R (2010) Ethylene and nitric oxide involvement in the upregulation of key genes related to iron acquisition and homeostasis in Arabidopsis. J Exp Bot 61:3885–3899

Garg BK (2003) Nutrient uptake and management under drought: nutrient moisture interaction. Curr Agric 27:1–8

Ghassemi F, Jakeman AJ, Nix HA (1995) Salinisation of land and water resources: human causes, extent, management and case studies. CAB international, Sydney, NSW

Gimenez C, Gallardo M, Thompson RB (2005) Plant water relations. In: Hillel D (ed) Encyclopedia of soils in the environment. Elsevier, Oxford, pp 231–238

Glenn MD (2012) The mechanisms of plant stress mitigation by kaolin-based particle films and applications in horticultural and agricultural crops. Hort Science 47:710–711

Golezani GK, Abriz FS (2018) Foliar sprays of salicylic acid and jasmonic acid stimulate H+ -ATPase activity of tonoplast, nutrient uptake and salt tolerance of soybean. Ecotoxicol Environ Saf 166:18–25

Goldberg D, Gornat B, Rimon DE (1976) Drip irrigation - principles, design and agricultural practices. Drip Irrigation Scientific Publications, Israel

Goreta S, Leskovar DI, Jifon JL (2007) Gas exchange, water status, and growth of pepper seedlings exposed to transient water deficit stress are differentially altered by antitranspirants. J Am Soc Hortic Sci 132(5):603–610

Grattan SR, Grieve CM (1992) Mineral element acquisition and growth response of plants grown in saline environments. Agric Ecosyst Environ 38:275–300

Guan XK, Song L, Wang TC, Turner NC, Li FM (2015) Effect of drought on the gas exchange, chlorophyll fluorescence and yield of six different-era spring wheat cultivars. J Agron Crop Sci 201:253–266. https://doi.org/10.1111/jac.12103

Gunes A, Pilbeam DJ, Inal A, Coban S (2008) Influence of silicon on sunflower cultivars under drought stress, I: Growth, antioxidant mechanisms, and lipid peroxidation. Commun Soil Sci Plant Anal 39:1885–1903

Guo J, Ling H, Wu Q, Xu L, Que Y (2014) The choice of reference genes for assessing gene expression in sugarcane under salinity and drought stresses. Sci Rep 4:7042. https://doi.org/10.1038/srep07042

Habibi G (2013) Exogenous salicylic acid alleviates oxidative damage of barley plants under drought stress. Acta Biol Szeged 56:57–63

Hafez EH, Abou El Hassan WH, Gaafar IA, Seleiman MF (2015) Effect of gypsum application and irrigation intervals on clay saline-sodic soil characterization, rice water use efficiency, growth, and yield. J Agric Sci 7:208–219

Hansel FD, Amado TJ, Ruiz Diaz DA, Rosso LH, Nicoloso FT, Schorr M (2017) Phosphorus fertilizer placement and tillage affect soybean root growth and drought tolerance. Agron J 109(6):2936–2944. https://doi.org/10.2134/agronj2017.04.0202

Hazrati S, Farahbakhsh M, Heydarpoor G, Besalatpour AA (2020) Mitigation in availability and toxicity of multi-metal contaminated soil by combining soil washing and organic amendments stabilization. Ecotoxicol Environ Saf 201:110807

Hojati M, Modarres-Sanavy S, Karimi M, Ghanati F (2011) Responses of growth and antioxidant systems in Carthamus tinctorius L. under water deficit stress. Acta Physiol Plant 33(1):105–112. https://doi.org/10.1007/s11738-010-0521-y

Horie T, Kaneko T, Sugimoto G, Sasano S, Panda SK, Shibasaka M et al (2011) Mechanisms of water transport mediated by PIP aquaporins and their regulation via phosphorylation events under salinity stress in barley roots. Plant Cell Physiol 52:663–675. https://doi.org/10.1093/pcp/pcr027

Hu Y, Schmidhalter U (1998) Spatial distributions and net deposition rates of mineral elements in the elongating wheat (Triticum aestivum L.) leaf under saline soil conditions. Planta 204(2): 212–219. https://doi.org/10.1007/s004250050249

Huang M, Zhang Z, Zhu C, Zhai Y, Lu P (2019) Effect of biochar on sweet corn and soil salinity under conjunctive irrigation with brackish water in coastal saline soil. Sci Hortic 250:405–413

Huang L, Zhang L, Zeng R, Wang X et al (2020) Brassinosteroid priming improves peanut drought tolerance via eliminating inhibition on genes in photosynthesis and hormone signaling. Genes 11(8):919. https://doi.org/10.3390/genes11080919

Huang L, Liu Y, Ferreira JFS, Wang M, Na J, Huang J, Liang Z (2022) Long-term combined effects of tillage and rice cultivation with phosphogypsum or farmyard manure on the concentration of salts, minerals, and heavy metals of saline-sodic paddy fields in Northeast China. Soil Tillage Res 215:105222

Hussain M, Farooq M, Lee DJ (2017) Evaluating the role of seed priming in improving drought tolerance of pigmented and non-pigmented rice. J Agron Crop Sci 203:269–276

ICAR–CSSRI (2015) ICAR–Central Soil Salinity Research Institute Vision 2050. Indian Council of Agricultural Research, New Delhi

Iyengar ERR, Reddy MP (1996) Photosynthesis in high salt-tolerant plants. In: Pesserkali M (ed) Hand book of photosynthesis. Marshal Dekar, Baten Rose, pp 56–65. 08247 9708

James RA, Blake C, Byrt CS, Munns R (2011) Major genes for Na+ exclusion, Nax1 and Nax2 (wheat HKT1;4 and HKT1;5), decrease Na+ accumulation in bread wheat leaves under saline and waterlogged conditions. J Exp Bot 62(8):2939–2947

Jatav KS, Agarwal RM, Tomar NS, Tyagi SR (2014) Nitrogen metabolism, growth and yield responses of wheat (Triticum aestivum L.) to restricted water supply and varying potassium treatments. J Indian Bot Sociol 93(4):177–189. ISSN: 0019-4468

Jia W, Qin W, Zhang Q, Wang X, Ma Y, Chen Q (2018) Evaluation of crop residues and manure production and their geographical distribution in China. J Clean Prod 188:954–965

Jisha KC, Vijayakumari K, Puthur JT (2013) Seed priming for abiotic stress tolerance: an overview. Acta Physiol Plant 35(5):1381–1396. https://doi.org/10.1007/s11738-012-1186-5

Jung JY, Shin R, Schachtman DP (2009) Ethylene mediates response and tolerance to potassium deprivation in Arabidopsis. Plant Cell 21:607–621

Kader MA, Lindberg S (2010) Cytosolic calcium and pH signaling in plants under salinity stress. Plant Signal Behav 5(3):233–238

Kang L, Yue S, Li S (2014) Effects of phosphorus application in different soil layers on root growth, yield, and water-use efficiency of winter wheat grown under semi-arid conditions. J Integr Agric 13(9):2028–2039

Karlowsky S, Augusti A, Ingrisch J, Akanda MKU, Bahn M, Gleixner G (2018) Drought-induced accumulation of root exudates supports post-drought recovery of microbes in mountain grassland. Front Plant Sci 9:1593

Kettlewell PS, Heath WL, Haigh IM (2010) Yield enhancement of droughted wheat by film antitranspirant application: rationale and evidence. Agric Sci 01:143–147. https://doi.org/10.4236/as.2010.13017

Khajeh-Hosseini M, Powell AA, Bimgham IJ (2003) The interaction between salinity stress and seed vigor during germination of soybean seeds. Seed Sci Technol 31:715–725

Khalaki MA, Moameri M, Lajayer BA, Astatkie T (2021) Influence of nano-priming on seed germination and plant growth of forage and medicinal plants. Plant Growth Regul 93:13–28. https://doi.org/10.1007/s10725-020-00670-9

Khan MG, Srivastava HS (1998) Changes in growth and nitrogen assimilation in maize plants induced by NaCl and growth regulators. Biol Plant 41:93–99

Killingbeck KT (2004) Nutrients in senesced leaves: keys to the search for potential resorption and resorption proficiency. Ecology 77:1716–1727

Kramer PJ, Boyer JS (1995) Water relations of plants and soils. Academic Press, New York, NY

Ladrera R, Marino D, Larrainzar E, Gonzalez EM, Arrese-Igor C (2007) Reduced carbon availability to bacteroids and elevated ureides in nodules, but not in shoots, are involved in the nitrogen fixation response to early drought in soybean. Plant Physiol 145:539–546. https://doi.org/10.1104/pp.107.102491

Leogrande R, Vitti C (2018) Use of organic amendments to reclaim saline and sodic soils: a review. Arid Land Res Manag 33:1–21

Lisar SY, Rahman IM, Hossain MM, Motafakkerazad R (2012) Water stress in plants: causes, effects and responses. IntechOpen, London

Lobell DB, Burke MB, Tebaldi C, Mastrandrea MD, Falcon WP, Naylor RL (2008) Prioritizing climate change adaptation needs for food security in 2030. Science 319(5863):607–610. https://doi.org/10.1126/science.1152339

Luo W, Zuo X, Ma W, Xu C, Li A, Yu Q et al (2018) Differential responses of canopy nutrients to experimental drought along a natural aridity gradient. Ecology 99:2230–2239

Lynch JP, Brown KM (2001) Topsoil foraging-an architectural adaptation of plants to low phosphorus availability. Plant Soil 237:225–237. https://doi.org/10.1023/A:1013324727040

Manonmani V, Begum MAJ, Jayanthi M (2014) Halo priming of seeds. Res J Seed Sci 7:1–13

Marchand M, Abd El Hadi H (2002) Long-term experiments comparing the impact on soils and field crops of potassium chloride vs. potassium sulfate. Acta Hortic 573:49–54

Martinez V, Cerda A (1989) Influence of N source on rate of Cl, N, Na and K uptake by cucumber seedlings grown in saline condition. J Plant Nutr 12(8):971–983

Martínez-Vilalta J, Garcia-Forner N (2017) Water potential regulation, stomatal behaviour and hydraulic transport under drought: deconstructing the iso/anisohydric concept. Plant Cell Environ 40:962–976

Maser P, Eckelman B, Vaidyanathan R, Horie T, Fairbairn DJ, Kubo M et al (2002) Altered shoot/root Na+ distribution and bifurcating salt sensitivity in Arabidopsis by genetic disruption of the Na+ transporter AtHKT1. FEBS Lett 531:157–161. https://doi.org/10.1016/S0014-5793(02)03488-9

McDonald MB (2000) Seed priming. In: Seed technology and its biological basis. Sheffield Academic Press, Sheffield, pp 287–325

Meena MD, Joshi PK, Narjary B, Sheoran P, Jat HS, Chinchmalatpure AR, Yadav RK, Sharma DK (2016) Effects of municipal solid waste compost, rice-straw compost and mineral fertilisers on biological and chemical properties of a saline soil and yields in a mustard–pearl millet cropping system. Soil Res 54:958–969

Mukarram M, Choudhary S, Kurjak D, Petek A, Khan MMA (2021) Drought: sensing, signaling, effects and tolerance in higher plants. Physiol Plant 172:1291–1300

Munns R (2002) Comparative physiology of salt and water stress. Plant Cell Environ 25:239–250. https://doi.org/10.1046/j.0016-8025.2001.00808.x

Munns R (2005) Genes and salt tolerance: bringing them together. New Phytol 167:645–663

Munns R, Gilliham M (2015) Salinity tolerance of crops–what is the cost? New Phytol 208(3):668–673. https://doi.org/10.1111/nph.13519

Munns R, James RA (2003) Screening methods for salinity tolerance: a case study with tetraploid wheat. Plant Soil 253:201–218. https://doi.org/10.1023/A:1024553303144

Munns R, Tester M (2008) Mechanisms of salinity tolerance. Annu Rev Plant Biol 59(1):651–681. https://doi.org/10.1146/annurev.arplant.59.032607.092911

Munns R, Schachtman DP, Condon AG (1995) The significance of the two-phase growth response to salinity in wheat and barley. Aust J Plant Physiol 13:143–160

Munns R, Husain S, Rivelli AR, Richard AJ, Condon AG, Megan PL, Evans SL, Schachtman DP, Hare RA (2002) Avenues for increasing salt tolerance of crops, and the role of physiologically based selection traits. Plant Soil 247:93–105

Nakano A, Uehara Y (1996) The effect of kaolin clay on cuticle transpiration in tomato. Acta Hortic 440:233–238

Nan J, Chen X, Wang X, Lashari MS, Wang Y, Guo Z, Du Z (2016) Effects of applying flue gas desulfurization gypsum and humic acid on soil physicochemical properties and rapeseed yield of a saline sodic cropland in the eastern coastal area of China. J Soils Sediments 16:38–50
Nawaz J, Hussain M, Jabbar A, Nadeem GA, Sajid M, Subtain M, Shabbir I (2013) Seed priming a technique. Int J Agric Crop Sci 6:1373–1381
Nielsen RL, Thomison PR, Brown GA, Halter AL, Wells J, Wuethrich KL (2002) Delayed planting effects on flowering and grain maturation of dent corn. Agron J 94(3):549–558
Nishimura T, Cha-um S, Takagaki M, Ohyama K (2011) Survival percentage, photosynthetic abilities and growth characters of two indica rice (Oryza sativa L. spp. indica) cultivars in response to isosmotic stress. Span J Agric Res 9:262–270
Nonami H (1998) Plant water relations and control of cell elongation at low water potentials. J Plant Res 111(3):373–382. https://doi.org/10.1007/BF0250780
O'Connell E (2017) Towards adaptation of water resource Systems to climatic and socio-economic Chang. Water Resour Manag 31:2965–2984
Oliveira MT, Medeiros CD, Frosi G, Santos MG (2014) Different mechanisms drive the performance of native and invasive woody species in response to leaf phosphorus supply during periods of drought stress and recovery. Plant Physiol Biochem 82:66–75
Osakabe Y, Osakabe K, Shinozaki K, Tran LSP (2014) Response of plants to water stress. Front Plant Sci 5:1–7
Palliotti A, Poni S, Berrios J, Bernizzoni GF (2010) Vine performance and grape composition as affected by earlyseasons our limitation induced with antitranspirants in two red Vitis viniferaL. cultivars. Aust J Grape Wine Res 16:426–433
Paparella S, Araújo SS, Rossi G, Wijayasinghe M, Carbonera D, Balestrazzi A (2015) Seed priming: state of the art and new perspectives. Plant Cell Rep 34:1281–1293
Paranychianakis NV, Chartzoulakis KS (2005) Irrigation of Mediterranean crops with saline water: from physiology to management practices. Agric Ecosyst Environ 106:171–187
Parida AK, Das AB (2005) Salt tolerance and salinity effects on plants: a review. Ecotox Environ Safe 60:324–349
Parker PS, Shonkwiler JS, Aurbacher J (2016) Cause and consequence in maize planting dates in Germany. J Agron Crop Sci 203:1–14. https://doi.org/10.1111/jac.12182
Passioura JB, Angus JF (2010) Improving productivity of crops in water-limited environments. In: Advances in agronomy, vol 106. Academic Press, Cambridge, MA, pp 37–75
Paul JW, Beauchamp EG (1993) Nitrogen availability for corn in soils amended with urea, cattle slurry, and solid and composted manures. Can J Soil Sci 73(2):253–266
Paul K, Pauk J, Deák Z, Sass L, Vass I (2016) Contrasting response of biomass and grain yield to severe drought in cappelle desprez and plainsman V wheat cultivars. PeerJ 4:e1708. https://doi.org/10.7717/peerj.1708
Pinkerton A, Simpson JR (1986) Interactions of surface drying and subsurface nutrients affecting plant growth on acidic soil profiles from an old pasture. Aust J Exper Agric 26(6):681–689
Prakash M, Ramachandran K (2000) Effects of moisture stress and antitranspirants on leaf chlorophyll. J Agron Crop Sci 184:153–156
Rahbarian R, Khavari-Negad R, Ganjeali A, Bagheri A, Najafi F (2011) Drought stress effects on photosynthesis, chlorophyll fluorescence and water relations in tolerant and susceptible chickpea (Cicer arientinum L.) genotypes. Acta Biol Cracov Ser Bot 53(1):47–56
Rahnama A, James RA, Poustini K, Munns R (2010) Stomatal conductance as a screen for osmotic stress tolerance in durum wheat growing in saline soil. Funct Plant Biol 37(3):255–263
Rengasamy P, North S, Smith A (2010) Diagnosis and management of sodicity and salinity in soil and water in the Murray Irrigation Region. The University of Adelaide, Adelaide, SA
Resnik ME (1970) Effect of mannitol and polyethylene glycol on phosphorus uptake by maize plants. Ann Bot 34(3):497–504
Richards R (1991) Crop improvement for temperate Australia: future opportunities. Field Crop Res 26:141–169

Ristvey AG, Belayneh BE, Lea-Cox JD (2019) A Comparison of irrigation-water containment methods and management strategies between two ornamental production systems to minimize water security threats. Water 11:2558

Roelfsema MRG, Hedrich R (2002) Studying guard cells in the intact plant: modulation of stomatal movement by apoplastic factors. New Phytol 153:425–431

Roy ST, Negrao S, Tester M (2014) Salt resistant crop plants. Curr Opin Biotechnol 26:115–124. https://doi.org/10.1016/j.copbio.2013.12.004

Sajjan AS, Badanur VP, Sajjanar GM (1999) Effect of external water potential on seed germination, seedling growth and vigor index in some genotypes of sunflower. In: Faroda SA, Joshi NL, Kathju S, Kar A (eds) Proceedings of a Symposium on Recent Advances in Management of Arid Ecosystem, pp 215–218

Salami M, Saadat S (2013) Study of potassium and nitrogen fertilizer levels on the yield of sugar beet in jolge cultivar. J Novel Appl Sci 2:94–100

Salehi-Lisar SY, Bakhshayeshan-Agdam H (2016) Drought stress in plants: causes, consequences, and tolerance. In: Drought stress tolerance in plants, vol 1. Springer, Berlin, pp 1–16

Saneoka H, Moghaieb RE, Premachandra GS, Fujita K (2004) Nitrogen nutrition and water stress effects on cell membrane stability and leaf water relations in Agrostis palustris Huds. Environ Exp Bot 52(2):131–138. https://doi.org/10.1016/j.envexpbot.2004.01.011

Sapeta H, Costa M, Lourenc T, Marocod J, Van-der Linde P, Oliveiraa MM (2013) Drought stress response in Jatropha curcas: growth and physiology. Environ Exp Bot 85:76–84

Sattar A, Cheema MA, Abbas T, Sher A, Ijaz M, Hussain M (2017) Separate and combined effects of silicon and selenium on salt tolerance of wheat plants. Russ J Plant Physiol 64:341–348

Schimel J, Balser TC, Wallenstein M (2007) Microbial stress response physiology and its implications for ecosystem function. Ecology 88:1386–1394. https://doi.org/10.1890/06-0219

Sen A, Puthur JT (2020a) Halo and UV-B priming influences various physiological and importantly yield parameters of Oryza sativa var. Vyttila 6. N Z J Crop Hortic Sci 49:1. https://doi.org/10.1080/01140671.2020.1844765

Sen A, Puthur JT (2020b) Influence of different seed priming techniques on oxidative and antioxidative responses during the germination of Oryza sativa varieties. Physiol Mol Biol Plants 26:551. https://doi.org/10.1007/s12298-019-00750-9

Shangguan Z et al (2000) Effects of nitrogen nutrition and water deficit on net photosynthetic rate and chlorophyll fluorescence in winter wheat. J Plant Physiol 156:46

Siddique MRB, Hamid A, Islam MS (2000) Drought stress effects on water relations of wheat. Bot Bull Acad Sin 41:35–39

Silla F, Escudero A (2006) Coupling N cycling and N productivity in relation to seasonal stress in Quercus pyrenaica Willd samplings. Plant Soil 282:301–311

Silva EC, Nogueira RJMC, Silva MA, Albuquerque M (2011) Drought stress and plant nutrition. Plant Stress 5(1):32–41

Singh KN, Chatrath R (2001) Salinity tolerance. In: Reynolds MP, Monasterio JIO, McNab A (eds) Application of physiology in wheat breeding. CIMMYT, Mexico, DF, pp 101–110

Singh DK, Sale PWG (1998) Phosphorus supply and the growth of frequently defoliated white clover (Trifolum repens L.) in dry soil. Plant Soil 205:155–168

Singh V, Pallaghy CK, Singh D (2006) Nutrition and tolerance of cotton to water stress I. Seed cotton yield and leaf morphology. Field Crop Res 96:191–198

Singhal RK, Bose B (2020) Wheat seedlings as affected by Mg (NO3)2 and ZnSO4 priming treatments. World Sci News 144:13–29

Siringam K, Juntawong N, Cha-um S, Kirdmanee C (2011) Salt stress induced ion accumulation, ion homeostasis, membrane injury and sugar contents in salt-sensitive rice (Oryza sajtiva L. spp. indica) roots under isosmotic conditions. Afr J Biotechnol 10:1340–1346

Siyal AA, Siyal AG, Abro Z (2001) Salt affected soils their identification and reclamation. Pak J Appl Sci 2:537–540

Souza RP, Machado EC, Silva JAB, Lagoa AMMA, Silveira JAG (2004) Photosynthetic gas exchange, chlorophyll fluorescence and some associated metabolic changes in cowpea (Vigna unguiculata) during water stress and recovery. Environ Exp Bot 51(1):45–56

Steudle E (2000) Water uptake by plant roots: an integration of views. Plant Soil 226:45–56. https://doi.org/10.1023/A:1026439226716

Sun YP, Yang JS, Yao RJ, Chen XB, Wang XP (2020) Biochar and fulvic acid amendments mitigate negative effects of coastal saline soil and improve crop yields in a three year field trial. Sci Rep 10:8946

Sutka M, Li G, Boudet J, Boursiac Y, Doumas P, Maurel C (2011) Natural variation of root hydraulics in Arabidopsis grown in normal and salt stressed conditions. Plant Physiol 155: 1264–1276. https://doi.org/10.1104/pp.110.163113

Taiz L, Zeiger E (1998) Plant physiology, 2nd edn. Sinauer Associates Publishers, Sunderland, MA

Taleshi K, Shokoh-far A, Rafiee M, Noormahamadi G, Sakinejhad T (2012) Safflower yield respond to chemical and biotic fertilizer on water stress condition. World Appl Sci J 20(11): 1472–1477

Tangpilig VC, Yambao EB, O'toole JC, De Datta SK (1987) Water stress effects on leaf elongation, leaf water potential, transpiration, and nutrient uptake of rice, maize, and soybean. Plant Soil 103(2):155–168

Terzi R, Kadioglu A (2006) Drought stress tolerance and antioxidant enzyme system in Ctenanthe setosa. Acta Biol Cracov Ser Bot 48:89–96

Thomas TTD, Puthur JT (2020) UV-B priming enhances specific secondary metabolites in Oryza sativa (L.) empowering to encounter diverse abiotic stresses. Plant Growth Regul 92:169–180. https://doi.org/10.1007/s10725-020-00628-x

Thomason WE, Battaglia ML (2020) Early defoliation effects on corn plant stands and grain yield. Agron J 112:1–9

Thornley JHM (1983) Crop yield and planting density. Ann Bot 52:257–259

Tiwari YK, Yadav SK (2020) Effect of high-temperature stress on ascorbate–glutathione cycle in maize. Agric Res 9:179–187

Turner NC, Wright GC, Siddique KHM (2001) Adaptation of grain legumes (pulses) to water-limited environments. Adv Agron 71:193–231. https://doi.org/10.1016/S0065-2113(01)71015-2

Ünyayar S, Keleþ Y, Ünal E (2004) Proline and ABA levels in two sunflower genotypes subjected to water stress. Bulg J Plant Physiol 30(3–4):34–47

USSLS (US Salinity Laboratory Staff) (1954) Diagnosis and improvement of saline and alkali soils. USDA Handbook No. 60. USDA, Washington, DC

Wang W, Vinocur B, Altman A (2003) Plant responses to drought, salinity and extreme temperatures: towards genetic engineering for stress tolerance. Planta 218(1):1–14

Wang M, Zheng Q, Shen Q, Guo S (2013) The critical role of potassium in plant stress response. Int J Mol Sci 14(4):7370–7390

Wen B (2015) Effects of high temperature and water stress on seed germination of the invasive species Mexican sunflower. PLoS One 10(10):e0141567

Wojtyla Ł, Lechowska K, Kubala S, Garnczarska M (2016) Molecular processes induced in primed seeds—increasing the potential to stabilize crop yields under drought conditions. J Plant Physiol 203:116–126

Wu FZ, Bao WK, Li FL, Wu N (2008) Effects of water stress and nitrogen supply on leaf gas exchange and fluorescence parameters of Sophora davidii seedlings. Photosynthetica 46(1): 40–48. https://doi.org/10.1007/s11099-008-0008-x

Wu L, Wei C, Zhang S, Wang Y, Kuzyakov Y, Ding X (2019) MgO-modified biochar increases phosphate retention and rice yields in saline-alkaline soil. J Clean Prod 235:901–909

Wu X, Peng J, Liu P, Bei Q, Rensing C, Li Y, Yuan H, Liesack W, Zhang F, Cui Z (2021) Metagenomic insights into nitrogen and phosphorus cycling at the soil aggregate scale driven by organic material amendments. Sci Total Environ 785:147329

Yadav S, Irfan M, Ahmad A, Hayat S (2011) Causes of salinity and plant manifestations to salt stress: a review. J Environ Biol 32:667–685

Zaki F (2011) The determinants of salinity tolerance in maize (Zea mays L.). University of Groningen, pp 11–15, 56

Zare M, Azizi MH, Bazrafshan F (2011) Effect of drought stress on some agronomic traits in ten barley (Hordeum vulgare) cultivars. Tech J Eng Appl Sci 1:57–62

Zhang T, Zhan X, Kang Y, Wan S, Feng H (2017) Improvements of soil salt characteristics and nutrient status in an impermeable saline–sodic soil reclaimed with an improved drip irrigation while ridge planting Lycium barbarum L. J Soils Sediments 17:1126–1139

Zheng M, Tao Y, Hussain S, Jiang Q, Peng S (2016) Seed priming in dry direct-seeded rice: consequences for emergence, seedling growth and associated metabolic events under drought stress. Plant Growth Regul 78:167–178

Zhu JK (2002) Salt and drought stress signal transduction in plants. Annu Rev Plant Biol 53:247–273

Zhu G, Gu Y, Shi Y (2021) Wang Plant hydraulic conductivity determines photosynthesis in rice under PEG-induced drought stress. Pak J Bot 53:409–417

Zlatev Z, Lidon FC (2012) An overview on drought induced changes in plant growth, water relations and photosynthesis. Emir J Food Agric 24:57–72

Multiple Stresses Are a Big Challenge for the Development of Tolerant Varieties: Shared and Unique Physiological Responses

2

Manorama Kesineni, Keerthi Sree Yama, Michael James Lindgow, and Nita Lakra

Abstract

Increased crop failures caused by unpredictability in the weather pose an immediate threat to livelihood systems dependent on agriculture that are already at threat from a significant increase in the number and intensity of different stress factors. Stress factors include combinations of abiotic and biotic stressors that simultaneously or sequentially impact plants and microbiomes, causing a significant decrease in plant growth, yield, and overall health. Recent research reveals that, with the increasing number and complexity of stressors simultaneously impacting the plant, growth and survival decline dramatically. New studies bring to light the fact that plant reactions to multiple stresses differ from individual stress reactions. Additionally, plants have regulated responses to both individual stress and combinations of stresses and make necessary physiological and molecular changes for its survival and sustainability of output. Plants exhibit shared responses which are common to individual and combination of stresses. Learning mechanisms are crucial for understanding how plants adapt under multiple stresses. The knowledge of these responses can be utilized for the development of crop varieties with broad spectrum stress tolerance.

Keywords

Drought · Salinity · Heat and stress tolerance

M. Kesineni · K. S. Yama · M. J. Lindgow · N. Lakra (✉)
College of Biotechnology, CCS Haryana Agricultural University, Hisar, Haryana, India

A. Kumar et al. (eds.), *Salinity and Drought Tolerance in Plants*,
https://doi.org/10.1007/978-981-99-4669-3_2

2.1 Introduction

Plants are simultaneously subjected to a variety of biotic and abiotic stressors in the field, which reduce crop yield and productivity. Numerous abiotic stressors, especially heat, cold, drought, and salinity, have a negative impact on plants (Zhu 2016; Qin et al. 2020). Stress combinations rather than individual stresses have been recognized as serious hazards faced by plants and caused high agroeconomic losses (Rizhsky et al. 2004; Mittler 2006; Kissoudis et al. 2014; Suzuki et al. 2014; Mahalingam 2015; Ramegowda and Senthil-kumar 2015; Lohani et al. 2020). Simultaneous occurrence of different abiotic stresses results in deployment of stress-adaptation strategies which are different and sometimes contrasting to those seen under individual stresses. For example, under combined drought and heat stress, *Arabidopsis thaliana* plants accumulate sucrose instead of proline (Rizhsky et al. 2004). The enhanced transpiration to cool leaf surface during heat stress aggravates the effects of concurrent drought and salinity because increased transpiration rate leads to more water loss and increased uptake of salts (Rizhsky et al. 2002; Mittler 2006). So, focus is diverted toward understanding of plant responses under combined stress conditions to develop plants with better adaptation under field conditions.

2.2 Prevailing Stress Responses in Plants

Stressors or complex stress events lead to the various stress-response phases such as the following:

1. **Response phase:** alarm reaction (beginning of stress)
   (a) Deviation of the functional norm
   (b) Decline of vitality
   (c) Catabolic processes exceeding anabolism
2. **Restitution phase:** stage of resistance (continuing stress)
   (a) Adaptation processes
   (b) Repair processes
   (c) Hardening (reactivation)
3. **End phase:** stage of exhaustion (long-term stress)
   (a) Stress intensity too high
   (b) Overcharge of the adaptation capacity
   (c) Chronic disease or death
4. **Regeneration phase:** partial or full regeneration of the physiological function, when the stressor is removed and the damage was not too high

The plants respond to the onset of stress by altering one or more physiological processes, such as photosynthesis, the movement of metabolites, and/or the uptake and translocation of ions. The plants diverge from their typical physiological standard as a result of this drop in metabolic activity, which lowers their vitality. Plants with low or no stress tolerance systems and consequently low resistance minima will experience acute damage and senescence quickly. However, most plants will start their stress coping mechanisms during this alarm period, including adjusting their metabolic fluxes, starting their repair processes, and making long-term metabolic and morphological changes. The general alarm syndrome (GAS), which is an optimal physiological stage under the influence of the stressor and corresponds to the plants' greatest resistance, will lead to a hardening of the plants by establishing a new physiological norm. The stage of exhaustion (end phase), in which physiology and vitality gradually deteriorate, appears when long-term stress and a stress dose overload the plants' stress coping mechanism. Cells will eventually die as a result of this extreme damage. However, the plants regenerate and adapt to new physiological norms when the stresses are removed at the appropriate moment before the senescence processes take over. The time and stage of exhaustion, at which the stressors are removed from the plant, determines to which new physiological standard, within the range of resistance minimum and maximum, the plants will follow. Plant resistance to stress can be subdivided into escape (successful reproduction before the onset of severe stress), avoidance (maintenance of a high plant water status during stress, e.g., caused by stomatal closure, trichomatous nature, reduced leaf area senescence of older leaves, etc. or by increased root growth), and tolerance (the maintenance of plant function at limited water availability and/or the recovery of plant water status and plant function after stress) that may involve osmotic adjustments but may also be the result of rigid cell walls or small cells. Water transport within a plant occurs under tension as determined by soil water availability, atmospheric vapor pressure deficit, and creating turgor pressure within cells. Physiological adjustments that maintain turgor pressure are important under changing environmental conditions. All of these factors are influenced by the activity of various transporters such as aquaporins, which are integral membrane proteins that function as channels to transfer select small solutes and water (Maurel et al. 2008; Vandeleur et al. 2014).

Different abiotic stresses such as cold, heat, drought, flooding, and salt can cause common cell disturbances and secondary stresses, including membrane damage, oxygen reactive species generation and damage, protein denaturation, and osmotic stress at a cellular level. The versatility of the response to different unfavorable conditions in plants is related to an intricate network that comprises different levels, such as cellular, physiological, and morphological defenses. Some authors suggest that the most generalized cellular responses in plants rely on cuticle, membrane lipid desaturation for membrane remodeling, activation of antioxidant enzymatic and nonenzymatic systems against ROS (reactive oxygen species), induction of molecular chaperones, and accumulation of osmolytes compatible with cell demand (He et al. 2018).

Most of the mechanisms for sensing and triggering adaptive responses to multiple abiotic stresses are associated with changes in proteins and lipids in biological membranes. Adverse conditions impose ultrastructural changes in biomolecules, which are sensed by receptors or specialized proteins, converging, typically, to Ca^{2+} accumulation in cytosol, as well as REDOX imbalances. These signals activate kinase cascades and other secondary events that stimulate phosphorylation/dephosphorylation cascades, culminating on TF activation and gene expression remodeling. Several enzymes involved in the biosynthesis of osmolytes, pigments, thermoprotectants, and other secondary metabolites are also activated, in addition to ROS detoxifying enzymes, which attempt to reestablish redox homeostasis in cells subjected to stress. At the systemic level, morphological changes are detected in the root and leaves, such as the proliferation of lateral roots in response to drought, and biochemical changes, such as the secretion of phytochelatins in response to heavy metals, as well as stomatal reclusion and closure, in addition to reduction of leaf area and abscission. If acclimation response mechanisms are not efficient enough to restore plant homeostasis, cells trigger early senescence, redirecting nutrient flow to reproductive tissues and seed generation through their stress-triggered cell death programs.

2.3 Plant Responses Under Individual Stresses

Plants generally apply unique strategies to counteract an individual stress. For example, under heat stress, transpiration increases to reduce the canopy temperature, which is attributed to increased leaf number and surface (Pandey et al. 2015). In contrast, under drought stress, plants undergo acclimatization by reducing water deprivation and increasing water uptake mediated by reduced vegetative growth and enhanced root system (Prasad et al. 2008; Pandey et al. 2015). However, under saline stress, plants have to counteract osmotic stress by accumulating/synthesizing compatible osmolytes or by compartmentalizing toxic ions such as sodium and chloride in vacuole (Kumar et al. 2017). It is naive to expect that plants under the combination of different stresses employ different adaptive mechanisms. Plants become acclimatized to drought stress through increasing the consumption and minimizing the loss of water. On the other side, adaptive strategy of heat tolerance involves rising transpiration to reduce canopy temperature, which is achieved through increased leaf number and area (Pandey et al. 2015). Multiple stresses resulted in the lowest stomatal conductance, followed by drought, salinity, and heat, respectively. These findings are partially in agreement with those of Rizhsky et al. (2002) who found that stomata of *Nicotiana tabacum* plant remained closed under either individual or multiple stresses of heat and drought. We observed that heat stress gave rise to higher value of stomatal conductance compared with those in all other stresses. Rizhsky et al. (2002) also found that heat shock increased stomatal conductance even higher than that of the control in the tobacco plant. Compared with the control, drought and multiple stresses improved water use efficiency, whereas heat and salinity stresses reduced it. Our findings are parallel to those of Alghamdi et al. (2015) who found that severe drought stress improved water use efficiency of *Vicia faba* genotypes.

2.4 Drought

Drought is an interval of time, in the order of months or years in duration, during which the actual moisture supply at a given time or place consistently falls short of the climatically appropriate moisture supply (Palmer 1965). The duration of drought appears to play a leading role in affecting the plant spikelet sterility, absence of dehiscence, inhibition of pollen tube elongation, reduced starch mobilization, inferior efficiencies of photosynthesis resulting from stomatal closure, and decreases in chlorophyll content and deficiencies in root growth explained in Table 2.1 (Zhou et al. 2007; Pirdashti et al. 2009). It will overcome by increased water uptake and conductance and decreases in water potential that arise from the accumulation of osmolytes (Degenkolbe et al. 2009; Price et al. 2002).

Table 2.1 Plant responses under individual stresses

Physiological parameter and processes	Individual stress		References
	Heat	**Drought**	
Chlorophyll fluorescence (F_v/F_m)	Decreased	Decreased	Abdelhakim et al. (2021)
Relative water content	Decreased	Decreased	
Water use efficiency	Decreased	Decreased	
Leaf gas exchange	Increased	Increased	
Transpiration rate (E)	Decreased	Decreased	
Difference between leaf and air cuvette temperature (ΔT)	Decreased	Decreased	
Vapor pressure deficit (VPD)	Decreased	Decreased	
	Heat	**Salinity**	
Growth (fresh weight of roots and shoots)	Slightly decreased	Decreased	Rivero et al. (2014)
Na^+ uptake (stem and leaves)	No change	Increased in stems and leaves	
Na^+ uptake (root)	No change	Increased	
K^+ uptake in stem, leaves, and roots	No change	Decreased	
Water potential	Decreased	Decreased	
CO_2 assimilation	Decreased	Decreased	
Transpiration rate	Increased	Decreased	
Photosynthetic efficiency (Fv/Fm)	Decreased	Decreased	
Photosynthetic efficiency (PS-II)	Decreased	Decreased	
	Drought	**Salinity**	
Plant growth	Decreased	Decreased	Ahmed et al. (2013)
Photosynthetic efficiency	Decreased	Decreased	
Chlorophyll content	Chl a, b reduced	Chl a, b reduced	
Na^+ uptake in root	–	Increased	

2.5 Salinity

Saline soils have a high content of chlorites and sulfates of magnesium, calcium, and sodium. The increase in Na^+ and Cl^- ions in guard cells, which induces a reduction in turgor pressure, is projected to be one of the primary causes of diminished stomatal conductance (increased resistance) (Moradi and Ismail 2007). Consequently, the net photosynthetic rate of the plant decreases due to the reduced conductance of CO_2 to photosynthetic tissue. A separate physiological change that is instigated due to salt stress is the increased efflux of reactive oxygen species (ROS). ROS-scavenging enzymes such as the superoxide dismutases (SODs) manifest reduced activity in rice seedlings during salt stress explained in Table 2.2. Due to the susceptibility of plant physiology to ROS with the oxidation of enzymatic systems including photosynthetic proteins and premature programmed cell death (PCD), the maintenance of pools of antioxidants such as ascorbate and glutathione is upregulated during salt stress, with the enhanced transcription of genes mediating the regeneration processes of key antioxidants (Moradi and Ismail 2007).

Table 2.2 Responses under combined stresses

Physiological parameters and processes	Combined stress	References
	Heat + drought	
Leaf expansion/size	Increased	Vile et al. (2012)
Leaf number	Decreased	
Leaf dry matter content	Increased	
Leaf insertion angle	Increased	
Stomatal density	Decreased	
Stomatal index	Decreased	
Biomass allocation	Decreased in vegetative, increased in roots and reproductive parts	
CO_2 assimilation	More than salt but less than heat	
Transpiration rate	Unaffected	
Chlorophyll fluorescence (F_v/F_m)	Decreased	Abdelhakim et al. (2021)
Relative water content	Decreased	
Water use efficiency	Decreased	
Leaf gas exchange	Increased	
Transpiration rate (E)	Decreased	
Difference between leaf and air cuvette temperature (ΔT)	Decreased	
Vapor pressure deficit (VPD)	Decreased	
	Drought + salinity	
Plant growth	Decreased	Ahmed et al. (2013)
Photosynthetic efficiency	Decreased	
Chlorophyll content	Chlorophyll b reduction	
Na^+ uptake in root	Increase and more than salt stress alone	

2.6 Heat

Heat stress is defined as the rise in soil and air temperature beyond a threshold level for a minimum amount of time such that permanent harm to plant growth and development occurs. Increased plasma membrane permeability, disturbed thylakoids, empty mitochondria, and spoiled PS-II were also observed (Zhang et al. 2005). The adaptive features are cell number as well as cell size reductions, reduction in stomatal aperture size and xylem vessels in roots and shoots, and enhanced density of trichomes on both the surfaces of the leaf (Anon et al. 2004).

2.7 Plant Responses Under Combined Stress

When two stresses occur concurrently, the adaptation strategy of plants to stress combination is governed by the interaction of two stresses which is conceived by plants as a new state of stress (Mittler 2006). Thus, adaptation strategies of plants to combined stress may be different from that of two individual stresses. The overall effect of stress combination on plants depends largely on the age of plant, the inherent stress resistant or susceptible nature of plant, and severity of two stresses involved. Plant responses to stress combination are primarily determined by the more severe stress (dominant stressor) such that the physiological and molecular processes of plants subjected to combined stress resemble with those observed under more severe individual stress. The shared responses under combined stresses constitute the generic morphophysiological and molecular events evoked by both stresses constituting stress combination. For example, drought, salinity, and chilling induce osmotic effect on plants resulting in induction of common physiological processes, one of which is accumulation of osmoprotectants. The other stress induced response shared by almost all abiotic stress conditions is the production of ROS. Heat and salt stress are known to commonly affect the transport and compartmentation of ions in plants. Drought and salinity stress evoke the generic response of creating a physiological water deficit in plants. Additionally, both stresses cause decreased CO_2 diffusion into chloroplast due to reduced stomatal opening leading to reduced carbon metabolism. Under drought, heat, and salinity, a decrease in root hydraulic conductivity, cuticular wax biosynthesis, and high potassium and sodium ratio, respectively, are the main adaptive features. Heat and drought stress differentially affect stomatal characteristics. Under combined heat and drought stress, stomata remained closed leading to increased leaf temperature of *Nicotiana tabacum* plants (Rizhsky et al. 2002). Plants under combined stress minimize leaf temperature in a unique way. Vile et al. (2012) reported that *A. thaliana* plants exposed to combined heat and drought stress adapt to heat stress by adjusting leaf orientation through increasing their leaf insertion angle. *A. thaliana* plants exposed to individual and combined heat and drought stresses increased stomatal density in response to drought stress which was reduced in response to heat stress. Under combined stress, stomatal density decreased (Vile et al. 2012). This suggests that in case of a stress combination constituting of two stresses differing in their severity, plant's physiological processes are apparently determined them or severe stress. The combined heat and drought stress led to higher leaf temperature in two genotypes of *T. aestivum*, Ofanto and Cappelli, which differ in water use efficiency (WUE). Cappelli is characterized by higher WUE and lower stomatal conductance compared to Ofanto. The combined stress led to a higher leaf temperature in Cappelli as compared to Ofanto (Aprile et al. 2013). This indicates that the effect of combined stress also varies among the genotypes of a particular plant species. The combined heat and drought stress have been shown to affect a number of physiological processes more severely than the individual stresses. Rizhsky et al. (2002) reported that *N. tabacum* plants exposed to simultaneous heat and drought stress led to greater suppression of photosynthetic activity as compared to individual stresses. Similarly, as compared to individual stresses, combined heat and drought stress lead to enhanced lipid peroxidation in *Lotus japonicas* (Sainz et al. 2010) and severe abnormalities in the ultrastructure of chloroplasts and mitochondria in *T. aestivum* (Szucs et al. 2010; Grigorova et al. 2012). The combined stress also led to a greater reduction in photosynthetic activity and enhanced production of ROS in *Populus yunnanensis* (Li et al. 2014) and greater diminution in root viability and photochemical efficiency of photosystem II (PS-II) in *Festuca arundinacea* (Jiang and Huang 2001). The reduction in photosynthetic activity is a response shared between the individual heat and drought stresses. However, photosynthesis is less affected by heat stress and only high temperatures (>40 °C) are known to be detrimental. Heat stress mediated reduction in photosynthesis mainly occurs due to enhanced photorespiration (Prasad et al. 2008), reduced RuBisCO activity (Salvucci and Crafts-Brandner 2004), and reduced PS-II activity. Heat stress did not reduce photosynthetic activity of tobacco plants, but drought stress and combined heat and drought stress led to more than 80% reduction in photosynthetic activity (Rizhsky et al. 2002). The RuBisCO activity in *Cicer arietinum* leaves was increased with heat stress and decreased with drought stress and combined stress (Fig. 2.1) (Awasthi et al. 2014).

Similarly, Sainz et al. (2010) reported significant disruption in PS-II function when *L. japonicas* plants were subjected to combined heat and drought stress. Jiang and Huang (2001) compared the response of *F. arundinacea* and *Lolium perenne* to combined heat and drought stress and observed that stress combination led to enhanced reduction in photochemical efficiency of PS-II, as compared to individual stresses. The modulation of mitochondrial respiration is also a shared response under drought and heat stress (Prasad et al. 2008). The rate of dark respiration increased with increasing temperatures, whereas drought stress reduced plant respiration rates (Bryla et al. 2001). Similar observations were made by Rizhsky et al. (2002) who found that drought stress led to suppression of respiration but heat and combined drought and heat stress led to enhancement of respiration in *N. tabacum* leaves.

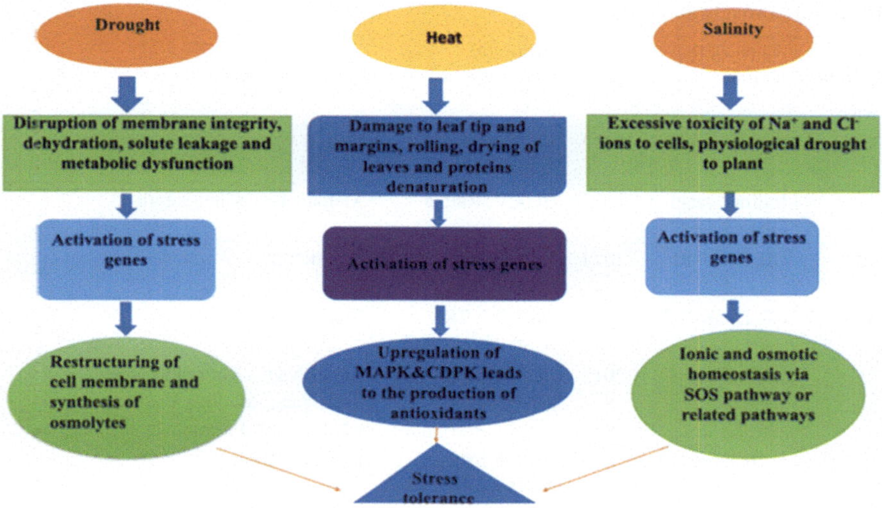

Fig. 2.1 Different stress responses

2.8 Crops Developed for Individual and Multiple Stresses

Simultaneous occurrence of different stresses results in deployment of stress-adaptation strategies which are different and sometimes contrasting to those seen under individual stresses. Now focus should be diverted toward understanding plant responses under combined stress conditions for better adaptation under field conditions or development of crop plants with multiple stress tolerance. The various research institutes are now involved and trying to develop crops for combined stress.

Different institutes developed the varieties for different crops for stresses listed in Tables 2.3, 2.4, and 2.5.

2.9 Challenges

Science and technology need to play an essential role in meeting the challenge of moving the globe into safe operating space where agriculture can satisfy the food needs of every single individual, especially those living in agriculturally marginal ecosystems. Although technological innovation has contributed significantly toward the development of stress tolerant varieties showing tolerance to different types of abiotic stresses, the imperative need is to ascertain strategic plans for the development of varieties possessing multiple stress tolerance. Moreover, the most important challenge for current system of agriculture and seed production is that the dissemination and adoption of stress tolerant varieties (STVs) is low mainly due to the lack of information and nonavailability of seeds among farmers (Zaidi et al. 2018; Emerick et al. 2016; Waza et al. 2016). The existing system of agricultural extension

Table 2.3 Drought tolerant crops

S. No.	Name of the variety	Year	Institute/ organization	Grown in states
Rice				
1.	Satyabhama	2012	ICAR-NRRI, Cuttack	Orissa
2.	Indira Barani Dhan 1	2012	IGKV, Raipur	Maharashtra, Chandigarh
3.	CR Dhan 40	2014	ICAR-NRRI, Cuttack	Jharkhand
4.	CR Dhan 202	2014	ICAR-NRRI, Cuttack	Jharkhand, Orissa
5.	DRR Dhan 42	2015	ICAR-IIRR, Hyderabad	Tamil Nadu, Andhra Pradesh, Telangana, Madhya Pradesh, Chandigarh, Jharkhand
6.	DRR Dhan 44	2015	ICAR-IIRR, Hyderabad	Haryana, Bihar
7.	Sabour Shree	2015	BAU, Ranchi	Bihar
8.	Purna	2017	ICAR-NRRI, Cuttack	Gujarat
9.	DRR Dhan 47	2018	ICAR-IIRR, Hyderabad	Telangana, Andhra Pradesh, Karnataka, Kerala, Pondicherry
10.	DRR Dhan 50	2018	ICAR-IIRR, Hyderabad	Andhra Pradesh, Telangana, Tamil Nadu, Karnataka, Bihar, Orissa, Chandigarh, Uttar Pradesh, Madhya Pradesh
11.	Tripura Khara 1	2018	Research Complex for NEH, Lembucherra	Tripura
12.	CR Dhan 801	2019	ICAR-NRRI, Cuttack	Andhra Pradesh, Telangana, Orissa, Uttar Pradesh, West Bengal
13.	CR Dhan 210	2020	ICAR-NRRI, Cuttack	Orissa
Wheat				
14.	Netravati	2011	MPKV, Rahuri	Maharashtra, Karnataka
15.	HD 2987	2011	IARI, New Delhi	Maharashtra, Karnataka, Andhra Pradesh, Goa, Tamil Nadu
16.	KRL-213	2012	ICAR-CSSRI, Karnal	Assam, Bihar, Delhi, Haryana, Uttar Pradesh, Rajasthan, Punjab, West Bengal, Uttarakhand
17.	HD 3043	2012	IARI, New Delhi	Haryana, Rajasthan, Uttar Pradesh, Delhi, Rajasthan, Uttarakhand, Madhya Pradesh
18.	PBW 644	2012	PAU, Ludhiana	Delhi, Haryana, Himachal Pradesh, Jammu and Kashmir, Punjab, Rajasthan, Uttar Pradesh, Uttarakhand
19.	WH 1142	2015	CCS, Hissar	Nwpz
20.	Sabour Nirjal	2017	BAU, Sabour	Bihar

(continued)

Table 2.3 (continued)

S. No.	Name of the variety	Year	Institute/ organization	Grown in states
21.	HUW 669	2018	BHU, Varanasi	Uttar Pradesh
22.	DBW 252	2020	ICAR-IIBWR, Karnal	Uttar Pradesh, Bihar, Jharkhand, Orissa, West Bengal, Assam
Maize				
23.	JM-216	2000	ZARS, Chhindwara	Karnataka
24.	Pratap Hybrid Makka-1	2003	AICMIP, Udaipur	Rajasthan
25.	Pratap Makka-3	2005	AICMIP, Udaipur	Gujarat, Madhya Pradesh, Rajasthan
26.	Pratap Makka-5	2006	AICMIP, Udaipur	Gujarat, Madhya Pradesh, Rajasthan
27.	HQPM-1	2007	AICRP	Chandigarh, Haryana
28.	Bajauramakka	2008	ARS, Himachal Pradesh	Himachal Pradesh, Assam, Jammu and Kashmir, Sikkim, Uttarakhand
29.	HQPM-5	2020	IARI, New Delhi	Andhra Pradesh, Karnataka, Gujarat, Telangana, Tamil Nadu, Madhya Pradesh, Uttar Pradesh, Rajasthan
Barley				
30.	RD-2592	2004	ARS, Durgapur	Rajasthan, Haryana, Uttar Pradesh
31.	JB-58	2005	JNKV, Jabalpur	Madhya Pradesh
32.	RD 2660	2006	ARS, Durgapur	Rajasthan, Haryana, Uttar Pradesh
Pearl millet				
33.	HHB 223	2010	CCSHAU, Hissar	Rajasthan, Gujarat, Haryana, Punjab, Delhi, Uttar Pradesh, Madhya Pradesh
34.	HHB 216	2010	CCSHAU, Hissar	Rajasthan, Gujarat, Haryana
35.	HHB 226	2011	CCSHAU, Hissar	Rajasthan, Gujarat, Haryana, Madhya Pradesh, Punjab
36.	RHB 177	2011	AICPMIP, Durgapura	Aicpmip, Durgapura
37.	Bio 70	2012	Biore Pvt. Ltd. Hyderabad	Rajasthan, Gujarat, Haryana
38.	HHB 234	2013	CCSHAU, Hissar	Rajasthan, Gujarat, Haryana
39.	Punjabh 306	2017	PrabhatAgri Biotech Ltd	Maharashtra, Karnataka, Andhra Pradesh, Telangana, Tamil Nadu
40.	Balwan	2018	Nuziveedu seeds Ltd. Hyderabad	Rajasthan
41.	NBH 4903	2018	Nuziveedu seeds Ltd. Hyderabad	Maharashtra, Karnataka, Andhra Pradesh, Telangana, Tamil Nadu
Sorghum				
42.	Parbanimoti	2002	VNMKV, Parbhani	Maharashtra

(continued)

Table 2.3 (continued)

S. No.	Name of the variety	Year	Institute/ organization	Grown in states
43.	Phule Vasudha	2007	MPKV, Rahuri	Maharashtra, Karnataka
44.	Phule Chitra	2008	MPKV, Rahuri	Maharashtra, Karnataka
45.	CSV-17	2009	AICRP, Sorghum	Rajasthan, Andhra Pradesh, Karnataka, Tamil Nadu
46.	Pant Chari 5	2010	GBPUAT, Uttarakhand	Uttarakhand
47.	Phule Panchami	2010	MPKV, Rahuri	Maharashtra, Karnataka
48.	Pant Chari 7	2011	GBPUAT, Uttarakhand	Uttarakhand
49.	DSV-2	–	AICRP, Sorghum	Cvrc, Karnataka
50.	PS-4	–	UAS, Bengaluru	Karnataka
51.	SIA-326	–	UAS, Bengaluru	Karnataka

and seed supply needs to be aligned in a way to successfully deliver the proven stress tolerant technology to farmers. Agricultural extension system needs reformations to enhance the diffusion and adoption of STVs. Research based strategies ought to be implemented so that STVs are adopted at a scale where it can achieve maximal benefits. Strategic and intentional collaborations should be ensured for enhancing the diffusion of STVs to farmers. These strategies have the potential to ensure food security especially under the climatically vulnerable agroecologies.

2.10 Conclusions and Prospects for Future

Plants have evolved sophisticated adaptive mechanisms to withstand diverse and complex abiotic stresses. With the advent of new technologies such as genomics and genetic transformation, significant progress has been made in understanding these complex traits in higher plants. Significant advancements in crop genome characterization and the optimization of genome editing technology in crops have and will continue to advance our understanding and capabilities toward development of stress tolerant crops. Ultimately, genome editing or transgenic approaches need to be combined with efforts using conventional and marker-assisted breeding activities to achieve the desired improved varieties. Further, it is important to take into account climate change models, which differ geographically, to guide breeding programs in target trait identification for selection and identification of new adapted germplasm for tolerance to multiple stresses (Harrison et al. 2014). These efforts will lead to tangible practical outcomes that may help mitigate the effects of climate change, especially with respect to drought, salinity, and heat stresses, and will contribute to improved crop productivity and food security.

Table 2.4 Salinity tolerant crops

S. No.	Name of the variety	Year	Institute/ organization	Grown in states
Rice				
1.	Luna Suvarna	2010	ICAR-NRRI, Cuttack	Orissa
	CR Dhan 403			
2.	Luna Sampad	2010	ICAR-NRRI, Cuttack	Orissa
	CR Dhan 402			
3.	DRR Dhan 39	2010	ICAR-IIRR, Hyderabad	Orissa, Kerala, Gujarat
4.	CSR 43	2011	ICAR-CSSRI, Karnal	Uttar Pradesh
5.	Luna Barial	2012	ICAR-NRRI, Cuttack	Orissa
	CR Dhan 406			
6.	Luna Sankhi	2012	ICAR-NRRI, Cuttack	Orissa
	CR Dhan 405			
7.	GNR-5	2018	NAU, Navsari	Gujarat
8.	CSR 46	2018	ICAR-CSSRI, Karnal	Uttar Pradesh
9.	CSR 60	2018	ICAR-CSSRI, Karnal	Uttar Pradesh, Punjab
10.	CSR 56	2018	ICAR-CSSRI, Karnal	Haryana, Uttar Pradesh
11.	CSR 52	2019	ICAR-CSSRI, Karnal	Uttar Pradesh
12.	Panvel 3	–	BSIV, Dapoli	Konkan Region of Maharashtra
13.	CARI Dhan 5	–	CARI, Port Blair	Andaman and Nicobar Islands
Wheat				
14.	KRL 213	2011	ICAR-CSSRI, Karnal	Bihar, Delhi, Haryana, Uttar Pradesh, Rajasthan, Punjab, West Bengal, Uttarakhand
15.	KRL 210	2012		

(continued)

Table 2.4 (continued)

S. No.	Name of the variety	Year	Institute/ organization	Grown in states
			ICAR-CSSRI, Karnal	Bihar, Delhi, Haryana, Uttar Pradesh, Rajasthan, Punjab, West Bengal, Uttarakhand
16.	KRL 283	2018	ICAR-CSSRI, Karnal	Uttar Pradesh

Table 2.5 Multiple stress tolerant crops

S. No.	Name of the variety	Year	Institute/organization	Tolerance
Rice				
1.	Kalinga 1	1973	ICAR-NRRI, Cuttack	Drought, Cold
2.	MTU 1010	2000	RARS, Maruteru	Submergence, drought
3.	DRR Dhan 47	2018	ICAR-IIRR, Hyderabad	High temperature, drought
4.	DRR Dhan 50	2018	ICAR-IIRR, Hyderabad	Drought, submergence
5.	DRR Dhan 52	2019	ICAR-IIRR, Hyderabad	High temperature, drought
6.	CR Dhan 801	2019	ICAR-NRRI, Cuttack	Drought, submergence
Wheat				
7.	A-9-30-1	1974	AAU ARS, Arnej	Drought, heat
8.	Ajanta	1983	CoA, Badnapur	Drought, heat
9.	KRL 19	2000	ICAR-CSSRI, Karnal	Salinity, water logging
10.	Netravati	2011	MPKV, Rahuri	Drought, lodging
11.	KRL 19	2011	ICAR-CSSRI, Karnal	Submergence, salinity
12.	KRL 210	2012	ICAR-CSSRI, Karnal	Submergence, salinity
13.	KRL-213	2012	ICAR-CSSRI, Karnal	Drought, salinity
14.	WH 1142	2015	CCS, Hissar	Drought, lodging
15.	KRL 283	2018	ICAR-CSSRI, Karnal	Submergence, salinity
Pear millet				
16.	GHB-538	2005	JAU, Jamnagar	Drought, lodging
17.	Rhb-177	2011	AICPMIP, Durgapura	Drought, lodging
18.	HHB-234	2013	CCSHAU, Hissar	Drought, lodging
Pigeon pea				
19.	UPAD-120	1976	GBPUA&T, Pantnagar	High temperature, salinity
20.	Bahar	1986	RAU, Pusa	High temperature, salinity
21.	MAL 13	2005	BHU, Varanasi	Drought, salinity
Chick pea				
22.	JG-315	1984	JNKV, Jabalpur	High temperature, drought
23.	JG-11	1999	JNKV, Jabalpur	High temperature, drought
24.	JG-6	2006	JNKV, Jabalpur	High temperature, drought
25.	GNG 1581	2008	ARS, Sriganganagar	Drought, water logging
26.	JG-14	2009	JNKV, Jabalpur	High temperature, drought
Lentil				
27.	Kota Masoor 3	2020	AU, Kota	Drought, high temperature

References

Abdelhakim LOA, Palma CFF, Zhou R, Wollenweber B, Ottosen CO, Rosenqvist E (2021) The effect of individual and combined drought and heat stress under elevated CO_2 on physiological responses in spring wheat genotypes. Plant Physiol Biochem 162:301–314

Ahmed IM, Cao F, Zhang M, Chen X, Zhang G, Wu F (2013) Difference in yield and physiological features in response to drought and salinity combined stress during anthesis in Tibetan wild and cultivated barleys. PLoS One 8:e77869

Alghamdi SS, Al-Shameri AM, Migdadi HM, Ammar MH, El-Harty EH, Khan MA, Farooq M (2015) Physiological and molecular characterization of faba bean (Viciafaba L.) genotypes for adaptation to drought stress. J Agron Crop Sci 201(6):401–409

Anon S, Fernandez JA, Franco JA et al (2004) Effects of water stress and night temperature preconditioning on water relations and morphological and anatomical changes of Lotus creticus plants. Sci Hortic 101:333–342

Aprile A, Havlickova L, Panna R, Mare C, Borrelli GM, Marone D et al (2013) Different stress responsive strategies to drought and heat in two durum wheat cultivars with contrasting water use efficiency. BMC Genomics 14:821

Awasthi R, Kaushal N, Vadez V, Turner NC, Jens B, Siddique KHM et al (2014) Individual and combined effects of transient drought and heat stress on carbon assimilation and seed filling in chickpea. Funct Plant Biol 41:1148–1167

Bryla DR, Bouma TJ, Hartmond U, Eissenstat DM (2001) Influence of temperature and soil drying on respiration of individual roots in citrus: integrating greenhouse observations into a predictive model for the field. Plant Cell Environ 24:781–790

Degenkolbe P, Do PT, Zuther E, Repsilber D, Walter D, Hincha K, Khol KI (2009) Expression profiling of rice cultivars differing in their tolerance to long-term drought stress. J Plant Mol Biol 69:133–153

Emerick K, de Janvry A, Sadoulet E, Dar MH (2016) Technological innovations, downside risk, and the modernization of agriculture. Am Econ Rev 106:1537–1561

Grigorova B, Vassileva V, Klimchuk D, Vaseva I, Demirevska K, Feller U (2012) Drought, high temperature, and their combination affect ultrastructure of chloroplasts and mitochondria in wheat (*Triticumaestivum L.*) leaves. J Plant Interact 7:204–213

Harrison MT, Tardieu F, Dong Z, Messina CD, Hammer GL (2014) Characterizing drought stress and trait influence on maize yield under current and future conditions. Glob Chang Biol 20:867–878

He M, He CQ, Ding NZ (2018) Abiotic stresses: general defenses of land plants and chances for engineering multistress tolerance. Front Plant Sci 9:1771

Jiang Y, Huang B (2001) Physiological responses to heat stress alone or in combination with drought: a comparison between tall fescue and perennial ryegrass. HortScience 36:682–686

Kissoudis C, van de Wiel C, Visser RGF, van der Linden G (2014) Enhancing crop resilience to combined abiotic and biotic stress through the dissection of physiological and molecular cross talk. Front Plant Sci 5:207

Kumar J, Singh S, Singh M, Srivastava PK, Mishra RK, Singh VP, Prasad SM (2017) Transcriptional regulation of salinity stress in plants: a short review. Plant Gene 11:160–169

Li X, Yang Y, Sun X, Lin H, Chen J, Ren J et al (2014) Comparative physiological and proteomic analyses of poplar (Populus yunnanensis) plantlets exposed to high temperature and drought. PLoS One 9:e107605

Lohani N, Jain D, Singh MB, Bhalla PL (2020) Engineering multiple abiotic stress tolerance in canola, Brassica napus. Front Plant Sci 11:3

Mahalingam R (2015) Consideration of combined stress: a crucial paradigm for improving multiple stress tolerance in plants. In: Combined stresses in plants. Springer, Cham, pp 1–25

Maurel C, Verdoucq L, Luu DT, Santoni V (2008) Plant aquaporins: membrane channels with multiple integrated functions. Annu Rev Plant Biol 59:595–624

Mittler R (2006) Abiotic stress, the field environment and stress combination. Trends Plant Sci 11: 15–19

Moradi F, Ismail AM (2007) Responses of photosynthesis, chlorophyll fluorescence and ROS-scavenging systems to salt stress during seedling and reproductive stages in rice. Ann Bot 99:1161–1173

Palmer WC (1965) Meteorological drought. Research Paper No. 45. Department of Commerce, Washington, DC. 58 pp

Pandey P, Ramegowda V, Senthil-Kumar M (2015) Shared and unique responses of plants to multiple individual stresses and stress combinations: physiological and molecular mechanisms. Front Plant Sci 6:723

Pirdashti H, Sarvestani ZT, Bahmanyar MA (2009) Comparison of physiological responses among four contrast rice cultivars under drought stress conditions. World Acad Sci Eng Technol 49:52

Prasad PVV, Staggenborg S, Ristic Z (2008) Impacts of drought and/or heat stress on physiological, developmental, growth, and yield processes of crop plants. In: Ahuja LR, Reddy VR, Saseendran SA, Yu Q (eds) Response of crops to limited water: understanding and modeling water stress effects on plant growth processes: advances in agricultural systems modeling. Series 1. ASA-CSSA-SSSA, Madison, WI, pp 301–356

Price AH, Cairns JE, Horton P, Jones HG, Griffiths H (2002) Linking drought resistance mechanisms to drought avoidance in upland rice using a QTL approach, progress and new opportunities to integrate stomatal and mesophyll responses. J Exp Bot 53:989–1004

Qin H, Li Y, Huang R (2020) Advances and challenges in the breeding of salt-tolerant rice. Int J Mol Sci 21(21):8385

Ramegowda V, Senthil-kumar M (2015) The interactive effects of simultaneous biotic and abiotic stresses on plants: mechanistic understanding from drought and pathogen combination. J Plant Physiol 176:47–54

Rivero RM, Mestre TC, Mittler R, Rubio F, Garcia Sanchez F, Martinez V (2014) The combined effect of salinity and heat reveals a specific physiological, biochemical and molecular response in tomato plants. Plant Cell Environ 37:1059–1073

Rizhsky L, Liang H, Mittler R (2002) The combined effect of drought stress and heat shock on gene expression in tobacco. Plant Physiol 130:1143–1151

Rizhsky L, Liang H, Shuman J, Shulaev V, Davletova S, Mittler R (2004) When defense pathways collide. The response of Arabidopsis to a combination of drought and heat stress. Plant Physiol 134:1683–1696

Sainz M, Diaz P, Monza J, Borsani O (2010) Heat stress results in loss of chloroplast Cu/Zn superoxide dismutase and increased damage to photosystem II in combined drought-heat stressed Lotus japonicus. Physiol Plant 140:46–56

Salvucci ME, Crafts-Brandner SJ (2004) Inhibition of photosynthesis by heat stress: the activation state of Rubisco as a limiting factor in photosynthesis. Physiol Plant 120:179–186

Suzuki N, Rivero RM, Shulaev V, Blumwald E, Mittler R (2014) Tansley review abiotic and biotic stress combinations. New Phytol 203:32–43

Szucs A, Jager K, Jurca ME, Fabian A, Bottka S, Zvara A et al (2010) Histological and microarray analysis of the direct effect of water shortage alone or combined with heat on early grain development in wheat (Triticum aestivum). Physiol Plant 140:174–188

Vandeleur RK, Sullivan W, Athman A, Jordans C, Gilliham M, Kaiser BN et al (2014) Rapid shoot-to-root signalling regulates root hydraulic conductance via aquaporins. Plant Cell Environ 37: 520–538

Vile D, Pervent M, Belluau M, Vasseur F, Bresson J, Muller B et al (2012) Arabidopsis growth under prolonged high temperature and water deficit: independent or interactive effects? Plant Cell Environ 35:702–718

Waza SA, Jaiswal HK, Sravan T, Bano DA, Priyanka K, Singh PK, Umesh U (2016) Heterosis for yield and quality traits in rice (*O. sativa L.*). J Appl Nat Sci 8:1510–1522

Zaidi NW, Singh M, Kumar S, Sangle UR, Singh R, Prasad R, Singh SS, Singh S, Yadav AK, Singh A et al (2018) Trichoderma harzianum improves the performance of stress tolerant rice varieties in rainfed ecologies of Bihar, India. Field Crop Res 220:97–104

Zhang JH, Huang WD, Liu YP et al (2005) Effect of temperature acclimation pretreatment on the ultrastructure of mesophyll cells in young grape plants (*Vitis vinifera L. cv. Jingxiu*) under cross temperature stresses. J Integr Plant Biol 47:959–970

Zhou Y, Lam HM, Zhang J (2007) Inhibition of photosynthesis and energy dissipation induced by water and high light stresses in rice. J Exp Bot 58:1207–1217

Zhu JK (2016) Abiotic stress signaling and responses in plants. Cell 167(2):313–324

Physio-Biochemical Approaches for Raising Drought Tolerance in Plants: Recent Progress and Future Perspectives

3

Sanjoy Shil

Abstract

Terrestrial plants are subject to a wide range of biotic and abiotic stress that drastically affects their growth, development, and ultimate productivity all over the world. Drought stress above all is one of the major constraints toward plant as they are sessile in nature. With the rapid increase in population, there is an urgent need to improve drought tolerance in plants. Drought tolerance have long been regarded as a complex mechanism between physio-morphological and biochemical strategies at cell, tissue, and whole plant level which includes improvement in root system (prolific and deep root penetration) to increase water uptake, increasing diffusive resistance to reduce water loss, leaf anatomical and morphological changes (leaf rolling, cuticle thickness, small and succulent leaves) to reduce transpirational water loss, phenotypic plasticity, osmotic accumulation and osmotic adjustment (OA) with low-molecular-weight osmolytes (glycine betaine, proline, and other amino acids), stomatal conduction and regulation, maintenance or alterations of relative water status and electrolytic leakage, hormonal balances and phytohormone signaling pathways, ROS scavenging (production of ROS and accumulation of free radicals), and application of exogenous chemicals, hormones and osmoprotectants etc. are the key mechanism to cope against drought stress. Although very recently transgenic approaches are one of the probabilistic tools that improve the level of drought tolerance in several plants, such approaches may be achieved only when there is a basic understanding as well as investigation of physio-biochemical mechanism of plants toward drought stress. Thus, an interdisciplinary approach including all mechanistic biological tools like breeding, physiology, biochemistry, molecular and gene engineering, etc. will be pertinent to raise the level of drought tolerance in plants. This present

S. Shil (✉)
Bidhan Chandra Krishi Viswavidyalaya (Bankura Campus), Bankura, West Bengal, India

A. Kumar et al. (eds.), *Salinity and Drought Tolerance in Plants*,
https://doi.org/10.1007/978-981-99-4669-3_3

chapter explores the recent advances of morphological as well as physio-biochemical approaches to perk up the level of drought tolerance in plants and also highlighted the probabilistic approaches that may be developed in the near future.

Keywords

Drought stress tolerance · Physio-biochemical strategies · Future prospects

3.1 Introduction

Generally, drought stress occurs when the available water in the soil is reduced and atmospheric conditions cause unceasing loss of water by evapotranspiration. Drought is a major environmental threat to agricultural production globally and is caused by inadequate rainfall and/or inconsistent rainfall patterns. Most of the crops exhibit a broad genotypic variability and a wide range of injury in response to drought stress affecting millions of hectares of rainfed field crops in South and Southeast Asia. For instance, in some states of India, severe drought can cause as much as 40% yield loss especially in rice that are extremely sensitive to drought (Lafitte et al. 2007) and use about 30% of the freshwater for agricultural crops worldwide where drought adversely affects more than half of the rice production area (Plaut et al. 2004).

Drought stress in plant leads to reduction in growth and development which depends on cell division, enlargement, and differentiation and involves genetic, morphophysiological, biochemical, and phenological and metabolic events and their complex interactions that ultimately reduce yield and quality of the crop. So, there is a need to develop the capacity of a plant to adapt to various environments determined by its genome (the blueprint of the genetic characteristics). Drought tolerance has long been regarded as a complex trait mainly related to various physio-biochemical parameters such as root traits, osmotic adjustment (OA), and maintenance of plant water status. QTL for these component traits and yield under drought have been identified and are being employed in molecular breeding.

3.2 Physiological Strategies Associated with Drought Tolerance in Plants

3.2.1 Root Morphological Traits and Plasticity

Roots often faces a substantial drought stress even though water is available deeper in the soil and at the same time plants have physio-biochemical mechanisms and genetic regulation for root adaptation to drought that help the plants to identify specific genes and biochemical pathways gene-based marker selection to develop plants with better roots (rooting depth/hardpan penetration, rooting thickness and rooting distribution pattern etc). Plants may develop longer taproot that helps to reach soil layer where sufficient water is available for avoiding dehydration (Taylor et al. 1978) and increases the number of lateral roots per unit of taproot length so that there is an increase of biomass partitioning to roots and increases the root/shoot ratio.

3.2.2 Shoot Traits

Plants are proficient in raising the level of drought tolerance by means of improved water conduction. They achieve this by increasing the cross section of the vessels, having densely veined leaves, reducing transport distances through the formation of shorter internodes, reducing transpiration by closing stomata and increasing cuticle thickness, covering leaves with trichomes, reducing the number and size of leaves, enabling the reversible folding or curling of the leaf blades, shedding of leaves, or developing succulence in leaves. Plants can also cease their metabolic activities (anabiosis) and synthesize and accumulate protective substances, such as dehydrins (proteins with a stable structure under cell dehydration) or carbohydrates that stabilize the phospholipids of the cell membranes.

3.2.3 Stomatal Conductance and Gaseous Exchange

Stomata are the entrance of water loss and CO_2 absorbability. Drought stress causes stomatal closure that results in declined rate of photosynthesis due to less CO_2 absorbability, and thus stomatal closure is one of the earliest responses to drought stress and causes restriction of gas exchange under drought concomitant with the reduced water potential and turgor associated with even a small decrease in RWC (Naithani et al. 2012). Stomatal conductance is affected by drought (Naithani et al. 2012; Flexas and Medrano 2002), and it is the key physiological shoot trait that influences leaf CO_2 diffusion, electron transport rate, water vapor exchange, carboxylation efficiency, WUE, respiration, transpiration, etc. Under drought stress, leaf water status always interacts with stomatal conductance, and there is a correlation between leaf water potential and stomatal conductance. Drought stress significantly decreases the relative leaf expansion rate, stomatal conductance, and leaf turgor which ultimately increase ABA content in the leaf and xylem (Liu et al. 2003).

Drought stress results in root suberization (Singh et al. 2012) and reduction of root meristematic activity, thus arresting root elongation. Atmospheric drought regulates stomatal closure, whereas the combination of atmospheric and surface drought regulates leaf transpiration (Naithani et al. 2012), and drought-induced stomatal closure is a limitation for photosynthesis. Under drought stress, stomatal closure and reduction in mesophyll conductance results in a reduction of CO_2 to the carboxylation site, and thus photosynthesis is affected (Tardieu et al. 2010; Wang et al. 2012).

3.2.4 Leaf Water Potential (PWP)

Leaf water potential (LWP) is the measure of whole plant water status and has long been predicted as an indicator of dehydration avoidance (Pantuwan et al. 2002). LWP for drought tolerance is always high that minimizes the effects of drought on spikelet sterility and thus on grain yields (Jongdee et al. 2002). When water deficit in leaves goes beyond a certain threshold level, the stomata closes by lowering the rate of transpiration. Stomatas help to regulate water loss when the tissue water status becomes too low, thereby minimizing the severity of water shortage in plants. Thus, higher LWP is maintained by stomatal closure, and varietal differences in stomatal response to water status have been reported (Jongdee et al. 1998).

3.2.5 Relative Water Content (RWC) and Water Use Efficiency (WUE)

RWC, leaf water potential, osmotic potential, pressure potential, and transpiration rate are the major attributes that are affected by drought (Liu et al. 2004). Water potential significantly declined in roots, leaves, and pods under drought, but root water potential dropped much earlier than in leaves and pods (Ouvrard et al. 1996). Genes controlling root traits and stomatal development and guarding cell movements strongly impact water use efficiency (WUE) and represent the best targets for molecular breeding programs. Very recently, plant varieties having higher level of drought tolerance and/or WUE have been developed through conventional breeding approaches using molecular assisted selection (Hu and Xiong 2014). Conventional breeding has used selection of the drought-resistant genotypes as major QTL donors for introgression into high-yielding/drought-susceptible varieties (Steele et al. 2006).

High WUE may enhance crop productivity under drought, and there is a positive correlation between WUE and total biomass yield under drought which suggests that improvement of the WUE of a crop plant should result in superior yield performance if a high HI can be maintained (Wright 1996). Therefore, understanding on the physiological basis for stomatal regulation and improved WUE in drought-stressed plants is required (Liu et al. 2005). In turn, a plant with higher WUE reduces the rate of photosynthesis due to reduced rate of transpiration and consequently slows the rate of plant growth (Condon et al. 2004). Currently, there is an extensive use of genotypes with increased WUE (Pereira et al. 2006).

Transpiration efficiency (assimilation or dry matter accumulation per unit of water transpired) is another related trait to WUE (Fischer 1981). TE is under genetic control and excludes amount of water lost by soil evaporation and hence should be considered as a potential trait.

3.2.6 Osmolyte Accumulation and Osmotic Adjustment

Osmotic adjustment is the active accumulation of solutes (organic and inorganic) that occurs in cytosol in response to drought maintaining the leaf turgor when tissue water potential declines (Rhodes and Samaras 1994). Drought stress altered leaf water relations by decreasing leaf water potential (Ψw), osmotic potential (Ψs), turgor osmotic potential ($Ψs^{100}$), and RWC but increasing osmotic adjustment (OA). OA has been shown to improve stomatal conductance for efficient intake of CO_2 (Kiani et al. 2007) and induce root's ability to uptake more water (Chimenti et al. 2006). It also maintains photosynthesis at lower water potentials, delayed leaf senescence and death, reduced flower abortion, improved root growth, and increased water extraction from the soil as water deficit develops (Turner et al. 2001).

Such osmotic adjustment is caused by accumulation of certain osmolytes (proline, sucrose, mannitol, glycine betaine, soluble carbohydrates, free amino acids, inorganic cations, organic acids, and other solutes) that improve the uptake of water from moisture deficit soil; however, these adjustments strongly depends on the intensity of water stress.

Among these osmolytes, proline accumulation and mobilization is the first response of plants exposed to water-deficit stress to raise drought tolerance. Proline acts as a signaling molecule to modulate mitochondrial functions, influence cell proliferation or cell death, and activate specific gene expression making the plant to recover from stress (Szabados and Savoure 2009). Increase in proline content and decrease in hydrogen peroxide and lipid peroxidation implied its protective role in tolerance against drought stress. Proline exhibits three main roles under stress, i.e., as a signaling molecule, an antioxidative defense molecule, and a metal chelator (Hayat et al. 2012). Thus, proline content can be used as a marker to screen for drought screening tolerance.

3.2.7 Root-Shoot Signaling

Drought stress reduces the root-shoot growth and alters their distribution that is an effect of root-shoot signaling (Novák and Lipiec 2012) where roots stimulate a signal cascade to the shoots via xylem. Phytohormones (ABA, cytokinins, and ethylene) act as signal molecules involved in the physiological growth regulation (Schachtman and Goodger 2008), and the closure of stomata is greatly regulated by abscisic acid, cytokinins, etc. (Daszkowska-Golec and Szarejk 2013), and thus they have been implicated in the root to shoot signaling through transpiration stream. Among these phytohormones, ABA is recognized as an important root-shoot stress signal (Schachtman and Goodger 2008) as during drought stress, ABA is synthesized by the roots and channeled in the xylem to the shoot where it suppresses leaf expansion leading to closure of stomata (Wang et al. 2000). ABA also promotes the efflux of K^+ from the guard cells that leads to loss of turgor pressure, and drought stress causes an increase in ABA level up to 50-fold due to deprivation of cell turgor or cell membrane disruption (Guerrero and Mullet 1986). Currently, it is understood that there is a drought-induced root-to-leaf signaling through the transpiration stream which results in stomatal closure.

3.2.8 High-Throughput Phenomics

High-throughput phenotyping is one of the basic tools that help to standardize specific collection of phenotypic data in plants. With the advancement of recent technologies like genomics, proteomics, metabolomics, and transcriptomics, phenotyping has become limiting (Salas Fernandez et al. 2017). Plant performance in the field is highly heterogeneous, and phenotyping at canopy level under stress environment explores more accurate data on plants. Recently, high-throughput phenotyping platforms (HTPPs) are used for successful phenotyping for physio-morphological traits under stress (Rutkoski et al. 2016) that makes a bridge between various omics technologies, and such phenotyping is easily accessible, more accurate, and more economic (Araus and Cairns 2014; Fahlgren et al. 2015; Bai et al. 2016; Zhang et al. 2017). Thus, there is a need for comprehensive high-throughput phenotyping of physio-morphological traits to improve the breeding efficiency of drought-tolerant crops.

3.3 Biochemical Strategies Associated with Drought Tolerance in Plants

3.3.1 Non-Stomatal Mechanisms

Non-stomatal mechanisms include alteration in chlorophyll and carotenoid biosynthesis and chloroplast structure; accumulation, transport, and distribution of assimilates; changes in chlorophyll stability index and membrane stability index; and membrane injury. Chlorophyll a and b and carotenoid content reduction under drought is a typical symptom of oxidative stress resulting in the photooxidation and degradation of the photosynthetic pigments. Such stress has the ability to reduce the tissue concentrations of chlorophylls and carotenoids (Kiani et al. 2008), primarily with the production of ROS in the thylakoids (Reddy et al. 2004).

3.3.2 Antioxidative Defense Mechanism and Drought Tolerance

Plant develops an antioxidative defense system that protects plant cells from oxidative damage by scavenging ROS while they are under drought (Fig. 3.1).

Such stress leads to an imbalance between antioxidant defenses and the amount of reactive oxygen species (ROS) resulting in oxidative stress by producing ROS that cause membrane injuries, protein degradation, and enzyme inactivation. The enhanced activities of different components of the antioxidant system include antioxidant enzymes (superoxide dismutase (SOD), peroxidase (POD/POX),

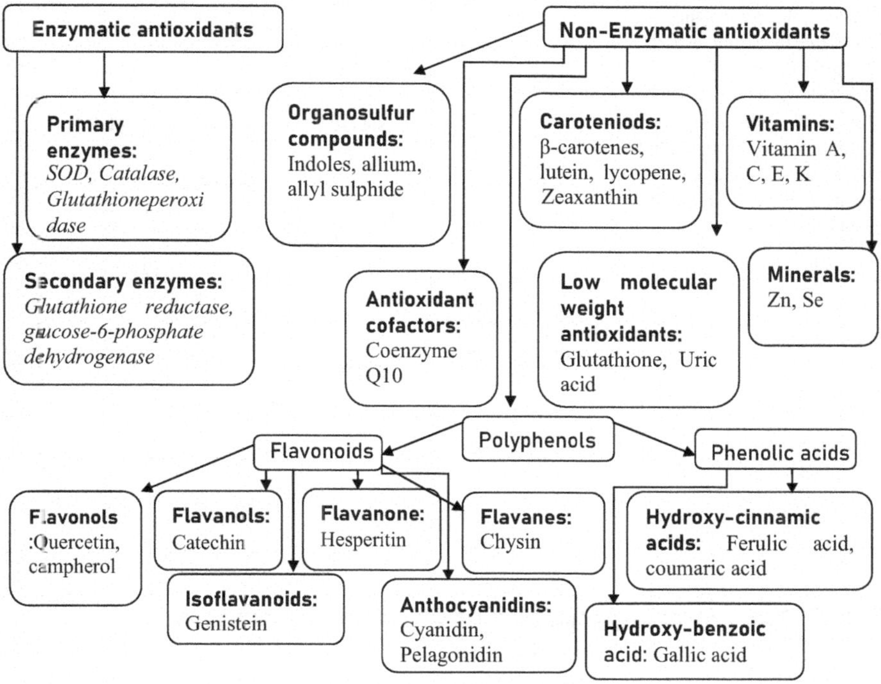

Fig. 3.1 Antioxidative defense strategies by various enzymatic and nonenzymatic antioxidants

catalase (CAT), ascorbate peroxidase (APX), glutathione reductase (GR), monodehydroascorbate reductase (MDHAR), dehydroascorbate reductase (DHAR), glutathione peroxidase (GPX), guaiacol peroxidase (GOPX), glutathione-S-transferase (GST), etc.) present in chloroplasts, mitochondria, peroxisomes, cytosol, and stroma, and nonenzymatic elements (ascorbic acid, glutathione (GSH), phenolic compounds, alkaloids, nonprotein amino acids, carotenoid, α-tocopherols, thioredoxin, polyamines (PAs), salicylates (SAs), vitamin E, compatible solutes such as proline, GB, and zeaxanthin, etc.) decrease oxidative damage caused by ROS production (Fig. 3.1) and develop resistance and improve the drought tolerance in plants (Cruz de Carvalho 2008; Yazdanpanah et al. 2011).

For example, glutathione peroxidase (GPx) is a group of selenium dependent enzymes, and it consists of cytosolic, plasma, phospholipid hydroperoxide, and gastrointestinal glutathione peroxidase. GPx (cellular and plasma) catalyzes the reaction of H_2O_2 by reduced glutathione (GSH), and as a result, oxidized glutathione (GSSG) is produced, and it is again recycled to its reduced form by glutathione reductase (GR) and reduced nicotinamide adenine dinucleotide phosphate (NADPH) (Fig. 3.2).

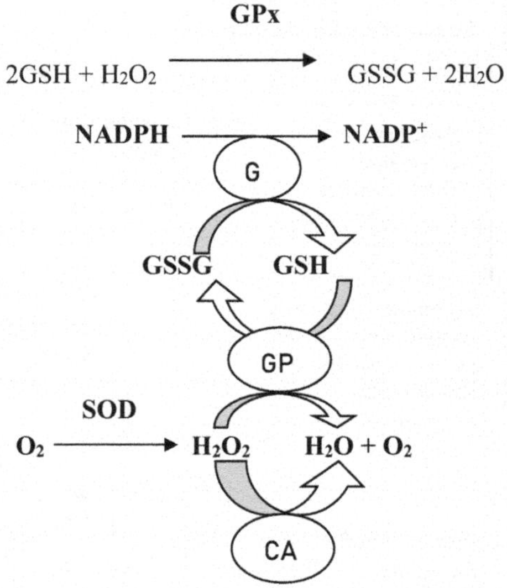

Fig. 3.2 Strategies in removing free radical by primary enzymatic antioxidants. *GR* glutathione reductase, *GPX* glutathione peroxidase, *SOD* superoxide dismutase, *CAT* catalase, *GSH* glutathione, *GSSG* oxidized glutathione

3.3.3 Exogenous Application of Chemical Substances

The combined management strategies such as nutrient management, exogenous use of growth regulators (gibberellic acids, jasmonic acid, brassinosteroids, salicylic acid, etc.), specific exogenous osmoprotectants (proline, glycine betaine, trehalose, etc.), antioxidants (ascorbic acid, GSH, tocopherols, etc.), signaling molecules (NO, H_2O_2, etc.), polyamines (spermidine, spermine, putrescine, etc.), and trace elements (Se, Si, etc.) help in improving drought tolerance, and they are also found effective in mitigating drought-induced damage in plants. Among these, nutrient management is regarded as a quick and more effective and better strategy to tackle drought stress. Exogenous application of brassinolide, uniconazole, and methyl jasmonate improved the drought tolerance with increased activities of SOD, CAT and APX, ABA, and total improved carotenoid contents in maize (Li et al. 1998), while methyl jasmonate brought about a threefold increase in the β-carotene synthesis as well as degradation of the chlorophyll contents in the epidermal peels (Pérez et al. 1993).

3.3.3.1 Application of Hydrogels

Hydrogels are super polymeric absorbent that retains considerable amount of water, and their exogenous application causes increase in water availability, water retention capacity of soil, and water uptake and decreases nutrient loss by percolation and leaching. It may also enhances the this way soil aeration, soil permeability, infiltration rates, reducing irrigation frequency and water loss, decreasing soil erosion, and drainage and a faster rate of plant root and shoot growth. In general, cellulose, pectin, chitin, and carboxymethyl cellulose (CMC) are the natural macromolecules having higher potential to absorb water to form hydrogels.

3.3.3.2 Nutrient Application

Nutrients N, P, K, Ca, Mg, Zn, Cu, and Si play an important role in developing drought tolerance in plant by regulating water relation; maintaining charge balance, electron carriers, structural components, and enzyme activation; and providing osmoticum for turgor and growth regulation and metabolism. N, K, Zn, Mg, Ca, and Si reduce the toxicity of ROS by concomitant increase in the concentration of antioxidants like superoxide dismutase (SOD), catalase (CAT), and peroxidase (POD) in plant and thereby improve drought tolerance in plants. P, K, Mg, and Zn improve the root growth that increases the water intake and thereby regulate water relation and improve stomatal conductance. Ca also helps to maintain high tissue water potential under drought condition and improve drought tolerance by osmotic adjustment. K^+ greatly improves in stomata regulation under drought by increasing extra-chloroplastic K^+ concentrations with surplus application of K^+ that could prevent photosynthesis inhibition under moisture stress (Egilla et al. 2005).

3.3.3.3 Application of Growth Hormones and Osmoprotectants

Osmoprotectants (proline, glycinebetaine (GB), polyamines, brassinolides, etc.) and phytohormones (salicylic acid, ABA, GA_3, etc.) have been found to improve drought tolerance with elevated osmotic adjustment to maintain turgor and antioxidant accumulation to detoxify ROS to maintain stability of membrane structures, enzymes, and other macromolecules under drought stress (Anjum et al. 2011). Proline as an osmoregulatory solute acts as osmoprotectant under drought stress, and its concentration is increasing in stressed plants due to stimulation of proline biosynthesis. Increase in proline content and decrease in hydrogen peroxide and lipid peroxidation implied its protective role against drought stress and improve tolerance. Exogenous GA application can enhance leaf photosynthetic capacity, stomata conductance, and transpiration rate of cotton crop under stress (Lichtfouse et al. 2009). SA is a secondary metabolite that induces drought tolerance in plants by regulating several physiological processes through signaling (Senaratna et al. 2000; Singh and Usha 2003). Exogenous application of methyl salicylic acid on leaves of water-stressed perennial plants promotes leaf senescence, as programmed leaf senescence in drought-stressed plants contributes to nutrient remobilization, thus allowing the rest of the plant to benefit from the nutrients accumulated during the life span of the leaf (Abreu and Munne-Bosch 2008). Polyamines are associated with the response of plants to drought via signaling (Bae et al. 2008). Silicon (Si) application promotes plant growth under drought stress with notable improvement in nutrient uptake and antioxidants and other osmoprotectants (Hattori et al. 2005; Gunes et al. 2008; Shen et al. 2010).

3.3.3.4 Application of Antitranspirants and Other Chemicals

Antitranspirants (kaolin, PMA, ABA, waxes, and silicon oil) help to defend against drought stress. KCl, KH_2PO_4, $CaCl_2$, or thiourea also helps in seed hardening, and Cycocel (CCC) improves drought tolerance. Biofertilizers, viz., *Azospirillum* or phosphobacteria, are also useful to ameliorate drought stress.

3.4 Conclusion and Future Perspectives

Prolonged droughts cause production of ROS in chloroplasts, mitochondria, and peroxisomes and attack cell biomolecules (DNA, lipids, proteins, and carbohydrates) and ultimately lead to cell death under severe stress. Plants can mechanize themselves to tolerate under such undesirable situation through enzymatic and nonenzymatic defense system, by accumulating osmolytes (organic or inorganic) and osmoprotectants (proline and glycine betaine). In recent era, plant biologists are trying to develop stress-tolerant plants without compromising yield and productivity through transgenic approaches exhibiting differential capabilities for ROS production and elimination and look for molecular markers and gene manipulation for stress tolerance where antioxidants, ROS, and osmolytes play a functional network to develop stress tolerance. To induce drought resistance in plants, the mechanism of action and optimal concentrations of exogenous chemicals needs to be understood along with the appropriate plant developmental stage. The role of H_2O_2 as a signaling molecule and identification of regulatory components in the signaling pathways are fundamental clues for future research on stress management. Even though physiological strategies of drought tolerance in plants are relatively well understood, further studies to resolve physiological basis of assimilate partitioning from source to sink, phenotypic flexibility, and high-throughput phenotyping are preferred arena of research. Modern omics tools like genomics, proteomics, metabolomics, and ionomics will be helpful to understand molecular basis of plant response to drought stresses and their tolerance. Thus, an integrated approach is essential to improve drought tolerance level in plants, and researchers must use the latest genomic resources in combination with physio-biochemical techniques and bioinformatics with an ecophysiological understanding of the interactions between plant genotypes and environment.

References

Abreu ME, Munne-Bosch S (2008) Salicylic acid may be involved in the regulation of drought induced leaf senescence in perennials: a case study in field-grown *Salvia officinalis* L. plants. Environ Exp Bot 64:105–112

Anjum SA, Wang C, Farooq M, Hussain M, Xue L (2011) Brassinolide application improves the drought tolerance in maize through modulation of enzymatic antioxidants and leaf gas exchange. J Agron Crop Sci 197:177–185. https://doi.org/10.1111/j.1439-037X.2010.00459.x

Araus JL, Cairns JE (2014) Field high-throughput phenotyping: the new crop breeding frontier. Trends Plant Sci 19:52–61. https://doi.org/10.1016/j.tplants.2013.09.008

Bae H, Kim S-H, Kim MS, Sicher RC, Lary D, Strem MD, Natarajan S, Bailey BA (2008) The drought response of *Theobroma cacao* (cacao) and the regulation of genes involved in polyamine biosynthesis by drought and other stresses. Plant Physiol Biochem 46:174–188

Bai G, Ge Y, Hussain W, Baenziger PS, Graef G (2016) A multi-sensor system for high throughput field phenotyping in soybean and wheat breeding. Comput Electron Agric 128:181–192. https://doi.org/10.1016/j.compag.2016.08.021

Chimenti CA, Marcantonio M, Hall AJ (2006) Divergent selection for osmotic adjustment results in improved drought tolerance in maize (*Zea mays* L.) in both early growth and flowering phases. Field Crops Res 95:305–315

Condon AG, Richards RA, Rebetzke GJ, Farquhar GD (2004) Breeding for high water-use efficiency. J Exp Bot 55:2447–2460

Cruz de Carvalho MH (2008) Drought stress and reactive oxygen species: production, scavenging and signaling. Plant Signal Behav 3:156–165. https://doi.org/10.4161/psb.3.3.5536

Daszkowska-Golec A, Szarejk OI (2013) Open or close the gate—stomata action under the control of phytohormones in drought stress conditions. Plant Cell Biol 4:1–16

Egilla JN, Davies FT, Boutton TW (2005) Drought stress influences leaf water content, photosynthesis, and water-use efficiency of *Hibiscus rosa-sinensis* at three potassium concentrations. Photosynthetica 43:135–140. https://doi.org/10.1007/s11099-005-5140-2

Fahlgren N, Gehan MA, Baxter I (2015) Lights, camera, action: high-throughput plant phenotyping is ready for a close-up. Curr Opin Plant Biol 24:93–99. https://doi.org/10.1016/j.pbi.2015.02.006

Fischer RA (1981) Optimizing the use of water and nitrogen through breeding of crops. Plant Soil 58:249–278

Flexas J, Medrano H (2002) Drought inhibition of photosynthesis in C3 plants: stomatal and non-stomatal limitations revisited. Ann Bot 89:183–189

Guerrero F, Mullet JE (1986) Increased abscisic acid biosynthesis during plant dehydration requires transcription. Plant Physiol 80:588–591

Gunes A, Kadioglu YK, Pilbeam DJ, Inal A, Coban S, Aksu A (2008) Influence of silicon on sunflower cultivars under drought stress, II: essential and non essential element uptake determined by polarized energy dispersive X-ray fluorescence. Comm Soil Sci Plant Anal 39:1904–1927

Hattori T, Inanaha S, Araki H, An P, Morita S, Luxova M, Lux A (2005) Application of silicon enhanced tolerance in *Sorghum bicolor*. Physiol Plant 123:459–466

Hayat S, Hayat Q, Alyemeni MN, Wani AS, Pichtel J, Ahmad A (2012) Role of proline under changing environments: a review. Plant Signal Behav 7(11):1456–1466

Hu H, Xiong L (2014) Genetic engineering and breeding of drought-resistant crops. Ann Rev Plant Biol 65:715–741

Jongdee B, Fukai S, Cooper M (1998) Genotypic variation for grain yield of rice under water-deficit conditions. In: Michalk DL, Pratley JE (eds) Agronomy, growing a greener future. Presented at the proceedings of 9th Australian agronomy conference, Wagga, pp 403–406

Jongdee B, Fukai S, Cooper M (2002) Leaf water potential and osmotic adjustment as physiological traits to improve drought tolerance in rice. Field Crop Res 76:153–163

Kiani SP, Talia P, Maury P, Grieu P, Heinz R, Perrault A, Nishinakamasu V, Hopp E, Gentzbittel L, Paniego N, Sarrafi A (2007) Genetic analysis of plant water status and osmotic adjustment in recombinant inbred lines of sunflower under two water treatments. Plant Sci 172:773–787

Kiani SP, Maury P, Sarrafi A, Grieu P (2008) QTL analysis of chlorophyll fluorescence parameters in sunflower (*Helianthus annuus* L.) under well-watered and water-stressed conditions. Plant Sci 175:565–573

Lafitte HR, Yongsheng G, Yan S, Li ZK (2007) Whole plant responses, key processes, and adaptation to drought stress: the case of rice. J Exp Bot 58:169–175

Lichtfouse E, Navarrete M, Debaeke P, Souchère V, Alberola C, Ménassieu J (2009) Agronomy for sustainable agriculture. A review. Agron Sustain Dev 29:1–6. https://doi.org/10.1051/agro:2008054

Li L, Van Staden J, Jager AK (1998) Effects of plant growth regulators on the antioxidant system in seedlings of two maize cultivars subjected to water stress. Plant Growth Regul 25:81–87. https://doi.org/10.1023/A:1010774725695

Liu FL, Jensen CR, Andersen MN (2003) Hydraulic and chemical signals in the control of leaf expansion and stomatal conductance in soybean exposed to drought stress. Funct Plant Biol 30:65–73

Liu F, Jensen CR, Andersen MN (2004) Drought stress effect on carbohydrate concentration in soybean leaves and pods during early reproductive development: its implication in altering pod set. Field Crops Res 86:1–13

Liu F, Andersen MN, Jacobsen S-E, Jensen CR (2005) Stomatal control and water use efficiency of soybean (*Glycine max* L. Merr) during progressive soil drying. Environ Exp Bot 5433–5440

Naithani KJ, Ewers BE, Pendall E (2012) Sap flux-scaled transpiration and stomatal conductance response to soil and atmospheric drought in a semi-arid sagebrush ecosystem. J Hydrol 464–465:176–185

Novák V, Lipiec J (2012) Water extraction by roots under environmental stresses. In: Halasi-Kun J, Stekauerová V, Fodor I, Nagy V, Sinóros-Szabó B, Pinto RL (eds) Pollution and water resources, columbia university seminar proceedings: impact of anthropogenic activity and climate changes on the environment of central europe and USA. Slovak Academy of Sciences—Hungarian Academy of Sciences—Columbia University, San Francisco, CA

Ouvrard O, Cellier F, Ferrare K, Tousch D, Lamaze T, Dupuis JM, Casse-Delbart F (1996) Identification and expression of water stress and abscisic acid-regulated genes in a drought tolerant sunflower genotype. Plant Mol Biol 31:819–829

Pantuwan G, Fukai S, Cooper M, Rajatasereekul S, O'Toole JC (2002) Yield response of rice (Oryza sativa L.) genotypes to drought under rainfed lowlands: 2. Selection of drought resistant genotypes. Field Crop Res 73:169–180

Pereira JS, Chaves MM, Caldeira MC, Correia AV (2006) Water availability and productivity. In: Morison JIL, Morecroft MD (eds) Plant growth and climate change. Blackwell Publishing Ltd, Oxford, pp 118–145

Pérez HP, Acosta DE, Padilla RS, Acosta GJ (1993) Effect of drought on seed quality of common beans (*Phaseolus vulgaris* L.). Agric Técn Méx 25:107–114

Plaut Z, Butow BJ, Blumenthal CS, Wrigley CW (2004) Transport of dry matter into developing wheat kernels and its contribution to grain yield under post anthesis water deficit and elevated temperature. Field Crops Res 86:185–198

Reddy AR, Chaitanya KV, Vivekanandan M (2004) Drought induced responses of photosynthesis and antioxidant metabolism in higher plants. J Plant Physiol 161:189–1202

Rhodes D, Samaras Y (1994) Genetic control of osmoregulation in plants. In: Strange K (ed) In cellular and molecular physiology of cell volume regulation. CRC Press, Boca Raton, pp 347–361

Rutkoski J, Poland J, Mondal S, Autrique E, Pérez LG, Crossa J et al (2016) Canopy temperature and vegetation indices from high-throughput phenotyping improve accuracy of pedigree and genomic selection for grain yield in wheat. G3 (Bethesda) 6(9):2799–2808. https://doi.org/10.1534/g3.116.032888

Salas Fernandez MG, Bao Y, Tang L, Schnable PS (2017) A high-throughput, field-based phenotyping technology for tall biomass crops. Plant Physiol 174:2008–2022. https://doi.org/10.1104/pp.17.00707

Schachtman DP, Goodger JQD (2008) Chemical root to shoot signaling under drought. Trends Plant Sci 13:281–287

Senaratna T, Touchell D, Bunn E, Dixon K (2000) Acetyl salicylic acid (aspirin) and salicylic acid induce multiple stress tolerance in bean and tomato plants. Plant Growth Regul 30:157–161

Shen X, Zhou Y, Duan L, Li Z, Eneji AE, Li J (2010) Silicon effects on photosynthesis and antioxidant parameters of soybean seedlings under drought and ultraviolet-B radiation. J Plant Physiol 167:1248–1252

Singh B, Usha K (2003) Salicylic acid induced physiological and biochemical changes in wheat seedlings under water stress. Plant Growth Regul 39:137–141

Singh CM, Kumar B, Mehandi S, Chandra K (2012) Effect of drought stress in Rice: a review on morphological and physiological characteristics. Trends Biosci 5(4):261–265

Steele KA, Price AH, Shashidhar HE, Witcombe JR (2006) Marker-assisted selection to introgress rice QTLs controlling root traits into an Indian upland rice variety. Theor Appl Genet 112:208–221

Szabados L, Savoure A (2009) Proline: a multifunctional amino acid. Trends Plant Sci 15:89–97
Tardieu F, Parent B, Simonneau T (2010) Control of leaf growth by abscisic acid: hydraulic or non-hydraulic processes? Plant Cell Environ 33:636–647
Taylor HM, Burnett E, Booth GD (1978). Taproot elongation rates of soybeans Z Acker Pfl anzenbau Bd. 146
Turner NC, Wright GC, Siddique KHM (2001) Adaptation of grain legumes (pulses) to water limited environments. Adv Agron 71:193–123
Wang Z, Xiong Y, Wang M (2000) Study on expert system of irrigation forecast and decision making for water saving. Institute of Agricultural Soil Water Engineering of North West Science and Technology University of Agriculture Forestry, Yangling, Shaanxi
Wang ZX, Chen L, Ai J, Qin HY, Liu YX, Xu PL, Jiao ZQ, Zhao Y, Zhang QT (2012) Photosynthesis and activity of photosystem IIin response to drought stress in Amur grape (Vitis amurensis Rupr.). Photosynthetica 50:189–196
Wright G (1996) Review of ACIAR selection for water use efficiency in legumes project recommends further research. ACIAR Food Legume Newslett:2–3
Yazdanpanah S, Baghizadeh A, Abbassi F (2011) The interaction between drought stress and salicylic and ascorbic acids on some biochemical characteristics of *Satureja hortensis*. Afr J Agron 6:798–807
Zhang X, Huang C, Wu D, Qiao F, Li W, Duan L et al (2017) High-throughput phenotyping and QTL mapping reveals the genetic architecture of maize. Plant Physiol 173:1554–1564. https://doi.org/10.1104/pp.16.01516

Physiological, Biochemical, and Molecular Responses to Salt Stress and Seed Priming Approach to Enhance Salt Tolerance in Bread Wheat

4

Parul, Arvind Kumar, Shiksha Chourasia, Raj Kumar, Geeta Devi, Ashwani Kumar, Anita Mann, Parvender Sheoran, and Satish Kumar Sanwal

Abstract

Soil and water salinization induced by climate change and anthropogenic activities is a major threat to the wheat production, globally. Wheat is a major source of calories for two-thirds of the global population. Salt stress adversely affects seed germination, plant growth and development, and physiological processes through reduction in chlorophyll content, hormonal imbalance, alteration in metabolic activities, and modification in household gene expression, which ultimately decreases the yield of wheat. Therefore, detailed understanding of the effects of salt stress in wheat crop is essential to devise the remedial measures for reducing its adverse effect of the crop. In the past, various approaches including management of saline soils, selection of salt-tolerant germplasm, and development of salt-tolerant cultivars through conventional and molecular breeding approaches have been applied to a limited extent. However, various crop management practices, such as seed priming, nutrient management, and exogenous applications of organic (phytohormones, osmolytes) and inorganic chemicals, were found to be more effective in managing the adverse effects of salinity in wheat crop as the biological approaches are very costly, time consuming, and labor intensive. Therefore, in this chapter, we reviewed from available literature the salt stress induced effects on physiological, biochemical,

Parul · A. Kumar (✉) · S. Chourasia · R. Kumar · A. Kumar · A. Mann · S. K. Sanwal
ICAR – Central Soil Salinity Research Institute, Karnal, Haryana, India
e-mail: arvind.kumar2@icar.gov.in

G. Devi
Rashtriya Kisan Post Graduate College, Shamli, Uttar Pradesh, India

P. Sheoran
ICAR – Central Soil Salinity Research Institute, Karnal, Haryana, India

ICAR – Agricultural Technology Application Research Institute, Ludhiana, Punjab, India

A. Kumar et al. (eds.), *Salinity and Drought Tolerance in Plants*,
https://doi.org/10.1007/978-981-99-4669-3_4

and molecular responses of wheat and the advantage of seed priming with chemical in plant growth and grain yield improvement of wheat under salt stress.

Keywords

Wheat · Salinity · Abiotic stress

4.1 Introduction

Bread wheat (*Triticum aestivum* L.) is a major cereal crop of Poaceae family and is grown all over the world. During 2020–2021, ~736 million tons of wheat grain was harvested from ~217 million hectares of land, worldwide (FAO 2020). It is the important source of dietary fibers and provides nearly 20% of the daily total protein and calorie requirements in human diet (Shewry and Hey 2015). Globally, wheat production is hampered by various abiotic (drought, high or low temperature and salinity, etc.) and biotic factors, resulting in the adverse effect of the crop quality and productivity. Among them, soil salinity is one of the major limiting factors that results in the low yield and poor quality of grains. Climate change and poor quality irrigation water are the major factors responsible for soil salinization which leads to degradation and desertification of valuable land resources (Rubio et al. 2009). Moreover soil salinization has been identified as a major cause for decline in the productivity of irrigated and rain-fed lands across the world (Kumar et al. 2022). Worldwide, more than 20% of soils are salt-affected, and the anthropogenic

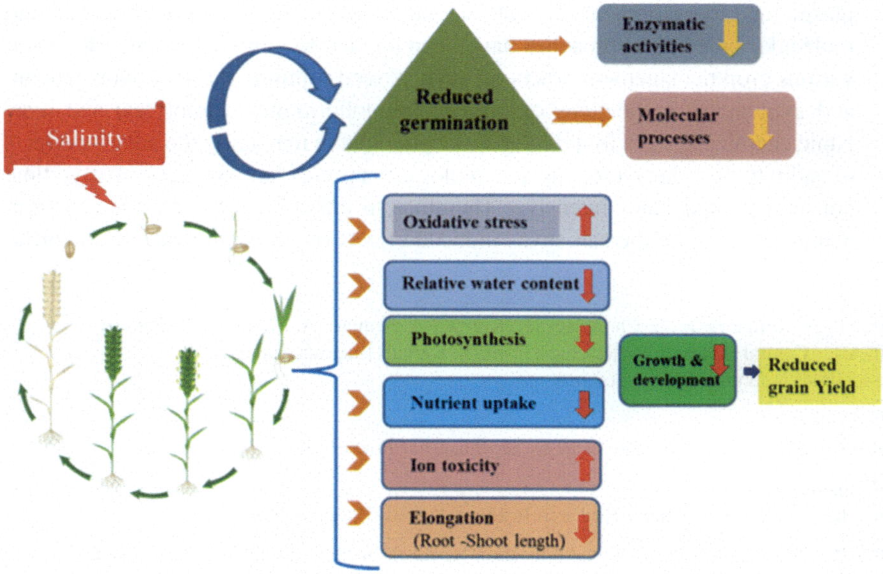

Fig. 4.1 Effect of salt stress on wheat

disturbances and changing climate are further increasing the extent of area under such soils (Ding et al. 2021). In salinity soils, EC above 6–8 dS m^{-1} adversely affects the growth and yield of wheat crops, leading to a decline in the overall productivity of wheat crop (Royo and Abió 2003) (Fig. 4.1). According to the estimate of Food and Agriculture Organization (FAO), salt stress has devastated 397 million hectares of wheat cultivation worldwide (World Health Organization 2019). Effects of salt stress depend on the severity and duration of stress, and the developmental stages of crop, which ultimately alters the various physiological and metabolic processes in plants (James et al. 2011). Initially, soil salinity-induced osmotic stress suppresses the early plant growth, which is followed by the development of ion generated toxicity in tissues. Once plants are exposed to salt stress, the osmotic stress is detected by the plants, which ultimately results in the deficiency of water and solutes in plants (hyperosmotic stress). It also causes rapid stomatal closure, which reduces the plant's ability to assimilate CO_2 and ultimately inhibits the photosynthesis in plants. Additionally, osmotic stress develops several physiological changes, such as interruption of membranes and deficiency of essential nutrients, and impairs the ability to detoxify reactive oxygen species (ROS), and alterations in antioxidant enzymes activities (Munns and Tester 2008). On the other hand, excess accumulation of Na^+ and Cl– ions in the plant tissues causes ionic stress (hyper-ionic stress) which leads to premature leaf senescence and death of plants in extreme cases. Na^+ toxicity results in the inhibition of enzymatic activities, thereby resulting in the impairing of plant metabolism, such as Calvin cycle. Presence of high level of Na^+ in the cytoplasm restricts uptake and transport of potassium (K^+) as well as other macro- and micronutrients, including N, P, Ca^{2+}, and Zn^{2+}. Moreover, osmotic and ionic stresses caused by the soil salts also result in the accumulation of reactive oxygen species (such as singlet oxygen, superoxide, hydroxyl radical, and hydrogen peroxide), which reflects adverse impacts on cellular structures and macromolecules like DNA, lipids, and enzymes that interrupt the vital cellular functions of plants.

Salt tolerance is a complex polygenic trait governed by several genetic factors (Arzani and Ashraf 2016). Rahman et al. (2005) reported that under certain conditions, crop shows improvement in their growth in salt stress such as increase in K^+, decrease of Na^+, optimization of the Na^+ and K^+ ion ratio, improvement in transpiration efficiency, regulation of osmotic potential, and development of antioxidant system of the exposed plants. Wheat, in general, is more sensitive to salinity, and it inhibits plant growth and development, resulting in the low productivity or even crop failure under extreme salinity conditions. Therefore, understanding the physiological basis of stress tolerance in plants is essential for devising the selection and breeding programs. Underlying mechanisms of wheat responses to salt stress at critical developmental stages are essential for improving breeding techniques and developing genetically modified salt-tolerant varieties. According to recent studies, the modification in leaf and stem anatomical features in different wheat genotypes is also an important adaptation trait under salt stress (Nassar et al. 2020). For instances, pigmented wheat genotypes with high anthocyanin content can maintain significantly higher dry matter production under salinity conditions, suggesting the significant role of phenolic compounds in imparting salinity tolerance in the crop (Mbarki et al. 2018). Moreover, worldwide, various approaches have been applied to improve crop performance under salt stress, including the introduction of desirable genes, genotype screening, genotype selection, and conventional and molecular breeding techniques (Hassan et al. 2018a, b; Chattha et al. 2020). Additionally, other methods, such as the use of osmoprotectants, nutrient management, seed priming, and hormone application, have also been found effective in alleviating salt stress in crops (Hasanuzzaman et al. 2017; Erdal et al. 2010). Therefore, the primary goal of this chapter is to examine recent research advancement in complex physiological, biochemical, and molecular mechanisms involved and amelioration approaches applied to increase salt stress tolerance in the bread wheat.

Effect of salt stress on plants: In this process, mainly two phases of response occurs in plant, i.e., osmotic effect and ion-specific effect.

4.2 Osmotic Effect

Once plants are exposed to salt stress, the concentration of neutral salts surrounding the plant roots increases which decreases the water potential and increases the osmotic pressure of soil, resulting in a reduction in the transportation of water from soil via root to plant shoot. This decrease in water uptake negatively affects the plant growth, reduces the leaf expansion, and slows down the emergence of new leaves in early plant growth stages. Another response of plants to salinity was early stomatal closure as a consequence of the osmotic effect (Munns and Tester 2008).

4.3 Ion-Specific Effect

The NaCl is the most neutral salt in soil solution that affects the plant growth. The ion-specific effect begins when salt levels in older leaves reach to a critical level causing cells to die (Munns et al. 2006; Munns and Tester 2008). It was observed that Na^+ accumulates more in leaves than in roots, and a high proportion of Na^+ in leaf causes metabolic disruption in plants, owing to toxicity of Na^+ and competition of Na + with K^+ at the binding sites, which is a cofactor that regulates the activity of over 50 enzymes in the cytoplasm of leaf cells (Tester and Davenport 2003). The activity of several enzymes in leaf cells is also inhibited when the Na^+ level in the leaf increased, which results in dead leaves (Fig. 4.2).

4.4 Mechanisms of Salinity Tolerance in Plants

Generally, plants possess four major strategies/mechanisms to cope up with salt stress, i.e., salt exclusion, salt extrusion, salt dilution, and compartmentation of ions. Salt-tolerant plants exhibit one or more of the mechanisms to counter the salt stress (Fig. 4.3).

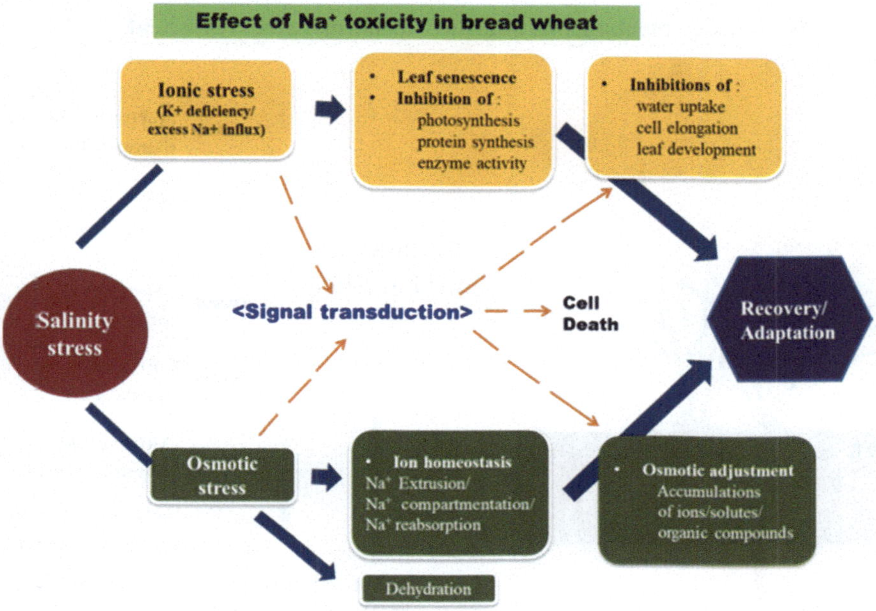

Fig. 4.2 Effects of sodium toxicity in bread wheat

4.5 Salt Exclusion

Salt exclusion is a process by which plants can maintain low levels of potentially harmful ions such as sodium by avoiding uptake in the leaves during salt stress conditions. Wheat plant maintains a significantly lower concentration of Na^+ and Cl– ions in the shoot than in the root, which is the key feature of salt tolerance in the species known as salt exclusion (Horie et al. 2009). At low level of salt stress, Na^+ ion exclusion mechanisms can provide tolerance in plants to a certain extent. Alternatively, in order to tolerate high-salinity levels, plants must be able to tolerate Na^+ toxicity as well as maintain a high cellular K^+ concentration (Zhang et al. 2008). To counteract the negative effects of salinity on plants, some ion transporters must be overexpressed or modified to reduce Na^+ import. The *HKT* gene found in the *Nax2* locus is responsible for the salt exclusion mechanism in wheat. For instance, Demidchik et al. (2002) observed that a large amount of Na^+ is pumped back into the soil solution from the root cells across the plasma membrane via a Na^+/H^+ antiporter. Lauchli et al. (2008) proposed that the efflux of Na^+ from root cells could be mediated by the *SOS1* gene, which codes for a protein similar to the plasma membrane Na^+/H^+ antiporter. Pardo et al. (2006) also observed that a large family of *CHX* genes may also be involved in Na^+ efflux from the root cells.

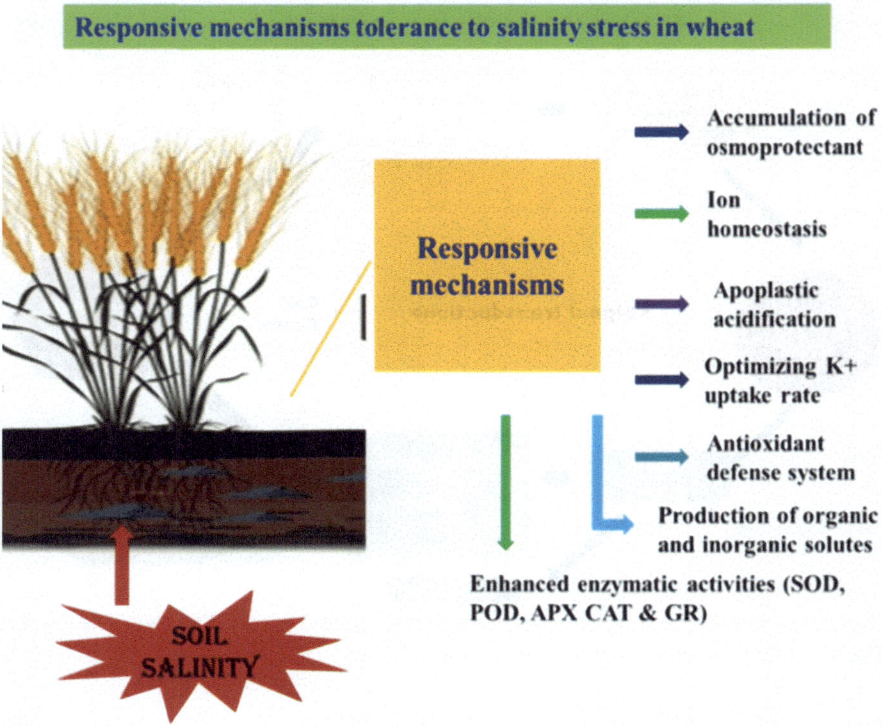

Fig. 4.3 Mechanisms of salt stress alleviation in bread wheat

4.6 Plant Level Compartmentalization

Once the concentration of salts in older leaves of plants increases above the critical limit of tolerance, at that point, the compartmentalization of salt in cell vacuole becomes crucial to prevent ion induced toxicity cytoplasm, and a similar tolerance mechanism exists in the halophytes (Munns 2005). Moreover, few halophytes accumulate Na^+ and Cl− ions in the root and root-stem junctions, resulting in lower salinity levels in the shoots than in roots. For example, eu-halophytes can survive in high-salinity environments as these can store ions in their delicate leaves or stems and dilute salt concentration in cells leading to the reduced ion toxicity in the cells.

4.7 Salt Excretion

Halophytic plants remove excess salt ions through unique tissue modifications, such as salt bladders, salt hairs, and salt glands. Halophytes can survive in low levels of sodium chloride (200 mM) and can withstand higher salt levels in the soils. Even under high salt concentrations, they can maintain normal reproduction and growth (Meng et al. 2018). It has been established that the salt bladders and salt glands are not found in all halophytes and only half of them use this type of mechanism to stabilize salt ion concentration in the plant tissue.

4.8 Cellular Level Compartmentalization

Under high-salinity condition, plant fails to restrict the entry of sodium ions in cells, resulting in the disruption of ion homeostasis, deficiency of essential ions such as K^+, and excess of harmful Na^+ in the cytoplasm. To cope up with the ion toxicity, plants compartmentalize ions in their cells, through various ion transporters and ion channels. Cell wall modification, vesicle trafficking, vacuolar compartmentalization, ROS detoxification, K^+ and NO_3- ion homeostasis, osmotic control of transport proteins, and accumulation of appropriate osmolytes are all important processes at the cellular level involved in adapting to higher level of salt stress. To protect cells from toxic effects of excess sodium ions, plants typically adsorb and retain these toxic ions inside the vacuole via tonoplast transporters through Na^+/H^+ anti-transporters (*NHX*), as those trap the excess amount of cytoplasmic sodium ions in vacuoles. The vacuolar H^+-ATPases and H^+-pyrophosphatases (H^+- ATPases) generate a proton motive force, which is used by anti-transporters (*NHX*) to sequester ions within the vacuole (Silva and Geros 2009). Further, the Na^+/H^+ anti-transporter, a part of SOS signaling pathways, reduces the sodium ions in the cytoplasm by compartmentalizing excess sodium ions into vacuoles, resulting in the improved salt tolerance in plants (Mahajan and Tuteja 2005).

4.9 Effects of Salt Stress on Germination, Growth, and Yield

In complete life cycle of a wheat plant, the seed germination is a highly sensitive phase that occurs through imbibition of water (Kumari and Kaur 2018). Salinity-induced osmotic stress suppresses seed germination via reducing water uptake or causing ionic toxicity. These effects collectively inhibit cell division and expansion and also stimulate the activities of some specific enzymes, leading to the reduced utilization of seed reserves (El-Hendawy et al. 2019). According to recent studies, wheat seedlings are moderately sensitive to salt stress, and at higher salt stress intensity, the decreased growth and development of plants and subsequent seedling death have been reported (Saddiq et al. 2021). Additionally, yield and yield associated attributes such as number of spikelets per spike, number of productive tillers per plant, grain weight, and biomass yield are also affected by the higher salt stress The physiological, such as, osmotic stress, ionic imbalance, and oxidative stress are the major factor hindering crop yield under salt stress. Osmotic stress triggers an increase in salt accumulation in cell sap and tissues, which manifests as leaf burn and wilting. This ionic imbalance causes nutrient imbalance, resulting in the reduced germination and disruption of various metabolic processes (Hussain et al. 2019). Besides that, reactive oxygen species generated oxidative stress causes lipid peroxidation and nucleic acid disruption, which reduce the consistency and overall grain yield (Dehnavi et al. 2020; Kumari and Kaur 2020). Hence, seed priming with natural or synthetic growth regulators to increase seed germination in saline environments may be a better option to enhance the salinity tolerance in wheat crop.

4.10 Effects on Root Architecture

Root is the first plant organ which responds to salt stress immediately. Under salt stress conditions, root system uptakes water and excludes salts. Few researchers have observed a symbiotic association between performance of plant and changes in root system architecture (Faiyue et al. 2010; Ristova and Busch 2014). Robin et al. (2016) concluded that wheat genotypes grown under the hydroponics condition showed nearly about 25–40% reduction in the length, density, and surface area of roots under salt stress.

4.11 Physiological Responses to Salt Stress

In order to survive at higher soil salinity levels, plants develop a variety of physiological and biochemical mechanisms. Assessing alterations in these parameters under saline environment and their role in imparting salinity tolerance to plants has important implications. Salt stress affects water balance, photosynthesis and respiration, turgor pressure, enzyme efficiency, cell membrane degeneration, and mineral nutrient intake and assimilates storage in plant tissues (Jamil et al. 2007). According to Sairam et al. (2005) and Kumar et al. (2017), the tolerant and sensitive salt wheat genotypes differently responded for various physio-biochemical parameters (membrane stability index, relative water content, chlorophyll content, total sugar and soluble starch content, ascorbic acid content, and activity of ROS scavenging enzymes) under salt stress.

4.12 Effects on Photosynthesis

Plant's survival depends on optimal photosynthetic activity, which is strongly influenced by the environmental factors (Badawy et al. 2021). Presence of higher amount of salt in soil solution inhibits root water uptake, resulting in the development of drought stress in plants. In order to cope drought stress, plant closes their stomata to conserve the water resulting in the reduction of CO_2 availability for performing the photosynthesis. Photosynthesis is also hindered by both accumulation of ions (Na^+ and Cl−) in chloroplast and decreased water potential in plants as a consequence of salt stress (Hasanuzzaman et al. 2013).

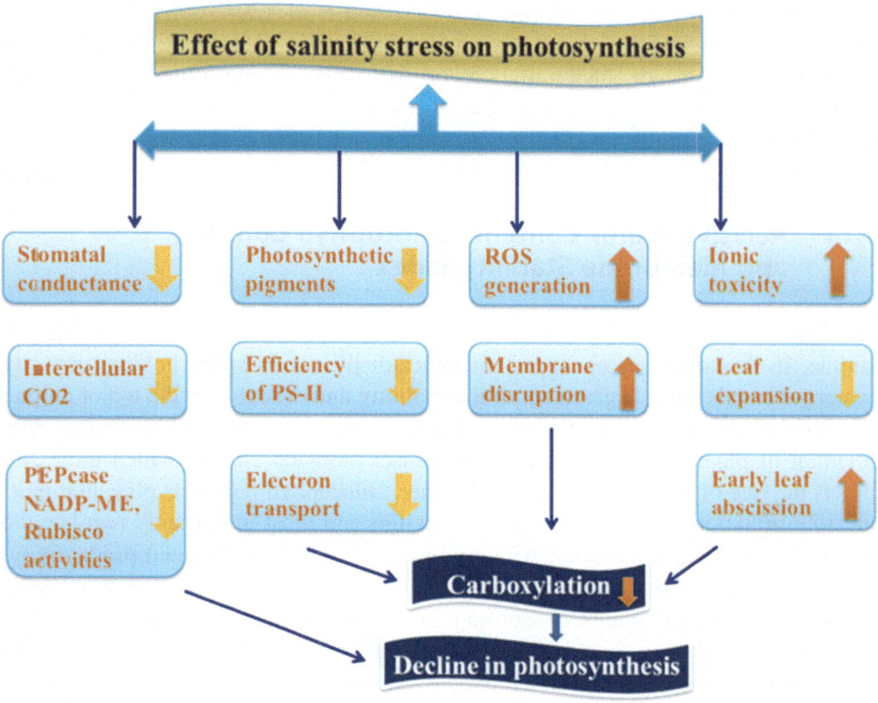

Fig. 4.4 Effects of salt stress on wheat photosynthesis

Additionally, salt stress substantially decreases photosynthetic pigments in the chloroplast (Afzal et al. 2008) to adversely affect photosynthetic efficiency that results in an overall decrease in growth and productivity of plants. The salt stress induced imbalance in ROS and antioxidant species increases ionic toxicity, reduces leaf growth, and induces early leaf senescence which affects carboxylation and thus reduces the photosynthetic activity. Furthermore, salt stress also reduces the efficiency of PS-II, stomatal conductance, intercellular CO_2, and electron transport resulting in an overall decrease in photosynthesis of plant species (Fig. 4.4).

4.13 Oxidative Damage

Salt stress induced stomatal closure prevents CO_2 entry into the leaves. This process inhibits CO_2 fixation and allows chloroplast to stimulate massive amounts of energy, resulting in the development of reactive oxygen species (ROS) (Wahid et al. 2007; Parida and Das 2005; Hasanuzzaman et al. 2011, 2013). ROS damages the major cell molecules such as lipids, proteins, and nucleic acids and causes peroxidation of polyunsaturated fatty acids in the membranes (Smirnoff 2005). The production of ROS increases under salt stress, resulting in cellular toxicity in the various crop plants. Under such conditions, salt-sensitive wheat cultivars display higher H_2O_2 and lipid peroxidation than salt-tolerant cultivars. Zou et al. (2016) observed that salt stress (100 mMNaCl) elevated the malondialdehyde (MDA) levels in wheat seedlings by 35% or 68% after 5 or 10 days of exposure, respectively. ROS-mediated membrane damage has also been identified as a major cause of salinity-induced cellular toxicity in rice, tomato, and citrus crops.

4.14 Relative Water Content, Chlorophyll Content, and Membrane Stability Index

RWC (relative water content) is a measure of the amount of water in a fresh leaf sample. It also serves as an important indicator of plant's water holding capacity (water potential) under a given condition. Salinity and drought are the major abiotic stresses that cause osmotic stress via decreasing water uptake and lowering water potential of plants. Consequently, RWC causes stomatal closure which indirectly affects the rate of carbon assimilation. Further, chlorophyll pigments play an important role in various plant physiological activities and most importantly regulate the plants' photosynthetic capacity, which ultimately influences the plant productivity. Under salt stress, the decrease in chlorophyll content is more in sensitive plants compared to tolerant plants. According to Jeranyama and DeMoranville (2008), decrease in plant growth and productivity under salt stress was attributed to the reduction in photosynthetic activity. Moreover, higher salt stress intensity adversely affects the cell and plasma membrane, and membrane integrity, resulting in the electrolyte leakage which is a measure of membrane stability index. Kumar et al. (2017) highlighted that salt stress tolerant wheat genotypes showed a lower decrease in RWC, MSI, and chlorophyll content than sensitive genotypes.

4.15 Ion Homeostasis

Ionic homeostasis regulates ion flux by maintaining low Na^+ and high K^+ concentration in the plant tissues (Farooq et al. 2015). Maintenance of Na^+/K^+ ratio in the cytoplasm through ion homeostatic balance necessitates the participation of Na^+ and K^+ transporters as well as channels which is the key mechanism of salinity tolerance in wheat (Assaha et al. 2017). Plants exclude excess soluble salt from the cytosol via primary and secondary active transports (Li et al. 2006a, b) and accumulate positively charged ions in the plasma and tonoplast membranes to maintain ion homeostasis during salt stress. Additionally, higher concentration of Na^+ interferes with K^+ accumulation, and stomata regulation is compartmentalized in the vacuole to protect the cytosol from the toxic effects of Na^+. The high level of Na^+ and Cl− in roots adversely affects metabolic path, photosynthetic efficiency, and growth of the crop plants (Deinlein et al. 2014). Murthy (1979) demonstrated that the decreased K^+ and increased Na^+ contents in leaves under salt stress reduce the grain yield of wheat genotype (UP-301).

4.16 Biochemical Responses

In salt-affected areas, the osmotic pressure of soil exceeds the osmotic pressure of root cell sap, thereby decreasing the water absorption potential of plants. Under such circumstances, a plant employs adaptive osmoregulation process to mitigate the negative effects of osmotic stress (Hasanuzzaman et al. 2013b; Farooq et al. 2015). The electrically neutral organic solutes called osmolytes (proline, glycine betaine, sorbitol, mannitol, and polyamine) are produced in the cell cytoplasm to regulate osmotic potential of ions in the vacuole under stress condition. These also maintain the osmotic balance of the cells and their surroundings and use an adaptive mechanism called osmotic adjustment to keep the plant cells turgid, enabling the plant to sustain under salt and moisture stress conditions. Ashraf and Foolad (2007) observed a clear relationship between the stress tolerance and accumulation of osmolytes under various stress conditions.

4.17 Proline

Proline accumulation is regarded as an important precautionary biochemical response in wheat against the salt stress. Increasing proline content in plant cells prevents negative effects of osmotic stress at the cell level. Even in adverse conditions, proline accumulation aids the plant's growth and development. It acts as an antioxidant, a signaling molecule, and a metal chelator in plants under stress condition (Hayat et al. 2012). Proline can be produced by plants in glutamate or ornithine pathways. However, under salt stress, glutamate is most important precursor molecule in the synthesis of proline. Pyrrolinecarboxylic acid reductase (P5CR) and pyrroline carboxylic acid synthase (P5CS) enzymes played an important catalytic role in proline biosynthesis. The expression of genes involved in proline biosynthesis increases during salt stress, while the gene encoding proline dehydrogenase involved in proline breakdown decreases in salt-tolerant wheat genotypes (Tavakoli et al. 2016).

4.18 Glycine Betaine

Glycine betaine, a nontoxic water-soluble organic metabolite, plays a significant role in osmoregulation in plants under abiotic stresses (Wani et al. 2013). It is synthesized from choline with a two-step reaction, involving two or more enzymes, i.e., choline monooxygenase and betaine aldehyde dehydrogenase. Both of these enzymes activated in the presence of salt stress. Glycine betaine protects the cells through osmotic adjustment and protein stabilization and also protects photosynthetic apparatus from stress (Cha-Um and Kirdmanee 2010) and reduction of ROS (Saxena et al. 2013a, b). Ahmad et al. (2013) reported that the foliar spray of glycine betaine on stressed plants resulted in the pigment stabilization as well as increase in photosynthetic rate and growth of the wheat plant.

4.19 Antioxidant Regulation

Biotic and abiotic stresses cause disruption of electron transport chains (ETC) in chloroplasts and mitochondria. Under such conditions, molecular oxygen (O_2-) acts as an electron acceptor, resulting in the ROS accumulation. Reactive oxygen species (ROS) are modified and activated forms of atmospheric oxygen that are an inevitable by-product of aerobic metabolism that occurs naturally in plants. Singlet oxygen ($^1O_2$), hydroxyl radical ($\cdot OH^-$), superoxide radical ($\cdot O_2^-$), and hydrogen peroxide (H_2O_2) are all highly oxidizing compounds that are harmful to cell integrity (F. Groß et al. 2013). These ROS molecules end up causing lipid peroxidation and denaturation of proteins and nucleic acids in the plant cells (Ahmad and Sharma 2008). In response to environmental stresses, crop plants typically build up ROS (reactive oxygen species) molecules in various subcellular compartments. The accumulation of ROS is the main cause of oxidative stress, which leads to serious injury in plants resulting in reduced crop growth and yield. ROS production and detoxification are continuous processes inside the plant cells, and thereby they maintain an equilibrium called redox potential (Caverzan et al. 2016) which is regulated by enzymatic and nonenzymatic antioxidants (Gondim et al. 2013). Antioxidant metabolism involves activation of enzymatic and nonenzymatic compounds, essential for detoxifying ROS caused by the salt stress. The activity of antioxidant enzymes, such as superoxide dismutase (SOD), catalase (CAT), glutathione peroxidase (GPX), ascorbate peroxidase (APX), and glutathione reductase (GR), as well as the deposition of nonenzymatic antioxidant compounds is positively associated with salt stress tolerance (Gupta et al. 2005). Gill et al. (2013) and Tuteja et al. (2013) identified a couple of helicase proteins (e.g., DESD-box helicase and OsSUV3 dual helicase) that contribute in plant salinity tolerance by enhancing photosynthesis and antioxidant machinery. The antioxidant role of anthocyanin and flavonoid has also been extensively studied in plants under salt stress. It has been reported that the anthocyanin accumulation increases once plants are exposed to salt stress. However, Van Oosten et al. (2013) isolated the anthocyanin-impaired response-1(*air1*) mutant, which was incapable of accumulating anthocyanins under salt stress. The identification and characterization of *AIR1*gene open up new opportunities for investigating the links between abiotic stress and antioxidant accumulation in the form of flavonoids and anthocyanins. Additionally, ascorbate is one of the most important antioxidants in a cell, and exogenous application of ascorbate reduces the negative effects of salt stress in a variety of plant species and encourages plant recovery from the stresses (Agarwal and Shaheen 2007; Munir and Aftab 2011). Glutathione is another important antioxidant that aids in stress reduction, because it can interact with free radicals like hydrogen peroxide, superoxide, and hydroxyl radicals. According to Aly-Salama and Al-Mutawa (2009) research, the exogenous application of glutathione protects the plasma membrane permeability and cell viability throughout the salt stress in *Allium cepa*.

4.20 Polyamines

Polyamines (PA) are tiny, low-molecular-weight polycationic molecules and play a wide range of functions in plant growth and development, including the control of cell proliferation, differentiation and morphogenesis, dormancy breaking, somatic embryogenesis and seed germination, development of flowers and fruit, and senescence (Knott et al. 2007; Gupta et al. 2013a, b). It is essential for abiotic stress tolerance, including salinity, and the increase in polyamine levels is associated with plant stress tolerance in wheat (Gupta et al. 2013a, b; Kov'acs et al. 2010).

4.21 Molecular Responses to Salt Stress

4.21.1 ROS Homeostasis

Salt stress disrupts the availability of CO_2 in leaves and the electron transport chain (ETC) in mitochondria and chloroplast as a result of the production and accumulation of ROS (Navarro-Yepes et al. 2014). Chloroplast (photosystem I and II), mitochondria (respiratory complex I and III), and peroxisomes (glycolate oxidase) are the primary sites of ROS generation. Accumulation of higher ROS concentration is extremely harmful to plants. Irregularities in the ETC cause lipid peroxidation, enzyme inhibition, nucleic acid damage, and activation of the programmed cell death pathway (Hossain et al. 2021). Lipid peroxidation increases electrolyte leakage, disturbs membrane permeability, and activates protein oxidation. For instance in wheat, MDA content was found to be increased up to 35% at 100 mM and 73% at 300 mM NaCl salt stress (Hasanuzzaman et al. 2013). Most of the plants have a natural defense system such as an antioxidant to detoxify the harmful effect of ROS. The antioxidant enzymes such as CAT, SOD, POX, APX, GR, GPX, DHAR and MDHAR, while non-enzyme antioxidants such as glycine betaine, and proline, sugars (sucrose and raffinose), polyols (mannitol and sorbitol) (Munns and Tester 2008; Wang et al. 2001) protect the cells by osmotic adjustment and stabilize the tertiary protein structure and shielded photosynthetic apparatus from ROS damage. Several genes regulate the biosynthesis of these compatible solutes and enhance the salt tolerance of the crop plants. For example, the rate-limiting enzyme, i.e., pyrroline-5-carboxylate synthetase (*P5CS*) gene, is involved in proline biosynthesis to enhance the salt tolerance in *Arabidopsis thaliana*. Both halophytes and glycophytes sequester Na^+ ion which is an important mechanism of tissue tolerance to salt stress (Munns and Tester 2008; Tester and Davenport 2003). Previous studies reported that the increased concentration of catalase in both salt-tolerant and salt sensitive wheat genotypes enhances the salt stress tolerance (Sairam et al. 2002). Mandhania et al. (2006) reported enhanced activity of SOD and CAT in bread wheat, which is involved in the detoxification of H_2O_2 to H_2O and O_2 under salt stress. Salt-tolerant wheat genotypes produced higher concentration of AsA and CAT to counter the effect of salt stress in comparison with sensitive genotypes (Athar et al. 2007).

4.21.2 Effect on Ion Homeostasis

Ionic imbalances are the most severe impacts caused by salt stress. The combinations of multiple ions in saline environment determine the nutrient imbalance which leads to either deficiency or excessiveness of a particular nutrient. Under moderate to high-salt stress, the nutrient imbalance is prevented by ion transportation, which tends to create influx of ions into the cell from the plasma membrane of the vacuole (Blumwald et al. 2000). Whenever the influx rate is higher, anions (Cl−) and cations (Na^+, Mg^{2+}, Ca^{2+}) accumulate in cytosol, matrix, and stroma of the cell instead of the vacuole, which results in nutrient imbalance, especially the deficiency of K^+ in plant tissues. To counter ionic imbalance, the apoplast continues to accumulate ions resulting in ion transport inhibition from apoplast to symplast (Speer and Kaiser 1991). The accumulation of salt in leaf apoplast disrupts cell water relations, leading to wilting and inhibition of enzymatic activity, which interferes with the protein surface charges, besides destabilizing the molecular level interactions (Flowers et al. 1991). Compartmentation of excess Na^+ in vacuole is tolerance mechanism of plants against the ion toxicity (Cordovilla et al. 1995). High Na^+ interferes with K^+ accumulation and stomata regulation in plants. Therefore, to optimize the K^+ uptake ratio, plants restrict the Na^+ entry and exclude the Na^+ from the cell to maintain the K^+/Na^+ ratio in the cytosol and to increase plant survival under salt stress (Li et al. 2006a, b). A higher concentration of Na^+ and Cl− disturbs the uptake of K^+ and Ca^{2+} causing lesions on tissues and reduction in shoot dry weight of plants. Similarly, antagonistic effect of Cl− has also been observed with NO^3- and $PO_4{}^3-$ that reduced the yield of wheat (Abdelgadir et al. 2005; Massa et al. 2009). Na^+ is a major cause of creating both ionic toxicity and osmotic stress during the salt stress; thus, maintaining ion homeostasis is very crucial for proper growth and development of wheat under salt stress. Under higher level of salt stress conditions, plants fail to exclude Na^+ from shoot tissue leading to accumulation of Na^+ in vacuoles via cytoplasm, which creates a detrimental effect on various stages of plant growth. Therefore, to regulate the Na^+ sequestration, several channels/transporters such as vacuolar NHX (Na^+/H^+) transporter transport Na^+ from cytosol into the vacuole. The two types of H^+ pumps are present in vacuoles, i.e., H-ATPase (V-ATPase) and vacuolar pyrophosphatase (V-PPase) (Wang et al. 2001), and V-ATPase is known as dominant due to playing a role in maintaining solute homeostasis through facilitating vesicle fusion by using secondary transport. Previous studies on *Vigna unguiculata* indicated that the activity of the V-ATPase pump increased, while the activity of the V-PPase pump was inhibited in seedlings under salt stress (Otoch et al. 2001). The functional activity of the NHX (Na^+/H^+) transporter depends upon tonoplast membrane-localized H^+-ATPase (V-ATPase) and H^+-ATPase (V-PPase). Overexpression of vacuolar AtNHX1 or AVP1 enhances the salinity tolerance in plants (Ishitani et al. 2000).

4.21.3 Ion Transporters

Plants maintain the ionic balance of Na^+ and Cl− inside the cell by removing excess salts via primary and secondary transport systems by compartmentalizing into vacuoles (Li et al. 2006a, b). Na^+ exclusion in plants can be achieved by different channels/transporters present in the cell membrane and by maintaining a K^+/Na^+ ratio. Previous studies reported the presence of low Na^+ content with high K^+/Na^+ ratio in the roots and shoots of wheat plant (Yadav et al. 2018). Higher Na^+ content in leaves creates metabolic toxicity that causes death of leaves due to inhibition of various enzymatic activities and reduction in photosynthetic ability in the leaf cell. K^+ works as cofactor in the activation of more than 50 enzymes in the cytosol of leaves, which are necessary for protein synthesis. Several previous studies have conducted the detailed and comprehensive analysis of Na^+ exclusion and stress induced tolerance mechanism in various crop species (Munns et al. 2000; Genc et al. 2007). In contrast, in few crops like soybean, citrus, and grapevine, the Cl− ion induced more toxicity compared to Na^+. To prevent the entry of Na^+ in photosynthetically active tissues, the Na^+ exclusion mechanisms have been identified in cereal crops such as rice, bread wheat, and barley. The mechanism generally involves up- or downregulations of specific ion channels/transporters allowing control of Na^+ transport in cytoplasm throughout the plant's life. Further, two identified transporter genes were found to be involved in Na^+ exclusion: SOS (salt overly sensitive) and HKT (high-affinity potassium transporter) family. The SOS gene includes three proteins; SOS1 encodes functional Na^+/H^+ antiporter which is controlled by generated proton gradient through ATPase activity localized in the plasma membrane. The studies suggested that SOS1 (Na^+/H^+ antiporter) is involved in Na^+ exclusion back into the soil, and overexpression of this SOS1 gene confers salt tolerance in transgenic plants. The second SOS2 gene encodes a protein serine/threonine kinase and consists N-terminal (catalytic) and C-terminal (regulatory) domain involved in Ca^+ ion signals (Conde et al. 2011; Liu et al. 2000). The third gene SOS3, Ca^+ ion binding protein contains a myristonylation site at N-terminal, conferring salt tolerance (Ishitani et al. 2000). In wheat, two upregulated SOS genes were also identified, i.e. *TaSOS1* (a transmembrane Na^+/H^+ antiporter) and *TaSOS4* (a cytoplasmic pyridoxal (PL) kinase), and their expression was measured in cultivated and wild wheat by using qRT-PCR technique under salt stress. High-affinity potassium transporter (HKT) gene family plays a crucial role in Na^+ exclusion via minimizing the entry of Na^+ into the roots from the soil (Tester and Davenport 2003). The *Nax1* and *Nax2* genes belonging to the HKT gene family were initially identified in durum wheat, which excludes the Na^+ from xylem tissues and maintains its low concentration in the leaf tissue. The *Nax* genes have already been utilized in the crop improvement program for developing salt-tolerant wheat genotypes. The presence of the *Nax2* gene in the durum wheat produced 25% more grain yield, while the presence of both the *Nax1* and *Nax2* gene in bread wheat reduces Na^+ concentration by 60% in leaves under salt stress conditions (James et al. 2006; Singh et al. 2015). The HKT includes two classes known for controlling Na^+ ion distribution in plants: class I or Na^+ selective transport and class II or Na^+/K^+

cotransport. The *Arabidopsis* gene *AtHKT1; 1* is localized in root stele and the overexpression of this gene enhances salt tolerance in plants by decreasing Na^+ transport to the shoot via the transpiration stream (Ramezani et al. 2013). Salt tolerance in wheat genotypes can be achieved by downregulation of *TaHKT 2* gene (Singh et al. 2015). Further, the genes including dehydrin, expansins, and transcription factors were involved in increasing salt tolerance in wheat such as *DHN-5* (Brini et al. 2007), *TaEXPA2* (Chen et al. 2017), *TaEXPB23* (Han et al. 2012), *TaSTRG* (Zhou et al. 2009), *TaMYB2A* (Mao et al. 2011), and *TaNAC29* (Xu et al. 2015). Overall, Na^+ exclusion is the most effective approach for improving salt tolerance in cereal crops by preventing the entry of Na^+ ions into the shoot.

4.22 Approaches to Enhance Salt Tolerance Through Seed Priming and Exogenous Application of Organic Chemicals

Seed germination is the most critical stage affected by salt stress which needs sui management for obtaining higher germination and vigorous seedlings. The priming process enhances resistance of seed against the salt stress. Among the different priming techniques, chemical priming is a prominent approach for managing salt stress in plants. For instance, H_2O_2 treated wheat seed showed 56% higher germination rate than hydroprimed (He and Gao 2009), suggesting the effectiveness of chemicals over hydropriming. Seed priming with SA improved pre- and postemergence salt stress tolerance of wheat, seed germination, tissue water status, and root-shoot ratio (Amin and Abolfazl 2017 ; Razzaq et al. 2013). It was also observed that the seed treated with 300 μmol L^{-1} of melatonin enhanced radicle length and number, plumule length, and lysine (Lys) content in grain which was indicative of a positive correlation with a morphological index of germination (Li et al. 2000).

4.23 Pre-Sowing Seed Treatment

The low water potential and higher osmotic stress drastically reduce the seed germination rate and total seed germination of wheat crop in salt-affected soils. The reduced water uptake in the saline soils alters various structural, physiological, and biochemical parameters, leading to reduction in seed germination (Poljakoff-Mayber et al. 1994). The seed germination and early-seedling establishment are the most sensitive growth phases under salt stress (Foolad 2004). However, rapid and uniform seed germination and early-seedling growth are of vital importance for satisfactory wheat production in saline soils. Therefore, it is important to alleviate the adverse effects of salts at the initial growth stages for improving the establishment of a crop in saline soils (Ashraf and Foolad 2005; Ashraf et al. 2003). Many studies have shown that the treating seed with water or solutions of inorganic or organic salts before sowing improves the salt tolerance of plants (Babaeva et al. 1999; Chang-Zheng et al. 2002). During pre-sowing seed treatment, seeds are immersed in a solution which is referred to as seed priming. Priming creates external water potential that is low enough to prevent germination but allows some pre-germinative physiological and biochemical phenomena to occur or for a duration that is insufficient for germination to pass a critical point (Bradford 1986). It has been observed that the primed seeds usually germinate more rapidly than unprimed seeds once placed in an appropriate germination environment. Over the last two decades, the seed priming approach has been used extensively to enhance the rate and uniformity of seed germination and the emergence of crop plants grown under both saline and nonsaline conditions. Seed priming as a technique was adopted to improve the rate and synchrony of seed germination and can be subdivided into hydropriming, halopriming, osmopriming, and hormone-priming.

4.23.1 Hydropriming

Hydropriming involves soaking the seeds in water before sowing (Pill and Necker 2001) and may or may not be followed by air-drying of the seeds. Hydropriming may enhance seed germination and seedling emergence under both saline and nonsaline conditions. For example, Roy and Srivastava (1999) found that soaking wheat kernels in water improved their germination rate under saline conditions.

4.23.2 Halopriming

Halopriming refers to soaking of the seed in inorganic salt solution. A large number of studies have demonstrated that the seed halopriming treated seed significantly improves germination, seedling emergence and establishment of seedlings, and final crop yield in salt-affected soils. For example, presoaking *Echinacea purpurea* seed in either 0.1% $MnSO_4$ or 0.05% $ZnSO_4$ solution increased germination percentage by 35% or 38%, respectively, and field emergence by 27–41% (Babaeva et al. 1999). The halopriming not only promotes seed germination but may also stimulate subsequent growth of plants, thereby enhancing the final crop yield (Eleiwa 1989; Sallam 1999). Under salt stress, presoaking seed in inorganic salts (NaCl, $CaCl_2$, and Na_2SO_4) solutions improves grain yield in wheat (*Triticuma aestivum*). For instance, seed priming with $CaCl_2$ was more effective than NaCl; however, soaking in 3% Na_2SO_4 was the most effective treatment to alleviate the salt stress in wheat (Mehta et al. 1979).

4.23.3 Hormone-Priming

Pretreating seed with optimal concentrations of plant growth regulators (PGRs) can effectively improve germination as well as growth and yield of crops under both normal and stress conditions (Lee et al. 1998). Several growth regulators are commonly used for seed priming, including auxins (IAA, IBA, and NAA), gibberellins (GA), gibberellin antagonists, kinetin, abscisic acid, polyamines

(PAs), ethylene, brassinolide, salicylic acid (SA), triacontanol, and ascorbic acid. For instance, soaking seed with IAA, NAA, or GA improves seed germination of wheat (Balki and Padole 1982). Similarly, priming of seeds with a moderate concentration of kinetin (150 mg L^{-1}) was found equally effective in improving growth and grain yield of two wheat cultivars, suggesting the beneficial effects of kinetin on photosynthetic capacity and water use efficiency under saline conditions (Iqbal and Ashraf 2005). Moreover, exogenous application of IAA in the form of foliar spray increases the grain yield and yield components in a wheat cultivar cv. Shahkar-92 (Hegazi et al. 1995) grown under saline conditions. Similarly, exogenous application of salicylic acid (SA) as a foliar spray counteracted the deleterious effects of NaCl on wheat (Sakhabutdinova et al. 2003). In the same crop, application of 100 mg L^{-1} SA proved to be effective in alleviating the adverse effects of salt stress on wheat seedlings (Al-Hakimi and Hamada 2001). Wheat seed pretreated with different concentrations of gibberellic acid (GA) also showed varying positive results in terms of germination under saline conditions, with presoaking in 50 mg L^{-1} of GA having the best result (Parashar and Varma 1988). In another study, adverse effects of salt stress on seed germination were mitigated by soaking seed in varying concentrations of IAA, IBA, or gibberellic acid (GA3) (Gulnaz et al. 1999). For example, 200 mg L^{-1} GA3 resulted in 100% seed germination at 13.1 dS m^{-1}, while IAA and IBA increased germination at 8.4 dS m^{-1}.

4.24 Conclusion

Climate change induced salt stress is adversely affecting the quality and yield of wheat crop, globally. Wheat plants exhibit variable responses to salt stress for physiological, biochemical, and molecular functions. Salt stress negatively influences photosynthesis, nutrient uptake, and water relationships and also alters hormonal and ionic imbalance through salinity-induced oxidative stress. The salt stress tolerance mechanisms in wheat have been extensively studied and delineated. However, detailed understanding of salinity sensitive biochemical and molecular parameters is required at seed germination and seedling stages. Additionally, the studies on root architecture, root-shoot signaling, and their cross talk with nutrient and water uptake under salt stress are still required to be documented for better understanding of salinity tolerance mechanism with the aim to develop salinity tolerant cultivars in wheat. Despite the extensive application of various conventional, molecular, and transgenics approaches for improving salinity tolerance in wheat, the only very few salt-tolerant wheat varieties were developed, globally. Further, the use of exogenous chemical (inorganic salt, plant hormones) for seed priming and foliar spray was found much more convenient in enhancing plant growth under saline environment. In wheat seed, priming treatment has been established as a promising practice to enhance salinity tolerance. The different kinds of chemicals are used for seed priming, and they create contrasting response in plant metabolic activities as few chemicals improve plant hormones and enzymes, while others enhance the activity of plant growth regulators. For instance, seed priming with the chemicals like plant hormones (IAA, NAA, or GA) and inorganic salts (Zn, Mn, and Na) improves the seed germination, pigment contents, membrane stability, water holding capacity, membrane stability, water use efficiency, membrane stability, yield-related traits, and biological and grain yields, resulting in an overall favorable increase in physiological processes in wheat. However, there is still lack of understanding about the potential of chemical molecules under salt stress which require the greater attention. Besides this, the efficiency of already tested chemicals as well as the effect of chemicals individually or in combination on the various developmental stages is required to be standardized for ease of use. Therefore, integrated approach that encompasses plant breeders, physiologists, biotechnologists, and agronomists is essential for enhancing salinity tolerance and producing sustainable yield in wheat under salt-affected agroecosystems.

References

Abdelgadir EM, Oka M, Fujiyama H (2005) Characteristics of nitrate uptake by plants under salinity. J Plant Nutr 28:33–46. https://doi.org/10.1081/PLN-200042156

Afzal I, Rauf S, Basra SMA, Murtaza G (2008) Halopriming improves vigor, metabolism of reserves and ionic contents in wheat seedlings under salt stress. Plant Soil Environ 54:382–388. https://doi.org/10.17221/408-PSE

Agarwal S, Shaheen R (2007) Stimulation of antioxidant system and lipid peroxidation by abiotic stresses in leaves of Momordicacharantia. Braz J Plant Physiol 19(2):149–161. https://doi.org/10.1590/S1677-04202007000200007

Ahmad P, Sharma S (2008) Salt stress and phyto-biochemical responses of plants. Plant Soil Environ 54(3):89–99; https://www.researchgate.net/publication/242580345

Ahmad R, Lim CJ, Kwon SY (2013) Glycine betaine: a versatile compound with great potential for gene pyramiding to improve crop plant performance against environmental stresses. Plant Biotechnol Rep 7:49–57. https://doi.org/10.1007/s11816-012-0266-8

Al-Hakimi AMA, Hamada AM (2001) Counteraction of salinity stress on wheat plants by grain soaking in ascorbic acid, thiamin or sodium salicylate. Biol Plant 44:253–261

Aly-Salama KH, Al-Mutawa MM (2009) Glutathione-triggered mitigation in salt-induced alterations in plasmalemma of onion epidermal cells. Int J Agric Biol 11(5):639–642

Amin N, Abolfazl B (2017) Consequences of seed priming with salicylic acid and hydro priming on smooth vetch seedling growth under water deficiency. J Agric Sci 9(12):259. https://doi.org/10.5539/jas.v9n12p259

Arzani A, Ashraf M (2016) Smart engineering of genetic resources for enhanced salinity tolerance in crop plants. Crit Rev Plant Sci 35(3):146–189. https://doi.org/10.1080/07352689.2016.1245056

Ashraf M, Foolad MR (2005) Pre-sowing seed treatment a shotgun approach to improve germination, plant growth, and crop yield under saline and non-saline conditions. Adv Agron 88:223–271

Ashraf M, Foolad MR (2007) Roles of glycine betaine and proline in improving plant abiotic stress resistance. Environ Exp Bot 59(2):206–216. https://doi.org/10.1016/j.envexpbot.2005.12.006

Ashraf M, Kausar A, Ashraf MY (2003) Alleviation of salt stress in pearl millet (*Pennisetumglaucum* (L.) R. Br.) through seed treatments. Agronomie 23:227–234

Assaha DV, Ueda A, Saneoka H, Al-Yahyai R, Yaish MW (2017) The role of Na^+ and K^+ transporters in salt stress adaptation in glycophytes. Front Physiol 8:509. https://doi.org/10.3389/fphys.2017.00509

Athar H, Khan A, Ashraf M (2007) Exogenously applied ascorbic acid alleviates salt-induced oxidative stress in wheat. Environ Exp Bot 63:224–231. https://doi.org/10.1016/j.envexpbot.2007.10.018

Babaeva EY, Volobueva VF, Yagodin BA, Klimakhin GI (1999) Sowing quality and productivity of Echinacea purpurea in relation to soaking the seed in manganese and zinc solutions. Izvestiya Timiryazevskoi Sel'skokhozyaistvennoi Akademii 4:73–80

Badawy SA, Zayed BA, Bassiouni SMA, Mahdi AHA, Majrashi A et al (2021) Influence of nano silicon and nano selenium on root characters, growth, ion selectivity, yield, and yield components of rice (*Oryza sativa* L.) under salinity conditions. Plan Theory 10:1657. https://doi.org/10.3390/plants10081657

Balki AS, Padole VR (1982) Effect of pre-soaking seed treatments with plant hormones on wheat under conditions of soil salinity. J Indian Soc Soil Sci 30:361–365

Blumwald E, Aharon GS, Apse MP (2000) Sodium transport in plant cells. Biochim Biophys Acta 1465:140–151. https://doi.org/10.1016/S0005-2736(00)00135-8

Bradford KJ (1986) Manipulation of seed water relations via osmotic priming to improve germination under stress conditions. HortScience 21(5):1105–1112. https://doi.org/10.21273/HORTSCI.21.5.1105

Brini F, Hanin M, Mezghani I, Berkowitz GA, Masmoudi K (2007) Overexpression of wheat Na^+/H^+ antiporter *TNHX1* and H^+-pyrophosphatase TVP1 improve salt-and drought-stress tolerance in Arabidopsis thaliana plants. J Exp Bot 58:301–308. https://doi.org/10.1093/jxb/erl251

Caverzan A, Casassola A, Brammer SP (2016) Antioxidant responses of wheat plants under stress. Genet Mol Biol 39(1):1–6. https://doi.org/10.1590/1678-4685-GMB-2015-0109

Chang-Zheng H, Jin H, Zhi-Yu Z, Song-Lin R, Wen-Jian S (2002) Effect of seed priming with mixed-salt solution on germination and physiological characteristics of seedling in rice (*Oryza sativa* L.) under stress conditions. J Zhejiang Univ (Agric Life Sci) 28:175–178

Chattha MU, Hassan MU, Khan I, Chattha MB, Aamer M et al (2020) Impact of planting methods on biomass production, chemical composition and methane yield of sorghum cultivars. Pak J Agric Sci 57:43–51. https://doi.org/10.21162/PAKJAS/20.7112

Cha-Um S, Kirdmanee C (2010) Effect of glycine betaine on proline, water use, and photosynthetic efficiencies, and growth of rice seedlings under salt stress. Turk J Agric For 34(6):517–527

Chen Y, Han Y, Kong X, Kang H, Ren Y, Wang W (2017) Ectopic expression of wheat expansin gene $TaEXPA_2$ improved the salt tolerance of transgenic tobacco by regulating Na^+/K^+ and antioxidant competence. Physiol Plant 159(2):161–177. https://doi.org/10.1111/ppl.12492

Conde A, Chaves MM, Gerós H (2011) Membrane transport, sensing and signaling in plant adaptation to environmental stress. Plant Cell Physiol 52(9):1583–1602. https://doi.org/10.1093/pcp/pcr107

Cordovilla MP, Ocana A, Ligero F, Lluch C (1995) Salinity effects on growth analysis and nutrient composition in four grain legumes-*rhizobium* symbiosis. J Plant Nutr 18:1595–1609. https://doi.org/10.1080/01904169509365006

Dehnavi AR, Zahedi M, Ludwiczak A, Cardenas Perez S, Piernik A (2020) Effect of salinity on seed germination and seedling development of sorghum (Sorghum bicolor (L.) Moench) genotypes. Agronomy 10:859. https://doi.org/10.3390/agronomy10060859

Deinlein U, Stephan AB, Horie T, Luo W, Xu G, Schroeder JI (2014) Plant salt-tolerance mechanisms. Trends Plant Sci 19(6):371–379. https://doi.org/10.1016/j.tplants.2014.02.001

Demidchik V, Davenport R, Tester M (2002) Nonselective cation channels in plants. Annu Rev Plant Biol 53:67–107. https://doi.org/10.1146/annurev.arplant.53.091901.161540

Ding Z, Kheir AS, Ali OA, Hafez E, Elshamey EA et al (2021) A vermicompost and deep tillage system to improve saline-sodic soil quality and wheat productivity. J Environ Manag 277:111–388. https://doi.org/10.1016/j.jenvman.2020.111388

Eleiwa ME (1989) Effect of prolonged seed soaking on the organic and mineral components of immature pods of soybeans. Egypt J Bot 32:149–160

El-Hendawy S, Elshafei A, Al-Suhaibani N, Alotabi M, Hassan W, Dewir YH et al (2019) Assessment of the salt tolerance of wheat genotypes during the germination stage based on germination ability parameters and associated SSR markers. J Plant Interact 14:151–163. https://doi.org/10.1080/17429145.2019.1603406

Erdal S, Aydın M, Genisel M, Taspınar MS, Dumlupinar R et al (2010) Effects of salicylic acid on wheat salt sensitivity. Afr J Biotechnol 10:5713–5718. https://doi.org/10.5897/AJB10.1550

Faiyue B, Vijayalakshmi C, Nawaz S, Nagato Y, Taketa S, Ichii M, Al-Azzawi MJ, Flowers TJ (2010) Studies on sodium bypass flow in lateral rootless mutants lrt1 and lrt2, and crown rootless mutant crl1 of rice (*Oryza sativa* L.). Plant Cell Environ 33(5):687–701. https://doi.org/10.1111/j.1365-3040.2009.02077.x

FAO (2020) India at a glance; http://www.fao.org/india/fao-in-india/india-at-a-glance/en/

Farooq M, Hussain M, Wakeel A, Siddique KHM (2015) Salt stress in maize effects resistance mechanisms and management: a review. Agron Sustain Dev 35:461–448. https://doi.org/10.1007/s13593-015-0287-0

Flowers TJ, Hajibagherp MA, Yeo AR (1991) Ion accumulation in the cell walls of rice plants growing under saline conditions: evidence for the Oertli hypothesis. Plant Cell Environ 14:319–325. https://doi.org/10.1111/j.1365-3040.1991.tb01507.x

Foolad MR (2004) Recent advances in genetics of salt tolerance in tomato. Plant Cell Tiss Org Cult 76:101–119

Genc Y, Mcdonald GK, Tester M (2007) Reassessment of tissue Na^+ concentration as a criterion for salinity tolerance in bread wheat. Plant Cell Environ 30(11):1486–1498. https://doi.org/10.1111/j.1365-3040.2007.01726.x

Gill SS, Tajrishi M, Madan M, Tuteja N (2013) A DESDbox helicase functions in salinity stress tolerance by improving photosynthesis and antioxidant machinery in rice (*Oryza sativa* L. cv. PB1). Plant Mol Biol 82(1–2):1–22

Goncim FA, Miranda RDS, Gomes-Filho E, Prisco JT (2013) Enhanced salt tolerance in maize plants induced by H_2O_2 leaf spraying is associated with improved gas exchange rather than with non-enzymatic antioxidant system. Theor Exp Plant Physiol 25(4):251–260. https://doi.org/10.1590/S2197-00252013000400003

Groß F, Durner J, Gaupels F (2013) Nitric oxide, antioxidants and prooxidants in plant defence responses. Front Plant Sci 4:419. https://doi.org/10.3389/fpls.2013.00419

Gulnaz A, Iqbal J, Azam F (1999) Seed treatment with growth regulators and crop productivity. II Response of critical growth stages of wheat (*Triticumaestivum* L.) under salinity stress. Cereal Res Commun 27:419–426

Gupta KJ, Stoimenova M, Kaiser WM (2005) In higher plants, only root mitochondria, but not leaf mitochondria reduce nitrite to NO, in vitro and in situ. J Exp Bot 56(420):2601–2609. https://doi.org/10.1093/jxb/eri252

Gupta K, Dey A, Gupta B (2013a) Plant polyamines in abiotic stress responses. Acta Physiol Plant 35(7):2015–2036. https://doi.org/10.1007/s11738-013-1239-4

Gupta K, Dey A, Gupta B (2013b) Polyamines and their role in plant osmotic stress tolerance. Climate change and plant abiotic stress tolerance. Wiley-VCH Verlag Gmb H & Co, Weinheim, pp 1053–1072. https://doi.org/10.1002/9783527675265.ch40

Han Y, Xiu Li A, Li F, Rong ZM, Wang W (2012) Characterization of a wheat (Triticumaestivum L.) expansin gene, Ta*EXPB23*, involved in the abiotic stress response and phytohormone regulation. Plant Physiol Biochem 54:49–58. https://doi.org/10.1016/j.plaphy.2012.02.007

Hasanuzzaman M, Hossain MA, Fujita M (2011) Nitric oxide modulates antioxidant defense and the methylglyoxal detoxification system and reduces salinity-induced damage of wheat seedlings. Plant Biotechnol Rep 5:353–365. https://doi.org/10.1007/s11816-011-0189-9

Hasanuzzaman M, Nahar K, Fujita M, Ahmad P, Chandna R et al (2013) Enhancing plant productivity under salt stress: relevance of poly-omics. In: Ahmad P, Azooz MM, Prasad MNV (eds) Salt stress in plants: omics, signaling and responses, pp 113–156. https://doi.org/10 1007/978-1-4614-6108-1_6

Hasanuzzaman M, Nahar K, Fujita M (2013b) Plant response to salt stress and role of exogenous protectants to mitigate salt-induced damages. In: Ecophysiology and responses of plants under salt stress. Springer, New York, NY, pp 25–87. https://doi.org/10.1007/978-1-4614-4747-4_2

Hasanuzzaman M, Nahar K, Rahman A, Anee TI, Alam MU et al (2017) Approaches to enhance salt stress tolerance in wheat. Intech Open Limited 5 Princes Gate Court, London, pp 151–187. https://doi.org/10.5772/67247

Hassan MU, Chattha MU, Mahmood A, Sahi ST (2018a) Performance of sorghum cultivars for biomass quality and biomethane yield grown in semi-arid area of Pakistan. Environ Sci Pollut Res 25:12800–12807. https://doi.org/10.1007/s11356-018-1575-4

Hassan MU, Chattha MU, Chattha MB, Mahmood A, Sahi ST (2018b) Bio-methane production from sorghum elite lines under the climatic conditions of Pakistan. Maydica 63:1–8

Hayat S, Hayat Q, Alyemeni MN, Wani AS, Pichtel J, Ahmad A (2012) Role of proline under changing environments: a review. Plant Signal Behav 7(11):1456–1466. https://doi.org/10.4161/psb.21949

He L, Gao Z (2009) Pretreatment of seed with H2O2 enhances drought tolerance of wheat (Triticum aestivum L.) seedlings. Afr J Biotechnol 8(22):6151–6157. https://doi.org/10.5897/AJB09.490

Hegazi AM, El-Gaaly FM, El-Din NMN (1995) Effect of some growth regulators on yield and yield components of wheat grown under saline conditions. Ann Agric Sci 33:709–717

Horie T, Hauser F, Schroeder JI (2009) HKT transporter-mediated salinity resistance mechanisms in Arabidopsis and monocot crop plants. Trends Plant Sci 14(12):660–668. https://doi.org/10.1016/j.tplants.2009.08.009

Hossain MA, Hoque T, Zaid A, Wani SH, Mostofa MG, Henry R (2021) Targeting the ascorbate-glutathione pathway and the glyoxalase pathway for genetic engineering of abiotic stress-tolerance in rice. In: Molecular breeding for rice abiotic stress tolerance and nutritional quality. https://doi.org/10.1002/9781119633174.ch21

Hussain S, Shaukat M, Ashraf M, Zhu C, Jin Q, Zhang J (2019) Salinity stress in arid and semi-arid climates: effects and management in field crops. In: Climate change and agriculture. IntechOpen. https://doi.org/10.5772/intechopen.87982

Iqbal M, Ashraf M (2005) Presowing seed treatment with cytokinins and its effect on growth, photosynthetic rate, ionic levels and yield of two wheat cultivars differing in salt tolerance. J Integr Plant Biol 47:1315–1325

Ishitani M, Liu J, Halfter U, Kim CS, Shi W, Zhu JK (2000) SOS3 function in plant salt tolerance requires N-myristoylation and calcium binding. Plant Cell 12(9):1667–1677. https://doi.org/10.1105/tpc.12.9.1667

James RA, Davenport RJ, Munns R (2006) Physiological characterization of two genes for Na^+ exclusion in durum wheat, *Nax1* and *Nax2*. Plant Physiol 142:1537–1547. https://doi.org/10.1104/pp.106.086538

James RA, Blake C, Byrt CS, Munns R (2011) Major genes for Na^+exclusion, N*ax1* and N*ax2* (wheat HKT1;4 and HKT1;5), decrease Na^+accumulation in bread wheat leaves under saline and waterlogged conditions. J Exp Bot 62(8):2939–2947. https://doi.org/10.1093/jxb/err003

Jamil M, Lee KJ, Kim JM, Kim HS, Rha ES (2007) Salinity reduced growth PS2 photochemistry and chlorophyll content in radish. Sci Agric 64:111–118. https://doi.org/10.1590/S0103-90162007000200002

Jeranyama P, DeMoranville CJ (2008) Gas exchange and chlorophyll content of cranberry under salt stress. In: IX international vaccinium symposium, vol 810, pp 753–758. https://doi.org/10.17660/ActaHortic.2009.810.100

Knott JM, Romer P, Sumper M (2007) Putative spermine synthases from *Thalassiosirapseudonana*and *Arabidopsis thaliana* synthesize thermospermine rather than spermine. FEBS Lett 581(16):3081–3086. https://doi.org/10.1016/j.febslet.2007.05.074

Kov'acs Z, Simon-Sarkadi L, Szucs A, Kocsy G (2010) Differential effects of cold, osmotic stress and abscisic acid on polyamine accumulation in wheat. Amino Acids 38(2):623–631. https://doi.org/10.1007/s00726-009-0423-8

Kumar S, Beena AS, Awana M, Singh A (2017) Physiological, biochemical, epigenetic and molecular analyses of wheat (*Triticumaestivum*) genotypes with contrasting salt tolerance. Front Plant Sci 8:1151. https://doi.org/10.3389/fpls.2017.01151

Kumar R, Singh A, Bhardwaj AK, Kumar A, Yadav RK, Sharma PC (2022) Reclamation of salt-affected soils in India: progress, emerging challenges, and future strategies. Land Degrad Dev 33(13):2169–2180. https://doi.org/10.1002/ldr.4320

Kumari A, Kaur R (2018) Evaluation of benzyl-butyl phthalate induced germination and early growth vulnerability to barley seedlings (Hordeumvulgare L.). Indian J Ecol 45:174–177

Kumari A, Kaur R (2020) A review on morpho-physiological traits of plants under phthalates stress and insights into their uptake and translocation. Plant Growth Regul 91:327–347. https://doi.org/10.1007/s10725-020-00625-0

Lauchli A, James RA, Huang CX, McCully M, Munns R (2008) Cell-specific localization of Na+ in roots of durum wheat and possible control points for exclusion. Plant Cell Environ 31: 1565–1574

Lee SS, Kim JH, Hong SB, Yuu SH, Park EH (1998) Priming effect of rice seedson seedling establishment under adverse soil conditions. Korean J Crop Sci 43:194–198

Li J, Li X, Guo L, Lu F, Feng X, He K et al (2006a) A subgroup of MYB transcription factor genes undergoes highly conserved alternative splicing in Arabidopsis and rice. J Exp Bot 57:1263–1273. https://doi.org/10.1093/jxb/erj094

Li WYF, Wong FL, Tsai SN, Phang TH, Shao G et al (2006b) Tonoplast located GmCLC1 and GmN*HX1* from soybean enhance NaCl tolerance in transgenic bright yellow (BY)-2 cells. Plant Cell Environ 29:1122–1137. https://doi.org/10.1111/j.1365-3040.2005.01487.x

Li P, Zhu Y, Song X, Song F (2020) Negative effects of long-term moderate salinity and short-term drought stress on the photosynthetic performance of Hybrid Pennisetum. Plant Physiol Biochem 155:93–104. https://doi.org/10.1016/j.plaphy.2020.06.033

Liu J, Ishitani M, Halfter U, Kim CS, Zhu JK (2000) The Arabidopsis thaliana SOS2 gene encodes a protein kinase that is required for salt tolerance. Proc Natl Acad Sci 97(7):3730–3734. https://doi.org/10.1073/pnas.97.7.3730

Mahajan S, Tuteja N (2005) Cold, salinity and drought stresses: an overview. Arch Biochem Biophys 444(2):139–158. https://doi.org/10.1016/j.abb.2005.10.018

Mandhania S, Madan S, Sawhney V (2006) Antioxidant defense mechanism under salt stress in wheat seedlings. Biol Plant 50:227–231. https://doi.org/10.1007/s10535-006-0011-7

Mao X, Jia D, Li A, Zhang H, Tian S, Zhang X et al (2011) Transgenic expression of TaMYB2A confers enhanced tolerance to multiple abiotic stresses in Arabidopsis. Funct Integr Genom 11: 445–465. https://doi.org/10.1007/s10142-011-0218-3

Massa D, Mattson NS, Lieth HJ (2009) Effects of saline root environment (NaCl) on nitrate and potassium uptake kinetics for rose plants: a Michaelis–Mentenmodelling approach. Plant Soil 318:101–115. https://doi.org/10.1007/s11104-008-9821-z

Mbarki S, Sytar O, Zivcak M, Abdelly C, Cerda A, Brestic M (2018) Anthocyanins of coloured wheat genotypes in specific response to salt stress. Molecules 23:1518. https://doi.org/10.3390/molecules23071518

Mehta PC, Puntamkar SS, Seth SP (1979) Effect of pre-soaking of seeds in different salts with varying concentration on the germination and yield of wheat grown on salinized soil. New Agric 6:73–76

Meng X, Zhou J, Sui N (2018) Mechanisms of salt tolerance in halophytes: current understanding and recent advances. Open Life Sci 13(1):149–154. https://doi.org/10.1515/biol-2018-0020

Munir N, Aftab F (2011) Enhancement of salt tolerance in sugarcane by ascorbic acid pretreatment. Afr J Biotechnol 10(80):18362–18370. https://doi.org/10.5897/AJB11.2919

Munns R (2005) Genes and salt tolerance: bringing them together. New Phytol 167(3):645–663. https://doi.org/10.1111/j.1469-8137.2005.01487.x

Munns R, Tester M (2008) Mechanisms of salinity tolerance. Annu Rev Plant Biol 59(1): 651–681.10.1146/arplant.2008.59.issue-1. https://doi.org/10.1146/annurev.arplant.59.032607.092911

Munns R, Hare RA, James RA, Rebetzke GJ (2000) Genetic variation for improving the salt tolerance of durum wheat. Aust J Agric Res 51(1):69–74. https://doi.org/10.1071/AR99057

Munns R, James RA, Lauchli A (2006) Approaches to increasing the salt tolerance of wheat and other cereals. J Exp Bot 57:1025–1043. https://doi.org/10.1093/jxb/erj100
Murthy ASP (1979) Effect of saline water irrigation on sodium and potassium uptake in UP 301 wheat (*Triticum aestivum* L.). Ann Arid Zone 18, 62–67
Nassar R, Kamel HA, Ghoniem AE, Alarcón JJ, Sekara A, Ulrichs C et al (2020) Physiological and anatomical mechanisms in wheat to cope with salt stress induced by seawater. Plan Theory 9: 237. https://doi.org/10.3390/plants9020237
Navarro-Yepes J, Burns M, Anandhan A, Khalimonchuk O, Del Razo LM, Quintanilla-Vega B et al (2014) Oxidative stress, redox signaling, and autophagy: cell death versus survival. Antioxid Redox Signal 21:66–85. https://doi.org/10.1089/ars.2014.5837
Otoch MDLO, Sobreira ACM, de Aragão MEF, Orellano EG, Lima MDGS, de Melo DF (2001) Salt modulation of vacuolar H^{+}-ATPase and H^{+}-pyrophosphatase activities in Vignaunguiculata. J Plant Physiol 158(5):545–551. https://doi.org/10.1078/0176-1617-00310
Parashar A, Varma SK (1988) Effect of presowing seed soaking in gibberellic acid, duration of soaking, diVerent temperatures and their interaction on seed germination and early seedling growth of wheat under saline conditions. Plant Physiol Biochem (New Delhi) 15:189–197
Pardo J, Cubero B, Leidi E, Quintero F (2006) Alkali cation exchanges: roles in cellular homeostasis and stress tolerance. J Exp Bot 57:1181–1199. https://doi.org/10.1093/jxb/erj114
Parida AK, Das AB (2005) Salt tolerance and salinity effect on plants: a review. Ecotoxicol Environ Saf 60:324–349. https://doi.org/10.1016/j.ecoenv.2004.06.010
Pill WG, Necker AD (2001) The effects of seed treatments on germination and establishment of Kentucky bluegrass (*Poapratense* L.). Seed Sci Technol 29:65–72
Poljakoff-Mayber A, Somers GF, Werker E, Gallagher JL (1994) Seeds of *Kosteletzkyavirginica* (Malvaceae): their structure, germination and salt tolerance. Am J Bot 81:54–59
Rahman MA, Chikushi J, Yoshida S, Yahata H, Yasunaga E (2005) Effect of high air temperature on grain growth and yields of wheat genotypes differing in heat tolerance. J Agric Meteorol 60: 605–608. https://doi.org/10.2480/agrmet.605
Ramezani A, Niazi A, Abolimoghadam AA, Babgohari MZ, Deihimi T, Ebrahimi M, Akhtardanesh H, Ebrahimie E (2013) Quantitative expression analysis of TaSOS1 and TaSOS4 genes in cultivated and wild wheat plants under salt stress. Mol Biotechnol 53(2): 189–197. https://doi.org/10.1007/s12033-012-9513-z
Razzaq A, Ali Q, Qayyum A, Mahmood I, Ahmad M, Rasheed M (2013) Physiological responses and drought resistance index of nine wheat (*Triticum aestivum* L.) cultivars under different moisture conditions. Pak J Bot 45:151–155
Ristova D, Busch W (2014) Natural variation of root traits: from development to nutrient uptake. Plant Physiol 166(2):518–527
Robin AHK, Matthew C, Uddin MJ, Bayazid KN (2016) Salinity-induced reduction in root surface area and changes in major root and shoot traits at the phytomer level in wheat. J Exp Bot 67(12): 3719–3729. https://doi.org/10.1093/jxb/erw064
Roy NK, Srivastava AK (1999) Effect of presoaking seed treatment on germination and amylase activity of wheat (*Triticumaestivum* L.) under salt stress conditions. Rachis 18:46–51
Royo A, Abió D (2003) Salt tolerance in durum wheat cultivars. Span J Agric Res 1:27–35. https://doi.org/10.5424/sjar/2003013-32
Rubio JS, Garcia-Sanchez F, Rubio F, Martinez V (2009) Yield, blossom-end rot incidence and fruit quality in pepper plants under moderate salinity are affected by K^{+} and Ca^{2+} fertilization. Sci Hortic 119:79–87. https://doi.org/10.1016/j.scienta.2008.07.009
Saddiq MS, Iqbal S, Hafeez MB, Ibrahim AM, Raza A et al (2021) Effect of salinity stress on physiological changes in winter and spring wheat. Agronomy 11(6):1193. https://doi.org/10.3390/agronomy11061193
Sairam RK, Rao KV, Srivastava GC (2002) Differential response of wheat genotypes to long term salinity stress in relation to oxidative stress, antioxidant activity and osmolyte concentration. Plant Sci 163:1037–1046. https://doi.org/10.1016/S0168-9452(02)00278-9

Sairam RK, Srivastava GC, Agarwal S, Meena RC (2005) Differences in antioxidant activity in response to salinity stress in tolerant and susceptible wheat genotypes. Biol Plant 49(1):85–91. https://doi.org/10.1007/s10535-005-5091-2

Sakhabutdinova AR, Fatkhutdinova DR, Bezrukova MV, Shakirova FM (2003) Salicylic acid prevents the damaging action of stress factors on wheat plants. Bulg J Plant Physiol 21:114–319

Sallam HA (1999) Effect of some seed-soaking treatments on growth and chemical components on faba bean plants under saline conditions. Ann Agric Sci (Cairo) 44:159–171

Saxena SC, Kaur H, Verma P, Petla BP, Andugula VR, Majee M (2013a) Osmoprotectants: potential for crop improvement under adverse conditions. In: Tuteja N, Singh Gill S (eds) Plant acclimation to environmental stress. Springer, New York, NY. https://doi.org/10.1007/978-1-4614-5001-6_9

Saxena SC, Kaur H, Verma P, Prakash B, Rao V, Majee M (2013b) Osmoprotectants: potential for crop improvement under adverse conditions. In: Tuteja N, Gill SS (eds) Plant accimilation to environmental stress, 1st edn. Springer, pp 197–232

Shewry PR, Hey SJ (2015) The contribution of wheat to human diet and health. Food Energy Secur 4(3):178–202. https://doi.org/10.1002/fes3.64

Silva P, Geros H (2009) Regulation by salt of vacuolar H^+-ATPase and H^+- pyrophosphatase activities and Na^+/H^+ exchange. Plant Signal Behav 4(8):718–726. https://doi.org/10.4161/psb.4.8.9236

Singh A, Bhushan B, Gaikwad K, Yadav OP, Kumar S, Rai RD (2015) Induced defence responses of contrasting bread wheat genotypes under differential salt stress imposition. Indian J Biochem Biophys 52:75–85

Smirnoff N (2005) Antioxidants and reactive oxygen species in plants. Blackwell Publishing, New York, NY. https://doi.org/10.1002/9780470988565

Speer M, Kaiser WM (1991) Ion relations of symplastic and apoplastic space in leaves from Spinaciaoleracea L. and Pisumsativum L. under salinity. Plant Physiol 97:990–997. https://doi.org/10.1104/pp.97.3.990

Tavakoli M, Poustini K, Alizadeh H (2016) Proline accumulation and related genes in wheat leaves under salinity stress. J Agric Sci Technol 18(3):707–716

Tester M, Davenport R (2003) Na^+ tolerance and Na^+ transport in higher plants. Ann Bot 9:503–227. https://doi.org/10.1093/aob/mcg058

Tuteja N, Sahoo RK, Garg B, Tuteja R (2013) OsSUV3 dual helicase functions in salinity stress tolerance by maintaining photosynthesis and antioxidant machinery in rice (*Oryza sativa* L. cv. IR64). Plant J 761:115–127. https://doi.org/10.1111/tpj.12277

Van Oosten MJ, Sharkhuu A, Batelli G, Bressan RA, Maggio A (2013) The *Arabidopsis thaliana* mutant air1 implicates SOS3 in the regulation of anthocyanins under salt stress. Plant Mol Biol 83:405–415. https://doi.org/10.1007/s11103-013-0099-z

Wahid A, Perveen M, Gelani S, Basra SMA (2007) Pretreatment of seed with H_2O_2 improves salt tolerance of wheat seedlings by alleviation of oxidative damage and expression of stress proteins. J Plant Physiol 164:283–294. https://doi.org/10.1016/j.jplph.2006.01.005

Wang B, Lüttge U, Ratajczak R (2001) Effects of salt treatment and osmotic stress on V-ATPase and V-PPase in leaves of the halophyte Suaeda salsa. J Exp Bot 52(365):2355–2365. https://doi.org/10.1093/jexbot/52.365.2355

Wani SH, Singh NB, Haribhushan A, Mir JI (2013) Compatible solute engineering in plants for abiotic stress tolerance-role of glycine betaine. Curr Genomics 14(3):157–165. https://doi.org/10.2174/1389202911314030001

World Health Organization (2019) Preventing disease through healthy environments: exposure to highly hazardous pesticides: a major public health concern (No. WHO/CED/PHE/EPE/19.4.6). World Health Organization, Geneva

Xu Z, Wang C, Xue F, Zhang H, Ji W (2015) Wheat NAC transcription factor TaNAC29 is involved in response to salt stress. Plant Physiol Biochem 96:356–363. https://doi.org/10.1016/j.plaphy.2015.08.013

Yadav S, Solanki YPS, Singh V, Yashveer S, Kesh H (2018) Assessment of diversity for yield and its component traits in F2 population of wheat (Triticumaestivum L.) under salinity conditions. Appl Biol Res 20:256–261. https://doi.org/10.5958/0974-4517.2018.00035.6

Zhang GH, Su Q, An LJ, Wu S (2008) Characterization and expression of a vacuolar Na+/H+ antiporter gene from the monocot halophyte Aeluropuslittoralis. Plant Physiol Biochem 46(2): 117–126. https://doi.org/10.1016/j.plaphy.2007.10.022

Zhou W, Li Y, Zhao BC, Ge RC, Shen YZ, Wang G et al (2009) Overexpression of TaSTRG gene improves salt and drought tolerance in rice. J Plant Physiol 166:1660–1671. https://doi.org/10.1016/j.jplph.2009.04.015

Zou P, Li K, Liu S, He X, Zhang X et al (2016) Effect of sulfated chitooligosaccharides on wheat seedlings (*Triticumaestivum* L.) under salt stress. J Agric Food Chem 64:2815–2821. https://doi.org/10.1021/acs.jafc.5b05624

Drought and Salinity Stress: An Overlapping Osmotic Resistance

5

Norah Johal and Prinka Goyal

Abstract

The present scenario sounds an alarm over the increasing global mean temperature that is coupled with reduced water availability. The present review highlights the osmotic changes that occur on account of water deficit under drought and salinity. Reduced water potential ameliorates the tension in the xylem tissue, thus ultimately leading to xylem cavitation and embolism. The sustained deficit leads to the production of increased reactive oxygen species (ROS) that is not comprehended by usual antioxidative defense system. The upregulation of antioxidative system, variation in root architecture, presence of trichomes, and stomatal regulation alleviate the overall water use efficiency (WUE) under drought and saline conditions. These mitigation strategies are coupled with expression of dehydrins and salt overly sensitive genes (SOS) under drought and salinity, respectively, at the molecular level to provide resistance against the same.

Keywords

Water deficit · Xylem cavitation · Reactive oxygen species (ROS) · Water use efficiency (WUE)

N. Johal (✉)
Department of Agronomy, Punjab Agricultural University, Ludhiana, India
e-mail: norah-cobsbot@pau.edu

P. Goyal
Department of Botany, Punjab Agricultural University, Ludhiana, India

A. Kumar et al. (eds.), *Salinity and Drought Tolerance in Plants*,
https://doi.org/10.1007/978-981-99-4669-3_5

5.1 Introduction

Drought and salinity are ever-increasing abiotic stresses that not only affect the crop yield in general but also disturb the soil ecological balance and shift the cropping pattern at a non-mitigating rate. The proportionate increase in the rate of carbon emissions and depletion of water table in a non-sustainable way has added up to an increase in the mean global temperature of 1 °C ultimately creating drought-like conditions (IPCC 2018). Further on, it is predicted that at this rate, the temperature might go up by 1.5 °C by 2050 (IPCC 2018). These prolonged droughts bring the salts and ions on the upper strata of soil affecting the rhizosphere zone of crop plants ending up in saline or sodic soils. However, both drought and salinity have a common node of lowered water potential due to unavailability or non-absorption of water. The response of plants toward decline in water potential depends upon the exposure of abiotic stress—gradual or abrupt (Ruehr et al. 2019). The gradual stress offers the plant to mitigate and cope up by upregulating the defense mechanisms physiologically, biochemically, and ultimately at the molecular level. The cross talk between these three mechanisms offers the plant an innate level resistance toward stress. However, in the latter case where the stress is severe and abrupt, plant's machinery reaches a level of no point of return in a short span.

Prolonged drought and salinity stress reduce water potential, negate xylem hydraulic conductance, and also interfere with transpiration rate and stomatal conductance (Sevanto et al. 2018). The abovementioned changes in the xylem tissue ultimately affect the rate of photosynthesis and thus affect the crop yield. The coping mechanism of plants comprises of an osmotic balance comprising of compatible solutes that tend to maintain the ionic homeostasis. Solutes, viz., proline, glycine-betaine, mannitol, sorbitol, trehalose, and hexose sugars, not only act as compatible solutes but also as signaling molecules to notify the changes under water deficit or increased salt concentration to the nucleus (Arif et al. 2020). Further changes in the molecular machinery produce stress-related proteins, viz., SOS proteins. This review will discuss changes in water potential under drought and salinity, xylem hydraulic conductance, homeostasis against drought and salinity, and mitigation strategies and future perspectives of drought and salinity resistant cultivars/varieties of major crops.

5.2 Water Potential Under Drought and Salinity

The reduction in water potential (Ψ_w) is a common node on account of salinity and drought stress. The unavailability of water around the rhizosphere of the plant creates tension in the xylem sap ultimately reflecting in lower xylem water potential values (Ψ_x). In salinity stress, the presence of ions in high concentration further adds to the lowering of turgor potential and relative water content (RWC). The reduction in the Ψ_w is not an individual entity but an amalgamation of various dependent factors, viz., stomatal conductance, photosynthetic rate (A_n), and carbon source allocation (C_i). The strain on the xylem conductance is much more worse when the stress is prolonged over a larger period of time. During mild and moderate stress water deficit exposure, the setback to the photosynthesis rate (A_n) is only in terms of stomatal closure. When the stress subsides, the stomata reopen and plant regains the normal photosynthetic rate. However, during severe stress conditions, WUE declines even after the stomatal closure and tends to cross a threshold barrier beyond which the damage to the photosynthetic apparatus is irreversible. When the drought and salinity stress conditions revert back to the optimal WUE environment, the carbon sources are diverted back to the repair of photosynthetic requisites. The tolerance capacity of species from temperate and semiarid biomes against low water availability has been attributed to enhanced biochemical capacity and faster repair of photosynthetic machinery. Leaf cooling and heat exposure are critical factors that need further consideration. In the leaf that is equipped with well-watered conditions, leaf transpiration imparts an optimal cooling strategy so as to prevent leaf heating (Drake et al. 2018). The temperature difference of 2–3 °C prevails between well-watered and water limited conditions, and this difference can cause substantial damage and also affects the gas exchange recovery (Birami et al. 2018). However, when water supply becomes limiting, stomata close and leaves can heat up. Water use efficiency (WUE) is one such trait which determines the sustenance of tolerant genotypes under water deficit condition (Table 5.1). Deep rooted genotypes maintain a constant value of WUE under physical and physiologically drought prone areas. The deep root system draws its water requirement making use of groundwater table, thus supporting drought avoidance strategy. This is evident from karst ecosystem under drought depicting contrasting behavior of leaf water potential and deeper root system. The species with small xylem diameter has a lower capacity for water uptake and depicts more negative water potentials and ultimately lower water use efficiency (WUE) ending up in drought-induced canopy defoliation and mortality. The species with large-diameter xylem vessels with deeper root system possess high water uptake capacity ultimately. Such species have hgh leaf-level WUE and thus stand high survival rates under drought with modertae canopy defoliation (Ding et al. 2021).

5.3 Xylem Hydraulic Conductance

The cohesive balance between the transpiration and water potential in plants especially woody trees is disturbed when the latter are exposed to the water deficit conditions. The negative values of xylem potential under drought and salinity stress reaches a threshold level below which the process of cavitation starts. The closure of stomata on account of water deficit also is determined by these threshold values; however, it is accounted that the stomata closure occurs much early before the plant hits threshold value (Choat et al. 2018). This might be an escape strategy by the plants to overcome the water deficit conditions. The xylem under tension tends to avoid embolism by prior stomatal closure, and this stomatal closure causes decline in transpiration and photosynthetic rates and ultimately disturbs the translocation of photosynthates (Fig. 5.1). In another study in two tomato varieties (Abdalla et al.

Table 5.1 Differential response of crop species under drought and salinity

Species	Stress	Treatment	Type of experiment	Response	Reference
Apple trees (*Malus domestica* Borkh.)	Drought	Drought stress for 60 days	Field	Reduction in leaf water potential Reduced hydraulic conductance	Bhusal et al. (2019)
11 co-occurring woody species in karst ecosystem	Drought	135-day rainfall exclusion experiment	Field	Negative water potential Reduced WUE Moderate canopy defoliation	Ding et al. (2021)
17 Californian woody species	Drought	Extreme to exceptional drought	Field	Stem xylem vulnerability Xylem cavitation Reduced water potential with stomatal closure	Pivovaroff et al. (2018)
Soybean genotypes	Drought	Progressive soil drying and rewetting	Pot experiment	Tolerant genotype exhibits higher WUE by water translocation and ionic homeostasis	Mesquita et al. (2020)
Two native (*Quercus pubescens* [Qp] and *Prunus mahaleb* [pm]) and two invasive (*Robinia pseudoacacia* [Rp] and *Ailanthus altissima* [aa]) woody angiosperms	Drought	3-month-old plants subjected to reduced leaf conductance, i.e., 10% in comparison with control plants for further 2 months	Pot experiment	Stomatal closure Xylem embolism Carbon starvation Hydraulic failure	Savi et al. (2016)
Portulaca oleracea L. cv. Grene	Salinity	25-day-old plants exposed to 100 mM and 300 mM NaCl	Growth experiment	Exposure to 300 mM NaCl reduced water potential, increased MDA content, transition from C4 to CAM pathway	Hnilickova et al. (2021)
T. Ramoisissima, *S. pectinate*, and *D spicata*	Salinity	0, 15, and 30 g L^{-1} NaCl for 2 weeks	Greenhouse experiment	Low water potential, low photosynthesis	Betzen et al. (2019)
Cucumis melo cv. *Huaghemi*	Salinity	0.8 g^{-1}, 2 g^{-1}, and 5 g^{-1} NaCl	Field experiment	Reduced specific leaf area (SLA), leaf area ratio (LAR), net assimilation rate (NAR), water use efficiency (WUE)	Tedeschi et al. (2017)

Alfaalfa variety	Salinity	NaCl (0, 75, 150, 225 mM)	Greenhouse experiment	Reduction in leaf water potential at 150 and 225 mM	Sun et al. (2016)
Wheat genotypes	Salinity	NaCl (4, 6, 8, and 10 dSm^{-1})	Field Experiment	Reduced evapotranspiration, yield, and soil water potential	Abedinpour (2017)
Populus euphratica	Salinity	NaCl (50, 100, 150, and 200 mM)	Laboratory experiment	Reduced stomatal area aperture, loss of hydraulic conductivity	Rajput et al. (2017)

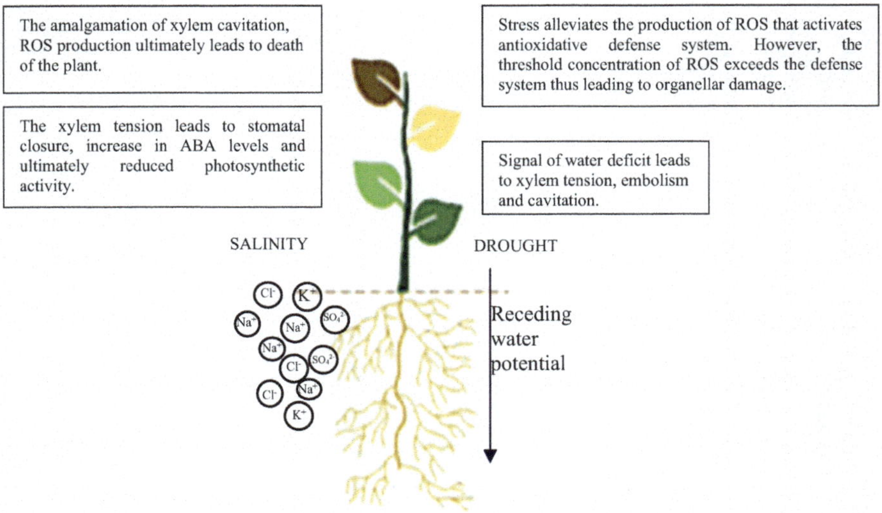

Fig. 5.1 Diagrammatic representation of lowered water potential under drought and salinity

2021), an interesting observation was postulated. Under well-watered condition, there was a direct correlation between leaf water potential (Ψ_l) and stomatal conductance; however, as the water levels receded, the relation became nonlinear one. It was observed that under water deficit conditions, the cross talk between lower-ground parts, viz., root architecture comprising of root length and root area and soil hydraulic conductance and stomatal conductance, played major role for plants to cope up with drought stress. Similarly, in *Vitis vinifera* L., the 4-year recurrent water stressed plants possessed xylem hydraulic memory depicting enhanced water use efficiency (WUE) in comparison with well water vines (Tombesi et al. 2018). In addition to this, there is a disturbance in the ionic homeostasis leading to the generation of reactive oxygen species (ROS). The generated ROS further damage the organelle structures and also cellular membrane. The increased threshold levels of ROS especially in the meristematic tissues are an important biomarker to assess the low levels of water availability. Water deficit in amalgamation with increased ROS leads to more serious consequence of cavitation, cytorrhysis, cell death, and ultimately tree mortality (Mantova et al. 2022). While discussing the xylem hydraulic conductance, the role of aquaporins cannot be neglected. Aquaporins are set of intrinsic membrane proteins that play a pivotal role in maintain ionic and osmotic homeostasis (Maurel et al. 2015). It is pointed that aquaporins improve osmotic balance by preventing membrane damage and leakiness, improving root architecture, viz., growth area and length, and by the upregulation of ROS scavenging constituent mechanisms (Patel and Mishra 2021).

5.4 Water Use Efficiency (WUE) Under Drought and Salinity

An important aspect to counteract water deficit is to have optimal water use efficiency (WUE) under such conditions. It is postulated that WUE is the most important factor that contributes to ultimate yield (Panda et al. 2021). Under the rainfed conditions, "more crop per drop" is considered as the future conservation strategy for enhanced production under water deficit conditions (Sharma et al. 2015). It is defined as the ratio of yield vs. evapotranspiration (WUE = yield/ET) (Zhao et al. 2014). Water use efficiency is internally dependent on stomatal size aperture (Bertolino et al. 2019), root architecture (Xu et al. 2016), and transpiration efficiency (Sinclair 2018). In another experiment, wheat and bean cultivars were exposed to two salinity levels (more than and less than 8 dSm^{-1}). The increased shoot/root ratio improved WUE under higher salinity levels (Khataar et al. 2018).

In another case study, the drought tolerance of pearl millet was attributed to leaf area development and deep root system under water deficit conditions (Zegada-Lizarazu and Iijima 2005). Apart from the root architecture and stomatal conductance, an interesting observation that is contributing to improve WUE is the trichome density. Trichome forms an extra layer on the leaf surface that reduces the rate of transpiration, and this reduction under water deficit conditions leads to improved WUE that was documented in *Solanum lycopersicum* (Galdon-Armero et al. 2018).

5.5 Mitigation Strategies Against Drought and Salinity

Three main strategies that sense the lowered water potential (Ψ) are the stress avoidance, stress tolerance, and stress escape. The avoidance and tolerance strategy work in a cohesive fashion with each other. The avoidance mechanism involves the maintenance of optimal levels of water potential so that the plants should not reach the critical threshold levels. Once the threshold limit is reached, the damage can be recorded in terms of xylem embolism and cavitation leading to damage to xylem vessels. The avoidance strategy involves alteration in stomatal number, trichome hairs and leaf area to cope up with the external water deficit environment (Kebede et al. 2019). In salinity, an association of root with arbuscular mycorrhizal fungi (AMF) is an ameliorating approach against salt stress. The AMF induce the exclusion of toxic Na^+ and K^+ ions to maintain a favorable osmotic gradient (Evelin et al. 2019). Tolerance strategy primarily involves the contribution of enzymatic and nonenzymatic constituents to counter the ROS (reactive oxygen species) that are produced on account of water deficit stress under drought and salinity. Shortening of life cycle, rapid plant development, and self-reproduction are some of the responses of escape strategy (Seleiman et al. 2021). However, every mechanistic approach toward reduced water potential is very elaborative which seems to be beyond the scope of this chapter. However, two stress-specific proteins Dehydrins and Salt-overly sensitive (SOS) deserve a special mention against drought and salinity which help to mitigate the plants against reduced water potential.

Dehydrins (DHS) are set of stress-regulated hydrophilic proteins that depict an upregulated expression against drought stress. They are ubiquitous in nature, i.e., present in every compartment of cell, viz., chloroplast, mitochondria, nucleus, cytoplasm, and endoplasmic reticulum (Riyazuddin et al. 2022). They impart drought tolerance in the cell by maintaining the osmotic potential by production of osmolytes, detoxification of ROS, protecting the photosynthetic machinery, and maintaining optimal water potential levels in the system. The former mechanism retains water levels in the system under water deficit conditions. The SOS pathway provides a specific mechanism for the extrusion of Na^+ ions to establish an osmotic balance. Under non-uniform saline conditions, plants tend to redirect their root growth by differential concentration of auxin hormone in order to avoid salt stress (Li and Zhang 2008; Sun et al. 2008). However, in a uniform saline environment, plant roots tend to reduce the number and length of root hairs so as to reduce the exposure of surface area to external Na^+ ions. Recent results revealed that the SOS signaling pathway coupled with avoidance of saline environment plays a role in the plastic development of root hairs under salt stress (Wang and Li 2008; Wang et al. 2008).

References

Abdalla M, Ahmed MA, Cai G, Wankmuller F, Schwartz N, Litig O, Javaux M, Carminati A (2021) Stomatal closure during water deficit is controlled by below ground hydraulics. Ann Bot 129: 161–170

Abedinpour M (2017) Wheat water use and yield under different salinity of irrigation water. J Water Land Dev 33:3–9

Arif Y, Singh P, Siddiqui H, Bajguz A, Hayat S (2020) Salinity induced physiological and biochemical changes in plants: an omic approach towards salt stress tolerance. Plant Physiol Biochem 156:64–77

Bertolino LT, Caine RS, Gray JE (2019) Impact of stomatal density and morphology on water-use efficiency in a changing world. Front Plant Sci 10:1–225

Betzen BM, Smart CM, Maricle KL, Maricle BR (2019) Effects of increasing salinity on photosynthesis and plant water potential in Kansas salt marsh species. Trans Kans Acad Sci 122:49–58

Bhusal N, Han SG, Yoon TM (2019) Impact of drought stress on photosynthetic response, leaf water potential, and stem sap flow in two cultivars of bi-leader apple trees (malus× domestica Borkh.). Sci Hortic 246:535–543

Birami B, Gattmann M, Heyer AG, Grote R, Arneth A, Ruehr NK (2018) Heat waves alter carbon allocation and increase mortality of Aleppo pine under dry conditions. Front For Glob Change 1: 1–8

Choat B, Brodribb TJ, Brodersen CR, Duursma RA, López R, Medlyn BE (2018) Triggers of tree mortality under drought. Nature 558:531–539

Ding Y, Nie Y, Chen H, Wang K, Querejeta JI (2021) Water uptake depth is coordinated with leaf water potential, water-use efficiency and drought vulnerability in karst vegetation. New Phytol 229:1339–1353

Drake JE, Tjoelker MG, Varhammar A, Medlyn BE, Reich PB, Leigh A, Pfautsch S, Blackman CJ, Lopez R, Aspinwall MJ (2018) Trees tolerate an extreme heat wave via sustained transpirational cooling and increased leaf thermal tolerance. Glob Chang Biol 24:2390–2402

Evelin H, Devi TS, Gupta S, Kapoor R (2019) Mitigation of salinity stress in plants by arbuscular mycorrhizal symbiosis: current understanding and new challenges. Front Plant Sci 10:1–470

Galdon-Armero J, Fullana-Pericas M, Mulet PA, Conesa MA, Martin C, Galmes J (2018) The ratio of trichomes to stomata is associated with water use efficiency in *Solanum lycopersicum* (tomato). Plant J 96:607–619

Hnilickova H, Kraus K, Vachova P, Hnilicka F (2021) Salinity stress affects photosynthesis, malondialdehyde formation, and proline content in *Portulaca oleracea* L. Plan Theory 10:1–845

Intergovernmental Panel on Climate Change (IPCC) (2018) Global warming of 1.5°C. An IPCC special report on the impacts of global warming of 1.5°C above pre-industrial levels and related global greenhouse gas emission pathways, in the context of strengthening the global response to the threat of climate change, sustainable development, and efforts to eradicate poverty

Kebede A, Kang MS, Bekele E (2019) Advances in mechanisms of drought tolerance in crops, with emphasis on barley. Adv Agron 156:265–314

Khataar M, Mohammadi MH, Shabani F (2018) Soil salinity and matric potential interaction on water use, water use efficiency and yield response factor of bean and wheat. Sci Rep 8:1–13

Li X, Zhang W (2008) Salt-avoidance tropism in *Arabidopsis thaliana*. Plant Signal Behav 3:351–353

Mantova M, Herbette S, Cochard H, Torres-Ruiz JM (2022) Hydraulic failure and tree mortality: from correlation to causation. Trends Plant Sci 27:335–345

Maurel C, Boursiac Y, Luu DT, Santoni VR, Shahzad Z, Verdoucq L (2015) Aquaporins in plants. Physiol Rev 95:1321–1358

Mesquita RO, Coutinho FS, Vital CE, Nepomuceno AL, Williams TCR, de Oliveira Ramos HJ, Loureiro ME (2020) Physiological approach to decipher the drought tolerance of a soybean genotype from *Brazilian savana*. Plant Physiol Biochem 151:132–143

Panda A, Rangani J, Parida AK (2021) Physiological and metabolic adjustments in the xero-halophyte *Haloxylon salicornicum* conferring drought tolerance. Physiol Plant 172:1189–1211

Patel J, Mishra A (2021) Plant aquaporins alleviate drought tolerance in plants by modulating cellular biochemistry, root architecture and photosynthesis. Physiol Plant 172:1030–1044

Pivovaroff AL, Cook VM, Santiago LS (2018) Stomatal behaviour and stem xylem traits are coordinated for woody plant species under exceptional drought conditions. Plant Cell Environ 41:2617–2626

Rajput VD, Yaning C, Ayup M, Minkina T, Sushkova S, Mandzhieva S (2017) Physiological and hydrological changes in *Populus euphratica* seedlings under salinity stress. Acta Ecol Sin 37: 229–235

Riyazuddin R, Nisha N, Singh K, Verma R, Gupta R (2022) Involvement of dehydrin proteins in mitigating the negative effects of drought stress in plants. Plant Cell Rep 41:519–533

Ruehr NK, Grote R, Mayr S, Arneth A (2019) Beyond the extreme: recovery of carbon and water relations in woody plants following head and drought stress. Tree Physiol 39:1285–1299

Savi T, Casolo V, Luglio J, Bertuzzi S, Gullo MAL, Nardini A (2016) Species-specific reversal of stem xylem embolism after a prolonged drought correlates to endpoint concentration of soluble sugars. Plant Physiol Biochem 106:198–207

Seleiman MF, Al-Suhaibani N, Ali N, Akmal M, Alotaibi M, Refay Y, Battaglia ML (2021) Drought stress impacts on plants and different approaches to alleviate its adverse effects. Plan Theory 10:1–259

Sevanto S, Ryan M, Dickamn LT, Derome D, Patera A, Defraeye T, Pangle RE, Hudson PJ, Pockman WT (2018) Is desiccation tolerance and avoidance reflected in xylem and phloem anatomy of two coexisting arid-zone coniferous trees? Plant Cell Environ 41:1551–1654

Sharma B, Molden D, Cook S (2015) Water use efficiency in agriculture: Measurement, current situation and trends (No. 612-2016-40604)

Sinclair TR (2018) Effective water use required for improving crop growth rather than transpiration efficiency. Front Plant Sci 9:1–1442

Sun F, Zhang W, Hu H, Li B, Wang Y, Zhao Y, Li K, Liu M, Li X (2008) Salt modulates gravity signalling pathway to regulate growth direction of primary roots in *Arabidopsis*. Plant Physiol 146:178–188

Sun J, Yang G, Zhang W, Zhang Y (2016) Effects of heterogeneous salinity on growth, water uptake, and tissue ion concentrations of alfalfa. Plant Soil 408:211–226

Tedeschi A, Zong L, Huang CH, Vitale L, Volpe MG, Xue X (2017) Effect of salinity on growth parameters, soil water potential and ion composition in *Cucumis melo* cv. Huanghemi in North-Western China. J Agron Crop Sci 203:41–55

Tombesi S, Frioni T, Poni S, Palliotti A (2018) Effect of water stress 'memory' on plant behavious during subsequent drought stress. Environ Exp Bot 150:106–114

Wang Y, Li X (2008) Salt stress-induced cell reprogramming, cell fate switch and adaptive plasticity during root hair development in Arabidopsis. Plant Signal Behav 3:436–438

Wang YN, Zhang WS, Li KX, Sun FF, Han CK, Wang Y, Li X (2008) Salt-induced plasticity of root hair development is caused by ion disequilibrium in *Arabidopsis thaliana*. J Plant Res 121: 87–96

Xu C, Tao H, Tian B, Gao Y, Ren J, Wang P (2016) Limited-irrigation improves water use efficiency and soil reservoir capacity through regulating root and canopy growth of winter wheat. Field Crops Res 196:268–275

Zegada-Lizarazu W, Iijima M (2005) Deep root water uptake ability and water use efficiency of pearl millet in comparison to other millet species. Plant Prod Sci 8:454–460

Zhao H, Wang RY, Ma BL, Xiong YC, Qiang SC, Wang CL, Liu CA, Li FM (2014) Ridge furrow with full plastic film mulching improves water use efficiency and tuber yields of potato in a semiarid rainfed ecosystem. Field Crop Res 161:137–148

Physiological and Postharvest Quality Changes of Horticultural Crops Under Salt Stress

6

Mukesh Kumar Mehla, Akshay, Rajkumar, Ashwani Kumar, Satish Kumar Sanwal, Anita Mann, and Rajender Kumar Yadav

Abstract

Salt stress is a significant abiotic stress that can adversely affect the growth and productivity of horticultural crops. Salt stress develops due to the accumulation of excess salt in the soil, leading to osmotic and ion imbalances within the plant affecting its growth and development. Salt stress also alters the physiological characteristics of horticultural crops. Salt stress affects the physiological parameters such as photosynthesis, respiration rate, water relations, oxidative stress, nutrient uptake and utilization, plant hormones, fruit development, ripening and senescence, etc. This in turn results in alteration and deterioration the postharvest quality of horticulture crops. Further research in these contexts will enhance our understanding of these physiological parameters and their interactions with environmental factors. This chapter summarizes the current knowledge of changes in postharvest quality of horticultural crops resulting from salt stress. The different mechanisms underlying these changes and strategies to mitigate the negative effects of salt stress on postharvest quality of horticultural crops are also discussed.

Keywords

Salt stress · Horticulture · Postharvest quality · Physiology and biochemical changes

M. K. Mehla · Akshay · Rajkumar (✉) · A. Kumar · S. K. Sanwal · A. Mann · R. K. Yadav
ICAR – Central Soil Salinity Research Institute, Karnal, Haryana, India

A. Kumar et al. (eds.), *Salinity and Drought Tolerance in Plants*,
https://doi.org/10.1007/978-981-99-4669-3_6

6.1 Introduction

Salinity stress is a major threat to global agriculture, with approximately 20% of total cultivated and 33% of irrigated agricultural lands being affected by it. The expansion of salinized areas is on the rise annually, with a rate of 10%. This increase can be attributed to several factors, including low levels of precipitation, high surface evaporation, weathering of indigenous rocks, irrigation with saline water, and substandard cultural practices. The projection is that by 2050, over half of the cultivable land will be affected by salinization (Jamil et al. 2011).

Horticultural crops are important sources of food, nutrition, and income for millions of people worldwide. However, the production and postharvest quality of these crops are often threatened by various environmental stresses, including salinity stress. Salinity stress, which is caused by high levels of salt in the soil and water, can negatively affect the growth, development, and productivity of horticultural crops. Furthermore, the impact of salt stress on the postharvest quality of horticultural crops has been a topic of growing concern. The physiological changes induced by salt stress during plant growth and development can result in alterations to the composition and quality of harvested produce. These changes can include decreased water content, changes in color and flavor, reduced nutritional value, and increased susceptibility to decay and spoilage. The precise mechanisms by which salt stress affects postharvest quality are not fully understood, but it is believed that this relates to changes in various physiological processes and regulation mechanisms, such as enzymatic activity, hormone signaling, and membrane integrity. Therefore, understanding the physiological changes that occur in horticultural crops grown under salt stress conditions is critical for developing effective mitigation strategies to maintain postharvest quality and reduce food waste.

The negative impact of salinity stress on horticultural crops is well documented. Salinity stress affects plants by altering their physiological processes, including photosynthesis, respiration, and transpiration, leading to reduced growth and development (Munns et al. 2006). The high salt concentration in the soil and water can also cause osmotic stress, leading to water deficit and ion toxicity in plants (Flowers and Colmer 2008). Salt stress can induce changes in the cell membrane structure and composition, leading to increased permeability and lipid peroxidation, which can result in reduced cell viability and membrane integrity during postharvest storage (Hassan et al. 2019; Rastogi et al. 2018). Salt stress can also affect the antioxidant defense mechanisms of horticultural crops, resulting in oxidative stress and damage to cellular components such as proteins and DNA (Rai et al. 2021). In addition, salt stress can alter the expression of genes and proteins involved in various metabolic pathways, such as photosynthesis, respiration, and carbohydrate metabolism, which can impact the postharvest quality of horticultural crops (Hao et al. 2020; Wang et al. 2017). Salinity stress can lead to reduced fruit size and weight, decreased sugar content, and altered flavor and aroma (Parida and Das 2005). In addition, salinity stress can also cause physical damage to the crops, such as fruit cracking and wilting, which can further reduce their market value and shelf life (Singh et al. 2015). These physiological changes can have significant impacts on the yield and postharvest quality of horticultural crops.

Several factors can influence the extent and severity of physiological changes in horticultural crops under salt stress conditions during postharvest storage. These include the crop species, cultivar, stage of maturity, duration and severity of salt stress exposure, storage temperature and humidity, and postharvest treatments such as washing, packaging, and refrigeration (Kamal et al. 2021; Li et al. 2018; Zhang et al. 2020). Understanding the physiological changes that occur in horticultural crops under salt stress conditions is crucial for developing effective mitigation strategies to reduce the negative impacts of salinity stress on crop yield and postharvest quality. This chapter provides an overview of the physiological changes that occur in horticultural crops due to salinity stress, focusing on their effects on postharvest quality and potential strategies for mitigating these effects.

6.2 Impact of Salt Stress on Horticultural Crops

Salt stress is a significant abiotic stress factor that limits the growth, productivity, and quality of horticultural crops. It occurs when the soil or irrigation water contains high levels of salt, which leads to ion toxicity and osmotic stress in plants. Salt stress affects the plant's water relations, nutrient uptake, and photosynthesis, leading to reduced growth and yield. It also affects the postharvest physiology and biochemical changes of crops, leading to reduced shelf life and quality. Salt stress affects the nutritional and medicinal properties of harvested crops, which can result in economic losses for farmers and decreased access to healthy food for consumers.

Salt stress affects the physiological responses of horticultural crops, including respiration rate, ethylene production, and chlorophyll content. Salt stress leads to an increase in respiration rate, which can lead to a decrease in postharvest quality due to a decrease in shelf life (Jiang et al. 2019). Ethylene production is also affected by salt stress, which can lead to premature ripening and decreased shelf life (Adams et al. 2019). Chlorophyll content is also affected by salt stress, leading to a decrease in the photosynthetic capacity of horticultural crops, resulting in a decrease in yield and quality (Shukla et al. 2018). Study by Hayat et al. (2012) revealed that the application of salt stress reduced the rate of respiration in tomato fruits. The reduction in respiration was associated with the suppression of enzymatic activities, which resulted in lower ATP synthesis. Salt stress also caused a decrease in ethylene production in cucumber fruits, which could be attributed to the impairment of ethylene biosynthesis (Khan et al. 2015). In contrast, salt stress resulted in the elevation of ethylene production in some crops, such as tomato and bell pepper, as reported by Hernández-Montiel et al. (2017) and Dukare et al. (2020), respectively (Fig. 5.1).

Salt stress has a significant impact on the physiological parameters of horticultural crops. The effect of salt stress on different parameters varies among different crops and depends on the severity and duration of the stress. Salt stress can induce oxidative stress in plants, which can lead to a decrease in photosynthesis rate,

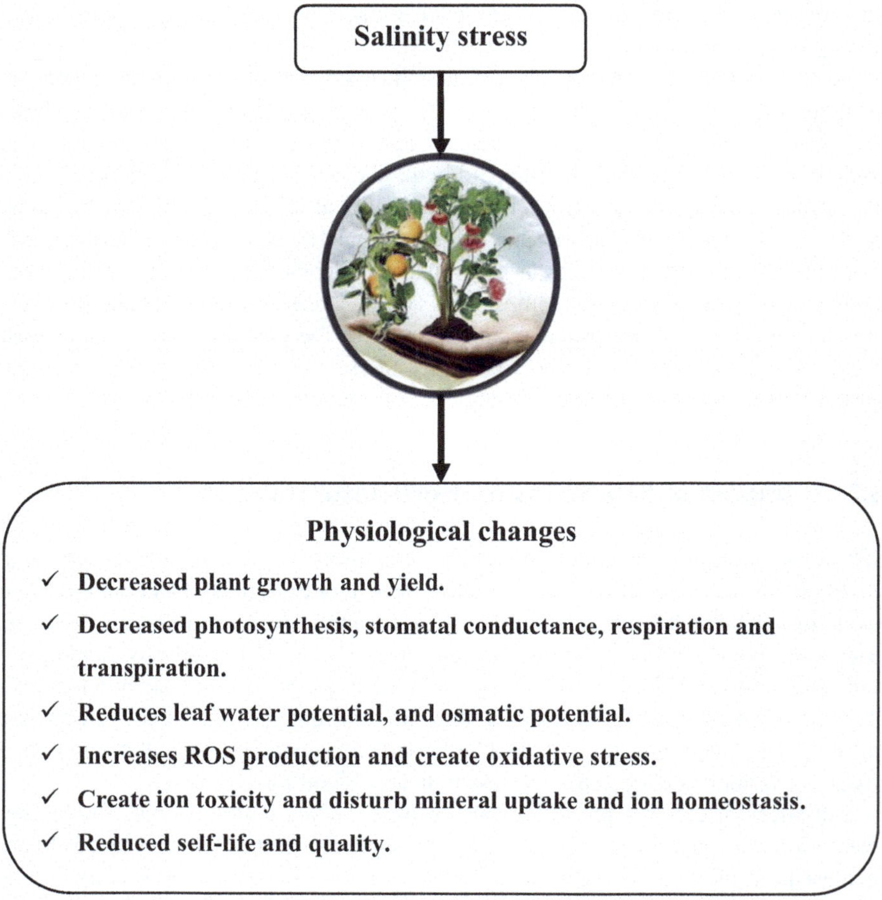

Fig. 6.1 Various physiological characteristics influenced by salt stress

chlorophyll content, and sugar metabolism. However, plants also respond to salt stress by increasing the activity of antioxidant enzymes and enhancing the accumulation of phenolic compounds, flavonoids, and other antioxidants.

6.3 Impact of Salt Stress on Physiological Parameters of Horticultural Crops

Horticultural crops are influenced by various physiological parameters that play a crucial role in plant growth and development. Understanding these parameters and how they interact with environmental factors is crucial in optimizing crop productivity and quality. The physiological parameters discussed further includes photosynthesis, respiration, transpiration, water use efficiency, nutrient uptake and utilization, stress tolerance, plant hormones, fruit development and ripening, and flower development and senescence.

Salt stress can lead to disturbance of water relations on horticultural crops resulting in decreased water uptake and increased transpiration, resulting in water deficit and ion toxicity. Accumulation of toxic ions such as Na^+ and Cl^- in plant tissues interferes with nutrient uptake, resulting in nutrient imbalances and deficiencies (Munns and Tester 2008). Additionally, salt stress disrupts the integrity of the cell membrane, leading to electrolyte leakage and loss of cellular turgor pressure (Ashraf and Foolad 2007). These changes can have a significant impact on the postharvest quality of horticultural crops, including reduced fruit size, weight, and quality.

Salt stress also affects the process of photosynthesis in horticultural crops. The high salinity levels reduce the efficiency of photosystem II (PS-II) in the thylakoid membrane, resulting in decreased chlorophyll content and reduced photosynthetic activity (Parida and Das 2005). The decrease in photosynthesis leads to a reduction in the production of carbohydrates, which are essential for plant growth and fruit development (Munns and Tester 2008). Additionally, salt stress can induce oxidative stress, which damages the photosynthetic machinery and reduces the efficiency of photosynthesis (Gill and Tuteja 2010).

Furthermore, salt stress affects mineral uptake in horticultural crops, leading to nutrient imbalances and deficiencies. The high levels of Na^+ and Cl^- ions in the soil reduce the uptake of essential nutrients such as K^+, Ca^{2+}, and Mg^{2+} (Munns et al. 2012). These imbalances can have negative effects on the quality of horticultural crops, including reduced fruit size, color, and flavor.

Another significant effect of salt stress on horticultural crops is the activation of antioxidant defense systems. Salt stress can lead to the accumulation of reactive oxygen species (ROS), which are highly reactive molecules that can cause oxidative damage to plant cells (Hasanuzzaman et al. 2013). To counter this oxidative stress, plants activate their antioxidant defense systems, including enzymatic and nonenzymatic antioxidants (Gill and Tuteja 2010). The activation of these systems helps to maintain plant health and productivity under salt stress conditions. Effects of salt stress on various physiological parameters are summarized in Table 6.1.

6.4 Photosynthesis

Photosynthesis is a process by which green plants use sunlight to synthesize food from carbon dioxide and water. In horticultural crops, it is a vital process for plant growth and development playing an important role in yield and quality of produce. This process occurs in the chloroplasts of plant cells, and the end products are oxygen and glucose. The rate of photosynthesis in horticultural crops is affected by several factors, including light intensity, temperature, and carbon dioxide concentration (Hasanuzzaman et al. 2013). In general, higher light intensity and carbon dioxide concentration and lower temperatures increase the rate of photosynthesis in horticultural crops. Salt stress can affect photosynthesis by reducing the efficiency of

Table 6.1 Various physiological changes induced by salt stress

Physiological changes	Important findings	Reference
Biochemical characteristics	Salt stress reduces the concentration of vitamin C and soluble sugars but increases the concentration of proline and phenolic compounds in tomato and cucumber	Ahmad et al. (2016), Hasanuzzaman et al. (2018)
Growth and development	Salt stress can affect plant growth and development by reducing cell expansion and division, altering root morphology, and inducing premature senescence	Hasegawa et al. (2000), Munns and Tester (2008)
Ion homeostasis	Plants have developed different mechanisms to maintain ion homeostasis under salt stress conditions, such as compartmentalization, exclusion, and sequestration	Munns and Tester (2008), Shabala and Cuin (2008)
Molecular mechanisms	Salt stress alters the expression of genes related to stress response and metabolism in tomato and cucumber	Ahmad et al. (2016)
	Salt stress can induce changes in gene expression, protein synthesis, and signal transduction pathways in plants, leading to the activation of stress-responsive genes and proteins	Munns and Tester (2008)
Nutritional quality	Drought stress can enhance the concentration of antioxidants, phenolic acids, and carotenoids in some horticultural crops	Khalid et al. (2018), Pereira et al. (2018)
Osmotic balance	Plants can maintain osmotic balance under salt stress conditions by accumulating compatible solutes, such as proline and glycine betaine, and adjusting the osmotic potential of the cell	Ashraf and Foolad (2007), Shabala and Cuin (2008)
Physical characteristics	Salt stress causes a decrease in fruit size, weight, and firmness in tomato and cucumber	Ahmad et al. (2016)
Water uptake	Salt stress can reduce water uptake in crops by increasing the soil osmotic potential and reducing the water potential gradient	Munns and Tester (2008), Shabala and Cuin (2008)

photosystem II, the main photosynthetic machinery in plants, resulting in a decrease in photosynthetic rate and leading to a reduction in yield (Koyro 2006). Salt stress can reduce the rate of photosynthesis by affecting various components of the photosynthetic apparatus, such as chlorophyll content, stomatal conductance, and the efficiency of the photosynthetic electron transport system (Parida and Das 2005). High salinity can also induce oxidative stress and damage the photosynthetic machinery by generating reactive oxygen species (ROS). However, some studies have reported that certain horticultural crops can adapt to salt stress by modifying their photosynthetic processes and enhancing their ability to scavenge ROS (Munns and Tester 2008). This adaptation mechanism involves the activation of several enzymes, such as superoxide dismutase (SOD), catalase (CAT), and peroxidase (POD), which play a crucial role in the detoxification of ROS (Yang et al. 2012; Zhang and Kirkham 1996). Additionally, the use of exogenous substances, such as plant growth regulators and osmo-protectants, can enhance the tolerance of horticultural crops to salt stress by improving the efficiency of their photosynthetic machinery. Therefore, it is important to understand the impact of salt stress on photosynthesis and the mechanisms underlying plant adaptation to salt stress in order to develop effective strategies to improve crop productivity in saline environments.

6.5 Respiration

Respiration is a crucial process of energy production and plant growth. It is the process by which plants convert sugars and oxygen into energy, carbon dioxide, and water. This process occurs in the mitochondria of plant cells, and the end products are carbon dioxide and water. The rate of respiration in horticultural crops is influenced by several factors, including temperature, oxygen concentration, and the availability of energy substrates. In general, higher temperatures and oxygen concentrations increase the rate of respiration in horticultural crops. However, excessive respiration can lead to a reduction in crop productivity and quality. Salt stress can have detrimental effects on various metabolic processes in plants which includes respiration. Studies have demonstrated that salt stress can lead to a decrease in the respiration rate of crops, which can affect their overall growth and development (Koyro 2006). The reduction in respiration rates may be attributed to several factors, including the inhibition of enzymes involved in respiration, such as cytochrome oxidase, as well as a decrease in the availability of oxygen, which is required for aerobic respiration (Zhang and Kirkham 1996). Moreover, salt stress can also lead to the accumulation of toxic ions such as Na^+ and Cl^- in the plant cells, which can impair mitochondrial function and decrease energy production (Maathuis and Amtmann 1999). However, some crops, such as halophytes, have developed unique mechanisms to cope with salt stress, which allows them to maintain their respiration rates even under high salinity conditions (Sairam and Tyagi 2004). Therefore, further research is required to develop strategies to enhance the resilience of horticultural crops to salt stress and to mitigate the negative impact of salt stress on their respiration rates. This could include the identification and selection of salt-tolerant cultivars, as well as the use of exogenous substances such as plant growth regulators or organic amendments to enhance the plant's ability to cope with salt stress (Munns and Tester 2008). In conclusion, salt stress is a significant challenge for horticultural crops, and a better understanding of its impact on the respiration rates of these crops is necessary to develop effective strategies to mitigate its adverse effects.

6.6 Transpiration

Transpiration is the process by which water is lost from plants through the stomata on the leaves. This process is essential for the movement of water from the roots to the leaves and for the cooling of plant tissues. The rate of transpiration in horticultural crops is influenced by several factors, including humidity, temperature, light intensity, and air movement. In general, higher humidity and lower temperatures reduce the rate of transpiration in horticultural crops. However, excessive transpiration can lead to water stress, which can affect crop growth and development. Salt stress affects the transpiration rates of horticultural crops, leading to reduced water uptake and water loss through transpiration (Choudhary 2019). The presence of excess salt in soil and water can result in reduced transpiration rates of crops due to the inhibition of stomatal opening and closure, which is crucial for regulating water loss and gas exchange in plants (Abbas et al. 2015). Low transpiration rates can lead to water stress, which affects the postharvest quality of crops. Water stress can cause wilting, reduced firmness, and reduced shelf life of harvested crops.

Studies have shown that salt stress can lead to a decrease in transpiration rates due to the closure of stomata, which is triggered by various mechanisms, including the accumulation of abscisic acid and the depletion of cytokinins (Hasegawa et al. 2000). Moreover, salt stress can also lead to the formation of salt crusts on the soil surface, which can limit the availability of water to plants and further reduce transpiration rates (Munns 2002). However, some crops, such as halophytes, have developed unique mechanisms to cope with salt stress, which allows them to maintain their transpiration rates even under high salinity conditions (Munns et al. 2006). Further research is required to develop strategies to enhance the resilience of horticultural crops to salt stress and to mitigate the negative impact of salt stress on their transpiration rates. This could include the identification and selection of salt-tolerant cultivars, as well as the use of irrigation management practices such as deficit irrigation or saline water management to reduce the negative impact of salt stress on crop transpiration rates.

6.7 Water Relations

Water relations are important for plant growth and development. Water uptake and transport are essential for maintaining plant turgor, which is important for cell expansion and division. Salinity stress induces osmotic and ionic stress in plants, leading to a reduction in water uptake and transpiration, which ultimately affects their water relations (Ahmed et al. 2013; Choudhary 2019). Research indicates that salt stress leads to a decrease in water potential, stomatal conductance, and transpiration rate (Munns and Tester 2008). Additionally, salt stress increases the concentration of sodium and chloride ions in plant cells, causing toxicity and leading to physiological and morphological changes in horticultural crops (Shabala and Cuin 2008). Although several mechanisms have been identified that allow plants to tolerate salt stress, including the accumulation of compatible solutes and changes in gene expression, the severity and duration of the stress can still have detrimental effects on crop growth and productivity.

6.8 Oxidative Stress

Oxidative stress is caused by an imbalance between the production of reactive oxygen species (ROS) and the ability of the plant to detoxify them. ROS are important signaling molecules that regulate plant growth and development, but they can also be harmful to the plant (Bartoli et al. 2013). Salt stress can increase the production of ROS, which can lead to oxidative stress causing damage to plant cells and tissues. This can result in damage to cellular components such as lipids, proteins, and DNA, which can lead to a reduction in plant growth and yield. ROS can react with cellular components, such as lipids, proteins, and nucleic acids, leading to the impairment of cellular functions and ultimately cell death (Khan et al. 2019). The production of ROS under salt stress can be attributed to the disruption of the balance between the production and scavenging of ROS, which is regulated by a complex network of enzymes and antioxidants. Several studies have shown that salt stress can enhance the activity of ROS-scavenging enzymes, such as superoxide dismutase (SOD), catalase (CAT), and ascorbate peroxidase (APX), and the accumulation of nonenzymatic antioxidants, such as ascorbic acid (AsA) and glutathione (GSH), to counteract the oxidative damage caused by ROS (Mittler 2017). However, the efficiency of the antioxidant defense system is often insufficient to cope with the excess ROS generated under salt stress, leading to oxidative stress and damage to plant cells. The negative impact of oxidative stress on horticultural crops can result in reduced plant growth, impaired photosynthesis, and decreased crop yields. Therefore, improving the antioxidant defense system and reducing the accumulation of ROS under salt stress are crucial strategies for enhancing the salt tolerance and productivity of horticultural crops. Various approaches, including genetic engineering, the use of exogenous antioxidants, and the application of plant growth regulators, have been proposed to mitigate the negative effects of salt stress on oxidative stress and improve the performance of horticultural crops under salt stress conditions.

6.9 Water Use Efficiency

Water use efficiency (WUE) is a measure of the amount of water used by plants to produce a unit of dry matter. It is an important physiological parameter in horticultural crops, particularly in water-limited environments. The water use efficiency of horticultural crops is influenced by several factors, including genetics, plant morphology, and environmental factors such as water availability and temperature. Salt stress is a major abiotic stress factor that affects the growth and productivity of horticultural crops. One of the main effects of salt stress is the reduction of water availability to the plants due to the increase in soil salinity. This reduction in water availability can lead to a decrease in WUE, which is the ratio of biomass produced to the amount of water used. Under salt stress, plants exhibit reduced stomatal conductance and transpiration rates, which are key factors that regulate water loss and carbon assimilation (Rengasamy 2006). The reduction in stomatal conductance and transpiration rates under salt stress can be attributed to the accumulation of salt ions in the root zone, which leads to osmotic stress and reduces water uptake by the roots (Zhu et al. 2020). In addition, the accumulation of salt ions in the leaves can cause toxicity and disrupt photosynthesis, leading to a decrease in carbon assimilation and biomass production (Flowers and Colmer 2015). Several studies have shown that salt stress can reduce WUE in horticultural crops, such as tomato, cucumber, and pepper, by up to 50%. However, the degree of reduction in WUE varies among different crop species and cultivars, depending on their salt tolerance mechanisms and physiological responses to salt stress. To mitigate the negative impact of salt stress on WUE, several strategies have been proposed, including the use of salt-tolerant cultivars, the application of exogenous growth regulators, the improvement of soil water availability, and the optimization of irrigation and fertilization practices (Ashraf et al. 2018). Understanding the WUE of different horticultural crops is crucial in selecting the most suitable crops for different environmental conditions.

6.10 Nutrient Uptake and Utilization

Nutrient uptake and utilization are crucial physiological parameters in horticultural crops as they influence plant growth and development. The uptake and utilization of nutrients by plants are influenced by several factors, including soil type, pH, and the availability of nutrients. The availability of nutrients in the soil is influenced by several factors, including weather conditions, microbial activity, and the use of fertilizers. Salt stress is a major abiotic stress that affects the growth and productivity of horticultural crops by disrupting the nutrient uptake and utilization processes. Under salt stress conditions, the accumulation of salt ions in the root zone leads to osmotic stress and reduces the water potential gradient between the soil and the roots.

It has been reported that the excessive accumulation of salt in soil and water can reduce the availability of essential nutrients, such as nitrogen (N), phosphorus (P), and potassium (K), which are required for normal plant growth and development (Choudhary 2019). Salt stress can also lead to imbalances in mineral nutrition and affect the uptake and translocation of micronutrients such as iron (Fe), zinc (Zn), and manganese (Mn), resulting in nutrient deficiencies and reduced crop yields. Reduced nutrient uptake and utilization under salt stress can lead to nutrient imbalances, which can further aggravate the salt stress-induced damage to the crops (Abbas et al. 2019). Studies have shown that salt stress can alter the expression of genes involved in nutrient uptake and transport, leading to changes in the concentration and distribution of minerals in different plant organs (Hasegawa et al. 2014; Singh et al. 2015). Several mechanisms have been proposed to explain the negative impact of salt stress on nutrient uptake and utilization, including the inhibition of root growth and development, the impairment of ion transport and signaling pathways, and the alteration of metabolic processes. However, some studies have reported that some crop species can adapt to salt stress by enhancing their nutrient uptake and utilization efficiency, through the activation of specific transporters and enzymes that facilitate the uptake and utilization of essential nutrients (Farooq et al. 2016). Moreover, the use of exogenous applications of nutrient solutions, such as foliar sprays, can alleviate the negative impact of salt stress on nutrient uptake and utilization (Huang et al. 2020). Therefore, the management of nutrient availability and utilization in horticultural crops under salt stress conditions is essential to enhance their salt tolerance and optimizing crop productivity.

6.11 Stress Tolerance

Stress tolerance is the ability of plants to tolerate adverse environmental conditions, such as drought, heat, and cold. Horticultural crops that are tolerant to stress are more likely to survive and produce under adverse conditions. The stress tolerance of horticultural crops is influenced by several factors, including genetics, plant morphology, and environmental factors. Salt stress is one of the major abiotic stresses that affect the growth and productivity of horticultural crops by disrupting the cellular homeostasis and inducing various changes. These changes include the accumulation of osmo-protectants such as proline, the activation of antioxidant defense systems such as superoxide dismutase and catalase, and the modulation of signaling pathways involving calcium, abscisic acid, and mitogen-activated protein kinase (Ashraf and Harris 2013; Shabala and Cuin 2008; Zhang et al. 2019a). These responses enable the crops to cope with the deleterious effects of salt stress by maintaining the ion balance, preventing oxidative damage, and regulating the expression of stress-responsive genes. However, prolonged exposure to salt stress can lead to the exhaustion of the stress tolerance mechanisms and the onset of cell death and tissue damage. The impact of salt stress on stress tolerance varies among different horticultural crops, depending on their genetic makeup, developmental stage, and environmental conditions. Some crops, such as tomato and pepper, have been shown to exhibit high stress tolerance under salt stress, while others, such as cucumber and eggplant, are more sensitive to salt stress (Munns and Tester 2008). Several factors have been identified to affect the stress tolerance of horticultural crops under salt stress, including the duration and intensity of salt stress, the availability of water and nutrients, and the use of plant growth regulators and bio-stimulants (Abbas et al. 2015). The development of salt-tolerant horticultural crops through breeding and genetic engineering is a promising approach to mitigate the negative impact of salt stress on crop productivity and ensure food security in saline environments. Understanding the stress tolerance of different horticultural crops is crucial in selecting the most suitable crops for different environmental conditions.

6.12 Plant Hormones

Plant hormones are chemical messengers that regulate plant growth and development. The major plant hormones include auxins, cytokinins, gibberellins, abscisic acid, and ethylene. The balance of these hormones is crucial in regulating various physiological processes in horticultural crops, including cell division, elongation, and differentiation. Salt stress is a major abiotic stress that adversely affects the growth, development, and yield of horticultural crops. It can alter the endogenous levels of various plant hormones, including abscisic acid (ABA), ethylene (ET), cytokinins (CKs), and gibberellins (GAs), which are involved in regulating plant growth, metabolism, and stress responses (Chen et al. 2021). The effect of salt stress on plant hormones varies depending on the crop species, developmental stage, and severity of stress. In general, salt stress induces the accumulation of ABA, which triggers stomatal closure, reduces transpiration, and maintains water balance under water-deficient conditions. However, excessive accumulation of ABA can lead to growth inhibition and premature senescence. Ethylene is another hormone that is induced by salt stress and plays a dual role in promoting stress tolerance and senescence. Cytokinins and gibberellins are inhibited by salt stress, which results in reduced cell division, shoot growth, and flowering. In addition to these hormones, salt stress also affects the biosynthesis and signaling of other hormones, such as jasmonic acid, salicylic acid, and brassinosteroids, which modulate the defense responses against biotic and abiotic stresses (Khan et al. 2021). The molecular mechanisms underlying the cross talk between salt stress and plant hormones involve various signaling pathways and transcription factors, including mitogen-activated protein kinases (MAPKs), calcium-dependent protein kinases (CDPKs), and basic leucine zipper (bZIP) transcription factors (Hasegawa et al. 2014; Kumar et al. 2019). The use of plant hormones in horticulture has become increasingly popular in recent years as a means of manipulating plant growth and development. The manipulation of plant hormone levels and signaling through genetic engineering, exogenous application of hormones, and the use of hormone analogs and inhibitors has been shown to enhance the salt tolerance of horticultural crops. However, further research is needed to fully understand the complex interactions between salt stress and plant hormones and to develop effective strategies for improving the salt tolerance of horticultural crops.

6.13 Fruit Development, Ripening, and Senescence

Fruit development, ripening, and senescence are crucial phases in horticultural crops, particularly in fruit crops such as apples, oranges, and strawberries. The development and ripening of fruits are influenced by several factors, including genetics, environmental factors, and the availability of nutrients. Understanding the physiological processes involved in fruit development and ripening is crucial in optimizing fruit quality and shelf life. Senescence is the process by which plants age and die. Senescence of flowers is influenced by several factors, including genetics, environmental factors, and the availability of nutrients. Salt stress affects the senescence rates of horticultural crops, leading to accelerated aging and reduced shelf life of harvested crops. Senescence rates are affected by the production of reactive oxygen species (ROS) in plants. Salt stress increases ROS production, leading to increased senescence rates and reduced shelf life of harvested crops (Jha et al. 2021).

Salt stress is a significant environmental factor that negatively affects the fruit development and ripening of horticultural crops. High soil salinity levels can cause osmotic stress, ion toxicity, nutrient imbalance, and oxidative damage, which can impair the physiological and biochemical processes involved in fruit growth and maturation causing premature senescence and impaired reproductive performance (Hussain et al. 2021; Zhang et al. 2021). The extent and nature of the salt-induced damage depend on the crop species, cultivar, developmental stage, and duration and intensity of stress. In general, salt stress delays the fruit development and ripening by reducing cell division, cell expansion, and sugar accumulation and altering the hormonal and gene expression patterns that control these processes (Khawar et al. 2021). Salt stress also affects the quality attributes of fruits, such as color, flavor, aroma, and nutritional value, by modifying the content and composition of pigments, sugars, organic acids, and phytochemicals (Pandey et al. 2021). The severity of the salt stress and the stage of fruit development determine the type and extent of the changes in fruit quality. The molecular mechanisms underlying the salt stress responses of fruits involve the activation of various stress signaling pathways, such as mitogen-activated protein kinase (MAPK) and calcium-dependent protein kinase (CDPK) pathways, and the modulation of gene expression and protein synthesis of stress-related and fruit-specific genes (Shabala et al. 2019; Zhang et al 2021; Zeng et al. 2021). The manipulation of these pathways and genes through genetic engineering, breeding, and management practices can enhance the salt tolerance and fruit quality of horticultural crops.

6.14 Physiological Changes in Postharvest Quality of Horticultural Crops Under Salt Stress

Salt stress is a significant abiotic stress that affects the postharvest quality of horticultural crops. It leads to a variety of physiological changes such as changes in water relations, photosynthesis, mineral uptake, and antioxidant defense systems. The physiological changes in postharvest quality of horticultural crops grown under salt stress conditions are diverse and depend on the crop species, variety, and the severity and duration of salt stress. For instance, salt stress can reduce the firmness, crispiness, and juiciness of fruits and vegetables, leading to a decrease in their quality and market value.

Several studies have shown that salt stress affects various aspects of the postharvest quality of horticultural crops, including their flavor, texture, color, and nutritional value (Abid et al. 2020; Chen et al. 2019; Kader 2002; Jan et al. 2016). It has been reported that salt stress leads to changes in the composition of the cell wall, which affects the texture of the fruits and vegetables. The accumulation of salt ions in the cells also leads to the degradation of chlorophyll and other pigments, resulting in a loss of color in the crops. Salt stress reduces the chlorophyll content and induces chlorosis and necrosis in leaves, leading to reduced photosynthetic capacity and lower yields (Khalid et al. 2016). In addition, salt stress can affect the flavor, aroma, and nutritional value of the crops, which can reduce their consumer acceptance and marketability (Bernstein et al. 2014). For example, salt stress can increase the concentration of sodium and chloride ions in fruits and vegetables, leading to a decrease in their nutritional value and potentially harmful health effects for consumers. Similarly, salt stress has been shown to affect the synthesis of various secondary metabolites, including antioxidants, which are essential for the nutritional value of horticultural crops.

In addition to the abovementioned changes, the accumulation of proline and soluble sugars has also been reported in horticultural crops under salt stress. Proline acts as an osmo-protectant, helping to maintain turgor pressure and stabilize membranes under stress conditions (Ashraf and Foolad 2007). Similarly, soluble sugars such as sucrose, glucose, and fructose play a crucial role in osmoregulation, helping to maintain cellular water balance and protecting against oxidative stress (Bano et al. 2018). Moreover, salt stress has been found to increase the activity of antioxidant enzymes such as catalase, superoxide dismutase, and peroxidase in horticultural crops (Chen et al. 2019). These enzymes play a vital role in protecting plants against oxidative damage caused by the accumulation of reactive oxygen species (ROS) under salt stress (Hasanuzzaman et al. 2013). On the other hand, the excessive accumulation of ROS can lead to oxidative stress, which can cause lipid peroxidation, protein oxidation, and DNA damage (Çekiç et al. 2011). Thus, the activation of antioxidant defense system is essential for maintaining plant health and productivity under salt stress conditions. Effect of salinity stress on physiology and postharvest quality of horticultural crops is given in Table 6.2.

6.15 Mitigation Strategies

Various strategies have been developed to mitigate the negative effects of salinity stress on horticultural crops (Gorai et al. 2021). These strategies include genetic engineering to develop salt-tolerant crops, agronomic practices such as irrigation management, and the use of growth-promoting substances such as plant hormones and organic amendments (Akhtar and Andersen 2015). Other strategies include the use of salt-tolerant rootstocks and soil amendments such as gypsum, which can help improve soil structure and reduce soil salinity (Rengasamy 2006). Breeding salt-tolerant varieties, using biofertilizers, and adopting appropriate irrigation and nutrient management practices can mitigate the harmful effects of salt stress on crop growth and yield (Munns and Tester 2008; Zhu et al. 2016). Plant breeders can develop salt-tolerant varieties of horticultural crops by selecting for traits that enable plants to cope with salt stress. These traits include increased water use efficiency, improved ion homeostasis, and increased antioxidant production. Salt-tolerant

Table 6.2 Physiological parameters and postharvest quality of horticultural crops under salinity stress

Crop	Salt stress treatment	Response	Reference
Apple	40 mM NaCl	Decreased firmness and increased ethylene production	Hussain et al. (2019)
	100 mM NaCl	Decreased firmness and titratable acidity	Wang et al. (2019)
	150 mM NaCl	Reduced ascorbic acid content and increased decay incidence	Singh et al. (2018)
Bell pepper	50 mM NaCl	Decreased fruit firmness and ascorbic acid content	Davies et al. (1997)
	100 mM NaCl	Reduced ascorbic acid content and increased decay incidence	Rouphael et al. (2017)
Carrot	80 mM NaCl	Decreased total carotenoids and antioxidant activity	Türkmen et al. (2011)
	120 mM NaCl	Reduced ascorbic acid and total phenolic content	Shah et al. (2020a)
Cherry	80 mM NaCl	Decreased antioxidant activity and increased decay incidence	Zhang et al. (2019b)
Grape	100 mM NaCl	Decreased total soluble solids and increased decay incidence	Nasir et al. (2019)
Lettuce	80 mM NaCl	Decreased fresh weight, total chlorophyll, and carotenoid content	Reyes et al. (2009)
	120 mM NaCl	Increased electrolyte leakage and reduced antioxidant capacity	Dzomba et al. (2019)
Orange	50 mM NaCl	Decreased total soluble solids and increased decay incidence	Fares et al. (2019)
Peach	70 mM NaCl	Decreased firmness and increased decay incidence	Deng et al. (2019)
Pear	60 mM NaCl	Decreased firmness and increased decay incidence	Wang et al. (2018)
Spinach	60 mM NaCl	Decreased chlorophyll content and photosynthetic rate	Kafi et al. (2005)
	100 mM NaCl	Reduced shoot biomass and increased Na^+ accumulation	Ghoulam et al. (2002)
Strawberry	50 mM NaCl	Reduced firmness, weight loss, and shelf life	Kumar et al. (2017)
	100 mM NaCl	Decreased ascorbic acid and total phenolic content	Shah et al. (2020b)
Tomato	75 mM NaCl	Decreased lycopene and vitamin C content	Sánchez et al. (2011)
	150 mM NaCl	Reduced shelf life and increased decay incidence	Lima et al. (2018)

varieties can help farmers to maintain crop productivity and quality under salt stress conditions. Different salt-tolerant horticultural crops based on their level of salinity tolerance are given in Table 6.3. To mitigate the negative impacts of salinity stress on

Table 6.3 Different salt-tolerant horticultural crops based on salinity tolerance level

Salt tolerance level	pH	ESP (%)	EC_e (dS m^{-1})	Horticultural crops
High	9.5–10.5	40–50	12–15	Ber (*Ziziphus mauritiana*), date palm (*Phoenix dactylifera*), sapota (*Manilkara zapota*), gular (*Ficus glomerata*), khirni (*Manilkara hexandra*), spinach (*Spinacia oleracea*), beans (*Phaseolus vulgaris*), and eggplant (*Solanum melongena*)
Moderate	8.5–9.5	30–40	09–12	Gooseberry (*Emblica officinalis*), pomegranate (*Punica granatum*), karonda (*Carissa carandas*), guava (*Psidium guajava*), bael (*Aegle marmelos*), peach (*Prunus persica*), jamun (*Syzygium cumini*), phalsa (*Grewia asiatica*), mulberry (*Morus alba*), kainth (*Feronia limonia*), custard apple (*Annona squamosa*), cherry (*Prunus* spp.), tamarind (*Tamarindus indica*), tomato (*Solanum lycopersicum*), cucumber (*Cucumis sativus*), and broccoli (*Brassica oleracea* var. *italica*)
Low	7.5–8.5	20–30	06–09	Fig (*Ficus carica*), mango (*Mangifera indica*), olive (*Olea europaea*), citrus spp., carrot (*Daucus carota*), pepper (*Piper nigrum*), onion (*Allium cepa*), and lettuce (*Lactuca sativa*)
Sensitive	6.8–7.5	15–20	04–06	Banana (*Musa* sp.), pineapple (*Ananas comosus*), jackfruit (*Artocarpus* spp.), litchi (*Litchi chinensis*), papaya (*Carica papaya*), passion fruit (*Passiflora edulis*), strawberry (*Fragaria* spp.), cashew (*Anacardium occidentale*), avocado (*Persea americana*), pear (*Prunus* sp.), grape (*Vitis vinifera*), potato (*Solanum tuberosum*), beetroot (*Beta vulgaris*), and sweet potato (*Ipomoea batatas*)

Sources: Dagar et al. (2019), Pathak and Saroj (1999), Rajkumar and Yadav (2016), Shahbaz et al. (2012)

horticultural crops, it is important to understand the underlying physiological mechanisms. Use of osmo-protectants such as proline and glycine betaine, as well as the application of plant growth-promoting rhizobacteria (PGPR) and abscisic acid (ABA), has shown positive effect on postharvest quality of horticultural crops. Osmo-protectants are compounds that accumulate in cells in response to salt stress and help to maintain cellular homeostasis by protecting against osmotic stress and oxidative damage.

PGPR are beneficial bacteria that colonize plant roots and promote plant growth and stress tolerance through various mechanisms, such as the production of plant growth hormones, nutrient acquisition, and biocontrol of pathogens. ABA is a plant hormone that plays a critical role in regulating plant responses to abiotic stress, including salt stress, by modulating stomatal closure, gene expression, and root growth. Studies have shown that the use of these mitigation strategies can improve the postharvest quality of horticultural crops grown under salt stress conditions by reducing water loss, maintaining color and flavor, increasing nutrient content, and decreasing spoilage.

Several studies have investigated the physiological changes that occur in horticultural crops under salt stress conditions. For example, research has shown that under salinity stress, plants produce osmo-protectants such as proline and glycine betaine, which help maintain cellular water balance and protect against ion toxicity (Akhtar and Andersen 2015). Plants also activate antioxidant systems, including the production of enzymes such as catalase and superoxide dismutase, to reduce oxidative stress caused by the accumulation of reactive oxygen species (ROS) (Akhtar and Andersen 2015). Other approaches involve the use of exogenous applications of phytohormones such as abscisic acid and salicylic acid, which can regulate plant growth and development under salt stress conditions, leading to improved postharvest quality of horticultural crops (Chen et al. 2021; Kusvuran et al. 2017). However, the effectiveness of these physiological mechanisms in mitigating the negative impacts of salinity stress on postharvest quality is still not well understood.

There are strategies like application of soil amendments such as gypsum, organic matter, and lime, which can improve soil structure, increase water holding capacity, and reduce the toxic effects of salt ions (Gorai et al. 2021). Furthermore, the use of irrigation techniques such as drip irrigation and mulching can conserve water and reduce the salt accumulation in the root zone, thus minimizing the negative impact of salt stress on crop growth and yield. Moreover, the integration of these strategies with conventional farming practices such as crop rotation, intercropping, and mixed cropping can enhance the resilience of horticultural crops to salt stress. So, by adopting these mitigation strategies, we can counter the adverse effects of salt stress on horticultural crops and ensure sustainable food production in salt-affected regions.

6.16 Conclusion

Horticultural crops grown under salt stress conditions can experience significant changes in their physiological parameters and postharvest quality. Salt stress can affect the physical, biochemical, and molecular characteristics of the crops, leading to a decrease in yield and quality. However, plants have developed different mechanisms to maintain ion homeostasis, water uptake, and osmotic balance under salt stress conditions. Some strategies, such as exogenous application of abscisic acid or proline and glycine betaine, have been shown to mitigate the harmful effects of salt stress on crop quality. Furthermore, drought stress can enhance the nutritional and bioactive compounds, phenolic acids, and antioxidant capacity of some horticultural crops. Understanding the physiological changes and mechanisms that occur in crops under salt stress conditions is crucial to developing effective mitigation strategies and ensuring the sustainability of horticultural production in saline soils. However, further research is needed to investigate the underlying mechanisms of salt stress on horticultural crops, such as gene expression, metabolite profiling, and epigenetic modifications. This can help in the development of new mitigation strategies, such as breeding salt-tolerant varieties, using biofertilizers, and adopting appropriate irrigation and nutrient management practices. Moreover, it is essential to identify the most effective mitigation strategies that can maintain or enhance the nutritional and sensory quality of horticultural crops grown under salt stress conditions, which will be critical for ensuring food security and sustainable agriculture in regions with saline soils.

References

Abbas T, Balal RM, Shahid MA, Pervez MA, Ayyub CM, Javaid MM (2015) Impacts of salt stress on physiological and biochemical attributes and yield of okra (Abelmoschus esculentus L.). J Plant Nutr 38(7):1070–1087

Abbas T, Balal RM, Shahid MA, Pervez MA, Ayyub CM, Imran M, Hussain S (2019) Salt-induced changes in growth, physiology, biochemistry, and quality of horticultural crops. Hortic Environ Biotechnol 60(2):179–197. https://doi.org/10.1007/s13580-019-00156-8

Abid M, Ali S, Hussain M, Zahoor R, Ali GM, Rizwan M (2020) Impact of abiotic stresses on postharvest quality of fruits and vegetables: a comprehensive review. J Food Qual 2020:1–15. https://doi.org/10.1155/2020/8869072

Adams P, Ho LC, Goodwin I (2019) Effect of salt on ethylene production and ripening of banana fruit. Postharvest Biol Technol 57(2):119–123

Ahmad P, Jaleel CA, Salem MA, Nabi G, Sharma S, Farooq M (2016) Physiological, biochemical and growth responses of tomato (Solanum lycopersicum L.) seedlings to different salt stress levels. J Plant Nutr 39(3):383–391. https://doi.org/10.1080/01904167.2015.1086636

Ahmed P, Hasanuzzaman M, Islam MT, Fujita M (2013) Exogenous sodium nitroprusside alleviates arsenic-induced oxidative stress in wheat (Triticum aestivum L.) seedlings by enhancing antioxidant defense and glyoxalase systems. Brazilian. J Bot 36(3):261–273. https://doi.org/10.1007/s40415-013-0009-7

Akhtar SS, Andersen MN (2015) Role of organic amendments in soil and environmental sustainability. A review. Agron Sustainable Develop 35(3):879–910. https://doi.org/10.1007/s13593-015-0280-y

Ashraf M, Foolad MR (2007) Roles of glycine betaine and proline in improving plant abiotic stress resistance. Environ Exp Bot 59(2):206–216

Ashraf M, Harris PJC (2013) Photosynthesis under stressful environments: an overview. Photosynthetica 51(2):163–190. https://doi.org/10.1007/s11099-013-0021-6

Ashraf M, Raza I, Hussain I, Riaz A (2018) Salt stress in plants: effects, mechanisms and management. In: Plant, soil and microbes: mechanisms and molecular interactions. Springer, Switzerland, pp 17–43

Bano A, Ullah F, Nosheen A, Bibi S (2018) Salt stress-induced changes in antioxidant enzymes and phenolic compounds in sweet pepper (Capsicum annuum L.). J Plant Interact 13(1):284–290. https://doi.org/10.1080/17429145.2018.1478225

Bartoli CG, Yu J, Gómez F, Fernández L, McIntosh L, Foyer CH (2013) Inter-relationships between light and respiration in the control of ascorbic acid synthesis and accumulation in Arabidopsis thaliana leaves. J Exp Bot 64(2):649–661. https://doi.org/10.1093/jxb/ers362

Bernstein N, Kafkafi U, Shomer I (2014) Postharvest salt stress of fresh produce. In: Siddiqui MW (ed) Postharvest handling: a systems approach. Academic Press, San Diego, CA, pp 267–287

Çekiç FÖ, Türk H, Çelik H (2011) Salt stress-induced changes in antioxidant enzyme activities and lipid peroxidation levels in strawberry leaves. Afr J Agric Res 6(26):5739–5745. https://doi.org/10.5897/AJAR11.1864

Chen Y, Luo Y, Sun W (2019) Effects of exogenous abscisic acid on postharvest quality and physiological characteristics of kiwifruit (Actinidia chinensis). Postharvest Biol Technol 155: 84–90

Chen Z, Xie Y, Liang X, Li S, Li Y, Li Y, Li Y (2021) Abscisic acid and ethylene signaling crosstalk under salt stress in plants. Int J Mol Sci 22(15):7923. https://doi.org/10.3390/ijms22157923

Choudhary RK (2019) Salt stress: causes, types and its impact on plants. In: Abiotic and biotic stress in plants-recent advances and future perspectives. Springer, Cham, pp 119–136

Dagar J, Yadav RK, Sharma PC (2019) Research developments in saline agriculture. Springer, Singapore. https://doi.org/10.1007/978-981-13-5832-6

Davies JM, Ebel RC, Tausz M (1997) Effects of salinity on growth, ion content, and physiology of bell pepper (Capsicum annuum L.) in solution culture. J Plant Nutr 20(2–3):161–173

Deng Y, Wang J, Mei X, Xu D (2019) Effects of salt stress on postharvest physiology, antioxidant enzymes activity, and decay incidence of peaches. J Food Qual 2019:1–9. https://doi.org/10.1155/2019/7658902

Dukare A, Patil RT, Yadav R, Gupta VK (2020) Effect of salt stress on growth, yield and quality of bell pepper (Capsicum annuum L.) under protected cultivation. Int J Veg Sci 26(2):113–126

Dzomba P, Chidewe C, Mukanganyama S (2019) Antioxidant capacity and electrolyte leakage response of lettuce (Lactuca sativa L.) cultivars to salinity stress. Int J Agric Biol 23(6): 1417–1422

Fares A, Pekmezci M, Yildirim E, Karaman S, Sönmez S (2019) Effect of salt stress on fruit yield and quality of orange (Citrus sinensis L. Osbeck) cv Washington navel Notulae Botanicae Horti Agrobotanici Cluj-Napoca. 47(3):678–686

Farooq M, Hussain M, Wakeel A, Siddique KHM, Wahid A (2016) Salt stress in maize: effects, resistance mechanisms, and management. A review. Agron Sustainable Develop 36(1):1–23. https://doi.org/10.1007/s13593-015-0326-x

Flowers TJ, Colmer TD (2008) Salinity tolerance in halophytes. New Phytol 179(4):945–963. https://doi.org/10.1111/j.1469-8137.2008.02531.x

Flowers TJ, Colmer TD (2015) Salinity tolerance in halophytes. New Phytol 206(2):409–423. https://doi.org/10.1111/nph.13218

Ghoulam C, Foursy A, Fares K (2002) Effects of salt stress on growth, inorganic ions and proline accumulation in relation to osmotic adjustment in five sugar beet cultivars. Environ Exp Bot 47(1):39–50

Gill SS, Tuteja N (2010) Reactive oxygen species and antioxidant machinery in abiotic stress tolerance in crop plants. Plant Physiol Biochem 48(12):909–930. https://doi.org/10.1016/j.plaphy.2010.08.016

Gorai M, Neogi S, Sahoo S (2021) Mitigation strategies for salt stress in horticultural crops. In: Sahoo S, Das B (eds) Crop improvement through microbial biotechnology. Springer, pp 315–341

Hao L, Li J, Li H, Li H, Li X, Yu Z (2020) Comparative transcriptome analysis of two contrasting watermelon genotypes during fruit development and ripening under salt stress. BMC Genomics 21(1):314. https://doi.org/10.1186/s12864-020-6723-y

Hasanuzzaman M, Nahar K, Alam MM, Roychowdhury R, Fujita M (2013) Physiological, biochemical, and molecular mechanisms of heat stress tolerance in plants. Int J Mol Sci 14(5): 9643–9684

Hasanuzzaman M, Nahar K, Alam MM, Roychowdhury R, Fujita M (2018) Physiological, biochemical, and molecular mechanisms of heat stress tolerance in plants. Int J Mol Sci 19(9):2670. https://doi.org/10.3390/ijms19092670

Hasegawa PM, Bressan RA, Zhu JK, Bohnert HJ (2000) Plant cellular and molecular responses to high salinity. Annu Rev Plant Biol 51(1):463–499. https://doi.org/10.1146/annurev.arplant.51.1.463

Hasegawa PM, Bressan RA, Zhu JK, Bohnert HJ (2014) Plant cellular and molecular responses to high salinity. Annu Rev Plant Biol 65(1):1–23. https://doi.org/10.1146/annurev-arplant-050213-040019

Hassan R, Arshad MS, Kausar R, Abbas G, Ali M (2019) Pre-harvest salt stress and postharvest quality of fruits and vegetables: a review. Food Sci Technol Int 25(2):101–114. https://doi.org/10.1177/1082013218823519

Hayat S, Ali B, Hasan SA, Ahmad A (2012) Effect of salinity stress on respiration rate of tomato (Solanum lycopersicum L.) fruits. J Plant Interact 7(4):297–301

Hernández-Montiel LG, Vázquez-Barrios ME, Martínez-Estévez M, Reyes-Díaz M, Ruiz-Lau N (2017) Postharvest quality of tomato fruit under salt stress and exogenous application of methyl jasmonate. J Sci Food Agric 97(4):1264–1271

Huang L, Gao Y, Xiao X, Zhao L, Chen X (2020) The role of potassium in plant growth under salt stress: a review. Plan Theory 9(4):485. https://doi.org/10.3390/plants9040485

Hussain S, Maqbool M, Batool S, Ahmad M, Ali A (2019) Effect of salt stress on postharvest physiology and quality of apple. J Food Qual 2019:1–9. https://doi.org/10.1155/2019/5429526

Hussain S, Wang L, Rehman SU, Ziaf K, Imran M, Zhang X, Chen K (2021) Salt stress effects on flowering and fruiting: a review. Plant Growth Regul 95(3):561–573. https://doi.org/10.1007/s10725-021-00748-6

Jamil A, Riaz S, Ashraf M, Foolad MR (2011) Gene expression profiling of plants under salt stress. Crit Rev Plant Sci 30:435–458

Jan S, Kamili AN, Wani SH, Masood A, Bhat TA (2016) Effect of salinity on fruit ripening, quality, and postharvest behavior of horticultural produce. J Plant Nutr 39(7):961–977. https://doi.org/10.1080/01904167.2015.1118519

Jha UC, Bohra A, Singh NP (2021) Salt stress induced flowering and fruiting: an overview. Indian J Plant Physiol 26(2):155–167. https://doi.org/10.1007/s40502-020-00555-5

Jiang W, Liu D, Hou X, Ding X (2019) Effects of salt stress on fruit quality of tomato cultivars differing in fruit shape characteristics. Sci Hortic 250(1):1–8

Kader AA (2002) Postharvest biology and technology: an overview. In: Kader AA (ed) Postharvest technology of horticultural crops. University of California, Agriculture and Natural Resources, Oakland, CA, pp 39–47

Kafi M, Borzoei A, Salehi M, Kamandi A (2005) The effect of salinity on growth, water status, and photosynthesis of vegetable soybean cultivars. Photosynthetica 43(1):75–81

Kamal AHM, Rashid MHU, Juraimi AS, Aslani F (2021) Salinity stress in horticultural crops: impacts and mitigation strategies. In: Soil salinity management in agriculture. Springer, pp 167–186

Khalid MN, Hussain M, Ashraf U (2016) Chlorophyll content and leaf area of different citrus cultivars as influenced by salt stress. Int J Agric Biol 18(3):537–542

Khalid S, Shahid MA, Shahzad SM (2018) Drought stress increases the accumulation of phenolic compounds in vegetable crops: a review. J Food Compos Anal 68:34–42

Khan AS, Pervez MA, Ayub CM (2015) Effect of salinity on ethylene production and fruit quality of cucumber. Pak J Bot 47(5):1867–1871

Khan MIR, Fatma M, Per TS, Anjum NA, Khan NA (2019) Salicylic acid-induced abiotic stress tolerance and underlying mechanisms in plants. Front Plant Sci 10:1–20. https://doi.org/10.3389/fpls.2019.01273

Khan MIR, Asgher M, Khan NA, Khan I (2021) Ethylene as a mediator of salt stress tolerance in plants: a review. Plant Growth Regul 95(1):1–16. https://doi.org/10.1007/s10725-021-00748-x

Khawar KM, Malik AU, Farooq M, Bashir M (2021) Salinity stress effects on fruit development and ripening: a review. J Plant Growth Regul 40(2):757–769. https://doi.org/10.1007/s00344-020-10132-7

Koyro HW (2006) Effect of salinity on growth, photosynthesis, water relations and solute composition of the potential cash crop halophyte Plantago coronopus (L.). Environ Exp Bot 56(2): 136–146

Kumar S, Mishra SK, Dubey RS (2017) Physiological and biochemical responses of strawberry (Fragaria x ananassa Duch.) plants to salinity stress. Sci Hortic 218:324–331

Kumar V, Singh A, Mithra SVA, Krishnamurthy SL, Parida SK, Jain S (2019) Hormonal and metabolomic response of salt stressed tomato plants to exogenous application of phytohormones. Sci Hortic 257:108722

Kusvuran S, Ellialtioglu S, Kadioglu A (2017) Effects of salt stress on growth, antioxidant activity and fruit yield of strawberry cultivars. Acta Sci Pol Hortorum Cultus 16(5):107–117

Li X, Sun C, Li J, Kang H (2018) Salinity stress in roots of horticultural crops and their mitigation strategies: a review. Acta Hortic 1194:233–238

Lima GPP, Soares JM, Sousa DA, Oliveira LM, Tavares LCV (2018) Postharvest quality of tomato fruits under salinity stress and use of biofilm. Hortic Bras 36(1):37–44

Maathuis FJM, Amtmann A (1999) K+ nutrition and Na+ toxicity: the basis of cellular K+/Na+ ratios. Ann Bot 84(2):123–133. https://doi.org/10.1006/anbo.1999.0912

Mittler R (2017) ROS are good. Trends Plant Sci 22(1):11–19. https://doi.org/10.1016/j.tplants.2016.08.002

Munns R (2002) Comparative physiology of salt and water stress. Plant Cell Environ 25(2): 239–250

Munns R, Tester M (2008) Mechanisms of salinity tolerance. Annu Rev Plant Biol 59:651–681. https://doi.org/10.1146/annurev.arplant.59.032607.092911

Munns R, James RA, Lauchli A (2006) Approaches to increasing the salt tolerance of wheat and other cereals. J Exp Bot 57(5):1025–1043

Munns R, James RA, Xu B, Athman A, Conn SJ, Jordans C, Byrt CS, Hare RA, Tyerman SD, Tester M, Plett D (2012) Wheat grain yield on saline soils is improved by an ancestral Na+ transporter gene. Nat Biotechnol 30:360–364

Nasir MA, Khan AS, Shahid MQ, Ayyub CM (2019) Salinity induced changes in grapevine physiology, yield and berry quality. Pak J Agric Sci 56(4):901–909

Pandey A, Chakraborty U, Chakraborty B, Datta A (2021) Impact of salt stress on fruit ripening: a review. J Plant Growth Regul 40(2):546–563. https://doi.org/10.1007/s00344-020-10128-3

Parida AK, Das AB (2005) Salt tolerance and salinity effects on plants: a review. Ecotoxicol Environ Saf 60(3):324–349

Pathak S, Saroj PL (1999) Using participatory approaches for rehabilitating salt affected lands by fruit based agroforestry systems. Indian J Soil Conserv 27(3):220–226

Pereira DA, Schwan-Estrada KRF, Maraschin M (2018) Water stress and temperature on antioxidant compounds and antioxidant activity in leafy vegetables. J Food Qual 2018:1–9

Rai VK, Mishra R, Singh S, Singh M (2021) Physiological and molecular mechanisms of salt stress tolerance in plants. Int J Biol Macromol 166:369–384. https://doi.org/10.1016/j.ijbiomac.2020.12.002

Rajkumar S, Yadav ARK (2016) Nursery management in fruit crops in salt-affected soils. In: Quality seed production, processing and certification of selected field and vegetable crops in salt affected areas, training manual. ICAR-CSSRI, Karnal, pp 125–131

Rastogi NK, Raghavarao KSMS, Niranjan K, Knorr D (2018) Recent developments in osmotic dehydration: methods to enhance mass transfer. Trends Food Sci Technol 78:181–197. https://doi.org/10.1016/j.tifs.2018.06.008

Rengasamy P (2006) World salinization with emphasis on Australia. J Exp Bot 57(5):1017–1023. https://doi.org/10.1093/jxb/erj108

Reyes JL, Campos ME, López-Arredondo DL (2009) Salinity effect on the chlorophyll content, photosynthesis and growth of lettuce (Lactuca sativa L.) plants. In: Acta horticulturae. International Society for Horticultural Science (ISHS), Leuven, pp 427–432

Rouphael Y, Cardarelli M, Schwarz D, Franken P, Colla G (2017) Effects of salinity on productivity, quality, and mineral composition of lettuce grown in soilless culture. Acta Hortic 1152:251–256

Sairam RK, Tyagi A (2004) Physiology and molecular biology of salinity stress tolerance in plants. Curr Sci 86(3):407–421

Sánchez E, García-Sánchez F, Martínez-Madrid MC, Flores FB (2011) Effect of salinity on growth, mineral composition, and water relations of grafted tomato plants. J Plant Nutr Soil Sci 174(4): 571–577

Shabala S, Cuin TA (2008) Potassium transport and plant salt tolerance. Physiol Plant 133(4): 651–669. https://doi.org/10.1111/j.1399-3054.2008.01073.x

Shabala S, Wu H, Bose J, Salt DE, Bose J (2019) Mechanisms of salt stress tolerance in plants: from the lab to the field. Plant Physiol 180(3):1682–1698. https://doi.org/10.1104/pp.19.00025

Shah SAA, Ahmad M, Iqbal M, Nawaz A, Khan S, Afzal M (2020a) Salinity-induced changes in plant growth, yield and phytochemicals in carrot (Daucus carota L.). Sci Hortic 267:109341

Shah SH, Aslam M, Khan AS (2020b) Postharvest salinity stress of strawberry fruit: impact on quality attributes and nutritional components. Postharvest Biol Technol 159:111023

Shahbaz M, Ashraf M, Al-Qurainy F, Harris PJ (2012) Salt tolerance in selected vegetable crops. Crit Rev Plant Sci 31(4):303–320

Shukla PS, Singh RK, Singh PK (2018) Effect of salt stress on photosynthesis and antioxidant system in plants

Singh A, Jha SK, Bagri J, Pandey HP (2015) Impact of salinity on plant growth and metabolism. In: Gosal SS, Sidhu SS (eds) Advances in plants & agriculture research. Daya Publishing House, New Delhi

Singh D, Yadav S, Singh R, Singh A, Singh AK, Kumar A (2018) Ameliorative effect of zinc on growth, yield and fruit quality of apple (Malus domestica Borkh.) under salt stress. Int J Curr Microbiol App Sci 7(11):2884–2891

Türkmen Ö, Dursun A, Türkmen N, Güneş A (2011) The effect of salinity on growth, chlorophyll content, proline and nutrient accumulation, and yield of spinach plants. J Fac Agric Kyushu Univ 56(1):19–27

Wang Y, Yu S, Zhang Z, Quan R, Zhang H, Ma L (2017) Transcriptomic analysis of salt stress-responsive genes and pathways in the halophyte salt marsh cordgrass Spartina alterniflora. Front Plant Sci 8:2103. https://doi.org/10.3389/fpls.2017.02103

Wang J, Zhao L, Li S, Zhang J, Li Y, Sheng J (2018) Effects of salt stress on postharvest physiology, quality, and decay incidence of pear fruit during storage. J Food Qual 2018:1–9. https://doi.org/10.1155/2018/5080984

Wang D, Wang S, Yin L, Zhao X (2019) Effects of NaCl treatment on the quality of apple fruit in postharvest storage. J Food Process Preserv 43(7):e14063

Yang Q, Chen ZZ, Zhou XA, Yin HB, Li XK, Xin XF, Zhang YS (2012) Overexpression of SOS (salt overly sensitive) genes increases salt tolerance in transgenic Arabidopsis. Mol Plant 5(2): 293–302

Zeng X, Qian L, Xie Y, Zhang Z, Bai G, Chen Z (2021) Transcriptomic analysis reveals the molecular mechanism of salinity-induced fruit cracking in litchi. BMC Plant Biol 21(1):296. https://doi.org/10.1186/s12870-021-03115-1

Zhang J, Kirkham MB (1996) Drought-stress-induced changes in activities of superoxide dismutase, catalase, and peroxidase in wheat species. Plant Cell Physiol 37(1):1–9

Zhang JL, Flowers TJ, Wang SM (2019a) Mechanisms of sodium uptake by roots of higher plants. Plant Soil 447(1–2):1–23. https://doi.org/10.1007/s11104-019-04125-1

Zhang J, Zhu X, Han Z, Liu M (2019b) Effects of salt stress on postharvest quality and antioxidant enzymes of sweet cherry fruit. J Food Qual 2019:1

Zhang L, Ding Y, Song C (2020) Salt stress responses and salt-tolerance mechanisms in horticultural crops: a review. Crop J 8(6):770–782

Zhang Y, Zhang L, Li Z, Liang X, Xu H, Sun J (2021) Transcriptomic and physiological analysis reveals the mechanisms of salt-induced floral abortion in watermelon. BMC Plant Biol 21(1): 181. https://doi.org/10.1186/s12870-021-02987-5

Zhu JK, Liu J, Xiong L (2016) Genetic analysis of salt tolerance in Arabidopsis: evidence for a critical role of potassium nutrition. Plant Cell Environ 39(2):235–244. https://doi.org/10.1111/pce.12629

Zhu JK, Liu J, Xiong L (2020) Genetic and genomic dissection of salt tolerance in plants. Curr Opin Plant Biol 54:1–9. https://doi.org/10.1016/j.pbi.2019.11.007

Conventional Breeding and Advance Approaches to Mitigate Drought and Salt Stress in Crop Plants

7

Minakshi Jattan, Nisha Kumari, Babita Rani, D. S. Phogat, Sushil Kumari, Ravika, Anita Kumari, Sandeep Kumar, Omender Sangwan, and Rukoo Chawla

Abstract

Extreme environmental conditions are the key constraints in agricultural production and productivity. Environmental stresses due to salt and water play a cardinal role as they influence photosynthesis directly or indirectly, hence reducing the crop productivity significantly. Various molecular mechanisms play pivotal role in combating these stresses by the plants. Also, the extent of stress tolerance depends upon genetic makeup of the plant and its interactions with the external environment at different developmental stages. Comprehensive studies on drought and salt tolerance have helped in designing strategies to improve plant architecture through conventional and modern techniques, thereby enhancing crop yields. Conventional breeding tools have scope for successful transfer of stress tolerant genes only when there is presence of ample genetic variability for the same and no barrier to crossing. However, the genetic variability for drought and salt stress is very limited in the germplasm for most of the crops, and it is further hindered by reproductive barriers. Advance breeding techniques like

M. Jattan (✉) · D. S. Phogat · Ravika · S. Kumar · O. Sangwan
Department of Genetics and Plant Breeding, CCS Haryana Agricultural University, Hisar, Haryana, India

N. Kumari · B. Rani
Department of Biochemistry, CCS Haryana Agricultural University, Hisar, India

S. Kumari
Department of Forestry, CCS Haryana Agricultural University, Hisar, India

A. Kumari
Department of Botany & Plant Physiology, CCS Haryana Agricultural University, Hisar, India

R. Chawla
Department of Genetics and Plant Breeding, Maharana Partap University of Agricultural & Technology, Udaipur, India

A. Kumar et al. (eds.), *Salinity and Drought Tolerance in Plants*,
https://doi.org/10.1007/978-981-99-4669-3_7

marker-assisted selection, quantitative trait loci (QTL) mapping, genetic engineering, genome editing, etc. have enormous potential to develop stress tolerant crop cultivars. The role of conventional as well as advance approaches in breeding for the development of drought and salt stress tolerant crop varieties and their future scope has been described in this chapter.

Keywords

Advance breeding · Conventional · Germplasm · Variation · Drought stress · Salt stress

7.1 Introduction

Increasing demand for food, diminishing cultivable land available, and environmental stressors are the most challenging problems that are faced globally and have encouraged new studies to examine their impact on food production. In an undulating environment, plants get affected to operate, and typical external fluctuations are balanced by internal dynamics without disturbing its normal functioning. Abiotic stress causes physical injury to the crop plants as a result of severe or persistently unfavorable environmental conditions. Abiotic stress is due to different factors like salinity, mineral toxicity, drought, cold or heat stress, etc. These factors provide important environmental constraints on global agricultural output (Fahad et al. 2017). Among these, drought and salinity stress are the serious threats to the whole agricultural sector and significantly hamper plant growth and agricultural productivity. Drought is one of the most frequent climate-related disasters occurring across large portions of the globe. It is generally induced by water unavailability due to uneven rainfalls or inadequate irrigation. However, it may also be provoked by other factors like elevated soil or air temperature and salinity or physical conditions of the soil. Over half of the world's arable land is predicted to be salinized by 2050 (Jamil et al. 2011). Both drought and salt stress have affected greater than 10% of arable land, resulting in speedy increase in desertification and soil salinization worldwide. Subsequently, average yields of major crops have been reduced by more than 50% (Bray et al. 2000). Hence, there is a paramount demand for new crop cultivars having enormous potential in terms of yield under abiotic adversities.

During the past, progress made in this area is meager because of great controversy among plant physiologists, breeders, and molecular biologists on how to improve stress tolerance in plants (Munns and Gilliham 2015). Since mechanisms showing drought and salt tolerance are very complex, varying from species to species and changing due to changes in environmental factors, investigation of mechanism of salt and drought tolerance using physiological and molecular based techniques is necessary. Under drought stress conditions, different morphophysiological, biochemical, and molecular attributes of the plants get impaired (Kumari et al. 2018). On the other hand, if soil salinity levels increase, plants extract water less easily from soil, aggravating water stress conditions. It also causes nutrient imbalances, limiting plant growth (Yadav et al. 2020). Plants are more sensitive to high salinity during seedling stages and when subjected to other stresses (e.g., disease, insect, nutrient).

Plant breeding has a significant role in addressing the challenges of food security at global level. To ameliorate harsh effects of drought and salt stress and improve the stress tolerance in crop plants, various breeding strategies, i.e., conventional as well as advance breeding, are deployed. The advancements have been made to improve stress tolerance in different crop plants through conventional and advance breeding approaches like marker-assisted breeding (MAB), genetic engineering, genome editing, etc. Even though scientists have made significant progress in increasing crop capacity to withstand stress, there continues to be a tremendous deal of room for improvement through conventional and molecular approaches. These two approaches have been discussed here.

7.2 Drought and Salt Tolerance

Drought is inadequacy of water availability, either due to low precipitation or soil moisture storage capacity. Dry soil and the loss of water through a high transpiration rate make the plant experience drought stress which leads to the loss of turgor. As a result, the development and growth rates are compromised. Ultimately, this water shortage lowers the viability of crop plant. On the other hand, salts are a common and necessary component of soil. Many salts (e.g., nitrates and potassium) are essential plant nutrients. Salts originate from mineral weathering, inorganic fertilizers, soil amendments (e.g., gypsum, composts, and manures), and irrigation waters. It is only when salts are present in relatively high amounts that plant growth is adversely affected. Hence, high concentration of soluble salts in soil, high enough to affect plant growth, is known as salinity. Salt stress is one of most detrimental environmental stresses, which causes ionic toxicity, osmotic, and oxidative stress simultaneously (Tanveer and Shabala 2018).

Both the terms, i.e., tolerance and resistance, are used to explain plant's capability to fight any stress. The difference between tolerance and resistance is that the former is the ability of a plant to grow despite acceptable yield losses whereas the latter is complete immunization to stress such that it does not lead to yield losses. In drought tolerance, plants uphold a certain level of physiological activity over the regulation and fine-tuning of numerous genes and metabolic pathways to decrease the damage (Mitra 2001). When subjected to drought circumstances, a plant's capacity to retain a satisfactory water requirement and turgidity is known as drought resistance. There are different mechanisms or adaptations (morphological, physiological, anatomical, and molecular) which are responsible for the resistance mechanisms. This generally includes stomatal mechanism, increased photosynthetic efficiency, lower rate of cuticular transpiration, reduced leaf area, leaves with thick cuticle, presence of awns, stomatal density, etc. The capacity of crops to endure in a salty environment while still producing growth and yields that are adequate is known as salt tolerance. In nature, different crops have different built-in capabilities to survive in saline conditions (Alkharabsheh et al. 2021). This classifies them in either of four classes, i.e., highly tolerant (barley as grain crop, cotton, Bermuda grass), moderately tolerant (barley as forage crop, sugar beet, wheat, sorghum, oat, soybean), moderately sensitive (rice, corn, peanut, alfalfa, sweet clover, sweet potato), or highly sensitive to saline conditions (bean, potato, tomato, pineapple, onion, strawberry).

7.3 Impact of Stress on Crop Plants

Under drought stress, plants exhibit frequent stomata closure, reduced leaf water potential, and turgor pressure which significantly hamper plant growth and development. It also influences ion homeostasis, nutrient metabolism, chlorophyll synthesis, and respiration. Drought stress exposure results in abated root activity, causing poor root hydraulic conductance. The reproductive stage of plants is more sensitive to drought stress in comparison with pre- or post-reproductive stages which severely affect yield and related attributes. Yadav et al. (2004) reported that in pearl millet (*Pennisetum glaucum* L.), exposure to drought stress at flowering stage results in complete sterility that may be due to disturbed assimilate movement to the developing ear.

The damage due to salt stress varies from species to species and also depends on environmental factors like light intensity and soil texture. One of the early reverberations of salt stress on plants is the reduction in total fresh or dry weight. Further, it causes an increased accumulation of phytohormones, i.e., abscisic acid and cytokinin, and decreased indole-3-acetic acid level which limits the plant growth, development, and reproduction (Kumari et al. 2019). Salts in the soil water may inhibit plant growth due to two reasons. First, the presence of salt in the soil water reduces water uptake by the plant, and this leads to growth rate reductions. This is known as water or osmotic deficit effect of salt stress. Second, when excessive salts enter the plants through transpiration stream, it will cause injury to the plant cells in the transpiring leaves. This is salt-specific or ion-excess effect of salt stress (Greenway and Munns 1980). Salt tolerance is defined as the percent biomass production in saline soil relative to biomass production in nonsaline soil, when plants are grown for an extended period of time.

7.4 Distinctive Traits Associated with Drought and Salt Tolerance

In order to survive under abiotic stress conditions, plants have to activate various physiological, biochemical, and molecular responses like modifications in leaf anatomy, morphology, enhanced synthesis of phytohormones and antioxidative enzymes, and upregulation of defense-responsive genes. The various traits associated with drought or salt tolerance in crop plants serve as important breeding tools in identifying stress tolerant genotypes and in introducing such tolerance traits into cultivated genotypes. The traits conferring drought or salt tolerance helps plant to adapt to such conditions through different mechanisms. These traits and mechanisms vary from one plant to another plant species. The plants growing under salt stress exhibit higher root proportion which results in retention and translocation of toxic ions in plants. Ultimately, this will affect survival/adaptation of the plants under salinity. García and Medina (2013) found that in sugarcane sodium ion (Na^+), sequestration in the xylem parenchyma of mature roots regulated the transport of this ion to the shoot. Higher Na sequestration might have correlation with the higher salinity tolerance. However, in sorghum which is a highly sensitive species toward chloride ion, its ability to exclude chloride ion from mesophyll and accumulate in epidermis and leaf sheath has direct relation with salt tolerance (Boursier and Lauchli 1990).

The root related traits for drought tolerance are root length, diameter, area, and root hair density. The leaf characteristics such as reduction in leaf size, leaf movement, or leaf rolling are also involved in plant defense to drought stress. According to Kashiwagi et al. (2015), in chickpea, better shoot biomass production and greater harvest index appear to be equally imperative for better survival under drought stress via a prolific and deep root system. Saglam et al. (2014) found that leaf rolling is an important mechanism which protects process of photosynthesis and reduces yield loss under drought stress environment in maize. The leaf rolling helps in maintaining the leaf moisture, prevents loss of photosynthetic pigments, and conserves Rubisco activity.

7.5 Breeding for Drought and Salt Tolerance

The use of stress tolerant plants on drought hit or salt-affected land is a more convenient approach to combat drought and salinity. It utilizes various methodologies through which source of drought or salt stress tolerance/resistance in crop plants is identified firstly and then utilized to develop drought or salt tolerant crop varieties. These methodologies include screening of available germplasm for drought/salt tolerance, conventional breeding and advance breeding approaches like molecular breeding, genomic selection, development of transgenic crops, genome editing, etc. (Fig. 7.1). The success has been achieved to develop salt or drought tolerance through conventional as well as advance approaches in various crop species, e.g., in cereal crops (wheat, barley, rice, maize), pulses (chickpea, soybean), fiber crops (cotton), oilseed crops (sunflower, groundnut), vegetable crops (tomato, potato), etc. (Table 7.1).

7.6 Conventional Breeding Approaches

In conventional breeding, the first step involves the identification of genetic variability for drought/salt tolerance among crop cultivars or sexually compatible plant species, and the genetic variation so identified can be introduced through different systems of mating into cultivars/lines with good agronomic characteristics. Conventional breeding approach for tolerance to abiotic stresses is complicated as

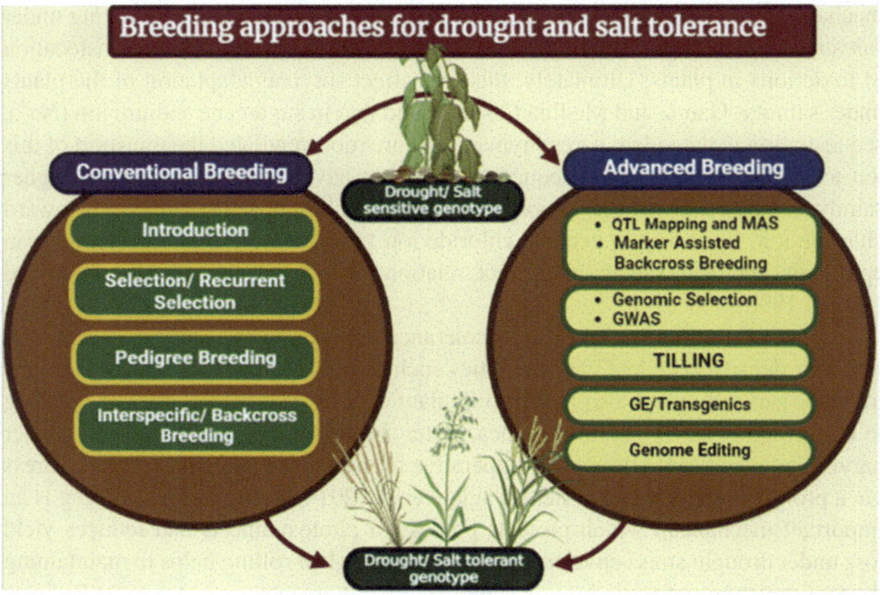

Fig. 7.1 Different breeding approaches for developing drought and salt tolerance in crop plants

compared to breeding for other traits (Fita et al. 2015). Despite this, efforts have been made in the past to develop crop cultivars with enhanced tolerance/resistance to drought/salt stress. A long chain of efforts has been done for introduction, selection, and recombination events like intergeneric and interspecific crosses, which led to the development of improved plant material for drought or salt tolerance. Breeding strategies for salt tolerance in cross-pollinating species by cycles of recurrent selection have been described long ago (Dewey 1962).

The pedigree selection is the widely used and one of the oldest breeding methods used for abiotic stress tolerance in crop plants. However, breeders generally prefer recurrent selection in most of self-pollinating species over pedigree selection for stress tolerance. An enormous amount of genetic diversity is exhibited by the landraces depicting different survival mechanisms and large variation in terms of responses to stress conditions (Reynolds et al. 2007). Blum and Sullivan (1986) identified superior genotypes in sorghum and millet during their studies for drought tolerance in dry and humid regions. Dencic et al. (2000) reported wheat genotypes with high yields under optimal and drought stress conditions in a 2-year experiment with cultivars and landraces. Yadav (2014) stated that the pearl millet landraces that evolved in dry areas as a consequence of natural and manmade selection over thousands of years revealed better adaptability to drought stress.

Different sources of drought and salt tolerance have been identified by the scientists in different crop species, e.g., in chickpea, wheat, maize, oat, and potato for drought tolerance and in rice, pigeon pea, common bean, and tomato for salt tolerance. The introduction of genes from wild salt tolerant species has been

Table 7.1 Conventional and advance breeding approaches utilized in different crops for drought and salt tolerance

Crop	Method	Gene/target action	Resistance	References
Wheat	Interspecific backcross	Gene transfer from *Aegilops tauschii*	Drought tolerance	Sohail et al. (2011)
Wheat	Genetic engineering	*TdPIP2*	Drought and salt tolerance	Ayadi et al. (2019)
Rice	Marker-assisted backcross	Four roots (QTL) from the tropical japonica rice variety Azucena into the Indian upland rice variety Kalinga III	Drought tolerance	Steele et al. (2007)
Barley	QTL (SSR markers)	*bPb-1278* and *bPb-8437*	Salt tolerance	Xue et al. (2010)
Maize	CRISPR/Cas9	*ARGOS8*	Drought tolerance	Shi et al. (2017)
Maize	Genetic engineering	*ZmVPP1*	Drought tolerance	Jia et al. (2020)
Soybean	QTL (SSR markers)	Eight QTL	Salt tolerance	Chen et al. (2008)
Groundnut	TILLING	*PLD*	Drought tolerance	Guo et al. (2015)
Cotton	Genetic engineering	*AVP1*	Salt and drought tolerance	Pasapula et al. (2011)
Tomato	CRISPR/Cas9	*SlMAPK3*	Drought tolerance	Wang et al. (2017)

explored for tomato (Perez Alfocea et al. 1994), wheat (King et al. 1997), and pigeon pea (Subbarao et al. 1990). *Helianthus anomalus*, a diploid annual sunflower species of hybrid origin that is endemic to active desert dunes, was successfully used in sunflower breeding with tolerance to drought stress (Ludwig et al. 2004). Similarly, *Oryza coarctata*, which is a halophytic relative of wild rice, has been studied for decades for salt tolerance to identify a way in which it can be utilized in salt tolerance improvement of cultivated rice (Garg et al. 2013). Recently, the Central Soil Salinity Research Institute (CSSRI) has developed six salt tolerant rice varieties (CSR46, CSR49, CSR52, CSR56, CSR60, and CSR76) through conventional breeding to increase productivity in salt-affected soils (Krishnamurthy et al. 2022). The International Maize and Wheat Improvement Center (CIMMYT), Mexico, developed drought tolerant maize in the 1970s. In 1997, CIMMYT extended its maize breeding programs to drought hit areas of southern Africa with an aim to develop maize for these areas. This center produced several maize hybrids, which were superior to maize hybrids developed by private enterprises for drought prone regions (Bänziger and Diallo 2004). In wheat crop also, the research reports from early 1970s are there, related to work based on directed selection for improved performance under water stress. The use of backcrossing method has led to the development of many drought tolerant varieties in rice. In rice, *Oryza glaberrima* has been used in interspecific backcrossing to improve the drought resistance in *Oryza sativa* (Ndjiondjop et al. 2010).

7.7 Constraints in Conventional Breeding for Drought and Salt Tolerance

Conventional breeding approaches for drought/salt stress tolerance are time-consuming and challenging because stress tolerance traits are controlled by numerous genes which are additive in expression. Also, the magnitude of variation present in the gene pool of most of the crops is low. While introgressing desired genes from one species to other through the conventional plant breeding, a number of undesired genes are also transferred, and the reproductive barrier among the donor species and cultivated variety further complicates the procedure. To achieve the desired gain through this approach, a number of breeding and selection cycles are essential. Thus, there is a need to look for more efficient approaches for enhanced drought/salt tolerance in crop plants.

7.8 Advance Breeding Approaches

Recent advancements made in the area of molecular biology and bioinformatics have offered substantial opportunities for enhancing conventional plant breeding programs. These emerging techniques have an enormous efficiency to boost the crop improvement programs.

7.8.1 Quantitative Trait Loci (QTL) Mapping and Marker-Assisted Selection (MAS)

The loci on chromosomes housing polygenes with additive effects in their expression are referred to as quantitative trait loci (QTL). QTL mapping and marker-assisted selection are a technique that has many advantages over phenotypic screening as a selection tool. In this approach, initially the molecular markers having association with the desired trait are identified. Ashraf et al. (2008) have accounted a variety of DNA markers such as RFLP (restriction fragment length polymorphism), RAPD (randomly amplified polymorphic DNA), CAPS (cleaved amplified polymorphic sequences), PCR indels, AFLPs (amplified fragment length polymorphisms), microsatellites (SSRs), and single nucleotide polymorphisms (SNPs) which are used frequently for inheritance studies for stress tolerance. These marker systems are used in marker-assisted breeding as they map close to specific genes or QTL associated with the trait of interest and help in identification of the individuals with desirable alleles. The marker-trait associations are recognized in

QTL mapping through reliable, dense molecular markers and accurate trait evaluation. Being a complex trait, the drought/salt stress tolerance QTL play a significant role in understanding the stress responses and thereby generating stress tolerant crop plants (Gorantla et al. 2005). Several QTL have been reported in different crop species (e.g., rice, barley, wheat, etc.), involved in the salt stress responses. Plant breeders have transferred whole *Saltol* QTL in to the elite rice varieties through marker-assisted backcrossing and marker-assisted selection (Waziri et al. 2016). Similarly, QTL mapping for the drought tolerance trait has been done in cotton, sorghum, and barley. Nakhla et al. (2021) identified a total of 23 and 46 QTL for the assayed traits associated with salt tolerance at the germination and seedling stages, respectively, in a rice backcross inbred line (BIL) population derived from a backcross of Africa rice ACC9 as donor and *indica* cultivar Zhenshan97 (ZS97) as the recurrent parent.

7.8.2 Genomic Selection (GS)

The conventional breeding strategies to select superior genotypes based on phenotypic traits have limited success due to low heritability, gene interactions, genotype-environment interactions, and polygenic inheritance of stress tolerance. With the advent of emerging genomic techniques, breeders have paved a way to select superior genotypes and speed up the efficiency of breeding programs. Genomic selection is one of these techniques which is economical, viable alternative to MAS and holds huge potential to predict the genotype performance (Shikha et al. 2017). It has accelerated crop improvement programs in cereals and several other crops. It is based on breeding values which are estimated from genomic dataset that explores the genetic variances within each individual (Heffner et al. 2009). It reduces the selection time as compared to phenotypic selection, e.g., in maize, barley, and *Arabidopsis*. It has contributed substantial genetic gain for drought stress tolerance in tropical maize germplasm (Beyene et al. 2015). Research efforts through this approach are on course in other crops, i.e., sugarcane, legumes, and wheat.

7.8.3 Genome-Wide Association Studies

Genome-wide association studies (GWAS) are an alternative approach to the QTL mapping. It relies on the principle of linkage disequilibrium (LD) which detects the association between the DNA marker and desired trait (Gomez et al. 2011). Shahzad et al. (2021) identified a total of 139 SNPs (single nucleotide polymorphisms) associated with the water loss ratio (WLR) using a genome-wide association study among 264 *B. napus* accessions, with the largest number of SNPs at chromosome A10, and 13 SNPs significantly were associated with the WLR. Furthermore, four putative candidate genes (*BnaC09.RPS6*, *BnaC09.MATE*, *BnaA10.PPD5*, and *BnaC09.Histone*) were found involved in drought tolerance. Quamruzzaman et al. (2022) conducted experiments to find the genotypic variation present among

328 wheat varieties for tolerance to salt at vegetative stage. The GWAS identified QTL for salt tolerance based on mixed linear model. Marker-trait associations (MTAs) were identified, and markers were found to be located on the wheat chromosomes 1B, 2B, 2D, 3A, 4B, and 5A.

7.8.4 Targeting-Induced Local Lesions in Genomes (TILLING)

It is a high-throughput reverse genetics approach which is based on screening target regions in the genome using polymerase chain reaction (PCR). It is an economical method for the identification of SNPs and mutations in the gene of interest in a mutant population. It can screen quickly induced mutations (physical or chemical mutagen induced) in the population and analyze missense, nonsense, or knockout mutations in the population. It has been used in wheat, maize soybean, and rice. Large number of proteins including transcription factors is expressed during drought stress in plants. The altered gene expression of some transcription factors has been reported to affect drought and salt tolerance (Hu et al. 2006). The drought tolerance of plants can be enhanced through the overexpression or suppression of these genes. In rice, zinc finger gene *DST* controlling leaf morphology enhanced drought tolerance through suppression (Huang et al. 2009).

7.8.5 Transgenic Approach

Recently, genetic engineering has played a pivotal role in generating stress tolerant crop plants as there is no reproductive barrier and risk of transfer of other undesirable traits. Gene pyramiding can also be achieved through this approach. Genetic engineering has been implemented worldwide as a potential technology for developing abiotic stress tolerant plants. Transgenic lines in different crops have been produced for tolerance to drought and salt stress. In maize, a potential inbred line DH4866 was transformed using *E. coli betA* gene expressing for choline dehydrogenase enzyme. The transformed line showed higher tolerance to drought as compared to wild-type plants when tested at the initial growth stages (Quan et al. 2004). Proline plays a multifaceted role in counteracting drought stress. Therefore, the gene for proline biosynthesis key enzyme, Pyrroline-5-carboxylate synthetase (P5CR), has been engineered in soybean (De Ronde et al. 2004). The transgenic lines showed enhanced expression for proline and hence more drought tolerance. In *Brassica napus*, Dalal et al. (2009) assessed the role of a group 4 LEA protein in stress tolerance. It was observed that the expression of *LEA4-1* gene in leaf tissues was induced by different stresses (e.g., ABA, salinity, low temperature, and drought). A gene stacking approach using two transcription factors, DREB1A and rice PHYTOCHROME-INTERACTING FACTOR-LIKE 1 (OsPIL1), was used by Kudo et al. (2017) to enhance plant growth under drought condition.

7.8.6 Genome Editing

Genome editing, a revolutionary and accurate genetic engineering technology that can modify specific target genes of the organism genome, is increasingly used in crop breeding. Recently, genome-editing technologies, especially CRISPR tools, have been adeptly applied to reveal how plants encounter drought and saline stress (Shelake et al. 2022). The CRISPR/Cas9 system precisely introduces mutation(s) in the target gene(s), resulting in enhanced tolerance to various stresses. Maize *ARGOS8* is a negative regulator of ethylene responses. Translocation of the GOS2 promoter to *ARGOS8* locus has been done in maize using this technique. It resulted in elevated expression of *ARGOS8* locus, and the *ARGOS8* variants had improved maize grain yield under field drought stress conditions as compared to wild type (Shi et al. 2017). *OsERA1* is a negative regulator of drought stress responses in rice. Mutants for this gene were created using CRISPR/Cas9 in rice. The mutants exhibited significantly lower relative stomatal conductance rates than those of the wild-type plants (Ogata et al. 2020). CRISPR/Cas9-mediated genome editing was done by Santosh Kumar et al. (2020) in *indica* rice cv. MTU1010 to generate different mutant alleles of *DST* gene. Badhan et al. (2021) utilized CRISPR/Cas9 gene editing of drought tolerance associated genes in chickpea protoplasts. *They selected two genes*, i.e., *4-coumarate ligase* (*4CL*) and *Reveille 7* (*RVE7*), associated with drought tolerance for gene editing. The results of knockout of these genes would be helpful in understanding drought mechanism in chickpea and development of new varieties with higher tolerance to drought stress. *AITRs* (ABA-induced transcriptional repressors) regulate plant responses to abiotic stresses. Using CRISPR/Cas9 genome editing, transgene-free *Gmaitr* mutants were generated in soybean. The seeds of these mutants showed a higher germination rate than the wild type under salt stress (Wang et al. 2021).

7.9 Conclusion and Future Perspective

Drought and salinity tolerance are based on complex genetic systems influenced by genome-environment interactions which affect plant growth, development, and yield potential. Developing drought/salinity resilient crop varieties is one of the most efficient and promising solutions to tackle these problems. Breeders have developed several stress tolerant varieties/cultivars using conventional breeding methods. However, these breeding procedures are less efficient and time-consuming. Also, the breeding for stress tolerance is challenging in terms of how to measure stress tolerance. For quick and effective transfer of tolerant genes, improvement in terms of stable inheritance and practical feasibility of these methods is essential. Modern breeding tools and strategies such as marker-assisted selection, genome-wide association studies, and genetic engineering are all rapid advancements in genomic research that can be utilized to develop drought/salinity tolerant crop varieties. Novel breeding technique, i.e., genome editing, is a pioneer technique and boon to elucidate gene functions and plant responses to stress tolerance. This technique has enabled precise and efficient genome engineering in an organism by adding, removing, or modifying specific regions in the genome. In the near future, CRISPR will prove to be a valuable resource and game changer to engineer crop plants to become resilient to climate change.

References

Alkharabsheh HM, Seleiman MF, Hewedy OA, Battaglia ML, Jalal RS, Alhammad BA, Schillaci C, Ali N, Al-Doss A (2021) Field crop responses and management strategies to mitigate soil salinity in modern agriculture: a review. Agronomy 11:2299

Ashraf M, Athar HR, Harris PJC, Kwon TR (2008) Some prospective strategies for improving crop salt tolerance. Adv Agron 97:45–110

Ayadi M, Brini F, Masmoudi K (2019) Overexpression of a wheat aquaporin gene, *Td* PIP2; 1, enhances salt and drought tolerance in transgenic durum wheat cv. Maali. Int J Mol Sci 20: 2389

Badhan S, Ball AS, Mantri N (2021) First report of CRISPR/Cas9 mediated DNA-free editing of *4CL* and *RVE7* genes in chickpea protoplasts. Int J Mol Sci 22:396

Bänziger M, Diallo AO (2004) Progress in developing drought and stress tolerant maize cultivars for eastern Africa. In: Friesen DK, Palmer AFE (eds) Integrated approaches to higher maize productivity in the new millennium: Proceedings of the 7th Eastern and Southern Africa Regional Maize Conference February 5–11, 2002. CIMMYT and Kenya Agriculture Research Institute (KARI), Nairobi, pp 189–194

Beyene Y, Semagn K, Mugo S, Tarekegne A, Babu R, Meisel B, Sehabiague P, Makumbi D, Magorokosho C, Oikeh S, Gakunga J, Vargas M, Olsen M, Prasanna BM, Banziger M, Crossa J (2015) Genetic gains in grain yield through genomic selection in eight bi-parental maize populations under drought stress. Crop Sci 55:154–163

Blum A, Sullivan CY (1986) The comparative drought resistance of landraces of sorghum and millet from dry and humid regions. Ann Bot 57:835–846

Boursier P, Lauchli A (1990) Growth responses and mineral nutrient relations of salt-stressed sorghum. Crop Sci 30:1126–1233

Bray EA, Bailey-Serres J, Weretilnyk E (2000) Responses to abiotic stresses. In: Gruissem W, Buchannan B, Jones R (eds) Biochemistry and molecular biology of plants. American Society of Plant Physiologists, Rockville, MD, pp 1158–1249

Chen H, Cui S, Fu S, Gai J, Yu D (2008) Identification of quantitative trait loci associated with salt tolerance during seedling growth in soybean (*Glycine max* L.). Crop Past Sci 59:1086–1091

Dalal M, Tayal D, Chinnusamy V, Bansal KC (2009) Abiotic stress and ABA-inducible Group 4 LEA from *Brassica napus* plays a key role in salt and drought tolerance. J Biotechnol 139: 137–145

De Ronde JA, Cress WA, Kruger GH, Strasser RJ, Van Staden J (2004) Photosynthetic response of transgenic soybean plants, containing an *Arabidopsis P5CR* gene, during heat and drought stress. J Plant Physiol 161:1211–1224

Dencic S, Kastori R, Kobiljski B, Duggan B (2000) Evaluation of grain yield and its components in wheat cultivars and landraces under near optimal and drought conditions. Euphytica 113:43–52

Dewey PR (1962) Breeding crested wheatgrass for salt tolerance. Crop Sci 2:403–407

Fahad S, Bajwa AA, Nazir U, Anjum SA, Farooq A, Zohaib A, Sadia S, Nasim W, Adkins S, Saud S, Ihsan MZ, Alharby H, Wu C, Wang D, Huang J (2017) Crop production under drought and heat stress: plant responses and management options. Front Plant Sci 8:1147. https://doi.org/10.3389/fpls.2017.01147

Fita A, Rodríguez-Burruezo A, Boscaiu M, Prohens J, Vicente O (2015) Breeding and domesticating crops adapted to drought and salinity: a new paradigm for increasing food production. Front Plant Sci 6:978. https://doi.org/10.3389/fpls.2015.00978

García M, Medina E (2013) Effect of salt stress on salt accumulation in roots and leaves of two sugarcane genotypes differing in salinity tolerance. J Trop Agric 51:15–22

Garg R, Verma M, Agrawal S, Shankar R, Majee M, Jain M (2013) Deep transcriptome sequencing of wild halophyte rice, *Porteresia coarctata*, provides novel insights into the salinity and submergence tolerance factors. DNA Res 21:69–84

Gomez G, Alvarez MF, Mosquera T (2011) Association mapping, a method to detect quantitative trait loci: statistical bases. Agron Colomb 29:367–376

Gorantla M, Babu PR, Reddy Lachagari VB, Feltus FA, Paterson AH, Reddy AR (2005) Functional genomics of drought stress response in rice: transcript mapping of annotated unigenes of an *indica* Rice (*Oryza sativa* L. cv. Nagina 22). Curr Sci 89(3):496–514

Greenway H, Munns R (1980) Mechanisms of salt tolerance in nonhalophytes. Annu Rev Plant Physiol 131:149–190

Guo Y, Abernathy B, Zeng Y, Ozias-Akins P (2015) TILLING by sequencing to identify induced mutations in stress resistance genes of peanut (*Arachis hypogaea*). BMC Genomics 16:1–13

Heffner EL, Sorrells ME, Jannink J (2009) Genomic selection for crop improvement. Crop Sci 49: 1–12

Hu H, Dai M, Yao J, Xiao B, Li X, Zhang Q, Xiong L (2006) Overexpressing a NAM, ATAF, and CUC (NAC) transcription factor enhances drought resistance and salt tolerance in rice. Proc Natl Acad Sci U S A 103:12987–12992

Huang XY, Chao DY, Gao JP, Zhu MZ, Shi M, Lin HX (2009) A previously unknown zinc finger protein, DST, regulates drought and salt tolerance in rice via stomatal aperture control. Genes Dev 23:1805–1817

Jamil A, Riaz S, Ashraf M, Foolad MR (2011) Gene expression profiling of plants under salt stress. CRC Crit Rev Plant Sci 30:435–458

Jia TT, Li JJ, Wang LF, Cao YY, Ma J, Wang H, Zhang DF, Li HY (2020) Evaluation of drought tolerance in ZmVPP1-overexpressing transgenic inbred maize lines and their hybrids. J Integr Agric 19:2177–2187

Kashiwagi J, Krishnamurthy L, Purushothaman R, Upadhyaya HD, Gaur PM, Gowda CLL, Ito O, Varshney RW (2015) Scope for improvement of yield under drought through the root traits in chickpea (*Cicer arietinum* L.). Field Crop Res 170:47–54

King IP, Forster BP, Law CC, Cant KA, Orford SE, Gorham J, Reader S, Miller TE (1997) Introgression of salt-tolerance genes from *Thinopyrum bessarabicum* into wheat. New Phytol 137:75–81

Krishnamurthy SL, Lokeshkumar BM, Rathor S, Warraich AS, Yadav S, Gautam RK, Singh RK, Sharma PC (2022) Development of salt-tolerant rice varieties to enhancing productivity in salt-affected environments. Environ Sci Proc 13:30

Kudo M, Kidokoro S, Yoshida T, Mizoi J, Todaka D, Fernie AR, Shinozaki K, Yamaguchi-Shinozaki K (2017) Double overexpression of DREB and PIF transcription factors improves drought stress tolerance and cell elongation in transgenic plants. Plant Biotechnol J 15:458–471

Kumari N, Avtar R, Kumari A, Sharma B, Rani B, Sheoran RK (2018) Antioxidative response of Indian mustard subjected to drought stress. J Oilseed Brass 9:40–44

Kumari N, Malik K, Rani B, Jattan M, Sushil AR, Devi S, Arya SS (2019) Insights in the physiological, biochemical and molecular basis of salt stress tolerance in plants. In: Giri B, Varma A (eds) Microorganism in saline environments: strategies and function. Soil biology, vol 56. Springer Nature, Basel, pp 353–374

Ludwig F, Rosenthal DM, Johnston JA, Kane NC, Gross BL, Lexer C, Dudley SA, Rieseberg LH, Donovan LA (2004) Selection on leaf ecophysiological traits in a desert hybrid *Helianthus* species and early-generation hybrids. Evolution 58:2682–2692

Mitra J (2001) Genetics and genetic improvement of drought resistance in crop plants. Curr Sci 80: 758–763

Munns R, Gilliham M (2015) Salinity tolerance of crops-what is the cost? New Phytol 208:668–673

Nakhla WR, Sun W, Fan K, Yang K, Zhang C, Yu S (2021) Identification of QTLs for salt tolerance at the germination and seedling stages in rice. Plants 10:428

Ndjiondjop MN, Manneh B, Cissoko M, Drame NK, Kakai RG, Bocco R, Baimey H, Wopereis M (2010) Drought resistance in an interspecific backcross population of rice (*Oryza* spp.) derived from the cross WAB56-104 (*O. sativa*)× CG14 (*O. glaberrima*). Plant Sci 179:364–373

Ogata T, Ishizaki T, Fujita M, Fujita Y (2020) CRISPR/Cas9-targeted mutagenesis of OsERA1 confers enhanced responses to abscisic acid and drought stress and increased primary root growth under nonstressed conditions in rice. PLoS One 15:e0243376

Pasapula V, Shen G, Kuppu S, Valencia JP, Mendoza M, Hou P, Chen J, Qiu X, Zhu L, Zhang X, Auld D, Blumwald E, Zhang H, Gaxiola R, Payton P (2011) Expression of an *Arabidopsis* vacuolar H^+-pyrophosphatase gene (AVP1) in cotton improves drought and salt tolerance and increases fibre yield in the field conditions. Plant Biotechnol J 9:8899

Perez Alfocea E, Guerrier G, Estan MT, Bolarin MC (1994) Comparative salt responses at cell and whole plant levels of cultivated and wild tomato species and their hybrid. J Hortic Sci 69:639–644

Quamruzzaman M, Manik SMN, Shabala S, Cao F, Zhou M (2022) Genome-wide association study reveals a genomic region on 5AL for salinity tolerance in wheat. Theor Appl Genet 135:709–721

Quan R, Shang M, Zhang H, Zhao Y, Zhang J (2004) Engineering of enhanced glycine betaine synthesis improves drought tolerance in maize. Plant Biotechnol J 2:477–486

Reynolds M, Dreccer F, Trethowan R (2007) Drought-adaptive traits derived from wheat wild relatives and landraces. J Exp Bot 58:177–186

Saglam A, Kadioğlu A, Demiralay M, Terzi R (2014) Leaf rolling reduces photosynthetic loss in maize under severe drought. Acta Bot Croat 73:315–332

Santosh Kumar VV, Verma RK, Yadav SK, Yadav P, Watts A, Rao MV, Chinnusamy V (2020) CRISPR-Cas9 mediated genome editing of drought and salt tolerance (OsDST) gene in *indica* mega rice cultivar MTU1010. Physiol Mol Biol Plants 26:1099–1110

Shahzad A, Qian M, Sun B, Mahmood U, Li S, Fan Y, Chang W, Dai L, Zhu H, Li J, Qu C, Lu K (2021) Genome-wide association study identifies novel loci and candidate genes for drought stress tolerance in rapeseed. Oil Crop Sci 6:12–22

Shelake RM, Kadam US, Kumar R, Pramanik D, Singh AK, Kim JY (2022) Engineering drought and salinity tolerance traits in crops through CRISPR-mediated genome editing: targets, tools, challenges, and perspectives. Plant Commun 3:100417

Shi J, Gao H, Wang H, Lafitte HR, Archibald RL, Yang M, Hakimi SM, Mo H, Habben JE (2017) *ARGOS8* variants generated by CRISPR-Cas9 improve maize grain yield under field drought stress conditions. Plant Biotechnol J 15(2):207–216

Shikha M, Kanika A, Rao AR, Mallikarjuna MG, Gupta HS, Nepolean T (2017) Genomic selection for drought tolerance using Genome-Wide SNPs in maize. Front Plant Sci 8:550

Sohail Q, Inoue T, Tanaka H, Eltayeb AE, Matsuoka Y, Tsujimoto H (2011) Applicability of *Aegilops tauschii* drought tolerance traits to breeding of hexaploid wheat. Breed Sci 61:347–357

Steele KA, Virk DS, Kumar R, Prasad SC, Witcombe JR (2007) Field evaluation of upland rice lines selected for QTLs controlling root traits. Field Crop Res 101:180–186

Subbarao GV, Johansen C, Kumar Rao JVDK, Jana MK (1990) Salinity tolerance in F_1 hybrids of pigeon pea and a tolerant wild relative. Crop Sci 30:785–788

Tanveer M, Shabala S (2018) Targeting redox regulatory mechanisms for salinity stress tolerance in crops. In: Salinity responses and tolerance in plants, vol 1. Springer, New York, NY, pp 213–234

Wang L, Chen L, Li R, Zhao R, Yang M, Sheng J, Shen L (2017) Reduced drought tolerance by CRISPR/Cas9-mediated SlMAPK3 mutagenesis in tomato plants. J Agric Food Chem 65(39): 8674–8682

Wang T, Xun H, Wang W, Ding X, Tian H, Hussain S, Dong Q, Li Y, Cheng Y, Wang C, Lin R, Li G, Qian X, Pang J, Feng X, Dong Y, Liu B, Wang S (2021) Mutation of *GmAITR* genes by CRISPR/Cas9 genome editing results in enhanced salinity stress tolerance in soybean. Front Plant Sci 12:779598. https://doi.org/10.3389/fpls.2021.779598

Waziri A, Kumar P, Purty RS (2016) Saltol QTL and their role in salinity tolerance in rice. Austin J Biotechnol Bioeng 3(3):1067
Xue D, Huang Y, Zhang G, Wei K, Westcott S, Li C, Chen M, Zhang X, Lance R (2010) Identification of QTLs associated with salinity tolerance at late growth stage in barley. Euphytica 169(2):187–196
Yadav OP (2014) Developing drought-resilient crops for improving productivity of drought-prone ecologies. Indian J Genet Plant Breed 74:548–552
Yadav RS, Hash CT, Bidinger FR, Devos KM, Howarth CJ (2004) Genomic regions associated with grain yield and aspects of post flowering drought tolerance in pearl millet across environments and tester background. Euphytica 136:265–277
Yadav S, Modi P, Dave A, Vijapura A, Patel D, Patel M (2020) Effect of abiotic stress on crops. Sustainable crop production. IntechOpen, London, pp 1–21

Strategies for Improving Tolerance to the Combined Effect of Drought and Salinity Stress in Crops

8

Himanshu Sharma, Archana Kumari, Om Prakash Raigar, Gaurav Augustine, Vikas Verma, Chayanika Lakhar, Arshvir Kaur Boparai, Hitashi Aggarwal, Aman Kumar, and Ashutosh K. Srivastava

Abstract

Abiotic variables such as water, light, nutrients, and temperature play a crucial role in optimizing growth and development of plants. The interplay of these variables results in significant yield losses in major agronomic crops, thereby putting global food security at risk. In recent times, the abiotic stress factors have been intensified by climate change, resulting in decreased crop yield and making global food security an implausible dream for agriculturists, policy makers, and researchers. Drought and salinity are two of the most important environmental factors hampering agricultural productivity worldwide. The concurrence of abiotic stress causes morphological, physiological, biochemical, and metabolic changes in plants, affecting their growth, development, and productivity. The responses of crops to this combined stress are complicated and are influenced by the severity and duration of these abiotic and other factors. In the present chapter, synergistic effects of drought and salinity stress on plant physiological processes, growth, and behaviour have been explained. Moreover, we have focused on the morphological, physiological, biochemical, and metabolic responses of plants to these combined stresses, underlying mechanisms that confer abiotic stress tolerance and key strategies for developing climate resilient crop.

H. Sharma · A. Kumari · O. P. Raigar · C. Lakhar · A. K. Boparai
Department of Plant Breeding and Genetics, Punjab Agricultural University, Ludhiana, India

G. Augustine · V. Verma · H. Aggarwal · A. Kumar
School of Agricultural Biotechnology, Punjab Agricultural University, Ludhiana, India

A. K. Srivastava (✉)
Department of Basic Science, Rani Lakshmi Bai Central Agricultural University, Jhansi, Uttar Pradesh, India

A. Kumar et al. (eds.), *Salinity and Drought Tolerance in Plants*,
https://doi.org/10.1007/978-981-99-4669-3_8

Keywords

Abiotic stress · Physiology · Salinity · Drought

8.1 Introduction

Environmental factors that affect plants growing in the field include salt, drought, and extremes in temperature. Plants might be subjected to two or more stressors at any same moment (e.g., drought and salinity). Drought, which affects almost 40% of the world's geographical area, is the stress that has the greatest impact on global agriculture. According to Trenberth et al. (2014), it is anticipated that global climate change would result in high temperatures and more severe, persistent drought in several regions of the world, which will have a significant influence on agricultural growth and output. A significant factor reducing agricultural output in addition to drought is salt. According to estimates, salinity has an impact on around 7% of the planet's geographical surface. The combined areas of soils impacted by salt and drought are 10.5 million km^2 and 60 million km^2, respectively (Rozema and Flowers 2008). These territories encompass all places impacted by salt and drought across the world, including those that aren't necessarily arable. Maintaining global food supplies for the expanding population has proven to be extremely difficult due to the loss of farmable land caused by either salinity or drought. Thus, there is an urgent need to develop and apply key strategies for improving tolerance to combined effect of drought and salinity stress in crops.

8.1.1 Physiological Response of Drought and Salinity Stress

Constant exposure of plants to the fluctuating environmental conditions especially when these conditions are extreme leads to stress in plants. In many crops, these stresses affect the plant growth, biomass, ion toxicity, osmotic stress, and net photosynthetic rate. The accumulation of salt not only inhibits the plant growth but also reduces water and nutrient uptake, resulting in osmotic or water deficit stress in plants. In some crops such as wheat, it was observed that due to combined stress of drought and salinity, the plants become necrotic resulting in significant reduction of root length, plant height, chlorophyll content, and biomass. Similarly, drought stress in sorghum leads to chlorophyll content reduction and promotes leaf senescence, which further leads to the lower grain yield. The growth of panicle exertion and peduncle length were also found to be affected due to the stressed conditions. Under ambient temperature, the drought stress in plants occurs due to low soil and atmospheric humidity following deficiency of water for normal growth of plants, and in response to this, the plants adapt their water balance to improve water uptake or to reduce water losses. In response to the drought stress, plant reacts in these ways: avoidance, tolerance, and escape (Fig. 8.1) (Ngara and Ndimba 2014).

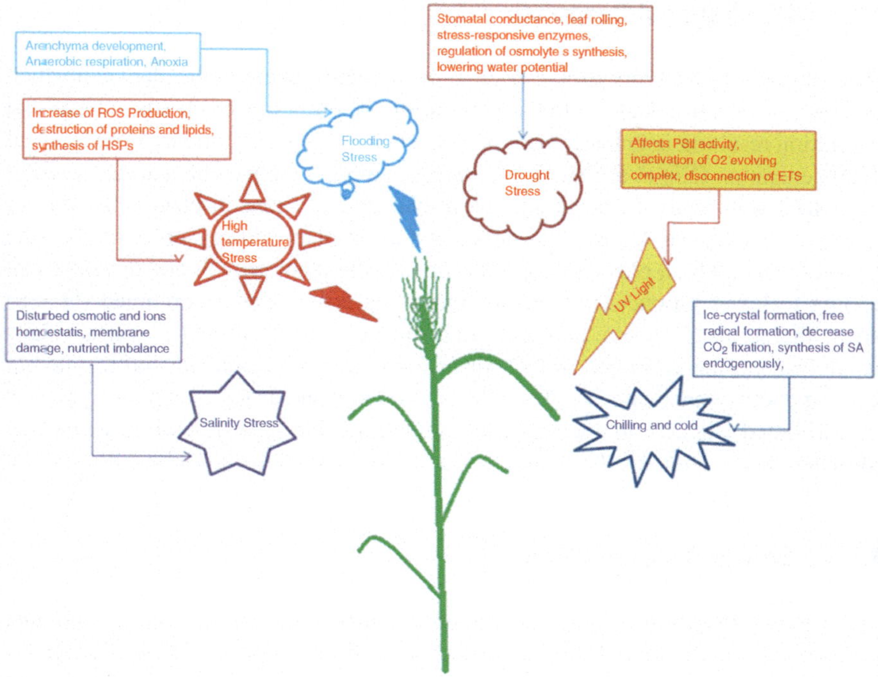

Fig. 8.1 Diverse abiotic stresses affect the plant and their physiological responses

Stress escape is the situation in which plant avoids the drought period by early maturing or completing life cycle before severe water deficient conditions, and this attribute is suitable for late-season drought stress environments. These varieties have early flowering, high photosynthetic capacity, lower total evapotranspiration, and lower yield potential. Avoidance is the plant ability to maintain higher hydration level/leaf water content under atmospheric or soil water stress conditions. These plants are characterized with small leaf area and reduce transpiration by stomatal closure, leaf rolling, and wax accumulation on leaf surface in water deficient conditions. In some cases, in response to water deficient conditions, the roots of the plant start synthesizing ABA which is transported through xylem from roots to leaves. To avoid water stress conditions, ABA plays an important role by affecting the stomata closure, leaf expansion reduction, and enhancing root growth. The stress tolerance ability of the plants means that as compared to the other plants under stress conditions, the tolerant plants will experience significantly lower level of changes. These plants attain tolerance by improving ability of osmotic adjustment and maintaining tissue turgidity by increasing cell wall elasticity.

8.1.2 Biochemical Response

These drought and salinity stresses lead to the oxidative stress in the plant cell which increases the electron leakage toward O_2 during the processes of photosynthesis and respiration resulting in enhancement of ROS such as H_2O_2, O_2, and hydroxyl radical (HO) generation (Xie et al. 2019). The salinity stress in plants which are exposed to high NaCl soils leads the accumulation of Na^+ and Cl^- ions in plant cells causing severe ion imbalance. Higher Na^+ concentration in plants during stress inhibits the K^+ ion uptake, which plays an important role in the plant growth and development leading to lower productivity and possibly plant death. Under combined (D + S) stress, the stomatal closure and leaf conductance were decreased in wheat plants which leads to the inhibition of CO_2 diffusion to carboxylation site and resulted in reduced photosynthetic uptake. The effects of salinity and drought on photosynthesis are attributed directly to the stomatal limitations for diffusion of gases which ultimately alters photosynthesis and the mesophyll metabolism (Dubey 2018).

8.1.3 Solute Accumulation

Under stress conditions, the concentration of other amino acids such as cysteine, arginine, and methionine decreases, whereas the proline concentration rises. Intracellular accumulation of proline during the salinity stress along with tolerance to the stress also serves as reservoir for organic nitrogen during the recovery under stress. For example, in rice, it was observed that the pretreated seeds with 1 mM proline produced rice seedlings which exhibited improved growth during salt stress. Glycine betaine (nontoxic cellular osmolyte) accumulation is found in large number of plants during stress (Hussain Wani et al. 2013). It plays important role under stress conditions, by raising the cell osmolarity, cell protection by osmotic adjustment, ROS reduction, and protecting photosynthetic apparatus from damages due to stress. Accumulation of glycine betaine is found in a wide variety of plants belonging to different taxonomical background.

In some plants, the accumulation of carbohydrates like sugars (fructose, sucrose, trehalose, and glucose) and starch under salt stress was also found to mitigate the stress by carbon storage, osmoprotectants, and scavenging of ROS (Goyal et al. 2023). During salt stress, accumulation of trehalose in the plant cell protects the plant from several physical and chemical damages (Sadak 2019). In rice roots, the content of starch under drought and salinity stress was observed to be low, while in the shoot, its concentration was not that much affected (Beckles and Thitisaksakul 2014).

8.1.4 Molecular Response

8.1.4.1 Genetic Basis of Drought and Salinity Stress Resistance

The traits responsible for drought and salinity stress tolerance such as yield stability, leaf rolling, root growth, ABA accumulation, osmotic adjustment, resistance to flower shedding, seedling recovery, and sustained formation of pod under stress are controlled by various major as well as minor genes. The characters of the leaf like waxy leaves and traits such as accumulation of ABA (wheat), proline accumulation (barley), and pod formation (French bean) are under the control of oligogenes (Rauf et al. 2016). On the other hand, the other remaining traits for drought or salinity stress are considered to be determined by the polygenes.

8.1.4.2 QTL Identification

There are many QTLs responsible for the abiotic stress tolerance. In many crop species, these QTLs for abiotic stress tolerance were identified and transferred to susceptible lines using linkage mapping and molecular markers linked to these QTLs through marker-assisted selection (MAS) such as QTL for osmotic adjustment in rice; for proline accumulation in *Brassica* species; for drought tolerance in rice, barley, and maize; or for salt tolerance in rice (Sahebi et al. 2018). However, due to some problems like linkage drag and the effect of environment, the transfer of marker linked QTL is limited. But new genomic approaches like next-generation sequencing (NGS), genome wide association study (GWAS), and CRISPR/Cas9 make it easier to identify and transfer these QTLs by providing genetic maps with better resolution and precisely locating the QTL position on the chromosome (Fiaz et al. 2019). The combination of genomic studies and transcriptomic analysis enhances the precision of identifying new genes and regulatory systems and their positions.

8.1.4.3 Transgenics

The transgenic plants with the overexpressing ion transporters, transformed with tissue-specific AtNHX1 gene from same species encoding for Na^+/K^+ antiporter, showed improved tolerance to NaCl and accumulated salt in leaves in many crops like *Arabidopsis thaliana* and tomato (Li et al. 2017). The manipulation of specific metabolic pathways and enhancing the cellular levels of particular osmolytes were also done in some crops to improve the abiotic stress tolerance. For example, in rice, the plants were transformed with choline oxidase gene, i.e., *codA*, resulting in higher levels of glycine betaine and simultaneously increasing tolerance to salinity (Bhowal et al. 2021). Similarly, in tobacco, the expression of 1-pyrroline-5-carboxilase synthetase leads to higher proline levels improving drought and salinity tolerance.

8.1.4.4 Activation of Antioxidant Systems

In response to abiotic stress conditions, transgenic plants with enhanced stress tolerance activates the antioxidant system by expressing different antioxidant enzymes like glutathione *S*-transferase/glutathione peroxidase in tobacco or superoxide dismutase in alfalfa and by expressing specific proteins to confer tolerance against abiotic stresses such as heat shock proteins (HSPs), LEA proteins, or transcription factors (Srivastava et al. 2019). There is another approach to increase the stress tolerance used in GM plants which involves overexpression of stress target proteins. In *A. thaliana*, two genes encoding the SR-like splicing factors were isolated from its cDNA library, and their expression in these transgenic plants of *A. thaliana* resulted in salt and drought tolerance (Naranjo et al. 2006).

8.2 Traits Which Are Crucial for Combined Salinity and Drought Stress Tolerance

Plants are frequently exposed to one or more abiotic stresses, including combined salinity and drought, which significantly lower plant growth. The two main abiotic stressors that interfere with plant growth and productivity are salinity and drought (Gull et al. 2019). Due to climate change, both salinity and drought stresses are gradually becoming more severe in many places, mainly in arid or semiarid areas (Manneh et al. 2007). Drought is frequently associated with salinity stress in coastal, arid, and semiarid regions. When the soil water evaporates, the salts become concentrated in the soil solution, resulting in combined drought and salinity (Angon et al. 2022). In general, the co-occurrence of many abiotic stresses is worse for crop output than a single stress event. For instance, the combined effects of salinity and drought on yield are more detrimental than the effects of each stress alone, as observed in barley (Yousfi et al. 2010), wheat (Yousfi et al. 2012), and potato. However, several studies to date have addressed the effects of single stresses on plant, and little is known about the physiological and molecular mechanisms underlying the acclimation of plants to a combination of salinity and drought (Zhao et al. 2020; Wu et al. 2013).

Adapting to unfavorable environment for growth and development is a common task for plants. A promising way of doing so is to identify measurable traits/parameters that are affected by combined stress conditions, which can be modified positively to improve crop productivity under combined stress conditions (Stokes and Inman-Bamber 2014). For example, a plant has evolved morphological and physiological adaptations, as well as signaling pathways that trigger biochemical and molecular mechanisms, to tolerate these extreme climatic circumstances. Some plants evolved with adapted characteristics to survive in harsh conditions including marshes, tundra, and deserts. By making morphophysiological adjustments, such as regulating stomata, reducing the number of leaves or their area, accumulating wax, improving the root system's ability to absorb available water, and using less water overall, it is possible to prevent droughts (Shelake et al. 2022). In the section below, we focus some important traits that can be used for crop improvement under combined salinity and drought stress.

8.2.1 Root System Architecture

Root system architecture has a dominant role in crop plants' response to abiotic stresses. Since roots grow underground, they are the first to sense abiotic stresses and adjust their genetic program for postembryonic development to survive the stress (Guo et al. 2009). Different abiotic stresses affect RSA in various ways. Deeper roots are associated with increased acquisition of water and mobile nutrients like N that may leach to lower soil layers (Lynch and Wojciechowski 2015). To prevent dehydration under mild or severe drought stress, the roots may alter their architecture and resource allocation (Smith and De Smet 2012; Hasibeder et al. 2015). However, under extreme dry stress, the roots shrivel, and the photosystem II in the leaf malfunctions (Fathi and Tari 2016).

8.2.2 Leaf Rolling and Stomatal Aperture

Leaf rolling is one of the consequences of turgor pressure adjustment observed in diverse plants when they are exposed to limited water environments. Leaf rolling during the stressful time decreases the leaf area, which in turn decreases the area that may be used for evapotranspiration and hence represents a drought acclimation response. Leaf rolling is a classic indication of soil water deficiency in wheat, as it is in other cereals. Leaf rolling is a classic indication of soil water deficiency in wheat, as it is in other cereals (Kadioglu et al. 2012). Early reactions to salinity and drought are quite similar because both cause water stress, which slows development, reduces stomatal aperture, and results in nutrient deficiencies (such K^+ and Ca^{2+}) (Pirasteh-Anosheh et al. 2016).

8.2.3 Canopy Temperature

According to Mason and Singh (2014), canopy temperature is a physiological trait that reflects crop water status, which is a recognized proxy for stomatal conductance (Deery et al. 2019). Canopy temperature has the potential to be an extremely helpful tool for the indirect selection of genotypes that are tolerant to heat and drought stress (Reynolds et al. 2010). Genotypes that retain cooler canopies are more likely to flourish in such conditions. It has been noted that canopy temperature is typically lower in deep-rooted genotypes because the crop may draw moisture from a deeper soil depth (López-Urrea et al. 2012). Canopy temperature and stomatal density may be effective measure of drought tolerance of wheat cultivars and yield improvement under drought stress conditions (Srivastava et al. 2017)

8.2.4 Osmotic Adjustment

Osmolytes or compatible solutes are small molecules sometimes known as osmoprotectants or osmolytes, have low molecular weights, and are water soluble and nontoxic at molar concentrations that enhance the cell potential to maintain turgor potential without hampering the normal physiological processes. Thus, damaging effects of drought are minimized by accumulation of solutes in cellular cytoplasm and vacuole (Taiz and Zeiger 2006). Osmotic regulation, antioxidant metabolism, and preservation of cell membrane stability are main defense strategy against abiotic stress. Plant cells enthusiastically accumulate solutes when cellular dehydration occurs as part of osmotic adjustment. Under salt, drought, and cold stresses, metabolic-compatible substances can accumulate in greater quantities and are key drivers of osmotic adjustment. They also aid in stabilizing membranes, protecting proteins' quaternary structures, and neutralizing poisonous substances which are the main tactics used by numerous creatures to overcome environmental stress (Abid et al. 2018). The most popular compatible substances are sugars (sucrose and trehalose), polyols (mannitol and sorbitol), polyamines (putrescine, spermine, and spermidine), amino acids (glutamine and proline), and quaternary amines and compatible solutes (glycine betaine and choline-O-sulfate) (Singh et al. 2015; Zulfiqar et al. 2021). Plants accumulate Osmoprotectants such proline, glycine betaine, and sugars as a result of an induced change in osmolarity. Increased proline levels in the niger showed an effective osmotic adjustment, which is consistent with findings in salt-stressed tobacco, sorghum, and *Sesuvium portulacastrum* (Çelik and Atak 2012; Kahrizi et al. 2012; Rajaravindran and Natarajan 2012; Deinlein et al. 2014).

8.3 Different Reactive Oxygen Species (ROS) Produced During Drought and Salinity Stress

The generation of ROS is a fundamental process in higher plants to transmit cellular signaling information in response to the changing environmental conditions. Different ROS are produced by the unavoidable leakage of electrons on to O_2 from the electron transport systems of chloroplasts, mitochondria, and plasma membranes or as a consequence of various metabolic pathways. Osmotic effects caused by salinity and drought change fundamental metabolic and enzymatic functions, increasing the production of reactive oxygen species (ROS). The main ROS include singlet oxygen ($^1O_2$), hydroxyl radicals (OH), superoxide anion (O_2), hydrogen peroxide (H_2O_2), and the latter two each of which have a distinct half-life and oxidizing potential (Apel and Hirt 2004; Miller et al. 2010; Hasanuzzaman et al. 2021).

Reactive $^1O_2$ can be efficiently quenched by small lipophilic compounds such as β-carotene and tocopherol or by a scavenging action of protein D1 protein present in photosystem II (Krieger-Liszkay 2005). However, minor portion of $^1O_2$ is able to diffuse to a smaller distance, where it can react with diverse biomolecules and potentially mediate its signaling pathways (Dogra et al. 2018). The superoxide radical ($O_2^{\bullet-}$) is formed mainly in the thylakoid-localized PSI during noncyclic electron transport. Its production is one of the responses to the pathogens that burst rapidly and transiently. It is mediated by NADPH oxidases, belonging to the respiratory burst oxidase homolog (RBOH) family (Sang and Macho 2017). Usually, H_2O is generated when cytochrome c oxidase interacts with O_2, and sometimes, it reacts with the different components of ETC to give $O_2^{\bullet-}$. Reactive $O_2^{\bullet-}$ often undergo further reactions to generate $_1O^2$, H_2O_2, and •OH (Janků et al. 2019). Several recent studies have demonstrated that H_2O_2 is involved in stress signaling pathways, which can activate multiple responses that strengthen resistance to various biotic and abiotic stresses (Petrov and Van Breusegem 2012; Molassiotis et al. 2016). Highly reactive •OH promotes oxidative cleavage of pectin and xyloglucans, thus causing cell elongation by loosening the cell wall. Drought stress increases Fe and Cu availability for Fenton reactions and could cause increased •OH production. There is potential of •OH for an impact on atmospheric methane levels, and after the plant's death, •OH contributes to its recycling into soil organic matter. Because cells have no enzymatic mechanism to eliminate •OH, its excess production can eventually lead to cell death (Pinto et al. 2003; Demidchik 2015; Richards et al. 2015).

8.3.1 Antioxidative Defense

If prolonged to a certain extent, drought stress or salinity stress alone or in combination inevitably resulted in oxidative damage due to the overproduction of reactive oxygen species. Reactive oxygen species (ROS), also called active oxygen species (AOS) or reactive oxygen intermediates (ROI), are the result of the partial reduction of atmospheric O_2. Both salt and drought can cause the production of reactive oxygen species (ROS) like superoxide, singlet oxygen, hydrogen peroxide, and hydroxyl radicals inside of plant cells. The overproduction of ROS causes oxidative damage to mainly cell membranes, proteins, lipids, and nucleic acids, causing oxidative stress (Gill and Tuteja 2010). The upregulation of antioxidant enzymes represents an important marker for drought stress (Laxa et al. 2019). In order to cope with the oxidative stress, plants usually rely on effective enzymatic antioxidant defense system, including catalase (CAT), peroxidase (POD), ascorbate peroxidase (APX), and superoxide dismutase (SOD) as well as nonenzymatic antioxidant defense system, including ascorbate, flavonoids, glutathione, tocopherols, and phenolics, involved in removing high levels of ROS. Some plants such as *Calendula officinalis*, *Solanum lycopersicum*, *Jatropha curcas*, and *Zea mays* have been discovered to upregulate antioxidant activity under drought and salinity stresses (Chaparzadeh et al. 2004; Mittova et al. 2004; Gao et al. 2008; Anjum et al. 2017). Similarly, by strengthening the resistance of host plants to oxidative damage, maintaining a high level of antioxidative enzyme activity significantly reduces the effects of drought or salt stress.

8.4 Strategies to Improve Combined Stress Tolerance

8.4.1 Breeding Approaches for Developing Combined Stress Tolerance

Drought and salinity are major abiotic stresses that severely affect agricultural yields. Combined stress of drought and salinity severely affects physiological, biochemical, and metabolic activities/processes of plant as compared to independent either drought or salinity. Earlier breeders concern about the improvement of either drought or salinity tolerance using classical breeding techniques, due to ever changing weather climatic changes and degradation of soil and water quality by extensive use of groundwater for agriculture. Degrading soil quality and poor groundwater conditions along with uneven distribution of rainfall create salt accumulation on top most soil layer (for crop growth and development) combined with drought. Therefore, to address the negative impact of combined drought and salinity stress on crops, there is a need to adopt new classical techniques to advance breeding strategies/approaches by using wild relatives and landraces, introducing elite breeding lines, inducing mutants, transferring beneficial quantitative trait loci (QTLs) genes, and altering gene metabolites into a commercially adopted well-known cultivars/varieties of crop plants (Kaur et al. 2023). Different breeding strategies that can be utilized for combined drought stress tolerance for traits viz., earliness, lesser number of stomata per unit area on leaves, thick and small leaves, leaf and stem waxiness, deep root system, determinate growth habit with salinity stress tolerance traits viz., osmoprotectant proteins, osmolytes, metabolites, stress protein transport, stomatal conductance, cell wall modification and root architectural changes. We can develop genotypes for both stresses using evaluation of large set of germplasm through artificial intelligence with molecular markers (SSR and SNPs), mutation breeding, haploid induction, mapping of QTLs/genes, marker-assisted selection (MAS)/marker-assisted backcross selection (MABC), advanced backcross QTL mapping, gene introgression, and gene pyramiding (Fig. 8.2). Now a days recent advances in modern plant breeding that exploits for generation of targeted mutations at desired loci/genic regions in plant genomes (Shelake et al. 2019) using genome editing tools that enable fast-track introgression of novel trait of interest into crop varieties (Shelake et al. 2019, 2022) represented in Fig. 8.2.

8.4.2 Searching for Drought and Salinity Stress Tolerance Germplasm

A lot of efforts have been invested by plant breeders and geneticists after the rediscovery of Mendel's laws, for the identification of desirable germplasm for drought tolerance, salinity tolerance, and both using accessions from primary center of origin to hot spots of crop diversity, landraces, and wild relatives for different crops. Earlier breeders use large-scale screening of germplasm using morphophysiological parameters for drought and salinity separately. These are very laborious and time-consuming and require experienced personnel. With the advancement of evaluation techniques, genetic inheritance, molecular markers, modification of germplasm, and mapping techniques also facilitate the improvement in crop plants for salt and drought tolerance (Deshmukh et al. 2014). A molecular marker increases the selection intensity for a trait of interest. Nowadays, advances in artificial intelligence (AI) along with molecular data of the genotype will increase selection efficiency by reducing the environmental variation for quantitative traits (salinity and drought tolerance). So through molecular techniques along with AI, we can easily identify simultaneously tolerant germplasm for drought and salinity in large population.

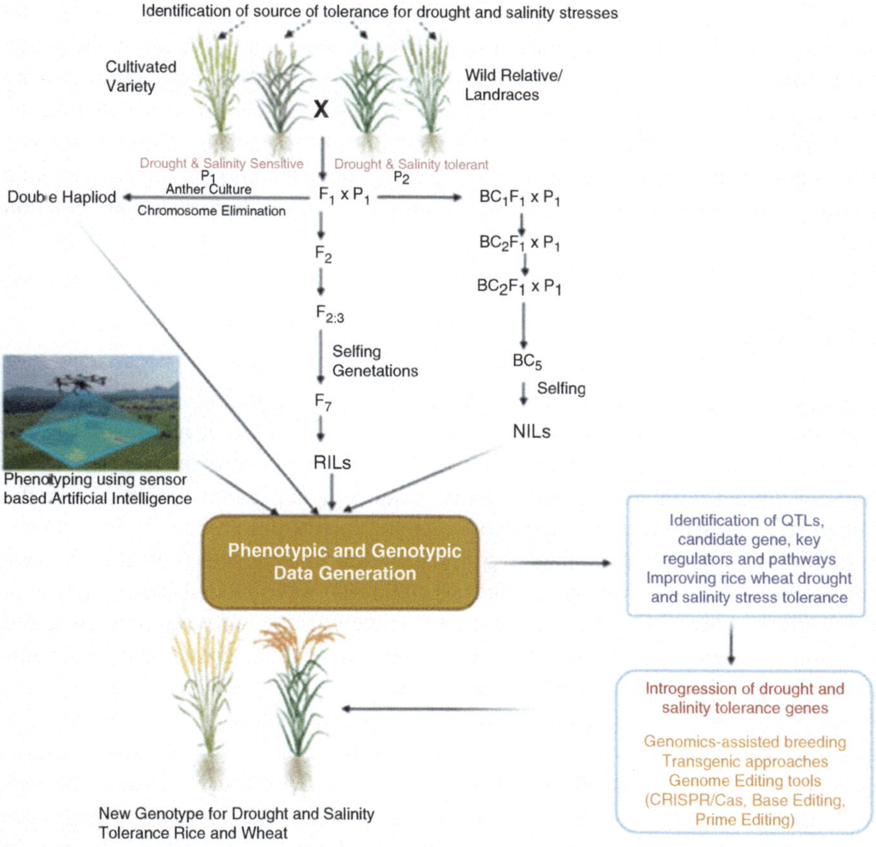

Fig. 8.2 Advanced breeding tools for development of drought and salinity tolerant genotype

8.4.3 Double Haploid

Double haploid (DH) production through anther culture, ovary culture, and chromosome elimination technique emerged as a new breeding tool having advantage by reducing the breeding cycle, homozygosity fixation, high selection efficiency, and expression of recessive alleles suitable for breeding. Guha and Maheshwari (1964) introduced the first haploid embryos and plantlets through anther culture of *Datura innoxia*. In this tool, we first developed haploid plantlets than diploidization of haploids artificially using genome doubling agents (i.e., colchicine treatment). DH technology can be utilized for fixation of desirable trait combination in an elite cultivar, and it will be used for the mapping of QTLs/genes controlling quantitative traits such as salt and drought tolerance. DH can be deployed for development of drought and salinity by using anther culture (rice, wheat, maize), ovary culture (wheat), and chromosome elimination technique, i.e., wheat × *H. bulbosum* (Inagaki 1985), wheat × maize (Laurie and Bennett 1988; Bains et al. 1998; Campbell et al.

1998), wheat × *Imperata cylindrica* (Chaudhary et al. 2005), *Hordeum vulgare* × *H. Bulbosum* (Kasha and Kao 1970), and inducer Stock 6 in maize (Lashermes and Beckert 1988). These techniques of haploid production can be utilized for the production of haploid from the hybrids of wild relative/landraces and simultaneously produce haploids having both drought and salinity tolerance traits within a shortened time periods using cytogenetic techniques such as genomic in situ hybridization (GISH) and FISH (fluorescence in situ hybridization) (Sharma et al. 2022a).

8.4.4 Mutation

Mutation is used to create genetic variation; if such variations are not present in population, then these mutant variants can be deployment for development of new varieties. A major key point in mutation breeding is identification of desirable mutation linked with trait of interest involving two steps: first screening out the mutant and then second confirmation of mutants (Forster and Shu 2012). In mutation breeding, the seeds/plant propagating materials are treated with mutagen (X-rays, gamma rays, chemical mutagens, and fast neutrons) with desired concentration at which plant shows LD 50% than grown for segregation and recombination and selection of salinity and drought tolerance plants and further grow for next generation to stabilization of mutant alleles.

Multiyear and location trials are conducted for evaluation of stable mutant for release of a new variety with trait of interest (Oladosu et al. 2016). Salt oversensitivity is achieved through mutation in barley that enhanced salinity tolerance through ion homeostasis (i.e., M4-73-30 and its wild-type cultivar). These two genotypes showed differential expression of HVA, HvSOS3, HvSOS2, and HvSOS1 genes in the roots of barley. And there is more Na^+ accumulation in shoot tissues of wild-type than mutant because sodium ion transfer from root to shoot is in the mutant type, i.e., salt tolerant (Yousefirad et al. 2018). Lethin et al. (2020) developed a mutant of wheat variety BARI Gom-25 using EMS (ethyl methanesulfonate) that showed semi-tolerance to salinity.

8.4.5 Wild Relative Exploitation

Wild relatives of crops are utilized as a source of biotic and abiotic stress tolerance using interspecific hybridization. Hybrid from wild relatives had lot of embryo and endosperm growth and developmental associated barriers. So it requires specialized methods/techniques for embryo rescue from those hybrids. Salt tolerant wild relatives of rice, *Oryza coarctata* (halophytic relative), was identified as a source which can be exploited for transferring salt tolerance QTL/genes into elite cultivated lines through marker-assisted backcrosses (Mishra et al. 2016; Garg et al. 2013).

8.4.6 QTL Mapping

Identification of genomic regions associated with drought and salinity tolerance can be done through different types of mapping populations, molecular marker, genomic sequencing, and statistical approaches. Zang et al. (2008) identified salt tolerance QTLs at seedling and tillering stages using introgression lines derived from IR64 (indica) × Binam (japonica). Yun et al. (2012) use BC2F8 advanced backcross introgression lines derived from Teqing × Binam in Teqing background for identification of both drought and salt stress tolerance QTL governing traits, i.e., days to seedling survival, score on salt toxicity symptoms on leaves, shoot K^+ concentration, shoot Na^+ concentration at seedling stage, panicle number per plant, thousand grain weight, seed fertility, plant height, and grain yield per plant.

8.4.7 Gene Introgression and Marker-Assisted Breeding

Introgression is the transfer of genes between crop species, which is mediated primarily by backcrossing. Introgressiomics is an extensive systematic improvement of plant genomes and populations through introgressions of fragments of genome from crop wild relatives into the genetic background of prevailing crops in order to develop new cultivars having desirable traits. Due to introgression, rampant gene flow may occur between species where their distribution patterns overlap and they interact so much that introgression has been described as an "invasion of the genome" (Mallet 2005). About 25% of plant species are known to hybridize with at least one other species (Mallet 2005; Schwenk et al. 2008). Greater genomic plasticity can be obtained in a crop using exotic genetic material that was previously nonexistent within the genome through introgression. This strategy can have myriad effects, and though it is most commonly thought to be deleterious, introgression may also provide the raw genetic materials for adaptation and speciation (Arnold and Martin 2009; Suarez-Gonzalez et al. 2018). However, a major limitation when assessing introgression is the availability of genetic resources to accurately estimate interspecific gene flow.

Changes in climatic conditions led to the emergence of various biotic and abiotic stresses which are becoming a major threat limiting crop production and productivity (Gautam et al. 2013). Thus, broadening the crop genetic base will serve as armor against these rising challenges under climate change. The crop wild relatives (CWRs) are known to possess useful alien alleles and cryptic genetic variation, which are introgressed and expressed in cultivated gene pool (Pratap and Gupta 2009). Recent advances in breeding and genomic tools and techniques provide an opportunity to introgress useful alleles left behind in the secondary and tertiary gene pool into the elite background useful for breeders. For crop improvement, genetic engineering strategies are relatively faster than traditional breeding programs, as well as cloning of genes responsible for imperative traits and introgression into plants.

Drought and salinity are the major abiotic stresses that dramatically cause threat to the food security in the world. *Triticum dicoccoides* and *Hordeum spontaneum*, the progenitors of cultivated wheat and barley, have adapted to a broad range of environments and developed rich genetic diversities for drought and salt tolerances. Drought- and salt-tolerant genes and quantitative trait loci (QTLs) have been identified in *T. dicoccoides* and *H. spontaneum* and have great potential in wheat and barley improvement. Advanced backcross QTL analysis, the introgression libraries based on wild wheat and wild barley as donors, and positional cloning of natural QTLs will play critical roles in elucidating the molecular control of drought and salt tolerance. Combining tolerant genes and QTLs in crop breeding programs can be aimed at improving tolerance to drought and salinity (Nevo and Chen 2010).

Root architecture traits are an important component for improving water stress and salinity adaptation. Considerable genetic variation exists for root-related traits in major crops such as wheat, rice, and maize (*Zea mays*) in both the domesticated species and wild relatives (Ribaut et al. 2009; Gowda et al. 2011). The presence of the rye-wheat translocation (1RS) reduced central metaxylem diameter, a trait that has been associated with drought tolerance in wheat (Richards and Passioura 1989; Sharma et al. 2010). Several *Agropyron* species and wheat 3 *A. elongatum* lines have been reported to enhance tolerance to abiotic stresses, including drought, waterlogging, and salinity (Dvorák et al. 1988; McDonald et al. 2001). The physiological and molecular consequences of introgressing an alien chromosome segment (7DL) from a wild wheat relative species (*Agropyron elongatum*) into cultivated wheat (*Triticum aestivum*) were studied by Placido et al. (2013). The wheat translocation line had improved water stress adaptation and higher root and shoot biomass compared with the control genotypes. The transcriptome analysis identified candidate genes associated with root development, and two of the candidate genes mapped to the site of translocation on chromosome 7DL based on single-feature polymorphism analysis.

8.4.8 Use of High-Throughput and Automated Phenotyping Techniques

The slow pace, high costs, and inconsistencies associated with trait quantification and data management using traditional phenotyping techniques still limit the progress of drought tolerance improvement. This could, also, have been contributing to the complexities of understanding the genetic and physiological basis of drought tolerance both at the phenotypic and the genomic level (Xu and Crouch 2008). The utilization of sophisticated, nondestructive, high-throughput phenotyping technologies with automated systems for capturing, storage, and statistical analysis of large volumes of data allows for fast and precise large-scale quantification and monitoring of various phenotypic traits (Araus and Cairns 2014). Ground and remote sensing techniques based on near- or far-infrared reflectance digital sensors, thermometers, and cameras are reported to precisely measure various phenotypic traits (Araus and Cairns 2014; Honsdorf et al. 2014). These include automated camera systems which are connected to computers for monitoring complex root architectural traits through periodic image capturing. Some of these tools are graced with image processing and analysis software. These advanced phenotyping technologies may create local databases for easy management of the vast amounts of data that will be generated.

8.4.9 Omics: Genomics, Phenomics, Metabolomics, Achievements

Plant ecology is a very complicated and interconnected system. For an edge in crop development, it is crucial to comprehend the fine-tuning and integration of various signals produced by environmental interactions in the plants. Multiple biotic and abiotic environmental challenges must be overcome by a plant. The outcome of the plant-environmental interaction may be controlled by prioritizing physiological pathways in plants; many of these are similar to stress responses while others have negative effects. Multiple stress causes result in complicated defensive signals in plants. The molecular and cellular processes behind abiotic stress adaptation in cereal crops have been extensively studied. The control of genes affects a plant's response to drought, high temperatures, and salt (upregulation or downregulation). In this context, multi-omics techniques can be used to tackle the difficult issue of figuring out changes in plants at the genomic, proteomic, or metabolomic level in order to unravel the causes.

8.4.10 Genomics

Crop development projects have greatly benefited from the use of genomics-based technology. The use of molecular markers, QTL mapping, genetic mapping, comparative genomics, the intricate interactions between plants and their environment, and expression quantitative trait loci have all made genomic and systems biology methods to the discovery of stress tolerant phenotypes possible. These methods can offer a useful method for locating candidate genes implicated in the ability to withstand abiotic stress. Thus, a large number of genomic information in the form of sequenced genomes and expression profiles are driving the development of breeding techniques for stress reduction (Fig. 8.3) (Tomar et al. 2014).

The discovery of restriction fragment length polymorphisms (RFLPs) and the subsequent correlation of these variants with a number of major agronomically significant variables presaged the promises of genomics to increase genetic variation per unit time in the 1980s. Later, numerous different marker systems were employed to map quantitative trait loci, most notably microsatellite or simple sequence repeat (SSR) markers (QTLs). However, despite their excellent usage in finding the polymorphisms and in applied breeding, these methods had limited throughput and were time- and cost-inefficient. For instance, there were only 279 loci on the first SSR map of wheat (Röder et al. 1998). According to Koussevitzky et al. (2008), the cytosolic ascorbate peroxidase 1 gene Apx1 is especially needed for *Arabidopsis* to be tolerant to drought and heat stress. Ectoine, a suitable osmolyte, is in charge of *Halomonas elongata* OUT30018's ability to tolerate salt. Tobacco plant (*Nicotiana*

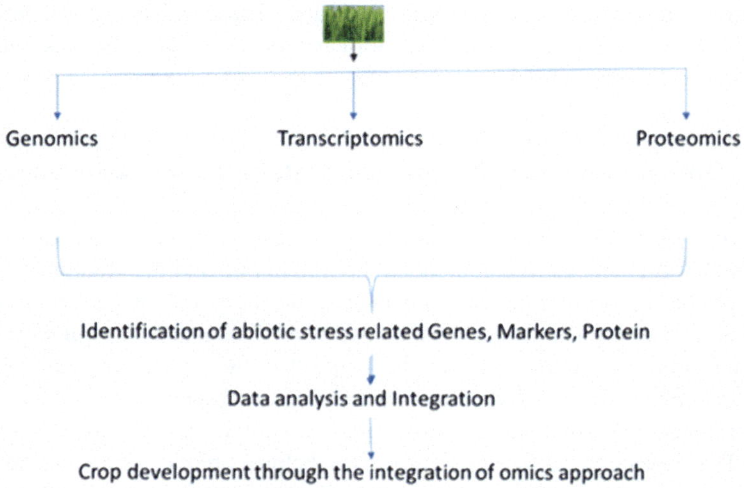

Fig. 8.3 Omics approaches for stress tolerance in plants

tabacum L.) cv. Bright Yellow 2 (BY2) was given three cloned ectoine biosynthesis genes, which increased resistance to hyperosmotic shock by accumulating ectoine and allowed it to develop normally in such circumstances (Nakayama et al. 2000). More than 15 million ha of rainfed lowland rice are under stress because of submergence in various areas of Asia (Neeraja et al. 2007). Problems with floods or anoxia affect 13% of the world's total geographical area (Cramer et al. 2011). In rice, a single main quantitative trait locus (QTL) located on chromosome 9 controls submergence tolerance (Toojinda et al. 2003). Molecular markers for the Sub1 gene were employed by Neeraja et al. (2007) in a backcross breeding effort using the recurrent parent, named Swarna. This Sub1 offers tolerance in giant varieties that are delicate. It has now been established that Sub1A is the main factor influencing submergence tolerance (Septiningsih et al. 2009). The development of submergence tolerance by marker-assisted backcrossing (MAB) has been greatly facilitated by this QTL.

Single nucleotide polymorphisms (SNPs), the smallest type of DNA variation, can be found, giving researchers the chance to examine potentially millions of DNA locations. As a result, it has become a popular site for markers. Large-scale germplasm assessments in many cereals have been made possible by high-throughput, high-efficiency, repeatability, and cheap cost per data point, which has led to the practically total replacement of RFLPs or SSR marker platforms. Both genotyping-by-sequencing and array-based genotyping are important techniques for SNP detection in grains. Both types of SNP identification may be accomplished using a variety of sequencing technologies. For the purpose of establishing the groundwork for whole-genome sequencing and determining the genetic basis of characteristics that are economically significant, high-density SNP genotyping is crucial.

Linkage mapping is the process of locating quantitative trait loci (QTLs) in segregating populations that are mostly artificially produced. Many features of economic significance are of a quantitative type, meaning that they are controlled by the coordinated action of numerous genetic loci, such as grain yield, stress tolerance, and disease resistance (Bernardo 2010). To analyze the genetic basis of a quantitative characteristic, segregating populations including practically hundreds of individuals is necessary. The development of different segregating populations, including the F2 population, recombinant inbred lines (RILs), doubled haploid (DH) population, heterogeneous inbred family (HIF), near-isogenic lines (NIL), advanced intercross recombinant inbred lines (AI-RIL), backcross inbred lines, and multiparent advanced generation intercross (MAGIC), is primarily driven by the resources available and the goals of the research. Through QTL, which can be found at the Gramene database (http://www.gramene.org/qtl/, accessed on October 5, 2022), the major cereal crops are mapped. It has QTLs for abiotic stressors such nutrient deficiency, heavy metal toxicity, salt, cold, and severe temperature. The Sorghum QTL Atlas is available online at https://aussorgm.org.au/sorghum-qtl-atlas/ containing 150 QTL and GWAS research that have been published in sorghum from 1995 to the present (accessed on October 6, 2022). It is beneficial for sorghum crop development and comparative genomics (Mace et al. 2019). The ability to identify natural differences in all recombination events that take place during the evolutionary processes of a variety of organisms makes GWAS an effective tool. SNP identification is the primary goal of the GWAS (Marees et al. 2018). To detect genetic differences based on complex variables like growth rate, blooming period, and yield, WGAS analyses complete samples using databases that contain reference plant genome sequences and genetic maps.

8.4.11 Transcriptomics

Transcriptomics is the study of gene expression patterns that can identify the genes causing abiotic stress. For abiotic stress to be managed successfully, understanding such reactions may be essential. Investigating the functional characterization of specific genes in stress tolerance systems is made possible by transcriptome profiling. For better plant genomic resources, more next-generation sequencing (NGS) and RNA sequences for short RNAs have been produced recently (Imran et al. 2021).

The high-throughput transcriptome sequencing dataset provides an efficient strategy for gene discovery, molecular marker development, and marker-assisted breeding (Munaweera et al. 2022). Typical plant responses to abiotic stress involve a network of molecules. ABA (abscisic acid) is a major phytomolecule that plays an important role in responding to a variety of stresses such as drought, heat stress, high salinity, heavy metal stress, low temperature, etc. They are also involved in various developmental processes, including seed germination, seed dormancy, and closure of stomata. ABA-dependent pathways appear to recruit antioxidant and osmoprotectant mechanisms, whereas ABA-independent pathways generally involve protective proteins (Rane et al. 2021; Khan et al. 2021).

Under abiotic conditions including drought, salt, and cold, different miRNAs in rice, *Medicago*, *Phaseolus*, *Arabidopsis*, and other plants have a regulatory function (Trindade et al. 2010). Noncoding RNAs called miRNAs, which have a length of 19–23 nucleotides, have a regulatory function in a number of biological processes (Budak et al. 2015). *Arabidopsis* overexpressing osaMIR393 exhibits tolerance to salt overload, and miR393 regulates this tolerance (Gao et al. 2011). According to Zhao et al. (2009), miR169 reduces the stress caused by salt and drought in rice via altering the expression of a nuclear transcription factor called YA (NF-YA). Drought resistance is conferred by tomato plants that overexpress the miR169c gene, which regulates the expression of the gene(s) involved in stomatal activity. WD-repeat proteins, which are essential for cucumber stress tolerance, are regulated by Bvu-miR13 (Li et al. 2015). MiRNAs target stress signaling pathways, which are involved in root development, leaf morphogenesis, and stress response, in addition to regulating TFs (Curaba et al. 2014). In silico analysis was used to identify 13 mature miRNAs in *B. vulgaris* plants (Li et al. 2015). Superoxide dismutase SOD1 and SOD2 mRNA activity is targeted by miR398 that plays a function in decreasing ROS and subsequent consequences of abiotic stress (Kantar et al. 2011). Distinct kinds of miRNAs regulate different cellular responses to stress and metabolic functions such transcriptional control, auxin homeostasis, ion transport, and apoptosis (Li et al. 2010). Additionally, it has been discovered that plants' responses to aluminum stress are regulated by miRNA (Lima et al. 2011). MiRNA expression in two distinct rice subspecies, japonica and indica, with variable levels of aluminum tolerance, was compared. The use of RT-qPCR identified 16 different types of miRNA responses, demonstrating a complicated response to aluminum stress.

The impacts of water stress, salinity stress, and combination stress were disclosed by Osthoff et al. (2019). Transcriptome sequencing was carried out in this study 6 and 24 h following the stress treatment. The gene expression responses that were shown at 6 h of stress are still present at 24 h. However, compared to a 6-h stress treatment, hundreds of genes were further altered after 24 h. Nine hundred fifty-three genes showed differential expr bHLH (basic Helix-Loop-Helix), ERF (Ethylene Response Factors), and HSF (Heat Shock Factors) transcription factors that have all been enriched at each time point under the three shocks (6 and 24 h). Additionally, the bZIP was enriched following both combined stress and water deficiencies at both time periods. After 24 h of treatment, salt stress, bZIP, G2-like, and HD-ZIP TFs were enriched, while 1802 genes did so following a 24-h salt therapy. bHLH (basic Helix-Loop-Helix), ERF (Ethylene Response Factors), and HSF (Heat Shock Factors) transcription factors have all been enriched at each time point under the three shocks (6 and 24 h). Additionally, the bZIP was enriched following both combined stress and water deficiencies at both time periods. After 24 h of treatment, salt stress, bZIP, G2-like, and HD-ZIP TFs were enriched.

The differently expressed genes in bread wheat across the salt-tolerant and sensitive cultivars were also discovered by comparative transcriptome analysis following a 150 mM salinity treatment. Five thousand one hundred and twenty-eight genes were differently expressed between salt-treated and control plants, according to the findings. One thousand nine hudred and ninety-five genes among them showed an upregulation, and 3133 genes showed a downregulation. In addition, 109 and 210 genes, respectively, were discovered to express themselves exclusively in salt-treated and control plants (Amirbakhtiar et al. 2019).

8.4.12 Proteomics

The development of genomic technologies has altered the way we study biological systems by giving us a blueprint for potential gene products. The lack of a link between mRNA and protein quantity owing to posttranslational modifications, protein function, and localization results from the static nature of the genome. Therefore, it is crucial to research protein interactions and structure in order to understand how they affect plant growth and development. A comprehensive, high-throughput method called proteomics allows for the thorough identification and study of protein expression in a cell, tissue, or organelle of an organism at a certain time and under particular circumstances (Tan et al. 2013). Proteome profile study gives in-depth understanding of many metabolic processes and how they interact with distinct regulatory pathways in a biological system. Proteomics is a potent tool that, compared to other methods, including genomics technologies, offers a robust and superior description of how a cell functions. The identification of proteins and their expression profiles, posttranslational modifications (PTMs), and protein-protein interactions under stressed and normal conditions have been made possible recently by advances in quantitative proteomics research using high-resolution and mass accuracy instruments (Agregán et al. 2021). Systematic high-throughput methods, such as 2D-PAGE (two-dimensional polyacrylamide gel electrophoresis), MALDI-TOF/MS/MS, and LC-MS/MS, are used to conduct plant proteome analyses. Utilizing protein mapping, PTM characterization, protein-protein interaction, and systems biology techniques, the elucidation of protein activities and functional protein networks is assessed in plant metabolic and signaling pathways (Holman et al. 2013).

Proteomic methodologies offer fresh research techniques for a deeper comprehension of the molecular causes of drought stress in rice. There were 900 proteins found in rice roots under drought stress. Among these, 38% had more changes than the untreated group. Tubulins and proteins linked to transport were reduced, but chitinases and redox proteins implicated in disease were elevated. The proteomes of drought resistant (IRAT109) and susceptible (Zhenshan97B) cultivars were compared. In contrast to the sensitive variety (two proteins in increased abundance; 15 proteins in decreased abundance), the tolerant variety had higher protein abundance (14 proteins in increased abundance; six proteins in decreased abundance). Proteins implicated in the action of the enzyme's superoxide dismutase and dehydroascorbate reductase shielded IRAT109 (Ji et al. 2012).

Three distinct wheat cultivars were tested for their capacity to produce grain under drought stress using an iTRAQ-based quantitative proteome analysis:

RAC875 (drought-tolerant), Excalibur (drought-tolerant), and Kukri (drought-intolerant). At various stages of drought stress, these cultivars displayed distinctive physiological reactions, with equally drought-tolerant types (RAC875, Excalibur) displaying variations in their protein responses. In contrast to Excalibur, which lacked substantial proteins, RAC875 proteins demonstrated considerably abundant proteins. The ability of a protein involved in oxidative stress metabolism and ROS scavenging to boost SOD and CAT enzyme activity was present in all three cultivars at the same time. However, by reducing the proteins necessary for photosynthesis and the Calvin cycle, both drought-tolerant cultivars are implicated in the prevention of ROS.

iTRAQ and 2-DGE investigations on the maize leaf proteome were carried out to examine the impact of drought on the proteome. This research revealed that protective and stress-related proteins were upregulated in both genotypes (chaperones and dehydrins) (Benešová et al. 2012). Out of the 220 proteins that were identified as having differential expression, it was discovered that 106 were upregulated in the tolerant genotype and downregulated in the sensitive genotype. The preventive and detoxifying proteins ascorbate peroxidase, superoxide dismutase, glutathione reductase, and catalase were upregulated in the drought-tolerant genotype CE704. Additionally, the levels of the translation initiation factor eIF3 and the mitochondrial translation elongation factor EF-TuM were both increased. Comparative proteome analysis also discovered distinct protein accumulation locations in maize between the genotypes that are resistant (YE8112) and susceptible (MO17) to drought stress (Zenda et al. 2019). They discovered 721 proteins that were differently abundant (DAPs). Among them, 107 distinct DAPs shared by drought-sensitive and drought-tolerant lines following drought treatment, 84 distinct DAPs exclusive to MO17, and 13 particular DAPs peculiar to YE8112 were found. The tolerant genotype's (YE8112) elevated proteins promoted chaperons like the ASR1 protein, photosynthesis (PSII), and lipid metabolism.

Rice's responses to salt (150 mM NaCl) stress were examined by Xu et al. (2015) utilizing a proteomic method. A total of 56 proteins were discovered to have undergone substantial alteration, 16 of which improved the photosynthesis, antioxidant, and oxidative phosphorylation pathways by upregulating peroxiredoxin Q and photosystem I subunit D. The abundances of thioredoxin x, thioredoxin peroxidase, glutathione *S*-transferase F3, PSI component H, light-harvesting antenna complex I subunits, vacuolar ATP synthase subunit H, chloroplast chaperonin, and ATP synthase delta chain were significantly decreased in response to salt exposure. The phosphor proteomic variations between a salt-tolerant and a salt sensitive maize cultivar under short-term salt stress were discovered by Zhao et al. (2016). The proteome analysis was carried out utilizing the iTRAQ technique after salt treatment with 200 mM NaCl in two different rice cultivars, Pokkali (tolerant) and IR64 (sensitive). When compared to the sensitive cultivar IR64, the tolerant cultivar Pokkali had a greater concentration of proteins. Ascorbate peroxidase, superoxide dismutase, peptidyl-prolyl cis-trans isomerases, glyoxalase II, and oxygen evolving enhancer proteins OEE1 and OEE3, as well as PsbP, are among the proteins that have been discovered as being involved in stress tolerance (Lakra et al. 2019).

8.4.13 Metabolomics

In order to comprehend complicated biological processes and decode plant metabolomes, one relatively new "-omics" method called metabolomics was developed. A cell, tissue, organ, or organism's tiny molecule (1500 Da) may be thoroughly profiled and compared using metabolomics (Deborde et al. 2017). A biological system's metabolites must be identified and quantified in order to study their compositions and interactions with the environment (Khakimov et al. 2014). Furthermore, metabolomics focuses on examining biological activity as opposed to genomes, transcriptomics, and proteomics, making it comparatively simpler to link to the phenotype (Ibarra-Estrada et al. 2016). Metabolomics has been extensively utilized to study how plants respond to stress in adaptive ways. It is crucial to study the production of certain metabolites under diverse stressors in order to comprehend how plants respond to adverse environmental conditions (Sharma et al. 2023). New substances and unique metabolic pathways that accumulate under various stress circumstances are discovered by metabolomic research (Wen et al. 2014). Additionally, metabolomic research contributes to a better knowledge of already known metabolic pathways. Numerous metabolome studies have been carried out over the past 10 years to look at how metabolite concentrations alter in response to various biotic and abiotic stress conditions.

In response to abiotic stressors, plants accumulate several metabolites such as trehalose, glycine betaine, IAA, etc. According to Allen et al. (2009), the sheer accumulation of a particular chemical does not determine stress tolerance; rather, tolerance is determined by the flux's adaptation to several routes for defense and development. Stoichiometry and metabolic modification have been described as methods for maintaining plants' optimal fitness (Rivas-Ubach et al. 2012). Time-series studies with the plant *Arabidopsis thaliana* revealed that metabolic activities react to abiotic changes more quickly than transcriptional activities (Caldana et al. 2011).

8.5 Antioxidative Enzyme-Based Approach for Combined Stress Tolerance

8.5.1 Superoxide Dismutase

Superoxide dismutase (SOD) is the first detoxification enzyme that acts as a component of first line defense system against reactive oxygen species. SODs are a class of metalloenzymes that catalyze the dismutation of two molecules of $O_2^{\bullet-}$ into O_2 and H_2O_2. In plants, there are three main groups of SODs that have been reported: Cu/Zn-SODs, Mn-SODs, and Fe-SODs. SOD removes $O_2^{\bullet-}$ and hence decreases the risk of OH formation via the metal catalyzed Haber Weiss-type reaction (Gill and Tuteja 2010). Together with other antioxidant enzymes like CAT, peroxidases (POX), and APX, SOD is able to detoxify the resultant H_2O_2 and stop the creation of additional damaging ROS like OH•. According to the previous studies, overexpression of Cu/Zn SOD and Mn-SOD in rice, sweet potato, *Arabidopsis*, alfalfa, and tobacco (Prashanth et al. 2008) improved the capacity of drought tolerance and recovery in plants.

8.5.2 Catalase (CAT)

Catalase (CAT), also known as a H_2O_2 oxidoreductase, is a heme-containing enzyme that catalyzes the dismutation of H_2O_2 into H_2O and O_2 mainly in peroxisomes (Vellosillo et al. 2010). Numerous studies demonstrated that expression abundance of CATs increased under various stresses for scavenging of ROS (Vandenabeele et al. 2004). The features that distinguish CAT from other H_2O_2 metabolizing enzymes such as ascorbate peroxidases (APX), peroxiredoxins (PRX), glutathione/thioredoxin peroxidases (GPX), and glutathione *S*-transferases (GST) are that CAT do not require a reductant. When compared to control plants, the activities of SOD, POD, and CAT in the roots of salt-stressed barley plants were greatly increased at Day 2, but they significantly decreased from Day 4 onward (Liang et al. 2003). Combined stress of high temperatures and salt resulted in a decrease of CAT activity at higher levels in durum than bread wheat genotypes, while salt stress alone significantly increased CAT activity in all genotypes under normal temperature.

8.5.3 Peroxidase

Peroxidases (POXs) are oxidoreductases that transform a variety of compounds via a free radical mechanism into oxidized or polymerized products. These POXs scavenge the H_2O_2 produced during oxidative stress. POXs are reported from all plants, animals, and microbes and are essential for living systems. POXs are involved in many physiological and biochemical processes, including the cross-linking of molecules in the cell wall and lignin and suberin formation by oxidation of cinnamyl alcohol.

8.5.4 Ascorbate Peroxidase (APX)

Ascorbic acid-glutathione (AsA-GSH) cycle is essential metabolic pathway for protection against ROS and regulation of the cellular level of H_2O_2 in plants. This pathway includes ascorbate peroxidase (APX), dehydroascorbate reductase (DHAR) and glutathione reductase (GR), and monodehydroascorbate reductase (MDHAR) with antioxidant metabolites ascorbic acid (AsA) and reduced glutathione (GSH). It plays an important role of maintaining redox homeostasis in plants to protect them from oxidation damage (Foyer and Noctor 2016). Increased activity of APX was reported under drought stress in maize and soybean (Jiang and Zhang 2001). Under combined stress (drought and salinity), antioxidant enzymes, for example, guaiacol peroxidase activity was upregulated as compared to control (Attia et al. 2020). The drought-tolerant wheat variety's APX, CAT, and SOD activity peaked at the end of flowering, milky ripeness, and wax ripeness, respectively; the drought-sensitive cultivar, on the other hand, had the lowest SOD and GR activities at waxy maturity and the least amount of APX activity at milky ripeness (Huseynova 2012).

8.5.5 Glutathione Reductase (GR)

Glutathione reductase (GR), also known as glutathione disulfide reductase (GSR), is a flavoprotein belonging to the family of NADPH-dependent oxidoreductases. It catalyzes the reduction of GSSG to GSH and plays an important role in cell defense against ROS (Hernández et al. 2017). An increase in GR activity in plants results in the accumulation of GSH and ultimately confers stress tolerance in plants. Various studies have demonstrated that drought stress increased GR activity in tobacco, barley, maize, wheat, rice, and pea (Sharma and Dubey 2005). Liang et al. (2003) reported that GR activity in the roots of salt-stressed plants at second day was unchanged, but at Day 4 and later, it drastically decreased in barley. Under normal (24/16 °C) temperature conditions, drought stress activated GR activity, and under high temperature water stress conditions, it significantly increased in all genotypes.

8.5.6 Monodehydroascorbate Reductase (MDHAR)

Monodehydroascorbate reductase (MDHAR) is one of the key enzymes in the conversion of oxidized ascorbate back to reduced ascorbate in plants (Kavitha and Murugan 2017). The exposure of plants to environmental stress conditions like high light leads to very quick oxidation of AsA to MDHA in chloroplast. Reddy et al. (2004) found that the enzyme activity of MDHAR was significantly high in the water-stressed leaves of mulberry cultivars, while contrary to this finding, reported that the activity of MDHAR decreased significantly in the heat and drought treatments in Crofton weed. MDHAR activity was enhanced under various other stresses such as salt, high light, and UV radiations. Sharma and Dubey (2005) reported that the activities of MDHAR, DHAR, and GR were higher in drought stressed rice seedlings.

8.5.7 Dehydroascorbate Reductase (DHAR)

Dehydroascorbate reductase (DHAR) is a monomeric enzyme, which is a member of the GSHS-transferase superfamily which maintains AsA in its reduced form by catalyzing the reduction of dehydroascorbate (DHA) to AsA using GSH as a reducing substrate (Hasanuzzaman et al. 2021). AsA recycling through MDHAR or DHAR is critical for maintaining the AsA level and redox state in the adaptation of plants to environmental conditions. Knockout mutants of *Arabidopsis* DHAR were unable to show any significant differences in total AsA content until challenged with the abiotic stress, which showed the importance of DHAR in reducing the DHA under stress conditions.

8.5.8 Glutathione *S*-Transferase (GST)

Glutathione *S*-transferases (GSTs) are multigene superfamily with cytosolic, mitochondrial, and microsomal localization in plants (Kumar and Trivedi 2018). GSTs are involved in safeguard of the cells against biotic and abiotic stresses and provide tolerance by catalyzing S-conjugation between the thiol group of GSH and electrophilic moiety in the hydrophobic and toxic substrate (Nianiou-Obeidat et al. 2017). Enzymatic activity of GSTs is induced by various fungal elicitors, wounding as well as by cold, drought, oxidative stress, heavy metals, and salt (Srivastava et al. 2019). Drought-tolerant and drought-sensitive sorghum genotypes showed efficient H_2O_2 scavenging mechanisms with significantly higher activities of GSTs (Jogeswar et al. 2006). According to the Rahdari et al. (2012), the ability of antioxidant enzymes to scavenge ROS and reduce the damaging influences is closely related to plant drought or salinity stress resistance. Drought or salinity not only adversely affects numerous subcellular compartments, for instance, vacuole, cytoplasm, and nucleus, and also affects the cell organs, and whole plant level, subsequently affecting plant biomass and health.

8.6 Secondary Metabolite Enzymes and Compounds

Phenolic compounds are derived from the phenylpropanoid pathway, mainly from aromatic amino acid phenylalanine in most plants or tyrosine in few cases. Enzymes involved in the phenylpropanoid biosynthetic pathway include lyases, transferases, ligases, oxygenases, and reductases, many of which are encoded by gene superfamilies.

8.6.1 Phenylalanine Ammonia-Lyases (PAL)

In plants, phenylalanine ammonia-lyases (PAL) catalyze the first step of the phenylpropanoid pathway, i.e., conversion of L-Phe to trans-cinnamic acid, which supplies the precursors for many phenolic compounds (Sharma et al. 2022b). PALs are ubiquitous in plants and also found in fungi but have not yet been detected in animals. Studies have shown that PAL activity responds to various stresses, such as wounding, drought, salinity, heavy metals, and infection by viruses, bacteria, or fungi (MacDonald and D'Cunha 2007).

8.6.2 Tyrosine Ammonia-Lyases (TAL)

Tyrosine ammonia-lyases (TAL) convert L-tyrosine to ammonia and *p*-coumaric acid (Sharma et al. 2022b). TAL has been studied less than PAL, and it remains unclear whether TAL activity is due to a capability of PAL to accept tyrosine as a substrate or due to the activity of a specific enzyme (Jendresen et al. 2015). The rarity of TAL is likely a reflection of its specialized role in 4-coumaric acid biosynthesis, which is used as the cofactor for photoactive yellow protein (PYP) in *Rhodobacter* and initiates the conversion of tyrosine to *N*-(*m*,*p*-dihydroxy cinnamoyl) taurine moiety of Saccharomycin antibiotics.

8.6.3 Polyphenol Oxidase (PPO)

Polyphenol oxidase (PPO) is widely distributed in bacteria, animals, plants, and fungi (Boeckx et al. 2015). PPO has been involved in the formation of pigments, oxygen scavenging, defense against plant pathogens, and herbivores. During various abiotic stresses (cold, heat, and drought), there was significant increase in phenolic compounds, and oxidation of these accumulated phenolics was proposed to be inhibited by significant decreases in PPO. This decrease in PPO activity, following abiotic stress, was associated with improved antioxidant capacity of plant under stress (Sofo et al. 2005).

8.6.4 Phenolic Compounds

Phenols are plant secondary metabolites that hold an aromatic ring bearing at least one hydroxyl group which are synthesized by the shikimate-phenylpropanoid biosynthetic pathway. Secondary metabolites are generally classified into two groups: nitrogen compounds (alkaloids, nonprotein amino acids, amines, alkamides, cyanogenic glycosides, and glucosinolates) and non-nitrogen compounds (monoterpenes, diterpenes, triterpenes, tetraterpenes, sesquiterpenes, saponins, flavonoids, steroids, and coumarins). Structurally, phenolic compounds are composed of the aromatic ring bonded directly to at least one (phenol) or more (polyphenol) hydroxyl groups (-OH) and other substituents, such as methoxyl or carboxyl groups, which cause the polar character of the compounds and allow dissolution in water. Phenolic compounds are usually divided into two groups—simple phenols and more complex derivatives—often containing several aromatic rings linked together (Sharma et al. 2019). Phenolic compounds have the ability to donate electron; hence, their hydroxyl groups can directly contribute to antioxidant action by stimulating the synthesis of endogenous antioxidant molecules in the cell (Bendary et al. 2013). Phenolic compounds chelate iron and copper ions due to the presence of suitable functional groups—hydroxyl and carboxyl—while some phenolic compounds also inhibit membrane lipid peroxidation by "catching" alkoxyl radicals. These activities of phenolic compounds are dependent on the structure of molecules and the number and position of hydroxyl groups. Under drought stress, increase in the number of flavonoids and phenolic acids was reported (Akula et al. 2018).

A previous study documented that phenylpropanoid biosynthetic pathway is one of the most stimulated pathways under salinity condition. The activation of phenylpropanoid biosynthetic pathway results in the improved production of various phenolic compounds, such as hydroxybenzoic acids (e.g., gallic acid, vanillic acid, syringic acid, *p*-hydroxybenzoic acid, and ellagic acid), hydroxycinnamic acids (e.g., caffeic acid, chlorogenic acid, *p*-coumaric acid, *m*-coumaric acid, ferulic acid, sinapic acid, and trans-cinnamic acid), and flavonoids (e.g., quercetin and iso-quercetin, rutin, luteolin and luteolin-7-O-glycoside, apigenin, kaempferol, and luteolin) (Sharma et al. 2019).

8.6.5 Flavonoids

They are synthesized through the phenylpropanoid pathway. Chalcone synthase (CHS), chalcone isomerase (CHI), and flavanone 3-hydroxylase (F3H) are key enzymes in this pathway (Hodaei et al. 2018). There are different classes of flavonoids depending upon the level of oxidation and pattern of substitution of the C ring, while individual compounds within a class differ in the pattern of substitution of the A and B rings. Important flavonoids in plants are flavones, flavanones, isoflavones, flavonols, flavanonols, flavan-3-ols, chalcones, and anthocyanidins. Until now, more than 5000 different flavonoids have been described which are classified into six major subclasses, such as flavones, flavonols, flavanones, flavanols, anthocyanidins, and isoflavones (Ross and Kasum 2002). Their structural variation in each subgroup is partly due to the degree and their hydroxylation, methoxylation, prenylation, and glycosylation pattern. Flavonoids can directly scavenge ROS by donating hydrogen atom, thus inactivating ROS (Sharma et al. 2023). Flavonoids themselves get converted to phenoxyl radical which can further react with other free radicals to form stable quinone structure (Treml and Šmejkal 2016).

8.7 Photosynthetic Machinery

8.7.1 Metabolism (Photosynthesis and Respiration)

The green leaves of plants indicate its capability for photosynthetic activity and offers important clues about agricultural productivity, as well as the physiological and phenological state of the plant. One of the primary processes impacted by salinity and drought is photosynthesis (Chaves et al. 2009). However, long-term salt stress exposure also causes plants to experience ionic stress, which impairs photosynthesis and causes leaf senescence, both of which have a detrimental impact on growth (Chaves et al. 2009). When plants are stressed by prolonged drought stress, their roots continue to elongate, which may be explained by the demand for groundwater on the part of the plant (Brunner et al. 2015). Long-term salinity stress might cause heavier roots to store more chloride. The photosynthetic components are negatively impacted by an excessive accumulation of ions, primarily Na^{2+}, which lowers enzyme activity and pigment synthesis.

Plants use stomata for two crucial processes: controlling CO_2 entry into the leaf and regulating transpiration, which supplies plants with nutrients and regulates leaf temperature. These stressful circumstances reduce the rate of CO_2 assimilation, and the excess light received that is not utilized by the plant may increase the generation of reactive oxygen species (ROS), which in turn causes oxidative stress. The reduction of the stomata aperture width (stomatal closure) prevents the loss of water to the atmosphere, and this protection mechanism is considered an adaptation response of plants to the onset of drought conditions (Saradadevi et al. 2017). Plants with salt exclusion mechanisms can either prevent salt from entering plant cells or reduce the amount of salt present in the cytoplasm by compartmentalizing salt in the vacuoles. In addition to lowering water potential and stomatal conductance, the combined stress also decreased plant photosynthetic efficiency.

8.8 Accumulation of Stress Protein

During water deficit and salt stress conditions, plants not only unregulated the production of osmoprotectants (e.g., proline and glycine betaine), antioxidant enzymes, but also enhance the expression of several drought- and salt stress-responsive genes and proteins such as late embryogenesis abundant proteins (LEA) and dehydrins (Fahad et al. 2015; Wani et al. 2016; Ullah et al. 2018).

8.9 Conclusion

Various studies regarding the drought and salinity stress response in plants have suggested the diverse adaptation mechanism to combat such conditions. These studies have improved our knowledge regarding the strategies and mechanisms involved to improve the tolerance in plants. Plants mitigate the combined stresses by regulating the genomic, proteomic, and metabolic responses. Plants have built-in adaptation mechanism to tackle complex abiotic conditions. Technological advancements can help to develop plant species with resistance toward the combined drought and salinity stresses. These advancements include breeding strategies, omics approaches, enzyme engineering, and other metabolic adjustments. These strategies further may be helpful to tackle the stress conditions for plant by maintaining the stress-responsive signaling. In this way, these approaches can be the coordinated efforts to develop resistance plants/varieties which can be helpful in maintaining food and nutritional security.

References

Abid M, Tian Z, Zahoor R, Ata-Ul-Karim ST, Daryl C, Snider JL, Dai T (2018) Pre-drought priming: a key drought tolerance engine in support of grain development in wheat. Adv Agron 152:51–85.4

Agregán R, Echegaray N, López-Pedrouso M, Aadil RM, Hano C, Franco D, Lorenzo JM (2021) Proteomic advances in cereal and vegetable crops. Molecules 26:4924

Akula R, Gill SS, Ravishankar GA (2018) Protective role of Indoleamines (Serotonin and Melatonin) during abiotic stress in plants. In: Metabolic adaptations in plants during abiotic stress. CRC Press, pp 221–228

Allen DK, Libourel IG, Shachar-Hill Y (2009) Metabolic flux analysis in plants: coping with complexity. Plant Cell Environ 32:1241–1257. https://doi.org/10.1111/j.1365-3040.2009.01992.x

Amirbakhtiar N, Ismaili A, Ghaffari MR, Firouzabadi FN, Shobbar ZS (2019) Transcriptome response of roots to salt stress in a salinity-tolerant bread wheat cultivar. PLoS One 14:e0213305

Angon PB, Tahjib-Ul-Arif M, Samin SI, Habiba U, Hossain MA, Brestic M (2022) How do plants respond to combined drought and salinity stress?—a systematic review. Plants 11(21):2884

Anjum SA, Ashraf U, Tanveer M, Khan I, Hussain S, Shahzad B, Wang LC (2017) Drought induced changes in growth, osmolyte accumulation and antioxidant metabolism of three maize hybrids. Front Plant Sci 8:69

Apel K, Hirt H (2004) Reactive oxygen species: metabolism, oxidative stress, and signaling transduction. Annu Rev Plant Biol 55:373

Araus JL, Cairns JE (2014) Field high-throughput phenotyping: the new crop breeding frontier. Trend Plant Sci 19(1):52–61

Arnold ML, Martin NH (2009) Adaptation by introgression. J Biol 8:82

Attia EA, Singh BP, Dashora K, Abdel-Azeem AM (2020) A potential antimicrobial, extracellular enzymes, and antioxidants resource: endophytic fungi associated with medicinal plants. Int J Biosci 17(1):119–132

Bains NS, Mangat GS, Singh K, Nanda GS (1998) A simple technique for the identification of embryo carrying seeds from wheat 9 maize crosses prior to dissection. Plant Breed 117:191–192

Beckles DM, Thitisaksakul M (2014) How environmental stress affects starch composition and functionality in cereal endosperm. Starch 66(1–2):58–71

Bendary E, Francis RR, Ali HMG, Sarwat MI, El Hady S (2013) Antioxidant and structure–activity relationships (SARs) of some phenolic and anilines compounds. Ann Agric Sci 58(2):173–181

Benešová M, Hola D, Fischer L, Jedelský PL, Hnilička F, Wilhelmová N, Rothova O, Kočová M, Prochazkova D, Honnerova J (2012) The physiology and proteomics of drought tolerance in maize: early stomatal closure as a cause of lower tolerance to short-term dehydration? PLoS One 7:e38017

Bernardo R (2010) Breeding for quantitative traits in plants. Stemma Press, Woodbury

Bhowal B, Chandra P, Saxena SC (2021) Engineering glycine betaine biosynthesis in alleviating abiotic stress effects in plants. In: Compatible solutes engineering for crop plants facing climate change. Springer, Cham, pp 97–127

Boeckx T, Winters AL, Webb KJ, Kingston-Smith AH (2015) Polyphenol oxidase in leaves: is there any significance to the chloroplastic localization? J Exp Bot 66(12):3571–3579

Brunner I, Herzog C, Dawes MA, Arend M, Sperisen C (2015) How tree roots respond to drought. Front Plant Sci 6:547

Budak H, Kantar M, Bulut R, Akpinar BA (2015) Stress responsive miRNAs and isomiRs in cereals. Plant Sci 235:1–13. https://doi.org/10.1016/j.plantsci.2015.02.008

Caldana C, Degenkolbe T, Cuadros-Inostroza A, Klie S, Sulpice R, Leisse A et al (2011) High-density kinetic analysis of the metabolomic and transcriptomic response of *Arabidopsis* to eight environmental conditions. Plant J 67:869–884. https://doi.org/10.1111/j.1365-313X.2011.04640.x

Campbell AW, Griffin WB, Conner AJ, Rowarth JS, Burritt DJ (1998) The effects of temperature and light intensity on embryo numbers in wheat doubled haploid production through wheat × maize crosses. Ann Bot 82:29–33

Çelik Ö, Atak C (2012) The effect of salt stress on antioxidative enzymes and proline content of two Turkish tobacco varieties. Turk J Biol 36(3):339–356

Chaparzadeh N, D'Amico ML, Khavari-Nejad RA, Izzo R, Navari-Izzo F (2004) Antioxidative responses of Calendula officinalis under salinity conditions. Plant Physiol Biochem 42(9): 695–701

Chaudhary HK, Sethi GS, Singh S, Pratap A, Sharma S (2005) Efficient haploid induction in wheat by using pollen of Imperata cylindrica. Plant Breed 124:96–98

Chaves MM, Flexas J, Pinheiro C (2009) Photosynthesis under drought and salt stress: regulation mechanisms from whole plant to cell. Ann Bot 103(4):551–560

Cramer GR, Urano K, Delrot S, Pezzotti M, Shinozaki K (2011) Effects of abiotic stress on plants: a systems biology perspective. BMC Plant Biol 11:163. https://doi.org/10.1186/1471-2229-11-163

Curaba J, Singh MB, Bhalla PL (2014) miRNAs in the crosstalk between phytohormone signalling pathways. J Exp Bot 65:1425–1438. https://doi.org/10.1093/jxb/eru002

Deborde C, Moing A, Roch L, Jacob D, Rolin D, Giraudeau P (2017) Plant metabolism as studied by NMR spectroscopy. Prog Nucl Magn Reson Spectrosc 102:61–97

Deery DM, Rebetzke GJ, Jimenez-Berni JA, Bovill WD, James RA, Condon AG et al (2019) Evaluation of the phenotypic repeatability of canopy temperature in wheat using continuous-terrestrial and airborne measurements. Front Plant Sci 10:875

Deinlein U, Stephan AB, Horie T, Luo W, Xu G, Schroeder JI (2014) Plant salt-tolerance mechanisms. Trends Plant Sci 19(6):371–379

Demidchik V (2015) Mechanisms of oxidative stress in plants: from classical chemistry to cell biology. Environ Exp Bot 109:212–228

Deshmukh R, Sonah H, Patil G, Chen W, Prince S, Mutava R, Vuong T, Valliyodan B, Nguyen HT (2014) Integrating omic approaches for abiotic stress tolerance in soybean. Front Plant Sci 5:244

Dogra V, Rochaix JD, Kim C (2018) Singlet oxygen-triggered chloroplast-to-nucleus retrograde signalling pathways: an emerging perspective. Plant Cell Environ 41(8):1727–1738

Dubey RS (2018) Photosynthesis in plants under stressful conditions. In: Handbook of photosynthesis. CRC Press, Boca Raton, FL, pp 629–649

Dvorák J, Edge M, Ross K (1988) On the evolution of the adaptation of Lophopyrum elongatum to growth in saline environments. Proc Natl Acad Sci U S A 85:3805–3809

Fahad S, Hussain S, Matloob A, Khan FA, Khaliq A, Saud S, Hassan S, Shan D, Khan F, Ullah N, Faiq M (2015) Phytohormones and plant responses to salinity stress: a review. Plant Growth Regul 75:391–404

Fathi A, Tari DB (2016) Effect of drought stress and its mechanism in plants. Int J Life Sci 10(1): 1–6

Fiaz S, Ahmad S, Noor MA, Wang X, Younas A, Riaz A et al (2019) Applications of the CRISPR/Cas9 system for rice grain quality improvement: perspectives and opportunities. Int J Mol Sci 20(4):888

Forster BP, Shu QY (2012) Plant mutagenesis in crop improvement: basic terms and applications. In: Shu QY, Forster BP, Nakagawa H (eds) Plant mutation breeding and biotechnology. CABI, Wallingford, pp 9–20

Foyer CH, Noctor G (2016) Stress-triggered redox signalling: what's in pROSpect? Plant Cell Environ 39(5):951–964

Gao X, Starmer J, Martin ER (2008) A multiple testing correction method for genetic association studies using correlated single nucleotide polymorphisms. Genet Epidemiol 32(4):361–369

Gao P, Bai X, Yang L, Lv D, Pan X, Li Y et al (2011) osa-MIR393: a salinity and alkaline stress-related microRNA gene. Mol Biol Rep 38:237–242. https://doi.org/10.1007/s11033-010-0100-8

Garg R, Verma M, Agrawal S, Shankar R, Majee M, Jain M (2013) Deep transcriptome sequencing of wild halophyte rice, Porteresia coarctata, provides novel insights into the salinity and submergence tolerance factors. DNA Res 21:69–84

Gautam HR, Bhardwaj ML, Kumar R (2013) Climate change and its impact on plant diseases. Curr Sci 105:1685–1691

Gill SS, Tuteja N (2010) Reactive oxygen species and antioxidant machinery in abiotic stress tolerance in crop plants. Plant Physiol Biochem 48(12):909–930

Gowda VRP, Henry A, Yamauchi A, Shashidhar HE, Serraj R (2011) Root biology and genetic improvement for drought avoidance in rice. Field Crop Res 122:1–13

Goyal M, Kumari A, Kumari A, Sharma H, Vasmatkar P, Gupta N (2023) Oxidative stress and antioxidant defense in mitigating abiotic stresses in forage crops: a physiological and biochemical perspective. In: Molecular interventions for developing climate-smart crops: a forage perspective. Springer Nature Singapore, Singapore, pp 109–135

Guha S, Maheshwari SC (1964) In vitro production of embryos from anthers of Datura. Nature 204: 497

Gull A, Lone AA, Wani NUI (2019) Biotic and abiotic stresses in plants. In: Abiotic and biotic stress in plants. IntechOpen, London, pp 1–19

Guo D, Liang J, Li L (2009) Abscisic acid (ABA) inhibition of lateral root formation involves endogenous ABA biosynthesis in Arachis hypogaea L. Plant Growth Regul 58(2):173–179

Hasanuzzaman M, Raihan MRH, Masud AAC, Rahman K, Nowroz F, Rahman M et al (2021) Regulation of reactive oxygen species and antioxidant defense in plants under salinity. Int J Mol Sci 22(17):9326

Hasibeder R, Fuchslueger L, Richter A, Bahn M (2015) Summer drought alters carbon allocation to roots and root respiration in mountain grassland. New Phytol 205(3):1117–1127

Hernández JA, Barba-Espín G, Diaz-Vivancos P (2017) Glutathione-mediated biotic stress tolerance in plants. In: Glutathione in plant growth, development, and stress tolerance, pp 309–329

Hodaei M, Rahimmalek M, Arzani A, Talebi M (2018) The effect of water stress on phytochemical accumulation, bioactive compounds and expression of key genes involved in flavonoid biosynthesis in Chrysanthemum morifolium L. Ind Crops Prod 120:295–304

Holman JD, Dasari S, Tabb DL (2013) Informatics of protein and posttranslational modification detection via shotgun proteomics. In: Proteomics for biomarker discovery. Springer, Berlin, pp 167–179

Honsdorf N, March TJ, Berger B, Tester M, Pillen K (2014) High-throughput phenotyping to detect drought tolerance QTL in wild barley introgression lines. PLoS One 9(5):e97047

Huseynova IM (2012) Photosynthetic characteristics and enzymatic antioxidant capacity of leaves from wheat cultivars exposed to drought. Biochimica et Biophysica Acta (BBA)-Bioenergetics 1817(8):1516–1523

Hussain Wani S, Brajendra Singh N, Haribhushan A, Iqbal Mir J (2013) Compatible solute engineering in plants for abiotic stress tolerance-role of glycine betaine. Curr Genom 14(3): 157–165

Ibarra-Estrada E, Soto-Hernández RM, Palma-Tenango M (2016) Metabolomics as a tool in agriculture. In: Metabolomics - fundamentals and applications. IntechOpen, London, pp 148–168

Imran QM, Falak N, Hussain A, Mun B-G, Yun B-W (2021) Abiotic stress in plants; stress perception to molecular response and role of biotechnological tools in stress resistance. Agronomy 11:1579

Inagaki M (1985) Chromosome doubling of the wheat haploids obtained from crosses with Hordeum bulbosom L. Jpn J Breed 35:193–195

Janků M, Luhová L, Petřivalský M (2019) On the origin and fate of reactive oxygen species in plant cell compartments. Antioxidants 8(4):105

Jendresen CB, Stahlhut SG, Li M, Gaspar P, Siedler S, Förster J, Maury J, Borodina I, Nielsen AT (2015) Highly active and specific tyrosine ammonia-lyases from diverse origins enable enhanced production of aromatic compounds in bacteria and Saccharomyces cerevisiae. Appl Environ Microbiol 81(13):4458–4476

Ji K, Wang Y, Sun W, Lou Q, Mei H, Shen S, Chen H (2012) Drought-responsive mechanisms in rice genotypes with contrasting drought tolerance during reproductive stage. J Plant Physiol 169:336–344

Jiang M, Zhang J (2001) Effect of abscisic acid on active oxygen species, antioxidative defence system and oxidative damage in leaves of maize seedlings. Plant Cell Physiol 42(11):1265–1273

Jogeswar G, Pallela R, Jakka NM, Reddy PS, Venkateswara Rao J, Sreenivasulu N, Kavi Kishor PB (2006) Antioxidative response in different sorghum species under short-term salinity stress. Acta Physiol Plant 28:465–475

Kadioglu A, Terzi R, Saruhan N, Saglam A (2012) Current advances in the investigation of leaf rolling caused by biotic and abiotic stress factors. Plant Sci 182:42–48

Kahrizi S, Sedighi M, Sofalian O (2012) Effect of salt stress on proline and activity of antioxidant enzymes in ten durum wheat cultivars. Ann Biol Res 3(8):3870–3874

Kantar M, Lucas SJ, Budak H (2011) miRNA expression patterns of Triticum dicoccoides in response to shock drought stress. Planta 233:471–484. https://doi.org/10.1007/s00425-010-1309-4

Kasha KJ, Kao KN (1970) High frequency haploid production in barley (Hordeum vulgare L.). Nature 225:874–876

Kaur M, Malik P, Devi U, Mukta A, Kaur A, Dhillon GS, Padhy AK, Sharma H, Sharma A, Kaur S (2023) Wheat biofortification: a molecular breeding outlook. In: QTL mapping in crop improvement. Academic Press, pp 163–201

Kavitha CH, Murugan K (2017) Modulating role for antioxidant system in desiccation tolerance of Dicranopteris linearis. Indian J Sci Res:1–13

Khakimov B, Bak S, Engelsen SB (2014) High-throughput cereal metabolomics: current analytical technologies, challenges and perspectives. J Cereal Sci 59:393–418

Khan MIR, Palakolanu SR, Chopra P, Rajurkar AB, Gupta R, Iqbal N, Maheshwari C (2021) Improving drought tolerance in rice: ensuring food security through multi-dimensional approaches. Physiol Plant 172:645–668

Koussevitzky S, Suzuki N, Huntington S, Armijo L, Sha W, Cortes D et al (2008) Ascorbate peroxidase 1 plays a key role in the response of *Arabidopsis thaliana* to stress combination. J Biol Chem 283:34197–34203. https://doi.org/10.1074/jbc.M806337200

Krieger-Liszkay A (2005) Singlet oxygen production in photosynthesis. J Exp Bot 56(411): 337–346

Kumar S, Trivedi PK (2018) Glutathione S-transferases: role in combating abiotic stresses including arsenic detoxification in plants. Front Plant Sci 9:751

Lakra N, Kaur C, Singla-Pareek SL, Pareek A (2019) Mapping the 'Early Salinity Response' triggered proteome adaptation in contrasting rice genotypes using ITRAQ approach. Rice 12:3

Lashermes P, Beckert M (1988) Genetic control of maternal haploidy in maize (Zea mays L.) and selection of haploid inducing lines. Theor Appl Genet 76:405–410

Laurie DA, Bennett MD (1988) The production of haploid plants from wheat 9 maize crosses. Theor Appl Genet 76:393–397

Laxa M, Liebthal M, Telman W, Chibani K, Dietz KJ (2019) The role of the plant antioxidant system in drought tolerance. Antioxidants 8(4):94

Lethin J, Shakil SS, Hassan S, Sirijovski N, Töpel M, Olsson O, Aronsson H (2020) Development and characterization of an EMS-mutagenized wheat population and identification of salt-tolerant wheat lines. BMC Plant Biol 20:18

Li YF, Zheng Y, Addo-Quaye C, Zhang L, Saini A, Jagadeeswaran G et al (2010) Transcriptome-wide identification of microRNA targets in rice. Plant J 62:742–759. https://doi.org/10.1111/j.1365-313X.2010.04187.x

Li JL, Cui J, Cheng DY (2015) Computational identification and characterization of conserved miRNAs and their target genes in beet (*Beta vulgaris*). Genet Mol Res 14:9103–9108. https://doi.org/10.4238/2015.August.7.19

Li N, Wang X, Ma B, Du C, Zheng L, Wang Y (2017) Expression of a Na+/H+ antiporter RtNHX1 from a recretohalophyte Reaumuria trigyna improved salt tolerance of transgenic Arabidopsis thaliana. J Plant Physiol 218:109–120

Liang Y, Chen QIN, Liu Q, Zhang W, Ding R (2003) Exogenous silicon (Si) increases antioxidant enzyme activity and reduces lipid peroxidation in roots of salt-stressed barley (Hordeum vulgareL.). J Plant Physiol 160(10):1157–1164

Lima JC, Arenhart RA, Margis-Pinheiro M, Margis R (2011) Aluminum triggers broad changes in microRNA expression in rice roots. Genet Mol Res 10:2817–2832. https://doi.org/10.4238/2011.November.10.4

López-Urrea R, Montoro A, Mañas F, López-Fuster P, Fereres E (2012) Evapotranspiration and crop coefficients from lysimeter measurements of mature 'Tempranillo' wine grapes. Agric Water Manag 112:13–20

Lynch JP, Wojciechowski T (2015) Opportunities and challenges in the subsoil: pathways to deeper rooted crops. J Exp Bot 66(8):2199–2210

MacDonald MJ, D'Cunha GB (2007) A modern view of phenylalanine ammonia lyase. Biochem Cell Biol 85(3):273–282

Mace E, Innes D, Hunt C, Wang X, Tao Y, Baxter J, Hassall M, Hathorn A, Jordan D (2019) The sorghum QTL atlas: a powerful tool for trait dissection, comparative genomics and crop improvement. Theor Appl Genet 132:751–766

Mallet J (2005) Hybridization as an invasion of the genome. Trends Ecol Evol 20:229–237

Manneh B, Kiepe P, Sie M, Ndjiondjop M, Drame NK, Traore K et al (2007) Exploiting partnerships in research and development to help African rice farmers cope with climate variability. SAT eJournal 4:1

Marees AT, de Kluiver H, Stringer S, Vorspan F, Curis E, Marie-Claire C, Derks EM (2018) A tutorial on conducting Genome Wide Association studies: quality control and statistical analysis. Int J Methods Psychiatr Res 27:e1

Mason RE, Singh RP (2014) Considerations when deploying canopy temperature to select high yielding wheat breeding lines under drought and heat stress. Agronomy 4(2):191–201

McDonald MP, Galwey NW, Ellneskog-Staam P, Colmer TD (2001) Evaluation of Lophopyrum elongatum as a source of genetic diversity to increase the waterlogging tolerance of hexaploid wheat (Triticum aestivum). New Phytol 151:369–380

Miller GAD, Suzuki N, Ciftci-Yilmaz SULTAN, Mittler RON (2010) Reactive oxygen species homeostasis and signalling during drought and salinity stresses. Plant Cell Environ 33(4):453–467

Mishra S, Singh B, Misra P, Rai V, Singh NK (2016) Haplotype distribution and association of candidate genes with salt tolerance in Indian wild rice germplasm. Plant Cell Rep 35:2295

Mittova V, Guy M, Tal M, Volokita M (2004) Salinity up-regulates the antioxidative system in root mitochondria and peroxisomes of the wild salt-tolerant tomato species Lycopersicon pennellii. J Exp Bot 55(399):1105–1113

Molassiotis A, Job D, Ziogas V, Tanou G (2016) Citrus plants: a model system for unlocking the secrets of NO and ROS-inspired priming against salinity and drought. Front Plant Sci 7:229

Munaweera TIK, Jayawardana NU, Rajaratnam R, Dissanayake N (2022) Modern plant biotechnology as a strategy in addressing climate change and attaining food security. Agric Food Secur 11:26

Nakayama H, Yoshida K, Ono H (2000) Ectoine, the compatible solute of *Halomonas elongata*, confers hyperosmotic tolerance in cultured tobacco cells. Plant Physiol 122:1239–1247. https://doi.org/10.1104/pp.122.4.1239

Naranjo MA, Forment J, Roldan M, Serrano R, Vicente O (2006) Overexpression of Arabidopsis thaliana LTL1, a salt-induced gene encoding a GDSL-motif lipase, increases salt tolerance in yeast and transgenic plants. Plant Cell Environ 29(10):1890–1900

Neeraja CN, Maghirang-Rodriguez R, Pamplona A, Heuer S, Collard BCY, Septiningsih EM et al (2007) A marker-assisted backcross approach for developing submergence-tolerant rice cultivars. Theor Appl Genet 115:767–776. https://doi.org/10.1007/s00122-007-0607-0

Nevo E, Chen G (2010) Drought and salt tolerances in wild relatives for wheat and barley improvement. Plant Cell Environ 33(4):670–685

Ngara R, Ndimba BK (2014) Understanding the complex nature of salinity and drought-stress response in cereals using proteomics technologies. Proteomics 14(4–5):611–621

Nianiou-Obeidat I, Madesis P, Kissoudis C, Voulgari G, Chronopoulou E, Tsaftaris A, Labrou NE (2017) Plant glutathione transferase-mediated stress tolerance: functions and biotechnological applications. Plant Cell Rep 36:791–805

Oladosu Y, Rafii MY, Abdullah N, Hussin G, Ramli A, Rahim HA, Usman M (2016) Principle and application of plant mutagenesis in crop improvement: a review. Biotech Equip 30(1):1–16

Osthoff A, Donà dalle Rose P, Baldauf JA, Piepho HP, Hochholdinger F (2019) Transcriptomic reprogramming of barley seminal roots by combined water deficit and salt stress. BMC Genomics 20:1–14

Petrov VD, Van Breusegem F (2012) Hydrogen peroxide—a central hub for information flow in plant cells. AoB Plants 2012:pls014

Pinto E, Sigaud-kutner TC, Leitao MA, Okamoto OK, Morse D, Colepicolo P (2003) Heavy metal–induced oxidative stress in algae 1. J Phycol 39(6):1008–1018

Pirasteh-Anosheh H, Saed-Moucheshi A, Pakniyat H, Pessarakli M (2016) Stomatal responses to drought stress. Water Stress Crop Plant Sustain Appr 1:24–40

Placido DF, Campbell MT, Folsom JJ, Cui X, Kruger GR, Baenziger PS, Walia H (2013) Introgression of novel traits from a wild wheat relative improves drought adaptation in wheat. Plant Physiol 161(4):1806–1819

Prashanth SR, Sadhasivam V, Parida A (2008) Over expression of cytosolic copper/zinc superoxide dismutase from a mangrove plant Avicennia marina in indica rice var Pusa Basmati-1 confers abiotic stress tolerance. Transgenic Res 17:281–291

Pratap A, Gupta SK (2009) Biotechnological interventions in host plant resistance. In: Peshin R, Dhawan AK (eds) Integrated pest management: innovation, dissemination and impact. Springer, Dordrecht, pp 183–207

Rahcari P, Hosseini SM, Tavakoli S (2012) The studying effect of drought stress on germination, proline, sugar, lipid, protein and chlorophyll content in purslane (Portulaca oleracea L.) leaves. J Med Plants Res 6(9):1539–1547

Rajaravindran M, Natarajan S (2012) Effects of salinity stress on growth and biochemical constituents of the halophyte Sesuvium portulacastrum. Int J Res Biol Sci 2(1):18–25

Rane J, Singh AK, Tiwari M, Prasad PVV, Jagadish SVK (2021) Effective use of water in crop plants in dryland agriculture: implications of reactive oxygen species and antioxidative system. Front Plant Sci 12:778270

Rauf S, Al-Khayri JM, Zaharieva M, Monneveux P, Khalil F (2016) Breeding strategies to enhance drought tolerance in crops. In: Advances in plant breeding strategies: agronomic, abiotic and biotic stress traits. Springer, Cham, pp 397–445

Reddy AR, Chaitanya KV, Vivekanandan M (2004) Drought-induced responses of photosynthesis and antioxidant metabolism in higher plants. J Plant Physiol 161(11):1189–1202

Reynolds MP, Hays D, Chapman S (2010) Breeding for adaptation to heat and drought stress. Clim Change Crop Prod 1:71–91

Ribaut JM, Betran J, Monneveux P, Setter T (2009) Handbook of maize: its biology. Springer, New York, NY

Richards R, Passioura J (1989) A breeding program to reduce the diameter of the major xylem vessel in the seminal roots of wheat and its effect on grain yield in rain-fed environments. Aust J Agric Res 40:943–950

Richards S, Aziz N, Bale S, Bick D, Das S, Gastier-Foster J et al (2015) Standards and guidelines for the interpretation of sequence variants: a joint consensus recommendation of the American

College of Medical Genetics and Genomics and the Association for Molecular Pathology. Genet Med 17(5):405–423
Rivas-Ubach A, Sardans J, Perez-Trujillo M, Estiarte M, Penuelas J (2012) Strong relationship between elemental stoichiometry and metabolome in plants. Proc Natl Acad Sci U S A 109: 4181–4186. https://doi.org/10.1073/pnas.1116092109
Röder MS, Korzun V, Wendehake K, Plaschke J, Tixier M-H, Leroy P, Ganal MW (1998) A microsatellite map of wheat. Genetics 149:2007–2023
Ross JA, Kasum CM (2002) Dietary flavonoids: bioavailability, metabolic effects, and safety. Annu Rev Nutr 22(1):19–34
Rozema J, Flowers T (2008) Crops for a salinized world. Science 322:1478–1480
Sadak MS (2019) Physiological role of trehalose on enhancing salinity tolerance of wheat plant. Bull Natl Res Cent 43(1):1–10
Sahebi M, Hanafi MM, Rafii MY, Mahmud TMM, Azizi P, Osman M et al (2018) Improvement of drought tolerance in rice (Oryza sativa L.): genetics, genomic tools, and the WRKY gene family. Biomed Res Int 2018:3158474
Sang Y, Macho AP (2017) Analysis of PAMP-triggered ROS burst in plant immunity. In: Plant pattern recognition receptors. Humana Press, New York, NY, pp 143–153
Saradadevi R, Palta JA, Siddique KH (2017) ABA-mediated stomatal response in regulating water use during the development of terminal drought in wheat. Front Plant Sci 8:1251
Schwenk K, Brede N, Streit B (2008) Introduction. Extent, processes and evolutionary impact of interspecific hybridization in animals. Philos Trans R Soc Lond Ser B Biol Sci 363:2805–2811
Septiningsih EM, Pamplona AM, Sanchez DL, Neeraja CN, Vergara GV, Heuer S et al (2009) Development of submergence-tolerant rice cultivars: the Sub1 locus and beyond. Ann Bot 103: 151–160. https://doi.org/10.1093/aob/mcn206
Sharma P, Dubey RS (2005) Modulation of nitrate reductase activity in rice seedlings under aluminium toxicity and water stress: role of osmolytes as enzyme protectant. J Plant Physiol 162(8):854–864
Sharma S, Demason DA, Ehdaie B, Lukaszewski AJ, Waines JG (2010) Dosage effect of the short arm of chromosome 1 of rye on root morphology and anatomy in bread wheat. J Exp Bot 61: 2623–2633
Sharma H, Chawla N, Dhatt AS (2019) Nutraceutical content and free radical scavenging capacity of brinjal (Solanum melongena L.) genotypes. Sci Hort 244:294–303
Sharma P, Chaudhary HK, Kapoor C, Manoj NV, Singh K, Sood VK (2022a) Molecular cytogenetic analysis of novel wheat-rye translocation lines and their characterization for drought tolerance and yellow rust resistance. Cereal Res Commun 50(4):655–665
Sharma H, Chawla N, Dhatt AS (2022b) Role of phenylalanine/tyrosine ammonia lyase and anthocyanidin synthase enzymes for anthocyanin biosynthesis in developing Solanum melongena L. genotypes. Physiol Plant 174(5):e13756
Sharma H, Singh S, Shamshad M, Padhy AK, Kaur R, Kashyap L, Srivastava P, Mavi GS, Kaur S, Sharma A, Sohu VS (2023) Variability in iron, zinc, phytic acid and protein content in pre-breeding wheat germplasm under different water regimes. Plant Growth Regul 100:531–543. https://doi.org/10.1007/s10725-022-00943-5
Shelake RM, Pramanik D, Kim JY (2019) Evolution of plant mutagenesis tools: a shifting paradigm from random to targeted genome editing. Plant Biotechnol Rep 13:423–445
Shelake RM, Kadam US, Kumar R, Pramanik D, Singh AK, Kim JY (2022) Engineering drought and salinity tolerance traits in crops through CRISPR-mediated genome editing: targets, tools, challenges, and perspectives. Plant Commun 3:100417
Singh A, Meena M, Kumar D, Dubey AK, Hassan MI (2015) Structural and functional analysis of various globulin proteins from soy seed. Crit Rev Food Sci Nutr 55(11):1491–1502
Smith S, De Smet I (2012) Root system architecture: insights from Arabidopsis and cereal crops. Phil Trans R Soc B Biol Sci 367(1595):1441–1452
Sofo A, Dichio B, Xiloyannis C, Masia A (2005) Antioxidant defences in olive trees during drought stress: changes in activity of some antioxidant enzymes. Funct Plant Biol 32(1):45–53

Srivastava A, Srivastava P, Sharma A, Sarlach RS (2017) Canopy temperature an effective measure of drought stress tolerance in RIL population of wheat. Vegetos 30:1. https://doi.org/10.4172/2229-4473.1000212

Srivastava D, Verma G, Chauhan AS, Pande V, Chakrabarty D (2019) Rice (Oryza sativa L.) tau class glutathione S-transferase (OsGSTU30) overexpression in Arabidopsis thaliana modulates a regulatory network leading to heavy metal and drought stress tolerance. Metallomics 11(2): 375–389

Stokes CJ, Inman-Bamber NG (2014) Climate ready sugarcane: traits for adaptation to high CO2 levels. CSIRO, Townsville, QLD

Suarez-Gonzalez A, Lexer C, Cronk QCB (2018) Adaptive introgression: a plant perspective. Biol Lett 14:20170688

Taiz L, Zeiger E (2006) Secondary metabolites and plant defense. Plant Physiol 4:315–334

Tan L, Chen S, Wang T, Dai S (2013) Proteomic insights into seed germination in response to environmental factors. Proteomics 13:1850–1870

Tomar RSS, Deshmukh RK, Naik K, Tomar SMS (2014) Development of chloroplast-specific microsatellite markers for molecular characterization of alloplasmic lines and phylogenetic analysis in wheat. Plant Breed 133:12–18. https://doi.org/10.1111/pbr.12116

Toojinda T, Siangliw M, Tragroonrung S, Vanavichit A (2003) Molecular genetics of submergence tolerance in rice: QTL analysis of key traits. Ann Bot 91:243–253. https://doi.org/10.1093/aob/mcf072

Treml J, Šmejkal K (2016) Flavonoids as potent scavengers of hydroxyl radicals. Compr Rev Food Sci Food Saf 15(4):720–738

Trenberth KE, Dai A, Van der Schrier G, Jones PD, Barichivich J, Briffa KR, Sheffield J (2014) Global warming and changes in drought. Nat Clim Chang 4:17–22

Trindade I, Capitao C, Dalmay T, Fevereiro MP, Santos DM (2010) miR398 and miR408 are up-regulated in response to water deficit in Medicago truncatula. Planta 231:705–716. https://doi.org/10.1007/s00425-009-1078-0

Ullah A, Manghwar H, Shaban M, Khan AH, Akbar A, Ali U, Ali E, Fahad S (2018) Phytohormones enhanced drought tolerance in plants: a coping strategy. Environ Sci Pollut Res 25:33103–33118

Vandenabeele S, Vanderauwera S, Vuylsteke M, Rombauts S, Langebartels C, Seidlitz HK, Zabeau M, Van Montagu M, Inzé D, Van Breusegem F (2004) Catalase deficiency drastically affects gene expression induced by high light in Arabidopsis thaliana. Plant J 39(1):45–58

Vellosillo T, Vicente J, Kulasekaran S, Hamberg M, Castresana C (2010) Emerging complexity in reactive oxygen species production and signaling during the response of plants to pathogens. Plant Physiol 154(2):444–448

Wani SH, Kumar V, Shriram V, Sah SK (2016) Phytohormones and their metabolic engineering for abiotic stress tolerance in crop plants. Crop J 4(3):162–176

Wen W, Li D, Li X, Gao Y, Li W, Li H, Liu J, Liu H, Chen W, Luo J (2014) Metabolome-based genome-wide association study of maize kernel leads to novel biochemical insights. Nat Commun 5:1–10

Wu QS, Srivastava AK, Zou YN (2013) AMF-induced tolerance to drought stress in citrus: a review. Sci Hortic 164:77–87

Xie X, He Z, Chen N, Tang Z, Wang Q, Cai Y (2019) The roles of environmental factors in regulation of oxidative stress in plant. Biomed Res Int 2019:9732325

Xu J, Lan H, Fang H, Huang X, Zhang H, Huang J (2015) Quantitative proteomic analysis of the rice (Oryza sativa L.) salt response. PLoS One 10:e0120978

Xu Y, Crouch JH (2008) Marker-assisted selection in plant breeding: from publications to practice. Crop Sci 48(2):391–407

Yousefirad S, Soltanloo H, Ramezanpour SS, Zaynalinezhad K, Shariati V (2018) Salt oversensitivity derived from mutation breeding improves salinity tolerance in barley via ion homeostasis. Biol Plant 62:775–785

Yousfi N, Slama I, Ghnaya T, Savouré A, Abdelly C (2010) Effects of water deficit stress on growth, water relations and osmolyte accumulation in Medicago truncatula and M. laciniata populations. Comptes Rendus Biologies 333(3):205–213

Yousfi M, Eichwald O, Merbahi N, Jomaa N (2012) Analysis of ionization wave dynamics in low-temperature plasma jets from fluid modeling supported by experimental investigations. Plasma Sources Sci Technol 21(4):045003

Yun W, Jinping Z, Yong S, Jauhar A, Jianlong X, Zhikang L (2012) Identification of genetic overlaps for salt and drought tolerance using simple sequence repeat markers on an advanced backcross population in rice. Crop Sci 52(4):1583–1592

Zang J, Sun Y, Wang Y et al (2008) Dissection of genetic overlap of salt tolerance QTLs at the seedling and tillering stages using backcross introgression lines in rice. Sci China Ser C 51:583–591. https://doi.org/10.1007/s11427-008-0081-1

Zenda T, Liu S, Wang X, Liu G, Jin H, Dong A, Yang Y, Duan H (2019) Key maize drought-responsive genes and pathways revealed by comparative transcriptome and physiological analyses of contrasting inbred lines. Int J Mol Sci 20:1268

Zhao MG, Chen L, Zhang LL, Zhang WH (2009) Nitric reductase-dependent nitric oxide production is involved in cold acclimation and freezing tolerance in *Arabidopsis*. Plant Physiol 151: 755–767. https://doi.org/10.1104/pp.109.140996

Zhao F, Zhang D, Zhao Y, Wang W, Yang H, Ta F, Li C, Hu X (2016) The difference of physiological and proteomic changes in maize leaves adaptation to drought, heat, and combined both stresses. Front Plant Sci 7:1471

Zhao A, Yu Q, Feng L, Zhang A, Pei T (2020) Evaluating the cumulative and time-lag effects of drought on grassland vegetation: a case study in the Chinese Loess Plateau. J Environ Manag 261:110214

Zulfiqar F, Chen J, Younis A, Abideen Z, Naveed M, Koyro HW, Siddique KH (2021) Biochar, compost, and biochar–compost blend applications modulate growth, photosynthesis, osmolytes, and antioxidant system of medicinal plant Alpinia zerumbet. Front Plant Sci 12:707061

Stress Protectants in Amelioration of Abiotic Stress: Mechanism of Action and Potential Role in Sustaining Oilseed and Pulses Productivity

9

Jogendra Singh, Vijayata Singh, M. L. Jakhad, Vikash Dutt, Pooja K. Shekhawat, Manish Kumar Suthar, and Parvin Kumar Bagaria

Abstract

To meet the demand for oilseed and pulse crops, it is important to either increase agricultural productivity or make better use of underutilized land by using abiotic stress-tolerant types. With the help of recombination breeding and diverse germplasms could enhance productivity as well as yield but it has proven to be more challenging using conventional approaches because of the quantitative nature of inheritance to tolerant against stress. To mitigate the abiotic stress, plants employ different approaches through generating the ample amount of stress protectants such as amino acids, proteins and enzymes, organic acids, and various metabolites controlled by specific genes. Thus, there is a crucial need to establish the abiotic stress tolerant varieties which can persist under prolonged stress while keeping the exemplary levels of yield and productivity in upward trend. In keeping view of these aspects, this chapter discusses the breeding

J. Singh · V. Singh (✉) · V. Dutt
ICAR-Central Soil Salinity Research Institute, Karnal, Haryana, India
e-mail: vijayata.singh@icar.gov.in

M. L. Jakhad
Sri Karan Narendra Agriculture University, Jobner, Rajasthan, India

P. K. Shekhawat
ICAR-Central Soil Salinity Research Institute, Karnal, Haryana, India

Sri Karan Narendra Agriculture University, Jobner, Rajasthan, India

M. K. Suthar
ICAR-Directorate of Medicinal & Aromatic Plant Research, Anand, Gujarat, India

P. K. Bagaria
ICAR-Indian Institute of Maize Research, Ludhiana, Punjab, India

A. Kumar et al. (eds.), *Salinity and Drought Tolerance in Plants*,
https://doi.org/10.1007/978-981-99-4669-3_9

schema to enhance tolerance level of stress by using conventional methods, omics approach, high throughput phenotyping (HTP), gender responsive approaches for enhancing the adoption of released tolerant variety.

Keywords

Conventional breeding · Omics approaches · *Brassica juncea* · Lentil

9.1 Introduction

Total population of world is expected to reach 9.2×10^9 people by 2050. To meet out the demand for feeding the population is required to increase either crop productivity or use unutilized area which are affected by salinity and alkalinity. As compared to other stress, salt stress is directly related with soil and water. Along with salt stress, physiological drought is always come. Therefore, if we concentrate on salt stress, we can screen the drought tolerant germplasm along with salt tolerant germplasm. In keeping view of the current agricultural production is totally dependent on large amounts of inputs and good quality water for irrigation which are major limiting factor. Taking all these things into account, oilseed and pulses could be played potential role for unutilized area for agriculture because oilseed and pulses need low input and less water.

Plants exposed to diverse abiotic stresses experience morphological and anatomical damage such as changes in osmotic balance, fluctuation in ionic homeostasis, and defects in photosynthetic activity, which consequently reduces the growth and productivity. To mitigate the abiotic stress, plants employ different approaches through generating the ample amount of stress protectants such as amino acids, proteins and enzymes, organic acids, and various metabolites controlled by specific genes (Singh et al. 2022). Thus, there is a crucial need to establish the abiotic stress tolerant varieties which can persist under prolonged stress while keeping the exemplary levels of yield and productivity in upward trend. Recent implementations of system biology for crop improvement by detecting the crucial stress protectants have increased our vast knowledge towards multigenic traits, responses, and defense mechanism of crops towards abiotic stress (Singh et al. 2022). We will discuss the various modern and traditional approaches used to identify the stress protectants and their mechanism of action to avoid abiotic stress in oilseed and pulses crop. In keeping view of these aspects, this chapter discusses the breeding schema to enhance tolerance level of stress by using conventional methods, omics approach, high throughput phenotyping (HTP), gender responsive approaches for enhancing the adoption of released tolerant variety.

9.2 Breeding Scheme to Enhance Abiotic Stress Especially Salt Stress Tolerance by Using Conventional Methods

In brassica Juncea (oilseed) and Lentil (pulses) productivity improved and has had tremendous success over the years due to the availability of genetic resources and breeding knowledge (Ranalli and Cubero 1997). Although the yield could be increased by employing different germplasm and recombination breeding, conventional methods have been shown to be more difficult owing to the quantitative character of heredity and resistance to salt stress. The nature of variation for different traits in a crop can be distinguished in two forms: first one, discrete or discontinuous variation, i.e., showing distinct classes with well-defined phenotypic expression. Second one is, non-discrete or continuous variation, i.e., manifesting indistinguishable or varying degree of phenotypic expression. Varietal adaptability to salt tolerance is important for the stabilization of crop production both over regions and years. Adaptability of salt tolerance is the ability of a genotype to exhibit relatively stable performance in different levels of salt stress. It is measured in terms of phenotypic stability of a genotype over several environments. Stability of the tolerant genotype is the suitability of the genotype for a general cultivation over a wide range of environments (Fig. 9.1). In keeping view of all these perspective regarding breeding schema to develop salt tolerance variety of *Brassica juncea* and lentil, we developed work flow which are as follows:

After selecting the donors (which high tolerant levels for abiotic stress) and recipient parents with the help of phenotyping. In order to improve the genetics of quantitative features that are directly related to salt tolerance, we apply biometrical

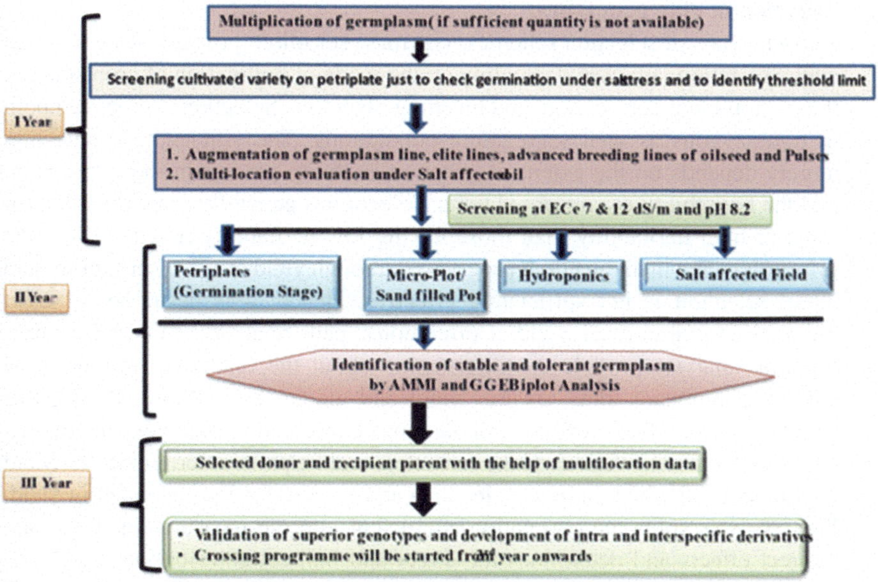

Fig. 9.1 Breeding schema to identify donors and recipient parents for salt tolerance

methodologies to aid in the selection of appropriate parents for hybridization as well as in the decision of breeding methods. There are numerous biometrical strategies that can be used to create materials for salt tolerance in four main ways, which are as follows:

(a) **Evaluation of polygenic variation:**
The efficiency of selection largely depends upon the magnitude of genetic variability present in the existing germplasm. Thus, the success of genetic improvement in any character depends on the nature of variability present in the gene pool for that character. Hence, an insight into the magnitude of variability present in the gene pool of *Brassica juncea* and lentil is of utmost importance to a plant breeder for starting a judicious plant breeding program. The degree of these components for different traits is determined by the simple measures of variability, particularly the coefficient of variation, which divides the total variance into phenotypic, genotypic, and environmental components. The study of components of genetic variance helps in further partitioning of genetic variance into additive, dominance, and epistatic components. A magnitude of these components is a measure of the type of gene action involved in the expression of various traits. Information about gene action helps in deciding a breeding procedure for the genetic important of a trait. The D^2 statistics evaluates large number of germplasm lines for genetic diversity and helps in the identification of genetically divergent parents for their exploitation in hybridization programs. Metroglyph analysis is a simple technique for the evaluation of phenotypic variability in large number of germplasm lines at a time.

(b) **Selection of elite genotypes:**
Various types of selection schemes, viz. mass selection, progeny selection, and cyclic selection, are used depending on the mode of pollination of crop species, the predominant gene action, and breeding objective. Selection is practiced both in homozygous as well segregating populations. The efficiency of selection largely depends on the extent of genetic variability present in the population, and the heritability of concerned traits. Selection is generally more effective for traits of high heritability than those having low heritability. High yield is the prime objective in all breeding programs, although yield has low heritability and direct selection is not sufficiently effective. Hence, it is desirable to select indirectly for improved yield. Correlations, path analysis, and discriminant function analysis provide the information about the relative contribution of various component traits towards yield and aid in the selection of superior genotypes from the breeding populations. Correlation measures the mutual relationship among various plant characters and helps in determining the yield components on which indirect selection can be based for improvement in yield. Path analysis splits the correlation coefficient into the measures of direct and indirect effects and determines the direct and indirect contribution of various characters towards yield. Discriminant function helps in the identification of

traits combination having high selection efficiency than direct selection for yield.

(c) **Choice of Parents and Breeding procedures:**
Hybridization is the most potent technique for breaking yield barriers and evolving varieties having built-in high yield potential. The selection of suitable parents for hybridization is one of the most important steps in a breeding program. Selection of parents on the basis of phenotypic performance alone is not a sound procedure since phenotypically superior lines may yield poor recombinants in the segregating generation. It is, therefore, essential that parents should be chosen on the basis of their genetic varieties or strains in terms of their genetic makeup. Three biometrical techniques, viz., generation mean analysis, biparental cross analysis, and triple test cross analysis are used for the analysis of individual cross in terms of components of genetic variance. Another three biometrical techniques, namely diallel cross, partial diallel cross, and line x tester analysis are used for the evaluation of several single crosses and commonly used in the selection of parents for hybridization. Tri-allele analysis is used for evaluating several three-way crosses and quadric-allele analysis is adopted for the simultaneous evaluation of several double crosses. The parents are chosen on the basis of their combining ability and the breeding procedure is decided on the basis of gene action involved in the expression of various quantitative characters.

(d) **Evaluation of varietal adaptability**
Salt tolerant varietal adaptability to salt fluctuations is important for the stabilization of crop production both over locations and years. Estimation of phenotypic stability, which involved regression analysis, has proven to be a valuable technique in the assessment of varietal adaptability. Stability analysis is useful in the identification of adaptable genotypes and in predicting the response of various genotypes over changing environments. It is generally agreed that the more stable genotypes can somehow adjust their phenotypic responses to provide some measures of uniformity in spite of salt fluctuations. The buffering ability of the segregating populations seems to be directly related with the homeostatic responses (buffering capacity) of the parental lines. Therefore, it is feasible to develop phenotypically stable high potential genotypes by incorporation of homeostatic genotypes in the hybridization program.

9.3 Omics Approach to Enhance Salt Tolerance

Both oilseeds and pulses are reported to be highly affected by variety of abiotic stress (Choudhary et al. 2018; Joshi et al. 2021; Singh et al. 2022). Under prolonged abiotic stress (temperature, metals toxicity, drought, water, or salt), elevated levels of reactive oxygen species (ROS) such as $^1O_2$, $O_2^{\bullet-}$, H_2O_2, and $OH^{\bullet}$ induce damage to photosynthetic machinery and biomolecules such as DNA, protein, and lipids, in response to this, plants mitigate the damage through anti-oxidant defense mechanism (Sultana et al. 2014). The main organelles supporting the anti-oxidative machinery include chloroplast, mitochondria, and peroxisomes which scavenges the ROS through enzymatic and non-enzymatic anti-oxidants. Enzymatic antioxidants include superoxide dismutase (SOD), catalase (CAT), peroxidase (POX), ascorbate peroxidase (APX), mono-dehydroascorbate reductase (MDHAR), dehydroascorbate reductase (DHAR), and glutathione reductase (GR) and non-enzymatic antioxidants are reduced glutathione (GSH), ascorbate (AA), carotenoids, and tocopherols (Gill and Tuteja 2010). In case of pulses, report suggested that tolerant chickpea genotypes displayed higher activities of CAT and POX when exposed to drought conditions (Choudhary et al. 2018). Report on L-glutamic acid treated Lentil (*Lens culinaris* Medik.) exposed to salt stress, suggested increased accumulation of photosynthetic pigments (Chlorophyll, Carotenoids), osmolytes (Proline, Glycine betaine) works synergistically with anti-oxidants (non-enzymatic and enzymatic) by reducing the H_2O_2 and malonaldehyde contents along with increased seedling growth and yield per plant (Fardus et al. 2021).

Under the influence of heat stress, tolerant lentil variety PDL-2 displayed high glycine betaine content along with increased activity of SOD, APX, and glutathione peroxidase (GPX) compared to JL-3 (sensitive). Transcriptome analysis of both the genotypes resulted in higher number of contigs in the tolerant PDL-2 compared to JL-3. Similar trend was observed in SNPs, where highest number of SNPs was detected in the tolerant control followed by sensitive treated, tolerant treated, and sensitive control samples (Singh et al. 2019a, b). Multiple members of heat shock protein (HSPs) family protect the cellular protein structures and remove the faulty or denatured polypeptides from plant under stress. The role of HSPs as stress protectants was confirmed through the detection of several HSP-encoding transcripts by RNA-seq analysis of Lentil under different abiotic stresses (Khorshidvand et al. 2021). Higher accumulation of protein such as proline having osmoregulatory and anti-oxidative properties provided drought tolerance to chickpea genotypes (Phule G 09103, Phule G 2008-74 and Digvijay) (Ulemale et al. 2013). Another stress protectant, namely Thiourea enhanced the salt, drought, heat, and cold resistance in mung bean by regulating the proline content along with pentose phosphate pathway (Waqas et al. 2019). Glycine Betaine is another stress marker providing osmoprotection to plants in response to osmotic stress induced by drought, salinity, and temperature fluctuations. Exogenous treatment of betaine and proline solution to drought stressed lentil varieties resulted in increased activity of glutathione-*S*-transferase (GST), an anti-oxidative enzyme which catalyzes the reduction of free radicals and peroxides generated via environmental stresses (Molla et al. 2014).

Oilseed crops commonly grown throughout the globe such as soybean, sunflower, canola, peanut, cottonseed, sesame, and mustard are vulnerable to various abiotic stresses. Various studies evaluated the impact of major abiotic stress on oilseed crops and found significant reduction in germination percentage, oil, protein, and crude fiber content (Singh et al. 2014a, b; Singh and Sharma 2016; Muhammad et al. 2021). Substantial decline in net photosynthetic activity, stomatal conductance, water use efficiency, transpiration rate, CO_2 assimilation rate was recorded in *Brassica* genotypes under saline conditions (EC_{iw} 15 dS/m). Higher ionic content in root and shoot samples further supported the detrimental impact of salt stress on plant growth (Singh et al. 2019a, b). However, pretreatment of the seeds with H_2O_2 stimulated the drought tolerance in *Brassica juncea* by reducing endogenous H_2O_2 production along with increased activities of DHAR and GPX stress marker (Dat et al. 1998).

Report identified and characterized significant SOD genes in *Brassica juncea* and *Brassica rapa* through genome wide identification and chromosome location mapping, indicating their important role to avoid abiotic stress. Verma and group further identified stress responsive SOD genes in both the genotypes via Fragments per kilobase permillion reads (FPKM) analysis of sequence read archive (SRA) database available for drought, heat, and salt stress (Verma et al. 2019). Singh et al. found highly expressed genes (SOS1, SOS2, SOS3, ENH1, and NHX1) against the salt stress in tolerant genotypes (CS 54 and CS 52-SPS-1-2012). These genes specifically regulated the ionic homeostasis in mustard plants facing saline conditions. Another set of genes (APX1, APX4, DHAR1, and MDHAR3) circumvented the oxidative stress due to salinity in the tolerant mustard varieties (Singh et al. 2019a, b). Through high throughput sequencing of 588 *Brassica napus* samples, around 5,294,158 single nucleotide polymorphisms (SNPs) and 1,307,151 indels were generated. Additionally, 60 loci identified via GWAS were considered to be associated with abiotic stress tolerance and other traits, which may provide valuable information for crop improvement (Lu et al. 2019). Furthermore, a non-targeted metabolome analysis conducted on mustard varieties CS 60 and CS 245 under the influence of salt stress found higher number of key organic acids, organosulphates, and lipid derivatives in salt tolerant CS 60 compared to sensitive genotype CS 245. High levels of amino acid accumulation (Glutamyl-cysteine, *N*-acetyl-L-glutamate 5-phosphate) prohibited the oxidative load induced by salt stress in CS 60 genotype (Singh et al. 2022). Transcript analysis showed that thiourea modulated the expression and activity of mitochondrial homeostasis and ATPase of *Brassica juncea* under salinity stress (Srivastava et al. 2009).

9.4 High Throughput Phenotyping (HTP) to Enhance Abiotic Stress Tolerance

Major abiotic stresses, viz. Salt stress, Draught, Mineral toxicity/deficiency, cold stress are challenging to the crop production worldwide ultimately to the food security. Various approaches from screening techniques to management are available but most effective, inexpensive, and manageable option is the development of tolerant crop cultivars (Kim et al. 2020). Significant progress has been achieved in crop genomics with continuously evolution of new technologies, however, crop improvement using convenient data is far away from satisfactory due to the lack of high-quality phenotypic data. Plant phenotyping is a corner stone in breeding program, i.e., quantitative description of plant's anatomical, ontological, physiological, and biochemical properties (Walter et al. 2015).

Consequently, there is a need to collect highthroughput, efficient, and complete trait data like plant development, architecture, plant photosynthesis, growth, biomass productivity, physiological, biochemical data in order to unlock the huge amount of plant information (Mir et al. 2019). This led to the invention of High Throughput Phenotyping (HTP) (Rutkoski et al. 2016). HTP is based on spectrum, sensors, and computation to accelerate the phenotypic data generation. It can potentially be cost effective as well as breeding cycle can be significantly shortened. It includes various types of approaches like spectroscopy (spectrum pattern of absorption, transmission, and reflection of photons are primarily determined by plant pigments, constituents, and structure), thermograpgy (plant temperature acts as indicator of physiological status of plant), photogrametry (stand count and ground cover), RGB imaging for phenology and envirotyping (soil characterization, weather characterization, and field management). Nowadays sensors and different types of platforms are also available for aerial phenotyping, e.g., drones, NDVI and vegetation indices, thermography, RGB, and Hyperspectral. Manned aircraft have been also developed where different types of multispectral and thermal cameras are attached like NDVI, MCARI, chlorophyll index, LAI, CAB, FAPAR, QPAR, etc. Nowadays focus is also given on the development of ground rover phenotyping based on deep learning platforms. HTP approaches are noninvasive, multiple spatio-temporal dimensions, fast tracking, systemic data collection with reduced errors and allows automation (Kim et al. 2020).

9.5 High Throughput Phenotyping Approaches for Abiotic Stress

9.5.1 RGB (Red, Green Blue) Images

Imaging technologies differs for healthy plants to diseased/stressed plants in electromagnetic radiation interaction such as absorbance, transmission, reflectance, emission, fluorescence, etc. RGB as most easily accessible sensor works on sensing visible wavelength ranging from 400 to 700 nm. Traits such as vegetation indices, plant height, plant structure, growth rates, and morphological traits can be obtained (Kim et al. 2018; Crimmins and Crimmins 2008; Deery et al. 2014; Liu et al. 2017), e.g., Green seeker for NDVI and IR Thermometer for canopy temperature (Reynolds et al. 2020).

9.5.2 Hyperspectral Imaging

Vegetation, leaf growth, quality grains, level of phytochemicals, water indices, soil cover status, level of phytochemicals, and photosynthetic traits can be obtained by sensing thousands of bands per pixel by hyperspectral imaging sensors such as hyperspectral camera, thermal camera, and spectrometer (Stagakis et al. 2010; Zhao et al. 2013; El-Hendawy et al. 2019). Hyperspectral imaging can measure at wide range with improved prediction accuracy as well as identification of specific plant stress response. Plant species such as arabidopsis, wheat, rice, and rye are phenotyped using this technology (Jangra et al. 2021).

9.5.3 Thermal Imaging

LAI, disease severity, insect infestation to seed, stomatal conductance, canopy temperature, transpiration rates, water stress responses, etc. can be measured by thermal imaging (suitable to image temperature changes) that works based on sensing emitted radiation of object that increases with the object temperature above absolute zero (700 ~ 10^6 nm). Thermal imaging is applicable in both controlled as well as field conditions but it needs soil background corrections and affected by wind and transient cloudiness (Baluja et al. 2012; Berni et al. 2009; Gago et al. 2015; Leinonen et al. 2006).

9.5.4 Fluorescence Imaging

Chlorophyll conductance, photosynthetic rates, pigment composition, quantum yield, health status of leaf as well as non-photochemical quenching can be estimated by fluorescence sensor (180–800 nm) that works by sensing fluorescence emitted by short wave light absorption of susceptible molecule. Major advantage of it is that it can detect stress before appearance of visual symptoms but major limitations are on robustness, reproducibility, and data analysis and need to follow strict protocols (Chaerle et al. 2006).

9.5.5 NIR (Near Infrared) Images

NIR sensors detect highest reflectance of green area in between 700 and 1300 nm (chlorophyll conductance and vegetation indices) and more absorbance by water than visible spectrum at above 1300 nm (Bei et al. 2011; Bendig et al. 2015; Yang et al. 2017), for example, drones for IR and spectral images and Phenocart (Reynolds et al. 2020), FT-NIR for oil quality parameters (Singh et al. 2014a, b).

9.5.6 Light Detection and Ranging (LiDAR)

It is the new remote sensing technology that measures various parameters such as leaf area index (LAI), nitrogen status, vegetation cover, Canopy, leaves, and height by scanning surface of target object and analyzing reflected light as distance (Lefsky et al. 2002; Lin 2015; Eitel et al. 2014; Madec et al. 2017; Omasa et al. 2006).

9.5.7 Other HTP Technologies

NMR (nuclear magnetic resonance) play important role in measuring the plant responses on plant leaves exposed to dehydration or to osmotic stresses (Capitani et al. 2009). MRI (magnetic resonance imaging) as powerful 3D-imaging tool of structures (Van As and Van Dusschoten 1997) can image both soil water contents, morphometric parameters, metabolic study and root architectures (Pohlmeier et al. 2008). 3D imaging based on stereo cameras and time of flight cameras can phenotype canopy and shoot structure, plant height and root structure with high accuracy with low noise level, e.g., sorghum, maize, pepper, rye, etc. Positron emission tomography (PET) can generate both quantitative and spatial information for water transport, flow speed, and sectorality but till date it has limited application in plants such as barley, sugarbeet, etc. X-Ray digital and computed sensors have been also applied for phenotyping in crop like wheat, rice for grain quality, monomorphic parameters, flow speed, etc. (Jangra et al. 2021).

9.5.8 HTTP Platforms

Innovations in sensing approaches for aerial as well as ground phenotyping have opened the door for investigation studies and significant phenotyping, e.g., robotics, aeronautics computing, drones, phenomobiles, phenoliner, etc. Others are like Shoelomics to study root architecture (Trachsel et al. 2011), Phenoscope for large scale experiment (Tisné et al. 2013), Breed vision as multisensor platform (Busemeyer et al. 2013), Rhizomes for paper based high throughput root phenotyping (Le Marié et al. 2014), plant accelerator (Neilson et al. 2015), Microphenotron (Burrell et al. 2017), Phenocart developed at CIMMYT (Crain et al. 2016), Terresttrial 3D laser (Friedli et al. 2016), Phenoboot 1.0 (Salas Fernandez et al. 2017), GPhenoVision (Jiang et al. 2018), Phenofield for characterization of abiotic stress, variety tolerance evaluation and other physiological analysis, etc. (Beauchene et al. 2019).

9.6 Gender Responsive Approaches for Enhancing the Adoption of Salt Tolerant Variety of Mustard and Lentil

9.6.1 Approach Ameliorating Feminization in Agriculture

Salinity aggravating the hardship in agriculture culminating in migration of men (Chen 2018) for better job opportunities led female farmers to take up farming (Kawarazuka et al. 2022) in toto. Although feminization is already taking place, research has shown that spatially specified gendered preferences and adoption techniques evolve with adaption choices. Therefore, efforts can be made to improve the same in the issue soil as well (Ravera et al. 2016). Climate change has already diminished the income generating opportunities of women farmers due to gendered norms (ADB 2013). Hence reclamation efforts through salt tolerant lentil (Panuccio et al. 2021) and mustard varieties (Singh et al. 2020) are promising in the salt affected areas/zones, as both crops are low input intensive and fulfill nutrient and forage requirement. This would be a right step in providing an opportunity to the female to derive income from their salt affected land. Eventually a sense of achievement which was hitherto missing in them would set an example to lead in successful adoption of the salt tolerant varieties in similar salt affected areas.

9.6.2 Nudging on Familial Food and Nutritional Security

Albeit being provider to the family, most men and women farmers are ignorant on food and nutritional security. If they can be nudged on, especially the woman of the family on the proper soil health management there can be a drastic change in the familial nutritional security (IFPRI 2020). Adoption of salt tolerant mustard and lentil cropping can release the stress a small holder farmer feels in feeding his/her family, as in it leads to augmenting farm income through effective soil management, hence better yield and eventually fulfilling the nutritional security of the family.

9.6.3 Alleviating Drudgery Approach

Due to unpredictable crop yields and longer travel lengths for fuel, fodder, and water for their livestock, work burdens increased for women in rainfed agricultural households in Maharashtra, India (World Bank 2022). World Bank report even says installation of solar based irrigation systems have benefitted women to grow high value climate smart crops and earn higher incomes in problem areas. This is an indication on how drudgery alleviating approach can reduce the drudgery already existing in the problematic soil through salt tolerant variety of lentil and mustard, which is already in the cropping practice of Indo Gangetic salt affected areas in India. Studies corroborated the fact that women can lead adaptation strategies efficiently if interruption through social norms don't supersede (Md et al. 2022).

9.6.4 Participatory Development Stakeholder Approach

Actively participating in decision making, adoption practices have proven to encourage people to commit to successful ventures. Farming is not an exception, but women are pushed to background in all the decision-making scenarios in a rural agrarian Indian household. The problem is grave especially when their land under cultivation is problematic or salt affected. Encouraging women to make a participatory stakeholder in the adoption of salt tolerant varieties of mustard and lentil crops can yield results in both tangible and intangible manner. Former being in the form increased yield due to participation of women with better information on salt tolerant varieties and latter being feeling a responsible stakeholder in the augmented farm income. Study conducted by Galiè (2013) depicted how participative plant breeding can assert identity for women as farmers. Albeit women performing majority roles of farming in salt affected areas their role as farmers are not positioned due to dominant norms.

9.7 Conclusion

To mitigate the abiotic stress, plants employ different approaches through generating the ample amount of stress protectants such as amino acids, proteins and enzymes, organic acids, and various metabolites controlled by specific genes. Therefore, there is a crucial need to develop and release the tolerant varieties for abiotic stress by using conventional breeding methods or modern breeding methods.

References

Asian Development Bank (2013) Gender equality and food security—women's empowerment as a tool against hunger Mandaluyong City. Asian Development Bank, Philippines

Baluja J, Diago MP, Balda P, Zorer R, Meggio F, Morales F, Tardaguila J (2012) Assessment of vineyard water status variability by thermal and multispectral imagery using an unmanned aerial vehicle (UAV). Irrig Sci 30:511–522

Beauchene K, Leroy F, Fournier A, Huet C, Bonnefoy M, Lorgeou J, de Solan B, Piquemal B, Thomas S, Cohan J-P (2019) Management and characterization of abiotic stress via PhénoField®, a high throughput field phenotyping platform. Front Plant Sci 10:904

Bei R, Cozzolino D, Sullivan W, Cynkar W, Fuentes S, Dambergs R, Pech J, Tyerman S (2011) Non-destructive measurement of grapevine water potential using near-infrared spectroscopy. Aust J Grape Wine Res 17:62–71

Bendig J, Yu K, Aasen H, Bolten A, Bennertz S, Broscheit J, Gnyp ML, Bareth G (2015) Combining UAV-based plant height from crop surface models, visible, and near infrared vegetation indices for biomass monitoring in barley. Int J Appl Earth Obs Geoinf 39:79–87

Berni JAJ, Zarco-Tejada PJ, Suarez L, Fereres E (2009) Thermal and narrow and multispectral remote sensing for vegetation monitoring from an unmanned aerial vehicle. IEEE Trans Geosci Remote Sens 47:722–738

Burrell T, Fozard S, Holroyd GH, French AP, Pound MP, Bigley CJ, James Taylor C, Forde BG (2017) The Microphenotron: a robotic miniaturized plant phenotyping platform with diverse applications in chemical biology. Plant Methods 13:10

Busemeyer L, Mentrup D, Möller K, Wunder E, Alheit K, Hahn V, Maurer HP, Reif JC, Würschum T, Müller J, Rahe F, Ruckelshausen A (2013) Breed Vision—a multi-sensor platform for non-destructive field-based phenotyping in plant breeding. Sensors 13:2830–2847

Capitani D, Brilli F, Mannina L, Proietti N, Loreto F (2009) In situ investigation of leaf water status by portable unilateral nuclear magnetic resonance. Plant Physiol 149:1638–1647

Chaerle L, Leinonen I, Jones HG, Van Der Straeten D (2006) Monitoring and screening plant populations with combined thermal and chlorophyll fluorescence imaging. J Exp Bot 58:773–784

Chen JJ (2018) Climate change is making soil saltier, forcing farmers worldwide to find new livelihoods. Scrollin. https://scroll.in/article/904032/climate-change-is-making-soil-saltier-forcing-farmers-worldwide-to-find-new-livelihoods. Accessed 25 Dec 2022

Choudhary AK, Sultana R, Vales MI, Saxena KB, Kumar RR, Ratnakumar P (2018) Integrated physiological and molecular approaches to improvement of abiotic stress tolerance in two pulse crops of the semi-arid tropics. Crop J 6(2):99–114

Crain JL, Wei Y, Barker J, Thompson SM, Alderman PD, Reynolds M, Zhang N, Poland J (2016) Development and deployment of a portable field phenotyping platform. Crop Sci 56:965

Crimmins MA, Crimmins TM (2008) Monitoring plant phenology using digital repeat photography. Environ Manag 41:949–958

Dat JF, Foyer CH, Scott IM (1998) Changes in salicylic acid and antioxidants during induced thermotolerance in mustard seedlings. Plant Physiol 118:1455–1461

Deery D, Jimenez-Berni J, Jones H, Sirault X, Furbank R (2014) Proximal remote sensing buggies and potential applications for field-based phenotyping. Agronomy 4:349–379

Eitel JUH, Magney TS, Vierling LA, Brown TT, Huggins DR (2014) LiDAR based biomass and crop nitrogen estimates for rapid, non-destructive assessment of wheat nitrogen status. Field Crop Res 159:21–32

El-Hendawy SES, Al-Suhaibani N, Elsayed SEM, Alotaibi M, Hassan WM, Schmidhalter U (2019) Performance of optimized hyperspectral reflectance indices and partial least squares regression for estimating the chlorophyll fluorescence and grain yield of wheat grown in simulated saline field conditions. Plant Physiol Biochem 144:300–311

Fardus J, Hossain MS, Fujita M (2021) Modulation of the antioxidant defense system by exogenous l-glutamic acid application enhances salt tolerance in lentil (Lens culinaris Medik.). Biomol Ther 11(4):587

Friedli M, Kirchgessner N, Grieder C, Liebisch F, Mannale M, Walter A (2016) Terrestrial 3D laser scanning to track the increase in canopy height of both monocot and dicot crop species under field conditions. Plant Methods 12:9

Gago J, Douthe C, Coopman RE, Gallego PP, Ribas-Carbo M, Flexas J, Escalona J, Medrano H (2015) UAVs challenge to assess water stress for sustainable agriculture. Agric Water Manag 153:9–19

Galiè A (2013) The empowerment of women farmers in the context of participatory plant breeding in Syria: towards equitable development for food security. Dissertation

Gill SS, Tuteja N (2010) Reactive oxygen species and antioxidant machinery in abiotic stress tolerance in crop plants. Plant Physiol Biochem 48(12):909–930

International Food Policy Research Institute (2020) Why gender matters *for soil health as part of sustainable food systems*. https://www.ifpri.org/blog/why-gender-matters-soil-health-part-sustainable-food-systems. Accessed 26 Dec 2022

Jangra S, Chaudhary V, Yadav RC, Yadav NR (2021) High-throughput phenotyping: a platform to accelerate crop improvement. Phenomics 1:31–53

Jiang Y, Li C, Robertson JS, Sun S, Xu R, Paterson AH (2018) GPhenoVision: a ground mobile system with multi-modal imaging for field-based high throughput phenotyping of cotton. Sci Rep 8:1213

Joshi R, Ramawat N, Jha J, Durgesh K, Singh M, Talukdar A, Tomar SMS, Singh D (2021) Salt stress in pulses: a learning from global research on salinity in crop plants. Indian J Genet Plant Breed 81(2):159–185

Kawarazuka N, Doss CR, Farnworth CR, Pyburn R (2022) Myths about the feminization of agriculture: implications for global food security. Glob Food Sec 33:100611. https://doi.org/10.1016/j.gfs.2022.100611

Khorshidvand M, Ismaili A, Sohrabi SS (2021) Identification and expression pattern of lentil's HSPs under different abiotic stresses. Plant Biotechnol Rep 15:609–625

Kim DW, Yun HS, Jeong SJ, Kwon YS, Kim SG, Lee WS, Kim HJ (2018) Modeling and testing of growth status for Chinese cabbage and white radish with UAV-based RGB imagery. Remote Sens 10:563–587

Kim J, Kim KS, Kim Y, Chung YS (2020) A short review: comparisons of high-throughput phenotyping methods for detecting drought tolerance. Sci Agric 78:4

Le Marié C, Kirchgessner N, Marschall D, Walter A, Hund A (2014) Rhizoslides: paper-based growth system for non-destructive, high throughput phenotyping of root development by means of image analysis. Plant Methods 10:13

Lefsky MA, Cohen WB, Parker GG, Harding DJ (2002) Lidar remote sensing for ecosystem studies: lidar, an emerging remote sensing technology that directly measures the three dimensional distribution of plant canopies, can accurately estimate vegetation structural attributes and should be of particular interest to forest, landscape, and global ecologists. Bioscience 52:19–30

Leinonen I, Grant OM, Tagliavia CPP, Chaves MM, Jones HG (2006) Estimating stomatal conductance with thermal imagery. Plant Cell Environ 29:1508–1518

Lin Y (2015) Lidar: an important tool for next-generation phenotyping technology of high potential for plant phenomics? Comput Electron Agric 11:61–73

Liu S, Baret F, Andrieu B, Burger P, Hemmerlé M (2017) Estimation of wheat plant density at early stages using high resolution imagery. Front Plant Sci 8:1–10

Lu K, Wei L, Li X, Wang Y, Wu J (2019) Whole-genome resequencing reveals Brassica napus origin and genetic loci involved in its improvement. Nat Commun 10:1154

Madec S, Baret F, de Solan B, Thomas S, Dutartre D, Jezequel S, Hemmerlé M, Colombeau G, Comar A (2017) High-throughput phenotyping of plant height: comparing unmanned aerial vehicles and ground lidar estimates. Front Plant Sci 8:2002–2015

Md A, Gomes C, Dias JM, Cerdà A (2022) Exploring gender and climate change nexus, and empowering women in the South Western coastal region of Bangladesh for adaptation and mitigation. Climate 10:172. https://doi.org/10.3390/cli10110172

Mir RR, Reynolds M, Pinto F, Khan MA, Bhat MA (2019) High-throughput phenotyping for crop improvement in the genomics era. Plant Sci 282:60–72

Molla MR, Ali MR, Hasanuzzaman M, Al-Mamun MH, Ahmed A, Nazim-ud-Dowla MAN, Rohman MM (2014) Exogenous proline and betaine-induced upregulation of glutathione transferase and glyoxalase I in lentil (Lens culinaris) under drought stress. Not Bot Horti Agrobot Cluj Napoca 42:73–80

Muhammad A, Ahmad WE, Milan S, Saddam H, Usman Z, Zohaib AM, Muhammad HR, Marian B, Disna R, Laura LT, Ibrahim AA, Ayman ELS (2021) Adaptation strategies to improve the resistance of oilseed crops to heat stress under a changing climate: an overview. Front Plant Sci 12:767150

Neilson EH, Edwards AM, Blomstedt CK, Berger B, Moller BL, Gleadow RM (2015) Utilization of a high-throughput shoot imaging system to examine the dynamic phenotypic responses of a C4 cereal crop plant to nitrogen and water deficiency over time. J Exp Bot 66:1817–1832

Omasa K, Hosoi F, Konishi A (2006) 3D lidar imaging for detecting and understanding plant responses and canopy structure. J Exp Bot 58:881–898

Panuccio MR, Mallamaci C, Attinà E, Muscolo A (2021) Using digestate as fertilizer for a sustainable tomato cultivation. Sustainability 13:1574. https://doi.org/10.3390/su13031574

Pohlmeier A, Oros-Peusquens A, Javaux M, Menzel MI, Vanderborght J, Kaffanke J, Romanzetti J, Lindenmair H, Shah NJ (2008) Changes in soil water content resulting from *Ricinus* root uptake monitored by magnetic resonance imaging. Vadose Zone J 7:1010–1017

Ranalli P, Cubero JI (1997) Bases for genetic improvement of grain legumes. Field Crop Res 53: 69–82

Ravera F, Martín-López B, Pascual U et al (2016) The diversity of gendered adaptation strategies to climate change of Indian farmers: a feminist intersectional approach. Ambio 45(Suppl 3): 335–351. https://doi.org/10.1007/s13280-016-0833-2

Reynolds M, Chapman S, Crespo-Herrera L, Molero G, Mondal S, Pequeno DN, Pinto F, Pinera-Chavez FJ, Poland J, Rivera-Amado C, Saint Pierre C (2020) Breeder friendly phenotyping. Plant Sci 295:110396

Rutkoski J, Poland J, Mondal S, Autrique E, Pérez LG, Crossa J, Reynolds M, Singh R (2016) Canopy temperature and vegetation indices from high-throughput phenotyping improve accuracy of pedigree and genomic selection for grain yield in wheat. G3 (Bethesda) 6:2799–2808

Salas Fernandez MG, Bao Y, Tang L, Schnable PS (2017) A highthroughput, field-based phenotyping technology for tall biomass crops. Plant Physiol 174:2008–2022

Singh J, Sharma PC (2016) Comparative effects of soil and water salinity on oil quality parameters of *Brassica juncea*. J Oilseed Brassica 7(1):29–37

Singh M, Rathore S, Raja P (2014a) Physiological and stress studies of different rapeseed-mustard genotypes under terminal heat stress. Int J Genet Eng Biotechnol 5:133–142

Singh J, Sharma PC, Sharma SK, Rai M (2014b) Assessing the effect of salinity on the oil quality parameters of Indian mustard (*Brassica juncea* L. Czern & Coss) using Fourier transform near-infrared reflectance (FT-NIR) spectroscopy. Grasas y Aceites 65:e009

Singh J, Singh V, Vineeth TV, Kumar P, Kumar N, Sharma PC (2019a) Differential response of Indian mustard (Brassica juncea L., Czern and Coss) under salinity: photosynthetic traits and gene expression. Physiol Mol Biol Plants 25(1):71–83

Singh D, Singh CK, Taunk J (2019b) Genome wide transcriptome analysis reveals vital role of heat responsive genes in regulatory mechanisms of lentil (Lens culinaris Medikus). Sci Rep 9:12976

Singh J, Sharma PC, Singh V (2020) Breeding mustard for salt tolerance: problems and prospects. In: Islam AKMA, Hossain MA, Islam AKMM (eds) Brassica breeding and biotechnology. IntechOpen, London. https://doi.org/10.5772/intechopen.94551

Singh J, Singh V, Dutt V, Walia N, Kumawat G, Jakhar ML, Yadava DK, Sharma PC (2022) Insights into salt tolerance of mustard (Brassica juncea L. Czern & Coss): a metabolomics perspective. Environ Exp Bot 194:104760

Srivastava AK, Ramaswamy NK, Mukopadhyaya R, Jincy MG, D'Souza SF (2009) Thiourea modulates the expression and activity profile of mtATPase under salinity stress in seeds of Brassica juncea. Ann Bot 103:403–410

Stagakis S, Markos N, Sykioti O, Kyparissis A (2010) Monitoring canopy biophysical and biochemical parameters in ecosystem scale using satellite hyperspectral imagery: an application on a Phlomis fruticosa Mediterranean ecosystem using multiangular CHRIS/PROBA observations. Remote Sens Environ 114:977–994

Sultana R, Choudhary AK, Pal AK, Saxena KB, Prasad BD, Singh R (2014) Abiotic stresses in major pulses: current status and strategies. In: Gaur R, Sharma P (eds) Approaches to plant stress and their management. Springer, New Delhi

Tisné S, Serrand Y, Bach L, Gilbault E, Ben Ameur R, Balasse H, Voisin R, Bouchez D, Durand-Tardif M, Guerche P, Chareyron G, Da Rugna J, Camilleri C, Loudet O (2013) Phenoscope: an automated large-scale phenotyping platform offering high spatial homogeneity. Plant J 74:534–544

Trachsel S, Kaeppler SM, Brown KM, Lynch JP (2011) Shovelomics: high throughput phenotyping of maize (*Zea mays* L.) root architecture in the field. Plant Soil 341:75–87

Ulemale CS, Mate SN, Deshmukh DV (2013) Physiological indices for drought tolerance in chickpea (*Cicer arietinum* L.). World J Agric Sci 9:123–131

Van As H, Van Dusschoten D (1997) NMR methods for imaging of transport processes in micro-porous systems. Geoderma 80:389–403

Verma D, Lakhanpal N, Singh K (2019) Genome-wide identification and characterization of abiotic-stress responsive SOD (superoxide dismutase) gene family in Brassica juncea and B. rapa. BMC Genomics 20(1):227

Walter A, Liebisch F, Hund A (2015) Plant phenotyping: from bean weighing to image analysis. Plant Methods 11:1–11

Waqas MA, Kaya C, Riaz A, Farooq M, Nawaz I, Wilkes A, Li Y (2019) Potential mechanisms of abiotic stress tolerance in crop plants induced by Thiourea. Front Plant Sci 10:1336

World Bank Blogs (2022). https://blogs.worldbank.org/climatechange/gender-smart-agriculture-only-way-forward-women-and-climate. Accessed 25 Dec 2022

Yang G, Liu J, Zhao C, Li Z, Huang Y, Yu H, Xu B, Yang X, Zhu D, Zhang X, Zhang R, Feng H, Zhao X, Li Z, Li H, Yang H (2017) Unmanned aerial vehicle remote sensing for field-based crop phenotyping: current status and perspectives. Front Plant Sci 8:1111–1136

Zhao K, Valle D, Popescu S, Zhang X, Mallick B (2013) Hyperspectral remote sensing of plant biochemistry using Bayesian model averaging with variable and band selection. Remote Sens Environ 132:102–119

Insights into Drought and Salinity Stress Tolerance in Crop Plant Through Various Genomic Approaches Under Changing Climate

10

Amit Sharma, Yadav Rajesh, Ravika Sheoran, Minakshi Jattan, Deepak Kaushik, Navreet Kaur Rai, Naresh, and Ashok

Abstract

Anthropogenic climate change is inducing the dangerous and sweeping disruption in nature while touching the billions of lives across the globe through food security. Extreme temperatures, low water availability, salinity, mineral deficiency, and toxicity are major abiotic factors that affect crop's productivity and sustainability worldwide. To mitigate the demand of growing population, there is a need to increase the production and productivity of crop under abiotic challenges. Despite the great efforts for abiotic tolerance in crops has been done by conventional breeding as well as by biotechnological method but results so far obtained have been rather scarce due to the complexity of the response to salt and drought stress. Therefore, novel techniques are required for isolate and functionally characterization of gene that involved in salinity and drought tolerance. In this chapter, we focus on the genomics tools, viz., traditional QTL, MAS, transcriptomic analysis, association mapping, GWAS, NGSs, TILLING and EcoTILLING, RNAi, CRISPR/Cas, and other gene tagging approaches for genetically and functionally dissection of these complex traits. Confidently, the collaborative use of above-mentioned genomics tools can help in deep insights into the genetic and physiological mechanisms of tolerance to drought and salinity in crop plants, thus empowering the accepted schematic of future breeding programs.

Keywords

Abiotic stress · Drought · Genomic tools · Salinity

A. Sharma · Y. Rajesh (✉) · R. Sheoran · M. Jattan · D. Kaushik · N. K. Rai · Naresh · Ashok
Department of Genetics and Plant Breeding, CCS Haryana Agricultural University, Hisar, India

A. Kumar et al. (eds.), *Salinity and Drought Tolerance in Plants*,
https://doi.org/10.1007/978-981-99-4669-3_10

10.1 Introduction

In a sustainable ecosystem, there should be enough food to feed the world's constantly growing population. However, the challenge posed by the prospect of climate change and unforeseen natural extremes has increased that would lead to food insecurity (Abberton et al. 2016). In the fields, crop plants are frequently subjected to one or more biotic and/or abiotic stresses that drastically impair agricultural production and productivity, creating a huge gap between demand and supply. Among the different abiotic stresses, drought and salinity most burtal environmental factors that remarkably impacted on agricultural output worldwide (Chantre Nongpiur et al. 2016; Fathi and Tari 2016). According to one report, about 10.5 and 60 million km^2 of land are plagued by salinity and drought, respectively (Zhang et al. 2014). It is become more and more difficult to sustainably feed the world's population in an acceptable and healthy manner. These challenges have been exacerbated in part by past human activity. Global temperatures have risen by an average of 1 °C from the pre-industrial era due to the 150 years of fast economic expansion and the associated increase in greenhouse gas (GHG) emissions. If things continue as they are, the average global warming between 2030 and 2050 is predicted to be 1.5 °C (Fróna et al. 2021). As per one climatic report of UN, rising temperatures varnished the Himalayan glaciers till by 2035 that is water resources of Asia's largest rivers (Ganges, Brahmaputra, Mekong, Indus, Yangtze, Salween, and Yellow). As a consequence, by the end of the twenty-first century, the melting of glaciers will have accelerated sea level up to 59 cm (Hossain et al. 2021).

Drought causes severe damage during all the stages of crop growth from seedling to maturity. In case of wheat, occurrence of drought at early growth stage resulting in poor seedling establishment and less number of tillers per unit area. Whereas the production of dry matter, effective tillers, and grains per plant are decreased by the occurrence of drought during the middle stage of growth. Similarly, terminal drought resulting in poor assimilates production, poor fertility, and low grain weight (Tiwari et al. 2015). In another example of tomato number of seeds, the size of the seeds and the quality of the seeds are typically reduced by drought stress during vegetative or early reproductive growth (Pervez et al. 2009). Similarly, plant growth and development are severely inhibited by salt stress, which also damages membranes, causes ion imbalances due to the accumulation of Na^+ and Cl^-, increases lipid peroxidation, and increases the production of reactive oxygen species such as superoxide radicals, hydrogen peroxide, and hydroxyl radicals (Shiferaw et al. 2011; Rasool et al. 2013). As population of the world is 7.75 billon till 2020 and by 2050 it will reach up to 9 billion (Manning 2015) which will continue worsen the current problems of global food insecurity. Whereas food demand of growing population will be increased by two folds till 2050 (Rosegrant et al. 2009). Although a large number of high-yielding varieties of field crops were developed over the past two to three decades yet they could not overcome the hurdles in production improvement due their vulnerability to biotic and abiotic stresses (Warkentin et al. 2015; Duc et al. 2015). Therefore, to cope with the demand of growing population there is need to increase the production and productivity of crop with challenging environment.

However, conventional breeding has absolutely dominated since the beginnings of plant domestication for developing crops and varieties that have assisted in the formation of modern societies (Fedoroff 2010). The global increase in food production is about 32 metric tonnes per year through improved agriculture practices and conventional breeding, which is 12 metric tonnes less than what is needed to meet the target of the Declaration of the World Summit on Food Security, which aims to acquire 70% more food by 2050 (Tester and Langridge 2010). However, due to intricate networks of plant response mechanisms against these stress, conventional breeding approaches become difficult for improvement (Sinclair 2011). After that another one strategy used to cope with climatic change is transgenic technology. However, it is still controversial in most countries due to political and moral issues (Chantre Nongpiur et al. 2016). But results so far obtained have been rather scarce due to the complexity of the response to salt and drought stress. Therefore, novel techniques are required for isolate and functionally characterization of gene that involved in salinity and drought tolerance. In this chapter, we focus on the genomics tools, viz., traditional QTL, MAS, transcriptomic analysis, association mapping, GWAS, NGSs, TILLING and EcoTILLING, RNAi, ZFNs, CRISPR/Cas, and other gene tagging approaches for genetically and functionally dissection of these complex traits. Confidently, the collaborative use of above-mentioned genomics tools can help in deep insights into the genetic and physiological mechanisms of tolerance to drought and salinity in crop plants, thus empowering the accepted schematic of future breeding programs.

10.2 Identification of Salt and Drought Stress Responsive Genes

For the development of salt and drought tolerant varieties, we must first have a comprehensive understanding of the complex mechanisms by which plants respond to salt and drought stress. Therefore, the first crucial step to achieving the required results is to identify the genes/QTLs implicated in salinity stress response that would be highly helpful to the breeder in order to create new drought and salt resistant cultivars (Dilnur et al. 2019). Effective genomic approaches that increase stress tolerance will be based on the discovery of novel genes, the characterization of their expression patterns in response to drought and salt stress, and a better understanding of their roles in stress adaption (Cushman and Bohnert 2000). Various approaches for gene discovery available are briefly described as follows.

10.2.1 Quantitative Trait Loci (QTL) Analysis

Tolerance to salinity and drought is multiplex phenomenon that is influenced by both genotypes of crops as well as by environment. Therefore, for the development of salt and drought resistant varieties there is a need to understand the genetic basis of underlying stress mechanism (Bizimana et al. 2017; Kamoshita et al. 2008). Quantitative trait locus (QTL) is location in DNA that is linked to a particular phenotypic

Table 10.1 QTLs (quantitative trait loci) associated with crop plant production under salinity and drought stresses

Stresses	Crop	Identified genes	References
Salinity and drought	Rice	Identified QTLs associated with salinity and drought tolerance at seedling stage	Kanjoo et al. (2011)
Salinity and drought	Barley	Two QTL for drought tolerance and one QTL for salinity tolerance	Fan et al. (2015)
Drought	Wheat	Detection of QTLs associated with yield and its components under drought	Rehman Arif et al. (2020)
Salinity	Soyabean	Salt tolerance QTL confers a large dominant effect over salt sensitivity	Hamwieh and Xu (2008)
Salinity	Wheat	Identified QTL associated with salinity tolerance in wheat	Díaz De León et al. (2011)

trait, which can be attributed to polygenic effects, i.e., is governed by two or more genes. These QTLs are often found on different chromosomes. On the other hand, quantitative trait locus (QTL) analysis is a statistical technique that is used to explain the genetic dissection of variation in quantitative traits by correlating their phenotypic data and genotypic data (Falconer and Mackay 1996; Kearsey 1998). Chromosomal mapping is used for localization of chromosomal region that significantly influence the variation of quantitative traits in a population. It assists in the identification of the genes that responsible for variation and to understand that how many numbers of QTL have a significant contribution to trait variation in a population (Zeng 2001). The insights of QTL mapping through marker-assisted breeding can assist to improve the genetic potential of crops. A well explained example as in case of rice for drought and salinity tolerance, where marker-assisted backcrossing was used to create chromosomal segment substitution lines (CSSL) of the KDML105 having QTLs for drought tolerance (DT) on chromosomes 1, 3, 4, 8, and 9 (Toojinda et al. 2011). Table 10.1 illustrates various instances of QTLs (quantitative trait loci) for improving crop production under saline and drought conditions.

10.2.2 Marker-Assisted Selection (MAS)

Marker-assisted selection (MAS) involves the use of molecular markers that are closely linked with agronomic traits of interest which is either controlled by individual genes or QTL. Molecular markers aids in selecting the parents having contrasting traits for improvement of new targeted traits (Monneveux et al. 2003). For the development of better hybrids, donor parents with good heritability and combining ability for drought tolerance-related traits should be used. The quality of molecular markers varies based on the ability to identify closeness and distant relationships among accessions. For example, restriction fragment length polymorphism (RFLP), amplified fragment length polymorphism (AFLP), and simple sequence repeat (SSR) markers have polymorphism and random amplified polymorphic DNA (RAPD) markers are simple, time-saving due to their rapidity, and require small amounts of

DNA. The advantages of SSR are that they are co-dominant, polymorphic, highly informative, easy to apply with relative high resolution, and less expensive for molecular biotechnology studies. Among the number of molecular markers used for crop genotyping, single nucleotide polymorphisms (SNP) markers are currently at the peak due to high abundance, ability to identify the polymorphism at single base level and can accommodate the whole genome with high throughput and high resolution compared to others.

Marker-assisted backcrossing (MABC) is used as the tool for selection of the best traits by the plant breeders that contain alleles with high recovery of recurrent parent genome. The use of marker-assisted backcrossing can exploit traits which have been difficult to exploit through conventional methods from one variety to another in a number of crops. MABC helps to identify tightly linked QTLs associated with traits of interest (Gill et al. 2017). They are transferred to the potential lines through marker-assisted backcrossing, involving foreground and background selection, thus reducing the breeding cycle and linkage drag as compared to a conventional approach. MAS utilizes novel sources of genetic materials provided by allelic variations in the gene of interest and can create, design, and develop a cheap molecular marker for salt tolerance (Rahman et al. 2008). MAS is the most promising and successful method for developing new salt-tolerant rice lines (Singh et al. 2016). Molecular marker studies have also been successfully used to deploy drought tolerance in sorghum through mapping of QTLs conferring drought tolerance. By applying MABC, Saltol QTLs was introgressed into a popular elite variety in different countries: Saltol QTLs was introgressed into two Bangladeshi varieties, BR11 and BRRI dhan28 (Huyen et al. 2012), AS996 and BT7 in Vietnam (Bimpong et al. 2016), Rassi in West Africa (Singh et al. 2011), and Pusa Basmati 1121 and PB6 in India (Chukwu et al. 2019; Geetha et al. 2017). In India, a multi-institutional program on the introgression of Saltol QTLs into elite mega-rice varieties is in progress (Shailani et al. 2021).

10.2.3 Transcriptomic Analysis

Transcriptomics is one of the branches of omics that is gaining wide acceptance in the field of crop improvement signifying its importance in the study of an entire set of transcripts (Piétu et al. 1999). All types of transcripts, including messenger RNAs (mRNAs), microRNAs (miRNAs), and different types of long noncoding RNAs (lncRNAs) come under Transcriptomics. Due to complex nature of abiotic stress induced responses, researcher's community had limited success in improving of salt or drought tolerance through conventional or transgenic breeding (Flowers 2004; Yamaguchi and Blumwald 2005). The use of high-throughput methods in modern transcriptomics has made possible to analyze the expression of diverse transcripts in different physiological or pathological conditions and widening our base of understanding of the relationships between the transcriptome and the phenotype. At transcriptional level large numbers of genes have been identified in different crops that display tolerance to various abiotic stresses. Large number of stress-inducible

Table 10.2 Identification of different TF families through transcriptome analysis relevant to salt and drought stress tolerance

Sr. no.	TF family	Crop	Target stress	References
1	ERF	Tomato, barley, rice, and wheat	Salt	Zhang et al. (2004), Jung et al. (2010), Xu et al. (2007)
2	WRKY	Maize	Salt	Li et al. (2013)
3	NAC	Melon	Salt	Wei et al. (2016)
4	AP2-EREBP	Chickpea	Drought	Garg et al. (2016)
5	bHLH	Chickpea	Drought	Garg et al. (2016)

genes has been identified by analyzing the transcript of those genes to improve the stress tolerance of transgenic plants (Umezawa et al. 2006). A total of 4954 and 5545 genes were found in drought tolerant and salinity-tolerant genotypes of chickpea, respectively, by comparative analysis of the transcriptomes at different development stages (Garg et al. 2016). Song et al. (2020) found a total of 7622 differentially expressed genes (DEGs) under salt stress, of which 4055 were upregulated by carrying out transcriptome analysis.

Transcription factors (TFs) are key regulators of stress tolerance mechanism and play important role in regulating the differential response of genotypes under stress conditions. The response of upland cotton under salinity stress was analyzed by comparative phenotypic and transcriptomic analysis and it was found that more TF genes were upregulated in stress tolerant than stress sensitive (Zhang et al. 2021). Recently, there are several reports (Rasheed et al. 2016; Takahashi et al. 2018; Nakaminami et al. 2018) deciphering the role of peptide hormones and small open reading frames (sORFs) in regulating the plant response under drought and salt stress. Also the importance of non-coding RNAs in mitigating the stress conditions has been extensively examined in recent years (Chekanova 2015). The role and example of various transcription families through transcriptome analysis relevant to salt and drought stress tolerance in different crops are discussed in Table 10.2.

10.2.4 Association Mapping (AM)

The high-resolution method of association mapping, which is based on the theory of linkage disequilibrium, has tremendous potential for the analysis of complex genetic characteristics. The size of the experimental population, the magnitude of the target allele effect, the density of markers used, the rate of LD decay between marker and target allele, errors in phenotyping and genotyping data, and the desired statistical significance level all influence the power of association studies. As a result of the development of the novel high-throughput genotyping and sequencing technologies, it is expected that this approach will be widely used in agricultural plants in the long term. Association mapping is a valuable tool for the detection of novel genes or QTLs of important field crops in regard of soil salinity and drought

Table 10.3 The association mapping for the sustainability of crop production under soil salinity and drought stresses

Crop	Stress	Major finding	References
Rice	Salinity	The association mapping strategy identified four loci significantly associated with salinity tolerance	Pushparajan et al. (2011)
Rice	Drought	Genomic characterization haplotype variation for potential use as molecular markers to combine drought avoidance and tolerance traits	Beena et al. (2021)
Barley	Drought	Association mapping identified 48 molecular markers that are significantly associated with 6 drought tolerance related traits	Abou-Elwafa (2016)
Cotton	Salanity	Among these eight marker loci, two (NAU2580-1 and NAU2580-2) were highly significantly associated with salt tolerance	Zhao et al. (2016)
Rice	Salinity	Two markers (Wn11463 and RM22418) showed significant association with salinity tolerance score	Emon et al. (2015)

stress. Soil salinity and drought are important abiotic stresses that affect crop productivity. Development of soil salinity and drought resistant cultivars are critical to reduce climate-related risk, maintain productivity, and enhance livelihood of crop growers. Certain accessions were found to have beneficial allele to improve traits, plant height, root length and spikelet fertility, which contribute to the grain yield under stress. On a collection of 107 barley accessions tested in both well-watered and drought-stressed conditions, association mapping was used to investigate the relationships between 76 SSR markers and 6 drought-related characteristics and revealed are significantly associated (Abou-Elwafa 2016). Eight marker loci were identified as being associated with salt tolerance in cotton, of which two loci (NAU2580-1 and NAU2580-2) accounts for the majority of the phenotypic variance in salt tolerance, indicating a substantial correlation between the marker NAU2580 and salt tolerance genes in cotton. Eight new STS markers based on three salt tolerance genes were developed, but only Wn11463,a marker for SKC1, was significantly associated with salt tolerance, suggested that STS marker Wn11463 would be helpful in MAS for rice salinity breeding (Emon et al. 2015). A detailed discussion on the association mapping for the sustainability of crop production in soil salinity and drought stress is available in Table 10.3.

10.2.5 Genome-Wide Association Mapping (GWAS)

The response of different plant species, to withstand in various environmental stresses is a complex phenomenon. Scientists all across the world are employing GWAS to identify the underlying molecular mechanism of such complicated traits. A large panel of diversified germplasm is typically used to give a high-resolution gene mapping method that is based on recombination events (Syed et al. 2020). Though the panel size is a big limitation of this study, however, several studies were carried out by many researchers on different crops such as cotton (Mahmood et al.

Table 10.4 Genome-wide association mapping for identifying QTLs under salinity and drought stresses

Stress	Crop	Major finding	References
Drought	Rice	Identified QTL containing candidate genes for drought	Hoang et al. (2019)
Salinity	Rice	Candidate genes can be identified by QTL	Kumar et al. (2015)
Drought	Chickpea	Identified candidate genes that can facilitate the fast development of drought tolerant varieties	Li et al. (2018)
Drought	Wheat	Identify genomic regions associated with seedling stage root architecture and shoot traits using GWAS	Thirunavukkarasu et al. (2014)
Drought	Wheat	Identify QTL associated with yield and yield components	Edae et al. (2014)
Drought and Salinity	Alfalfa	Identified alfalfa cultivars with enhanced resistance to drought and salt stresses	Yu et al. (2016)

2019), wheat (Cormier et al. 2014), and maize (Zaidi et al. 2016). In these genomic approaches by using cost-effective genome-wide DNA markers, whole genome exposure has been accomplished in two different crops. This approach has higher statistical power and resolution to identify significant QTLs to account for the broad phenotypic variability found when compared to a biparental mapping population. As in GWAS, the population size is larger, such that in most studies, single nucleotide polymorphisms (SNPs) have been used in genome-wide association studies to find a number of QTLs for abiotic stress, particularly drought and salinity tolerance. Common QTLs for traits linked to drought and salinity tolerance can be identified by employing meta-analyses (Abdelraheem et al. 2019). Details on genome-wide association mapping for identifying QTLs under salinity and drought stresses are discussed in Table 10.4.

10.2.6 New Generation Sequencing (NGS)

The NGS technologies has made a remarkable revolution in the field of DNA sequencing and greatly enhanced the progress of genome sequencing through use of robust, multitasking, and cost-effective technologies such as roche 454 sequencing, illumina, SOLiD, Ion torrent, oxford nanopore technology tec. Ability of this technology to detect large number of single nucleotide polymorphisms (SNPs) and insertions/deletions (InDels) has proved its potential in the domain of crop improvement (Huang et al. 2013; Varshney et al. 2009). Through the identification of large number of SNPs positions in the genome we are able to understand the genetic basis of phenotypic differences exhibited by different cultivars in response to various abiotic stresses, such as salinity and drought. Endophytes play important role in mitigation of salinity stress through induction of synthesis of antioxidants species and production of osmolytes, which help in overcoming the osmotic effect caused by stress factors (Barnawal et al. 2016). Therefore, NGS may hold great promise in understanding microbiome of both the epiphytic and endophytic (Verma et al. 2021). Tang et al. (2013) identified drought tolerant genes in *Populus euphratica* through the use of Roche's 454-GS FLX System. Illumina sequencing technology was applied to sequence a transcriptome of red clover (*Trifolium pratense* L.) for identification of genes that were responsible for drought tolerance. These genes were found to give three metabolites (pinitol, proline, and malate) the concentration of which increased under the influence of drought stress (Yates et al. 2014). Area of saline soil is increasing day by day in different part of the country and has negative impact on both quantity and quality so; large scale studies have been conducted deciphering the use of NGS to find genes responsible for salt tolerance. Wang et al. (2015) found genes responsible for copper tolerance through the application of Illumina technology and same technology was used by Le et al. (2011) in their study to explore the function of plant-specific NAC transcription factor family during development and dehydration stress in soybean. Both Roche and Illumina technologies were used to interrogate the transcriptome analysis of finger millet, for its potential in influencing the tolerance to salinity, drought, and diseases (Rahman et al. 2014). Juntawong et al. (2014) also found its calibre in elucidating molecular responses of Jatropha roots to waterlogging through transcriptome profiling. NGS has revolutionized the field of moleculars markers mining and identification and mapping of causal mutations within a short period at relatively low cost. Because of fast sequencing with less error, now it is possible to develop a large number of molecular markers specific to salinity and drought tolerance without using reference genomes, for example, in durum wheat (Trebbi et al. 2011) and common bean (Cortés et al. 2011). NGS technologies also have applications in the genome editing approaches to edit a specific gene related with tolerance to abiotic stresses as some examples given in Table 10.5.

Table 10.5 Identification of gene for salt and drought stress tolerance by different NGS technology

Sr. no	NGS based techniques/ approaches	Name of the gene(s)	Trait(s)	Crop/ species	References
1	De novo sequencing	GmCHX1	Salt stress	Wild soybean W05	Qi et al. (2014)
2	IlluminaHiSeq2000	XLOC_000635	Salt stress	Chickpea	Mahdavi et al. (2018)
3	IlluminaHiSeq2000	Ca_03175	Salt stress	Chickpea	Mahdavi et al. (2018)
4	MutMap	OsRR22	Salt tolerance	Rice	Papworth et al. (1996)
5	IlluminaHiSeq 2500	18,369 differently expressed genes	Drought tolerance	Lentil (*Lens culinaris*)	Singh et al. (2017)

10.2.6.1 TILLING and EcoTILLING

In genomics, the focus on functional genomics is increasing to deduce the function of a gene of interest. Reverse genetic techniques have been widely used to analyze the resulting phenotype from sequence data. Approaches such as site-directed mutagenesis, gene knockout, RNAi, and transposon tagging involve transgenic material, which is sometimes not feasible in case of many plant species. TILLING (Targeting Induced Local Lesions IN Genome) is a non-transgenic reverse genetics high-throughput approach that is applicable to all plant species which can be mutagenized, regardless of their pollinating system, ploidy level, or genome size. It focuses on identifying SNPs (single nucleotide polymorphisms) and/or INDELS (insertions/deletions) in a gene or genes of interest from a mutagenized population. This technique needs prior information related to DNA sequence and it takes benefits of a mismatch endonuclease to locate and detect induced mutations. Ultimately, it can provide an allelic series of silent, missense, nonsense, and splice site mutations to examine the effect of various mutations in a gene. TILLING has proven to be a practical, efficient, and an effective approach for functional genomic studies in numerous plant species. Ecotype TILLING (EcoTILLING), a variant of TILLING, analyses genetic variation naturally present in populations and has been successfully utilized in plants to discover SNPs including rare ones.

TILLING enables rapid identification of mutations in genes of interest from within a mutagenized population and has been applied to a wide range of crops including wheat, rice, maize, and sorghum in order to evaluate mutations in target genes. In cereals, several space-induced and EMS-induced mutant populations have been used to identify mutants with important traits including salinity tolerance, grain size, and recombinant crossovers via TILLING by sequencing (TbyS). Through EcoTILLING, allelic variation through identification of polymorphism and association analyses was found for resistance to drought in rice. A new durum wheat TILLING population has been developed by using ethyl methane sulfonate (EMS) with five target genes related to drought tolerance (DHN11; pTd27; DHNWWZ; RUBISCO; SNAC). The method of Ecotype Targeting Induced Local Lesions in Genomes (Ecotilling) was used to analyze the genes associated with drought tolerance level in rice. The study illuminated that association analysis aimed at Ecotilling diversity of natural groups could facilitate the isolation of rice and wheat genes related to complex quantitative traits (Table 10.6).

10.2.6.2 RNAi

The finding and discovery of RNAi was not systematic rather accidental when Jorgensen (1995) in an attempt for upregulation of activity of chalcone synthase for deep purple flower pigmentation in petunia found variegations and some cases of complete colour loss. This phenomenon was called as co-suppression. Later on research of Guo and Kemphues (1995), Mello and Fire (1995) established use of both sense and antisense strands was more effective in target gene silencing. This ended up being the turning point in RNAi research and the RNA interference phenomenon in drought and salinity tolerance. There are several miRNAs associated with drought responsiveness such as miR169, miR396, miR171, miR319, miR393,

Table 10.6 Identified gene for salt and drought stress tolerance by Tilling and EcoTilling

Species	Gene name	No. of genes	Trait	References
Oryza sativa	OSCP17	1	Salt tolerance	Negrao et al. (2011)
		19	Drought tolerance	Yu et al. (2012)
	osCPK17, osRMC, osNHX1, osHKTI; 5, SalT	5	Salt resistant	Negrao et al. (2013)
Durum wheat	DHN11; pTd27; DHNWWZ; RUBISCO; SNAC	5	Drought tolerance	El Yadini et al. (2021)

miR156, miR158, etc. (Younis et al. 2014; Liu et al. 2009a, b). Other miRNAs viz., miR159 in *Arabidopsis* germinating seeds involved in the cleavage of MYB101 and MYB33 transcripts induced by ABA or drought treatments (Reyes and Chua 2007; Abe et al. 2003). The plants overexpressing MYB101 and MYB33 were better osmotic stress tolerant and ABA hypersensitive. mi167 as a result to drought stress was found to be downregulated in rice seedings but upregulated in *Arabidopsis* (Liu et al. 2008, 2009a, b). In the event of drought stress, for necessary accumulation of Phospholipase D (PLD) the expression of miR167 was resultantly inhibited in maize (Ding et al. 2009; Wei et al. 2009). NFYA5 (a subunit of nuclear factor Y, NF-Y) known to play role in environmental stress responses was regulated by miR169. Over expression of NFYA5 lead to drought tolerant phenotypes whereas over expression of miR169 are very sensitive to drought stress (Li et al. 2008). Several others like miR393 targeting TIR1 (transport inhibitor response 1) were upregulated in *Arabidopsis* and rice (Dharmasiri and Estelle 2002) transgenic tobacco plant over expressing the miR396 improves drought stress tolerance by lowering the stomata density (Liu et al. 2009a, b; Yang and Yu 2009), etc. There were reports of several differentially regulated salt stress responsive miRNAs in plant like *Arabidopsis* such as miR156, miR158, miR159, miR319, miR393, miR394, miR396, miR397, etc., thus miRNAs play role in accordance response to salt stress. In rice during salt stress miR393 works in Auxin pathways by targeting the phytosulfokinne receptor (Liu et al. 2008). In response to salt stress, the genes miR2001, miR2004, and miR2006 that, respectively, target putative proteins (Protein GPR107 precursor), (EMB2745), and (exonuclease FAD binding domain containing protein) were increased. During salt stress, miR2003 targets a Heat repeat family protein that is produced and contains the ribosomal protein S11 (Jian et al. 2010). A number of miRNA are involved in both salt and drought stress. The *Arabidopsis thaliana* plant has miRNAs that are involved in both salt and drought conditions, including miR159, miR167, miR168, miR169, miR393, miR396, and miR397. Some miRNA have varied actions depending on the type of stress, for example, some miRNA are upregulated under drought stress whereas the same miRNA are downregulated under salt stress.

10.2.6.3 CRISPR/Cas9

CRISPR is a natural immune system of *Streptococcus pyogenes* against viruses. Cas9 protein cleaves out viral genome to prevent it from hijacking bacterial genome and causing disease, while CRISPR array saves a copy of the invading virus genome in the bacterial genome to thwart future viral attack. CRISPR/Cas9 falls under the bacterial-specific Class II division of the CRISPR/Cas system, which is further segmented into Classes I through III. This latest technology is used in drought and salinity stress tolerance. As, *SlLBD40* gene (LBD family), was knockout in tomato with reduced water loss from plants (Liu et al. 2020). Similarly, *SlARF4* (auxin response factor) knockout's leaves were able to stand upright even after wilting in comparisons to wild types tomato (Chen et al. 2021a, b). The OST2 gene, which codes for a plasma membrane H^+ ATPase involved in stomatal response, was knocked out through CRISPR/dCas9. This resulted in a high degree of stomatal closure with a low level of water loss, which improved *Arabidopsis*' ability to withstand drought (Osakabe et al. 2016). A key positive regulator of the drought stress response is AREB1/ABF2 (ABA-responsive element binding protein 1/ABRE binding factor). The histone acetyltransferase (HAT) from *Arabidopsis* improves chromatin relaxation and encourages gene expression. The promoter activity of AREB1 was significantly boosted by HAT coupled with dCas9 (Roca Paixão et al. 2019). OsERA1 mutant showed increased drought tolerance in rice as stomatal conductance reduced comparatively. As shown, multiple CRISPR/Cas9 alterations improved drought resistance (Shi et al. 2017).

The OsRR22 gene mutants perform better than wild types in 0.75% sodium chloride (NaCl) as compared to wild types as it performs in both cytokinin signal transduction and metabolism (Zhang et al. 2019). OsVDE in rice was found to be negatively regulating in salinity stress and its mutant had lower transpiration water loss because of enhanced stomatal closure and higher ABA content (Wang et al. 2021). A class of transcription factors called AITRs controls how plants react to abiotic stresses. There are six genes in *Arabidopsis* that produce AITRs. Using CRISPR/Cas9 to knock out any one of these genes resulted in increased tolerance to salt treatment, demonstrating that the complete family of AITR genes in *Arabidopsis* increases salinity tolerance (Chen et al. 2021a, b). TaHAG1 overexpressed (TaHAG1-OE) wheat plants had better root length, fresh weight, and less chlorotic leaves was demonstrated by CRISPR/Cas9 mutated seedlings (Zheng et al. 2021). Using CRISPR/Cas9 technology, a portion of the potato coilin gene that codes for the CTD was altered. The wild forms displayed rapid yellowing, leaf fall, and more severe root development inhibition under salt-stress conditions, demonstrating that potato coilin is implicated in the plant defense response to salinity (Makhotenko et al. 2019).

10.3 Conclusion

Ensuring present and future food security, agricultural production is biggest challenge in varying environment, especially in developing nations. Therefore, there is need to understand the genomic architecture of the plant in response to abiotic stresses, especially, salinity and drought. Due to the complexity nature of plant in response to salt and drought stress, traditional breeding methods are no longer effective ways to increase agricultural productivity. In this chapter we discussed about the genomic tools to understand the genetic dissection of plant in response to salt and drought stress tolerance that help in boosting up the crop production in adverse environment conditions. However, it should be noted that the primary function of genomics as a tool is to improve existing technologies, rather than its replacement. Finally, to develop a crops or varieties that are resilient to salinity and drought with high production, any two or all three genomics approaches with conventional breeding would be effective.

References

Abberton M, Batley J, Bentley A (2016) Global agricultural intensification during climate change: a role for genomics. Plant Biotechnol J 14:1095–1098

Abdelraheem A, Esmaeili N, O'Connell M, Zhang J (2019) Progress and perspective on drought and salt stress tolerance in cotton. Ind Crop Prod 130:118–129

Abe H, Urao T, Ito T, Seki M, Shinozaki K, Yamaguchi-Shinozaki K (2003) Arabidopsis AtMYC2 (bHLH) and AtMYB2 (MYB) function as transcriptional activators in abscisic acid signaling. Plant Cell 15:63–78

Abou-Elwafa SF (2016) Association mapping for drought tolerance in barley at the reproductive stage. C R Biol 339(2):51–59

Barnawal D, Bharti N, Tripathi A, Pandey SS, Chanotiya CS, Kalra A (2016) ACC-deaminase-producing endophyte Brachbacterium paraconglomeratum strain SMR20 ameliorates chlorophytum salinity stress via altering phytohormone generation. J Plant Growth Regul 35: 553–564

Beena R, Kirubakaran S, Nithya N, Manickavelu A, Sah RP, Abida PS, Siddique KH (2021) Association mapping of drought tolerance and agronomic traits in rice (Oryza sativa L.) landraces. BMC Plant Biol 21(1):1–21

Bimpong IK, Manneh B, Sock M, Diaw F, Amoah NKA, Ismail AM, Gregorio G, Singh RK, Wopereis M (2016) Improving salt tolerance of lowland rice cultivar 'Rassi'through marker-aided backcross breeding in West Africa. Plant Science 242:288–299

Bizimana JB, Luzi-Kihupi A, Murori RW, Singh RK (2017) Identification of quantitative trait loci for salinity tolerance in rice (Oryza sativa L.) using IR29/Hasawi mapping population. J Genet 96(4):571–582

Chantre Nongpiur R, Lata Singla-Pareek S, Pareek A (2016) Genomics approaches for improving salinity stress tolerance in crop plants. Curr Genomics 17(4):343–357

Chekanova JA (2015) Long non-coding RNAs and their functions in plants. Curr Opin Plant Biol 27:207–216

Chen S, Zhang N, Zhou G, Hussain S, Ahmed S, Tian H, Wang S (2021a) Knockout of the entire family of *AITR* genes in Arabidopsis leads to enhanced drought and salinity tolerance without fitness costs. BMC Plant Biol 21:137

Chen M, Zhu X, Liu X, Wu C, Yu C, Hu G, Chen L, Chen R, Bouzayen M, Zouine M (2021b) Knockout of auxin response factor SlARF4 improves tomato resistance to water deficit. Int J Mol Sci 22:3347

Chukwu SC, Rafii MY, Ramlee SI, Ismail SI, Oladosu Y, Okporie E, Jalloh M (2019) Marker-assisted selection and gene pyramiding for resistance to bacterial leaf blight disease of rice (*Oryza sativa* L.). Biotechnol Biotechnol Equip 33(1):440–455

Cormier F, Le Gouis J, Dubreuil P, Lafarge S, Praud S (2014) A genome-wide identification of chromosomal regions determining nitrogen use efficiency components in wheat (Triticum aestivum L.). Theor Appl Genet 127:2679–2693

Cortés AJ, Chavarro MC, Blair MW (2011) SNP marker diversity in common bean (Phaseolus vulgaris L.). Theor Appl Genet 123:827–845

Cushman JC, Bohnert HJ (2000) Genomic approaches to plant stress tolerance. Curr Opin Plant Biol 3(2):117–124

Dharmasiri S, Estelle M (2002) The role of regulated protein degradation in auxin response. Plant Mol Biol 49:401–409

Díaz De León JL, Escoppinichi R, Geraldo N, Castellanos T, Mujeeb-Kazi A, Röder MS (2011) Quantitative trait loci associated with salinity tolerance in field grown bread wheat. Euphytica 181(3):371–383

Dilnur T, Peng Z, Pan Z, Palanga KK, Jia Y, Gong W, Du X (2019) Association analysis of salt tolerance in Asiatic cotton (Gossypium arboretum) with SNP markers. Int J Mol Sci 20:2168

Ding D, Zhang L, Wang H, Liu Z, Zhang Z, Zheng Y (2009) Differential expression of miRNAs in response to salt stress in maize roots. Ann Bot 103:29–38

Duc G, Agrama H, Bao S, Berger J, Bourion V, De Ron AM, Zong X (2015) Breeding annual grain legumes for sustainable agriculture: new methods to approach complex traits and target new cultivar ideotypes. Crit Rev Plant Sci 34(1–3):381–411

Edae EA, Byrne PF, Haley SD, Lopes MS, Reynolds MP (2014) Genome-wide association mapping of yield and yield components of spring wheat under contrasting moisture regimes. Theor Appl Genet 127(4):791–807

El Yadini M, Guaadaoui A, Labhilili M, Bounimi S, Azeqour M (2021) Screening mutations in drought tolerance genes using simple method for tilling in durum wheat (Triticum turgidum L. subsp. durum (Desf.)). Plant Cell Biotechnol Mol Biol 22:131–142

Emon RM, Islam MM, Halder J, Fan Y (2015) Genetic diversity and association mapping for salinity tolerance in Bangladeshi rice landraces. Crop J 3(5):440–444

Falconer DS, Mackay TFC (1996) Introduction to quantitative genetics, 4th edn. Prentice Hall, London

Fan Y, Shabala S, Ma Y, Xu R, Zhou M (2015) Using QTL mapping to investigate the relationships between abiotic stress tolerance (drought and salinity) and agronomic and physiological traits. BMC Genomics 16(1):1–11

Fathi A, Tari DB (2016) Effect of drought stress and its mechanism in plants. Int J Life Sci 10:1–6

Fedoroff NV (2010) The past, present and future of crop genetic modification. New Biotechnol 27: 461–465

Flowers T (2004) Improving crop salt tolerance. J Exp Bot 55:307–319

Fróna D, Szenderák J, Harangi-Rákos M (2021) Economic effects of climate change on global agricultural production. J Nat Conserv 44:117

Garg R, Shankar R, Thakkar B (2016) Transcriptome analyses reveal genotype- and developmental stage-specific molecular responses to drought and salinity stresses in chickpea. Sci Rep 6:19228

Geetha S, Vasuki A, SeLvam PJ, Saraswathi R, Krishnamurthy SL, Dhasarathan M, Baskar M (2017) Development of sodicity tolerant rice varieties through marker assisted backcross breeding. Electron J Plant Breed 8(4):1013–1031

Gill MB, Zeng F, Shabala L, Zhang G, Fan Y, Shabala S, Zhou M (2017) Cell-based phenotyping reveals QTL for membrane potential maintenance associated with hypoxia and salinity stress tolerance in barley. Front Plant Sci 8:1941

Guo S, Kemphues KJ (1995) par-1, a gene required for establishing polarity in C. elegans embryos, encodes a putative Ser/Thr kinase that is asymmetrically distributed. Cell 81(4):611–620

Hamwieh A, Xu D (2008) Conserved salt tolerance quantitative trait locus (QTL) in wild and cultivated soybeans. Breed Sci 58(4):355–359

Hoang GT, Van Dinh L, Nguyen TT, Ta NK, Gathignol F, Mai CD, Jouannic S, Tran KD, Khuat TH, Do VN (2019) Genome-wide association study of a panel of Vietnamese rice landraces reveals new QTLs for tolerance to water deficit during the vegetative phase. Rice 12:1–20

Hossain A, Skalicky M, Brestic M, Maitra S, Ashraful Alam M, Syed MA, Islam T (2021) Consequences and mitigation strategies of abiotic stresses in wheat (Triticum aestivum L.) under the changing climate. Agronomy 11(2):241

Huang X, Lu T, Han B (2013) Resequencing rice genomes: an emerging new era of rice genomics. Trends Genet 29:225–232

Huyen LTN, Cuc LM, Ismail AM, Ham LH (2012) Introgression the salinity tolerance QTLs Saltol into AS996, the elite rice variety of Vietnam. Am J Plant Sci 3:981–987

Jian X, Zhang L, Li G, Zhang L, Wang X, Cao X, Fang X, Chen F (2010) Identification of novel stress-regulated microRNAs from Oryzasativa L. Genomics 95:47–55

Jorgensen RA (1995) Cosuppression, flower color patterns, and metastable gene expression states. Science 268(5211):686–691

Jung KH, Seo YS, Walia H, Cao P, Fukao T, Canlas PE, Amonpant F, Bailey Serres J, Ronald PC (2010) The submergence tolerance regulator Sub1A mediates stress-responsive expression of AP2/ERF transcription factors. Plant Physiol 152:1674–1692

Juntawong P, Sirikhachornkit A, Pimjan R, Sonthirod C, Sangsrakru D, Yoocha T, Tangphatsornruang S, Srinives P (2014) Elucidation of the molecular responses to waterlogging in Jatropha roots by transcriptome profiling. Front Plant Sci 5:658

Kamoshita A, Babu RC, Boopathi NM, Fukai S (2008) Phenotypic and genotypic analysis of drought-resistance traits for development of rice cultivars adapted to rainfed environments. Field Crops Res 109(1–3):1–23

Kanjoo V, Jearakongman S, Punyawaew K, Siangliw JL, Siangliw M, Vanavichit A, Toojinda T (2011) Co-location of quantitative trait loci for drought and salinity tolerance in rice. Genomics Genet 4(2):126–138

Kearsey MJ (1998) The principles of QTL analysis (a minimal mathematics approach). J Exp Bot 49:1619–1623

Kumar V, Singh A, Mithra SA, Krishnamurthy SL, Parida SK, Jain S, Tiwari KK, Kumar P, Rao AR, Sharma SK, Khurana JP (2015) Genome-wide association mapping of salinity tolerance in rice (Oryza sativa). DNA Res 22(2):133–145

Le DT, Nishiyama R, Watanabe Y, Mochida K, Yamaguchi-Shinozaki K, Shinozaki K, Tran LS (2011) Genome-wide survey and expression analysis of the plant-specific NAC transcription factor family in soybean during development and dehydration stress. DNA Res 18:263–276

Li WX, Oono Y, Zhu J, He XJ, Wu JM, Iida K, Lu XY, Cui X, Jin H, Zhu JK (2008) The Arabidopsis NFYA5 transcription factor is regulated transcriptionally and posttranscriptionally to promote drought resistance. Plant Cell 20:2238–2251

Li H, Gao Y, Xu H, Dai Y, Deng D, Chen J (2013) ZmWRKY33, a WRKY maize transcription factor conferring enhanced salt stress tolerances in Arabidopsis. Plant Growth Regul 70:207–216

Li Y, Ruperao P, Batley J, Edwards D, Khan T, Colmer TD, Pang J, Siddique KH, Sutton T (2018) Investigating drought tolerance in chickpea using genome-wide association mapping and genomic selection based on whole-genome resequencing data. Front Plant Sci 9:190

Liu HH, Tian X, Li YJ, Wu CA, Zheng CC (2008) Microarray-based analysis of stress-regulated microRNAs in Arabidopsis thaliana. RNA 14:836–843

Liu D, Song Y, Chen Z, Yu D (2009a) Ectopic expression of miR396 suppresses GRF target gene expression and alters leaf growth in Arabidopsis. Physiol Plant 136:223–236

Liu Q, Zhang YC, Wang CY, Luo YC, Huang QJ, Chen SY, Zhou H, Qu LH, Chen YQ (2009b) Expression analysis of phytohormone regulated microRNAs in rice, implying their regulation roles in plant hormone signaling. FEBS Lett 583:723–728

Liu L, Zhang J, Xu J, Li Y, Guo L, Wang Z, Zhang X, Zhao B, Guo YD, Zhang N (2020) CRISPR/Cas9 targeted mutagenesis of *SlLBD40*, a lateral organ boundaries domain transcription factor, enhances drought tolerance in tomato. Plant Sci 301:110683

Mahdavi MK, Garg V, Nasrollahnezhad Ghomi AA, Kudapa H, Chitikineni A, Zaynali Nezhad K et al (2018) RNASeq analysis revealed genes associated with drought stress response in kabuli chickpea (*Cicer arietinum* L.). PLoS ONE 13(6):e0199774. https://doi.org/10.1371/journal.pone.0199774

Mahmood T, Khalid S, Abdullah M, Ahmed Z, Shah MKN, Ghafoor A, Du X (2019) Insights into drought stress signaling in plants and the molecular genetic basis of cotton drought tolerance. Cell 9(1):105

Makhotenko AV, Khromov AV, Snigir EA, Makarova SS, Makarov VV, Suprunova TP, Kalinina NO, Taliansky ME (2019) Functional analysis of coilin in virus resistance and stress tolerance of potato *Solanum tuberosum* using CRISPR-Cas9 editing. Dokl Biochem Biophys 484:88–91

Manning DA (2015) How will minerals feed the world in 2050? Proc Geol Assoc 126(1):14–17

Mello C, Fire A (1995) DNA transformation. Methods Cell Biol 48:451–482

Monneveux P, Reynolds MP, Aguilar JG, Singh RP, Weber WE (2003) Effects of the 7DL. 7Ag translocation from Lophopyrum elongatum on wheat yield and related morphophysiological traits under different environments. Plant Breed 122(5):379–384

Nakaminami K, Okamoto M, Higuchi-Takeuchi M (2018) AtPep3 is a hormone-like peptide that plays a role in the salinity stress tolerance of plants. Proc Natl Acad Sci U S A 115(22): 5810–5815

Negrao S, Almadanim C, Pires I, McNally KL, Oliveira MM (2011) Use of EcoTILLING to identify natural allelic variants of rice candidate genes involved in salinity tolerance. Plant Genet Resour 9:300–304

Negrao S, Almadanim MC, Pires IS, Abreu IA, Maroco J, Courtois B, Gregorio GB, McNally KL, Oliveira MM (2013) New allelic variants found in key rice salt-tolerance genes: an association study. Plant Biotechnol J 11:87–100

Osakabe Y, Watanabe T, Sugano SS, Ueta R, Ishihara R, Shinozaki K, Osakabe K (2016) Optimization of CRISPR/Cas9 genome editing to modify abiotic stress responses in plants. Sci Rep 6(1):1–10

Papworth C, Bauer JC, Braman J (1996) QuikChange site-directed mutagenesis. Strategies 9:3–4

Pervez MA, Ayub CM, Khan HA, Shahid MA, Ashraf I (2009) Effect of drought stress on growth, yield and seed quality of tomato (Lycopersicon esculentum L.). Pak J Agric Sci 46(3):174–178

Piétu G, Mariage-Samson R, Fayein NA, Matingou C, Eveno E, Houlgatte R, Auffray C (1999) The Genexpress IMAGE knowledge base of the human brain transcriptome: a prototype integrated resource for functional and computational genomics. Genome Res 9(2):195–209

Pushparajan N, Krishnasamy V, Babu RC, Kannanbabu JR (2011) Association mapping of salinity tolerance in rice using molecular markers. Int J Bioresour Stress Manag 2(3):307–312

Qi X, Li MW, Xie M, Liu X, Ni M, Shao G, Lam HM (2014) Identification of a novel salt tolerance gene in wild soybean by whole-genome sequencing. Nat Commun 5:4340

Rahman M, Sun Z, McVetty PB, Li G (2008) High throughput genome-specific and gene-specific molecular markers for erucic acid genes in *Brassica napus* (L.) for marker-assisted selection in plant breeding. Theor Appl Genet 117(6):895–904. https://doi.org/10.1007/s00122-008-0829-9

Rahman H, Jagadeeshselvam N, Valarmathi R, Sachin B, Sasikala R, Senthil N, Sudhakar D, Robin S, Muthurajan R (2014) Transcriptome analysis of salinity responsiveness in contrasting genotypes of finger millet (Eleusine coracana L.) through RNA-sequencing. Plant Mol Biol 85: 485–503

Rasheed S, Bashir K, Nakaminami K (2016) Drought stress differentially regulates the expression of small open reading frames (sORFs) in Arabidopsis roots and shoots. Plant Signal Behav 11(8):e1215792. https://doi.org/10.1080/15592324.2016.1215792

Rasool S, Hameed A, Azooz MM, Siddiqi TO, Ahmad P (2013) Salt stress: causes, types and responses of plants. In: Ecophysiology and responses of plants under salt stress. Springer, New York, pp 1–24

Rehman Arif MA, Attaria F, Shokat S, Akram S, Waheed MQ, Arif A, Börner A (2020) Mapping of QTLs associated with yield and yield related traits in durum wheat (*Triticum durum* Desf.) under irrigated and drought conditions. Int J Mol Sci 21(7):2372

Reyes JL, Chua NH (2007) ABA induction of miR159 controls transcript levels of two MYB factors during Arabidopsis seed germination. Plant J 49:592–606

Roca Paixão JF, Gillet FX, Ribeiro TP, Bournaud C, Lourenço-Tessutti IT, Noriega DD, de Melo BP, de Almeida-Engler J, Grossi-de-Sa MF (2019) Improved drought stress tolerance in Arabidopsis by CRISPR/dCas9 fusion with a histone acetyltransferase. Sci Rep 9:8080

Rosegrant MR, Ringler C, Sulser TB, Ewing M, Palazzo A, Zhu T (2009) Agriculture and food security under global change: prospects for 2025/2050. International Food Policy Research Institute, Washington, DC

Shailani A, Joshi R, Singla-Pareek SL, Pareek A (2021) Stacking for future: pyramiding genes to improve drought and salinity tolerance in rice. Physiol Plant 172(2):1352–1362

Shi J, Gao H, Wang H, Lafitte HR, Archibald RL, Yang M, Hakimi SM, Mo H, Habben JE (2017) ARGOS8 variants generated by CRISPR-Cas9 improve maize grain yield under field drought stress conditions. Plant Biotechnol J 15:207–216

Shiferaw B, Prasanna BM, Hellin J, Bänziger M (2011) Crops that feed the world 6. Past successes and future challenges to the role played by maize in global food security. Glob Food Sec 3(3): 307–327

Sinclair TR (2011) Challenges in breeding for yield increase for drought. Trends Plant Sci 16:289–293

Singh AK, Gopalakrishnan S, Singh VP, Prabhu KV, Mohapatra T, Singh NK, Marathi B (2011) Marker assisted selection: a paradigm shift in basmati breeding. Indian J Genet Plant Breed 71(2):120

Singh R, Singh Y, Xalaxo S, Verulkar S, Yadav N, Singh S, Singh NK (2016) From QTL to variety-harnessing the benefits of QTLs for drought, flood and salt tolerance in mega rice varieties of India through a multi-institutional network. Plant Sci 242:278–287

Singh D, Singh CK, Taunk J, Tomar RSS, Chaturvedi AK, Gaikwad K, Pal M (2017) Transcriptome analysis of lentil (*Lens culinaris* Medikus) in response to seedling drought stress. BMC Genomics 18:206

Song Q, Joshi M, Joshi V (2020) Transcriptomic analysis of short-term salt stress response in watermelon seedlings. Int J Mol Sci 21(17):6036

Syed MA, Alam MA, Hossain A, Islam MR, Vemuri H, Jahan N (2020) Rice breeding and genomics approaches for improving water and nitrogen use efficiency. In: Roychoudhury A (ed) Rice research for quality improvement: genomics and genetic engineering. Springer, Singapore

Takahashi F, Suzuki T, Osakabe Y (2018) A small peptide modulates stomatal control via abscisic acid in long-distance signalling. Nature 556(7700):235–238. https://doi.org/10.1038/s41586-018-0009-2

Tang S, Liang H, Yan D, Zhao Y, Han X, Carlson JE, Xia X, Yin W (2013) Populus euphratica: the transcriptomic response to drought stress. Plant Mol Biol 83:539–557

Tester M, Langridge P (2010) Breeding technologies to increase. Science 327:818–822

Thirunavukkarasu N, Hossain F, Arora K, Sharma R, Shiriga K, Mittal S, Mohan S, Namratha PM, Dogga S, Rani TS, Katragadda S (2014) Functional mechanisms of drought tolerance in subtropical maize (Zea mays L.) identified using genome-wide association mapping. BMC Genomics 15(1):1–12

Tiwari R, Sheoran S, Rane J (2015) Wheat improvement for drought and heat tolerance. In: Shukla RS, Mishra PC, Chatrath R, Gupta RK, Tomar SS, Sharma I (eds) Recent trends on production strategies of wheat in India. Directorate of Wheat Research, Karnal, pp 39–58

Toojinda T, Siangliw JL, Punyawaew K, Kanjoo V (2011) Development of single QTL near isogenic lines (NILs) of KDML105 for drought tolerance. Research report. National Center for Genetic Engineering and Biotechnology, NSTDA, Pathum Tanee

Trebbi D, Maccaferri M, de Heer P, Sorensen A, Giuliani S, Salvi S, Sanguineti MC, Massi A, van der Vossen EAG, Tuberosa R (2011) High-throughput SNP discovery and genotyping in durum wheat (Triticum durum Desf.). Theor Appl Genet 123:555–569

Umezawa T, Fujita M, Fujita Y, Yamaguchi-Shinozaki K, Shinozaki K (2006) Engineering drought tolerance in plants: discovering and tailoring genes to unlock the future. Curr Opin Biotechnol 17(2):113–122

Varshney RK, Nayak SN, May GD, Jackson SA (2009) Next-generation sequencing technologies and their implications for crop genetics and breeding. Trends Biotechnol 27:522–530

Verma H, Kumar D, Kumar V, Kumari M, Singh SK, Sharma VK, Droby S, Santoyo G, White JF, Kumar A (2021) The potential application of endophytes in management of stress from drought and salinity in crop plants. Microorganisms 9:1729

Wang Y, Dong C, Xue Z, Jin Q, Xu Y (2015) De novo transcriptome sequencing and discovery of genes related to copper tolerance in Paeoniaostii. Gene 576:126–135

Wang X, Ren P, Ji L, Zhu B, Xie G (2021) OsVDE, a xanthophyll cycle key enzyme, mediates abscisic acid biosynthesis and negatively regulates salinity tolerance in rice. Planta 255:6

Warkentin TD, Smykal P, Coyne CJ, Weeden N, Domoney C, Bing D (2015) Pea (*Pisum sativum* L.). In: De Ron AM (ed) Grain legumes. Springer, New York, pp 37–83

Wei L, Zhang D, Xiang F, Zhang Z (2009) Differentially expressed miRNAs potentially involved in the regulation of defense mechanism to drought stress in maize seedlings. Int J Plant Sci 170: 979–989

Wei S, Gao L, Zhang Y, Zhang F, Yang X, Huang D (2016) Genome-wide investigation of the NAC transcription factor family in melon (Cucumis melo L.) and their expression analysis under salt stress. Plant Cell Rep 35:1827–1839

Xu ZS, Xia LQ, Chen M, Cheng XG, Zhang RY, Li LC, Zhao YX, Lu Y, Ni ZY, Liu L (2007) Isolation and molecular characterization of the Triticum aestivum L. ethylene-responsive factor 1 (TaERF1) that increases multiple stress tolerance. Plant Mol Biol 65:719–732

Yamaguchi T, Blumwald E (2005) Developing salt-tolerant crop plants: challenges and opportunities. Trends Plant Sci 10:615–620

Yang F, Yu D (2009) Overexpression of Arabidopsis miR396 enhances drought tolerance in transgenic tobacco plants. Acta Bot Yunn 31:421–426

Yates SA, Swain MT, Hegarty MJ, Chernukin I, Lowe M, Allison GG, Ruttink T, Abberton MT, Jenkins G, Skøt L (2014) De novo assembly of red clover transcriptome based on RNA-Seq data provides insight into drought response, gene discovery and marker identification. BMC Genomics 15:453

Younis A, Siddique MI, Kim CK, Lim KB (2014) RNA interference (RNAi) induced gene silencing: a promising approach of hi-tech plant breeding. Int J Biol Sci 10:1150–1158

Yu S, Liao F, Wang F, Wen W, Li J, Mei H, Luo L (2012) Identification of rice transcription factors associated with drought tolerance using the EcoTILLING method. PLoS One 7:e30765

Yu LX, Liu X, Boge W, Liu XP (2016) Genome-wide association study identifies loci for salt tolerance during germination in autotetraploid alfalfa (Medicago sativa L.) using genotyping-by-sequencing. Front Plant Sci 7:956

Zaidi PH, Seetharam K, Krishna G, Krishnamurthy L, Gajanan S, Babu R, Vivek BS (2016) Genomic regions associated with root traits under drought stress in tropical maize (Zea mays L.). PLoS One 11(10):e0164340

Zeng ZB (2001) QTL mapping. In: Maloy S, Hughes K (eds) Brenner's encyclopedia of genetics, 2nd edn. Academic, San Diego, pp 8–12

Zhang H, Huang Z, Xie B, Chen Q, Tian X, Zhang X, Zhang H, Lu X, Huang D, Huang R (2004) The ethylene, jasmonate, abscisic acid and NaCl responsive tomato transcription factor JERF1 modulates expression of GCC box-containing genes and salt tolerance in tobacco. Planta 220: 262–270

Zhang X, Lu G, Long W, Zou X, Li F, Nishio T (2014) Recent progress in drought and salt tolerance studies in Brassica crops. Breed Sci 64:60–73

Zhang A, Liu Y, Wang F, Li T, Chen Z, Kong D, Bi J, Zhang F, Luo X, Wang J (2019) Enhanced rice salinity tolerance via CRISPR/Cas9-targeted mutagenesis of the OsRR22 gene. Mol Breed 39:47

Zhang J, Zhang P, Huo X, Gao Y, Chen Y, Song Z, Zhang J (2021) Comparative phenotypic and transcriptomic analysis reveals key responses of upland cotton to salinity stress during postgermination. Front Plant Sci 12:639104

Zhao YL, Wang HM, Shao BX, Chen W, Guo ZJ, Gong HY et al (2016) SSR-based association mapping of salt tolerance in cotton (Gossypium hirsutum L.). Genet Mol Res 15(2):gmr-15027370

Zheng M, Lin J, Liu X, Chu W, Li J, Gao Y, An K, Song W, Xin M, Yao Y (2021) Histone acetyltransferase TaHAG1 acts as a crucial regulator to strengthen salt tolerance of hexaploid wheat. Plant Physiol 186:1951–1969

Strategies to Improve Drought and Salinity Tolerance in Some Cash Crops Through Proteomics Perspective

11

Monika Sihmar, Renu Yadav, and Adhini S. Pazhany

Abstract

Abiotic stresses particularly drought and salinity are the major cause of crop loss all over the world. Abiotic stresses are expected to rise due to changing climatic conditions driven by anthropogenic activities. In the current scenario, it is utmost important that we reduce these crop losses and enhance the production to meet the increasing food demands. Therefore, the development of more tolerant new varieties is important to ensure food security to feed the world population. Proteomics is a powerful analytical tool for the investigation of crop response towards stress. Advancement in proteomic techniques allows the identification and quantification of stress responsive proteins, their expression analysis and translational study followed by protein modification. A study of collective information of differentially expressed proteins is important to elucidate the molecular mechanisms acting behind stress tolerance mechanism. Proteomic techniques clarify the tolerance mechanisms under major abiotic stress, and thus helpful in the development of abiotic stress tolerant varieties. Through proteomics based functional analyses we can characterize the relevant traits for the crop breeding and genetic engineering programs for the production of next generation crops.

Keywords

Abiotic stress · Drought · Salinity · Proteomics · Tolerant crops

M. Sihmar
Maharishi Dayanand University, Rohtak, India

R. Yadav (✉)
Central University of Haryana, Mahendragarh, India

A. S. Pazhany
ICAR – Sugarcane Breeding Institute, Coimbatore, India

A. Kumar et al. (eds.), *Salinity and Drought Tolerance in Plants*,
https://doi.org/10.1007/978-981-99-4669-3_11

11.1 Introduction

Stress, either biotic or abiotic in crops, is one of the major problems of agriculture worldwide. Abiotic stress confers to the environment in which plant resides; it can be physical or chemical whereas biotic stress occurs when plant get affected by biological unit like insect, birds, etc. (Gull et al. 2019). Abiotic stresses not only affect the growth and development of crops but also result in their yield loss. There are various common abiotic stresses faced by crops such as drought, salinity, temperature fluctuation (high or low), mineral, and waterlogging (Sharma et al. 2019). Crops generally counteract stresses through two basic strategies, i.e., one is stress avoidance and the other is stress tolerance. Among abiotic stresses, drought and salinity play major role in yield loss of crops. Salinity is one of the major contributing factors which spoils soil, retards plant growth and productivity in arid as well as semi-arid areas. It decreases the area of cultivable area (Singh et al. 2015; Li et al. 2015). The salt affected area is increasing day by bay; in current status, more than 6% of the world's total land area is affected by salinity (Munns and Tester 2008). Salinity imparts negative impact on overall growth and development of all the major crops. It results in the development of osmotic stress due to high salt concentration in the outside SAP. It also generates ion toxicity, one common way is through accumulation of access of salts on the transpiring leaves (Acosta-Motos et al. 2017). It affects both qualitative and quantitative traits of the crop. It is well-known fact that about 90% of agricultural land is affected by abiotic stresses (Barkla et al. 2013). Mantri et al. (2012) reported that the compromised plant growth and substantial yield loss had been obtained because of abiotic stresses. Due to abiotic stress, genes exhibit altered levels of expression. This altered expression levels can be analyzed through proteomics and could be used for breeding and gene editing programs in the creation of crop plants with better resistance capacity towards abiotic stresses (Raza et al. 2021). Stress proteomics is the study of proteins in plants which encounters the abiotic stresses such as temperatures (cold, frost, and heat), drought, excessive watering (waterlogging/flooding), salinity (sodicity), minerals (heavy metal and metalloid), disease attack (by fungi, bacteria, parasites, viruses, and insects), and weeds. Stress proteomic studies allow us to identify the stress-associated genes, their regulation, and function and are therefore helpful for crop improvement. Approximately 50% yield loss was reported because of abiotic stresses (heat, cold, drought, salinity, nutrient, and heavy metals) (Kajla et al. 2015). Proteomics aims the study of complete set of proteins in a sample at a given time and specific conditions, and it offers itself as universal analysis tool for any alteration or change in protein and proves itself as an effective method over transcriptomics for assessing the role of stress-related proteins (Jensen 2006).

Plants are necessarily adapted to an environment where stresses are present, and it is controlled by a cascade of various cellular responses and eventually expressing the proteins or set of proteins that protect them from damage (Ramanjulu and Bartels 2002). Different strategies are being employed for the development of tolerant crops. In stress tolerance, various factors play important role such as expression of various genes and proteins, several transcription factors, molecular chaperons, histone-modifying enzymes, osmolytes, and many metabolites. Proteomics means the study of proteins in all aspects enabled to develop tolerant crops with high efficiency. The focus of this present article is on proteomic techniques that strengthen the tolerance mechanisms of crops under major abiotic stress (salinity and drought), and thus helpful in the development of abiotic stress tolerant varieties. Cash crops are the agricultural crops which are grown for economic purposes by the farmers. These crops differ from subsistence crops in view that these are cultivated for market commodities. The cash crops include almost all crops as most of the plants grown nowadays are for revenue such as cereals, vegetables, fruits, oil-producing crops, etc. (Sturgeon 2013).

11.2 Protein Profiling of Various Plants During Salinity Stress and Drought Stress

The wide range of salinity induced effects has been reported on crop plants including decrease in transpiration and photosynthesis rate, stomatal conductance, and relative water content (Marriboina et al. 2017). The comparative proteomic analysis of different crop plants under salinity stress leads to the identification of numerous salt-responsive proteins, which has specific role in important biological processes such as photosynthesis, antioxidant, oxidation, stress and defense, energy metabolism, ion homeostasis, metabolism (carbohydrate, nitrogen and protein) signal transduction, energy, cell growth and cell structure, protein synthesis and storage, transporters, redox regulation, disease/defense-related processes, transcription, ATP synthesis, etc. (Fig. 11.1).

The salt stress response has been comprehensively studied by Xu et al. (2015). They identified 56 differentially expressed proteins (DEP's) after salinity treatment to rice seedlings. They isolated and identified various proteins involved in different biological processes such as photosynthesis, transcription, transcription factors, oxidative phosphorylation, antioxidant activity, carbon and nutrients, IAA, signal transduction, protein synthesis, protein degradation, nucleosome, actin regulation, biosynthesis, metabolism, etc. Functional categorization of these differentially expressed proteins has also been extensively studied (Xu et al. 2015). Under salt stress, it is necessary to protect photosynthetic apparatus as it is important for the growth of the plant. The photosynthetic apparatus of the plant generally includes RuBisCO activase, and chlorophyll-binding proteins, which regulates light reaction, photosynthesis-related processes, and CO_2 fixation. Many studies reported increased expression of RuBisCO activase and ATPase in leaves of salt stressed plants (Chen et al. 2015). Numerous proteins were upregulated and downregulated after stress treatment in various plants, and conflicting results were also found in many studies. The proteins 40s ribosomal protein, abscisic acid, actin, aspartate aminotransferase, ATP synthase, ATPase, catalase, dehydroascorbate reductase, dihydrolipoyl dehydrogenase, enolase, fasciclin-like arabinogalactan protein 2,3,4,6,11,13, ferritin, fructokinase, fructose-bisphosphate aldolase, glutamine synthetase, glutathione-*S*-transferase, glyceraldehyde 3-phosphate dehydrogenase, histone proteins (h1, h2a,

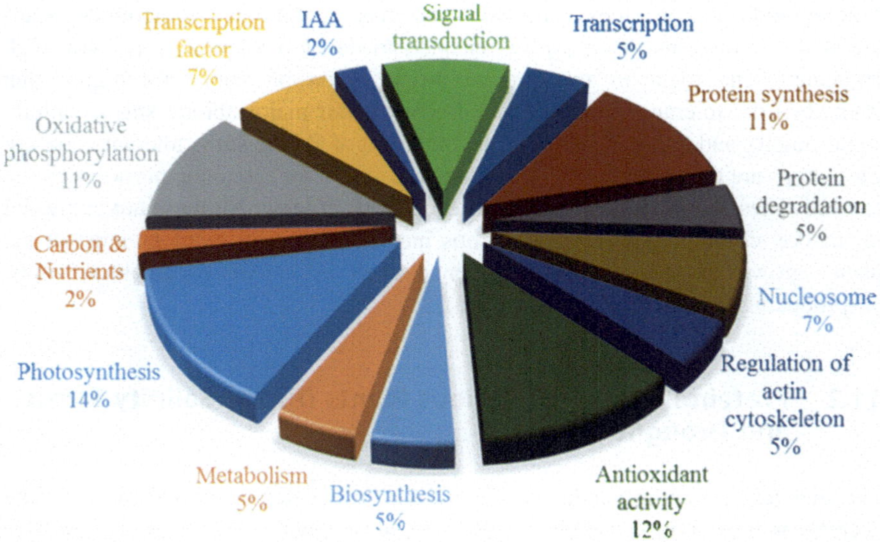

Fig. 11.1 Functional classification of differentially expressed proteins in rice under salt stress. The percentage for each process in which protein is involved is shown and represented in the Pie-chart (Xu et al. 2015)

h2b, h3, h4), iron deficiency-induced protein idi1, idi2, ids2, ids3, L-ascorbate peroxidase, lipoxygenase, malate dehydrogenase, monodehydroascorbate reductase, pathogenesis-related protein, peroxidase, phosphoenolpyruvate carboxykinase, phosphoglycerate kinase, ribosomal protein (s19, s4, s2, 119, l2, l3), rubisco large chain, rubisco small chain, succinate dehydrogenase, transketolase, ubiquitin are generally shows differential expression in plants during salt stress and are involved in a number of biological processes (Parker et al. 2006; Shen et al. 2017; Wu et al. 2014; Kim et al. 2005; Yousuf et al. 2017; Xu et al. 2015; Cheng et al. 2016; Sugimoto and Takeda 2009). Rubisco (large chain) is one of the most important proteins which helps in the process of photosynthesis that occurs inside the plant cell, it got upregulated during salt stress. Increase in the concentration of rubisco protein under salt stress is essential for plant so that plants can tolerate stress at low CO_2 levels (Acosta-Motos et al. 2017). For carboxylation of CO_2 at low CO_2 levels, a rise in the level of rubisco protein (large subunit) is beneficial. Ferritin protein also got upregulated during salt stress, these proteins can store iron atoms, and are employed for the regulation of iron atoms in the chloroplast, so that minimize the effect of iron toxicity. Hydroxyl radicals are one of the most potent ROS produced by the cell because of oxidative stress; its formation in the cell could be reduced by ferritin (Parker et al. 2006). Various proteins expression is different in different plants such as catalase shows upregulation in wheat while downregulation in barley (Witzel et al. 2009; Woodrow et al. 2017); dihydrolipoyl dehydrogenase shows upregulation in sugarcane while downregulation in barley and cotton (Passamani et al. 2017; Wu et al. 2014; Li et al. 2015); glutamine synthetase shows upregulation in mustard and wheat while downregulation in barley (Yousuf et al. 2017; Woodrow et al. 2017; Wu et al. 2014); glyceraldehyde 3-phosphate dehydrogenase shows upregulation in mustard and sugarcane while downregulation in cotton (Yousuf et al. 2017; Passamani et al. 2017; Cheng et al. 2016); monodehydroascorbate reductase shows upregulation in barley while downregulation in cotton (Li et al. 2015; Wu et al. 2014). Increase in the rate of photosynthesis may be related with the upregulation of ATP synthase because during salt stress plants need secondary transport mechanism which increases ATP synthesis in stressed plants (Parker et al. 2006; Shen et al. 2016; Wu et al. 2014; Kim et al. 2005; Yousuf et al. 2017). Various differentially expressed proteins under salinity in different cash crops are listed in Table 11.1.

Among abiotic stresses, drought is one of the primary yield limiting factors which reduces yield by reducing starch biosynthesis as enzymes involved in this process are also diminished. The stage of the plant determines the effect of drought stress on crop yield. Seed germination stage is the most affected stage by water deficiency. Drought stress affects both yield and quality during grain filling, and at flowering stage rate of photosynthesis decreases, which leads to the reduction in grain number and weight. It also promotes the accumulation of toxic metabolites in the plant. Drought stress is the most deleterious stress among all because it affects approximately every crop all over the world. Drought stress is not only the major problem in the arid and semi-arid areas but also in irrigated regions as a consequence of limited water supply in these areas (Sadaqat et al. 2003). As in other abiotic stresses, plants also respond towards drought stress by differentially expressing numerous proteins which help in adaptation to the changing environment and therefore provides tolerance against the stress. These are the most common differentially expressed protein in the plant during drought stress, 14–3–3 Protein, 40s ribosomal protein, actin, alcohol dehydrogenase, ATP synthase, beta-1,3-glucanase, caffeoyl-CoA *O*-methyltransferase, catalase, chaperone protein, citrate synthase, Cu-Zn superoxide dismutase, cytochrome b6-f complex iron-sulfur subunit, dehydroascorbate reductase, ferredoxin-NADP reductase, glutamine synthetase, glutathione-*S*-transferase, glutelin, glyceraldehyde 3 phosphate dehydrogenase, heat shock protein, lipoxygenase, malate dehydrogenase, NADH dehydrogenase, oxygen-evolving enhancer protein, phosphoglycerate kinase, ribosomal proteins, ribulose bisphosphate carboxylase large chain, rubisco small subunit protein, *S*-adenosylmethionine synthetase family protein, tubulin (Hao et al. 2015; Zadraznik et al. 2017; Faghani et al. 2015; Shi et al. 2014; Oh and Komatsu 2015; Koh et al. 2015; Ashoub et al. 2015).

ATP synthase, an enzyme responsible for ATP synthesis shows upregulation during drought stress as ATP synthesis process was known to be considerably affected by water stress so to overcome these stressed conditions overproduction of this protein occurs inside the cell. Many studies suggested that there is an upregulation of this protein under the stressed condition in tolerant lines while it shows downregulation in sensitive lines of the same plant. It happens because this enzyme can maintain ion homeostasis and energy production (Zadraznik et al. 2017; Shi et al. 2014; Ashoub et al. 2015; Faghani et al. 2015). A list of differentially

Table 11.1 A summary of the main results of differential-expression proteins under salt stress in conventional cash crops

Name of protein	Plant	Process involved	References
40S ribosomal protein	Wheat, cotton	Translation	Jiang et al. (2017), Li et al. (2015)
Abscisic acid	Rice, Maize, Tomato	Stress and defense	Xu et al. (2015), Gurmani et al. (2013), Zörb et al. (2013), Amjad et al. (2014)
Actin	Rice, Barley	Cell organization	Xu et al. (2015), Dermendjiev et al. (2021)
Aspartate aminotransferase	Cotton, Barley, Canola	Photosynthesis	Cheng et al. (2016), Wu et al. (2014), Wang et al. (2021a, b)
ATP synthase	Cotton, Barley, Mustard, Tomato	ATP biosynthesis	Shen et al. (2016), Wu et al. (2014), Yousuf et al. (2017), Vaishnav et al. (2020)
ATPase	Cotton, Barley	ATP synthesis	Cheng et al. (2016), Wu et al. (2014)
Catalase	Wheat, Onion	Redox regulation	Woodrow et al. (2017), Chaudhry et al. (2021)
Dehydroascorbate reductase	Barley, Rice	Oxidative stress	Shen et al. (2016), Kim et al. (2022)
Dihydrolipoyl dehydrogenase	Sugarcane, Barley, Cotton	Redox homeostasis	Passamani et al. (2017), Wu et al. (2014), Li et al. (2015)
Enolase	Cotton, Soybean	Carbohydrate metabolism	Cheng et al. (2016), Li et al. (2015), Liu et al. (2021)
Fasciclin-like arabinogalactan protein 2,3,4,6,11,13	Cotton	Cytoskeleton metabolism	Li et al. (2015), Jiang et al. (2017)
Ferritin	Rice	Chlorophyll metabolism	Xu et al. (2015)
Fructokinase	Cotton, mustard	Carbohydrate metabolism	Cheng et al. (2016), Yousuf et al. (2017)
Fructose-bisphosphate aldolase	Rice, Maize	Carbohydrate metabolism	Cheng et al. (2016), Araujo et al. (2021)
Glutamine synthetase	Mustard, Wheat, Barley	Nitrogen metabolism	Yousuf et al. (2017), Woodrow et al. (2017), Wu et al. (2014)
Glutathione-*S*-transferase	Cotton, Sugarcane, Wheat	Stress and defense	Li et al. (2015), Passamani et al. (2017), Hao et al. (2021)
Glyceraldehyde 3-phosphate dehydrogenase	Mustard, Sugarcane, Cotton	Carbohydrate metabolism	Yousuf et al. (2017), Passamani et al. (2017), Cheng et al. (2016)
Histone proteins (H1, H2A, H2B, H3, H4)	Wheat	Transcription	Jiang et al. (2017), Xu et al. (2015)

(continued)

Table 11.1 (continued)

Name of protein	Plant	Process involved	References
L-ascorbate per Oxidase	Mustard, Cotton	Stress and defense	Yousuf et al. (2017), Li et al. (2015)
Lipoxygenase	Sugarcane, Wheat	Energy metabolism	Passamani et al. (2017), Menga and Trono (2020)
Malate dehydrogenase	Mustard, Cotton	Carbohydrate metabolism	Yousuf et al. (2017), Cheng et al. (2016)
Monodehydroascorbate reductase	Barley, Cotton	Stress and defense	Li et al. (2015), Wu et al. (2014)
Pathogenesis-related protein	Barley, Wheat, Mustard, Sugarcane	Stress and defense	Passamani et al. (2017), Wu et al. (2014), Jiang et al. (2017), Yousuf et al. (2017)
Peroxidase	Cotton	Stress and defense	Cheng et al. (2016), Li et al. (2015)
Phosphoenolpyruvate carboxykinase	Cotton, Wheat	Carbohydrate metabolism	Cheng et al. (2016), Woodrow et al. (2017)
Phosphoglycerate kinase	Cotton, Barley	Photosynthesis	Cheng et al. (2016), Wu et al. (2014)
Ribosomal protein (S19, S4, S2, L19, L2, L3)	Mustard, Wheat	Translation	Yousuf et al. (2017), Xu et al. (2015), Jiang et al. (2017)
Rubisco large chain	Rice, Cotton, mustard	Photosynthesis	Kim et al. (2005), Cheng et al. (2016), Yousuf et al. (2017)
Rubisco small chain	Barley, Rice	Photosynthesis	Wu et al. (2014), Roy Choudhury et al. (2022)
Succinate dehydrogenase	Cotton	Stress and defense	Li et al. (2015)
Transketolase	Mustard	Carbohydrate Metabolism	Yousuf et al. (2017)
Ubiquitin	Mustard, Cotton	Energy metabolism	Yousuf et al. (2017), Li et al. (2015)
Vacuolar processing enzyme	Rice	Cell death	Kim et al. (2014)

expressed proteins under drought stress in different cash crops is given in Table 11.2. Plants differ in their extent of tolerance as different crop can tolerate stress to a different extent such as barley can tolerate salt and drought stress to a higher level than other crops; wheat is a moderately tolerant (Zeeshan et al. 2020). Above information regarding different differentially expressed proteins clearly shows that the abundance of same protein in crop under stress can be observed in increased or decreased state.

Table 11.2 A summary of the main results of differential-expression proteins under drought stress in conventional cash crops

Name of protein	Plant	Process involved	References
14–3–3 Protein	Wheat, Common bean	Signal transduction	Hao et al. (2015), Zadraznik et al. (2017), Faghani et al. (2015), Shi et al. (2014)
40S ribosomal protein	Common bean, Rice	Translation	Zadraznik et al. (2017), Hamzelou et al. (2020)
Actin	Bermudagrass, Soybean	Cell organization	Shi et al. (2014), Oh and Komatsu (2015)
Alcohol dehydrogenase	Common bean, Groundnut	Energy metabolism	Zadraznik et al. (2017), Jacob et al. (2022)
ATP synthase	Common bean, Bermudagrass, Barley, Wheat, Tomato	Photosynthesis	Zadraznik et al. (2017), Shi et al. (2014), Ashoub et al. (2015), Faghani et al. (2015), Yang et al. (2020)
Beta-1,3-glucanase	Wheat, Rice	Cell organization	Faghani et al. (2015), Wang et al. (2021a, b)
Caffeoyl-CoA *O*-methyltransferase	Wheat, Tobacco	Energy metabolism	Hao et al. (2015), Zhao et al. (2021a, b)
Catalase	Common bean, *B. napus*	Stress and defense	Zadraznik et al. (2017), Koh et al. (2015)
Chaperone protein	Common bean, Wheat, *B. napus*	Stress and defense	Zadraznik et al. (2017), Hao et al. (2015), Koh et al. (2015)
Citrate synthase	Common bean, Maize	Energy metabolism	Zadraznik et al. (2017), Hu et al. (2020)
Cu-Zn Superoxide dismutase	Wheat, Bermudagrass	Stress and defense	Faghani et al. (2015), Shi et al. (2014)
Cytochrome b6-f complex iron-sulfur subunit	Bermudagrass, Wheat	Photosynthesis	Shi et al. (2014), Faghani et al. (2015), Kamal et al. (2013)
Dehydroascorbate reductase	Wheat, Rice	Stress and defense	Hao et al. (2015), Melandri et al. (2020)
Ferredoxin-NADP reductase	Bermudagrass, Wheat, Barley	Amino acid synthesis	Ashoub et al. (2015), Faghani et al. (2015), Shi et al. (2014)
Glutamine synthetase	Barley, Bermudagrass	Amino acid synthesis	Ashoub et al. (2015), Shi et al. (2014)
Glutathione-*S*-transferase	Barley, Wheat, Soybean	Stress and defense	Oh and Komatsu (2015), Faghani et al. (2015)
Glutelin	Common bean, Barley	Storage	Zadraznik et al. (2017), Ye and Zhang (2020)
Glyceraldehyde 3 phosphate dehydrogenase	Soybean, Bermudagrass, Common bean, Wheat, *B. napus*	Carbohydrate metabolism	Oh and Komatsu (2015), Shi et al. (2014), Zadraznik et al. (2017), Faghani et al. (2015), Koh et al. (2015)
Heat shock protein	Bermudagrass, Common bean, Barley	Stress and defense	Shi et al. (2014), Zadraznik et al. (2017), Ashoub et al. (2015), Hao et al. (2015)

(continued)

Table 11.2 (continued)

Name of protein	Plant	Process involved	References
Lipoxygenase	Common bean, Soybean	Energy metabolism	Zadraznik et al. (2017), Oh and Komatsu (2015)
Malate dehydrogenase	Wheat, Bermudagrass, *B. napus*	Carbohydrate metabolism	Faghani et al. (2015), Shi et al. (2014), Koh et al. (2015)
NADH dehydrogenase	Bermudagrass, Common bean	Photosynthesis	Shi et al. (2014), Zadraznik et al. (2017)
Oxygen-evolving enhancer protein	Bermudagrass, Wheat, *B. napus*	Photosynthesis	Shi et al. (2014), Hao et al. (2015), Koh et al. (2015)
Phosphoglycerate kinase	Bermudagrass, Common bean	Carbohydrate metabolism	Shi et al. (2014), Zadraznik et al. (2017)
Ribosomal proteins	Common bean, Soybean	Translation	Zadraznik et al. (2017), Oh and Komatsu (2015)
Ribulose bisphosphate carboxylase large chain	Bermudagrass, *B. napus*, Wheat, Barley, Wheat, Common bean	Photosynthesis	Koh et al. (2015), Faghani et al. (2015), Ashoub et al. (2015), Hao et al. (2015), Shi et al. (2014), Zadraznik et al. (2017)
RuBisCO small subunit protein	Bermudagrass, *B. napus*, Wheat	Photosynthesis	Shi et al. (2014), Faghani et al. (2015), Koh et al. (2015)
S-adenosylmethionine synthetase family protein	Barley, Soybean, Wheat	Amino acid synthesis	Ashoub et al. (2015), Oh and Komatsu (2015), Faghani et al. (2015)
Tubulin	Common bean, Wheat	Cell organization	Zadraznik et al. (2017), Hao et al. (2015)

11.3 Drought and Salinity Tolerance Improving Strategies in Crops Through Proteomics Approach

The proteomic analysis of various crops has been done to elucidate the proteins involved after stress response in various crops that underlies salinity and drought stress. A number of different proteomic tools including basic to advanced have been used in the proteomic studies such as One dimensional Gel Electrophoresis (1-DE), Two dimensional Gel Electrophoresis (2-DE), Diagonal gel electrophoresis, Difference Gel electrophoresis (DIGE), Isobaric Tags for Relative and Absolute Quantitation (iTRAQ), Tandem Mass Tags (TMT), Stable Isotope Labeling by/with Amino acids in Cell culture (SILAC), Mass spectrometry based Matrix-assisted laser desorption/ionization (MS-MALDI), Electrospray ionization (ESI)—Liquid chromatography–mass spectrometry (LC-MS), hyphenated mass spectrometry technique (LC-ESI-MS/MS) to identify change in the abundance of proteins during salinity and drought stress.

The conventional breeding tools have limited success in the development of tolerant crops therefore new advanced omics strategies are gaining importance. The omics studies such as proteomics, transcriptomics, and metabolomics offer numerous advantages which are being used to decipher and understand the stress tolerance mechanism at molecular level. The use of proteomics-based markers can serve as an efficient selection tool over phenotype-based selection (Kumar et al. 2022). A combination of omics (genomics, proteomics, and metabolomics) approaches with bioinformatic tools can provide an all-inclusive understanding of the biological architecture of crops under stress. The molecular mechanisms of defense responses of the crop plant underlying stress condition regulated by changes in gene transcription level, cellular protein, and metabolite profile level. This deep insights of changes of proteome level as per stress conditions will thereby provide a better understanding for the development of stress tolerant crops (Dey et al. 2022). Current progressions in the plant proteomics specifically stress proteomics have led to the identification of numerous potential genes and proteins for the development of drought and salt-tolerant crops (Gupta et al. 2022). The differentially expressed proteins under stressed condition plays key role in tolerance and their elucidation led to the development of tolerant crops. In this chapter a list of differently abundant proteins has been provided under drought and salinity stress.

11.4 Conclusion and Future Prospective

Proteomic analysis on various common cash crops under different abiotic stresses unveiled important information on stress responsive proteins. There are numerous abiotic stresses, i.e., cold, heat, drought, waterlogging, salinity, ozone, inadequate or excessive light, irregularity in mineral nutrition, hypoxia, anoxia, heavy metals ion concentration, herbicide treatment, presence of radioactivity, mechanical stress (wounding) but cold, heat, drought, salinity, heavy metals ion concentration, and waterlogging stress which affects the plants. Plant growth adversely got affected by abiotic stresses, which results in very low crop productivity. To lessen the effect of these adverse conditions, plants change their various pathways so as to adapt according to the new conditions. There is numerous protein which shows differential expression during abiotic stresses and are involved in various biological processes. During these abiotic stresses plant responds by regulating their gene expression which directly means plant produces proteins which help them to acclimatize in the new environment. ROS are the main factors which affects plants during stress condition. Therefore, majority of proteins produced by the plants under stress conditions are proteins that are involved in the process of antioxidation. Other important proteins that take part in combating the effect of these stresses are involved in protein folding, photosynthesis, disease, and defense-related proteins (proteolytic enzymes and proteosomal factors). In this chapter we have discussed some common DEP's and their mechanism through which enable plants to tolerate stressed environmental conditions. The differential abundance of proteins under salinity and drought stress helps in the identification of proteins responsible for salinity and drought tolerance. The contributions of a number of antioxidants, cellular enzymes, plant hormones, and cell signaling molecules in stress adaptation by the plants have been revealed. Although a lot of research work has been done still several novel genes, proteins, and signaling molecules that have essential role in the tolerance mechanism yet to be identified and characterized. There is a need of continuous and further in-depth investigation for the acquisition of novel salinity and drought tolerance mechanisms of cash crops.

References

Acosta-Motos JR, Ortuño MF, Bernal-Vicente A, Diaz-Vivancos P, Sanchez-Blanco MJ, Hernandez JA (2017) Plant responses to salt stress: adaptive mechanisms. Agronomy 7(1):18

Amjad M, Akhtar J, Anwar-ul-Haq M, Yang A, Akhtar SS, Jacobsen SE (2014) Integrating role of ethylene and ABA in tomato plants adaptation to salt stress. Sci Hortic 172:109–116

Araujo GDS, Lopes LDS, Paula-Marinho SDO, Mesquita RO, Nagano CS, Vasconcelos FR, de Carvalho HH, Moura ADAAN, Marques EC, Gomes-Filho E (2021). H_2 O_2 priming induces proteomic responses to defense against salt stress in maize. Plant Mol Bio 106:33–48

Ashoub, Baeumlisberger M, Neupaertl M, Karas M, Brüggemann W (2015) Characterization of common and distinctive adjustments of wild barley leaf proteome under drought acclimation, heat stress and their combination. Plant Mol Biol 87(4–5):459–471

Barkla BJ, Vera-Estrella R, Pantoja O (2013) Progress and challenges for abiotic stress proteomics of crop plants. Proteomics 13(12–13):1801–1815

Chaudhry UK, Gökçe ZNÖ, Gökçe AF (2021) Drought and salt stress effects on biochemical changes and gene expression of photosystem II and catalase genes in selected onion cultivars. Biologia 76(10):3107–3121

Chen Y, Wang XM, Zhou L, He Y, Wang D, Qi YH, Jiang DA (2015) Rubisco activase is also a multiple responder to abiotic stresses in rice. PLoS One 10(10):e0140934

Cheng X, Deng G, Su Y, Liu JJ, Yang Y, Du GH, Chen ZY, Liu FH (2016) Protein mechanisms in response to NaCl-stress of salt-tolerant and salt-sensitive industrial hemp based on iTRAQ technology. Ind Crops Prod 83:444–452

Dermendjiev G, Schnurer M, Weiszmann J, Wilfinger S, Emanuel OTT, Gebert C, Weckwerth W, Verena IBL (2021) Tissue-specific proteome and subcellular microscopic analyses reveal the effect of high salt concentration on actin cytoskeleton and vacuolization in aleurone cells during early germination of barley. Int J Mol Sci 22(17):9642

Dey S, Malviya R, Gayen D (2022) Understanding abiotic stress tolerance in cereals through genomics and proteomics approaches. In: Omics approach to manage abiotic stress in cereals. Springer, Singapore, pp 73–102

Faghani E, Gharechahi, Komatsu S, Mirzaei M, Khavarinejad RA, Najafi F, Farsad LK, Salekdeh GH (2015) Comparative physiology and proteomic analysis of two wheat genotypes contrasting in drought tolerance. J Proteome 114:1–15

Gull A, Lone AA, Wani NUI (2019) Biotic and abiotic stresses in plants. In: Abiotic and biotic stress in plants. IntechOpen, London, pp 1–19

Gupta S, Gupta K, Nehra C, Gaur RK, Yadav D (2022) Biotechnological intervention for sugarcane improvement under salinity. Sugar Tech 25:15–31

Gurmani AR, Bano A, Ullah N, Khan H, Jahangir M, Flowers TJ (2013) Exogenous abscisic acid (ABA) and silicon (Si) promote salinity tolerance by reducing sodium (Na+) transport and bypass flow in rice ('Oryza sativa'indica). Aust J Crop Sci 7(9):1219–1226

Hamzelou S, Pascovici D, Kamath KS, Amirkhani A, McKay M, Mirzaei M, Atwell BJ, Haynes PA (2020) Proteomic responses to drought vary widely among eight diverse genotypes of rice (Oryza sativa). Int J Mol Sci 21(1):363

Hao P, Zhu J, Gu A, Lv D, Ge P, Chen G, Li X, Yan Y (2015) An integrative proteome analysis of different seedling organs in tolerant and sensitive wheat cultivars under drought stress and recovery. Proteomics 15(9):1544–1563

Hao Y, Xu S, Wang H, Kong L, Sun S (2021) Comparative analysis of the glutathione S-transferase gene family of four Triticeae species and transcriptome analysis of GST genes in common wheat responding to salt stress. Int J Genom 2021:6289174

Hu Y, Xie W, Chen B (2020) Arbuscular mycorrhiza improved drought tolerance of maize seedlings by altering photosystem II efficiency and the levels of key metabolites. Chem Biol Technol Agric 7(1):1–14

Jacob F, Mahatma M, Deshmukh Y, Kandoliya UK, Maraviya GV, Joshi M, Vala A (2022) Altered expression levels of transcripts of GNAC TFs during drought stress in susceptible and tolerant cultivars of groundnut. Plant Stress 3:100062

Jensen ON (2006) Interpreting the protein language using proteomics. Nat Rev Mol Cell Biol 7(6): 391–403

Jiang Q, Li X, Niu F, Sun X, Hu Z, Zhang H (2017) iTRAQ-based quantitative proteomic analysis of wheat roots in response to salt stress. Proteomics 17(8):1

Kajla M, Yadav VK, Khokhar J, Singh S, Chhokar RS, Meena RP, Sharma RK (2015) Increase in wheat production through management of abiotic stresses: a review. J Appl Nat Sci 7(2): 1070–1080

Kamal AHM, Cho K, Choi JS, Jin Y, Park CS, Lee JS, Woo SH (2013) Patterns of protein expression in water-stressed wheat chloroplasts. Biol Plant 57:305–312

Kim DW, Rakwal R, Agrawal GK, Jung YH, Shibato J, Jwa NS, Iwahashi Y, Iwahashi H, Kim DH, Shim IS, Usui K (2005) A hydroponic rice seedling culture model system for investigating proteome of salt stress in rice leaf. Electrophoresis 26(23):4521–4539

Kim Y, Wang M, Bai Y, Zeng Z, Guo F, Han N, Bian H, Wang J, Pan J, Zhu M (2014) Bcl-2 suppresses activation of VPEs by inhibiting cytosolic Ca2+ level with elevated K+ efflux in NaCl-induced PCD in rice. Plant Physiol Biochem 80:168–175

Kim YS, Park SI, Kim JJ, Shin SY, Kwak SS, Lee CH, Park HM, Kim YH, Kim IS, Yoon HS (2022) Over-expression of dehydroascorbate reductase improves salt tolerance, environmental adaptability and productivity in Oryza sativa. Antioxidants 11(6):1077

Koh J, Chen G, Yoo MJ, Zhu N, Dufresne D, Erickson JE, Shao H, Chen S (2015) Comparative proteomic analysis of Brassica napus in response to drought stress. J Proteome Res 14(8): 3068–3081

Kumar P, Choudhary M, Halder T, Prakash NR, Singh V, Sheoran S, Longmei N, Rakshit S, Siddique KH (2022) Salinity stress tolerance and omics approaches: revisiting the progress and achievements in major cereal crops. Heredity 128:1–22

Li J, Ban L, Wen H, Wang Z, Dzyubenko N, Chapurin V, Gao H, Wang X (2015) An aquaporin protein is associated with drought stress tolerance. Biochem Biophys Res Commun 459(2): 208–213

Liu A, Xiao Z, Wang Z, Lam HM, Chye ML (2021) Galactolipid and phospholipid profile and proteome alterations in soybean leaves at the onset of salt stress. Front Plant Sci 12:644408

Mantri N, Patade V, Penna S, Ford R, Pang E (2012) Abiotic stress responses in plants. In: Ahmad P, Prasad MNV (eds) Abiotic stress response in plants: metabolism, productivity and sustainability. Springer, New York, NY, p 2

Marriboina S, Sengupta D, Kumar S, Reddy AR (2017) Physiological and molecular insights into the high salinity tolerance of Pongamia pinnata (L.) pierre, a potential biofuel tree species. Plant Sci 258:102–111

Melandri G, AbdElgawad H, Riewe D, Hageman JA, Asard H, Beemster GT, Kadam N, Jagadish K, Altmann T, Ruyter-Spira C, Bouwmeester H (2020) Biomarkers for grain yield stability in rice under drought stress. J Exp Bot 71(2):669–683

Menga V, Trono D (2020) The molecular and functional characterization of the durum wheat Lipoxygenase TdLOX2 suggests its role in hyperosmotic stress response. Plants 9(9):1233

Munns R, Tester M (2008) Mechanisms of salinity tolerance. Annu Rev Plant Biol 59:651–681

Oh M, Komatsu S (2015) Characterization of proteins in soybean roots under flooding and drought stresses. J Proteome 114:161–181

Parker R, Flowers TJ, Moore AL, Harpham NV (2006) An accurate and reproducible method for proteome profiling of the effects of salt stress in the rice leaf lamina. J Exp Bot 57(5):1109–1118

Passamani LZ, Barbosa RR, Reis RS, Heringer AS, Rangel PL, Santa-Catarina C, Grativol C, Veiga CF, Souza-Filho GA, Silveira V (2017) Salt stress induces changes in the proteomic profile of micropropagated sugarcane shoots. PLoS One 12(4):e0176076

Ramanjulu S, Bartels D (2002) Drought-and desiccation-induced modulation of gene expression in plants. Plant Cell Environ 25(2):141–151

Raza A, Razzaq A, Mehmood SS, Hussain MA, Wei S, He H, Zaman QU, Xuekun Z, Yong C, Hasanuzzaman M (2021) Omics: the way forward to enhance abiotic stress tolerance in Brassica napus L. GM Crops Food 12(1):251–281

Roy Choudhury A, Roy SK, Trivedi P, Choi J, Cho K, Yun SH, Walitang DI, Park JH, Kim K, Sa T (2022) Label-free proteomics approach reveals candidate proteins in rice (Oryza sativa L.) important for ACC deaminase producing bacteria-mediated tolerance against salt stress. Environ Microbiol 24:3612

Sadaqat HA, Tahir MHN, Hussain MT (2003) Physiogenetic aspects of drought tolerance in canola (Brassica napus). Int J Agric Biol 5(4):611–614

Sharma JK, Sihmar M, Santal AR, Singh NP (2019) Impact assessment of major abiotic stresses on the proteome profiling of some important crop plants: a current update. Biotechnol Genet Eng Rev 35(2):126–160

Shen Q, Fu L, Dai F, Jiang L, Zhang G, Wu D (2016) Multi-omics analysis reveals molecular mechanisms of shoot adaption to salt stress in Tibetan wild barley. BMC genomics 17:1–15

Shen Q, Fu L, Qiu L, Xue F, Zhang G, Wu D (2017) Time-course of ionic responses and proteomic analysis of a Tibetan wild barley at early stage under salt stress. Plant Growth Regul 81(1): 11–21

Shi, Ye T, Chan Z (2014) Comparative proteomic responses of two bermudagrass (Cynodon dactylon (L). Pers.) varieties contrasting in drought stress resistance. Plant Physiol Biochem 82:218–228

Singh YP, Nayak AK, Sharma DK, Singh G, Mishra VK, Singh D (2015) Evaluation of Jatropha curcas genotypes for rehabilitation of degraded sodic lands. Land Degrad Dev 26(5):510–520

Sturgeon JC (2013) The cultural politics of ethnic identity in Xishuangbanna, China: tea and rubber as "cash crops" and "commodities". J Curr Chin Aff 41(4):109–131

Sugimoto M, Takeda K (2009) Proteomic analysis of specific proteins in the root of salt-tolerant barley. Biosci Biotechnol Biochem 73(12):2762–2765

Vaishnav A, Singh J, Singh P, Rajput RS, Singh HB, Sarma BK (2020) Sphingobacterium sp. BHU-AV3 induces salt tolerance in tomato by enhancing antioxidant activities and energy metabolism. Front Microbiol 11:443

Wang W, Pang J, Zhang F, Sun L, Yang L, Zhao Y, Yang Y, Wang Y, Siddique KH (2021a) Integrated transcriptomics and metabolomics analysis to characterize alkali stress responses in canola (Brassica napus L.). Plant Physiol Biochem 166:605–620

Wang Y, Liu M, Wang X, Zhong L, Shi G, Xu Y, Li Y, Li R, Huang Y, Ye X, Li Z (2021b) A novel β-1, 3-glucanase Gns6 from rice possesses antifungal activity against Magnaporthe oryzae. J Plant Physiol 265:153493

Witzel K, Weidner A, Surabhi GK, Börner A, Mock HP (2009) Salt stress-induced alterations in the root proteome of barley genotypes with contrasting response towards salinity. J Exp Bot 60(12): 3545–3557

Woodrow P, Ciarmiello LF, Annunziata MG, Pacifico S, Iannuzzi F, Mirto A, D'Amelia L, Dell'Aversana E, Piccolella S, Fuggi A, Carillo P (2017) Durum wheat seedling responses to simultaneous high light and salinity involve a fine reconfiguration of amino acids and carbohydrate metabolism. Physiol Plant 159(3):290–312

Wu D, Shen Q, Qiu L, Han Y, Ye L, Jabeen Z, Shu Q, Zhang G (2014) Identification of proteins associated with ion homeostasis and salt tolerance in barley. Proteomics 14(11):1381–1392

Xu J, Lan H, Fang H, Huang X, Zhang H, Huang J (2015) Quantitative proteomic analysis of the rice (Oryza sativa L.) salt response. PLoS One 10(3):e0120978

Yang X, Li Y, Chen H, Huang J, Zhang Y, Qi M, Liu Y, Li T (2020) Photosynthetic response mechanism of soil salinity-induced cross-tolerance to subsequent drought stress in tomato plants. Plants 9(3):363

Ye HONG, Zhang GP (2020) The influence of drought stress on malt quality traits of the wild and cultivated barleys. J Integr Agric 19(8):2009–2015

Yousuf PY, Ahmad A, Ganie AH, Sareer O, Krishnapriya V, Aref IM, Iqbal M (2017) Antioxidant response and proteomic modulations in Indian mustard grown under salt stress. Plant Growth Regul 81(1):31–50

Zadraznik T, Egge-Jacobsen W, Meglič V, Šuštar-Vozlič J (2017) Proteomic analysis of common bean stem under drought stress using in-gel stable isotope labeling. J Plant Physiol 209:42–50

Zeeshan M, Lu M, Sehar S, Holford P, Wu F (2020) Comparison of biochemical, anatomical, morphological, and physiological responses to salinity stress in wheat and barley genotypes deferring in salinity tolerance. Agronomy 10(1):127

Zhao D, Luan Y, Shi W, Zhang X, Meng J, Tao J (2021a) A Paeonia ostii caffeoyl-CoA O-methyltransferase confers drought stress tolerance by promoting lignin synthesis and ROS scavenging. Plant Sci 303:110765

Zhao S, Zhang Q, Liu M, Zhou H, Ma C, Wang P (2021b) Regulation of plant responses to salt stress. Int J Mol Sci 22(9):4609

Zörb C, Geilfus CM, Mühling KH, Ludwig-Müller J (2013) The influence of salt stress on ABA and auxin concentrations in two maize cultivars differing in salt resistance. J Plant Physiol 170(2): 220–224

Insights in Metabolomics Responses to Drought and Salinity Stress in Crop Plants

12

Sonia, Nisha Kumari, Hemanthkumar Manne, Minakshi Jattan, Babita Rani, Sushil, Ravika, Ram Avtar, Jyothi Duhan, Shweta, and Anubhuti Sharma

Abstract

The environmental stresses have no boundaries and without any warnings they hamper the plant biomass output and quality. Among these stresses, drought and salinity stress were the most significant in reducing the plant productivity. Nevertheless, because of its cumulative, subtle effects, and multifaceted nature, it negatively affects the morphological, physiological, biochemical, and molecular characteristics of plants and renders their ability to synthesize oxygen. In order to adapt to stress, plants have evolved a variety of intricate defense and adaptation mechanisms, including biochemical and physiological responses that vary depending on the species. We briefly highlight the complex adaption mechanisms and regularity network that enhance plant water stress tolerance and adaptability. The strategies used by plants under drought and saline conditions include modifications in growth pattern and structural dynamics, reduction of transpiration loss through changing stomatal conductivity, leaf rolling, root elongation, accumulation of compatible solutes, and enhancing the efficiency of transpiration, homeostatic balance between osmolytes, and delayed senescence. Approaches that are made to alleviate the stress include exogenous application

Sonia · N. Kumari · H. Manne (✉) · B. Rani · J. Duhan
Department of Biochemistry, CCS Haryana Agricultural University, Hisar, India

M. Jattan · Ravika · R. Avtar · Shweta
Oilseeds Section, Department of Genetics and Plant Breeding, CCS Haryana Agricultural University, Hisar, India

Sushil
Department of Entomology, CCS Haryana Agricultural University, Hisar, India

A. Sharma
ICAR – Directorate of Rapeseed-Mustard Research, Bharatpur, India

A. Kumar et al. (eds.), *Salinity and Drought Tolerance in Plants*,
https://doi.org/10.1007/978-981-99-4669-3_12

of potassium, secondary metabolites, and osmoprotectants. Advanced omics technologies include trans-genomics and metabolomics were performed to increase stress tolerance. Besides nanoparticles utilization that control antioxidant enzyme activity for stress adaption in plants, improving plant tolerance by restoration of cell homeostasis, and mitigating the negative impacts of water stress are very promising for agriculture.

Keywords

Drought stress · Oxidative stress · Salinity · Tolerance · Water relations

12.1 Introduction

Abiotic stress is one of the most significant factors that affect plant development and productivity; it has a significant effect on growth and is to blame for the greatest losses in crop output. In most plant species, growth decreases can reach up to 50% (Munns 2005). There are numerous environmental factors that have a serious impact on plants, including salinity, high temperatures (Rani et al. 2016), chilling, heat stress (Mohan et al. 2020), metal toxicity, and increasing carbon dioxide (CO_2) levels. Climate change has a notable influence on the development and productivity of different plants (Li et al. 2009). Various research studies have been done over the last few years on plants administered to abiotic stress. Plants that are subjected to various stresses also employ particular mechanisms to complete their life cycle. Different plant species respond differently according to habitat, physiology, and severity and duration of stress. The process by which plants fight drought also differs between plant species. Therefore, to adapt to harsh climatic condition like drought, plants use less resources and change their rate of development (Osakabe et al. 2014; Bielach et al. 2017). It is necessary to increase the plant's ability to withstand drought under particular evolving conditions. Fostering additional research on drought-tolerant plants and implementation of cost-effective and advantageous agricultural methods will be of utmost significance to fulfill future food demand. The impact of drought stress on plant growth and crop yield is discussed in this chapter. Different morphological, physiological, biochemical, and molecular mechanisms used by plants to mitigate the negative effects of drought on plant growth, such as decreased leaf size, decreased stem elongation, reduced root proliferation, increased water-use efficiency, closure of stomata, decreased turgor pressure, and increased ABA biosynthetic rate. We have summarized current molecular, physiological, and metabolic developments in understanding plant responses to salt and drought in this chapter. We also emphasize the significance of applying this knowledge to create crops that are resistant to the consequences of future expected climate change.

12.2 Drought and Salt Stress Combined Causes

Global climate change anticipated to intensify in future as a result of increasing temperatures and CO_2 levels, which eventually modifies rainfall distribution and pattern. Although scarce rainfall is typically the major cause of drought stress, high temperature events, intense light, and dry winds can also cause soil water to evaporate, which can exacerbate an already severe drought stress occurrence (Cohen et al. 2021). On a worldwide scale, drought stress conditions are often brought on by climate change and affect large regions. In addition to drought, salt stress is regarded as a major factor in plant water shortage (Mostofa et al. 2018). The presence of soluble salts in soil at larger quantities is termed as salinity. The two types of saline soils are sodic (or alkali) and saline soils. The presence of carbonate and insoluble sodium carbonate crust is what distinguishes sodic soils from other types. With an electrical conductivity of >4 dS/m, an electrical/ESP >15%, saline soils are common in dry locations, estuaries, and coastal margins. Around the world, nearly 20% of arable land is impacted by salinity, and this area is growing yearly. Plants are divided into halophytes and glycophytes based on how they respond to salt. High temperatures, hydrological and pedological processes, the raise of water-borne salts in soils, excessive fertilization, and overgrazing are some of the factors contributing to increasing salinization of soil (Szaboles 1994). Since drought conditions are likely to cause greater losses in crop yield than losses from any other abiotic stress, the relation between drought and salinity is necessary for crop productivity.

Elevated CO_2 levels reported a higher photosynthetic rate and produced greater yields (Brown et al. 2018). However, climate change has negative effects on agricultural and natural ecosystems. Glaciers may melt as a result of rising air temperatures. In reality, global warming has caused a drop in yearly cumulative precipitation in many rain-fed agricultural regions throughout the world (Warner and Afifi 2014). By the end of the twenty-first century, the predicted rises in temperature was roughly 2 °C higher than current levels materialize and nearly one-fifth of the global population may suffer from critical water scarcity (Ray et al. 2019). In comparison to places that get irrigation through canals, rivers, and water channels, more stress is anticipated under rainfall areas (Konapala et al. 2020). If current trends continue, it is predicted that summer precipitation amounts in rain-fed regions would decline by 70% by the start of the twenty-second century (Yu et al. 2013). Some researchers predict that increase in atmospheric CO_2 level would cause excessive rainfall and heavy crop loss (Guo et al. 2015; Reddy 2015). As a result, variations in monsoon rainfall intensity, incidence, and length have an impact on the moisture content of the rhizosphere, which in turn affects plant production in specific regions of the world.

12.3 Effects and Responses of Stress in Plants

Plants may experience a variety of challenges depending on the dynamics of their environment, which might negatively impact their growth and development (Battaglia et al. 2019; Bukhari et al. 2019). Due to climate change and global warming, the latter has lately caused drastic decreases in the output of grain crops and is only predicted to get worse (Bal and Minhas 2017; Hafez and Seleiman 2017). The symptoms of drought stress in plants are an increase in the rate of leaf senescence and drooping, scorching and limp leaves, leaf rolling and brittleness, closed flowers and flower sagging, etiolation, wilting, turgidity, premature fall, senescence, and yellowing of leaves (Khan et al. 2018; Ruehr et al. 2019). Excess water also has an influence on plant performance and hinders growth and ultimate production, whereas water deficit often has a significant negative impact (Zargar et al. 2017). Furthermore, variations in soil temperature, lack of oxygen, mechanical resistance, and salt can all have a significant effect on root structure. Root cells from salinized plants frequently have irregularly thickened and twisted cell walls (Shannon et al. 1994). In woody tree roots, salts frequently encourage the suberization of the hypodermis and endodermis, which produces a well-developed casparian strip closer to the root apex that is distinct from that observed in non-salinized roots (Walker et al. 1984). Additionally, certain plants morphology demonstrate how sensitive they are to salt. These changes restrict the spread of this crop to locations with high-quality irrigation water. High salt uptake ratio values promote sodicity, which improve soil resistance, inhibit root development, and inhibit water transport through the root through a reduction in hydraulic conductivity. Salinity inhibits plant growth by osmotic and toxic effects (Rengasamy and Olsson 1993).

12.4 Physiological Effects

An extensive deep root system is advantageous to uptake water from deeper layers. Water stress in *Catharanthus roseus* has been observed to cause an increase in root development (Jaleel et al. 2008). However, in maize, water stress did not significantly decrease root development (Sacks et al. 1997). The ratio of root to shoot in plants often rises under low water availability (Wu and Cosgrove 2000). When roots are exposed to drought stress, a signal cascade is triggered that travels through the xylem to the shoots and causes physiological changes that ultimately determine how well the plant will respond to the stress. The root-shoot signaling has been linked to ABA, cytokinins, ethylene, malate, and other unknown substances. The stomatal closure caused due to drought through the transpiration is a significant alteration to the fields under restricted water supply. The ABA encourages the guard cells to outflow K^+ ions, which lowers turgor pressure and causes stomata to close. Due to loss of cell turgor or cell membrane disruption, dehydration of plants has been found to result in an increase in ABA levels of up to 50-fold (Yang et al. 2021).

Environmental stress affects the photosynthetic apparatus directly by essentially interfering with all key processes involved in photosynthesis (Allen and Ort 2001).

Crop plants under severe drought stress had their gas exchange parameters severely hampered. This could be because of a reduction in leaf expansion, a malfunctioning photosynthetic system, early senescence of the leaves, and chloroplast lipids oxidation (Menconi et al. 1995). The entry of carbon dioxide and loss of water occurs at stomata. The closure of stomata deprives the leaves of CO_2, resulting in a reduction in photosynthetic carbon uptake and an increase in photorespiration. A common sign of oxidative stress is a reduction in chlorophyll content, which might be a result of chlorophyll breakdown and pigment photo-oxidation. Plants need photosynthetic pigments primarily to capture light and produce reducing agents. Depending on the length and intensity of the drought, numerous species have reported decreased or stable chlorophyll levels (Kyparissis et al. 1995). Besides, by using optical microscopy on semi-thin layers, Navarro et al. (2007) detected the anatomical alterations in the leaves of salt-stressed *Arbutus unedo*. There was no change in first layer of palisade cells, but second layer size considerably improved concurrently with salt levels (0, 52, and 105 mM NaCl). They also reported a significant decrease in the intercellular gaps inside the spongy mesophylls. This reduction had an impact on the conductance of CO_2 via the stomata and mesophyll. Fv/Fm, Y(II), qP, and the electron transport rate (ETR) often decrease in saline circumstances, but enhancements qN, NPQ, and Y(NPQ) was also presented (Acosta-Motos et al. 2015a, b; Moradi and Ismail 2007).

12.5 Biochemical Effects

An early stage of the plant's defensive response to stress is the formation of ROS, or the oxidative burst, which also serves as a supplementary stimulant to start the plants following defense response. As a by-product of oxygen regular metabolism, the generation of ROS occurs (free radicals, and peroxides) naturally and plays a significant role in cell signaling. However, when there is a drought, the levels of ROS rapidly rise, which causes oxidative damage to proteins, DNA, and lipids (Apel and Hirt 2004). The ROS may cause significant harm to plants because of their high reactivity. The damage is done by peroxidation of lipids, breakdown of proteins, cellular fragmentation and finally it leads to cell death. Lipid peroxidation is a direct result of the ROSs ability to target membrane lipids. These radicals include O_2^-, H_2O_2, and •OH as examples (Mittler 2002). Because of the excess ROS production caused by the drought, the amount of malondialdehyde (MDA) rises (Kumari et al. 2013; Naveen et al. 2021). A biomarker for peroxidation has been the amount of MDA in the cell body (Moller et al. 2007). The MDA is considered to be a noteworthy sign of membrane lipid peroxidation. A measure for the frequency of free radical reactions in tissues is lipid peroxidation. Additionally, these processes result in the production of reactive oxygen species, particularly H_2O_2, which is created during photorespiration in peroxisomes. The MDA and H_2O_2 concentrations in the water-stressed plants of *P. cathayana* enhanced by 88.9% and 99.7%, respectively, but they increased in *P. kangdingensis* about, 44% and 63.6%, respectively, according to Yang and Miao (2010).

12.6 Molecular Effects

Under drought stress, plants correctly detect changes in water deficit levels in the soil and acclimatize to the conditions by accumulating ABA. Additionally, both ABA-dependent and ABA-independent regulatory mechanisms control the stress responses brought on by drought (Nakashima et al. 2014). Hormone-like peptides have been implicated in new signaling functions that mediate responses to drought stress in the recent research (Takahashi et al. 2019). Plants detect the presence of a water shortage in their roots and share this information with other distant organs. The hydraulic change signal can move at a pace of 1 m/min through the vascular system. According to a previous study, lycophytes and ferns control stomatal closure without the use of ABA (Brodribb and McAdam 2011). To maintain stomatal closure, *Vitis vinifera* accumulates ABA under drought conditions (Tombesi et al. 2015).

The vasculature of dried leaves has significant expression of the NINE CIS EPOXYCAROTENOID DIOXYGENASE3 (NCED3) gene, which codes for a crucial ABA production enzyme (Iuchi et al. 2001). Additionally, the hydraulic tension brought on by turgor loss triggers the leaves ABA production (Christmann et al. 2007). Recent studies have demonstrated that the peptide CLAVATA3/EMBRYO-SURROUNDING REGION-RELATED25 (CLE25) act as a signal for ABA synthesis in the leaves (Christmann and Grill 2018). By the expression of NCED3 in the leaves, CLE25 travels from roots to leaves and improves accumulation of ABA. The BARELY ANY MERISTEM (BAM) 1 and BAM3 Receptor-like Protein Kinases (RLKs) are able to detect CLE25 in the leaves. Therefore, increase in the ABA which leads to stomatal closure and stress-inducing genes, are controlled by CLE25-BAM1 and BAM3 systems.

The plasma membrane is thought to be where extracellular Na^+ is detected, but the sensors associated with this process remain a mystery. The NaCl elevated concentration produces a Ca^{2+} influx into the cytosol through the plasma membrane and tonoplast, as evidenced by the fast rise of free Ca^{2+} in the cytosol (Sanders 2000; Tracy et al. 2008). Calcium sensors, namely SOS3 calcenerin of B like proteins, may contribute to an increase in the amount of free calcium in the cytosol. There is a hypothesis that dimerization of SOS3 and the successive interaction of the dimers with SOS2 calcinerin B-like protein kinase will result in an increase in the calcium concentration in the cytosol (CIPK24). The SOS1 is the Na^+/H^+ antiporter that is phosphorylated by the combination SOS3/SOS2 (Zhu 2002). SOS1 is thought to also serve as a sodium sensor, even though no osmosensor has yet been discovered (Shi et al. 2000).

12.7 Tolerance Mechanism

Plants use a variety of techniques and methods to mitigate the negative consequences of drought and salinity stress. Agriculturists employ a variety of methods to strengthen the ability of plants to withstand drought stress. Some of them are discussed below.

12.7.1 Morphological Modifications

Through evolution, many adaptation mechanisms have been developed by plants that help them to be more resilient to the negative consequences of stress (Batool et al. 2020). When plant is subjected to drought stress, three major coping mechanisms are stress avoidance, escape, and tolerance. The next sections go through the plant defense strategies against stress, including escape, avoidance, and tolerance.

12.7.1.1 Escape

To avoid the negative effects of drought, some plants use mechanisms including fast development and shortening the life cycle, self-reproduction, and seasonal growth before the start of the driest portion of the year (Alvarez et al. 2018). Although this process can sometimes result in a significant reduction in the duration of the plant growth cycle and the ultimate plant production (Blum 2011), early blooming is likely the finest potential escape adaptation mechanism in plants (Tekle and Alemu 2016).

12.7.1.2 Avoidance

A well-established deep root system checks the water scarcity by enhancing water uptake and reduces the transpiration losses (Dobra et al. 2010). Some other xeromorphic traits includes the presence of hairy cuticles and leaves limits the loss of water and preserves the water to present within the plant tissues (Brestic et al. 2018). Development of such structures, nevertheless, becomes burden to the plant in case of decreased production and smaller-than-average vegetative and reproductive components (Wasaya et al. 2018).

12.7.1.3 Resistance

On the other hand, a photosynthetic machinery level adaptive tolerance mechanism comprises restrictions on the growth of new leaves and decreases in plant leaf area. Similar to this, in the arid areas, the development of trichomes on the dorsal and ventral side of the leaves is an external trait which enables plant to withstand water shortages (Zhang et al. 2019). Thus, these specific physiological structures add an additional layer over the leaf and boost light reflection rate of leaf and provide an increased resistance over water loss. Thus, there is a decrease in the rate of water loss by leaf transpiration (Tiwari et al. 2020). However, it is widely acknowledged that adjustments to the root system, such as those to root size, length, expansion and growth rate, serve as primary method by which drought-tolerant plants counteract water shortages (Tzortzakis et al. 2020).

12.7.2 Ion Homeostasis

When plants are stressed by drought, potassium (K) treatment reduces the negative effects of the water shortage while maintaining plant productivity. The plants absorb more K for their intracellular regulatory mechanisms while under drought stress (Hasanuzzaman et al. 2018). Plants increased K levels result in oxidative damage, which might result in the formation of ROS during the photosynthesis process (Foyer 2018). In order to maintain photosynthetic carbon dioxide fixation, plants growing in stress have a high K requirement. The increase in ROS in plants during plant stress may be caused by a decrease in CO_2 (Seleiman et al. 2020a, b, c). When plants were cultivated under drought conditions, the generation of ROS disrupts the photosynthetic process and carbohydrate metabolism (Xin et al. 2018). By the application of a lower dosage of K than a greater dose, the low rate of photosynthetic activity was seen in plants growing under drought stress (Zahoor et al. 2017). Therefore, for plants to sustain their physiological activities, sufficient K is required. In the transport mechanisms that allow for plant absorption, Na^+ ions directly compete with K^+ ions. It should be emphasized that under conditions of high salinity, plants are significantly more successful at absorbing Na^+ than K^+. Due to their comparable physicochemical properties, Na^+ and K^+ ions compete against one another for the catalytic sites of metabolic enzymes that normally bind to K^+ cation. In cytoplasm, by maintaining a high ratios of K^+/Na^+ plants tolerate salt (Maathuis and Amtmann 1999; Zhu 2001). It has also been shown that applying K can improve the poor grain yield of crops cultivated in water-scarce conditions. For plants to function at their best physiologically, K can be applied topically or as a soil amendment (Brestic et al. 2018; Hawrylak-Nowak et al. 2018). As a result, K application is crucial for achieving optimal yield output of crops produced.

12.7.3 Osmotic Protection

Osmoprotectant signaling controls the wide spectrum of plant stressors that inhibit plant development and production. Under stressed conditions, osmoprotectants raise and are in charge of sustaining the physiological (Brito et al. 2019; Seleiman and Nahar 2019). Proline, fruton, mannitol, trehalose, and glycinebetaine are a few significant osmoprotectants found in plants under water stress (Zouari et al. 2019). These compounds enhance under stress and reduce the cells water potential, which encourages retention of water without inhibiting with regular metabolism. Osmotic adjustment under stress, preserves turgidity of cell and improves plant development (Bartels and Sunkar 2005; Krasensky and Jonak 2012). Additionally, when osmotic stress intensifies or lasts longer, these suitable solutes stabilize cellular structures (Hoekstra et al. 2001). Moreover, they also act as free radical scavengers, and thus by eliminating ROS they restore redox equilibrium of the cell (Miller et al. 2010). Exogenous supplementation of proline improves internal free proline levels and provides tolerance to drought (Semida et al. 2020). Finally, Sallam et al. (2019) reported that application of polyamines, such as spermidine, can effectively boost plant resistance to drought stress in crops like wheat and barley (*Hordeum vulgare* L.) 2019. Controlling the inflow of Na^+ and Cl^- ions at the interface between plant roots and soil is a key tactic for supporting essential activities when exposed to salt stress. Inorganic compound absorption is primarily regulated by halophytes in line with growth and osmotic homeostasis requirements (Flowers et al. 1977). The Na^+ ions enhancement in plant tissues, in contrast to halophytes, far surpasses the physiological requirements of glycophytes (Kronzucker et al. 2006). Therefore, by reducing Na^+ and Cl^- ion content, that enter glycophyte tissues might reduce the effects or symptoms of sodium stress.

12.7.4 Secondary Metabolites

Plant's ability to withstand drought can be increased by using synthetic and natural growth regulators (Ashraf et al. 2011). Since, water equilibrium under drought conditions is maintained by gibberellins (GA), it can be used to minimize the loss in length and weight of hypocotyl under stressed-seedlings (Javid et al. 2011). Following the administration of GA, the stomatal conductance, photosynthesis, and respiration rates of wheat, cotton, and maize all rose in water-scarce locations, leading to better grain yields than untreated-GA plants (Javid et al. 2011). The productivity of crops during droughts can also be increased by exogenous applications of ABA, uniconazole, brassinolide, and jasmonate (Fahad et al. 2015; Ahmad et al. 2018). It has also been investigated how polyamines (PAs) defend against salt stress. For instance, PA treatment enhanced glutathione and ascorbate as well as several other antioxidant enzymes, such as DHAR, GR, CAT, and GPX against salt stress (Nahar et al. 2016). In addition, researchers have looked into the potential role of PAs in grapevine (*Vitisvinifera* L.) plantlet response to salt stress.

Plant's ability to withstand drought is greatly improved by a growth hormone, salicylic acid (SA). By controlling growth, development, and defensive mechanisms, SA modifies how plants react to external challenges. The stimulation of SA inducing genes are PR1 and PR2 under drought stress provides more evidence for the function of SA in drought (Kang et al. 2013). According to Bandurska and Stroinski (2005), the SA treatment improved the stability of the membranes and proline levels and accumulation of ABA in water-stressed barley, gives an ability to plants to withstand stress. The SA treatment in wheat, withstanded the stress and improved AsA-GSH cycle (Singh and Usha 2003), besides it also reduced the significant water loss from wheat plants. Exogenous SA application increased antioxidative metabolism and reduced the detrimental effect of salt stress on photosynthesis. Additionally, under water stress, SA supply has improved *zea mays* defensive mechanisms (Saruhan et al. 2012). Foremost flavonoids including flavonols and anthocyanins provide tolerance (Nakabayashi et al. 2014). Additionally, these specific substances reduce oxidative stress in plants by acting as free radical scavengers.

12.7.5 Nanoparticles

The properties of nanoparticles (NPs) include particle shape, variable pore size, potential reactivity, and vast surface area. Specific contents are released when nanoparticles are directed into plant cellular organelles (Seleiman et al.

2020a, b, c). The SOD, CAT, and POD antioxidant enzyme activity was improved and regulated by the use of nanoparticles (Siddiqi and Husen 2016). For instance, the SOD activity in plants was increased when TIO_2 NPs were used (Ghasemlou et al. 2019). In order to boost the resistance of different plants to drought stress, several trace elements and their oxides of NPs were used in agriculture. It has been demonstrated that using silicon nanoparticles (Si-NPs) can lessen the harmful impacts of stress, including salinity, drought, chilling stress, and metal toxicity (Siddiqui et al. 2020). The improvement of osmoprotectants, sugar solubles, carbohydrates were reported by the usage of NPs like silica and ZnO nanoparticles (Seleiman et al. 2020a, b, c). Wheat plants resilience to drought was similarly improved by Si-NPs (Rizwan et al. 2019). Similar to this, using ZnO nanoparticles helped plants that were stressed by salt and drought (Seleiman et al. 2020a, b, c). The use of ZnO NPs during the early stages of development boosted sapling seed reserves and improved plant drought tolerance (Seydmohammadi et al. 2019). Additionally, it has been shown that ferrous and zinc work effectively together to increase a plants ability to withstand drought stress. The deleterious effects of drought stress on plants were decreased by using TIO_2 nanoparticles, which in turn activated a number of compounds (Movafeghi et al. 2018; Seleiman et al. 2020c). In lentil, the usage of silver and copper NPs were increased to provide resistance to stress. Nano-silica may increase the drought resilience of certain plant species (Seleiman et al. 2020a, b, c). Wheat resistance to drought was strengthened under zinc and copper NPs (Khan et al. 2019; Maswada et al. 2020).

12.7.6 Metabolic Engineering

The creation of drought-tolerant crops is one of the best ways to manage drought stress (Ullah et al. 2018). Therefore, improving drought tolerance without significantly lowering grain output is a difficult task. An ideal technique is to increase stress-related metabolites in plants and induce drought tolerance in them by metabolic engineering (Naghshbandi et al. 2019). One of the traditional approaches to provide resistance in arid and semi-arid areas is to successfully breed for drought resistance by engineering the biosynthetic pathway of raffinose. Galactinol synthase (AtGolS) gene with particular *AtGolS2* gene is triggered under drought, cold, and heat stress in particular causes the accumulation of raffinose and galactinol (Salvi et al. 2020). Heat-shock transcription factor A2 (*AtHsfA2*) also increased *AtGolS2* expression in response to oxidative stress (Nishizawa et al. 2008). Raffinose gene expression enhancement by *AtGolS2* expression, improved the plant's ability to withstand drought and defending them against oxidative stress. Through ROS scavengers and suitable solutes, galactinol and raffinose both have the capacity to protect cells from environmental stress (Salvi et al. 2020). This is because rice and soybean metabolome analyses revealed an increase in raffinose and galactinol levels, which represented a response to drought stress. To evaluate stress responses and phenotypes in dry field conditions, *AtGolS2* has been used to the transformation of rice and soybean. In transgenic rice and soybean, overexpressing *AtGolS2* boosted grain output in dry field conditions in addition to improving drought resistance (Honna et al. 2016). Crop plants are transformed by the application of *AtGolS2*, which triggers the plants' stress tolerance in dry regions. In transgenic *Arabidopsis* and *Brachypodium*, overexpression of *AtGolS2* increased the levels of galactinol and raffinose, improved tolerance against drought, and shielded plants suffering from oxidative damage (Taji et al. 2002; Himuro et al. 2014). These findings demonstrated that value of *AtGolS2* engineering as a biotechnological tool to minimize loss in grain yield during drought.

12.8 Conclusions

Drought and salinity stress were the abiotic stressors that is gaining attention since it has a negative impact on plant growth and development which significantly reduces plant biomass and production, contributing to global food insecurity. Certain plant processes that increase drought tolerance are adopted in order to combat the negative effects of drought stress on plants. The strategies that were adopted by plant under drought and saline conditions were growth in root length, minimization of transpiration loss by hindering the stomatal conductance, compatible solutes accumulation, production of more secondary metabolites, and delay in senescence. In saltwater environments, a healthy root system may guarantee water and nutrient intake and increase plant tolerance. Important processes for salt tolerance include salt exclusion, salt compartmentalization in vacuoles, and enhancement of osmoprotectants. Additionally, palisade parenchyma and interstitial gaps are increased in leaves while spongy tissue is decreased and whenever the stomatal aperture is smaller, CO_2 diffusion is aided by the parenchyma. Drought stress is regulated by the phytohormone ABA at the cellular and subcellular levels and imparts resistance. To provide resistance to the drought stress utilization of omics technology, i.e., transcriptomics, phenomics, metabolomics, and genomics are of great focus. Additional techniques including exogenous application of secondary metabolites, osmoprotectants, and potassium are worthwhile under water deficit conditions. Besides, application of nanoparticles and metabolic engineering adaptation improves resistance to drought stress by maintaining the cellular homeostasis and thus mitigating the negative effect of drought stress. In order to lessen the risk to the world's food security, these advanced technologies help us understand plants in dry conditions better and may boost their output.

References

Acosta-Motos JR, Díaz-Vivancos P, Alvarez S et al (2015a) Physiological and biochemical mechanisms of the ornamental *Eugenia myrtifolia* L. plants for coping with NaCl stress and recovery. Planta 242:829–846

Acosta-Motos JR, Díaz-Vivancos P, Alvarez S et al (2015b) NaCl-induced physiological and biochemical adaptative mechanism in the ornamental *Myrtus cummunis* L. plants. J Plant Physiol 183:41–51

Ahmad I, Kamran M, Ali S et al (2018) Seed filling in maize and hormones crosstalk regulated by exogenous application of uniconazole in semiarid regions. Environ Sci Pollut Res 25:33225–33239

Allen DJ, Ort DR (2001) Impact of chilling temperatures on photosynthesis in warm climate plants. Trends Plant Sci 6:36–42

Alvarez S, Rodríguez P, Broetto F et al (2018) Long term responses and adaptive strategies of *Pistacialentiscus* under moderate and severe deficit irrigation and salinity: osmotic and elastic adjustment, growth, ion uptake and photosynthetic activity. Agric Water Manag 202:253–262

Apel K, Hirt H (2004) Reactive oxygen species: metabolism, oxidative stress, and signal transduction. Annu Rev Plant Biol 55:373–399

Ashraf M, Akram NA, Al-Qurainy F et al (2011) Drought tolerance: roles of organic osmolytes, growth regulators, and mineral nutrients. In: Advances in agronomy, vol 111. Academic Press, London, pp 249–296

Bal SK, Minhas PS (2017) Atmospheric stressors: challenges and coping strategies. In: Abiotic stress management for resilient agriculture. Springer, Berlin

Bandurska H, Stroinski A (2005) The effect of salicylic acid on barley response to water deficit. Acta Physiol Plant 27:379–386

Bartels D, Sunkar R (2005) Drought and salt tolerance in plants. Crit Rev Plant Sci 24:23–58

Batool T, Ali S, Seleiman MF et al (2020) Plant growth promoting rhizobacteria alleviates drought stress in potato in response to suppressive oxidative stress and antioxidant enzymes activities. Sci Rep 10:16975

Battaglia M, Lee C, Thomason W et al (2019) Hail damage impacts on corn productivity: a review. Crop Sci 59:1–14

Bielach A, Hrtyan M, Tognetti VB (2017) Plants under stress: involvement of auxin and cytokinin. Int J Mol Sci 18:1427–1427

Blum A (2011) Plant water relations, plant stress and plant production. In: Plant breeding for water-limited environments. Springer, Berlin, pp 11–52

Brestic M, Zivcak M, Hauptvogel P et al (2018) Wheat plant selection for high yields entailed improvement of leaf anatomical and biochemical traits including tolerance to non-optimal temperature conditions. Photosynth Res 136:245–255

Brito C, Dinis LT, Moutinho-Pereira J et al (2019) Drought stress effects and olive tree acclimation under a changing climate. Plants 8:232

Brodribb TJ, Mcadam SA (2011) Passive origins of stomatal control invascular plants. Science 331(6017):582–585

Brown S, Nicholls RJ, Lázár AN et al (2018) What are the implications of sea-level rise for a 1.5, 2 and 3 C rise in global mean temperatures in the Ganges-Brahmaputra-Meghna and other vulnerable deltas? Reg Environ Chang 18:1829–1842

Bukhari SAH, Peerzada AM, Javed MH et al (2019) Growth and development dynamics in agronomic crops under environmental stress. In: Agronomic crops. Springer, Berlin, pp 83–114

Christmann A, Grill E (2018) Peptide signal alerts plants to drought. Nature 556(7700):178–179

Christmann A, Weiler EW, Steudle E et al (2007) A hydraulic signal in root to-shoot signalling of water shortage. Plant J 52(1):167–174

Cohen I, Zandalinas SI, Huck C et al (2021) Meta-analysis of drought and heat stress combination impact on crop yield and yield components. Physiol Plant 171:66–76

Dobra J, Motyka V, Dobrev P et al (2010) Comparison of hormonal responses to heat, drought and combined stress in tobacco plants with elevated proline content. J Plant Physiol 167:1360–1370

Fahad S, Nie L, Chen Y et al (2015) Crop plant hormones and environmental stress. In: Lichtfouse E (ed) Sustainable agriculture reviews. Springer, Cham, pp 371–400

Flowers TJ, Troke PF, Yeo AR (1977) The mechanism of salt tolerance in halophytes. Annu Rev Plant Physiol 28:89–121

Foyer CH (2018) Reactive oxygen species, oxidative signaling and the regulation of photosynthesis. Environ Exp Bot 154:134–142

Ghasemlou F, Amiri H, Karamian R et al (2019) Alleviation of the effects of on drought stress *Verbas cumnudicuale* by methyl jasmonate and titanium dioxide nanoparticles. Plant Physiol 9: 2911–2920

Guo HD, Zhang L, Zhu LW (2015) Earth observation big data for climate change research. Adv Clim Chang Res 6:108–117

Hafez EH, Seleiman MF (2017) Response of barley quality traits, yield and antioxidant enzymes to water-stress and chemical inducers. Int J Plant Prod 11:477–490

Hasanuzzaman M, Bhuyan MHM, Nahar K et al (2018) Potassium: a vital regulator of plant responses and tolerance to abiotic stresses. Agronomy 8(3):31

Hawrylak-Nowak B, Dresler S, Rubinowska K et al (2018) Selenium Biofortification enhances the growth and alters the physiological response of lambs lettuce grown under high temperature stress. Plant Physiol Biochem 127:446–456

Himuro Y, Ishiyama K, Mori F et al (2014) Arabidopsis galactinol synthase AtGolS2 improves drought tolerance in the monocot model *Brachypodium distachyon*. J Plant Physiol 171(13): 1127–1131

Hoekstra FA, Golovina EA, Buitink J (2001) Mechanisms of plant desiccation tolerance. Trends Plant Sci 6(9):431–438

Honna PT, Fuganti-Pagliarini R, Ferreira LC et al (2016) Molecular, physiological, and agronomical characterization, in greenhouse and in field conditions, of soybean plants genetically modified with AtGolS2 gene for drought tolerance. Mol Breed 36(11):157–157

Iuchi S, Kobayashi M, Taji T et al (2001) Regulation of drought tolerance by gene manipulation of 9-cisepoxycarotenoiddioxygenase, a key enzyme in abscisic acid biosynthesis in *Arabidopsis*. Plant J 27(4):325–333

Jaleel CA, Gopi R, Sankar B et al (2008) Differential responses in water use efficiency in two varieties of *Catharanthus roseus* under drought stress. Comp Rend Biol 331:42–47

Javid MG, Sorooshzadeh A, Moradi F et al (2011) The role of phytohormones in alleviating salt stress in crop plants. Aust J Crop Sci 5:726

Kang GZ, Li GZ, Liu GQ et al (2013) Exogenous salicylic acid enhances wheat drought tolerance by influence on the expression of genes related to ascorbate-glutathione cycle. Biol Plant 57: 718–724

Khan A, Pan X, Najeeb U et al (2018) Coping with drought: stress and adaptive mechanisms, and management through cultural and molecular alternatives in cotton as vital constituents for plant stress resilience and fitness. Biol Res 1:47–47

Khan ZS, Rizwan M, Hafeez M et al (2019) The accumulation of cadmium in wheat (Triticum aestivum) as influenced by zinc oxide nanoparticles and soil moisture conditions. Environ Sci Pollut Res 26:19859

Konapala G, Mishra AK, Wada Y et al (2020) Climate change will affect global water availability through compounding changes in seasonal precipitation and evaporation. Nat Commun 11:1–10

Krasensky J, Jonak C (2012) Drought, salt, and temperature stress induced metabolic rearrangements and regulatory networks. J Exp Bot 63(4):1593–1608

Kronzucker HJ, Szczerba MW, Darzi MMG et al (2006) The cytosolic Na^+: K^+ ratio does not explain salinity induced growth impairment in barley: a dual tracer study using 42K and 24Na. Plant Cell Environ 29:2228–2237

Kumari N, Ram A, Anitha K et al (2018) Antioxidative response of Indian mustard subjected to drought stress. J Oilseed Brassica 9(1):40–44

Kyparissis A, Petropoulun Y, Manetas Y (1995) Summer survival of leaves in a soft-leaved shrub (*Phlomis fruticose* L., Labiatae) under Mediterranean field conditions: avoidance of photo inhibitory damage through decreased chlorophyll contents. J Exp Bot 46:1825–1831

Li YP, Ye W, Wang M et al (2009) Climate change and drought: a risk assessment of crop-yield impacts. Clim Res 39:31–46

Maathuis FJM, Amtmann AK (1999) Nutrition and Na toxicity; the basis of cellular K^+/Na^+ ratios. Ann Bot 84:123–133

Maswada HF, Mazrou YS, Elzaawely AA et al (2020) Nanomaterials. Effective tools for field and horticultural crops to cope with drought stress: a review. Span J Agric Res 18:15
Menconi M, Sgherri C, Pinzino C et al (1995) Activated oxygen production and detoxification in wheat plants subjected to a water deficit programme. J Exp Bot 46:1123–1130
Miller G, Suzuki N, Ciftci-Yilmaz S et al (2010) Reactive oxygen species homeostasis and signalling during drought and salinity stresses. Plant Cell Environ 33(4):453–467
Mittler R (2002) Oxidative stress, antioxidants and stress tolerance. Trends Plant Sci 7:405–410
Mohan N, Kumari N, Jattan M et al (2020) Aftermath of terminal heat stress on Indian mustard (*Brassica juncea* L.): a brief review. J Oilseed Brassica 11(1):1–8
Moller IM, Jensen PE, Hansson A (2007) Oxidative modifications to cellular components in plants. Annu Rev Plant Biol 58:459–481
Moradi F, Ismail AM (2007) Responses of photosynthesis, chlorophyll fluorescence and ROS-scavenging systems to salt stress during seedling and reproductive stages in rice. Ann Bot 99:1161–1179
Mostofa MG, Ghosh A, Li ZG et al (2018) Methylglyoxal-a signaling molecule in plant abiotic stress responses. Free Radic Biol Med 122:96–109
Movafeghi A, Khataee A, Abedi M et al (2018) Effects of TiO2 nanoparticles on the aquatic plant Spirodela polyrrhiza: evaluation of growth parameters, pigment contents and antioxidant enzyme activities. J Environ Sci 64:130–138
Munns R (2005) Genes and salt tolerance: bringing them together. New Phytol 167:645–663
Naghshbandi MP, Tabatabaei M, Aghbashlo M et al (2019) Progress toward improving ethanol production through decreased glycerol generation in Saccharomyces cerevisiae by metabolic and genetic engineering approaches. Renew Sust Energ Rev 115:10935
Nahar K, Hasanuzzaman M, Rahman A et al (2016) Polyamines confer salt tolerance in mung bean (*Vigna radiata* L.) by reducing sodium uptake, improving nutrient homeostasis, antioxidant defense, and methylglyoxal detoxification systems. Front Plant Sci 7:1104–1104
Nakabayashi R, Mori T, Saito K (2014) Alternation of flavonoid accumulation under drought stress in Arabidopsis thaliana. Plant Signal Behav 9(8):e29518
Nakashima K, Yamaguchi-Shinozaki K, Shinozaki K (2014) The transcriptional regulatory network in the drought response and its crosstalk in abiotic stress responses including drought, cold, and heat. Front Plant Sci 5:170–170
Navarro A, Banon S, Olmos E et al (2007) Effects of sodium chloride on water potential components, hydraulic conductivity, gas exchange and leaf ultrastructure of *Arbutus unedo* plants. Plant Sci 172:473–480
Naveen N, Kumari N, Avtar A et al (2021) Evaluation of effect of brassinolide in *Brassica juncea* leaves under drought stress in field conditions. Horticulturae 7:514
Nishizawa A, Yabuta Y, Shigeoka S (2008) Galactinol and raffinose constitute a novel function to protect plants from oxidative damage. Plant Physiol 147(3):1251–1263
Osakabe Y, Osakabe K, Shinozaki K et al (2014) Response of plants to water stress. Front Plant Sci 5:86–86
Rani B, Kumari N, Jain P et al (2016) Antioxidative system as influenced by high temperatures stress in *Brassica juncea* (L.) Czern & Coss. Curr Trend Bio Pharm 10(2):118–125
Ray DK, West PC, Clark M et al (2019) Climate change has likely already affected global food production. PLoS One 14:217148
Reddy PP (2015) Impacts of climate change on agriculture. In: Climate resilient agriculture ensuring food security. Springer, Berlin, pp 43–90
Rengasamy P, Olsson KA (1993) Irrigation and sodicity. Aust J Soil Res 31:821–837
Rizwan M, Ali S, Ali B et al (2019) Zinc and iron oxide nanoparticles improved the plant growth and reduced the oxidative stress and cadmium concentration in wheat. Chemosphere 214:269–277
Ruehr NK, Grote R, Mayr S et al (2019) Beyond the extreme: recovery of carbon and water relations in woody plants following heat and drought stress. Tree Physiol 39:1285–1299

Sacks MM, Silk WK, Burman P (1997) Effect of water stress on cortical cell division rates within the apical meristem of primary roots of maize. Plant Physiol 114:519–527

Sallam A, Alqudah AM, Dawood MF et al (2019) Drought stress tolerance in wheat and barley: advances in physiology, breeding and genetics research. Int J Mol Sci 20:3137

Salvi P, Kamble NU, Majee M (2020) Ectopic over-expression of ABAresponsive Chickpea galactinol synthase (CaGolS) gene results in improved tolerance to dehydration stress by modulating ROS scavenging. Environ Exp Bot 171:103957

Sanders D (2000) Plant biology: the salty tale of *Arabidopsis*. Curr Biol 10:486–488

Saruhan N, Saglam A, Kadioglu A (2012) Salicylic acid pretreatment induces drought tolerance and delays leaf rolling by inducing antioxidant systems in maize genotypes. Acta Physiol Plant 34: 97–106

Seleiman MM, Nahar K (2019) Use of plant nutrients in improving abiotic stress tolerance in wheat. In: Hossain AM (ed) Wheat production in changing environments. Springer, Berlin, pp 481–495

Seleiman MF, Almutairi KF, Alotaibi M et al (2020a) Nano fertilization as an emerging fertilization technique: why modern agriculture can benefit from its use? Plants 10:2

Seleiman MF, Semida WM, Rady MM et al (2020b) Sequential application of antioxidants rectifies ion imbalance and strengthens antioxidant systems in salt-stressed cucumber. Plants 12:1783

Seleiman MF, Alotaibi M, Alhammad B et al (2020c) Effects of ZnO nanoparticles and biochar of rice straw and cow manure on characteristics of contaminated soil and sunflower productivity, oil quality, and heavy metals uptake. Agronomy 10:790

Semida WM, Abdelkhalik A, Rady MO et al (2020) Exogenously applied proline enhances growth and productivity of drought stressed onion by improving photosynthetic efficiency, water use efficiency and up-regulating osmoprotectants. Sci Hortic 272:109580

Seydmohammadi Z, Roein Z, Rezvanipour S (2019) Accelerating the growth and flowering of Eustoma grandiflorum by foliar application of nano-ZnO and nano-CaCO3. Plant Physiol Rep 25:140–148

Shannon MC, Grieve CM, Francois LE et al (1994) Whole-plant response to salinity. In: Plant-environment interactions. Springer, New York, NY, pp 199–244

Shi H, Ishatani M, Kim C et al (2000) The *Arabidopsis thaliana* salt tolerance gene SOSI encodes a putative Na^+/H^+ antiporter. Proc Natl Acad Sci U S A 97:6896–6901

Siddiqi KS, Husen A (2016) Engineered gold nanoparticles and plant adaptation potential. Nano-scale Res Lett 11:400

Siddiqui H, Ahmed KBM, Sami F et al (2020) Silicon nanoparticles and plants: current knowledge and future perspectives. In: Lichtfouse E (ed) Sustainable agriculture reviews, vol 41. Springer, Cham, pp 129–142

Singh B, Usha K (2003) Salicylic acid induced physiological and biochemical changes in wheat seedlings under water stress. Plant Growth Regul 39:137–141

Szabolcs I (1994) Handbook of plant and crop stress. Marcel Dekker, New York, NY

Taji T, Ohsumi C, Iuchi S et al (2002) Important roles of drought- and cold-inducible genes for galactinol synthase instress tolerance in *Arabidopsis thaliana*. Plant J 29(4):417–426

Takahashi F, Hanada K, Kondo T et al (2019) Hormone-like peptides and small coding genes in plant stress signaling and development. Curr Opin Plant Biol 51:88–95

Tekle AT, Alemu MA (2016) Drought tolerance mechanisms in field crops. World J Biol Med Sci 3:15–39

Tiwari P, Srivastava D, Chauhan AS et al (2020) Root system architecture, physiological analysis and dynamic transcriptomics unravel the drought responsive traits in rice genotypes. Ecotoxicol Environ Saf 207:111252

Tombesi S, Nardini A, Frioni T et al (2015) Stomatal closure is induced by hydraulic signals and maintained by ABA in drought-stressed grapevine. Sci Rep 5:12449

Tracy FE, Gilliham M, Dodd AN et al (2008) Cytosolic free Ca^{2+} in *Arabidopsis thaliana* are heterogeneous and modified by external ionic composition. Plant Cell Environ 31:1063–1073

Tzortzakis N, Chrysargyris A, Aziz A (2020) Adaptive response of a native Mediterranean grapevine cultivar upon short-term exposure to drought and heat stress in the context of climate Chang. Agronomy 10:249

Ullah A, Manghwar H, Shaban M et al (2018) Phytohormones enhanced drought tolerance in plants: a coping strategy. Environ Sci Pollut Res 25:33103–33118

Walker RR, Sedgley M, Blesing MA et al (1984) Ultrastructure and assimilate concentrations of roots of citrus genotypes differing in ability for salt exclusion. J Exp Bot 35:1481–1494

Warner K, Afifi T (2014) Where the rain falls: evidence from 8 countries on how vulnerable households use migration to manage the risk of rainfall variability and food insecurity. Clim Dev 6:1–17

Wasaya A, Zhang X, Fang Q et al (2018) Root phenotyping for drought tolerance: a review. Agronomy 8:241

Wu Y, Cosgrove DJ (2000) Adaptation of roots to low water potentials by changes in cell wall extensibility and cell wall proteins. J Exp Bot 51:1543–1553

Xin L, Zheng H, Yang Z et al (2018) Physiological and proteomic analysis of maize seedling response to water deficiency stress. J Plant Physiol 228:29–38

Yang F, Miao LF (2010) Adaptive responses to progressive drought stress in two poplar species originating from different altitudes. Silva Fennica 44:23

Yang X, Lu M, Wang Y, Wang Y, Liu z, Chen S (2021) Response mechanism of plants to drought stress. Horticulturae 7(3):50

Yu W, Yang YC, Savitsky A et al (2013) The Indus Basin of Pakistan: the impacts of climate risks on water and agriculture. The World Bank, Washington, DC

Zahoor R, Zhao W, Dong H et al (2017) Potassium improves photosynthetic tolerance to and recovery from episodic drought stress in functional leaves of cotton (*Gossypium hirsutum* L.). Plant Physiol Biochem 119:21–32

Zargar SM, Gupta N, Nazir M et al (2017) Impact of drought on photosynthesis: molecular perspective. Plant Gene 11:154–159

Zhang F, Wang P, Zou YN et al (2019) Effects of mycorrhizal fungi on root hair growth and hormone levels of taproot and lateral roots in trifoliate orange under drought stress. Arch Agron Soil Sci 65:1316–1330

Zhu J (2001) Plant salt tolerance. Trends Plant Sci 6:66–71

Zhu JK (2002) Sat and drought stress signal transduction in plants. Annu Rev Plant Biol 53:247–273

Zouari M, Hassena AB, Trabelsi L et al (2019) Exogenous proline mediated abiotic stress tolerance in plants: possible mechanisms. In: Osmoprotectant-mediated abiotic stress tolerance in plants. Springer, Berlin

Transcriptional Regulatory Network Involved in Drought and Salt Stress Response in Rice

13

Pushpendra Kumar, Pradeep Kumar, Suniti, Upendra Kumar, Avni, and Anita Mann

Abstract

Transcription factors (TFs) family intimately regulate gene expression in response to hormones, biotic and abiotic factors, symbiotic interactions, cell differentiation, stress signaling pathways in plants, and protective genome activities in response to water and salt stress conditions. TFs are specialized proteins which bind to specific DNA elements in gene promoters and modulate gene expression in response to various external and internal stimuli. Water and salt stress-responsive genes expression is regulated by a large number of common transcription factors (TFs). AP2/ERF, NAC, bZIP, HD-ZIP, and MYB family of transcription factor/genes are regulated by drought, salt, heat, cold, etc. DREB1/CBF, DREB2, and HD-ZIP TF/gene family control are not involved in the abscisic acid (ABA) dependent pathway of stress mitigation. OsAREBs/ABF, NAC, MYB, and MYC TF/genes are identified in ABA- dependent transcriptional networks in rice. TFs are crucial part of plant signal transduction pathway mediated by signal receptors, phytohormones, and other regulatory compounds also. The expression of downstream genes may produce a subset of TFs or regulate other functional proteins involved in physiological drought adaptation. Thus, the hierarchic regulations of TF activities, downstream gene expression, and protein–protein interaction comprise a complex regulatory network, which

P. Kumar (✉) · P. Kumar · Suniti
Department of Agricultural Biotechnology, SVP University of Agriculture & Technology, Meerut, UP, India

U. Kumar
Department of Molecular Biology and Biotechnology, CCS Haryana Agricultural University, Hisar, Haryana, India

Avni · A. Mann
ICAR – Central Soil Salinity Research Institute, Karnal, Haryana, India

A. Kumar et al. (eds.), *Salinity and Drought Tolerance in Plants*,
https://doi.org/10.1007/978-981-99-4669-3_13

participates in stress response and adaptation in rice crop. This chapter summarizes the basic mechanisms of water and salt stress response at plant tissue and cellular level through transcriptional factors with the integration and discussion of regulatory network based on scientific findings in last two decades in rice. But more insight is needed to find new tools for enhancing cereals' adaptation to abiotic stresses.

Keywords

Transcriptional factors · Abiotic stress · Rice · WRKY · MYB · bZIP

13.1 Introduction

All abiotic stress such as drought, high salinity, and high temperature due to the effect of climate change is most critical environmental factors that adversely affect crop yield and plant growth. These stresses induce different biochemical and physiological responses in plants. Plant adaptation to environmental stresses is dependent upon the activation of cascades of molecular networks involved in stress perception, signal transduction, and the expression of specific stress-related genes and metabolites (Vinocur and Altman 2005). These responses include stomatal closure, repression of cell growth, decrease the photosynthesis, activation of respiration, and upregulation and down regulation of transcription factors and genes. Plants respond and adapt to water deficit at both the cellular and molecular levels, for instance, by the accumulation of osmolytes and proteins specifically involved in stress tolerance. An assortment of genes with diverse functions are induced or repressed by these stresses (Shinozaki et al. 2003; Bartels and Sunkar 2005). Most of their gene products may function in stress response to tolerance at the cellular level. Significantly, the introduction of various stress-inducible genes via gene transfer resulted in improved stress tolerance and grain yield in rice (Zhang et al. 2004; Umezawa et al. 2006). Crop plants selected for their economic yield need to survive abiotic stress through mechanisms that maintain crop yield (Basu et al. 2016).

13.2 Morphological, Physiological, Biochemical, and Molecular Changes Under the Abiotic Stress

A schematic model of abiotic stress perception and responses at morphological, biochemical, physiological, and molecular level affecting yield attributes in rice has been presented in Fig. 13.1. Exhibition of a distinct or a combination of intrinsic changes ascertains the capacity of a plant to sustain itself under unfavorable environmental conditions (Farooq et al. 2009). This comprises a range of physiological and biochemical adjustments in plants including leaf wilting, leaf area reduction, leaf abscission, root growth stimulation, alterations in relative water content (RWC), electrolytic leakage (EL), production of reactive oxygen species (ROS), and

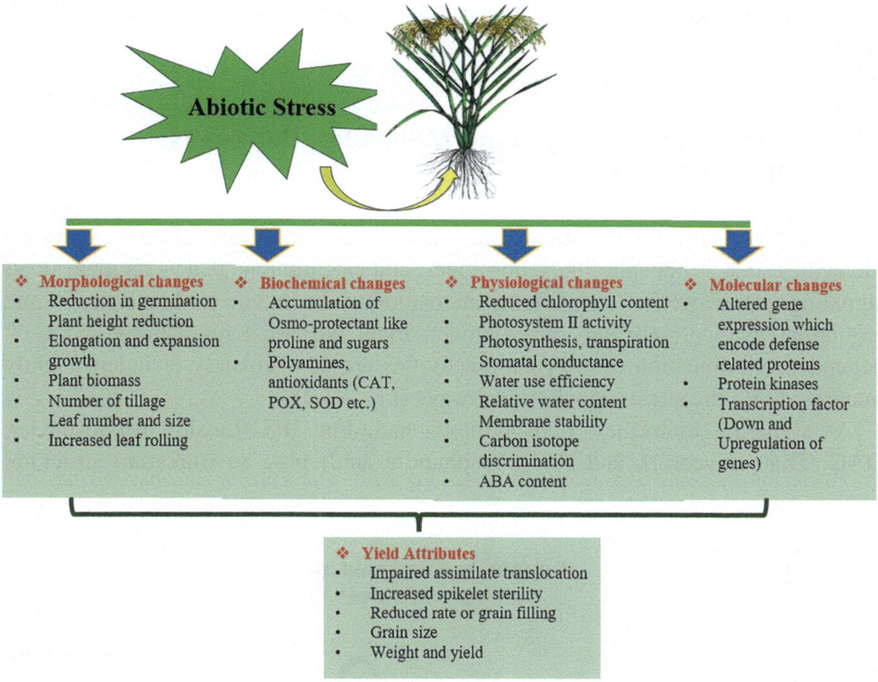

Fig. 13.1 A schematic model of abiotic stress perception and responses at morphological, biochemical, physiological, and molecular level affecting yield attributes in rice

accumulation of free radicals which disturb cellular homeostasis ensuing lipid peroxidation, membrane damage, and inactivation of enzymes thus influencing cell viability (Bartels and Sunkar 2005). Other than these, abscisic acid (ABA), a plant stress hormone, induces leaf stomata closure, thus reducing transpirational water loss and photosynthetic rate which improves the water-use efficiency (WUE) of the plant. Molecular responses to abiotic stress, on the other hand, include perception, signal transduction, gene expression, and ultimately metabolic changes in the plant thus providing stress tolerance (Agarwal et al. 2006)

Many genes respond to drought and salt stresses at the transcriptional level and the products of these genes function in the stress response and tolerance (Shinozaki and Yamaguchi-Shinozaki 2000; Bray et al. 2000; Mann et al. 2021). Transcriptome analyses have identified several genes that are induced by abiotic stresses, and these genes have been classified into two major groups (Shinozaki and Yamaguchi-Shinozaki 2000; Bray et al. 2000). One group encodes products that directly protect plant cells against stresses, whereas the products of the other group regulate gene expression and signal transduction in abiotic stress responses. Molecular and genomic analyses have shown that several different transcriptional regulatory systems are involved in stress-responsive gene induction (Mann et al. 2019; Shinozaki et al. 2003). Now, analyzing the functions of stress-inducible genes is critical to further our understanding of the molecular mechanisms governing plant stress response and tolerance, ultimately leading to enhancement of stress tolerance in rice through genetic manipulation.

13.3 Stress Signals, Signaling Molecules, Pathways, and ABA Role

Under dehydration (water stress and salt) stress conditions, diverse signaling molecules such as phospholipids, hormones, and calcium ions (Ca^{2+}) regulate stress signaling pathways for maintaining an osmotic adjustment or homeostasis and regulating plant growth and development. Plant hormonal regulations and Ca^{2+} dependent modification of enzymatic activities are co-ordinately or independently integrated into the stress signaling pathways (Fig. 13.2).

As signaling molecules, phospholipids including IP3 (Inositol triphosphate), DAG (Diacylglycerol), and PA (phosphatidic acid) play an important structural

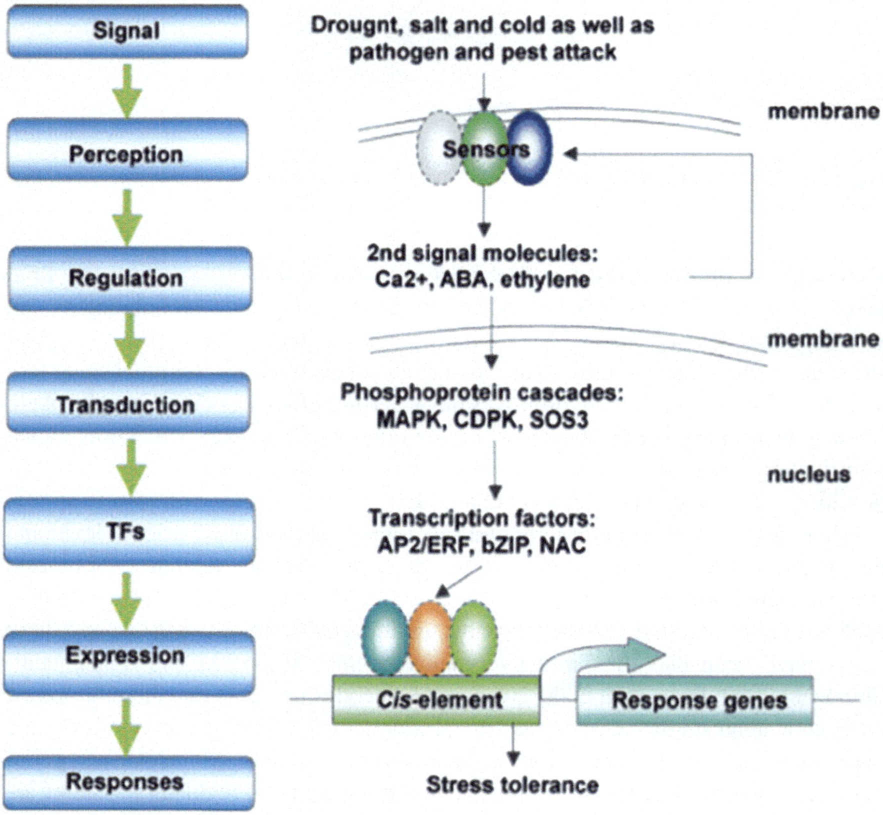

Fig. 13.2 Abiotic stress signal transduction and response pathway in plants (Source: Xu et al. 2011)

role during stress responses in inducing cytosolic Ca^{2+} spiking. Under stress conditions, PA and IP3 levels rapidly increase in rice, arabidopsis, and tobacco. Furthermore, several studies have shown that IP3 and its biosynthetic related genes rapidly increased in response to hyperosmotic stress and stress hormone ABA treatment (DeWald et al. 2001). The formation of the phospholipid-based signaling molecules is mainly regulated by phospholipase C and D (PLC/PLD). IP3 act as strong elicitors in mobilizing cytosolic Ca^{2+} levels in plants (Zhu et al. 2001). This implies the activation of phospholipid formations for salt stress induced cytosolic Ca^{2+} spiking possibly by membrane anchored salt signaling sensor proteins.

The phytohormones, ABA belongs to the class of isoprenoids (terpenoids) and involved in plant growth, development, and adaptation to various stress conditions (Verslues et al. 2006). A dynamic balance of biosynthesis and degradation determines the amount of available cellular ABA. In plants, these two processes are influenced by developmental and environmental factors such as light, salinity, and water stress. Indeed, the endogenous level of ABA and its biosynthetic genes in plant are rapidly increased by abiotic stresses including drought (Fig. 13.3) and salt stress (Fig. 13.4). The elevated ABA hormone aids plant to acclimate under lower water availability by closing guard cells and accumulating numerous proteins for osmotic adjustment. Interestingly, the expression of many ABA biosynthetic genes seems to be regulated by a stress-induced Ca^{2+}-dependent phosphorylation and its signaling pathways in rice (Saengngam et al. 2012). For example, overexpression of drought-responsive OsDSM2 (Drought-hypersensitive mutant2) and OsCam1-1 genes led to accumulation of ABA and tolerance to salt stress in rice. OsDSM2 and OsCam1-1 genes encode an ABA biosynthetic β-carotene hydrolase and a Ca^{2+}-binding calmodulin, respectively (Du et al. 2010; Saengngam et al. 2012). These results suggest that stress-activated Ca^{2+} spiking could provide the positive feedback loop for ABA biosynthesis, and this event might be critical for stress tolerance in rice.

Osmotic stress signaling is transduced via ABA-dependent or ABA-independent pathway. ABA- dependent pathway includes mitogen activated protein kinase (MAP Kinase) cascades, calcium-dependent protein kinases (CDPK), receptor-like kinases (RLK), SNF1-related protein kinases (SnRK), transcription factors (*OsRAB1*, *MYC/MYB,* and *OsNAC/SNAC*), and micro- RNAs. ABA-independent pathway includes transcription factors (*OsDREB1* and *OsDREB2*) and stress related genes (*OsPSY1*, *OsNCEDs*). Ionic stress does signaling via Ca^{2+}/PLC pathway and salt overly sensitive (SOS) pathway and Calmodulin (CaM) pathway. Ca^{2+} ions are sensed by Ca^{2+} sensor (OsCBL4) and the sensor activates calcineurin B-like protein kinase (OsCIPK24), which in turns activates Na^+/H^+ antiporter (OsSOS1), Na^+/H^+ antiporter (OsCAX1), vacuolar H^+/ATPase, vacuolar Na^+/H^+ exchangers (OsNHX1), and suppress K^+/Na^+ symporter (OsHKT1) to maintain ionic homeostasis under salt stress. Ca^{2+} ions activate calmodulin (OsMSR2) which further activates vacuolar Na^+/H^+ exchanger (OsNHX1) (Kumar et al. 2013).

An ABA-dependent Ca^{2+}-dependent protein kinases (CDPKs), OsCPK21, have been cloned and the OsCPK21-ox transgenic rice exhibited higher salt stress tolerance than wild type plant with enhanced expression of the ABA and salt-stress

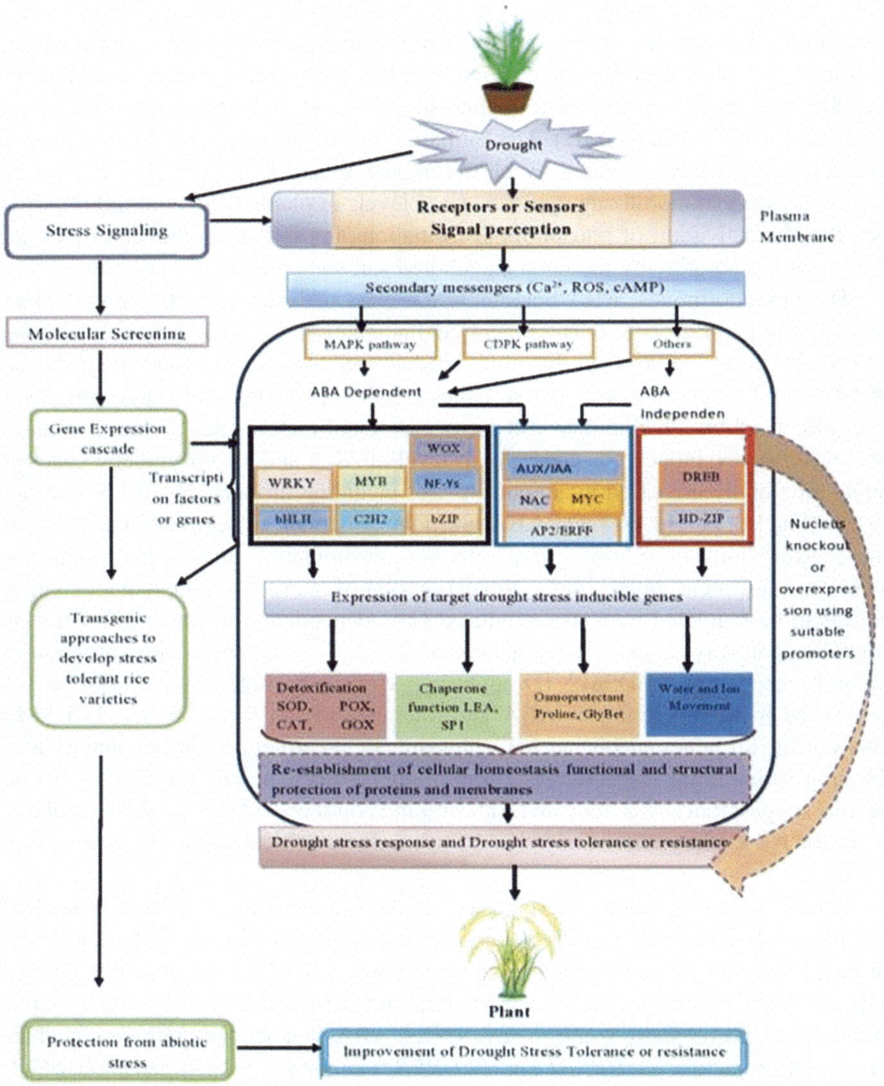

Fig. 13.3 Schematic diagram of transcriptional network involved in drought response

inducible genes such as OsNAC6 and Rab21 (Asano et al. 2011). An SNF1-related protein kinase (SnRK) functions in salt stress tolerance as well. In rice, ten members of SnRK2 family have been shown to be activated by hyperosmotic stress through phosphorylation (Kobayashi et al. 2004). Among them SAPK4 seems to play a role in the salt stress tolerance. SAPK4-ox transgenic rice revealed an improved salt tolerance with a reduced Na^+ accumulation in the cytosol. The vacuolar Na^+/H^+ antiporter gene, OsNHX1 is less expressed in the transgenic plants, indicating the

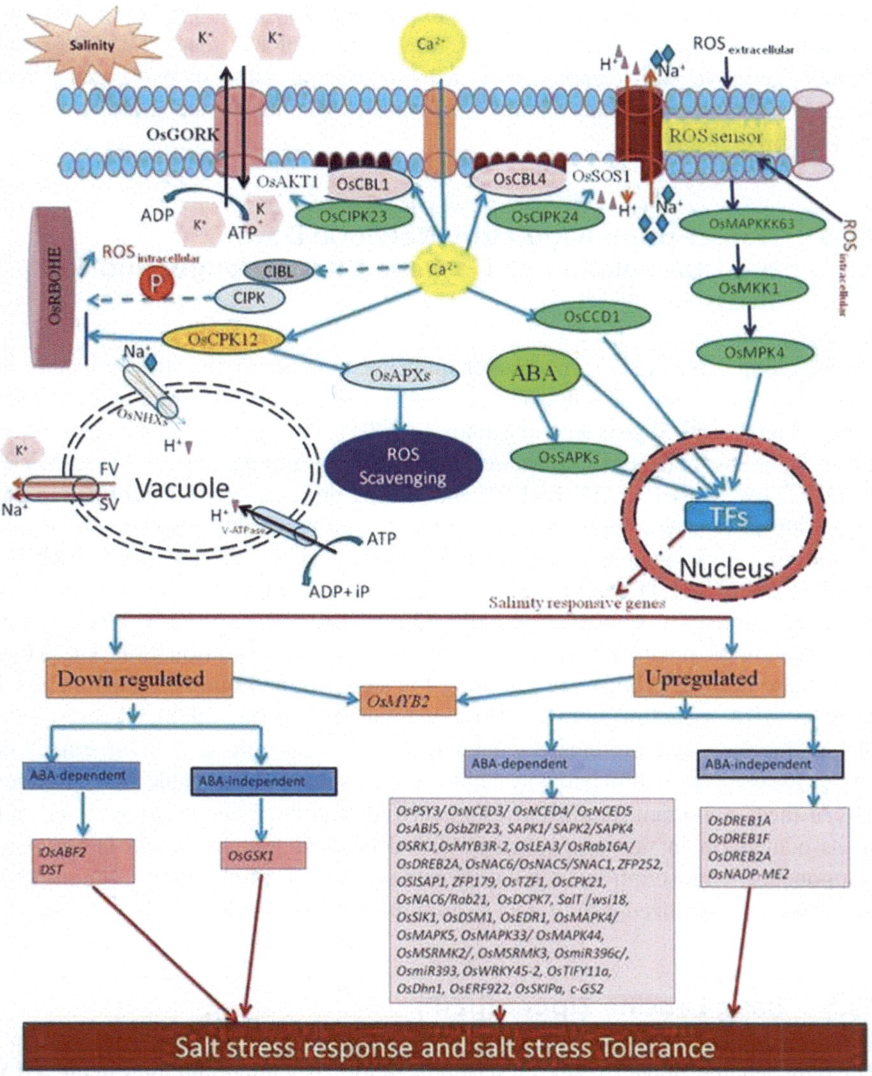

Fig. 13.4 Schematic diagram of salinity stress signal perception and response

reduced Na^+ accumulation due to cellular Na^+ exclusion rather than vacuolar sequestration of the ion (Diédhiou et al. 2008).

AREB/ABF-SnRK2 (sucrose non-fermenting-1 related protein kinase 2) pathway plays a crucial role in ABRE-mediated transcription in response to osmotic stress in plants (Fujita et al. 2013). Three members of the rice subclass III SnRK2 protein kinase family, *OsSAPK8*, *OsSAPK9*, and *OsSAPK10,* can be activated by ABA signal and hyperosmotic stress and directly phosphorylate a rice AREB/ABF, *OsTRAB1*, in response to ABA and hyperosmotic stress (Kobayashi et al. 2005).

OsSAPK9 positively regulates salt-stress tolerance and bacterial blight resistance by interacting with *OsSGT1* (Zhang et al. 2019). Two SnRK2 genes, *OsSAPK4* and *OSRK1*, improve salt tolerance in rice by regulating salt-responsive genes and functioning as the upstream regulators of stress signaling in rice (Diédhiou et al. 2008; Nam et al. 2012).

13.4 Transcription Regulatory Network: Down and Upregulation of TFs Gene Under Drought and Salt Stress

Transcription factors (TFs) have proven quite useful in improving stress tolerance in transgenic plants, through influencing expression of a number of stress-related target genes (Yamaguchi-Shinozaki and Shinozaki 2005). TFs are pivotal players in plant stress signaling and signal transduction pathways. Among the key TFs, MYB, WRKY, NAC, ZF-HD, AP2-EREBP, and bHLH proteins play crucial roles in the regulation of reprogramming the transcriptome and associated responses in stress. Considering this, genome-wide identification of NAC, ZF-HD, AP2-EREBP, WRKY, and bHLH TF families, performed in rice revealed the expression patterns and gene ontology of 17 NAC, 3 ZF-HD, 13 AP2-EREBP, 11 WRKY, 8 bHLH key genes having the putative novel variants in stress and signal transduction. These key players are needed to be studied in order to categorize and outline their functional roles in ABA signaling network (Muthuramalingam et al. 2018). Transcription factors families and proteins are tabulated with their role and involvement in respective gene regulation pathways under abiotic stress in rice (Table 13.1). Further TFs of these important families are summarized with their role in drought and salt stress adaptation in rice crop plants. ABA-dependent pathways and TFs as key components in transcriptional regulatory networks under salinity in rice. Figure 13.3 represents a generalized role of TFs in drought stress conditions.

13.5 Basic Leucine Zipper (bZIP)

The basic leucine zipper (bZIP) transcription factors play important roles in the ABA signaling pathway and could be involved in crosstalk among stress, hormone, and sugar signaling in rice. ABA, acting as a pivotal signal molecule in biotic and abiotic stress responses of plants, is required for full activation of AREB/ABF TFs (Yoshida et al. 2010). AREB/ABF is a bZIP (basic leucine zipper)-type transcription factor, which can bind to the cis-acting elements responding to ABA (Uno et al. 2000). AREB/ABF-homologous genes have been reported to be regulated by salt stress. Both OsABF1 and OsABF2 are positive regulators for abiotic stress responses and ABA-dependent signaling transduction pathways in rice (Hossain et al. 2010a, b). Many bZIP TFs, such as OsbZIP12 (Joo et al. 2014), OsbZIP23 (Xiang et al. 2008), OsbZIP46 (Tang et al. 2012), OsbZIP62 Yang et al. (2019), are positive regulator of ABA signaling and drought stress tolerance in rice and their expression are also

Table 13.1 Transcription factors with their function under drought and salt stress in rice and other plants

S. No.	Rice TF and gene	Type of stress	Play the role in metabolic cycle, ABA response	Trait regulated in crop	Transgenic plant/crop used	Reference
bZIP						
1	OsbZIP23	Drought, salt, PEG	Up regulated Yes	Photosynthesis, regulation of expression of genes involved in stress response and tolerance	Transgenic rice Plants	Xiang et al. (2008) Kim et al. (2017)
2	OsbZIP12	Drought	Yes	ABA signaling pathway, rapid stomatal closure	Transgenic rice Plants	Joo et al. (2014)
3	OsbZIP16	Drought	Yes	Seedling and tillering stages	Transgenic rice Plants	Chen et al. (2012)
4	OsbZIP45	Drought	Up regulated Yes	Different tissues such as root, leaves, flower, and growth stages of rice, TFs in response to drought	Transgenic rice Plants	Park et al. (2015)
5	OsbZIP46	Drought	Up regulated Yes	Interact with homologs of SnRK2 protein kinases that phosphorylate ABFs	Transgenic rice Plants	Zhu et al. (2012)
6	OsAREB1	Drought, heat	Yes	Positive regulator in drought/heat stresses response, but a negative regulator in flowering time in arabidopsis	Transgenic plants arabidopsis	Jin et al. (2010)
7	OsbZIP52, 71, 72/ RISBZ5	Drought, salinity	Yes	Flowers and the early stages of seed development	Transgenic plants, rice	Liu et al. (2012, 2014)
10	OsbZIP12	Drought	Yes	Involved in crosstalk among stress, hormone, and sugar signaling	Rice	Joo et al. (2014)

(continued)

Table 13.1 (continued)

S. No.	Rice TF and gene	Type of stress	Play the role in metabolic cycle, ABA response	Trait regulated in crop	Transgenic plant/crop used	Reference
11	OsbZIP 62	Drought	Yes	Promote with the ABA-activated protein kinases (SAPKs) under the stress conditions	Rice	Yang et al. (2019)
12	OsbZIP 66	Drought, salinity	Yes		Rice	Yoon et al. (2017)
13	OsABF2	Salt	Downregulated/ yes	Gene expressed modulation through an ABA-dependent pathway	Rice	Hossain et al. (2010a, b)
14	OsTF1L	Drought	Yes	Reproductive stage of growth	Rice	Bang et al. (2019)
15	OsABI5	Drought; salinity	Yes	Down regulated by cold and drought in seedlings low rice fertility	Tobacco	Zou et al. (2008)
MYB						
16	OsMYB4	Drought, freezing	NA	15% reduction in the RWC	Transgenic plants Tomato Apple	Vannini et al. (2004) Pasquali et al. (2008)
17	OsMYB3R-2	Drought, salinity, cold	Up regulated/yes	Regulation of stress-responsive gene expression	Transgenic plants arabidopsis/rice	Dai et al. (2007)
18	OsMYB2	Salinity	5 h up regulated and 24 h down regulated/yes	Over expression of OsMYB2 in rice seedling as positive regulator to salt, cold and dehydration stress tolerance	Transgenic rice	Yang et al. (2012)

19	*OsMYB48-1*	Drought, salinity	NA	Root, stem, sheath, leaf, and panicle, but mainly expressed in roots	Transgenic plants rice	Xiong et al. (2014)
20	MID1	Drought	NA	Improved pollen fertility and higher seed setting rate	Rice	Guo et al. (2016)
21	OsMYB1	Salt, drought, and heat stresses	NA	Positive and upregulation response	Transgenic plants rice	Deeba et al. (2017)
22	OsMYB6	Drought and salinity stress	–	Positive and upregulation response	Transgenic rice	Tang et al. (2019)
23	OsMYB1R1	Drought	–	Negatively regulates drought resistance	Transgenic rice	Peng et al. (2022)
24	OsMYB511	Drought, salt, heat stress	NA	Earlier development stage in rice panicles	Rice	Huang et al. (2015)
NAC						
24	OsNAC5 (ONAC009)	Drought, salt	Upregulated/yes	Roots, increases root diameter, regulation of OsLEA3 expression	Rice	Jeong et al. (2013) Takasaki et al. (2010)
25	SNAC1	Cold, drought, salt	Up regulated/yes	Play important role in enhancing salt stress tolerance	Transgenic rice	Hu et al. (2006)
26	OsNAC6	Drought, salinity	Up regulated NA	Regulation of stress-responsive gene expression OsCPK21 downstream gene	Transgenic rice	Nakashima et al. (2007) Asano et al. (2011)
27	ONAC045	Drought, salinity	NA	Photoassimilates are transported through phloem sieve element cells in plants	Rice	Zheng et al. (2009) Furuta et al. (2014)
28	SNAC2	Cold, drought, salt	NA	Overexpression in plants increased the sensitivity to ABA	Transgenic rice	Hu et al. (2008)

(continued)

Table 13.1 (continued)

S. No.	Rice TF and gene	Type of stress	Play the role in metabolic cycle, ABA response	Trait regulated in crop	Transgenic plant/crop used	Reference
29	EcNAC67	Drought, salinity	U	Increases RWC in leaves, delays leaf rolling symptoms, better stomatal regulation, maintains higher root and shoot biomass	Rice	Rahman et al. (2016)
30	ONAC002(SANC1/ OsNAC9) 003, OsNAC10, 022 122, 058, 066	Drought, salinity	Upr egulated/NA	Play important role in enhancing drought, and salt stress tolerance	Transgenic rice	Chen et al. (2014)
WRKY						
31	OsWRKY45	Drought, disease	Yes	Response to wounding, senescence, development dormancy and drought tolerance, solar ultraviolet-B radiation, metabolism, hormone signaling	Transgenic plants arabidopsis	Qiu and Yu (2009)
32	OsWRKY 3, 4, 8	Drought, cold, flood	Yes		Transgenic Plants rice and arabidopsis	Berri et al. (2009)
33	OsWRKY 11, 18, 22, 24, 42, 50, 53, 78, 84, 96, 100	Drought, heat	Up regulated	Response to wounding, senescence and drought tolerance, metabolism, hormone signaling pathways	Transgenic plants rice and arabidopsis	Berri et al. (2009)
34	OsWRKY45-2	Salt	Up regulated in after 4 days	Positively regulates ABA signaling and negatively regulates rice response to salt stress	Transgenic plants rice	Tao et al. (2011)
35	OsWRKY55	Drought	Up regulated	Plant height	Rice	Huang et al. (2021)

36	OSISAP1	Drought	NA	Delayed leaf rolling, lesser membrane damage, and lipid peroxidation under water deficit stress	Transgenic plants rice	Mukhopadhyay et al. (2004) Dansana et al. (2014)
37	*OsbHLH2*	Salt and osmotic stress	NO	Promotes OsTPP1 expression and increases trehalose content and resistance to chilling damage	Transgenic Plants rice	Zhou et al. (2009)
WUS/WOX						
38	OsWOX5	Drought	Up regulated/yes/no	Key regulator of root development in rice	Transgenic plants rice	Kawai et al. (2022)
39	OsWOX13	Drought	Up regulated/yes/no	Regulated spatially in vegetative organs but temporally in flowers and seeds	Transgenic plants rice	Minh-Thu et al. (2018)
40	*OsNS1/OsNS2, OsWOX3, OsWOX9A*	Drought	Up regulated/yes/no	Expressed in panicle and endosperm development	Transgenic plants rice	Cheng et al. (2014)
41	*OsWOX 4,9B, 9D, 10,12 A, B*	Drought	Up regulated/yes/no	Expressed in seeds (72 h after imbibition) during root emergence or growth; panicle development and embryogenesis	Transgenic Plants rice	Cheng et al. (2014)
42	OsWOX10	Drought	Upregulated Yes/No	Primordia formed from an early developmental stage	Transgenic plants rice	Kawai et al. (2022)
43	*WOX11*	Drought	Upregulated Yes/No	Controlling root hair formation and root system development	Transgenic Plants rice	Zhao et al. (2009)
C2H2						

(continued)

Table 13.1 (continued)

S. No.	Rice TF and gene	Type of stress	Play the role in metabolic cycle, ABA response	Trait regulated in crop	Transgenic plant/crop used	Reference
44	ZFP245	Drought, cold	Up regulated/yes	Exhibit growth retardation, but showed growth sensitivity against exogenous abscisic acid, increased free proline levels	Arabidopsis and tobacco	Huang et al. (2009)
45	ZFP182	Cold, salt, drought	Up regulated/no	ABA-induced antioxidant defense and the expression of *ZFP182* is regulated by rice MAPKs in ABA signaling	Arabidopsis and tobacco	Zhang et al. (2012)
46	ZFP179	Salt, PEG 6000, and ABA treatments	Yes	Highly expressed in immature spikes, and markedly induced in the seedlings	Arabidopsis and tobacco	Sun et al. (2010)
47	Oshox22	Drought, salt	Yes/NA	Auxin signaling and transport, Vascular development	Rice	Zhang et al. (2012)
48	OsHox4	Drought stress	Yes/NA	Regulation of apical embryo patterning, embryonic shoot meristem formation, organ polarity, vascular development and meristem function	Rice	Agalou et al. (2008)
49	OsIAA1	Drought stress	U	Auxin related development defects, including altered gravitropism and apical dominance	Transgenic rice, arabidopsis	Song et al. (2009)

50	OsIAA6	Drought stress	Yes	Apical dominance, lateral/ adventitious root formation, tropisms,	Transgenic rice, arabidopsis	Song et al. (2009)
51	OsIAA20	Drought, salinity	Yes		Transgenic rice, arabidopsis	Song et al. (2009)
52	DRO1	Drought	U	Influences root growth angle, induces root elongation and deeper rooting	Rice	Uga et al. (2013)
53	OsPYL/RCAR5	Drought, salinity	Yes	Stomatal closure, maintains the fresh weight of Leaf blade	Rice	Yadav et al. (2020)
54	OsPYL10	Dehydration and freezing	Yes	Maintained higher RWC, membrane stability index, chlorophyll content, and accumulated lower amount of MDA and H_2O_2 as compared with WT plants	Transgenic rice	Yadav et al. (2020)
55	OsPYL1, 4, 7, 8, 9, 11, 12, 13,	Drought,	Yes	Panicle among the tissues, seed (embryo, endosperm and caryopsis), shoots, flag leaf	Rice	Yadav et al. (2020)
56	OsPSY3/ OsNCED3/4/5	Drought, salt	Upregulated/ ABA-dependent	Regulate salt, dehydration stress tolerance through ABA biosynthesis	Transgenic rice	Welsch et al. (2008)
57	LEA3	Drought, salt	Up regulated/yes	MYB2 downstream gene	Transgenic rice	Duan and Cai (2012)
58	Rab16A/SalT/ wsi18	Drought, salt	Up regulated/yes	MYB2 downstream gene/ OsCDPK downstream gene	Transgenic rice	Yang et al. (2012) Saijo et al. (2000)
59	OsAL5, OsTMS5	Drought, heat	NA	Pollen, anther	Rice	Wen et al. (2021)

(continued)

Table 13.1 (continued)

S. No.	Rice TF and gene	Type of stress	Play the role in metabolic cycle, ABA response	Trait regulated in crop	Transgenic plant/crop used	Reference
AP2/ERF						
60	OsDREB1A, 1C, 1G, 2A	Drought, cold, salt, wounding	Up regulates/no	Enhanced stress tolerance through expression of stress-responsive genes in vegetative tissues and accumulation of osmolytes	Transgenic arabidopsis, tobacco, rice, wheat, soybeans	Dubouzet et al. (2003), Chen et al. (2008)
61	OsDREB2B	Drought, thermotolerance	NA	Leaf sheath, root tissues, increases root number and length	Transgenic plants arabidopsis	Matsukura et al. (2010)
62	OsDREB1F, 2A	Drought, salt, cold	Upregulation of ABA-independent signaling pathway	Almost all tissues, but higher in callus and panicle; Growth, fertility and development of rice and regulation of Salt stress tolerance	Transgenic plants arabidopsis, rice	Wang et al. (2008), Matsukura et al. (2010), Mallikarjuna et al. (2011), Cui et al. (2011)
63	DREB1, DREB2	Drought, salinity, cold, disease, and flooding stress	No	Function as *trans*-acting factors in two separate signal transduction pathways under low-temperature and dehydration conditions	Transgenic plants arabidopsis	Rashid et al. (2012), Bihani et al. (2011), Liu et al. (1998)
64	OsERF48, OsERF71	Drought	ABA independent	Enhances root growth, drought tolerance, and grain yield	Rice	Jung et al. (2017), Lee et al. (2017a, b)
65	OsABF2	Salt	Down regulated/ yes	Gene expressed modulation through an ABA-dependent pathway	Rice	Hossain et al. (2010a, b)

66	OsERF922	Salt	Down regulated/ yes	Negatively regulate the salt stress tolerance	Rice	Liu et al. (2012)
67	*OsDSM1*	Salt	Up regulated/yes	Early signaling component in regulating response to osmotic stress through scavenging of ROS	Rice	Ning et al. (2011)
68	*OsEDR1*	Salt	Up regulated/yes	An essential positive regulator of tolerance to salt stress	Rice	Kim et al. (2003)
69	*OsMAPK4/ OsMAPK5*	Salt	Up regulated/yes	An essential positive regulator of tolerance to salt stress	Rice	Xiong et al. (2003)
70	*OsMAPK33/ OsMAPK44*	Salt	Up regulated/yes	Play important role in enhancing salt stress tolerance through unfavorable ion homeostasis	Rice	Lee et al. (2011)
71	*OsMSRMK2/ OsMSRMK3*	Salt	Up regulated/yes	Play role in salt stress signaling pathway in rice	Rice	Agarwal et al. (2006)
72	*OsmiR396c/ OsmiR393*	Salt	Up regulated/yes	Overexpression shows sensitive phenotype in salt stress	Rice	Gao et al. (2011)
73	*OsTIFY11a*	Salt	Up regulated/yes	Regulation of seed germination	Rice	Ye et al. (2009)
74	*OsSK1Pa*	Salt, drought	Up regulated/yes	Play important role in enhancing salt and drought stress tolerance	Rice	Hou et al. (2020)
75	*cGS2*	Salt	Up regulated/yes	Accumulation of glutamine leads to salt stress tolerance	Rice	Hoshida et al. (2000)
76	*OtsA+OtsB*	Salt, drought	Up regulated/yes		*E. coli*	Garg et al. (2002)

(continued)

Table 13.1 (continued)

S. No.	Rice TF and gene	Type of stress	Play the role in metabolic cycle, ABA response	Trait regulated in crop	Transgenic plant/crop used	Reference
				Postively regulate salt and drought stress tolerance through the accumulation of trehalose		
77	*OsCPK12*	Salt	No	An essential positive regulator of tolerance to salt stress	Rice	Asano et al. (2012)
78	*Osrboh1/OsAPx2/ OsAPx8*	Salt	Regulated by *OsCPK12*/no	Play important role in H_2O_2 homeostasis	Rice	Asano et al. (2012)
79	*OsNADP-ME2*	Salt	Up regulated/no	Play important role in enhancing tolerance of plants to salt and osmotic stress	Rice	Liu et al. (2007)
80	*OsSOS1/ OsCIPK24/OsCBL4*	Salt	Up regulated/no	Important in Calcium signaling	Rice	Atienza Martínez et al. (2007)
81	*OsAKT1*	Salt	Up regulated/no	Play important role in enhancing salt tolerance in plants	Rice	Golldack et al. (2003)
82	*OsCLC1*	Salt	Up regulated/no	Act as proton chloride antiporters	Rice	Nakamura et al. (2006)

induced by drought, hydrogen peroxide, and abscisic acid (ABA) treatment. OsbZIP71 encodes a rice bZIP TF, an atomic-limited protein linked to the G-box theme but play a vital role in rice ABA-independent drought and salt tolerance and flowering time. OsbZIP72 plays a decisive role in drought resistance through ABA signaling and may help with drought tolerance in rice. OsbZIP72 is a critical regulator in abiotic stress reaction and ABA signaling transduction pathways (Liu et al. 2014). Upregulated expression of OsbZIP66 is significantly induced upon treatments of rice plants with drought, high salinity, and ABA-dependent manner (Yoon et al. 2017). TF OsbZIP46 directs ABA signaling-mediated drought tolerance in rice by regulating pressure-related genes (Tang et al. 2012). Expression of the OsbZIP23 gene causes an adverse effect on stress, including ABA, salt, and drought, while other stress-responsive genes of this family are slightly induced only by one or two of these stressors.

The expression of the OsABI5, a bZIP gene, is initiated by high salinity and ABA and down regulated by cold and drought in rice seedlings. Over expression of the OsABI5 gene in rice conserved high sensitivity to salt stress, and OsABI5 repression enhanced drought stress tolerance and resulted in low rice fertility (Zou et al. 2008). OsABF1 binds to DNA sequences containing an ACGT core motif. Overexpression of one of these genes, COR413-TM1, which encodes a putative thylakoid membrane protein, resulted in a drought tolerance phenotype without obvious side effects. In addition, OsABF1 directly regulates the expression of the protein phosphatase 2C (OsPP48 and OsPP108) and bZIP (OsbZIP23, OsbZIP46, and OsbZIP72) genes, thus forming a complex feedback circuit in the drought/abscisic acid signaling pathway (Zhang et al. 2017). OsbZIP23 transcription factor upregulated by drought are closely associated with the starch-sucrose pathway while those that are downregulated are involved in photosynthesis (Kim et al. 2017).

Rice (*Oryza sativa*) homeodomain-leucine zipper transcription factor gene, OsTF1L (*Oryza sativa* transcription factor 1-like), is a key regulator of drought tolerance mechanisms. Overexpression of the OsTF1L in rice significantly increased drought tolerance at the vegetative stages of growth and promoted both effective photosynthesis and a reduction in the water loss rate under drought conditions. Importantly, the OsTF1L overexpressing plants showed a higher drought tolerance at the reproductive stage of growth with a higher grain yield than non-transgenic controls under field drought conditions. Genome- wide analysis of OsTF1L overexpression plants revealed upregulation of drought-inducible, stomatal movement, and lignin biosynthetic genes (Bang et al. 2019). A b-ZIP TF OsGATA8 has been shown to contribute toward multiple stress tolerance and seed development in Arabidopsis and rice (Nutan et al. 2020).

13.6 MYB Factors

MYB transcription factors have been demonstrated to play key regulatory roles in plant growth, development, and abiotic stress response. The positive and upregulation response of OsMYB1 suggests that its overexpression in crop plants may help in providing protection to plants to grow under wounding, salt, drought, and heat stresses in rice (Deeba et al. 2017). OsMYB511 is a TF in rice that controls abiotic stress responses and has been activated by exogenous ABA, high temperature, and osmotic pressure. OsMYB511 gene showed high expression at an earlier development stage in rice panicles. A co-articulation investigation uncovered an extra two MYB qualities co-communicated with OsMYB511, suggesting that they coordinate direct pressure reactions in rice (Huang et al. 2015). OsMYB6-overexpressing plants showed increased tolerance to drought and salt stress compared with wild type plants, as are evaluated by higher proline content, higher CAT and SOD activities, lower REL and MDA content in transgenic plants under drought and salt stress conditions. These results indicate that OsMYB6 gene functions as a stress-responsive transcription factor which plays a positive regulatory role in response to drought and salt stress resistance (Tang et al. 2019). OsMYB48-1 functions as a novel MYB-related TF overexpressed in drought and salinity tolerance by regulating stress-induced ABA synthesis Xiong et al. (2014). MID1 (MYB Important for Drought Response1), encoding a putative R-R-type MYB-like transcription factor improved rice yield under drought. MID1 primarily expressed in root and leaf vascular tissues, with low level in the tapetum is induced by drought and other abiotic stresses. Compared with wild type, MID1-overexpressing plants were more tolerant to drought at both vegetative and reproductive stages and produced more grains under water stress (Guo et al. 2016). Expression of OsMYB1R1 was downregulated by drought stress. The OsMYB1R1- overexpression plants exhibited increased relative electrical conductivity (REC), increased malondialdehyde (MDA) content, and decreased proline content compared with the wild type, whereas lower REC and MDA content and higher proline content in the RNAi plants (Peng et al. 2022).

13.7 NAC Factors

The NAC (NAM, AFAT, and CUC) transcription factors play critical roles in rice development and stress regulation. NAC transcriptional factors constitute a large family with 158 members in *Oryza sativa* indica and several members of this family have been demonstrated to play crucial roles in rice abiotic stress response. NAC transcription factors are one of the regulatory proteins that are involved in stress signaling pathway. The involvement of NAC TFs in rice abiotic stress response is extensively explored. Over expression of ONAC002 (SANC1/OsNAC9), ONAC003, ONAC048 (SNAC2/OsNAC6), ONAC009 (OsNAC5), ONAC122 (OsNAC10), ONAC045, or ONAC058 (OsNAP) ONAC022, ONAC066 improved significantly the drought and salinity tolerance in transgenic rice (Hu et al. 2006, 2008; Nakashima et al. 2007; Zheng et al. 2009; Jeong et al. 2010, 2013; Takasaki et al. 2010; Song et al. 2011; Redillas et al. 2012; Chen et al. 2014; Liang et al. 2014; Fang et al. 2015; Hong et al. 2016; Yuan et al. 2019a, b) and some of these transgenic rice lines showed increased drought tolerance under severe drought stress conditions without any adverse effect on yield or even with yield increase.

Functional studies have identified at least 9 rice ONAC genes that play important roles in abiotic stress tolerance and these abiotic stress-related includes ONAC002 (SNAC1/OsNAC9), ONAC048 (SNAC2/OsNAC6), ONAC009 (OsNAC5), ONAC122 (OsNAC10), ONAC045, ONAC058 (OsNAP), ONAC022, ONAC095, and ONAC003 (SNAC3). It is found that overexpression of SNAC1, SNAC2 or ONAC022 significantly enhanced tolerance to dehydration, cold and salt stresses in transgenic rice plants, while transgenic rice plants overexpressing the root-specific OsNAC5, OsNAC6, OsNAC9, or OsNAC10 displayed significant improvement in drought tolerance. Root-specific overexpression of OsNAC10 enlarges roots, enhancing drought tolerance of transgenic plants, which increases grain yield significantly under field drought conditions (Jeong et al. 2010).

Overexpression of ONAC022; OsNAC2 is induced by drought, high salinity, and ABA-mediated pathways (Hong et al. 2016; Jiang et al. 2019). ONAC106 is induced by salt and cold stresses, indicating that this gene involves in abiotic stress response. The ABRE (Abscisic Acid Regulatory Element) cis-element is identified in the promoter region of ONAC106, suggesting that it may involve in the abscisic acid (ABA)-dependent signaling pathway (Basri et al. 2016). OsNAC2 directly binds the promoters of late embryogenesis abundant 3 (OsLEA3) and stress-activated protein kinases 1 (OsSAPK1), two marker genes in the abiotic stress and ABA response pathways, respectively. Rice OsNAC2 regulates both abiotic stress responses and ABA-mediated responses, and acts at the junction between the ABA and abiotic stress pathways (Shen et al. 2017a, b; Jiang et al. 2019).

Expression of ONAC095 is upregulated by drought stress and abscisic acid (ABA) but downregulated by cold stress. ONAC095 is found to have dual functions in drought and cold stress tolerance. The suppression of ONAC095 demonstrate that it plays opposite roles in drought and cold stress tolerance, acting as a negative regulator of drought response but as a positive regulator of cold response in rice (Huang et al. 2016). ONAC066 is a nucleus-localized transcription activator that can respond to multiple abiotic stress factors. Functional analyses using overexpression and RNAi-mediated suppression transgenic lines demonstrate that ONAC066 is a positive regulator of drought and oxidative stress tolerance in rice (Yuan et al. 2019a, b).

The rice ONAC022 over expression demonstrated higher survival ratio and less Na^- accumulation are observed in roots and shoots in response to drought and salt stress. Expression of ONAC022 is induced by drought, high salinity, and abscisic acid (ABA). ONAC022 functions as a stress-responsive NAC with transcriptional activator activity and plays a positive role in drought and salt stress tolerance through modulating an ABA-mediated pathway (Hong et al. 2016). Overexpression of OsNAC14 resulted in drought tolerance at the vegetative stage of growth. Field drought tests demonstrated that OsNAC14 overexpressing transgenic rice lines exhibited higher number of panicle and filling rate compared to non-transgenic plants under drought conditions. OsNAC14 mediates drought tolerance by recruiting factors involved in DNA damage repair and defense response resulting in improved tolerance to drought (Shim et al. 2018).

The OsNAC6 gene is one of the transcription factors in rice that regulate gene expression during stress conditions. The overexpression OsNAC6 transgene exhibited higher tolerance against drought and salinity stresses (Rachmat et al. 2014). OsNAC6-mediated root structural adaptations, including increased root number and root diameter, which enhanced drought tolerance. The OsNAC6 root-specific overexpressing transgenic rice lines are less affected by drought stress than nontransgenic controls. Genome-wide analyses of loss and gain of function mutants revealed that OsNAC6 upregulates the expression of direct target genes involved in membrane modification, nicotianamine (NA) biosynthesis, glutathione relocation, 30-phophoadenosine 50-phosphosulphate accumulation and glycosylation, which represent multiple drought tolerance pathways (Lee et al. 2017a, b). The knockout of OsNAC006 caused enhanced sensitivity to drought and heat tolerance in rice, which lowered chlorophyll levels, reduced SOD and POD enzyme activities, and increased MDA content. RNA sequencing (RNA-seq) transcriptome analysis revealed that OsNAC006 regulates the expression of genes mainly involved in response to stimuli, oxidoreductase activity, cofactor binding, and membrane-related pathways (Wang et al. 2020).

13.8 WRKY Factors

WRKY transcription factor (TF) is one of the largest TF families in plants and plays an important role in plant development and stress protection. WRKY transcription factors (TFs) have been reported to respond to biotic and abiotic stresses and regulate plant growth and development. OsWRKY11 is induced by pathogens, drought, and heat, suggesting a function in biotic and abiotic stress responses. It is a transcriptional activator that localized to the nucleus. Ectopic expression of OsWRKY11 resulted in enhanced resistance to a bacterial pathogen, Xanthomonas oryzae pv. ryzae. OsWRKY11 also bound directly to the promoter of a drought-responsive gene, RAB21, activating its transcription. OsWRKY11 integrates plant responses to pathogens and abiotic stresses by positively modulating the expression of biotic and abiotic stress-related genes (Lee et al. 2018). OsWRKY47 imparted drought stress tolerance. OsWRKY47 expression is caused by drought stress in plants, and their mutants showed higher susceptibility to drought and decreased yield, whereas overexpressing OsWRKY47 plants are more tolerant. Interestingly, a WRKY transcription factor named OsWRKY78 has been reported to be involved in regulation of grain size in rice under the drought conditions (Guo et al. 2019). OsWRKY55 is overexpressed in various tissues and plays a critical role in responses to drought stress and reduce plant height under normal conditions by decreasing the cell size in rice, further providing valuable information for crop improvement (Huang et al. 2021).

Upregulated WRKY57 transcription factor is able to confer drought tolerance to transgenic rice (Oryza sativa) plants. The enhanced drought tolerance of transgenic rice revealed lower water loss rates, cell death, malondialdehyde (MDA) contents, and relative electrolyte leakage while a higher proline content and reactive oxygen species-scavenging enzyme activities during stress conditions (Jiang et al. 2016). OsWRKY97, which positively regulates drought tolerance in rice, over expressed in various tissues and could be induced by various abiotic stresses and increased the accumulation of ABA and reduced water loss. Over expression of OsWRKY97 in plants achieved higher proline content and reduced levels of MDA and reactive oxygen species (ROS) (Hou et al. 2020). OsWRKY11, *OsWRKY56*, and *OsWRKY62*, suggesting a feedback control acting on the upregulation of WRKY transcription factors. Genes involved in the submergence stress and resulting aerenchyma development had a W-box in their promoter regions (Viana et al. 2018). OsWRKY5 negatively regulates drought tolerance, which is mainly expressed in developing leaves at the seedling and heading stages. Its expression is reduced by drought stress and by treatment with NaCl, mannitol, and abscisic acid (ABA). The loss of OsWRKY5 activity increased sensitivity to ABA, thus promoting ABA-dependent stomatal closure. OsWRKY5 functions as a negative regulator of ABA-induced drought stress tolerance, suggesting that inactivation of OsWRKY5 or manipulation of key OsWRKY5 targets could be useful to improve drought tolerance in rice cultivars (Lim et al. 2022).

13.9 Zinc Finger Proteins

The gene from the CCCHZF rice family, OsC3H10, primarily expressed in plants, consequently causes a rapid decline during seed imbibition; moreover, the expression of OsC3H10 was induced by drought high salinity and ABA. OsC3H10 regulated drought resistance by modulating stress-related gene expression involving various drought-tolerant pathways. However, root-specific overexpression of OsC3H10 was inadequate to cause drought tolerance, whereas the plant overall had increased drought tolerance (Seong et al. 2020). Several genes of zinc finger proteins are involved in playing essential roles in drought salt stress. Using CRISPR-Cas9 mediated genome editing in rice (OsDST), the DST gene increased drought and salinity stress tolerance and improved crop production. The DST mutant was first produced in rice, and stomatal density was associated with reducing stomatal development genes in the DST mutant (Kumar et al. 2020).

13.10 Basic Helix-Loop-Helix Protein (bHLH)

Basic helix-loop-helix protein (bHLH) is the most extensive class of transcription factors in eukaryotes, which can regulate gene expression through interaction with specific motif in target genes. bHLH transcription factor is not only universally involved in plant growth and metabolism, including photomorphogenesis, light signal transduction, and secondary metabolism, but also plays an important role in plant response to stress (Sun et al. 2018).

The rice OsbHLH068 gene is a part of the ABA-dependent pathway, and delayed seed germination and late flowering. OsbHLH068 overexpression in Arabidopsis resulted in late flowering, delayed seed germination, decreased salt-induced H2O2 accumulation, increased MDA, and promoted root elongation (Chen et al. 2017). The rice OsbHLH035 is involved in germinating seeds and enabling the recovery of seedlings from salt stress through the ABA-dependent and ABA-independent pathways. Over expression of the OsbHLH035 gene, seed germination was delayed, and the average growth of Arabidopsis seedlings recovered after salt stress (Ortolan et al. 2021). OsbHLH35 presented small and curved anthers, leading to a reduction of 72% on seed production. This regulation was also observed in planta through the analysis of transgenic plants over expressing OsGRF11 (OsGRF11 OE), confirming that OsGRF11 is a negative regulator of OsbHLH35 in rice. OsbHLH35 plays an essential role in anther development in rice and the fine control of its expression is crucial to ensure proper seed production (Ortolan et al. 2021).

It is well reported that bHLH TFs play essential roles in gene regulation in many plant species under various abiotic stressors. OsbHLH024 plays the role of a negative regulator of salt stress, which helps to understand better the molecular basis of rice production improvement under salt stress (Alam et al. 2022).

NH_4^+ is important for the growth and the yield production of paddy-soil grown rice. We further analyzed transcription factors respond to NH_4^+, and identified an NH_4^+-repressed bHLH transcription factor involved in NH_4^+ acquisition process. RNAi-mediated suppression of bHLH and bHLH overexpression (bHLH OX) resulted in the accumulation of lower and higher NH_4^+ contents in transgenic rice roots, respectively (Wang et al. 2017).

Salt, cold, and heavy metals are major factors limiting crop productivity and quality. OsSMP1 (stress membrane protein) gene of rice responds to cold, drought and heat treatment. The results of RT-PCR showed that the relative expression level of the OsSMP1 gene was upregulated in rice under ABA, high salt, and cold treatments. There were some cis-elements upstream of OsSMP1 that regulate ABA, low temperature, and drought. These results imply that OsSMP1 is a positive regulator of rice tolerance to salt, cold, and heavy metal via an ABA-dependent pathway and is potential for improving rice tolerance to abiotic stress (Zheng et al. 2021).

13.11 AP2/ERF

The AP2/ERF is one of the largest families of TFs, with 301 members in *Oryza sativa* crop plants, which regulate multiple responses such as stress, metabolism, and development in plants. The rice ABA-independent gene OsERF48 directly binds to the promoter of OsCML16 via AP2/ERF cis-acting regulatory elements, thereby activating its transcription. Overexpression of OsERF48 causes regulation of OsCML16, a calmodulin like protein gene that enhances root growth, drought tolerance, and grain yield and is involved in cell wall proteins, carbohydrate metabolism, and stress signaling in drought conditions in the field (Liu et al. 2020). The rice OsERF71 gene is an AP2/ERF TF involved in an ABA- independent pathway controlling drought resistance by regulating cell wall modifications. After OsERF71 overexpression, roots are sufficient for drought resistance phenotypes and increase yield under drought stress (Lee et al. 2017a, b).

AP2/ERF transcription factor HAIRY LEAF 6 (HL6) regulates auxin biosynthesis. HL6 controls the elongation of epidermis hair, and its regulatory role in the elongation of epidermis hair is mainly dependent on the function of OsWOX3B (Sun et al. 2017). GA is the main phytohormone for regulating plant height and its biosynthesis can be regulated by the AP2/ERF family gene. REDUCED PLANT HEIGHT1 (OsRPH1) negatively regulates plant height, internode length, and leaf sheath length by controlling the expression of GA-related genes. Exogenous GA3 treatment restores the defect of plant growth in overexpression lines. A recent study showed that a AP2 transcription factor, small organ size1 (SMOS1)/nitrogen-mediated tiller growth response 5 (NGR5) was a key element of gibberellin signaling pathway and interacted with gibberellin receptor GA-insenstive dwarf1 (GID1) protein. NGR5 can also interact with the protein complex of Polycomb Repressive Complex 2 (PRC2) to regulate the expression of target genes by mediating the methylation level of histone H3K27me3 (Wu et al. 2020). Ethylene is an important phytohormone for plant growth, development and stress tolerance, and its biosynthesis is regulated by the AP2/ERF transcription factor. The drought-responsive ERF gene, OsDERF1, negatively regulates ethylene synthesis by activating the transcription of OsERF3 and OsAP2-39, and plays a negative role in drought stress. SNORKEL1 and SNORKEL2 are strongly induced by ethylene and derived internode elongation through the GA pathway (Wan et al. 2011).

13.12 OsHBP1b Factors

Transcription factors (TFs), as the regulators of gene expression, are the key players contributing to stress tolerance and crop yield. Histone gene binding protein-1b (OsHBP1b) is a TF localized within the Saltol QTL in rice. Rice productivity is adversely affected by environmental stresses. Histone gene binding protein-1b (OsHBP1b) is a TF localized within the Saltol QTL in rice. Over-expression of the full-length gene encoding OsHBP1b in the homologous system (rice) has shown its contribution toward multiple stress tolerance and grain yield. HBP1b gene contributes to multiple abiotic stress tolerance through several molecular and physiological pathways and hence, may serve as an important gene for providing multiple stress tolerance and improving crop yield in rice (Das et al. 2019).

The gene OsHBP1b reported to be associated with salinity and drought tolerance in a model system tobacco. Overexpressing OsHBP1b exhibit better survival and favourable osmotic parameters under salinity stress than the wild type counterparts. These transgenic plants restricted reactive oxygen species accumulation by exhibiting high antioxidant enzyme activity (ascorbate peroxidase and superoxide dismutase), under salinity conditions. Additionally, these transgenic plants maintained the chlorophyll concentration, organellar structure, photosynthesis, and expression of photosynthesis and stress related genes even when subjected to salinity stress. Experiments conducted for other abiotic stresses such as drought and high temperature revealed improved tolerance in these transgenic plants with better root and shoot growth, better photosynthetic parameters, and enhanced antioxidant enzyme activity, in comparison with WT (Das et al. 2019).

13.13 Dehydration Responsive Element-Binding (DREB)

The DREB TF family belongs to the APETALA2/ethylene-responsive factor (AP2/ERF) super family of TFs. OsDREB2A is a key transcriptional activator that induces many heat and drought-responsive genes, increases tolerance to both heat and drought stress, and suppresses plant growth in rice. *DREB2A* expression is induced by stress, but stabilization of the DREB2A protein in response to stress is essential for activating the expression of downstream stress-inducible genes. Under non-stress growth conditions, an integral negative regulatory domain (NRD) destabilizes DREB2A, but the mechanism by which DREB2A is stabilized in response to stress remains unclear (Mizoi et al. 2019). Functionally characterized a DREB2-like gene, OsDRAP1 conferring drought tolerance (DT) in rice. OsDRAP1, containing many cis-elements in its promoter region, was expressed in all organs (mainly expressed in vascular tissues) of rice, and induced by a variety of environmental stresses and plant hormones. Overexpression of OsDRAP1 has a positive impact on maintaining water balance, redox homeostasis, and vascular development in rice plants under drought stress. OsDRAP1 interacted with many genes/proteins and could activate many downstream DT related genes, including important transcription factors such as OsCBSX3 to response drought stress, indicating the OsDRAP1-mediated pathways for DT involve complex genes networks (Huang et al. 2018). Co-expression of DREB2A and APX can provide enhanced drought tolerance in rice plants to combat climate change conditions Sandhya et al. (2021).

Molecular markers of DREB2A and BADH2 genes are also identified in 39 tested lines with approximately 250 and 2300 bp length, respectively. These lines have the potential to be developed on rainfed lowland rice or dry land because it has drought resistance Herawati et al. (2021). Over expression of *DREB1A* and rice *phytochrome-interacting factor-like 1* (*OsPIL1*) improve drought stress tolerance in various crops, although it also causes a severe dwarf phenotype. *OsPIL1* is a rice homologue of Arabidopsis *phytochrome-interacting factor 4* (*PIF4*), and it enhances cell elongation by activating cell wall-related gene expression (Kudo et al. 2017). All OsDREBs contained the AP2 domain and unique [K/R]GKKGPxN motif characteristic to DREB2 family. During rice growth and development, three *OsDREB2*s, namely *OsDREB2A*, *OsDREB2B,* and *OsABI4* are expressed and their expression is confined to embryo and endosperm tissues (Cui et al. 2011; Mallikarjuna et al. 2011). *OsDREB2A*, *OsDREB2B,* and *OsDREB2C* are expressed under abiotic stress conditions. *OsDREB2B* is expressed under drought, salinity, and cold stress conditions while *OsDREB2A* and *OsDREB2C* are expressed only under drought and salinity conditions (Herath 2016). Co-expression of DREB2A and APX were evaluated for drought tolerance during seed germination, vegetative, and reproductive stages and provide enhanced drought tolerance in rice plants to combat climate change conditions (Sandhya et al. 2021). This promising result suggests that transgenic approach can be a viable option for genetic enhancement of rice against abiotic stresses like drought Geda (2019).

Further studies suggested that *OsDREB1F* might also be involved in the ABA-dependent pathway (Wang et al. 2008). In addition to *DREB1s* and *DREB2s*, *OsDREB4-1* from the *DREB4* subgroup was found to be induced by high salt and was assumed to be a *trans*-acting factor in the DRE/DREB regulated stress-responsive pathway (Tian et al. 2005). *OsDREB6*, an A-6 type of DREB, was hypothesized to participate in stress responses in both ABA-dependent and ABA-independent signal transduction pathways (Ke et al. 2014). *SERF1*, belonging to group IIc ERFs of DREB subfamily, was reported to be a positive regulator of short- and long-term salt stress tolerance in rice (Schmidt et al. 2013) and could amplify the ROS-activated MAPK cascade signal in roots upon salt stress, which plays a dominant role in salt stress-induced root-to-shoot communication rather than ABA (Schmidt et al. 2013).

13.14 Other Transcription Factors, Genes, and MicroRNAs Play Crucial Role in Drought and Salt Stress

The *W USCHEL*-related homeobox (*WOX*) genes are important transcription regulators participated in plant development processes. The expressions of auxin- and cytokinins responsive genes were affected in *WOX11* overexpression and RNA interference transgenic plants (Zhao et al. 2009). Four genes (*OsWUS*, *OsNS1/OsNS2*, *OsWOX3*, and *OsWOX9A*) are expressed in panicle and endosperm development while six genes (*OsWOX5*, *OsWOX9B*, *OsWOX9D*, *OsWOX11*, *OsWOX12A*, and *OsWOX12B*) expressed in seeds during root emergence or growth (Chen et al. 2014)

WOX (WUSCHEL-related homeobox) is a plant TF linked to plant development and stress responses. The rice WOX13 gene belongs to the WOX subfamily of TFs and is ABA-responsive, essential for flower improvement, contains proteins, and is involved in drought and salinity stress. OsWOX13 was involved in the regulation of vegetative organs, flowers, and seeds. OsWOX13 caused early flowering and stress responses. OsWOX13 over expression resulted in early flowering and showed an extensive spectrum of effects on biological processes, such as abiotic and biotic stress, after drought and salinity stress (Minh-Thu et al. 2018). A transcriptome analysis revealed that OsWOX10 is highly upregulated in L-type lateral root primordium (LRPs). OsWOX10 overexpression in lateral root primordium (LRPs) increased the LR diameter in an expression-dependent manner. Conversely, the mutation in OsWOX10 decreased the L-type LR diameter under mild drought conditions (Kawai et al. 2022).

The ELIP and PPR genes are involved in chloroplast protection during dehydration and rehydration. ELIP3 may be involved in the regulation of the redox state of the cell and takes important role in protecting the photo-system under photo-oxidative stress in low temperatures. FL478 is a salt tolerant indica recombinant inbred line, which can be a good source of salt tolerance at the seedling stage in rice. Its employs more efficient mechanisms (especially in signal transduction of salt stress, influx and transport of k+, ionic and osmotic homeostasis, as well as ROS inhibition) to respond to the salt stress compared to its susceptible parent (Mirdar Mansuri et al. 2019). *SNF* and trehalose genes are known to be oxidant scavengers that protect the cell structure from the deleterious effect of drought. *TPP* and *TPS* genes were found in the starch and sucrose metabolism pathways, which are essential sugar-signaling metabolites regulating plant metabolism and other biological processes. *ABC-G* gene interacts with abscisic acid (ABA) phytohormone in the stomata opening during stress conditions. (Kwon et al. 2021).

The plant glycogen synthase kinase 3 (GSK3)-like kinases are highly conserved protein serine/threonine kinases that are constitutively active under normal growth conditions but become inactivated in response to diverse developmental and environmental signals (Mao et al. 2021). Grain size and shape are important determinants of grain weight and yield in rice and encodes by the OsSK41 (also known as OsGSK5), a member of the GLYCOGEN SYNTHASE KINASE 3/SHAGGY-like family. Rice near-isogenic lines carrying the loss-of-function allele of OsSK41 have increased grain length and weight. Co-expression of OsSK41 with OsARF4 increases the accumulation of OsARF4 in rice protoplasts. Loss of function of OsARF4 results in larger rice grains. RNA-sequencing analysis suggests that OsARF4 and OsSK41 repress the expression of a common set of downstream genes, including some auxin-responsive genes, during rice grain development (Hu et al. 2018). GSKs/SKs proteins is positively regulated by reactive oxygen species, whereas it is negatively regulated through ubiquitylation, deacetylation, and nitric oxide-mediated nitrosylation. GSKs/SKs proteins interact with proteins representing various signaling pathways, and on the basis of the complicated network of interactions the GSKs/SKs proteins differentially regulate various physiological, developmental, stress response, and yield-related processes (Zolkiewicz and Gruszka 2022).

Late embryogenesis abundant (LEA) proteins are involved in tolerance to drought, cold, and high salinity in rice. The *OsLEA4* gene in rice, which showed that *OsLEA4* was expressed in different organ tissues during different development stages of rice. The expression levels of *OsLEA4* in the leaves during the tillering stage and leaves and panicles during the heading stage, the filling stage, and the full ripe stage were dramatically increased. Moreover, based on seed germination, growth status, and physiological indices, the overexpression of *OsLEA4* in transgenic rice plants conferred increased resistance to drought, salt, and heavy metal stresses compared with the wild type (WT) plants (Hu et al. 2016).

MicroRNA (miRNA), a kind of small non-coding RNA, regulates gene expression at post-transcriptional levels. The verified targets of miRNAs encode a diverse range of regulatory proteins mostly target to TFs. The *miR164b* is found to target a rice TF *OsNAC2*. Rice plants overexpressing the *miR164b*-resistant form of *OsNAC2* showed enhanced salinity tolerance. Overexpression of *osa-MIR396c* and *osa-MIR393* in rice and *Arabidopsis* plants increased sensitivity to salinity stress via negatively mediating target growth-regulating factors and other regulatory proteins

(Gao et al. (2011). Further studies revealed that the downregulation of two auxin-receptor genes *transport inhibitor response 1* (*OsTIR1*) and *auxin signaling f-box 2* (*OsAFB2*) contributed to reduced salinity tolerance in *OsmiR393*-overexpressing rice plants (Xia et al. (2012). In addition, the expression of rice-specific *Osa-miR820* targeting *domain rearranged* methyltransferase 2 (*OsDRM2*) was regulated by salinity stress (Sharma et al. 2015). Several other salinity-responsive miRNAs have been identified and characterized in rice (Ganie et al. 2019). Undoubtedly, further identification and validation of drought and salinity-responsive miRNAs will bring more alternatives for crop improvement.

13.15 Conclusion and Future Aspects

Drought and salt stress conditions are perceived by specific proteins in cell membranes known as sensors. These sensors pass on the stress message through signaling molecules such as phospholipids, hormones, and calcium ions (Ca^{2+}) which regulate stress signaling pathways for maintaining signal transduction. These signals are finally perceived by transcription regulatory machinery in nucleus and regulatory network switch on to stress mitigation pathways. To some extent the mitigation process in drought and salt stress use common machinery. The ABA-dependent and ABA-i ndependent pathways are followed in mitigation of abiotic stress, which has numerous different TFs. In recent years new members of TFs families have been reported to be involved in drought and salt stress mitigation. ABA- dependent stress-responsive gene and transcription factors function in modulating plant response and tolerance to abiotic stress, focusing on bZIPs, bHLHs, NACs, AP2/ERFs, MYBs, and WRKYs. A major transcription system regulating ABA-independent gene expression in response to dehydration and cold stress includes a DRE/CRT cis-acting element and its DNA-binding protein, DREB/CBF. The mi-RNAs also play important role in regulation of stress related TFs. Furthermore, the studies on mi-RNA and SRA data availability of rice under different abiotic stress conditions shall provide more insights of TFs and co-expression regulatory network for mitigation of stress in rice crop.

References

Agalou A, Purwantomo S, Övernäs E, Johannesson H, Zhu X, Estiati A, Ouwerkerk PB (2008) A genome-wide survey of HD-Zip genes in rice and analysis of drought-responsive family members. Plant Mol Biol 66(1):87–103

Agarwal PK, Agarwal P, Reddy MK, Sopory SK (2006) Role of DREB transcription factors in abiotic and biotic stress tolerance in plants. Plant Cell Rep 25(12):1263–1274

Alam MS, Kong J, Tao R, Ahmed T, Alamin M, Alotaibi SS, Xu JH (2022) CRISPR/Cas9 mediated knockout of the OsbHLH024 transcription factor improves salt stress resistance in rice (Oryza sativa L.). Plants 11(9):1184

Asano T, Hakata M, Nakamura H, Aoki N, Komatsu S, Ichikawa H, Ohsugi R (2011) Functional characterisation of OsCPK21, a calcium-dependent protein kinase that confers salt tolerance in rice. Plant Mol Biol 75(1):179–191

Asano T, Hayashi N, Kobayashi M, Aoki N, Miyao A, Mitsuhara I, Ohsugi R (2012) A rice calcium dependent protein kinase OsCPK12 oppositely modulates salt stress tolerance and blast disease resistance. Plant J 69(1):26–36

Atienza Martínez J, Jiang X, Garciadeblas B, Mendoza I, Zhu JK, Pardo JM, Quintero FJ (2007) Conservation of the salt overly sensitive pathway in rice. Plant Physiol 143(2):1001–1012

Bang SW, Lee DK, Jung H, Chung PJ, Kim YS, Choi YD, Kim JK (2019) Overexpression of OsTF1L, a rice HD-Zip transcription factor, promotes lignin biosynthesis and stomatal closure that improves drought tolerance. Plant Biotechnol J 17(1):118–131

Bartels D, Sunkar R (2005) Drought and salt tolerance in plants. Crit Rev Plant Sci 24(1):23–58

Basri K, Sukiran NL, Zainal Z (2016) Cloning and characterization of the ONAC106 gene from Oryza sativa cultivar Kuku Belang. AIP Conf Proc 1784(1):020021

Basu S, Ramegowda V, Kumar A, Pereira A (2016) Plant adaptation to drought stress [version 1; referees: 3 approved]. F1000Research 5:1554

Berri S, Abbruscato P, Faivre-Rampant O, Brasileiro A, Fumasoni I, Satoh K, Piffanelli P (2009) Characterization of WRKYco-regulatory networks in rice and arabidopsis. BMC Plant Biol 9(1):1–22

Bihani P, Char B, Bhargava S (2011) Transgenic expression of sorghum DREB2 in rice improves tolerance and yield under water limitation. J Agric Sci 149(1):95–101

Bray E, Bailey-Serres J, Weretilnyk E (2000) Responses to abiotic stresses. In: Buchanan B, Gruissem W, Jones R (eds) Biochemistry and molecular biology of plants. American Society of Plant Biologists, Rockville, pp 1158–1203

Chen JQ, Meng XP, Zhang Y, Xia M, Wang XP (2008) Over-expression of OsDREB genes lead to enhanced drought tolerance in rice. Biotechnol Lett 30(12):2191–2198

Chen H, Chen W, Zhou J, He H, Chen L, Chen H, Deng XW (2012) Basic leucine zipper transcription factor OsbZIP16 positively regulates drought resistance in rice. Plant Sci 193:8–17

Chen X, Wang Y, Lv B, Li J, Luo L, Lu S, Zhang X, Ma H, Ming F (2014) The NAC family transcription factor OsNAP confers abiotic stress response through the ABA pathway. Plant Cell Physiol 55:604–619

Chen HC, Hsieh-Feng V, Liao PC, Cheng WH, Liu LY, Yang YW, Chang MC (2017) The function of OsbHLH068 is partially redundant with its homolog, AtbHLH112, in the regulation of the salt stress response but has opposite functions to control flowering in Arabidopsis. Plant Mol Biol 94(4):531–548

Cheng S, Huang Y, Zhu N, Zhao Y (2014) The rice WUSCHEL-related homeobox genes are involved in reproductive organ development, hormone signaling and abiotic stress response. Gene 549(2):266–274

Cui M, Zhang W, Zhang Q, Xu Z, Zhu Z, Duan F, Wu R (2011) Induced over-expression of the transcription factor OsDREB2A improves drought tolerance in rice. Plant Physiol Biochem 49(12):1384–1391

Dai X, Xu Y, Ma Q, Xu W, Wang T, Xue Y, Chong K (2007) Overexpression of an R1R2R3 MYB gene, OsMYB3R-2, increases tolerance to freezing, drought, and salt stress in transgenic Arabidopsis. Plant Physiol 143(4):1739–1751

Dansana PK, Kothari KS, Vij S, Tyagi AK (2014) OsiSAP1 overexpression improves water-deficit stress tolerance in transgenic rice by affecting expression of endogenous stress-related genes. Plant Cell Rep 33(9):1425–1440

Das P, Lakra N, Nutan KK, Singla-Pareek SL, Pareek A (2019) A unique bZIP transcription factor imparting multiple stress tolerance in rice. Rice 12(1):1–16

Deeba F, Sultana T, Javaid B, Mahmood T, Naqvi SMS (2017) Molecular characterization of a MYB protein from Oryza sativa for its role in abiotic stress tolerance. Braz Arch Biol Technol 60:352

DeWald DB, Torabinejad J, Jones CA, Shope JC, Cangelosi AR, Thompson JE, Hama H (2001) Rapid accumulation of phosphatidylinositol 4,5-bisphosphate and inositol 1,4,5-trisphosphate correlates with calcium mobilization in salt-stressed Arabidopsis. Plant Physiol 126(2):759–769

Diédhiou CJ, Popova OV, Dietz KJ, Golldack D (2008) The SNF1-type serine-threonine protein kinase SAPK4regulates stress-responsive gene expression in rice. BMC Plant Biol 8(1):1–13

Du H, Wang N, Cui F, Li X, Xiao J, Xiong L (2010) Characterization of the β-carotene hydroxylase gene DSM2 conferring drought and oxidative stress resistance by increasing xanthophylls and abscisic acid synthesis in rice. Plant Physiol 154(3):1304–1318

Duan J, Cai W (2012) OsLEA3-2, an abiotic stress induced gene of rice plays a key role in salt and drought tolerance. PLoS One 7(9):e45117

Dubouzet JG, Sakuma Y, Ito Y, Kasuga M, Dubouzet EG, Miura S, Yamaguchi-Shinozaki K (2003) OsDREB genes in rice, Oryza sativa L., encode transcription activators that function in drought-, high-salt-and cold-responsive gene expression. Plant J 33(4):751–763

Fang Y, Liao K, Du H, Xu Y, Song H, Li X, Xiong L (2015) A stress-responsive NAC transcription factor SNAC3 confers heat and drought tolerance through modulation of reactive oxygen species in rice. J Exp Bot 66(21):6803–6817

Farooq M, Wahid A, Kobayashi NSMA, Fujita DBSMA, Basra SMA (2009) Plant drought stress: effects, mechanisms and management. In: Sustainable agriculture. Springer, Dordrecht, pp 153–188

Fujita Y, Yoshida T, Yamaguchi-Shinozaki K (2013) Pivotal role of the AREB/ABF-SnRK2 pathway in ABRE-mediated transcription in response to osmotic stress in plants. Physiol Plant 147(1):15–27

Furuta T, Uehara K, Angeles-Shim RB, Shim J, Ashikari M, Takashi T (2014) Development and evaluation of chromosome segment substitution lines (CSSLs) carrying chromosome segments derived from Oryza rufipogon in the genetic background of Oryza sativa L. Breed Sci 63(5): 458–475

Ganie SA, Molla KA, Henry RJ, Bhat KV, Mondal TK (2019) Advances in understanding salt tolerance in rice. Theor Appl Genet 132:851–870

Gao P, Bai X, Yang L, Lyu D, Pan X, Li Y, Cai H, Ji W, Chen Q, Zhu Y (2011) osa-MIR393: a salinity- and alkaline stress-related microRNA gene. Mol Biol Rep 38:237–242

Garg AK, Kim JK, Owens TG, Ranwala AP, Choi YD, Kochian LV, Wu RJ (2002) Trehalose accumulation in rice plants confers high tolerance levels to different abiotic stresses. Proc Natl Acad Sci 99(25):15898–15903

Geda YF (2019) Determinants of teenage pregnancy in Ethiopia: a case–control study, 2019. Curr Med Issues 17(4):10–41

Golldack D, Quigley F, Michalowski CB, Kamasani UR, Bohnert HJ (2003) Salinity stress-tolerant and-sensitive rice (Oryza sativa L.) regulate AKT1-type potassium channel transcripts differently. Plant Mol Biol 51(1):71–81

Guo C, Yao L, You C, Wang S, Cui J, Ge X, Ma H (2016) MID1 plays an important role in response to drought stress during reproductive development. Plant J 88(2):280–293

Guo Y, Li P, Zou Y, Xie D, Lu J, Liu Q, Li Q (2019) Expression and functional analysis of rice OsWRKY78 transcription factor in response to salt stress. J Yangzhou Univ 40(2):18–24

Herath V (2016) Small family, big impact: in-silico analysis of DREB2 transcription factor family in rice. Comput Biol Chem 65:128–139

Herawati R, Alnopri A, Masdar M, Simarmata M, Sipriyadi S, Sutraati M (2021) Identification of drought tolerant and molecular analysis of DREB2A and BADH2 genes and yield potential of lines from single crossing bengkulu local rice varieties. Biodivers J Biol Divers 22:2

Hong Y, Zhang H, Huang L, Li D, Song F (2016) Overexpression of a stress-responsive NAC transcription factor gene ONAC022 improves drought and salt tolerance in rice. Front Plant Sci 7:4

Hossain MA, Cho JI, Han M, Ahn CH, Jeon JS, An G, Park PB (2010a) The ABRE-binding bZIP transcription factor OsABF2 is a positive regulator of abiotic stress and ABA signaling in rice. J Plant Physiol 167(17):1512–1520

Hossain A, Lee Y, Cho JI, Ahn CH, Lee SK, Jeon JS, Park PB (2010b) The bZIP transcription factor OsABF1 is an ABA responsive element binding factor that enhances abiotic stress signaling in rice. Plant Mol Biol 72(4):557–566

Hou D, Ye T, Zhang L, Fan J, Li C, Dong Y, Li L (2020) Overexpressing the WRKY transcription factor OsWRKY97 improves drought tolerance in rice. Res Square. https://doi.org/10.21203/rs.3.rs-17373/v1

Hoshida H, Tanaka Y, Hibino T, Hayashi Y, Tanaka A, Takabe T, Takabe T (2000) Enhanced tolerance to salt stress in transgenic rice that overexpresses chloroplast glutamine synthetase. Plant Mol Biol 43, 103–111

Hu H, Dai M, Yao J, Xiao B, Li X, Zhang Q, Xiong L (2006) Overexpressing a NAM, ATAF, and CUC (NAC) transcription factor enhances drought resistance and salt tolerance in rice. Proc Natl Acad Sci 103(35):12987–12992

Hu H, You J, Fang Y, Zhu X, Qi Z, Xiong L (2008) Characterization of transcription factor gene SNAC2 conferring cold and salt tolerance in rice. Plant Mol Biol 67(1):169–181

Hu T, Zhu S, Tan L, Qi W, He S, Wang G (2016) Overexpression of OsLEA4 enhances drought, high salt and heavy metal stress tolerance in transgenic rice (Oryza sativa L.). Environ Exp Bot 123:68–77

Hu Z, Lu SJ, Wang MJ, He H, Sun L, Wang H, Liu JX (2018) A novel QTL qTGW3 encodes the GSK3/SHAGGY-like kinase OsGSK5/OsSK41 that interacts with OsARF4 to negatively regulate grain size and weight in rice. Mol Plant 11(5):736–749

Huang J, Sun SJ, Xu DQ, Yang X, Bao YM, Wang ZF, Zhang H (2009) Increased tolerance of rice to cold, drought and oxidative stresses mediated by the overexpression of a gene that encodes the zinc finger protein ZFP245. Biochem Biophys Res Commun 389(3):556–561

Huang P, Chen H, Mu R, Yuan X, Zhang HS, Huang J (2015) OsMYB511 encodes a MYB domain transcription activator early regulated by abiotic stress in rice. Genet Mol Res 14:9506–9517

Huang L, Hong Y, Huijuan Zhang H, Li D, Song F (2016) Rice NAC transcription factor ONAC095 plays opposite roles in drought and cold stress tolerance. BMC Plant Biol 16:203

Huang L, Wang Y, Wang W, Zhao X, Qin Q, Sun F, Li Z (2018) Characterization of transcription factor gene OsDRAP1 conferring drought tolerance in rice. Front Plant Sci 9:94

Huang K, Wu T, Ma Z, Li Z, Chen H, Zhang M, Bian M, Bai H, Jiang W, Du X (2021) Rice transcription factor OsWRKY55 is involved in the drought response and regulation of plant growth. Int J Mol Sci 22:4337

Jeong JS, Kim YS, Baek KH, Jung H, Sun-Hwa H, Choi YD, Kim M, Reuzeau C, Kim JK (2010) Root-specific expression of OsNAC10 improves drought tolerance and grain yield in rice under field drought conditions. Plant Physiol 153:185–197

Jeong JS, Kim YS, Redillas MC, Jang G, Jung H, Bang SW, Kim JK (2013) OsNAC5 overexpression enlarges root diameter in rice plants leading to enhanced drought tolerance and increased grain yield in the field. Plant Biotechnol J 11(1):101–114

Jiang Y, Qiu Y, Hu Y, Yu D (2016) Heterologous expression of AtWRKY57 confers drought tolerance in Oryza sativa. Front Plant Sci 7:145

Jiang D, Zhou L, Chen W, Ye N, Xia J, Zhuang C (2019) Overexpression of a microRNA-targeted NAC transcription factor improves drought and salt tolerance in Rice via ABA-mediated pathways. Rice 12(1):1–11

Jin XF, Xiong AS, Peng RH, Liu JG, Gao F, Chen JM, Yao QH (2010) OsAREB1, an ABRE-binding protein responding to ABA and glucose, has multiple functions in Arabidopsis. BMB Rep 43(1):34–39

Joo J, Lee YH, Song SI (2014) Overexpression of the rice basic leucine zipper transcription factor OsbZIP12 confers drought tolerance to rice and makes seedlings hypersensitive to ABA. Plant Biotechnol Rep 8(6):431–441

Jung H, Chung PJ, Park SH, Redillas MCFR, Kim YS, Suh JW, Kim JK (2017) Overexpression of Os ERF 48 causes regulation of Os CML 16, a calmodulin-like protein gene that enhances root growth and drought tolerance. Plant Biotechnol J 15(10):1295–1308

Kawai T, Shibata K, Akahoshi R, Nishiuchi S, Takahashi H, Nakazono M, Inukai Y (2022) WUSCHEL-related homeobox family genes in rice control lateral root primordium size. Proc Natl Acad Sci 119(1):e2101846119

Ke YG, Yang ZJ, Yu SW, Li TF, Wu JH, Gao H, Luo LJ (2014) Characterization of OsDREB6 responsive to osmotic and cold stresses in rice. J Plant Biol 57(3):150–161

Kim JA, Agrawal GK, Rakwal R, Han KS, Kim KN, Yun CH, Jwa NS (2003) Molecular cloning and mRNA expression analysis of a novel rice (Oryzasativa L.) MAPK kinase kinase, OsEDR1, an ortholog of ArabidopsisAtEDR1, reveal its role in defense/stress signalling pathways and development. Biochem Biophys Res Commun 300(4):868–876

Kim SW, Lee SK, Jeong HJ, An G, Jeon JS, Jung KH (2017) Crosstalk between diurnal rhythm and water stress reveals an altered primary carbon flux into soluble sugars in drought-treated rice leaves. Sci Rep 7(1):1–18

Kobayashi Y, Yamamoto S, Minami H, Kagaya Y, Hattori T (2004) Differential activation of the rice sucrose nonfermenting1–related protein kinase2 family by hyperosmotic stress and abscisic acid. Plant Cell 16(5):1163–1177

Kobayashi Y, Murata M, Minami H, Yamamoto S, Kagaya Y, Hobo T, Hattori T (2005) Abscisic acid-activated SNRK2 protein kinases function in the gene-regulation pathway of ABA signal transduction by phosphorylating ABA response element-binding factors. Plant J 44(6):939–949

Kudo M, Kidokoro S, Yoshida T, Mizoi J, Todaka D, Fernie AR, Yamaguchi-Shinozaki K (2017) Double overexpression of DREB and PIF transcription factors improves drought stress tolerance and cell elongation in transgenic plants. Plant Biotechnol J 15(4):458–471

Kumar K, Kumar M, Kim SR, Ryu H, Cho YG (2013) Insights into genomics of salt stress response in rice. Rice 6(1):1–15

Kumar SVV, Verma RK, Yadav SK, Yadav P, Watts A, Rao MV, Chinnusamy V (2020) CRISPR-Cas9 mediated genome editing of drought and salt tolerance (OsDST) gene in indica mega rice cultivar MTU1010. Physiol Mol Biol Plants 26(6):1099–1110

Kwon E, Basnet P, Roy NS, Kim JH, Heo K, Park KC, Choi IY (2021) Identification of resurrection genes from the transcriptome of dehydrated and rehydrated Selaginella tamariscina. Plant Signal Behav 16(12):1973703

Lee SK, Kim BG, Kwon TR, Jeong MJ, Park SR, Lee JW, Park SC (2011) Overexpression of the mitogen-activated protein kinase gene OsMAPK33 enhances sensitivity to salt stress in rice (Oryza sativa L.). J Biosci 36(1):139–151

Lee DK, Chung PJ, Jeong JS, Jang G, Bang SW, Jung H, Kim YS, Ha SH, Choi YD, Kim J (2017a) The rice OsNAC6 transcription factor orchestrates multiple molecular mechanisms involving root structural adaptions and nicotianamine biosynthesis for drought tolerance. Plant Biotechnol J 15:754–764

Lee DK, Yoon S, Kim YS, Kim JK (2017b) Rice OsERF71-mediated root modification affects shoot drought tolerance. Plant Signal Behav 12(1):e1268311

Lee H, Cha J, Choi C, Choi N, Ji HS, Park SR, Hwang DJ (2018) Rice WRKY11 plays a role in pathogen defense and drought tolerance. Rice 11(1):1–12

Liang C, Wang Y, Zhu Y, Tang J, Hu B, Liu L, Chu C (2014) OsNAP connects abscisic acid and leaf senescence by fine-tuning abscisic acid biosynthesis and directly targeting senescence-associated genes in rice. Proc Natl Acad Sci 111(27):10013–10018

Lim C, Kang K, Shim Y, Yoo SC, Paek NC (2022) Inactivating transcription factor OsWRKY5 enhances drought tolerance through abscisic acid signaling pathways. Plant Physiol 188(4): 1900–1916

Liu Q, Kasuga M, Sakuma Y, Abe H, Miura S, Yamaguchi-Shinozaki K, Shinozaki K (1998) Two transcription factors, DREB1 and DREB2, with an EREBP/AP2 DNA binding domain separate two cellular signal transduction pathways in drought-and low-temperature-responsive gene expression, respectively, in Arabidopsis. Plant Cell 10(8):1391–1406

Liu S, Cheng Y, Zhang X, Guan Q, Nishiuchi S, Hase K, Takano T (2007) Expression of an NADP-malic enzyme gene in rice (Oryza sativa L.) is induced by environmental stresses; over-

expression of the gene in Arabidopsis confers salt and osmotic stress tolerance. Plant Mol Biol 64:49–58
Liu C, Wu Y, Wang X (2012) bZIP transcription factor OsbZIP52/RISBZ5: a potential negative regulator of cold and drought stress response in rice. Planta 235(6):1157–1169
Liu C, Mao B, Ou S, Wang W, Liu L, Wu Y, Wang X (2014) OsbZIP71, a bZIP transcription factor, confers salinity and drought tolerance in rice. Plant Mol Biol 84(1):19–36
Mallikarjuna G, Mallikarjuna K, Reddy MK, Kaul T (2011) Expression of OsDREB2A transcription factor confers enhanced dehydration and salt stress tolerance in rice (Oryza sativa L.). Biotechnol Lett 33(8):1689–1697
Mann A, Kumar N, Sharma C, Kumar A, Kumar A, Meena BL (2019) Functional annotation of differentially expressed genes under salt stress in Dichanthium annulatum. Indian J Plant Physiol 24:104–111. https://doi.org/10.1007/s40502-019-0434-8
Mann A, Kumar N, Kumar A et al (2021) de novo transcriptomic profiling of differentially expressed genes in grass halophyte Urochondra setulosa under high salinity. Sci Rep 11:1–14
Mao J, Li W, Liu J, Li J (2021) Versatile physiological functions of plant GSK3-like kinases. Gene 12(5):697
Matsukura S, Mizoi J, Yoshida T, Todaka D, Ito Y, Maruyama K, Yamaguchi-Shinozaki K (2010) Comprehensive analysis of rice DREB2-type genes that encode transcription factors involved in the expression of abiotic stress-responsive genes. Mol Gen Genomics 283(2):185–196
Minh-Thu PT, Kim JS, Chae S, Jun KM, Lee GS, Kim DE, Kim YK (2018) A WUSCHEL homeobox transcription factor, OsWOX13, enhances drought tolerance and triggers early flowering in rice. Mol Cells 41(8):781
Mirdar Mansuri R, Shobbar ZS, Babaeian Jelodar N, Ghaffari MR, Nematzadeh GA, Asari S (2019) Dissecting molecular mechanisms underlying salt tolerance in rice: a comparative transcriptional profiling of the contrasting genotypes. Rice 12(1):1–13
Mizoi J, Kanazawa N, Kidokoro S, Takahashi F, Qin F, Morimoto K, Yamaguchi-Shinozaki K (2019) Heat-induced inhibition of phosphorylation of the stress-protective transcription factor DREB2A promotes thermotolerance of Arabidopsis thaliana. J Biol Chem 294(3):902–917
Mukhopadhyay A, Vij S, Tyagi AK (2004) Overexpression of a zinc-finger protein gene from rice confers tolerance to cold, dehydration, and salt stress in transgenic tobacco. Proc Natl Acad Sci 101(16):6309–6314
Muthuramalingam P, Krishnan SR, Saravanan K, Mareeswaran N, Kumar R, Ramesh M (2018) Genome-wide identification of major transcription factor superfamilies in rice identifies key candidates involved in abiotic stress dynamism. J Plant Biochem Biotechnol 27(3):300–317
Nakamura A, Fukuda A, Sakai S, Tanaka Y (2006) Molecular cloning, functional expression and subcellular localization of two putative vacuolar voltage-gated chloride channels in rice (Oryza sativa L.). Plant Cell Physiol 47(1):32–42
Nakashima K, Tran LSP, Van Nguyen D, Fujita M, Maruyama K, Todaka D, Yamaguchi-Shinozaki K (2007) Functional analysis of a NAC-type transcription factor OsNAC6 involved in abiotic and biotic stress-responsive gene expression in rice. Plant J 51(4):617–630
Nam MH, Huh SM, Kim KM, Park WW, Seo JB, Cho K, Yoon II (2012) Comparative proteomic analysis of early salt stress-responsive proteins in roots of SnRK2 transgenic rice. Proteome Sci 10(1):1–19
Ning J, Zhang B, Wang N, Zhou Y, Xiong L (2011) Increased leaf angle1, a Raf-like MAPKKK that interacts with a nuclear protein family, regulates mechanical tissue formation in the Lamina joint of rice. Plant Cell 23:4334–4347
Nutan KK, Singla-Pareek SL, Pareek A (2020) The Saltol QTL-localized transcription factor OsGATA8 plays an important role in stress tolerance and seed development in Arabidopsis and rice. J Exp Bot 71(2):684–698
Ortolan F, Fonini LS, Pastori T, Mariath JE, Saibo NJ, Margis-Pinheiro M, Lazzarotto F (2021) Tightly controlled expression of OsbHLH35 is critical for anther development in rice. Plant Sci 302:110716

Park SH, Jeong JS, Lee KH, Kim YS, Do Choi Y, Kim JK (2015) OsbZIP23 and OsbZIP45, members of the rice basic leucine zipper transcription factor family, are involved in drought tolerance. Plant Biotechnol Rep 9(2):89–96

Pasquali G, Biricolti S, Locatelli F, Baldoni E, Mattana M (2008) Osmyb4 expression improves adaptive responses to drought and cold stress in transgenic apples. Plant Cell Rep 27(10): 1677–1686

Peng Y, Tang N, Zou J, Ran J, Chen X (2022) Rice MYB transcription factor OsMYB1R1 negatively regulates drought resistance. Plant Growth Regul 2022:1–11

Qiu Y, Yu D (2009) Over-expression of the stress-induced OsWRKY45 enhances disease resistance and drought tolerance in Arabidopsis. Environ Exp Bot 65(1):35–47

Rachmat A, Nugroho S, Sukma D, Hajrial Aswidinnoor H, Sudarsono (2014) Overexpression of OsNAC6 transcription factor from Indonesia rice cultivar enhances drought and salt tolerance. Emir J Food Agric 26(6):519–527. https://doi.org/10.9755/ejfa.v26i6.17672

Rahman H, Ramanathan V, Nallathambi J, Duraialagaraja S, Muthurajan R (2016) Over-expression of a NAC 67 transcription factor from finger millet (Eleusine coracana L.) confers tolerance against salinity and drought stress in rice. BMC Biotechnol 16(1):7–20

Rashid M, Guangyuan H, Guangxiao Y, Hussain J, Xu Y (2012) AP2/ERF transcription factor in rice: genome-wide canvas and syntenic relationships between monocots and eudicots. Evol Bioinforma 8:9369

Redillas MC, Jeong JS, Kim YS, Jung H, Bang SW, Choi YD, Kim JK (2012) The overexpression of OsNAC9 alters the root architecture of rice plants enhancing drought resistance and grain yield under field conditions. Plant Biotechnol J 10(7):792–805

Saengngam S, Takpirom W, Buaboocha T, Chadchawan S (2012) The role of the OsCam1-1 salt stress sensor in ABA accumulation and salt tolerance in rice. J Plant Biol 55(3):198–208

Saijo Y, Hata S, Kyozuka J, Shimamoto K, Izui K (2000) Over-expression of a single Ca2+-dependent protein kinase confers both cold and salt/drought tolerance on rice plants. Plant J 23(3):319–327

Sandhya J, Ashwini T, Manisha R, Vinodha M, Srinivas A (2021) Drought tolerance enhancement with co-Overexpression of DREB2A and APX in indica rice (Oryza sativa L.). American. J Plant Sci 12(2):234–258

Schmidt R, Mieulet D, Hubberten HM, Obata T, Hoefgen R, Fernie AR, Mueller-Roeber B (2013) Salt-responsive ERF1 regulates reactive oxygen species–dependent signaling during the initial response to salt stress in rice. Plant Cell 25(6):2115–2131

Seong SY, Shim JS, Bang SW, Kim JK (2020) Overexpression of OsC3H10, a CCCH-zinc finger, improves drought tolerance in rice by regulating stress-related genes. Plants 9(10):1298

Sharma N, Tripathi A, Sanan-Mishra N (2015) Profiling the expression domains of a rice-specific microRNA under stress. Front Plant Sci 6:333

Shen L, Hua Y, Fu Y, Li J, Liu Q, Jiao X, Wang K (2017a) Rapid generation of genetic diversity by multiplex CRISPR/Cas9 genome editing in rice. Sci China Life Sci 60(5):506–515

Shen J, Lv B, Luo L, He J, Mao C, Xi D, Ming F (2017b) The NAC-type transcription factor OsNAC2 regulates ABA-dependent genes and abiotic stress tolerance in rice. Sci Rep 7(1):1–16

Shim JS, Oh N, Chung PJ, Kim YS, Choi YD, Kim J-K (2018) Overexpression of OsNAC14 Improves Drought Tolerance in Rice. Front Plant Sci 9:310. https://doi.org/10.3389/fpls.2018.00310

Shinozaki K, Yamaguchi-Shinozaki K (2000) Molecular responses to dehydration and low temperature: differences and cross-talk between two stress signaling pathways. Curr Opin Plant Biol 3: 217–223

Shinozaki K, Yamaguchi-Shinozakiy K, Seki M (2003) Regulatory network of gene expression in the drought and cold stress responses. Curr Opin Plant Biol 6:410–417

Song Y, You J, Xiong L (2009) Characterization of OsIAA1 gene, a member of rice Aux/IAA family involved in auxin and brassinosteroid hormone responses and plant morphogenesis. Plant Mol Biol 70(3):297–309

Song SY, Chen Y, Chen J, Dai XY, Zhang WH (2011) Physiological mechanisms underlying OsNAC5-dependent tolerance of rice plants to abiotic stress. Planta 234(2):331–345

Sun SJ, Guo SQ, Yang X, Bao YM, Tang HJ, Sun H, Zhang HS (2010) Functional analysis of a novel Cys2/His2-type zinc finger protein involved in salt tolerance in rice. J Exp Bot 61(10): 2807–2818

Sun W, Gao D, Xiong Y, Tang X, Xiao X, Wang C, Yu S (2017) Hairy leaf 6, an AP2/ERF transcription factor, interacts with OsWOX3B and regulates trichome formation in rice. Mol Plant 10(11):1417–1433

Sun X, Wang Y, Sui N (2018) Transcriptional regulation of bHLH during plant response to stress. Biochem Biophys Res Commun 503(2):397–401

Takasaki H, Maruyama K, Kidokoro S, Ito Y, Fujita Y, Shinozaki K, Nakashima K (2010) The abiotic stress-responsive NAC-type transcription factor OsNAC5 regulates stress-inducible genes and stress tolerance in rice. Mol Gen Genomics 284(3):173–183

Tang N, Zhang H, Li X, Xiao J, Xiong L (2012) Constitutive activation of transcription factor OsbZIP46 improves drought tolerance in rice. Plant Physiol 158(4):1755–1768

Tang Y, Bao X, Zhi Y, Wu Q, Guo Y, Yin X, Liu K (2019) Overexpression of a MYB family gene, OsMYB6, increases drought and salinity stress tolerance in transgenic rice. Front Plant Sci 10: 168

Tao Z, Kou Y, Liu H, Li X, Xiao J, Wang S (2011) OsWRKY45 alleles play different roles in abscisic acid signalling and salt stress tolerance but similar roles in drought and cold tolerance in rice. J Exp Bot 62(14):4863–4874

Tian XH, Li XP, Zhou HL, Zhang JS, Gong ZZ, Chen SY (2005) OsDREB4 genes in rice encode AP2 containing proteins that bind specifically to the dehydration-responsive element. J Integr Plant Biol 47(4):467–476

Uga Y, Sugimoto K, Ogawa S, Rane J, Ishitani M, Hara N, Yano M (2013) Control of root system architecture by DEEPER ROOTING 1 increases rice yield under drought conditions. Nat Genet 45(9):1097–1102

Umezawa T, Fujita M, Fujita Y, Yamaguchi-Shinozaki K, Shinozaki K (2006) Engineering drought tolerance in plants: discovering and tailoring genes to unlock the future. Curr Opin Biotechnol 17(2):113–122

Uno Y, Furihata T, Abe H, Yoshida R, Shinozaki K, Yamaguchi-Shinozaki K (2000) Arabidopsis basic leucine zipper transcription factors involved in an abscisic acid-dependent signal transduction pathway under drought and high-salinity conditions. Proc Natl Acad Sci 97(21): 11632–11637

Vannini C, Locatelli F, Bracale M, Magnani E, Marsoni M, Osnato M, Coraggio I (2004) Overexpression of the rice Osmyb4 gene increases chilling and freezing tolerance of Arabidopsis thaliana plants. Plant J 37(1):115–127

Verslues PE, Agarwal M, Katiyar-Agarwal S, Zhu J, Zhu JK (2006) Methods and concepts in quantifying resistance to drought, salt and freezing, abiotic stresses that affect plant water status. Plant J 45(4):523–539

Viana VE, Marini N, Busanello C, Pegoraro C, Fernando JA, Da Maia LC, Costa de Oliveira A (2018) Regulation of rice responses to submergence by WRKY transcription factors. Biol Plant 62(3):551–560

Vinocur B, Altman A (2005) Recent advances in engineering plant tolerance to abiotic stress: achievements and limitations. Curr Opin Biotechnol 16:123–132

Wan L, Zhang J, Zhang H, Zhang Z, Quan R, Zhou S, Huang R (2011) Transcriptional activation of OsDERF1 in OsERF3 and OsAP2-39 negatively modulates ethylene synthesis and drought tolerance in rice. PLoS One 6(9):e25216

Wang Q, Guan Y, Wu Y, Chen H, Chen F, Chu C (2008) Overexpression of a rice OsDREB1F gene increases salt, drought, and low temperature tolerance in both Arabidopsis and rice. Plant Mol Biol 67(6):589–602

Wang L, Li Z, Lu M, Wang Y (2017) ThNAC13, a NAC transcription factor from Tamarix hispida, confers salt and osmotic stress tolerance to transgenic Tamarix and Arabidopsis. Front Plant Sci 8:635

Wang B, Zhong Z, Wang X, Han X, Yu D, Wang C, Song W, Zheng X, Chen C, Zhang Y (2020) Knockout of the OsNAC006 transcription factor causes drought and heat sensitivity in rice. Int J Mol Sci 21:2288. https://doi.org/10.3390/ijms21072288

Welsch R, Wust F, Bar C, Al-Babili S, Beyer P (2008) A third phytoene synthase is devoted to abiotic stress-induced abscisic acid formation in rice and defines functional diversification of phytoene synthase genes. Plant Physiol 147(1):367–380

Wen J, Zeng Y, Chen Y, Fan F, Li S (2021) Genic male sterility increases rice drought tolerance. Plant Sci 312:111057

Wu K, Wang S, Song W, Zhang J, Wang Y, Liu Q, Fu X (2020) Enhanced sustainable green revolution yield via nitrogen-responsive chromatin modulation in rice. Science 367(6478): eaaz2046

Xia K, Wang R, Ou X, Fang Z, Tian C, Duan J, Wang Y, Zhang M (2012) OsTIR1 and OsAFB2 downregulation via OsmiR393 overexpression leads to more tillers, early flowering and less tolerance to salt and drought in rice. PLoS ONE 7:e30039

Xiang Y, Tang N, Du H, Ye H, Xiong L (2008) Characterization of OsbZIP23 as a key player of the basic leucine zipper transcription factor family for conferring abscisic acid sensitivity and salinity and drought tolerance in rice. Plant Physiol 148(4):1938–1952

Xiong, Lizhong, Yang Y (2003) Disease resistance and abiotic stress tolerance in rice are inversely modulated by an abscisic acid–inducible mitogen-activated protein kinase. Plant Cell 15(3): 745–759

Xiong H, Li J, Liu P, Duan J, Zhao Y, Guo X, Li Z (2014) Overexpression of OsMYB48-1, a novel MYB-related transcription factor, enhances drought and salinity tolerance in rice. PLoS One 9(3):e92913

Xu ZS, Chen M, Li LC, Ma YZ (2011) Functions and application of the AP2/ERF transcription factor family in crop improvement. J Integr Plant Biol 53:570–585

Yadav SK, Santosh Kumar VV, Verma RK, Yadav P, Saroha A, Wankhede DP, Chinnusamy V (2020) Genome-wide identification and characterization of ABA receptor PYL gene family in rice. BMC Genomics 21(1):1–27

Yamaguchi-Shinozaki K, Shinozaki K (2005) Organization of cis-acting regulatory elements in osmotic-and cold-stress-responsive promoters. Trends Plant Sci 10(2):88–94

Yang A, Dai X, Zhang WH (2012) A R2R3-type MYB gene, OsMYB2, is involved in salt, cold, and dehydration tolerance in rice. J Exp Bot 63(7):2541–2556

Yang S, Xu K, Chen S, Li T, Xia H, Chen L, Luo L (2019) A stress-responsive bZIP transcription factor OsbZIP62 improves drought and oxidative tolerance in rice. BMC Plant Biol 19(1):1–15

Yang J, Liu S, Ji L, Tang X, Zhu Y, Xie G (2020) Identification of novel OsCML16 target proteins and differential expression analysis under abiotic stresses in rice. J Plant Physiol 249:153165

Ye H, Du H, Tang N, Li X, Xiong L (2009) Identification and expression profiling analysis of TIFY family genes involved in stress and phytohormone responses in rice. Plant Mol Biol 71(3): 291–305

Yoon S, Lee DK, Yu IJ, Kim YS, Choi YD, Kim JK (2017) Overexpression of the OsbZIP66 transcription factor enhances drought tolerance of rice plants. Plant Biotechnol Rep 11(1):53–62

Yoshida T, Fujita Y, Sayama H, Kidokoro S, Maruyama K, Mizoi J, Yamaguchi-Shinozaki K (2010) AREB1, AREB2, and ABF3 are master transcription factors that cooperatively regulate ABRE-dependent ABA signaling involved in drought stress tolerance and require ABA for full activation. Plant J 61(4):672–685

Yuan X, Wang H, Cai J, Bi Y, Li D, Song F (2019a) Rice NAC transcription factor ONAC066 functions as a positive regulator of drought and oxidative stress response. BMC Plant Biol 19: 278

Yuan X, Wang H, Cai J, Li D, Song F (2019b) NAC transcription factors in plant immunity. Phytopathol Res 1(1):1–13

Zhang JZ, Creelman RA, Zhu JK (2004) From laboratory to field. Using information from Arabidopsis to engineer salt, cold, and drought tolerance in crops. Plant Physiol 135(2):615–621

Zhang S, Haider I, Kohlen W, Jiang L, Bouwmeester H, Meijer AH, Ouwerkerk PB (2012) Function of the HD-Zip I gene Oshox22 in ABA-mediated drought and salt tolerances in rice. Plant Mol Biol 80(6):571–585

Zhang C, Li C, Liu J, Lv Y, Yu C, Li H, Liu B (2017) The OsABF1 transcription factor improves drought tolerance by activating the transcription of COR413-TM1 in rice. J Exp Bot 68(16): 4695–4707

Zhang F, Zeng D, Huang L, Shi Y, Chen T, Zhou Y (2019) Stress-activated protein kinase OsSAPK9 regulates tolerance to salt stress and resistance to bacterial blight in rice. Rice 12(1):1–15

Zhao Y, Hu Y, Dai M, Huang L, Zhou DX (2009) The WUSCHEL-related homeobox gene WOX11 is required to activate shoot-borne crown root development in rice. Plant Cell 21(3): 736–748

Zheng X, Chen B, Lu G, Han B (2009) Overexpression of a NAC transcription factor enhances rice drought and salt tolerance. Biochem Biophys Res Commun 379(4):985–989

Zheng S, Liu S, Feng J, Wang W, Wang Y, Yu Q, Chen R (2021) Overexpression of a stress response membrane protein gene OsSMP1 enhances rice tolerance to salt, cold and heavy metal stress. Environ Exp Bot 182:104327

Zhou J, Li F, Wang JL, Ma Y, Chong K, Xu YY (2009) Basic helix-loop-helix transcription factor from wild rice (OrbHLH2) improves tolerance to salt-and osmotic stress in Arabidopsis. J Plant Physiol 166(12):1296–1306

Zhu T, Budworth P, Han B, Brown D, Chang HS, Zou G, Wang X (2001) Toward elucidating the global expression patterns of developing Arabidopsis: parallel analysis of 8300 genes by a high-density oligonucleotide probe array. Plant Physiol Biochem 39:221–242

Zhu W, Zhang L, Zhang N, Xing Y, Jiang B (2012) The clone of wheat dehydrin-like gene wzy2 and its functional analysis in Pichia pastoris. Afr J Biotechnol 11(40):9549–9558

Zolkiewicz K, Gruszka D (2022) Glycogen synthase kinases in model and crop plants–from negative regulators of brassinosteroid signaling to multifaceted hubs of various signaling pathways and modulators of plant reproduction and yield. Front Plant Sci 13:939487

Zou M, Guan Y, Ren H, Zhang F, Chen F (2008) A bZIP transcription factor, OsABI5, is involved in rice fertility and stress tolerance. Plant Mol Biol 66(6):675–683

14 Advancement of Omics Approaches in Understanding the Mechanism of Salinity Tolerance in Legumes

Naresh Kumar, Gurpreet Kaur, Sunita Devi, Charu Lata, Hemant Dasila, Satish Kumar Sanwal, Ashwani Kumar, and Anita Mann

Abstract

Grain legume crops are important to meet global food security and crop rotation with legumes also enhance the soil fertility through symbiotic fixation of atmospheric nitrogen. Globally, legume crop production is challenged adversely by salinity stress, but the effect is variable from genotype to genotype and further in genotypes from stage to stage. Plants adapt to salinity induced changes through complex mechanisms and several omics-based studies have been conducted in salinity exposed legumes such as chickpea, soyabean, cowpea, etc., in an attempt to understand the complex molecular mechanism underlying salinity tolerance. Improvement of salt tolerance ability demands the understanding of role of genes and their products in contributing salinity stress tolerance. Comparative transcriptomic analysis of tolerant and sensitive genotypes had led to the understanding of various genes and their role in complex molecular mechanisms underlying salinity tolerance in legumes. Similarly, comparative proteomics had offered unexpected and poorly understood molecular resources associated with salt tolerance. Metabolomics studies had quantified several metabolites such as sugars, sugar acids, sugar phosphates, organic acids, amines, and amino acids in legumes bearing contrasting response toward salinity and the metabolite profiling had

N. Kumar (✉)
Eternal University, Rajgarh, Himachal Pradesh, India

ICAR – Central Soil Salinity Research Institute, Karnal, Haryana, India

G. Kaur · S. Devi · S. K. Sanwal · A. Kumar · A. Mann
ICAR – Central Soil Salinity Research Institute, Karnal, Haryana, India

C. Lata
ICAR-Indian Institute of Wheat and Barley Research, RRS, Shimla, Himachal Pradesh, India

H. Dasila
Eternal University, Rajgarh, Himachal Pradesh, India

A. Kumar et al. (eds.), *Salinity and Drought Tolerance in Plants*,
https://doi.org/10.1007/978-981-99-4669-3_14

provided new insights of legumes metabolic responses to salinity stress. This chapter emphasizes on the mechanisms underlying salinity stress tolerance in legumes and the omics approaches offer opportunities to make clear the molecular basis of salinity tolerance.

Keywords

Legumes · Salt stress · Tolerance · Transcriptomics · Proteomics

14.1 Introduction

With over 20,000 species, legumes are the most widely used plant family and one of the most significant crops in the world, having a significant impact on agriculture, the environment, and human and animal nutrition and health (Reddy et al. 2012; Mathesius 2022). In terms of global production, which makes up 27% of all primary crop production worldwide, legumes come in third place behind cereals and oilseeds. In addition to providing 33% of the dietary protein nitrogen (N) needs of humans (Chaudhary et al. 2022), grain legumes also serve as a source of revenue and livestock feed, making them an essential part of both human and animal diets. In descending order, the primary legumes in terms of their significance in the diet include common beans (*Phaseolus* spp.), peas (*Pisum sativum* L.), chickpeas (*Cicer arietinum* L.), broad beans (*Vicia faba* L.), pigeon peas (*Cajanus cajan* L.), cowpeas (*Vigna unguiculata* L.), and lentils (*Lens esculenta* L.). In addition, grain legumes particularly soybean (*Glycine max* L.) and peanut (*Arachis hypogeae* L.) are a significant source of processed vegetable oil, accounting for more than 35% of global production.

Abiotic stress is one of the main factors limiting crop yield in the semiarid tropic which includes regions of developing countries with a combined population of nearly 1.4 billion, and where grain legumes are primarily grown. The leading causes of agricultural losses globally are abiotic stresses, which includes drought stress, salinity stress, waterlogging stress, high temperatures stress, and more. Since the last decade, the overall production of legumes has declined due to these various abiotic stresses. In recent years, salt stress alone has caused a significant decline in the production of legume crops such as soyabean, mung bean, and chickpea (Mann et al. 2019; Nadeem et al. 2019). Salt stress primarily executes detrimental effects on legumes growth through imbalanced nutritional uptake and osmotic potential which leads to the secondary stress-oxidative stress. Moreover, salt stress causes the reduction in the chlorophyll content, stomatal conduction and disturbs the photosystem II activity which leads to the disturbance in the gaseous exchange.

Plants under salt stress show changes in cellular metabolism, growth, and development as a result of decreased CO_2 absorption, altered cytosolic enzymes, osmotic stress, ion toxicity (Na^+ and Cl^-), oxidative stress, and finally cell death (Shabala 2013; Sharma et al. 2013; Mann et al. 2021). In order to eliminate Na^+ and Cl^- ions from roots when their accumulation becomes toxic, plants respond to salinity stress by activating a variety of physiological, biochemical, and molecular mechanisms. These mechanisms can be divided into three main classes: ion exclusion, tissue tolerance, and osmotic tolerance and have been documented in soyabean, pea, and chickpea. Several plant biologists believe that biophysical and biochemical factors that control stress tolerance can be used to unravel the mechanisms governing the tolerance pathway by examining correlative evidence from various plant species. These discoveries provided conceptual frameworks for moving from phenotypic studies to analyses of proteins and enzymes, gene structure, and gene expression, which led to the development and assessment of transgenic and mutant plant species. In this chapter, we have outlined many multi-omic approaches that have been utilized to date to examine the function of various salt stress-responsive proteins in legume plants and to uncover genes and proteins that are involved in salt tolerance.

14.2 Omics: Solutions to Salt Stress in Legumes

During abiotic stresses, biotechnological approaches such as tissue culture, in vitro mutagenesis, marker-assisted breeding, and genetic transformation can overcome major issues of traditional breeding, however, it requires a deep biological knowledge of stressed species and mechanisms underlying tolerance to abiotic stresses. The mechanisms of tolerance can be estimated at various levels from the molecular level to the entire plant. Recent research has focused on the plant's molecular responses to salt stress, since responses are regulated by the plant genome (Hao et al. 2021). Till recent years, to study the plant stress responses was primarily focused on model plants like *Arabidopsis* rather than the other plants like legumes. Utilizing knowledge from *Arabidopsis*'s stress responses, research on legumes has leveraged it as an information source.

14.3 Transcriptomics

The transcriptome is the collection of all messenger RNAs (mRNAs) in a cell, tissue, or organism, and the study of mRNA populations is referred to as transcriptomics. Transcriptomics is a potent method for examining the role of genes in a variety of biological processes, finding potential candidates, and illuminating the molecular interactions between gene regulatory networks in response to abiotic stress. Transcriptome analysis has been progressively advancing our understanding of RNA-based gene regulatory network with the emergence of next-generation high-throughput sequencing technologies. In the post-genomic era, the transcriptomics has emerged as excited area of life sciences for many reasons such as transcriptome analysis shows how the genome expresses itself dynamically, it supports the proteomics research and additionally, non-coding RNA's functional and structural studies broaden the scope of transcriptomics. Abiotic stresses such as drought or salinity commonly activate about 50% of the genes in plants because they cause dehydration at the cellular level. Cellular dehydration mainly causes numerous changes in gene expression that ultimately determine how the plant reacts to a given environmental factor. The induced genes in response to cellular water deficit stress fall into a variety of functional categories, including metabolism, transport, signaling, transcription, hydrophilic proteins, and the unclassified, including the repression of genes related to photosynthesis and involved in plant growth and development. These genes are basically characterized on the basis of responding time such as some genes like protein kinases and transcription factors respond immediately within minutes or seconds after the abiotic stress while some genes such as ROS scavenger proteins, LEA proteins, heat shock proteins, etc. respond later in hours, days, or even weeks (Reddy et al. 2012).

The expression patterns of several salinity tolerance genes in chickpea have been studied using a variety of functional genomic resources, including cDNA-microarray (Mantri 2007), expressed sequence tags (ESTs) (Varshney et al. 2009) and deep super serial analysis of gene expression (SAGE) (Molina et al. 2011). Afterward, under salt stress, RNA-seq study of root tissue of chickpea at the vegetative and reproductive stages found 1376 and 3660 differentially expressed genes (DEGs), respectively. Most of these DEGs, in terms of gene function, were connected to cellular metabolic processes and cell redox homeostasis to ethylene hormone signaling under salt stress (Garg et al. 2016). Liao et al. (2003) studied that reactive oxygen species (ROS) scavenging proposed by expression analysis of the GmPAP3 gene upregulation during salt stress as a potential contributor to salt tolerance in soybean. Likewise, Lin et al. (2022) reported the comparative transcriptomic analysis of the leaves and roots of common vetch under salinity stress. A total of 6361 DEGs were identified in roots and leaves, with a predominant expression of genes related to Ca^{2+} transport in leaves, while in roots, the expressed genes were primarily associated with peroxidase activity. Salt stress adaption may be significantly influenced by the differential expression of the *alternative oxidase 1 (Aox1)* gene and higher stimulation of cellular antioxidant genes in root of *M. trancatula* (Mhadhbi et al. 2011, 2013). The differential expression of genes encoding transcription factors, including *WRKY*, *ERF/AP2*, *bZIP*, *bHLH*-type, *ZFP*, *YABBY*, and *HD-Zip*, is crucial for regulating the plant's response to salt stress (Chen et al. 2014; Deinlein et al. 2014; Zhao et al. 2017). Markedly, the higher expression of CBF4 in *M. truncatula* (Li et al. 2011) and higher expression of CBF4 and Zpt2-2 transcription factors in *Medicago* varieties (Mokhtari et al. 2017), making them more resistant to salt. Hiz et al. (2014) discovered differential expression of 441 salt-responsive TFs from 2678 putative TFs under salinity stress, 6422 and 4555 unigenes from leaf and root tissues, respectively, in common bean. In high salt concentration, during reproductive stage chickpea is highly sensitive. Based on this fact, Kaashyap et al. (2022) reported the comparative transcriptomic analysis in flowers of chickpea. The authors identified about 2022 DEGs in response to salt stress where genes such as FLOWERING LOCUS T (FT) and pollen development such as ABORTED MICROSPORES (AMS), rho-GTPase, and pollen-receptor kinase were found to be significantly differentially regulated suggesting the role in the salt tolerance. The transcription factor family "WRKY" is considered to have a wide range of functions in development and physiological processes, particularly in the drought and high salinity stress.

Table 14.1 Differentially expressed genes, transcription factors, and candidate genes in response to salinity stress in some legumes

S. No.	Legume species	DEGs/TFs/Genes	References
1.	Chickpea—JG11 (salt tolerant) and ICCV2 (salt sensitive)	2022 DEGs	Kaashyap et al. (2022)
2.	Chickpea—BG 1103, S7, DCP 92–3, ICCV 10, KWR 108, BG 256, K 850, JG 16, ICC 4463, CSG8962	*SOD, CAT, APX, MDHAR, DHAR, GR, POX, P5CS, P5CR, PD, HKT1*	Kaur et al. (2022)
3.	Chickpea–salt tolerant (ICCV 10, JG 11) and salt sensitive (DCP 92-3, Pusa 256)	21,698 DEGs (11,456 upregulated and 10,242 downregulated)	Kumar et al. (2021)
4.	Chickpea—Genesis 836, Hattrick, ICC12726, Rupali, Slasher and Yubileiny	*CaRab-GTP* gene	Sweetman et al. (2020)
5.	Soyabean—BB52 (salt tolerant) and N23674 (salt sensitive)	*GsCNGC20-d*	Pi et al. (2023)
6.	Soyabean—Qi Huang No.34 (Salt tolerant) and Dong Nong No.50 (Salt sensitive)	17,477 DEGs (6644 with distinct expression)	Hu et al. (2022)
7.	Soybean—JD19, LH3, and LD2	1482 DEGs (837 genes upregulated and 645 downregulated)	Jin et al. (2021)
8.	Soybean—Tianlong No. 1	1,235 differentially expressed genes	Liu et al. (2021)
9.	Soyabean	*GmCDPK3* Gene	Wang et al. (2019)
10.	Common bean	*SABATH* gene family	Aygören et al. (2022)
11.	Common beans—Syrian White, Jiyin1 and Landrace-60-Day Harvest	441 differentially expressed genes	Zhang et al. (2021)
12.	Common beans—Ispir (salt tolerant) and TR43477 (salt-susceptible)	71 differentially expressed genes	Niron et al. (2020)

Yu et al. (2016) reported the upregulation of 19*GmWRKYs* genes and 35 *GmWRKYs* genes with decreased expression in response to salt stress in soyabean. Similarly, in chickpea distinct expression was detected for WRKYs under salinity stress (Garg et al. 2016). A brief list of DEGs related to salinity stress tolerance in legumes are listed in Table 14.1.

Growing advances in sequencing technology have made it clear that non-coding RNA (ncRNA) molecules, such as long non-coding RNA (lncRNA) and short RNA (sRNA) or micro-RNA (miRNA), have a role in how well plants adapt to various abiotic stresses. Recently many studies have reported the function of miRNA in response to salt stress. It is now established that plants miRNAs affects several biochemical and physiological processes (Arshad et al. 2017). miR156 induced overexpression of different genes such as NHX1, SOS1, HC-ATPase is responsible for the improvement of salinity tolerance in alfalfa which leads to the increased biomass and reduced ion toxicity (Arshad et al. 2017).

14.4 Proteomics

Proteomics approach helps to understand the changes in the proteome. The study helps to understand the changes in the proteins in response to stress (Hossain 2020). A thorough understanding of the proteins present in the genome of the cell, tissue or an organism at a particular time (proteome) is necessary for understanding the biology of the cell or the organism (Reddy et al. 2012). Salinity badly affects the process of photosynthesis. Proteomic study in legumes investigated an alteration in the proteins involved in photosynthesis in plants grown under salinity. In alfalfa, the exposure to salt stress induced by NaCl led to the downregulation of three thylakoid membrane proteins: cytochrome b6-f complex iron-sulfur (Cyt b6/f), chlorophyll a/b binding protein (CAB), and chloroplast oxygen-evolving enhancer protein 1 (OEE1) (Xiong et al. 2017). These proteins play vital roles in the light reactions of photosynthesis, encompassing electron transfer, light harvesting and the light-induced oxidation of water. In chickpea, proteins related to photosynthesis, psbP domain-containing protein 1 (PPD1) (a nuclear encoded and thylakoid luminal protein, is an essential for PSI and PSII assembly and activity) and CAB got upregulated in the tolerant cultivar, both PPD1 and CAB transferred more excitation energy to the reaction center, where the accumulated plastocyanin can donate more electrons to photosystem I (PSI) to reduce $NADP^+$ to NADPH. OEE was downregulated in alfalfa (Xiong et al. 2017) but it got upregulated in the tolerant genotype of chickpea (Arefian et al. 2019) and soyabean (Awana et al. 2020). RuBisCO activase releases inhibitory sugar phosphates, such as ribulose-1,5-biphosphate, from the active RuBisCO sites so that CO_2 can activate the enzyme controlling carbamylation. RuBisCO activase got upregulated in alfalfa (Xiong et al. 2017) but its downregulation has been reported in soybean (Sobhanian et al. 2010). Proteomic study detected three subunits of Ribulose 1,5-bisphosphate carboxylase/oxygenase (RuBisCO), fixes the dissolved CO_2 to produce 3-phosphoglycerates, in NaCl treated alfalfa (Xiong et al. 2017) and upregulation of this enzyme was detected in the tolerant genotype of chickpea (Arefian et al. 2019) and authors suggested that the Calvin cycle in chickpea can be slowed by salinity but rapidly impaired in sensitive cultivar. Phosphoribulokinase (PRK) provides an immediate CO_2 acceptor and is considered the regulatory enzyme of the Calvin cycle, and upregulation under salt stress has been revealed in tolerant chickpea (Arefian et al. 2019) and soyabean (Awana et al. 2020) but in alfalfa NaCl treatment led to the downregulation of CP12 protein, a small nuclear encoded chloroplast protein, oligomerized with phosphoribulokinase (PRK) and $NADP^+$-GAPDH in the presence of NAD(H) to generate a PRK/CP12/GAPDH complex. Hence, the downregulation of CP12 appears to be associated with a decrease in photosynthesis activity (Xiong et al. 2017). Apart from these proteins, upregulation of alpha and beta subunits of ATP synthase (participate in photosynthesis) fructose-bisphosphate aldolase and transketolase (TKT) (involved in the regeneration phase of the Calvin cycle and the pentose phosphate pathway, thus influencing plant productivity) and magnesium chelatase (enzyme involved in the chlorophyll biosynthesis pathway) was found to be a salt tolerance factor in chickpea (Arefian et al. 2019). Photosystem I, II, cytochrome oxidase, and ATP synthase complex are thylakoid membrane-bound proteins involved in photosynthesis and these proteins were expressed under salt stress in pigeon pea (*Cajanus cajan*) (Jain et al. 2021).

Heat shock proteins (HSPs) are the high-temperature-inducible chaperones which regulate normal plant growth processes by helping with protein folding and preventing protein aggregation. HSPs are grouped into five families: HSP100s, HSP90s, HSP70s, HSP60s, and sHSPs (small HSPs) and a heat shock protein HSP70 got upregulated under NaCl stress in alfalfa seedlings (Xiong et al. 2017). Significant upregulation of three HSP70 molecular chaperones and one 20 kDachaperonin has been reported in salt tolerant chickpea (Arefian et al. 2019). Proteomic investigation in mung bean (*Vigna radiata* L.) under salt stress also detected the presence of HSP70. HSP70 maintains protein metabolism, proper protein folding, mitigates oxidative stress and helps in the proper development of legumes (Alharby and Hakeem 2021).

Plants utilize various groups of low molecular weight compounds, collectively known as osmoprotectants, to mitigate the osmotic stress imposed by salinity. These nontoxic compounds stabilize cellular structures and enzymes, act as metabolic signals, and scavenge reactive oxygen species produced under stressful conditions (Zulfiqar et al. 2020). A L-myo-inositol 1-phosphate synthase (MIPS) was upregulated by salt stress in the shoots of alfalfa and this enzyme catalyzes the first step of the pathway producing inositol containing substances and these substances act as osmolytes (Xiong et al. 2017). Late embryogenesis abundant (LEA) proteins, known as high-molecular osmolytes, function to protect the steady structure of proteins, chlorophyll, membranes, and cells. Upregulation with salinity has been reported in salt tolerant chickpea with salinity which is a sign of adaptation to saline conditions (Arefian et al. 2019). Level of two dehydrins and two late embryogenesis abundant (LEA) proteins increased with salt stress in soybean (Yin et al. 2018). Imino acid proline is a strong osmolyte and its presence was detected only under salt stress conditions in pigeonpea (*Cajanus cajan*) (Jain et al. 2021).

Reactive oxygen species (ROS) are important for the process of respiration, photorespiration, and photosynthesis under normal conditions but under saline conditions hike in their production leads to oxidative stress which causes cytotoxicity in plants. Plants avoid the oxidative stress by an enzymatic [superoxide dismutase (SOD), ascorbate peroxidase (APX), monodehydroascorbate reductase (MDHAR), dehydroascorbate reductase (DHAR), glutathione reductase (GR), catalase (CAT), and Peroxidase (POX)] and non-enzymatic (ascorbic acid, reduced glutathione, α-tocopherol, carotenoids, and flavonoids) antioxidant defense machineries. SOD enzyme removes superoxide ions while other antioxidative enzymes decompose the hydrogen peroxide (H_2O_2) generated by this enzyme and/or other reactions (García-Caparrós et al. 2019). A total of eight proteins (4 ascorbate peroxidases, 2 glutathione peroxidases, 1 ferritin protein (Fenton reaction (a reaction between ferrous ion and hydrogen peroxide) produces highly toxic hydroxyl and hydroxide ions and ferritin protein possess ferroxidase (conversion of ferrous (Fe^{2+}) to ferric (Fe^{3+}) activity) and lquinone reductase family protein (Maintains reduced forms of ubiquinones and α-tocopherolquinone, thus protects the cell organelles, especially plasma membrane, from oxidative stress)) related to antioxidative reactions got upregulated in alfalfa seedling roots and shoots experiencing salt stress (Xiong et al. 2017). Abundance of two SODs [Cu–Zn] increased by salt treatment in soybean. NaCl stress significantly enhanced the accumulation of POD (Q9XFI8) while decreasing the POD (C6TBQ4) abundance, suggesting that different POD members may have different functions in salt stress responses in soybeans (Yin et al. 2018). In chickpea, Cu/Zn superoxide dismutase (SOD) and ascorbate peroxidase (APX) were classified and upregulated in tolerant cultivar (Arefian et al. 2019). Glutathione S-transferase (GSTs), Phase 2 detoxification enzymes, functions in protecting the cellular macromolecules from the attack of reactive electrophiles. GSTs carry out the conjugation of glutathione (GSH) with a wide variety of endogenous and exogeneous electrophilic compounds. In pigeonpea (*Cajanus cajan*), proteomic study detected GSTs and APX under salt stress condition (Jain et al. 2021). A apart from these proteins, trypsin inhibitor got upregulated in soyabean under salt stress indicating its role ion H_2O_2 detoxification in soyabean seedlings (Sobhanian et al. 2010). Thiamine thiazole synthase and xanthoxin dehydrogenase, which are involved in the biosynthesis of thiamine and abscisic acid respectively, serve as defensive proteins that help protect seedlings against the harmful effects of reactive oxygen species. Apolipoprotein D (ApoD) is a small plasma membrane-associated protein known as lipocalin. Apolipoprotein D binds and scavenges peroxidated lipids, which helps to maintain membrane integrity. These defensive proteins were found in the tolerant cultivars of chickpea. Carbonic anhydrase (CA) also got upregulated in tolerant chickpea and it improves cell resistance to cytotoxic concentrations of H_2O_2 (Arefian et al. 2019).

Salt stress is first perceived by putative sensors in the root cell membranes and these signals are transmitted to the cellular machinery to regulate gene expression and changes in cellular metabolism designed to prevent or minimize the deleterious effects of stress. This signaling is mediated by different kinds of secondary messengers, such as Ca^{2+}. Two annexin proteins were identified in alfalfa (Xiong et al. 2017) and one in soyabean (Sobhanian et al. 2010). Annexins are a multigene, multifunctional family of Ca^{2+} dependent membrane binding proteins and mediate osmotic stress and abscisic acid signal transduction. A plasma membrane H^+-ATPase (PM H^+-ATPase) also got upregulated under salt stress in alfalfa. This protein is involved in the compartmentalization of toxic Na^+ within the vacuole. Hence, the enhanced presence of plant plasma membrane H^+-ATPase might assume a crucial function in alfalfa's ability to withstand salt stress (Xiong et al. 2017). In pigeonpea as well, effective sequestration of Na^+ into the vacuoles was associated with salt tolerance. This was attributed to the elevated expression of V-ATPase (TCONS_00029723) under salt stress, resulting in increased energy production to establish a robust proton gradient. This gradient facilitates the active transport of excessive cytoplasmic Na^+ into vacuoles, allowing strict regulation of ion compartmentalization by controlling the driving force of Na^+ transport. Moreover, MAPK (TCONS_00036358, TCONS_00036542, TCONS_00036806) were found in both the genotypes under salt stress, suggesting the involvement of MAPK signaling pathways to salt stress induced responses in the pigeonpea genotypes (Awana et al. 2020). Comparative proteomic analyses using contrasting genotypes under control and salt-stressed conditions provided the basis for revealing salt tolerance mechanisms in legumes. Proteomic study in salinity contrasting genotypes showed the change in abundance of the proteins related to energy metabolism, photosynthesis, carbon assimilation, photorespiration, salt stress-responsive proteins, antioxidative defense system and signaling. These findings highlight the significance of stress-responsive proteins in the adaptation of legumes to salinity stress.

14.5 Metabolomics

Genomics, transcriptomics, and proteomics are multi-omics approach that are frequently used in the expression study of genes and proteins but to study the expression of specific metabolites under any biotic or abiotic stress use of "metabolomics" study becomes prominent (Brunetti et al. 2018). Metabolomics brings about the physiological state of organisms as it specifies alternations in cell metabolome (Sumner et al. 2003). Therefore, metabolomics is considered to bring about the gap between phenotype and genotype and for complete understanding of both there is need to get integrative approach which compromises of metagenomics, transcriptomics, metabolomics, and proteomics (Dixon 2001). Metabolomics techniques are commonly employed to elucidate differential metabolic responses resulting from geographical and species-related factors. These techniques involve the analysis of variations in the content of small molecules. In recent studies, metabolomics has been extensively used to investigate and uncover the underlying resistance mechanisms in plants under abiotic stress conditions (Yue et al. 2020).

In case of legumes metabolomics approach has been widely used to determine the response against various stimuli as in case of *M. truncataula* where metabolites were analyzed via cell suspension (Bell et al. 2001). In order to gain insights into the mechanisms occurring in the root system of maize (*Zea mays*), a highly significant leguminous crop, comparative studies of metabolic profiling were conducted under NaCl stress (Yue et al. 2020). The main substances that are marked for metabolic profiling are fatty acids, organic acids, amino acids, sugars, polyhydric alcohols, and mineral elements under salt stress. Yue et al. (2020) reported to have a sharp rise in the content of betanin, sucrose, and fatty acids under NaCl stress in maize. In legumes application of metabolomics approach mostly confined to model legumes where significant number of changes has been observed in tricarboxylic acid and glycolysis cycle intermediates (Kumari et al. 2015; Jiao et al. 2018). Comparative analysis between cultivated soybean C08 and wild soybean W05 at metabolic level suggested that compounds like disaccharides and sugar alcohols, were higher in wild type as compared to cultivated soybean (Lu et al. 2013). Similar results were reported in wild type soybean (*G. soja*) where elevated level of metabolites (phenylalanine, citraconic acid, and other metabolites derived from glycolysis and TCA cycle) (Zhang et al. 2016). Li et al. (2017) reported significant amount of change in several metabolites such as isoleucine, phenylalanine, glutamic acid, asparagines, aspartic acid, 1-allothreonine, and some anti-antioxidants in soybean in response to salinity stress. Similar comparative metabolomics study in W1 (wild type soybean) and W2 (wild salinity tolerant soybean) revealed significant amount of accumulation of different TCA cycle intermediates, organic acids, and amino acids in wild type W2 which confers high tolerant ability to soybean under salinity stress than wild type W1 (Jiao et al. 2018). Thus, these metabolites can be used as a marker in distinguishing sensitive and salt tolerant genotype (Sanchez et al. 2010).

14.6 Genomics

Climate change and its impact on agriculture is major concern among the research to sustain the balance between food and feed demands. Legumes are considered as an essential diet due to its high protein, minerals, and vitamin content. With the advancement in technology, agriculture is also revolutionized. Therefore, technology advancement emerged with the legume production is a solution for mitigating stress effect in plants. Various OMICs approaches such as proteomics, genomics, metabolomics, and transcriptomics widely used to breed climate-resilient plants (Ali et al. 2022). Marker-assisted breeding has led to the identification of numerous Quantitative Trait Loci (QTL) and markers in cowpea and chickpea, facilitating the development of climate-resilient legumes. These QTLs and markers have proven instrumental in improving the resilience of legume crops to climate-related challenges. A successful transition to resilience and efficiency under abiotic stress conditions has been made possible by quick innovations and advancements in "omics" pertaining to the post-genomic epoch, such as molecular characterization, next-generation sequencing, modeling of various molecular and physiological knowledge, and association of these assertions with plant establishment (Pandey et al. 2021). As genomic size of legumes is large so model plants or system having short breeding season, enormous seed production and most importantly diploid in nature like *Medicago truncatula* and *Lotus japonicas* is used for genetics studies (Cook 1999; Cervantes et al. 2019; Stai et al. 2019). The focus of genomics is the physical integrity of the genome, with the intention of identifying, evaluating, and controlling genomic properties throughout the chromosomes. In diverse abiotic stress some genes are upregulated, downregulated, and remain silent so genomic information is mandatory to deal with transcriptomics, proteomic, and metabolomics, etc. (Chandrashekharaiah et al. 2021). Advancement in genomics tool for abiotic stress tolerant in legumes: genome sequence information is necessary for any type of genetic improvement in legumes under stress conditions. The combination of genomics techniques and markers assisted selection of legumes assist in easy identification of trait specific genes in population (Saade et al. 2016). Using massively multiplexed marker oligonucleotides like the Affymetrix GeneChip, SNPs are good for separating complicated traits because they are quick, high-throughput, co-dominant, abundant, affordable, and sequence-tagged (Missanga et al. 2021; Thudi et al. 2021). Axiom R SNP array genotyping at high-throughput is a productive and affordable technique for genotyping and the creation of high-density linkage maps. Recently, the "Axiom R CicerSNP Array" for genotyping recombinant inbred chickpea lines was created (Roorkiwal et al. 2016). Identification of QTL associated with stress tolerance is necessary for genetic and linkage mapping of that tolerant genomic regions. QTLs can identify genomic areas linked to the expression of the trait being researched (Kushwah et al. 2021). Instead of QTL, genome wide association mapping (GWAS) is with better precision, utility, viable, and cost effective. In case of legumes association mapping/GWAS is associated with abiotic stresses in legumes (Gondalia et al. 2022). Instead of these genomes editing in legumes with desired gene under abiotic stress is the recent research motive. Genome editing techniques are being used to investigate numerous potential genes linked to abiotic stress response. To date, numerous biotechnological tools are present to identify these genes. Somaclonal variations, tissue culture, marker-assisted breeding, mutagenesis, wide hybridization, double haploids, and genetic transformation are the main tools of genome editing (Aasim et al. 2018). Genome editing technology, encompassing cutting-edge methods, enables rapid and precise modification of agricultural genomes. This technology plays a crucial role in safeguarding plants against various hazards and enhancing crop yields. Genome editing techniques use site-specific endonucleases such as CRISPR-Cas9, transcription activator-like effector nucleases, and zinc-finger nucleases (ZFNs). Numerous investigations on genome editing have been carried out on important legume crops (Zhu et al. 2017). Badhan et al. (2021) first time used CRISPR/Cas9 and studied drought tolerance associated gene, Reveille 7 (RVE7), and 4-coumarate ligase for CRISPR/Cas9 editing in chickpea (*C. arietinum*) using protoplast. Recently various research studies are conducted on symbiotic nitrogen fixation (SNF) capability in cowpea. Ji et al. (2019) studied non-heritable cowpea mutations with hairy roots for genome editing using the CRISPR/Cas9 system, and reduced nodule formation was seen in the mutants with disrupted alleles. The work further continues by Juranić et al. (2020), three cowpea meiosis genes REC8 (encodes meiotic recombination protein), SPO11-1 (encodes SPO11 protein; an initiator of meiotic double-stranded breaks), and OSD1 (encodes Ophiostoma scytalone dehydratase protein promoting meiotic progression) were studied. These genes were used to induce the development of asexual seeds in cowpea plants. Due to biallelic mutations in exons 1 and 3 of the SPO11-1 gene, they discovered abnormalities in meiosis. Crisper/Cas9 also used in Mung bean and soyabean. However, numerous attempts have been undertaken to pinpoint the genetic areas connected to abiotic stressors. To speed up marker-assisted breeding, fewer attempts have been made to confirm the discovered genetic area. Therefore, the development of KASP markers for speed breeding is currently required (Ali et al. 2022). Legumes' genomic resources have recently made significant strides, laying the groundwork for novel breeding techniques including genome editing and genomic selection. When compared to grains, their efforts are still modest. Utilizing CRISPR/Cas9-based gene editing technology is crucial for the targeted enhancement of characteristics in legume crops.

14.7 Trans-genomics

As a complementary method for the genetic improvement of field crops, the use of transgenic technology or "trans-genomics" may provide a more focused gene-based approach for learning important information about the mechanisms governing stress tolerance, easing some of the major obstacles to crop productivity in developing countries. Grain legume tissue culture has generally been characterized as challenging (Pratap et al. 2018). It has been noted that this plant group's regeneration from both organogenesis and embryogenesis is refractory. It has been identified as a significant barrier to the creation of several transgenic legumes. Gene overexpression, gene suppression, promoter analysis, and T-DNA tagging are only a few examples of molecular genetics advancements that need for effective transformation systems (Kharb et al. 2021). Hence, implementing reliable regeneration techniques in legumes is a prerequisite for genetic transformation. A variety of genes in plants are activated in response to abiotic stress, increasing the levels of numerous osmolytes and proteins that may provide some degree of protection from these stresses.

Hence, to achieve a high level of resistance to drought or salt stress, it may be required to introduce many potentially helpful genes into the same plant. Through a variety of genetic transformation techniques, novel genes obtained from unusual sources of plants, animals, bacteria, and even viruses can be inserted into the crop with the possibility of regulating the timing, tissue specificity, and expression level of transferred genes for their optimum function. Nevertheless, utilizing *Agrobacterium tumefaciens*-mediated gene transfer has been a significant advance in the study of legume transgenics studies (Jha et al. 2019). Nonetheless, the recovery rate of transgenic lines in legume crops remains exceedingly low. All of the major legume crops, including *Vigna* species, *C. arietinum, C. cajan*, *Phaseolus* species, *Lupinus* species, *Vicia* species, *P. sativum*, and *soybean*, have undergone through genetic alteration to date (Reddy et al. 2012). Various genes responsible for several osmolytes have been engineered for overexpression in plants to create crops that can withstand stresses. Nevertheless, methods involving the transfer of a single functioning gene have not been very successful in enhancing plant tolerance beyond the short-term effects. While being essential to tropical agriculture, transgenic grain legumes, with the exception of soybean, have lagged behind as compared to their counterparts for example cereals, in moving from labs to major agricultural fields. Nevertheless, researchers performed some events of transgenic legumes successfully in major legume crops discussed in this chapter.

Chickpea is a lucrative crop, especially in low-input food production systems because of their high protein content, critical minerals, dietary fiber, and noteworthy capacity to fix atmospheric nitrogen. Activation of osmoregulatory genes in response to drought, salt, and high temperatures is one of the greatest strategies for abiotic stress tolerance (Soren et al. 2020). Sharmila et al. (2009) targeted the chloroplasts of chickpeas with a prokaryotic osmoregulatory choline oxidase gene (codA) to increase their ability to survive oxidative damage. When transgenic plant chloroplasts were tested for their ability to resist photo-inhibitory damage, the PS II activity of wild type plant chloroplasts exposed to high light intensity was much lower than that of transgenic chickpea chloroplasts. D1-pyrroline-5-carboxylate synthetase (P5CS) gene was introduced into chickpeas to produce excess of proline. The transgenic events demonstrated notable proline accumulation, particularly in the leaves, resulting in a reduction in the levels of free radicals, which are by-products of lipid peroxidation. Despite this, several events exhibited a significant increase in biomass production, indicating that the overexpression of proline had no beneficial impact on biomass accumulation. Increased amount of proline had minimal impact

Table 14.2 List of events on production of abiotic stress-tolerant transgenic legume crops

Gene	Cellular role	Host	Promoter used	Response	References
Gene *J*	Transcription factor	*Glycine max* (L.)	CaMV35S	Salt tolerance	Cheng et al. (2020)
GmNFYA13	ABA production	*Glycine max* (L.)	CaMV35S	Tolerance to abiotic stresses	Ma et al. (2020)
NTR1	Methyl jasmonate synthesis	*Glycine max* (L.)	CaMV35S	Abiotic stress tolerance	Xue and Zhang (2007)
p5cs	Proline biosynthesis	*C. arietinum*	CaMV35S	Drought tolerance	Bhatnagar-Mathur et al. (2009)
GmDREB1	Transcription factor	*M. sativa*	*A. thaliana* RD29A	Salt tolerance	Jin et al. (2010)
MDH	Malate dehydrogenase	*M. sativa*	CaMV35S	Tolerance to aluminum toxicity	Tesfaye et al. (2001)
Mn-sod	Dismutation of reactive oxygen intermediates in mitochondria	*M. sativa*	CaMV35S with a chloroplastic and mitochondrial transit peptide	Water and cold stress	Pandey et al. (2008)

on the yield architecture elements that are crucial for mitigating the detrimental impacts of drought stress in chickpea since the overexpression of the P5CSF129A gene only led to a minor increase in transpiration efficiency (TE). Large numbers of transgenic studies in chickpea are performed on regulatory genes or transcription factors such as DREB1A transcription factor from *Arabidopsis thaliana*, driven by a stress-inducible promoter from rd29A gene from *A. thaliana*. Prior research has established that regulatory genes or transcription factors, particularly those belonging to the AP2/ERF family, enhance stress tolerance by regulating the synchronized expression of numerous stress-related genes in transgenic plants of different origins.

In 2001, roundup-ready soybean was the only transgenic legume crop commercially cultivated in seven countries, namely the United States of America, Argentina, Canada, Mexico, Romania, Uruguay, and South Africa. Roundup ready soybean was the first transgenic soybean resistant to herbicide, commercially released in the USA in 1996 by Monsanto Company. Gene *J* positively regulated expression of *GmWRKY12*, *GmWRKY27*, *GmWRKY54*, *GmNAC*, and *GmSIN1* downstream salt stress response genes (Cheng et al. 2020). *GmNFYA13*, an NF-YA transcription factor in soybean, was strongly induced by salt, drought, ABA, and H_2O_2 stresses and enhances salt tolerance (Ma et al. 2020). Several successful attempts of transgenic legumes are given in Table 14.2.

Fig. 14.1 Integrated omics techniques for the salt tolerant cultivar development

14.8 Omics Integration to Develop Salt Tolerant Genotypes

Bioinformatics/computational biology is getting more attention as a result of the flood of high-throughput "omics" data in order to make better conclusions from the massive data sets. However, it is extremely difficult to resolve the complex features at the functional level by integrating this large "omics" data. Salinity is a complex attribute due to a complicated network of signal transduction pathways and several levels of regulation occurring at gene regulation, transcription, post-transcription, and post-translational levels. Hence, the "omics" including genomics, transcriptomics, proteomics, metabolomics, finally decides the phenotypic of trait contributing salinity tolerance. Several web-based resources have been built for various legumes to obtain information about genes, genomics, transcriptomics, proteomics, and other "omics," for example, Chickpea Transcriptome Database

(CTDB) in chickpea (Verma et al. 2015), Cowpea Genespace/Genomics Knowledge Base (CGKB) in cowpea (Chen et al. 2007), Soybean Knowledge Base in soybean (SKB, http://soykb.org), and Legume information system (LegumeInfo.org) (Dash et al. 2016). Using an efficient combination of multilayer "omics" research and breeding methods, it is possible to create tolerant cultivars in grain legumes as depicted in Fig. 14.1.

14.9 Conclusion

The use of biotechnological approaches to combat plant stresses has become more and more promising over time, although advancements in the field of legumes have been rather slow. For the successful application of "omics" to abiotic stress, it is crucial to comprehend the molecular aspects of stress response, including the mechanisms of gene expression for protein and metabolite production, as well as the subsequent impact on phenotypes. The "omics science" might be very useful in understanding plant salinity response to promote agricultural crop improvement programs. Additionally, the techniques such as selection and editing of genome using CRISPR-Cas9 and breeding techniques such as "speed breeding" could enhance the agricultural development efforts. The increased productivity of grain legume crops under salt stress will undoubtedly be aided by the greater understanding of plant salinity response and more effective breeding strategies.

References

Aasim M, Baloch FS, Nadeem MA et al (2018) Fenugreek (*Trigonella foenum-graecum* L.): an underutilized edible plant of modern world. In: Global perspectives on underutilized crops. Springer, Cham, pp 381–408

Alharby HF, Hakeem KR (2021) Proteomic investigations to assess the impact of salinity on *Vigna radiata* L. genotypes. Curr Proteomics 18:106–112

Ali A, Altaf MT, Nadeem MA et al (2022) Recent advancement in OMICS approaches to enhance abiotic stress tolerance in legumes. Front Plant Sci 13:952759

Arefian M, Vessal S, Malekzadeh-Shafaroudi S et al (2019) Comparative proteomics and gene expression analyses revealed responsive proteins and mechanisms for salt tolerance in chickpea genotypes. BMC Plant Biol 19:1–26

Arshad M, Gruber MY, Wall K, Hannoufa A (2017) An insight into microRNA156 role in salinity stress responses of Alfalfa. Front Plant Sci 8:356

Awana M, Jain N, Samota MK et al (2020) Protein and gene integration analysis through proteome and transcriptome brings new insight into salt stress tolerance in pigeonpea (*Cajanus cajan* L.). Int J Biol Macromol 164:3589–3602

Aygören AS, Güneş E, Muslu S et al (2022) Genome-wide analysis and characterization of SABATH gene family in phaseolus vulgaris genotypes subject to melatonin under drought and salinity stresses. Plant Mol Biol Report. https://doi.org/10.1007/s11105-022-01363-5

Badhan S, Ball AS, Mantri N (2021) First report of CRISPR/Cas9 mediated DNA-free editing of 4CL and RVE7 genes in chickpea protoplasts. Int J Mol Sci 22:396

Bell CJ, Dixon RA, Farmer AD et al (2001) The medicago genome initiative: a model legume database. Nucleic Acids Res 29:114–117

Bhatnagar-Mathur P, Vadez V, Jyostna Devi M et al (2009) Genetic engineering of chickpea (*Cicer arietinum* L.) with the P5CSF129A gene for osmoregulation with implications on drought tolerance. Mol Breed 23:591–606

Brunetti AE, Neto FC, Vera MC et al (2018) An integrative omics perspective for the analysis of chemical signals in ecological interactions. Chem Soc Rev 47:1574–1591

Cervantes E, Saadaoui E, Tocino Á, Gómez JJM (2019) Seed shape quantification in the model legumes: methods and applications. In: The model legume medicago truncatula. Wiley, Hoboken, pp 92–98

Chandrashekharaiah PS, Paul V, Kushwaha S et al (2021) Biotechnological approaches for enhancing stress tolerance in legumes. In: Sustainable agriculture reviews 51: legume agriculture and biotechnology, vol 2. Springer, Cham, pp 247–293

Chaudhary S, Dhanker R, Kumar R, Goyal S (2022) Importance of legumes and role of Sulphur oxidizing bacteria for their production: a review. Legum Res 45:275–284

Chen X, Laudeman TW, Rushton PJ et al (2007) CGKB: an annotation knowledge base for cowpea (*Vigna unguiculata* L.) methylation filtered genomic genespace sequences. BMC Bioinf 8:1–9

Chen X, Chen Z, Zhao H et al (2014) Genome-wide analysis of soybean HD-Zip gene family and expression profiling under salinity and drought treatments. PLoS One 9:e87156

Cheng Q, Gan Z, Wang Y et al (2020) The soybean gene J contributes to salt stress tolerance by up-regulating salt-responsive genes. Front Plant Sci 11:272

Cook DR (1999) Medicago truncatula–a model in the making! Curr Opin Plant Biol 2:301–304

Dash S, Campbell JD, Cannon EK et al (2016) Legume information system (LegumeInfo.org): a key component of a set of federated data resources for the legume family. Nucleic Acids Res 44: D1181–D1188

Deinlein U, Stephan AB, Horie T et al (2014) Plant salt-tolerance mechanisms. Trends Plant Sci 19: 371–379

Dixon RA (2001) Natural products and plant disease resistance. Nature 411:843–847

García-Caparrós P, Hasanuzzaman M, Lao MT (2019) Oxidative stress and antioxidant defense in plants under salinity. In: Reactive oxygen, nitrogen and sulfur species in plants: production, metabolism, signaling and defense mechanisms. Wiley, Hoboken, pp 291–309

Garg R, Shankar R, Thakkar B et al (2016) Transcriptome analyses reveal genotype-and developmental stage-specific molecular responses to drought and salinity stresses in chickpea. Sci Rep 6:1–15

Gondalia N, Vashi R, Barot V et al (2022) Genomic designing for abiotic stress tolerance in pea (*Pisum Sativum* L.). In: Genomic designing for abiotic stress resistant pulse crops. Springer, Cham, pp 45–113

Hao S, Wang Y, Yan Y et al (2021) A review on plant responses to salt stress and their mechanisms of salt resistance. Horticulturae 7:132

Hiz MC, Canher B, Niron H, Turet M (2014) Transcriptome analysis of salt tolerant common bean (*Phaseolus vulgaris* L.) under saline conditions. PLoS One 9:e92598

Hossain MS (2020) Proteomic studies: contribution to understanding plant salinity stress response. Glob J Bot Sci 8:1–10

Hu J, Zhuang Y, Li X et al (2022) Time-series transcriptome comparison reveals the gene regulation network under salt stress in soybean (*Glycine max*) roots. BMC Plant Biol 22:157

Jain N, Farhat S, Kumar R et al (2021) Alteration of proteome in germinating seedlings of piegonpea (*Cajanus cajan*) after salt stress. Physiol Mol Biol Plants 27:2833–2848

Jha UC, Bohra A, Jha R, Parida SK (2019) Salinity stress response and 'omics' approaches for improving salinity stress tolerance in major grain legumes. Plant Cell Rep 38:255–277

Ji J, Zhang C, Sun Z et al (2019) Genome editing in cowpea (*Vigna unguiculata*) using CRISPR-Cas9. Int J Mol Sci 20:2471

Jiao Y, Bai Z, Xu J et al (2018) Metabolomics and its physiological regulation process reveal the salt-tolerant mechanism in Glycine soja seedling roots. Plant Physiol Biochem 126:187–196

Jin T, Chang Q, Li W et al (2010) Stress-inducible expression of GmDREB1 conferred salt tolerance in transgenic alfalfa. Plant Cell Tissue Organ Cult 100:219–227

Jin J, Wang J, Li K et al (2021) Integrated physiological, transcriptomic, and metabolomic analyses revealed molecular mechanism for salt resistance in soybean roots. Int J Mol Sci 22:12848

Juranić M, Nagahatenna DS, Salinas-Gamboa R et al (2020) A detached leaf assay for testing transient gene expression and gene editing in cowpea (*Vigna unguiculata* [L.] Walp.). Plant Methods 16:1–17

Kaashyap M, Ford R, Mann A et al (2022) Comparative flower transcriptome network analysis reveals DEGs involved in chickpea reproductive success during salinity. Plan Theory 11:434

Kaur G, Sanwal SK, Sehrawat N et al (2022) Getting to the roots of *Cicer arietinum* L. (chickpea) to study the effect of salinity on morpho-physiological, biochemical and molecular traits. Saudi J Biol Sci 29:103464

Kharb P, Chaudhary R, Lakra N, Behl RK (2021) Transgenics for improving salt stress tolerance in legume crops chickpea and pigeon pea. Ekin J Crop Breed Genet 7:106–115

Kumar N, Soren KR, Bharadwaj C et al (2021) Genome-wide transcriptome analysis and physiological variation modulates gene regulatory networks acclimating salinity tolerance in chickpea. Environ Exp Bot 187:104478

Kumari A, Das P, Parida AK, Agarwal PK (2015) Proteomics, metabolomics, and ionomics perspectives of salinity tolerance in halophytes. Front Plant Sci 6:537

Kushwah A, Bhatia D, Rani U et al (2021) Molecular mapping of quantitative trait loci for ascochyta blight and botrytis grey mould resistance in an inter-specific cross in chickpea (*Cicer arietinum* L.) using genotyping by sequencing. Breed Sci 71:229–239

Li D, Zhang Y, Hu X et al (2011) Transcriptional profiling of *l* under salt stress identified a novel CBF transcription factor MtCBF4 that plays an important role in abiotic stress responses. BMC Plant Biol 11:1–19

Li M-W, Xin D, Gao Y et al (2017) Using genomic information to improve soybean adaptability to climate change. J Exp Bot 68:1823–1834

Liao H, Wong F-L, Phang T-H et al (2003) GmPAP3, a novel purple acid phosphatase-like gene in soybean induced by NaCl stress but not phosphorus deficiency. Gene 318:103–111

Lin X, Wang Q, Min X et al (2022) Comparative transcriptomic analysis of root and leaf transcript profiles reveals the coordinated mechanisms in response to salinity stress in common vetch. Int J Mol Sci 23:8477

Liu X, Yang X, Zhang B (2021) Transcriptome analysis and functional identification of GmMYB46 in soybean seedlings under salt stress. PeerJ 9:e12492

Lu Y, Lam H, Pi E et al (2013) Comparative metabolomics in Glycine max and Glycine soja under salt stress to reveal the phenotypes of their offspring. J Agric Food Chem 61:8711–8721

Ma X-J, Fu J-D, Tang Y-M et al (2020) GmNFYA13 improves salt and drought tolerance in transgenic soybean plants. Front Plant Sci 11:587244

Mann A, Kaur G, Kumar A, Sanwal SK, Singh J, SharmaPC. (2019) Physiological response of chickpea (*Cicer arietinum* L.) at early seedling stage under salt stress conditions. Legum Res 42(5):625–632

Mann A, Kumar N, Kumar A et al (2021) De novo transcriptomic profiling of differentially expressed genes in grass halophyte *Urochondra setulosa* under high salinity. Sci Rep 11:1–14

Mantri N (2007) Gene expression profiling of chickpea responses to drought, cold and high-salinity using cDNA microarray. PhD Thesis, RMIT University

Mathesius U (2022) Are legumes different? Origins and consequences of evolving nitrogen fixing symbioses. J Plant Physiol 276:153765

Mhadhbi H, Fotopoulos V, Mylona PV et al (2011) Antioxidant gene–enzyme responses in *Medicago truncatula* genotypes with different degree of sensitivity to salinity. Physiol Plant 141:201–214

Mhadhbi H, Fotopoulos V, Mylona PV et al (2013) Alternative oxidase 1 (Aox1) gene expression in roots of *Medicago truncatula* is a genotype-specific component of salt stress tolerance. J Plant Physiol 170:111–114

Missanga JS, Venkataramana PB, Ndakidemi PA (2021) Recent developments in Lablab purpureus genomics: a focus on drought stress tolerance and use of genomic resources to develop stress-resilient varieties. Legume Sci 3:e99

Mokhtari F, Rafiei F, Shabani L, Shiran B (2017) Differential expression pattern of transcription factors across annual Medicago genotypes in response to salinity stress. Biol Plant 61:227–234

Molina C, Zaman-Allah M, Khan F et al (2011) The salt-responsive transcriptome of chickpea roots and nodules via deepSuperSAGE. BMC Plant Biol 11:1–26

Nadeem M, Li J, Yahya M et al (2019) Grain legumes and fear of salt stress: focus on mechanisms and management strategies. Int J Mol Sci 20:799

Niron H, Barlas N, Salih B, Türet M (2020) Comparative transcriptome, metabolome, and ionome analysis of two contrasting common bean genotypes in saline conditions. Front Plant Sci 11: 599501

Pandey A, Chakraborty S, Datta A, Chakraborty N (2008) Proteomics approach to identify dehydration responsive nuclear proteins from chickpea (*Cicer arietinum* L.). Mol Cell Proteomics 7:88–107

Pandey AK, Rubiales D, Wang Y et al (2021) Omics resources and omics-enabled approaches for achieving high productivity and improved quality in pea (*Pisum sativum* L.). Theor Appl Genet 134:755–776

Pi B, Liu X, Huang Q et al (2023) Comparative transcriptomic analysis of Glycine soja and G. max and functional identification of GsCNGC20-d interacted with GsCDPK29 under salt stress. Environ Exp Bot 206:105185

Pratap A, Prajapati U, Singh CM et al (2018) Potential, constraints and applications of in vitro methods in improving grain legumes. Plant Breed 137:235–249

Reddy DS, Bhatnagar-Mathur P, Vadez V, Sharma KK (2012) Grain legumes (soybean, chickpea, and peanut): omics approaches to enhance abiotic stress tolerance. In: Improving crop resistance to abiotic stress. Wiley, Hoboken, pp 995–1032

Roorkiwal M, Rathore A, Das RR et al (2016) Genome-enabled prediction models for yield related traits in chickpea. Front Plant Sci 7:1666

Saade S, Maurer A, Shahid M et al (2016) Yield-related salinity tolerance traits identified in a nested association mapping (NAM) population of wild barley. Sci Rep 6:1–9

Sanchez DH, Szymanski J, Erban A et al (2010) Mining for robust transcriptional and metabolic responses to long-term salt stress: a case study on the model legume Lotus japonicus. Plant Cell Environ 33:468–480

Shabala S (2013) Learning from halophytes: physiological basis and strategies to improve abiotic stress tolerance in crops. Ann Bot 112:1209–1221

Sharma V, Kumar N, Verma A, Gupta VK (2013) Exogenous application of brassinosteroids ameliorates salt-induced stress in mung bean seedlings. Life Sci 2:7–13

Sharmila P, Phanindra MLV, Anwar F et al (2009) Targeting prokaryotic choline oxidase into chloroplasts enhance the potential of photosynthetic machinery of plants to withstand oxidative damage. Plant Physiol Biochem 47:391–396

Sobhanian H, Razavizadeh R, Nanjo Y et al (2010) Proteome analysis of soybean leaves, hypocotyls and roots under salt stress. Proteome Sci 8:1–15

Soren KR, Madugula P, Kumar N et al (2020) Genetic dissection and identification of candidate genes for salinity tolerance using Axiom® CicerSNP array in chickpea. Int J Mol Sci 21:5058

Stai JS, Yadav A, Sinou C et al (2019) Cercis: a non-polyploid genomic relic within the generally polyploid legume family. Front Plant Sci 10:345

Sumner LW, Mendes P, Dixon RA (2003) Plant metabolomics: large-scale phytochemistry in the functional genomics era. Phytochemistry 62:817–836

Sweetman C, Khassanova G, Miller TK et al (2020) Salt-induced expression of intracellular vesicle trafficking genes, CaRab-GTP, and their association with Na^+ accumulation in leaves of chickpea (*Cicer arietinum* L.). BMC Plant Biol 20:183

Tesfaye M, Temple SJ, Allan DL et al (2001) Overexpression of malate dehydrogenase in transgenic alfalfa enhances organic acid synthesis and confers tolerance to aluminum. Plant Physiol 127:1836–1844

Thudi M, Palakurthi R, Schnable JC et al (2021) Genomic resources in plant breeding for sustainable agriculture. J Plant Physiol 257:153351

Varshney RK, Hiremath PJ, Lekha P et al (2009) A comprehensive resource of drought-and salinity-responsive ESTs for gene discovery and marker development in chickpea (*Cicer arietinum* L.). BMC Genomics 10:1–18

Verma M, Kumar V, Patel RK et al (2015) CTDB: an integrated chickpea transcriptome database for functional and applied genomics. PLoS One 10:e0136880

Wang D, Liu Y-X, Yu Q et al (2019) Functional analysis of the soybean GmCDPK3 gene responding to drought and salt stresses. Int J Mol Sci 20:5909

Xiong J, Sun Y, Yang Q et al (2017) Proteomic analysis of early salt stress responsive proteins in alfalfa roots and shoots. Proteome Sci 15:1–19

Xue R, Zhang B (2007) Increased endogenous methyl jasmonate altered leaf and root development in transgenic soybean plants. J Genet Genomics 34:339–346

Yin Y, Qi F, Gao L et al (2018) iTRAQ-based quantitative proteomic analysis of dark-germinated soybeans in response to salt stress. RSC Adv 8:17905–17913

Yu Y, Wang N, Hu R, Xiang F (2016) Genome-wide identification of soybean WRKY transcription factors in response to salt stress. Springerplus 5:1–15

Yue JY, Wang LH, Dou XT et al (2020) Comparative metabolomic profiling in the roots of salt-tolerant and salt-intolerant maize cultivars treated with NaCl stress. Biol Plant 64:569–577

Zhang J, Yang D, Li M, Shi L (2016) Metabolic profiles reveal changes in wild and cultivated soybean seedling leaves under salt stress. PLoS One 11:e0159622

Zhang Q, Li M, Xia CY et al (2021) Transcriptome-based analysis of salt-related genes during the sprout stage of common bean (*Phaseolus vulgaris*) under salt stress conditions. Biotechnol Biotechnol Equip 35:1086–1098

Zhao S-P, Lu D, Yu T-F et al (2017) Genome-wide analysis of the YABBY family in soybean and functional identification of GmYABBY10 involvement in high salt and drought stresses. Plant Physiol Biochem 119:132–146

Zhu C, Bortesi L, Baysal C et al (2017) Characteristics of genome editing mutations in cereal crops. Trends Plant Sci 22:38–52

Zulfiqar F, Akram NA, Ashraf M (2020) Osmoprotection in plants under abiotic stresses: new insights into a classical phenomenon. Planta 251:1–17

Speed Breeding: A Budding Technique to Improve Crop Plants for Drought and Salinity Tolerance

15

Navreet Kaur Rai, Ravika, Rajesh Yadav, Minakshi Jattan, Karuna, Prashant Singh Rai, Nisha Kumari, Babita Rani, Amit Sharma, Sachin, and Sunaina Yadav

Abstract

The present dynamic climate conditions would escalate various abiotic stresses in crop plants such as drought, salinity, heat, mineral toxicity, etc. which ultimately affect the agricultural produce. Drought and salinity stress are the major concern affecting growth, metabolism, and development of plants, worldwide. Normally, the development of noble drought and salinity tolerant varieties through conventional approaches takes around 10–12 years due to long reproductive cycles of various crops and repetitive selection cycles. Due to the slow rate of generation advancement through conventional techniques, agricultural scientists, and researchers have put their efforts in adopting newer techniques for rapid generation advancement of three to nine generations in a single year, such as, speed breeding. Speed breeding is a collective technique which aims at reducing the reproductive period of crop plants by controlling environmental conditions. Temperature regulation, manipulation of the photoperiod regime, increasing CO_2 levels, regulating soil moisture, inducing plant growth and hormones, etc. enhance the growth in plants as well as the pace of shift between the vegetative and the reproductive phase which further helps crop plants to tolerate stress conditions. Moreover, rapid and multiple trait selection is possible through

N. K. Rai · Ravika (✉) · R. Yadav · M. Jattan · Karuna · A. Sharma · Sachin
Department of Genetics and Plant Breeding, CCS Haryana Agricultural University, Hisar, India
e-mail: ravika.sheoran@hau.ac.in

P. S Rai
Department of Biochemistry, Central University of Rajasthan, Ajmer, India

N. Kumari · B. Rani
Department of Biochemistry, CCS Haryana Agricultural University, Hisar, India

S. Yadav
Department of Genetics and Plant Breeding, IARI, New Delhi, India

A. Kumar et al. (eds.), *Salinity and Drought Tolerance in Plants*,
https://doi.org/10.1007/978-981-99-4669-3_15

speed breeding combined with the approaches of marker-assisted selection (MAS) and high-throughput phenotyping. This chapter deals with the novel concept of speed breeding, its methods and applications, its opportunities and challenges in developing elite varieties against drought and salinity stresses.

Keywords

Abiotic stresses · Drought stress · Salinity stress · Conventional · Speed breeding · Photoperiod · MAS · High-throughput phenotyping

15.1 Introduction

With the present count of around 7.8 billion, human population is expanding at faster rate and is expected to cross 9.9 billion by the mid of twenty-first century (IISD 2022). Feeding the world's growing population is a major challenge for mankind. Present crop improvement programs are seemingly insufficient to meet future demands due to dynamic climate fluctuations, or "Climate catastrophe," which is slowly but constantly making the Earth's biosphere warmer and drier, resulting devastating crop failures. Agricultural productivity depends on various environmental events occurring in the surroundings of crop plants and any unwanted change in these events may elevate the harmful effects of abiotic stresses, namely drought and salinity, on the plant's inherent potential. Abiotic or climatic stresses may influence plant growth and development by altering the plant's cellular mechanisms at physiological, morphological, biochemical, and metabolic levels. Drought and salinity have the major impact on agricultural produce. Agricultural economic losses related with drought alone, cost around $29 billion worldwide over the last decade (FAO 2019). The occurrence of drought and salinity stresses has direct effect on freshwater availability and water demand in agriculture.

Therefore, the development of stress tolerant cultivars of major crops is the most crucial step towards sustainable agricultural practices and stabilizing food safety issues. To cope with the unwanted and harsh drought and salinity effects, understanding the mechanism of plant responses to these stresses is crucial along with increasing the current rate of genetic gain. Lin et al. (2016) mentioned the urgency of enhancing the current genetic gain rate to buffer the global food security. Accelerating the plant breeding pipeline with decreased breeding cycle time, results into rapid production of improved crop varieties, in turn, increases the genetic gain per unit time.

Plant shows a variety of responses to external stress conditions. The development of drought and salt tolerance in plants requires manipulation of plant responses. During drought, plants abruptly close their stomata to reduce transpirational losses but this parallelly leads to the declined net photosynthetic efficiency and water use efficiency of the plant. Drought tolerance may be induced by altering the photosynthetic pigments and by increasing the root density. The mechanism of drought tolerance involves scavenging of ROS (reactive oxygen species) by enzymatic as well as non-enzymatic systems, maintaining stability of cell membrane and producing stress-responsive proteins. Osmotic adjustment of metabolites such as polyphones, amino acids, carbohydrates, etc. contributes for better drought adaptation. The plants followed a biphasic model for response to salinity, wherein responses similar to drought are induced in the early phase and ion toxicity developed in the long term. At the early stage, the decrease in water potential result in stomata closure, affecting plant growth and development, whereas, in later stage, ion accumulation, majorly Na^+, disturb the photosynthetic machinery leading to oxidative stress. On the onset of the first phase, there is a significant fall in plant growth along with the closure of stomata due to decline in water potential. In the second phase, ion accumulation, especially Na^+, results in increased oxidative stress as well as has effects on photosynthetic components including enzymes and pigments. Microbes can also save plants from the impact of stresses caused due to drought and salinity by nutrient solubilization (N, P, K, and Fe), producing phytohormones (IAA, cytokinin, GA, and ABA), and various other mechanisms.

Crop redesigning for drought and salinity stresses is crucially required. The redesigning of crops through conventional methods take a long route to reach its satisfactory goals. A series of events take place before the release and notification of an improved cultivar. A cultivar must go through several generation cycles and evaluation trials compiled with uncompromised screening for specific stress tolerant traits which ultimately increases the length of the crop improvement program. Breeding for new and high yielding varieties with market-oriented traits, previously lacking integrated pre-breeding program, may take more than a decade. The early stages of any breeding program, starting from initial parental crosses, demand an ample quantity of resources, time and space for targeted selection along with further genetic advancement. However, an alternate technique, named as "speed breeding," has the ability to produce multiple generations per year, eventually solving the concern related to accelerated genetic gain in major crops. Speed breeding helps breeders in improving crop varieties in short time and speed up the process of their development. Through the speed breeding channel, the problems associated with dynamic environmental conditions and hyper-competitive market demands for stress tolerant cultivars can be solved rapidly. Keeping the view of the urgent need for an alternate required for rapid development of stress tolerant varieties, this chapter deals with the special features of speed breeding, its techniques, opportunities, collaboration with other modern approaches, achievements in producing stress tolerance in plants and its challenges.

15.2 Speed Breeding: Time Saving Technique

Major goals of plant breeding comprise higher performance in terms of yield, nutritional quality, tolerance to biotic and abiotic stresses and adaptation. In conventional plant breeding, due to several phases of crossing, selection and evaluation required for the development of new crop varieties, it takes nearly one and half or two decades to achieve the goals of a new variety. Thus, the major bottleneck for the production of new varieties for stress tolerance in plants through conventional methods is the long duration of the seed-to-seed cycle. Basically, major events in any crop improvement program can be distinguished as following: the foremost important step is to select desirable parents that have useful traits to unite; performing hybridization among the selected individuals and developing their progenies; selecting the best progenies on the basis of the target traits and their further genetic advancement; multiple screening and evaluation of the best performing and stable progenies in different targeted environments; and new variety registration, seed multiplication and their farmers (Shimelis and Laing 2012). The lack of proper integrated pre-breeding approach, moreover, can slow the process of production and release of an enhanced variety through conventional breeding programs (Ahmar et al. 2020).

Dynamic climatic changes and the rapid rate of population growth are the major issues that need an urgent attention of crop breeders. Increasing demand for quality food in larger quantities made researchers and breeders to come up with better solutions. Earlier in the 1940s, many techniques had been used to anyhow manipulate the speed of the life cycle of plants such as single seed descent (Brim 1966) and shuttle breeding (Borlaug 1968). Current techniques can be used to reduce the period of each crop breeding cycle, namely, double haploid breeding and speed breeding (Dwivedi et al. 2015; Hickey et al. 2019). With the advent of modern techniques, manipulation of environmental conditions has become easier and cost effective. Recently, researchers have developed methods to manipulate growth conditions under controlled environment (CE) which further trim the plant's life period. All the techniques used to increase the cycle turnover are collectively known as speed breeding (SB) (Ghosh et al. 2018). The term "speed breeding" was initially coined by the researchers at the University of Queensland in 2003 to collectively describe all the techniques that are designed to speed up wheat breeding. Through traditional breeding methods, it takes nearly 8–10 years to develop a new variety, but with the advent of speed breeding, the breeding cycle can be condensed to up to twice or thrice times than the traditional methods (Fig. 15.1). Speed breeding can produce three to nine generations per annum whereas only one to two generations can be developed per annum with traditional methods. Major techniques under speed breeding involves accelerated form of single seed descent (aSSD) for instant production of homozygous populations; fast generation cycling (FGC) for more breeding cycles per annum involving unfavorable environment along with in vitro culturing of immature embryos; rapid generation cycling (RGC) that includes the application of DNA marker technology for more breeding cycles per annum, and rapid generation turnover (RGT) which involves seed harvest from immature plants and photoperiod response for increasing number of generations per annum.

Speed breeding techniques are currently being performed on a bigger scale to reduce time of breeding cycles and effectively grow multiple generations of a crop in a year. It is a collective method involving manipulation of various environmental conditions. The major goal of performing speed breeding in any crop improvement program is to speed up the vegetative, reproductive, and maturity phases so as to get the next crop as early as possible. This quick crop advancement, in turn, minimizes

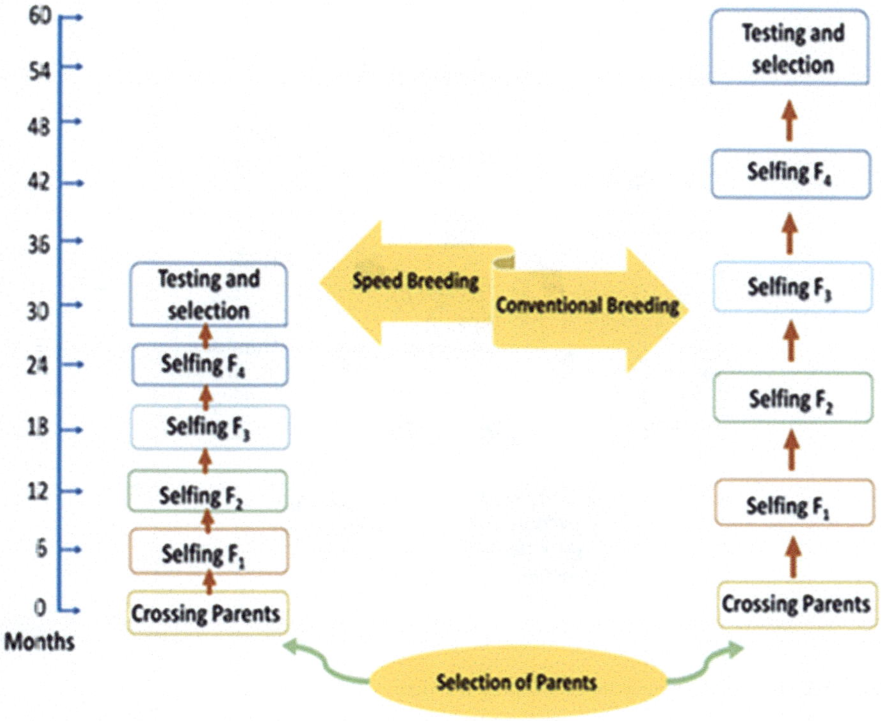

Fig. 15.1 The comparison of the time taken to produce a new variety under speed breeding and conventional breeding

the resources required, hence, making speed breeding a cost-effective approach for crop intensification. Speed breeding-specific growth chambers are used to speeds up crop generations and hence fasten up the research related to the generation of crosses, mutation, mapping populations, transformation, and adult plant phenotyping. Nowadays, researchers are giving emphasis in improving breeding efficiency by using an integrated approach, wherein, speed breeding is used with present plant breeding and biotechnological approaches. Speed breeding has been successfully applied to major model and pasture crops belonging to Fabaceae, Poaceae, and Brassicaceae families. Various protocols were developed for different species such as peanuts, rice, barley and wheat, soybeans, chickpea, and lentil.

15.3 Applications of Speed Breeding Techniques for Stress Tolerance

In plant breeding, developing a new cultivar is not an easy task. Altering the basic genotype of elite cultivar for the development of stress tolerant new varieties requires such gene combinations that do not segregate for several generations and

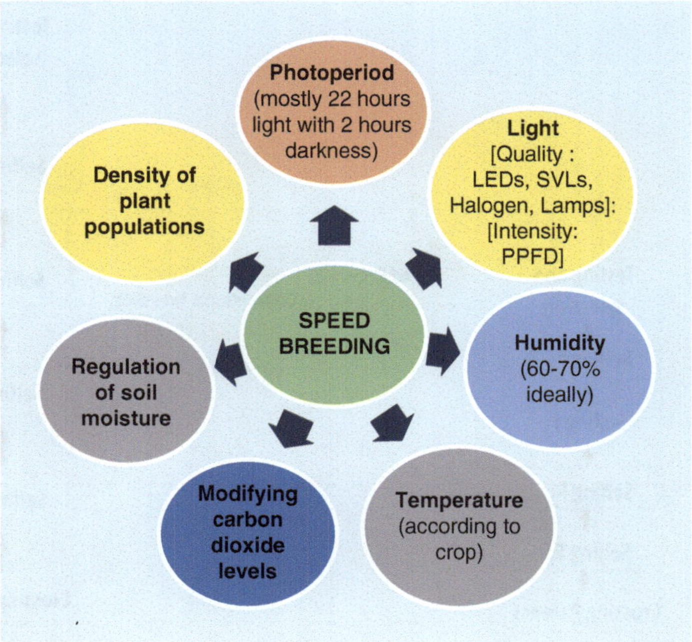

Fig. 15.2 Key altering components of speed breeding

remain pure in their actual form in the population. Fixation of such newly generated gene combinations is a necessity. Producing homozygous lines helps to fix the desired genes in the homozygous state which remains stable and does not segregate in further generations. Commonly, the single crop cycle of cereals and legumes ranges from 3 to 6 months/annum. The crops, like cassava, have a longer breeding cycle period of 15–18 months. Furthermore, in different agro-ecologies, diverse weather conditions prevail, for instance, acute temperature ranges, unpredictable and uneven rainfall pattern, day length, etc. that limits single generation cycle yearly.

Using speed breeding, the process of producing of homozygous lines accelerates through alteration of key components of the surrounding environment (Fig. 15.2) and hence, provides better opportunities for breeders to breed for stress tolerant varieties in lesser time and release superior early maturing varieties timely. Several crops are being used as model crops for the production of stable and more tolerant varieties through speed breeding. Moreover, the amenability of speed breeding with selection method has positive impact on its application in crop improvement. Major applicable areas of speed breeding are discussed below:

15.3.1 Developing Homozygous Lines for Accelerated Breeding

After the initial crossing between ideal parents having complementary traits, the speed breeding techniques have revolutionized the rapid production of homozygous lines. The techniques used majorly rely on the manipulation of photoperiodism, temperature, intensity of light, soil water content, soil nutrition status, and compact planting. All such techniques have been developed with the prime intentions of inducing early flowering as well as early seed set, manipulating the period of each generation cycle (Table 15.1). For accelerated breeding, the opportunities provided by speed breeding with the ease of producing three to nine generations per year is crucial. This situation is also ideal for population evaluation in targeted environments including major selection methods, namely, SSD, SPD, and SPS (El-Hashash and El-Absy 2019).

Speed breeding methods may directly affect the chromosomal structure of plants with the alteration of the key components responsible for the growth and development of the plants. Under controlled environment conditions, it is easier to adjust day and night temperature, photoperiod duration, light spectrum and intensity, and various other factors in view to reduce vegetative growth period and hasten the floral and embryo development stages for quick seed maturity. The speed of breeding cycle can be increased by enhancing the photosynthesis process by

Table 15.1 Rapid generation advancement using different techniques to alter the flowering period

Crop species	Techniques used	Modified days of flowering	Number of generations per annum	References
Arabidopsis Thaliana	Photoperiod; plant hormones; germination of immature seed	20–26	Ten	Ochatt and Sangwan (2008)
Hordeum vulgare	Temperature; photoperiod; soil fertility; germination of immature seed; embryo rescue	24–36	Nine	Zheng et al. (2013)
Canola	Photoperiod; soil moisture; light intensity; temperature; germination of immature seed	73	Four	Watson et al. (2018)
Cicer arietinum	Germination of immature seed; photoperiod	33	Seven	Samineni et al. (2020)
Arachis hypogaea	Temperature; photoperiod	25–27	Three	O'connor et al. (2013)
Cajanus cajan	Temperature; germination of immature seed; photoperiod	50–56	Four	Saxena et al. (2019)
Oryza sativa	Temperature; photoperiod; high-density planting	75–85	Four	Collard et al. (2017)
Glycine max	Temperature; photoperiod; germination of immature seed	23	Five	Jähne et al. (2020)
Triticum aestivum	Temperature; photoperiod; germination of immature seed; embryo rescue	28–41	Seven to eight	Zheng et al. (2013)

means of ideal light parameters, temperature ranges, and daytime length adjustment, ideally, 22 h light, 22 °C day and 17 °C night temperature, and high light intensity, followed by early harvesting of seeds to shift rapidly from seed to seed, minimizing breeding spans for some long day or day-neutral crops (Zhang et al. 2014). Rapid cycling can be achieved by improving the plant related factors (planting density, photoperiod, and temperature) by performing genetic manipulation aiming the pathways involved in flowering initiation, grafting immature plants to mature rootstocks, utilizing plant growth hormones, and harvesting immature seed (Richard et al. 2015). Speed breeding bank on intentional altering of number of parameters that are responsible for the normal growth and development of plants which are specified below.

15.3.1.1 Alteration of Photoperiod System

Photoperiod is defined as the duration of regular subjection of crops each day to planned light and dark periods for promotion of instant growth, development, early reproductive, and maturity stages (Vince-Prue 1994). For the early induction of flowering and seed set, there are variable requirements of photoperiod for every crop (Kouressy et al. 2008; Saito et al. 2009). Therefore, the proper understanding of the optimum parameters of light such as the quality of light, the intensity of light, and required duration of photoperiod that initiate flowering in different crops and genotypes is crucial for plant breeders. Manipulation of light parameters has given more importance due to the effect of light duration and the light quality on the time of flowering in plants. There is a direct effect of instantaneous and cumulative amount of light quality on plant growth, stomatal conductance, net photosynthetic rate, amount of intercellular CO_2, and rate of transpiration (Yang et al. 2017). The artificial environment under speed breeding provides continuous light to create extended daylight conditions that affects the photo-insensitive crop life cycles. The light parameters can control the phenomenon of flowering in plants and can help to produce early maturing varieties which can escape from drought conditions.

To facilitate speed breeding in various crops including barley, wheat, gram, pea, and canola, suitable light sources have been used with photosynthetic active radiation (PAR) ranging from 400 to 700 nm paired with an intensity ranging between 360 and 650 μmol per m^2 per second (Ghosh et al. 2018; Watson et al. 2018). Speed breeding protocols based on light parameters have the benefit of longer active photosynthesis during the entire light period, for the whole year (Bhatta et al. 2021). Artificial electric lamps are being used since a long time for enhancing plant growth and development. The extension of photoperiod has been widely applied for the manipulation of days required for flowering in many long day species. The discovery of advanced systems of LED lighting enables the adjustment of composition of light wavelength which effectively accelerated the lifecycle turnover of crop plants with the help of initiating light responses, like shade avoidance, and supporting quick initiation of flowering.

15.3.1.2 Manipulation of the Temperature Conditions

Temperature regulation of soil and air has direct effect on the germination of seeds and growth responses of plants, resulting in speedy growth and early flowering. Temperature ranging from lower to upper extremes initiates various effects on the development of plant changing the transition period between the pre-reproductive and the reproductive phases (McClung et al. 2016). For most crops, optimum temperatures for germination range from 12 and 30 °C, whereas for growth, reproductive, and maturity stages the optimum temperatures lie between 25 and 30 °C.

15.3.1.3 Maintenance of Soil Moisture and Soil Nutrition

Soil moisture and its nutritional status is very important for the growth of a healthy plant population, especially in areas affected by drought and salinity. Conserved soil moisture in the capillaries of soil can help the plants to tolerate drought. Sometimes, the accumulation of toxic nutrients leads to soil salinity conditions. Soil water stresses can induce remarkable variations in the development of crop plants by hampering days required for flowering, plant height, days required for seed setting and maturity (Hussain et al. 2018). In speed breeding, stress conditions caused by drought or flooding can be utilized to initiate early flowering and maturation. Creating drought stress conditions in crops such as, wheat, barley, and pearl millet, is the most common technique (Shavrukov et al. 2017).

15.3.1.4 Elevating Carbon Dioxide Levels

Varied range of responses is found in different crop species and their genotypes to elevated levels of CO_2. High carbon dioxide (CO_2) levels may enhance the speed required for the transition of plant from its vegetative state to reproductive state, ultimately leading to rapid plant growth (Jagadish et al. 2016).

15.3.1.5 Altering Planting Density

Cultivating crops in more dense condition as compared to the density needed to produce highest yield is termed as high-density planting. Planting crops in high density gives tall plants because of light competition, resulting in a rapid shift between the vegetative and the reproductive growth phases (Warnasooriya and Brutnell 2014). This approach is reliable in increasing the number of generations per annum by inducing early flowering and maturity.

15.3.1.6 Application of Plant Nutrients, Growth Regulators, and Organ Tissue Culture

Plant nutrients and growth hormones, along with in vitro immature seed germination, always plays an important role in accelerating growth and inducing flowering and early seed set (Bermejo et al. 2016). Under the conditions of controlled environments such as growth chambers and greenhouses, diverse reactions to applied plant growth regulators (PGRs) are found with closely monitored and controlled photoperiod and temperatures.

15.3.2 Model Crops

Cereal Crops To breed cereal varieties that are superior in every aspect, breeders have explored many novel approaches to produce homozygous lines resulting from hybridization in less possible time. For example, following the harvesting of immature seeds after 15–20 days of anthesis and treating of the seeds with H_2O_2 at a low temperature four to six generations of wheat were obtained (Mukade 1974). Speed breeding has been applied successfully for screening of various drought tolerance characters in barley (Ghosh et al. 2018; Zhang et al. 2015). Moreover, to reduce the breeding cycle time, direct germination of immature seeds and the embryo rescue technique has been utilized in sorghum (Rizal et al. 2014). In oats, increased photoperiod and application of a foliar mineral supplement has proved to significantly shorten the anthesis period for getting higher generation turnover (Tanaka et al. 2016).

Oilseed Crops In oilseeds, a standardized protocol was developed by Nagatoshi and Fujita (2019) for rapid generation advancement in high-quality, Japanese, soybean cultivar Enrej, that modified the crop period to 70 days from initial 102 to 132 days. Watson et al. (2018) worked on canola and finalized a speed breeding protocol which facilitates phenotyping of the pod-shattering trait along with the enhancement of the generation turnover. Dagustu et al. (2012) found a protocol while studying on sunflower in which short breeding period was achieved in breeding program using the embryo rescue technique.

Legume Crops In crops like, clover (*Trifolium subterraneum* L.), breeding cycle can be minimized by reducing the time to flowering with SSD technique assisted with in vitro conditions by the means of regulated temperature, an extended photoperiod, trimmed seed-filling phase, and embryo rescue. Due to the positive response of temperate pulses to photoperiod extension, protocols for rapid generation advancement have been produced in many legumes (Croser et al. 2016). For instance, in peanut, under greenhouse facilities with continuous light associated with optimal temperature and humidity, increased rate of plant growth was recorded (Jähne et al. 2020). In recent study of *Cicer arietinum*, the cycle time from seed to seed is reduced by inducing flowering earlier than normal and germinating the immature seeds (Samineni et al. 2020). The generation advancement method was found in pigeonpea where complete germination of immature seeds that were taken from 35-days-old plants supported novel opportunities for growing 3–4 generations in an annum (Saxena et al. 2017).

Fruit Crops Before flowering, long juvenile period is present in the number of fruit crops, in general, taking more than 20 years (Van Nocker and Gardiner 2014). To the rescue, various speed breeding methods have provided the opportunities of vigorous vegetative growth and early flowering in fruit crops, such as in apple the vegetative growth period is modified to nearly 10 months than the normal 5 years and in case of chestnut the period of vegetative phase has been modified to 2 years instead of initial

7 years (Baier et al. 2012). Using speed breeding techniques along with transgenic, marker-assisted breeding and early flowering plants, many new varieties with ideal traits were produced in apple (Flachowsky et al. 2011). Utilization of speed breeding is not limited to seed producing fruit crops as it has been used in several clonally propagated crops, like banana, mainly propagating through tubers and roots.

Vegetable Crops Shortening the generation intervals through extended photoperiod in vegetables that have effective response to increased daylight, like tomato, pepper, and amaranth, is becoming famous nowadays (Stetter et al. 2016). In tomato, immature seeds germination obtained from different plants with different maturity levels created new possibilities for getting five generations in a mean time where earlier only three were possible (Bhattaraj et al. 2009). In tomato and pepper, germination of immature embryos in in vitro conditions made it possible to obtain additional generations than the traditional breeding methods (Manzur et al. 2014). Moreover, the manipulation of photoperiod was observed valuable in grain amaranth, where synchronization in flowering among different germplasm lines was combined with DNA marker technology to develop and identify true hybrids, ultimately, speeding up the plant breeding program (Stetter et al. 2016).

15.4 Amenability of Speed Breeding and Selection Methods

Basically, speed breeding is applied for rapid generation production from one to next without any type of prior phenotypic selection. However, current techniques (e.g., high-throughput genotyping and phenotyping, marker-assisted selection, and genomic selection) provide reliable integration opportunities for successful targeted selection of desirable traits. The combination of speed breeding with proper selection methods can be used for the conservation of a productive breeding population, creating genetic diversity, and better yield production under the stressed environments that inhibits plant growth (Johnston et al. 2019). Traditional selection tools, namely, mass, bulk, pedigree, pure line, and recurrent methods, have requirements of genetically stable and superior plant for choosing desirably yielding genotypes. But the conventional tools are not suitable for speed breeding because of their time-consuming inbreeding and selection cycles. Therefore, some other selection methods that are appropriately suited with speed breeding are performed like, single seed descent (SSD), single pod descent (SPD), and single plant selection (SPS) methods. The brief discussion of these methods is given as follows.

15.4.1 Single Seed Descent Method

Under SSD method, the development of homozygous populations is geared with inbreeding in segregating population continuously by selecting one seed at a time from individual plant in F_2 and advancing these plants to the succeeding generation. The time invested in SSD for developing inbred lines is equal to that of the doubled haploid method (Yan et al. 2017). Key benefits offered by SSD selection method is the requirement of less area of land and less number of laborer for the maintenance of early generations. Single seed descent method gives the opportunities for the advancement of the selected progenies under small scale nurseries, growth chambers or greenhouses, with high-density plantings (Arbelaez et al. 2019; Funada et al. 2013). However, it has the drawback that SSD produces inferior progenies in advanced generations relatively more often than pure line selection, pedigree selection, and recurrent selection methods.

15.4.2 Single Pod Descent Method

Another selection method is the single pod descent (SPD) technique that involves the selection in each selected plant from F_2–F_4 generations involving one pod per plant unlike a single seed selection of single seed descent method. The single pod descent has a better role in maintaining each F_2 plants in advanced generations than single seed descent selection due to the presence of more seeds per pod in most of the crops. Another favorable aspect of SPD is that due to early selection of pods, advancement of a relatively smaller population is feasible. However, preliminary trials for the traits being selected are essential in speed breeding for determining the efficiency of SPD selection method.

15.4.3 Single Plant Selection Method

This method is based on the selection of all the harvested seeds of each selected plants and advancing them to further generations, where the advancement of future generation is done as plant-to-row. The major applications of single plant selection method are in barley where a modified backcross approach is used to produce introgression lines (ILs) in less than 2 years (Hickey et al. 2017). The single plant selection method has a major advantage over SPD which provides earlier selection of desirable individuals in a smaller population. Various studies have been done to produce superiorly resistant and better performing varieties with the help of the single plant selection method. The procedure of maintaining continuous light and temperature in advanced backcross generations helps in producing better responses towards stress tolerance.

15.5 Collaboration of Speed Breeding with Other Modern Techniques

Speed breeding does not require special labs and can be performed in basic settings, unlike tissue culture techniques where sophisticated lab facilities are among the basic requirements (Shivakumar et al. 2018). Speed breeding can be applied to a wide range of germplasm without requiring special in vitro culturing facilities. Various

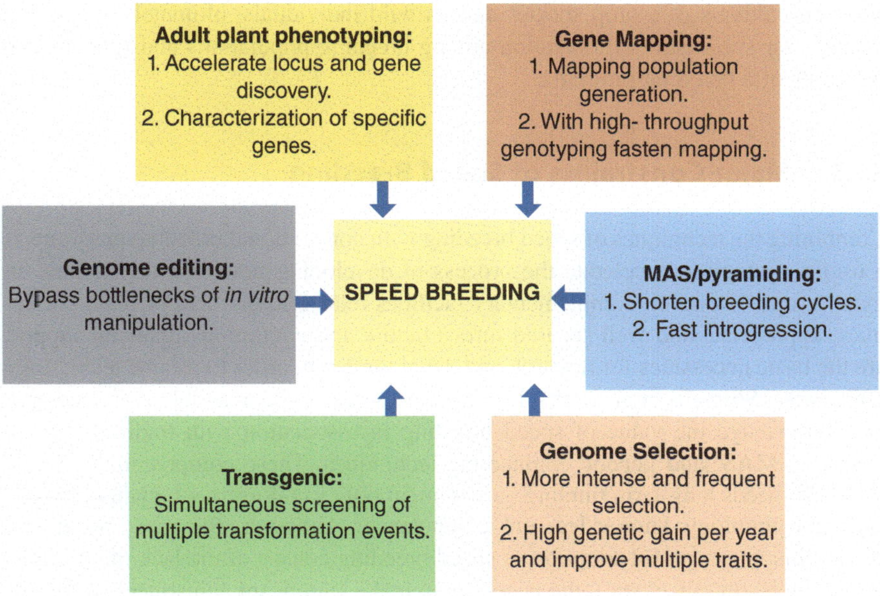

Fig. 15.3 Speed breeding collaboration with other modern techniques and its advantages

advantages offered by speed breeding over conventional methods for the development of stress tolerant varieties can be listed, involving the potential to expedite backcrossing, gene pyramiding, and transgenic pipelines. The art of this novel approach of speed breeding can be performed by researchers having lesser area by setting up small speed breeding units (Shimelis and Laing 2012). Integration of speed breeding with various other modern techniques speeds up the crop improvement program at incredible rates.

The flexible nature of speed breeding protocols make it easier to align and integrate it with numerous research purposes comprising multiple trait selection, phenotyping, genomic selection, population development, and genomic editing. To keep pace between the agricultural production and the changing climate as well as the rapidly expanding population, a variety of techniques can be used to achieve the integration of speed breeding with other techniques, namely, genotyping, high-throughput phenotyping, marker-assisted selection, genomic selection, gene editing, and re-domestication (Fig. 15.3).

Various phenotyping techniques facilitating speed breeding have been developed, permissing characterization and selection of important traits (Richard et al. 2015). Screening of important traits in early generations during population development is beneficial for breeders, saving time and funds by minimizing the expenditure of labor and field testing.

Synchronization of flowering among crop varieties and their distant relatives or wild species is a drawback in hybridization programs. Speed breeding methods can help to eliminate this drawback by aiding in the synchronization of the flowering between cultivars of a crop species and its wild individuals, ultimately enhancing chances for successful hybridization among breeding populations and speeding up the goals of a breeding program.

15.6 Major Constraints of Speed Breeding

Combining the techniques of speed breeding with conventional breeding programs is a useful strategy to accelerate the process of developing new tolerant varieties in major crops. However, complementary facilities related to the field of plant phonemics, appropriate and well secured infrastructure and continuous financial support are the basic necessities for research and development in order to use the technology effectively (Shimelis et al. 2019). For these resources to be available it is necessary to acknowledge the value of speed breeding in association with traditional plant breeding, MAS, and genetic engineering techniques. These comprehensive set of tools also needs long-term funding, government policy endorsement, plant breeding, and biotechnology knowledge and competence. Although, the most prominent constraints in successful adoption of speed breeding consist of the lack of appropriate facilities, untrained staff, the necessity to make significant adjustments to breeding programs' operations and design, and the lack of long-term finance. Here is a quick discussion of these difficulties.

15.6.1 Absence of Trained Manpower

A key limitation restraining the use of speed breeding in the government sector of developing countries is the absence of skilled and engaged plant breeders and the plant breeding technicians. The frequent turnover of plant breeding staff to non-governmental seed corporations and training institutes which provides more profit than public sector has a negative impact on the public sector breeding programs. The fact that only a small number of universities in developing nations provide postgraduate degrees in plant breeding is another reason why there are not many scientists with a focus on the respective field. In developing countries, the under developed statutory and administrative framework involved in managing the rights of plant breeders along with seed regulation affects the promotion of plant breeding to the extent that the benefit of value chain initiating from farmers and ending to consumers is greatly hampered and lacked in various resources (Tripp et al. 2007). As a consequence, there is an urgent need to pay attention and modify the policies and practices of developing countries dealing with the investments in the field of plant breeding via research, education, and personnel retention so that the feasibility of long-term crop improvement plans can be maintained and scientific advances can be achieved through the implementation of modern techniques such as speed breeding.

15.6.2 Scarce Infrastructure Facilities

To maintain environmental factors, mainly temperature, soil moisture, and photoperiod, sophisticated infrastructure is required for performing speed breeding. There is a very little institutional support to public plant breeding programs that restricts the large-scale application of the state-of- the-art breeding methods in many developing countries such as speed breeding and genetic engineering tools. Moreover, there is a limited stock of specialized equipment that are required for the selection of traits in initial stages of generation advancement (Ribaut et al. 2010). Additionally, an excessive dependence on donor organizations (a "donor mind-set") and a lack of coordination across regional breeding programs result in duplications of activities and resource expenditures. In order to effectively construct infrastructure, national and regional organizations must actively collaborate, and once infrastructure is in place, they must share resources and information. There is a chance to lower the price of building new infrastructure through the development of creative, local technology, including the usage of refurbished shipping containers outfitted with solar-powered equipment for automatically controlling temperature and light (Chiurugwi et al. 2018).

15.6.3 Inconsistent Electricity and Water Supplies

The alteration of several environmental factors, especially temperature, moisture, and photoperiod, in controlled conditions needs consistent electricity and water supplies. Speed breeding with indoor facilities needs sustainable, inexpensive and dependable energy for processes such as heating, cooling, and lighting. For example, in Queensland during winter, the price of temperature regulation found to be greater than half of the overall price accounted for plant management. The regulation of photoperiod and temperature for speed breeding in the plant breeding programs in public sector is severely hampered by unreliable electricity supplies. The activities such as preparation of land, irrigation, fertilization, and other common agronomic activities are necessary for growing crops in the field. These practices are expensive and call for significant infrastructural investments. Speed breeding have future in developing countries only when innovative solutions for the reliable supply of water and electricity are applied, for instance, using sustainable sources of energy, such as solar energy, to provide continuous supply of light to the areas that require controlled environmental conditions for speed breeding. Using already available technologies, a compact indoor kit for speed breeding fitted with LED lights and temperature controls can be used that are provided with a solar system backed up with battery system. As an alternative, semi-controlled systems based on field conditions supporting production of large populations with the management of high-density planting, along with moisture and nutritional stress at cheaper prices can be applied.

15.7 Conclusion

In the conclusion, it is justified to state that the advantages offered by speed breeding techniques in plant breeding are immense and play a crucial role in facing the problems of modern world. Drought and salinity related stress conditions can be overcome by the proper application of speed breeding. The earliness in breeding generation advancement will help to cope the adverse effects of climate catastrophes avoiding the increasing threats of crop failure and economic losses. The production of homozygous and stable tolerant varieties in less time will save the cost, labor, and time of the humans. Speed breeding provides better opportunities to breed and combine new genetic composition from different sources and ultimately develop novel cultivars. Moreover, the controlled environmental conditions help to check the infestation of pests and diseases in the crops which timely prevent the large scale infestation of such nuisance pests. To overcome the challenges of speed breeding, proper infrastructure and financial support is important in both public and private sector. The goal of sustainable agriculture production can be achieved with the incorporation of speed breeding with various other modern techniques.

References

Ahmar S, Gill RA, Jung KH, Faheem A, Qasim MU, Mubeen M, Zhou W (2020) Conventional and molecular techniques from simple breeding to speed breeding in crop plants: recent advances and future outlook. Int J Mol Sci 21(7):2590. https://doi.org/10.3390/ijms21072590

Arbelaez JD, Tandayu E, Reveche MY, Jarana A, van Rogen P, Sandager L, Stolt P, Ng E, Varshney RK, Kretzschmar T, Cobb J (2019) Methodology: Ssb-MASS: a single seed-based sampling strategy for marker-assisted selection in rice. Plant Methods 15(1):78. https://doi.org/10.1186/s13007-019-0464-2

Baier K, Maynard C, Powell WA (2012) Early flowering in chestnut species induced under high intensity, high dose light in growth chambers. J Am Chestnut Found 26:8–10

Bermejo C, Gatti I, Cointry E (2016) In vitro embryo culture to shorten the breeding cycle in lentil (Lens culinaris Medik). Plant Cell Tissue Organ Cult 127(3):585–590. https://doi.org/10.1007/s11240-016-1065-7

Bhatta M, Sandro P, Smith MR, Delaney O, Voss-Fels KP, Gutierrez L, Hickey LT (2021) Need for speed: manipulating plant growth to accelerate breeding cycles. Curr Opin Plant Biol 60: 101986. https://doi.org/10.1016/j.pbi.2020.101986

Bhattaraj SP, de la Pena RC, Midmore DJ, Palchamy K (2009) In vitro culture of immature seed for rapid generation advancement in tomato. Euphytica 167:23–30

Borlaug N (1968) Wheat breeding and its impact on world food supply. Canberra, Australian Academy of Sciences, pp 1–36

Brim CA (1966) A modified pedigree method of selection in soybeans 1. Crop Sci 6:220

Chiurugwi T, Kemp S, Powell W, Hickey LT, Powell W (2018) Speed breeding orphan crops. Theor Appl Genet 132(3):607–616

Collard BCY, Beredo JC, Lenaerts B, Mendoza R, Santelices R, Lopena V, Verdeprado H, Raghavan C, Gregorio GB, Vial L et al (2017) Revisiting rice breeding methods–evaluating the use of rapid generation advance (RGA) for routine rice breeding. Plant Prod Sci 20(4): 337–352. https://doi.org/10.1080/1343943X.2017.1391705

Croser JS, Pazos-Navarro M, Bennett RG, Tschirren S, Edwards K, Erskine W, Creasy R, Ribalta FM (2016) Time to flowering of temperate pulses in vivo and generation turnover in vivo–

in vitro of narrow-leaf lupin accelerated by low red to far-red ratio and high intensity in the far-red region. Plant Cell Tissue Organ Cult 127:591–599

Dagustu N, Bayram G, Sincik M, Bayraktaroglu M (2012) The short breeding cycle protocol effective on diverse genotypes of sunflower (Helianthus annuus L.). Turk J Field Crop 17:124–128

Dwivedi SL, Britt AB, Tripathi L, Sharma S, Upadhyaya HD, Ortiz R (2015) Haploids: constraints and opportunities in plant breeding. Biotechnol Adv 33(6):812–829. https://doi.org/10.1016/j.biotechadv.2015.07.001

El-Hashash EF, El-Absy KM (2019) Barley (Hordeum vulgare L.) breeding. In: Al-Khayri J, Jain S, Johnson DV (eds) Advances in plant breeding strategies: cereals. Springer, Cham, pp 1–45. https://doi.org/10.1007/978-3-030-23108-8_1

FAO (2019) Disasters causing billions in agricultural losses, with drought leading the way. FAO, Rome

Flachowsky H, Le Roux P-M, Peil A, Patocchi A, Richter K, Hanke M-V (2011) Application of a high-speed breeding technology to apple (Malus × domestica) based on transgenic early flowering plants and marker-assisted selection. New Phytol 192:364–377

Funada M, Helms TC, Hammond JJ, Hossain K, Doetkott C (2013) Single-seed descent, single-pod, and bulk sampling methods for soybean. Euphytica 192(2):217–226. https://doi.org/10.1007/s10681-012-0837-3

Ghosh S, Watson A, Gonzalez-Navarro OE, Ramirez-Gonzalez RH, Yanes L, Mendoza-Suárez M, Simmonds J, Wells R, Rayner T, Green P, Hafeez A, Hayta S, Melton RE, Steed A, Sarkar A, Carter J, Perkins L, Lord J, Tester M, Hickey LT (2018) Speed breeding in growth chambers and glasshouses for crop breeding and model plant research. Nat Protoc 13(12):2944–2963. https://doi.org/10.1038/s41596-018-0072-z

Hickey LT, Germán SE, Pereyra SA, Diaz JE, Ziems LA, Fowler RA, Platz GJ, Franckowiak JD, Dieters MJ (2017) Speed breeding for multiple disease resistance in barley. Euphytica 213:64. https://doi.org/10.1007/s10681-016-1803-2

Hickey LT, Hafeez N, Robinson H, Jackson SA, Leal-Bertioli SCM, Tester M, Gao C, Godwin ID, Hayes BJ, Wulff BBH (2019) Breeding crops to feed 10 billion. Nat Biotechnol 37(7):744–754. https://doi.org/10.1038/s41587-019-0152-9

Hussain HA, Hussain S, Khaliq A, Ashraf U, Anjum SA, Men S, Wang L (2018) Chilling and drought stresses in crop plants: implications, cross talk, and potential management opportunities. Front Plant Sci 9:393. https://doi.org/10.3389/fpls.2018.00393

IISD (2022) World population to reach 9.9 billion by 2050. https://sdg.iisd.org/news/world-population-to-reach-9-9-billion-by-2050/. Accessed 3 February 2022.

Jagadish SVK, Bahuguna RN, Djanaguiraman M, Gamuyao R, Prasad PVV, Craufurd PQ (2016) Implications of high temperature and elevated CO2 on flowering time in plants. Front Plant Sci 7:913. https://doi.org/10.3389/fpls.2016.00913

Jähne F, Hahn V, Würschum T, Leiser WL (2020) Speed breeding short-day crops by LED-controlled light schemes. Theor Appl Genet 133:2335–2342. https://doi.org/10.1007/s00122-020-03601-4

Johnston HR, Keats BJB, Sherman SL (2019) Population genetics. In: Pyeritz RE, Korf BR, Grody WW (eds) Emery and Rimoin's principles and practice of medical genetics and genomics foundations. Academic, New York, pp 359–373. https://doi.org/10.1016/B978-0-12-812537-3.00012-3

Kouressy M, Dingkuhn M, Vaksmann M, Heinemann AB (2008) Adaptation to diverse semi-arid environments of sorghum genotypes having different plant type and sensitivity to photoperiod. Agric For Meteorol 148(3):357–371. https://doi.org/10.1016/j.agrformet.2007.09.009

Lin Z, Cogan NOI, Pembleton LW, Spangenberg GC, Forster JW, Hayes BJ, Daetwyler HD (2016) Genetic gain and inbreeding from genomic selection in a simulated commercial breeding program for perennial ryegrass. Plant Genome 9:1–12

Manzur J, Oliva-Alarcón M, Rodríguez-Burruezo A (2014) In vitro germination of immature embryos for accelerating generation advancement in peppers (Capsicum annuum L.). Sci Hortic 170:203–210

McClung CR, Lou P, Hermand V, Kim JA (2016) The importance of ambient temperature to growth and the induction of flowering. Front Plant Sci 7:1266. https://doi.org/10.3389/fpls.2016.01266

Mukade K (1974) New procedures for accelerating generation advancement in wheat breeding. JARQ 8:1–5

Nagatoshi Y, Fujita Y (2019) Accelerating soybean breeding in a CO2-supplemented growth chamber. Plant Cell Physiol 60:77–84

O'Connor DJ, Wright GC, Dieters MJ, George DL, Hunter MN, Tatnell JR, Fleischfresser DB (2013) Development and application of speed breeding technologies in a commercial peanut breeding program. Peanut Sci 40(2):107–114. https://doi.org/10.3146/PS12-12.1

Ochatt SJ, Sangwan RS (2008) In vitro shortening of generation time in Arabidopsis thaliana. Plant Cell Tissue Organ Cult 93:133–137. https://doi.org/10.1007/s11240-008-9351-7

Ribaut JM, de Vicente MC, Delannay X (2010) Molecular breeding in developing countries: challenges and perspectives. Curr Opin Plant Biol 13(2):213–218. https://doi.org/10.1016/j.pbi.2009.12.011

Richard CA, Hickey LT, Fletcher S, Jennings R, Chenu K, Christopher JT (2015) Highthroughput phenotyping of seminal root traits in wheat. Plant Methods 11:13. https://doi.org/10.1186/s13007-015-0055-9. PMID: 25750658; PMCID: PMC4351910.

Rizal G, Karki S, Alcasid M, Montecillo F, Acebron K, Larazo N, Garcia R, Slamet-Loedin IH, Quick WP (2014) Shortening the breeding cycle of sorghum, a model crop for research. Crop Sci 54:520–529

Saito H, Yuan Q, Okumoto Y, Doi K, Yoshimura A, Inoue H, Teraishi M, Tsukiyama T, Tanisaka T (2009) Multiple alleles at Early flowering 1 locus making variation in the basic vegetative growth period in rice (Oryza sativa L.). Theor Appl Genet 119(2):315–323. https://doi.org/10.1007/s00122-009-1040-3

Samineni S, Sen M, Sajja SB, Gaur PM (2020) Rapid generation advance (RGA) in chickpea to produce up to seven generations per year and enable speed breeding. Crop J 8(1):164–169. https://doi.org/10.1016/j.cj.2019.08.003

Saxena K, Saxena RK, Varshney RK (2017) Use of immature seed germination and single seed descent for rapid genetic gains in pigeonpea. Plant Breed 136:954–957

Saxena KB, Saxena RK, Hickey LT, Varshney RK (2019) Can a speed breeding approach accelerate genetic gain in pigeonpea? Euphytica 215(12):1–7. https://doi.org/10.1007/s10681-019-2520-4

Shavrukov Y, Kurishbayev A, Jatayev S, Shvidchenko V, Zotova L, Koekemoer F, De Groot S, Soole K, Langridge P (2017) Early flowering as a drought escape mechanism in plants: how can it aid wheat production? Front Plant Sci 8:1950. https://doi.org/10.3389/fpls.2017.01950

Shimelis H, Laing M (2012) Timelines in conventional crop improvement: prebreeding and breeding procedures. Aust J Crop Sci 6(11):1542–1549

Shimelis H, Gwata ET, Laing MD (2019) Crop improvement for agricultural transformation in Southern Africa. In: Sikora RA, Terry ER, Vlek PLG, Chitja J (eds) Transforming agriculture in Southern Africa, 1st edn. Routledge, London, pp 97–103. https://doi.org/10.4324/9780429401701

Shivakumar M, Nataraj V, Kumawat G, Rajesh V, Chandra S, Gupta S, Bhatia VS (2018) Speed breeding for Indian agriculture: a rapid method for development of new crop varieties. Curr Sci 115:1241–1241

Stetter MG, Zeitler L, Steinhaus A, Kroener K, Biljecki M, Schmid KJ (2016) Crossing methods and cultivation conditions for rapid production of segregating populations in three grain amaranth species. Front Plant Sci 7:816

Tanaka J, Hayashi T, Iwata H (2016) A practical, rapid generation-advancement system for rice breeding using simplified biotron breeding system. Breed Sci 66:542–551

Tripp R, Louwaars N, Eaton D (2007) Plant variety protection in developing countries. A report from the field. Food Policy 32(3):354–371. https://doi.org/10.1016/j.foodpol.2006.09.003

Van Nocker S, Gardiner SE (2014) Breeding better cultivars, faster: applications of new technologies for the rapid deployment of superior horticultural tree crops. Hortic Res 1:14022

Vince-Prue D (1994) The duration of light and photoperiodic responses. In: Kendrick RE, Kronenberg GHM (eds) Photomorphogenesis in plants. Springer, Dordrecht. https://doi.org/10.1007/978-94-011-1884-2_17

Warnasooriya SN, Brutnell TP (2014) Enhancing the productivity of grasses under high-density planting by engineering light responses: from model systems to feedstocks. J Exp Bot 65(11): 2825–2834. https://doi.org/10.1093/jxb/eru221

Watson A, Ghosh S, Williams MJ, Cuddy WS, Simmonds J, Rey M-D, Hatta MAM, Hinchliffe A, Steed A, Reynolds D et al (2018) Speed breeding is a powerful tool to accelerate crop research and breeding. Nat Plants 4:23–29. https://doi.org/10.1038/s41477-017-0083-8

Yan G, Liu H, Wang H, Lu Z, Wang Y, Mullan D, Hamblin J, Liu C (2017) Accelerated generation of selfed pure line plants for gene identification and crop breeding. Front Plant Sci 8:1786. https://doi.org/10.3389/fpls.2017.01786

Yang L, Wang D, Xu Y, Zhao H, Wang L, Cao X, Chen Y, Chen Q (2017) A new resistance gene against potato late blight originating from Solanum pinnatisectum located on potato chromosome 7. Front Plant Sci 8:1729. https://doi.org/10.3389/fpls.2017.01729

Zhang F, Wen Y, Guo X (2014) CRISPR/Cas9 for genome editing: progress, implications and challenges. Hum Mol Genet 23:R40–R46

Zhang Z, Wei W, Zhu H, Challa GS, Bi C, Trick HN, Li W (2015) W3 is a new wax locus that is essential for biosynthesis of β-diketone, development of glaucousness, and reduction of cuticle permeability in common wheat. PLoS ONE 10:e0140524

Zheng Z, Wang HB, Chen GD, Yan GJ, Liu CJ (2013) A procedure allowing up to eight generations of wheat and nine generations of barley per annum. Euphytica 191(2):311–316. https://doi.org/10.1007/s10681-013-0909-z

Exploiting Integrated Breeding Strategies to Improve Salinity Tolerance in Crop Plants

16

Reetu Verma, Maharishi Tomar, Mitali Mahajan, Praveen Yadav, Anita Rana, and D. Seva Nayak

Abstract

Abiotic stress has emerged as a major threat to food security, accounting for the majority of crop and agricultural product losses worldwide. Salinity is one of the primary key variables that inhibit plant growth and productivity among other abiotic stresses. The growing negative effects of salinity stress (SS) are putting global food and nutritional security at jeopardy. Plants respond to high salinity stress by initiating a series of events and adapt by activating a number of stress-responsive genes. However, the complex and poorly known mechanism of salt tolerance (ST) are key roadblocks to breeding for improved ST. As a result, while making crop selections, the focus should be on assessing crop diversity and addressing adaptive/morpho-physiological traits. The quick and precise introgression of ST-related gene(s)/QTLs into salinity-susceptible cultivars to restore genotypes with improved ST is also important. Therefore, a sensible integration of molecular breeding, functional genomics, and transgenic technologies, as well as next-generation phenomics facilities, is required for the gradual tailoring of salinity-tolerant genotypes.

R. Verma (✉) · M. Tomar · P. Yadav · D. Seva Nayak
ICAR - Indian Grassland and Fodder Research Institute, Jhansi, Uttar Pradesh, India

M. Mahajan
CSIR - Institute of Himalayan Bioresource Technology, Palampur, Himachal Pradesh, India

A. Rana
ICAR-IARI, Regional Station, Katrain, Himachal Pradesh, India

A. Kumar et al. (eds.), *Salinity and Drought Tolerance in Plants*,
https://doi.org/10.1007/978-981-99-4669-3_16

16.1 Introduction

Soil salinity is one of the most important abiotic stresses especially in arid and semi-arid environments where it can lead to the degradation of arable soils (Akram et al. 2017). Over 800 million hectares of land throughout the world are salt-affected, either by salinity (397 million ha) or the associated condition of sodicity (434 million ha) (FAO 2015). By 2050, it is anticipated that 50% of cultivable land will be affected by salinity. Salinity is a soil condition characterized by high concentration soluble salts containing chlorides (Cl^-) and sulfates of sodium (Na_2SO_4), calcium ($CaSO_4$), and magnesium ($MgSO_4$). Among them, sodium chloride (NaCl) is most soluble, pervasive, and superabundant salt in the world (FAO 2008). Excess salt causes biochemical, morphological, physiological, and molecular alterations in plants, all of which have a negative impact on productivity and plant growth. Salinity poses two major threats to plant growth, i.e., osmotic stress and ionic stress (Flowers and Colmer 2015). One of the most detrimental effects of salinity stress is the accumulation of Na^+ and Cl^- ions in tissues of plants exposed to soils with high NaCl concentrations. High Na^+ concentration inhibits uptake of K^+ ions which is an essential element for growth and development that results into lower productivity and may even lead to death (Cao et al. 2018). As a consequences of salinity stress, alteration in plant productivity was noticed among different species worldwide. Therefore, increasing salinity tolerance in various crop species is emerging as an alarming issue especially in concern to the global food security. Furthermore, an attempt has been made to provide an overview of different "omics" approaches and their role in breeding strategies for developing salinity tolerance crops has been discussed in this chapter.

16.2 Crop Yield Losses Due to Salinity Stress Worldwide

Salinity stress adversely affects the agricultural production by about 30% for irrigated crops and 7% for dryland agriculture worldwide. Salinity stress declines plant growth and yield by altering physiological and biochemical processes (Tavakkoli et al. 2011). Also, the detrimental effects of salinity on plant have been caused mainly by increase in sodium and chlorine ions that leads to imbalance the other nutrients concentrations in soil solution, their uptake, transport, and partitioning within the plant parts (Grattan and Grieve 1999). The imbalance of ions creates the decisive conditions for plant survival by intercepting different plant mechanisms. Moderate soil salinity (8–10 dS m^{-1}) reduced yield by 15, 28, and 55% in cotton, wheat, and corn, respectively. Also, 55% yield loss in cotton was reported under high soil salinity of 18 dS m^{-1} (Satir and Berberoglu 2016). Hasan et al. (2015) observed decreased grains per spike, and seed yield in wheat cultivars when grown under salt stress (15 dS m^{-1}). Rice plants irrigated with 34.2 mM NaCl decreased about 9.8% panicle number, 21.6% grain number per panicle, 4.4% 1000-grain weight, and 20% yield compared with the control (without salt treatment) (Zhang et al. 2022). According to studies conducted by Farooq and Azam (2005) salinity levels especially beyond 150 mM of NaCl, significantly reduced the wheat grain yield. Salinity stress not only reduces the production of traditional crops but also alters the growth and yield of many medicinal and aromatic plants (MAPs). Salt stress shows negative effect on the growth, biomass, and essential oil yield of various MAPs, e.g., *Majoranahortensis* (Shalan et al. 2006), peppermint (*Mentha* × *piperita* L.), pennyroyal (*Mentha pulegium* L.), apple mint (*Menthasuaveolens*) (Aziz et al. 2008), *Thymus maroccanus*, *Ocimumbasilicum* (Said-Al Ahl and Mahmoud 2010), and spearmint (Khadhri et al. 2011). Mahajan et al. (2020) studied the effect of different concentrations of NaCl on *Stevia rebaudiana* and observed that plant irrigated with NaCl at 120 mM decreases dry leaf yield by about 23.0%.

16.3 Impact of Salinity Stress on Crops from Germination to Maturity

Seed germination is the vulnerable and most important stage in seedling establishment which defines the development of crop. Generally, seeds are poorly germinated when sown under saline conditions. Salinity causes various physiological and metabolic disturbances in plants. The alterations in physiological and biochemical characteristics are stage-specific and attribute to final yield potential. The crops are more susceptible to salinity at germination and early seedling stage than later development stages (Goldsworthy 1994; Shereen et al. 2005). Salt stress declines the endogenous level of phytohormones and delays seedling establishment. Seed germination reduction during salt stress was mainly linked with lowered osmotic potential of the soil solution, which leads to hinders water absorption by seeds. Also, sodium and chloride ion toxicity can inhibit the metabolism of dividing and expanding cells, and retards seeds germination (Khaje-Hosseini et al. 2003; Farsiani and Ghobadi 2009). These both consequences, i.e., osmotic stress and disturbed ions homoeostasis collectively inhibit cell division and modulate the key enzymes activity, resulting reduced seed reserves utilization (El-Hendawy et al. 2005). Furthermore, accelerated reactive oxygen species (ROS) induces oxidative stress that ultimately decreases overall yield of the affected seeds. Salinity reduces cell turgor and declines root and leaf elongation in vegetative plants (Fricke et al. 2006). Rice is more resistant to salt stress at reproductive and grain filling than vegetative stages. Various studies showed that salinity decreases rice stand density as well as seedling biomass production. Salinity decreases the amylase activity and negatively affect the rate of starch remobilization in wheat and rice. The roots are the most susceptible to salt stress and responsible for poor growth of other parts. Salinization in soil diminishes the shoot development by suppressing leaves extension, internodes development and causing leaf abscission. The first phase of salt stress strongly inhibited the shoot growth in maize (Pitann et al. 2009). In salt-affected soils, high concentrations of sodium and chloride ions in the rhizosphere reduce the uptake of nitrogen, potassium, calcium, magnesium, and iron, and causing water loss and necrosis in crops. In case of cereals, grain number and weight are important attributes for evaluating final grain yield. At the reproductive phase, salt stress negatively affects the yield potential through reducing photosynthetic rate, cell elongation, hastens the reproductive organ's senescence, reduction of flag leaf turgidity and leaf area, source-sink activity and assimilates synthesis. Salinization also antagonistically affects blossoming, fertilization, and natural product advancement. Several earlier studies revealed that during reproductive phase salt stress causes massive yield penalty in many crops, including maize, wheat (Kalhoro et al. 2016). In maize, salinity decreases grain number and weight, ultimately leads to reductions in grain yield during the reproductive phase (Kaya et al. 2013).

16.4 Genomics, Transcriptomics, and Proteomics Approach for Salinity Tolerance

Integrating numerous "omics" approaches such as genomics, transcriptomics, proteomics, and metabolomics in plants is the most prevalent means for dealing with plant salinity tolerance (Liu et al. 2019). Studies on "Omics" have enabled researchers to modulate traits, metabolomes, proteins, and genes that are indispensable in providing salinity tolerance to the plant (Jha et al. 2019).

16.5 Genomics Approach

A branch of "omics" called genomics studies a particular genome and provides insightful information on the biology of the organism (Gilliham et al. 2017). The researchers determine gene structures, gene, and intragenic sequences, and functional annotation of the genome are all aspects in the study of genomics. The entire process offers insightful information on the organism's genomic structure (El-Metwally et al. 2013; El-Metwally et al. 2014). For improving the salinity tolerance in plants, a comprehensive knowledge of the mechanisms by which plants counter salinity stress is imperative. Therefore, identifying the genes involved in salt stress response is a crucial initial step in gaining the knowledge needed to develop crops that are tolerant to salinity stress (Billah et al. 2021). However, locating the genes that respond to stress is just the first step; the next involves characterizing their functions and understanding how they work. The final step is using the obtained information to develop stress-tolerant crops via molecular breeding or genetic engineering (Singh et al. 2021a). Rice and Arabidopsis were the first two plant species with their genomes sequenced, thus a majority of plant science research is carried out using these models (Yadav et al. 2021).

Genomics-assisted breeding (GAB) is a technique that harnesses genomic resources, techniques, and tools, such as molecular markers, to expedite the process of selecting plants by using DNA markers linked to attractive attributes (Varshney et al. 2020; Varshney et al. 2021). Molecular or DNA markers are potent tools for increasing the effectiveness of plant breeding. Important markers used in breeding programs include single nucleotide polymorphism (SNP), microsatellites, amplified fragment length polymorphism (AFLP), random amplified polymorphic DNA

(RAPD), and restriction fragment length polymorphism (RFLP). For instance, BC3F4 rice (Indica-donor, japonica Italian varieties, recipient) lines were developed through introgression of salt tolerance using *Saltol* QTL, succeeded by marker-assisted backcrossing (MABC). Also pyramiding QTLs that govern tolerance were introgressed in the rice variety (Improved White Ponni) against several stressors, including salt stress, using the MABC breeding technique (Muthu et al. 2020). The substantial correlation between desirable attributes and markers suggests effective breeding, which can be assessed through gene mapping, recombination analysis, or quantitative trait loci (QTL) analysis.

Quantitative trait locus (QTL) is a term that is used for identifying genes that govern: molecular markers, phenotypic traits, and markers associated with these traits. The case studies that identify novel QTLs in crops under salinity stress are listed below, and Table 16.1 provides some more recent examples. Salinity stress was investigated among the recombinant inbred lines (RILs) population as a significant growth-limiting factor in rice. 935 QTLs described in rice over the previous 20 years were examined in recent meta-QTL research for several features that contribute to salt tolerance. These QTLs were derived from 13 different genetic background mapping populations, the bulk of which were RILs (F2:F4, F2, BILs, BC4F8, BC3F5, BC3F4, BC3F2, BC2F8, BC2F5, BC1F9, DHs, ILs, and RILs). These analyses suggest that the most promising genetic areas to increase salt tolerance are 63 meta-QTLs (Singh et al. 2021b). QTL analysis displayed a phenotypic variation including, root dry weight (RDW), shoot dry weight (SDW), root fresh weight (RFW), shoot fresh weight (SFW), root length (RL), shoot length (SL), found 21 stable QTLs. A major QTL for shoot length designated as qSL7, having a phenotypic variation of 7.5 and 6.8%, was detected against Na^+ and K^+ concentrations which can provide new avenues for rice salinity tolerance (Jahan et al. 2020). Seven additional multi-environmental QTLs for component characteristics, including spikelet degeneration, stress susceptibility index, and spikelet sterility, were discovered in the second study of rice under salt stress. Interactions between genotype and environment have a favorable influence on two significant QTLs (*qDEG-S-2-1* and *qSSI-STE-2-1*) (Chattopadhyay et al. 2021). For 15 agronomic traits in wheat, soil salinity-related QTLs were found, resulting in the discovery of 90 stable QTLs with phenotypic variation ranging from 2.34 to 32.43%.

Except for chromosomes 7D, 6B, and 4D these QTLs were found on all three genomes. In the QTL cluster, *QPh-4B* was also verified to be an allele of *Rht-B1*. Additionally, this research lays the groundwork for wheat salt-tolerant QTL cloning and allele-specific PCR markers that would aid MAS for salt-tolerant wheat breeding (Luo et al. 2021). Wheat has been mapped using biparental QTLs for salt-responsive characteristics at two distinct growth phases. About 22 overlapping loci and two unique candidate genes (*TaRN1* and *TaRN2*) with various expression patterns were found in roots during salt stress (Li et al. 2021). As a shoot ion-independent tolerance (QG) (1–5), (QCl.asl-3A) for Cl^- accumulation, and (QK: Na.asl-2DS2) for K^+: Na^+ DW, novel QTL for salt tolerance have been discovered in bread wheat. This research may facilitate the understanding of the genetic underpinnings of salt tolerance and expedite bread wheat sub-trait breeding (Asif et al. 2020). In chickpea,

Table 16.1 Summary of studies for identifying QTLs under salinity stress in different crops

Plant species	Stress	Approach	QTLs/ genes	No. of lines/ accessions used	Salient findings	References
Rice (*Oryza sativa*)	Saline water (8 dS m^{-1}); 40–45 days	Graphical GenoTyping and QTL IciMapping	QTLs (7)	BC_3F_5 generation (180 lines)	• Seven stable QTLs were found for component traits • *qDEG-S-2-1* and *qSSI-STE-2-1* QTLs were positively influenced by genotype × environment interactions QTLs	Chattopadhyay et al. (2021)
	From 6 dS m^{-1} to 12 dS m^{-1} NaCl with the increments of 2 dSm^{-1}/day; 14 days	R/qtl and qtlTools (Lovell)	QTLs (14)	(204) $F_{2:3}$ families From *IR29* × *Horkuch*	Fourteen QTLs for nine yield-related and physiological traits were found	Haque et al. (2020)
	100 and 50 mM NaCl; 14 days	MultiQTL package	QTLs (38)	RILs (132)	• Put of 38, 21 QTLs were found stable • *qSL7*, was identified as a novel QTL for shoot length having a phenotypic variation of 7.5 and 6.8%	Jahan et al. (2020)
	$MgSO_4$, $CaCl_2$, NaCl, in water with ECe ~40 mmol; until mature	Win QTL cartographer and Joinmap software	QTLs (39)	F2 population (140)	• 39 QTLs were identified during salinity stress for the reproductive stage • *qNa/KL-1.2*, *qKR-1*, qNa/KL-1.3, *qNaL-1.2* were present in roots and leaves • QTL *qGY-2* was identified for grain yield	Pundir et al. (2021)
Wheat (*Triticum aestivum*)	100 mM NaCl; 13 days	Illumina NextSeq 500, WinQTLCart, R/ASMap	QTLs (9)	128 RILs	QTLs for shoot ion-independent tolerance [QG (1-5) asl-7B], Cl^-	Asif et al. (2020)

					accumulation (QCl.asl-3A), and K^+: Na^+ DW (QK:Na.asl-2DS2)	
	1.2% (w/v) NaCl; 7 and 10 days	Wheat 660K SNP array and IciMapping	QTLs (91)	150 double haploid lines and 323 accessions	TaRN2 and TaRN1, were found as two novel candidate genes salt stress	Li et al. (2021)
	0.30% and 0.18% mM NaCl; 240 days	MapChart, IciMapping, Wheat 55K SNP array	QTLs (90)	F7 RILs including 350 lines	*QPh-4B* was established as an *Rht-B1* allele	Luo et al. (2021)
Chickpea (*Cicer arietinum*)	6 dS/m; 30, 60, and 90 days	Windows QTL Cartographer, Join Map, Axiom® *CicerSNP* array	QTLs (28)	(F_8) 201 RILs	QTL clusters governing yield and yield components were found on CaLG06 and CaLG03	Soren et al. (2020)
Barley (*Hordeum vulgare*)	75, 90, 120, and 150 mM NaCl; 14 days	Multiple QTL model (MQM), 50 diversity array technology (DArT), MapQTL software	QTLs (6)	85 barley germplasm, 103 DH lines	Shoot length (65.6% and 50.3%) showed the highest variation. QTL on 3H showed higher linkage	Mwando et al. (2021)

a comprehensive analysis found 28 QTLs for nine yield-related parameters, mostly on two salinity stress-related genomic regions, *CaLG03* and *CaLG06*. These QTL regions contain potential genes that encode kinases, including calcium-dependent protein kinases (CDPKs), MAPKs, histidine kinases (HKs), and kinases associated with sucrose non-fermenting (SnRK1).

16.6 Transcriptomics Approach

The field of transcriptomics studies the functional genome of lifeforms, including the vast array of transcripts present in a given cell and any post-transcriptional modifications (Hussain 2018). The extensive processes taking place inside a tissue cell are a major factor in how well a plant function. RNA sequencing (RNA-seq), microarray, and other sequencing techniques can all be used to conduct plant transcriptomic research. The good news is that improvements in RNA-seq technology enable us to investigate these extensive and widespread transcriptional changes inside particular plant tissues. With its enormous potential, RNA-seq enables researchers to decipher and take advantage of the intricate regulatory networks that are influenced by environmental factors or developmental changes in plant tissue (Rich-Griffin et al. 2020). Given its enormous potential, transcriptome analysis is currently a popular strategy for figuring out the complex molecular mechanisms driving various stressors, such as salinity (Table 16.2). The RNA-sequencing study of wild-type (WT-Zarjou) and salt-tolerant mutant (M4-73-30) cultivar of barley indicated specific genes that facilitate salinity tolerance by modulating ion transporters like SOS3/SOS2/SOS1, HAK, TPK1/KCO1, NHX1, and HKT for K^+, Na^+, Ca^+ transportation (Yousefirad et al. 2020). Numerous salinity-responsive transcription factors (TFs), including bZIP, HSF, MAD, CTR/DRE, NAC, AP2/ERF, and WRKY were discovered using RNA-sequence data. By using the fixed carbon and stored energy, the respiration and photosynthetic rates of the mutant lines were dramatically reduced, subsequently protecting the tissues from the harmful effects of salinity.

Additionally, ion transporter expression and transcripts associated with channels were elevated in mutant lines compared to the wild types for maintaining ion homeostasis (Yousefirad et al. 2020). Through RNA-sequence analysis, 73,401 genes in bread wheat that responded to salinity stress were discovered. Several early and late-stress responsive genes as well as genes involved in ROS scavenging, chaperons, and glucose metabolism were found among the identified genes. Enrichment analysis indicated that: pentose phosphate pathways, secondary metabolites, and carbohydrate metabolism showed higher enrichment during the salinity response (Ma et al. 2020).

Additionally, a recent study investigating the global transcriptomic profiling of the salt-sensitive cultivar TN6-18 of the model legume *Medicago truncatula* using microarray analysis. The study indicated a lower expression of many genes involved in stress signaling that were not previously associated with salinity and belonged to the TIR-NBS-LRR gene class (Filippou et al. 2021). In a different study, RNA-seq

Table 16.2 Recently transcriptomics studies under salinity stress in different crop plants

Plant species	Stress condition, Tissue, approach	Key findings	References
Cucumber (*Cucumis sativus*)	100 mM NaCl; 72 h, seeds, RNA-sequencing	Common enriched pathways were alpha-linolenic acid metabolism, secondary metabolites biosynthesis, unsaturated fatty acids biosynthesis, carbon fixation and metabolism, fatty acid degradation, dicarboxylate and glyoxylate metabolism, linoleic acid metabolism, photosynthesis, chlorophyll, and porphyrin metabolism	Du et al. (2021)
Oats (*Avena sativa*)	150 and 300 mM NaCl; 24 hours, Roots, RNA-sequencing	The expression Na^+/K^+ transporter genes were upregulated under salinity stress. Differentially expressed genes were enriched in glycolysis/ gluconeogenesis, galactose, sucrose, and starch metabolism	Xu et al. (2021)
Zoysia macrostachya	30 mM NaCl; 24 h, leaf, RNA-sequencing	Genes involved in ROS scavenging, ion homeostasis, and hormone signal transduction were identified	Wang et al. (2020)
Wheat (*Triticum aestivum*)	150 mM NaCl; 24 h, shoots and roots, RNA-sequencing	The identified genes were majorly involved in carbohydrate metabolism, chaperons, and ROS scavenging	Ma et al. (2020)
Tomato (*Solanum chilense*)	500 mM NaCl; 21 days, leaf, RNA-sequencing	Several differentially expressed genes involved in ethylene-, auxin, and Ca^{+2} signaling networks were recognized as key genes against salinity. Genes encoding stress and defence response, osmotic regulation, transporters, and ROS scavenging were upregulated under salinity	Kashyap et al. (2020)
Barley (*Hordeum vulgare*)	300 mM NaCl; 6 hours, shoots and roots, RNA-sequencing	Expression of channels-related and ion transporter genes was upregulated for maintaining ion homeostasis. Salinity stress-responsive different TFs were bZIP, HSF, NAC, AP2/EREBP, ERF, MIKC, MAD, AP2/ERF, CTR/DRE, and WRKY were identified	Yousefirad et al. (2020)

analysis was done comparing sensitive and tolerant genotypes of wheat at the osmotic stage (Zentos-tolerant and Syn86-susceptible) and ionic phase (Altay2000-tolerant and Bobur-susceptible). The initial upregulation of Ca^+-binding and cell wall synthesizing genes were found in the tolerant genotype in response to salt-associated osmotic stress and was thought to be a critical factor for improving salinity tolerance. As an alternative, the higher photosynthesis reserve at the osmotic stage was related to the downregulation of photosynthesis-associated and Ca^+-binding genes as well as the increased oxidative stress in the sensitive genotype.

The precise modulation of some Na^+/Ca^{+2} and ABC transporters during the ionic phase in the tolerant genotype identified their role in the regulation of Na^+ elimination and ion homeostasis (Duarte-Delgado et al. 2020). Two cultivars of castor bean (*Ricinus communis* L.), namely "Tongbi 5" and wild castor (Y) were used to better understand the mechanism underlying the ability of this crop to withstand salinity stress. The investigation produced a large number of differentially expressed genes, mostly from the bHLH TF, WRKY, NAC, ERF/AP2, and families (Lei et al. 2021). Both the wild and domesticated castor cultivars included hormone-related differentially expressed genes. The PP2C TFs family, GA and JA were the majority of the differentially expressed genes found in the cultivars. In response to salt, the GA signaling gene DELLA (GA signal suppressor) was downregulated in cultivars, suggesting that stress resistance or adaptation may be caused by the activation of other TFs, including NAC or WRKY. As previously documented, the DELLA gene controlled the trade-off between defence and growth, which reduced stress (Jusovic et al. 2018).

The well-known multi-stress resistor arbuscular mycorrhizal fungus (AMF) was typically utilized to promote plant development in unfavorable environments. In this context, AMF-induced salinity tolerance was mapped by using the transcriptomic approach in *Suaeda salsa* plant. *Funneliformis mosseae*, an AMF strain was used in the study. The research revealed 424 and 1306 differentially expressed genes in roots and shoots, respectively. The major differentially expressed genes in shoots were majorly involved in energy metabolism, carbohydrate, and photosynthesis, while those in root tissues were primarily related to starch and sucrose metabolism. The findings suggest that *Funneliformis mosseae* protects *Suaeda salsa* during salinity stress through the suppression of stress-induced ROS in the chloroplast (Diao et al. 2021).

16.7 Proteomics Approach

Proteomics deals with the localization, function, structure, and role, of functional proteins, also their implementation under stress or natural conditions. Post-transcriptional modifications and changes at the protein level can be studied in more detail through proteomics. As a result, proteomics is now a crucial method for locating important stress protein markers which can be useful in developing crops that are resistant to stress (Kosová et al. 2018). The key examples of proteomics studies under salinity stress in different crop plants are indicated in Table 16.3.

In a recent study, the proteome of salt- affected rice plants was analyzed. Two varieties of rice were used: cv. Jhelum, which is salt-sensitive and cv. Vytilla-4, which is very salt-tolerant. The investigation revealed a variety of distinct protein expression patterns and responses in rice plant tissues to salt stress. Particularly, under salinity stress in cv. Vytilla, photosynthesis-related proteins such as Chl a-b binding protein, carboxylase small chain, and ferredoxin were activated. The PS-II CP47, on the other hand, was suppressed in cv. Jhelum. Other carbon-fixation-related proteins (Cytochrome c oxidase subunit 5C, peroxisomal (S)-2-hydroxy-

Table 16.3 Recently proteomics studies under salinity stress in different crop plants

Plant species	Stress condition, tissue, approach	Key findings	References
Wheat (*Triticum aestivum*)	200 mM NaCl; 24, 48, 72 and 96 h, leaves, MALDI-TOF/TOF MS	Several differentially expressed proteins were significantly upregulated related to nitrogen, amino acid, and carbon metabolisms, transcription, translation, and Calvin cycle. Also proteins associated with vitamin, hormone synthesis, proteolysis, protein folding and plastoglobuli development were also upregulated under salinity stress	Zhu et al. (2021)
Cucumber (*Cucumis sativus*)	200 mM NaCl; 7 days, leaves, MALDI-TOF/TOF-MS	Differentially expressed proteins associated with signal transduction, cell defense, metabolic pathways were identified. Key proteins were peroxidase, protein disulfide-isomerase, glutathione S-transferase U25-like and cysteine synthase 1	Jiang et al. (2020)
Beet (*Beta vulgaris*)	300 mM NaCl; 3 weeks, Leaves, NanoLC-MS/MS	Differentially expressed proteins involved in signaling, ATP biosynthesis, cell wall modification, lipid metabolism were seen. Stress-related proteins, such as inorganic pyrophosphatase 2, heat shock proteins, chaperone proteins, and lipid transfer protein were significantly upregulated under salinity stress	Rasouli et al. (2020)
Alfalfa (*Medicago sativa*)	50, 100, 200 and 400 mM NaCl; 14 days, Leaf LC-MS/MS	Differentially expressed proteins involved in redox pathways, glutathione metabolism were upregulated. ROS metabolism and TCA cycle were identified as key pathways for enhancing salinity tolerance	Li et al. (2020)
Barley (*Hordeum vulgare*)	300 mM NaCl; 2, 4, and 6 days, roots and leaves and MALDI TOF-TOF	Proteins related to ROS scavenging, ATP synthase and photosynthesis were upregulated	Zhu et al. (2020)
Wheat (*Triticum aestivum*)	150 mM NaCl; 24 h, shoots and roots, iTRAQ	Major proteins enriched in pyruvate metabolism, cyanoamino acid metabolism, pentose phosphate pathway, purine metabolism, pyrimidine metabolism. Primarily, superoxide dismutase, phosphoenlpyruvate carboxylases, beta-glucosidases, and nucleoside diphosphate kinases were significantly upregulated under salt stress	Ma et al. (2020)

acid oxidase GLO1, and peroxisomal (S)-2-hydroxy-acid oxidase GLO5) that were essential for energy metabolism were markedly elevated in cv. Vytilla. This demonstrates unequivocally that these proteins may function as possible indicators in controlling the response of plants to salinity without impairing growth (Frukh et al. 2020). Hormone-related proteins in the salinity-stressed hulless barley were profiled. In this work, the germination of two cultivars (the salt-tolerant lk573 and the salt-sensitive landrace lk621) was examined under salinity stress. The lk573 cultivar showed proper germination under salinity when compared to the lower germination rate in lk621. The salt-tolerant cultivar showed 171 differentially expressed proteins. At 4 h following salt stress, several of them including proteins involved in terpenoid-quinone biosynthesis, ABC transporters, ascorbate and nitrogen metabolism were expressed (Lai et al. 2020).

In a different investigation, proteome profiles of wheat were analyzed under salinity stress. The focus of this study was to monitor the changes at a protein level in the chloroplast of wheat under salt stress. A total of 194 differentially expressed proteins were mapped in the chloroplast. These proteins were attributed to nitrogen, carbon metabolism, and Calvin cycle (Zhu et al. 2021). Salt-tolerant (T46 and T66) and salt-sensitive (N33 and N53) near-isogenic lines of barley were cultivated in soils treated with 300 mM NaCl. After performing a proteomics analysis, a set of proteins found in both tolerant and sensitive lines were identified. The majority of the dominantly expressed proteins in the tolerant lines were involved in photosynthesis, ROS scavenging, and ATP production (Zhu et al. 2020). As the ATP-mediated energy biosynthesis reaction primarily takes place in chloroplast, all of the aforementioned evidence pointed to the fundamental significance of chloroplast and photosynthesis-related activities. These ATP energy packets play a crucial role in preserving normal ion homeostasis, eliminating potentially damaging ROS, and possibly dictating the course of overall growth. Thus, the outlined proteins may aid in the development of functionally stress-resistant crops by the research community (Muthu et al. 2020).

16.8 Metabolomic Approach

Salinity stress can impose changes in a plant at a: biochemical, protein, and transcript levels. Fundamentally, the plant responds to stress only at the biochemical level without modulating the protein expression of the transcription process. These biological molecules are also known as metabolites, and metabolomics is the study of these metabolites (Razzaq et al. 2019; Raza 2022).

Because of their significant differences in structures and activities, metabolites research has recently gained popularity in the tech community. The key examples of metabolomic studies under salinity stress in different crop plants are indicated in Table 16.4. The primary metabolites involved in stress resistance were ascorbate, glutathione, and MDA were upregulated under salt stress during the early growth stage. Other stress-responsive metabolites, such as 3-ketoacyl-CoA synthase (KCS),

Table 16.4 Recently metabolomics studies under salinity stress in different crop plants

Plant species	Stress condition, tissue, approach	Key findings	References
Oats (*Avena sativa*)	150 and 300 mM NaCl; 24 h, roots, GC-TOF-MS	Several vital differentially accumulated metabolites were observed under salt stress, i.e., 3, 6-AnhydroD-galactose, melibiose, isomaltose, sophorose, and sucrose. Metabolites involved in organic acids, carbohydrates, and amino acids metabolisms and were also significantly accumulated	Xu et al. (2021)
Potato (*Solanum tuberosum*)	20 mM LiCl; 45 days, shoots, GC-MS	Organic acids, amines, fatty acids, alkanes, terpenes, and sugars were induced under salt stress	Hamooh et al. (2021)
Rimth Saltbush (*Haloxylon salicornicum*)	400 mM NaCl; 21 days, shoots, HPLC-DAD and GC-QTOF–MS	Most of the metabolites belonged to phytohormones, alkaloids, fatty acids, sugars, sugar alcohols, amines, organic acids, and amino acids. Enrichment analysis indicated threonine, serine, cysteine, phenylalanine, nucleotide, and sugar metabolism were substantially enriched under salinity stress	Panda et al. (2021)
Barley (*Hordeum vulgare*)	150 mM NaCl; 14 days, seeds, GC-MS	Under salinity stress, sugar acids, sugar alcohols, sugars, amino acids, and other derivatives acted as osmolytes. The identified metabolites governed TCA cycle, sugar, and amino acid metabolism	Derakhshani et al. (2020)
Tomato (*Solanum Lycopersicon*)	60 mM NaCl; 45 days, pericarp, GC-TOF-MS	Metabolites including nucleotides, organic acids, fatty acids, carbohydrates, alkylamines showed high accumulation. Glycolic acid, alpha-tocopherol, chlorogenic acid, D-galactose, trehalose, L-aspartic acid, L-valine, L-tryptophan, were induced by salt stress. The metabolites were involved in beta-alanine, aldarate, alanine, glutamate, ascorbate, aspartate metabolism, TCA cycle, arginine biosynthesis, glucuronate interconversions	Tang et al. (2020)

(continued)

Table 16.4 (continued)

Plant species	Stress condition, tissue, approach	Key findings	References
Bean (*Phaseolus vulgaris*)	125 mM NaCl; 3 days, leaf and roots, GC-CMS	Primarily, isoleucine, valine, and lysine were strongly induced under salinity stress. Amino acids and carbohydrate metabolisms were also boosted	Niron et al. (2020)

cinnamyl alcohol dehydrogenase (CAD), and cinnamoyl-CoA reductase (CCR) were identified as key lignin biosynthesis metabolites (Pan et al. 2020).

In Maize crop genotypes PH4CV (salt-susceptible) and PH6WC (salt-tolerant) were treated with 100 mM NaCl. The result indicated that a group of metabolites (alkaloids, organic acids, amino acids, and sugars) were induced more than twofold in the control treatment in the PH6WC (Yue et al. 2020). To better understand the metabolic response of cucumber and tomato to salt stress, a metabolomics study was conducted. Under 200 mM NaCl salt stress, the flavonoid levels of cucumber and tomato plants significantly increased. In comparison to the control treatment, cucumber and tomato showed an increase of 2% and 30%. Under salt stress (200 mM), saponin concentration in cucumber was also downregulated, but that of tomatoes was considerably increased (Abdel-Farid et al. 2020). Most plant developmental processes are primarily regulated by the metabolites found in the TCA cycle. The primary intermediate pathway, or TCA cycle, is what connects all other metabolic processes and maintains healthy plant growth. This is supported by a metabolomic investigation that involved stressing tomato plants with salinity. The study found that under salt stress, numerous metabolites, including carbohydrates and amino acids, accumulated in green and mature tomato fruit. Additionally, there was a considerable rise in the level of metabolites associated with the TCA cycle, which may act as a regulator of the tomato's response to NaCl stress (Tang et al. 2020). Overall, metabolomics is a special method that might enable us to comprehend the expanding field of metabolites and their alterations. Furthermore, one possible application for these stress-responsive metabolites is as a diagnostic. A further advantage of metabolomics is that it makes it easier for researchers to rewire the transcription factors (TFs) connected to proteins and metabolites and provides a clear, in-depth picture of how plants respond to stress on various levels.

References

Abdel-Farid IB, Marghany MR, Rowezek MM, Sheded MG (2020) Effect of salinity stress on growth and metabolomic profiling of Cucumis sativus and Solanum lycopersicum. Plan Theory 9(11):1626

Akram S, Siddiqui MN, Hussain BN, Al Bari MA, Mostofa MG, Hossain MA, Tran LSP (2017) Exogenous glutathione modulates salinity tolerance of soybean [*Glycine max* (L.) Merrill] at reproductive stage. J Plant Growth Regul 36:877–888

Asif MA, Garcia M, Tilbrook J, Brien C, Dowling K, Berger B et al (2020) Identification of salt tolerance QTL in a wheat RIL mapping population using destructive and non-destructive phenotyping. Funct Plant Biol 48(2):131–140

Aziz EE, Al-Amier H, Craker LE (2008) Influence of salt stress on growth and essential oil production in peppermint, pennyroyal, and apple mint. Int J Geogr Inf Syst 14:77–87

Billah M, Aktar S, Brestic M, Zivcak M, Khaldun ABM, Uddin MS, Bagum SA, Yang X, Skalicky M, Mehari TG, Maitra S, Hossain A (2021) Progressive genomic approaches to explore drought- and salt-induced oxidative stress responses in plants under changing climate. Plan Theory 10(9):1910

Cao D, Li Y, Liu B, Kong F, Tran LSP (2018) Adaptive mechanisms of soybean grown on salt-affected soils. Land Degrad Dev 29:1054–1064

Chattopadhyay K, Mohanty SK, Vijayan J, Marndi BC, Sarkar A, Molla KA, Chakraborty K, Ray S, Sarkar RK (2021) Genetic dissection of component traits for salinity tolerance at reproductive stage in rice. Plant Mol Biol Report 39(2):386–402

Derakhshani Z, Bhave M, Shah RM (2020) Metabolic contribution to salinity stress response in grains of two barley cultivars with contrasting salt tolerance. Environ Exp Bot 179:104229

Diao F, Dang Z, Cui X, Xu J, Jia B, Ding S, Zhang Z, Guo W (2021) Transcriptomic analysis revealed distinctive modulations of arbuscular mycorrhizal fungi inoculation in halophyte Suaeda salsa under moderate salt conditions. Environ Exp Bot 183:104337

Du C, Li H, Liu C, Fan H (2021) Understanding of the postgerminative development response to salinity and drought stresses in cucumber seeds by integrated proteomics and transcriptomics analysis. J Proteome 232:104062

Duarte-Delgado D, Dadshani S, Schoof H, Oyiga BC, Schneider M, Mathew B, Léon J, Ballvora A (2020) Transcriptome profiling at osmotic and ionic phases of salt stress response in bread wheat uncovers trait-specific candidate genes. BMC Plant Biol 20(1):428

El-Hendawy SE, Hu Y, Schmidhalter U (2005) Growth, ion content, gas exchange, and water relations of wheat genotypes differing in salt tolerances. Austral J Agric Res 56:123–134

El-Metwally S, Hamza T, Zakaria M, Helmy M (2013) Next-generation sequence assembly: four stages of data processing and computational challenges. PLoS Comput Biol 9(12):e1003345

El-Metwally S, Ouda OM, Helmy M (2014) First- and next-generations sequencing methods. In: El-Metwally S, Ouda OM, Helmy M (eds) Next generation sequencing technologies and challenges in sequence assembly. Springer, New York, pp 29–36

FAO (2008) FAOSTAT. FAO, Rome

FAO (2015) FAO land and plant nutrition management service. Food and Agriculture Organization of the United Nations, Rome

Farooq S, Azam F (2005) The use of cell membrane stability (CMS) technique to screen for salt tolerant wheat varieties. J Plant Physiol 163:629–637

Farsiani A, Ghobadi ME (2009) Effects of PEG and NaCl stress on two cultivars of corn (*Zea mays* L.) at germination and early seedling stages. World Acad Sci Eng Technol 57:382–385

Filippou P, Zarza X, Antoniou C, Obata T, Villarroel CA, Ganopoulos I, Harokopos V, Gohari G, Aidinis V, Madesis P, Christou A, Fernie AR, Tiburcio AF, Fotopoulos V (2021) Systems biology reveals key tissue-specific metabolic and transcriptional signatures involved in the response of Medicago truncatula plant genotypes to salt stress. Comput Struct Biotechnol J 19:2133–2147

Flowers TJ, Colmer TD (2015) Plant salt tolerance: adaptations in halophytes. Ann Bot 115:327–331

Fricke W, Akhiyarova G, Wei WX et al (2006) The short-term growth response to salt of the developing barley leaf. J Exp Bot 57:1079–1095

Frukh A, Siddiqi TO, Khan MIR, Ahmad A (2020) Modulation in growth, biochemical attributes and proteome profile of rice cultivars under salt stress. Plant Physiol Biochem 146:55–70

Gilliham M, Able JA, Roy SJ (2017) Translating knowledge about abiotic stress tolerance to breeding programmes. Plant J 90(5):898–917

Goldsworthy (1994) Calcium and salinity. Appl Biol 4:1–6

Grattan SR, Grieve CM (1999) Salinity-mineral nutrient relations in horticultural crops. Sci Hortic 78:127–157

Hamooh BT, Sattar FA, Wellman G, Mousa MAA (2021) Metabolomic and biochemical analysis of two potato (Solanum tuberosum L.) cultivars exposed to in vitro osmotic and salt stresses. Plan Theory 10(1):98

Haque T, Elias SM, Razzaque S, Biswas S, Khan SF, Jewel GMNA, Rahman MS, Juenger TE, Seraj ZI (2020) Natural variation in growth and physiology under salt stress in rice: QTL mapping in a Horkuch × IR29 mapping population at seedling and reproductive stages. 2020.03.01.971895

Hasan A, Hafiz HR, Siddiqui N, Khatun M, Islam R, Mamun AA (2015) Evaluation of wheat genotypes for salt tolerance based on some physiological traits. J Crop Sci Biotechnol 18:333–340

Hussain S (2018) Native RNA-sequencing throws its hat into the transcriptomics ring. Trends Biochem Sci 43(4):225–227

Jahan N, Zhang Y, Lv Y, Song M, Zhao C, Hu H, Cui Y, Wang Z, Yang S, Zhang A, Hu J, Ye G, Qian Q, Gao Z, Guo L (2020) QTL analysis for rice salinity tolerance and fine mapping of a candidate locus qSL7 for shoot length under salt stress. Plant Growth Regul 90(2):307–319

Jha UC, Bohra A, Jha R, Parida SK (2019) Salinity stress response and 'omics' approaches for improving salinity stress tolerance in major grain legumes. Plant Cell Rep 38(3):255–277

Jiang J, Ren X, Li L, Hou R, Sun W, Jiao C, Yang N, Dong Y (2020) H2S regulation of metabolism in cucumber in response to salt-stress through transcriptome and proteome analysis. Front Plant Sci 11:1283

Jusovic M, Velitchkova MY, Misheva SP, Börner A, Apostolova EL, Dobrikova AG (2018) Photosynthetic responses of a wheat mutant (Rht-B1c) with altered DELLA proteins to salt stress. J Plant Growth Regul 37(2):645–656

Kalhoro NA, Rajpar I, Kalhoro SA, Ali A, Raza S, Ahmed M et al (2016) Effect of salts stress on the growth and yield of wheat (*Triticum aestivum* L.). Am J Plant Sci 7:2257

Kashyap SP, Prasanna HC, Kumari N, Mishra P, Singh B (2020) Understanding salt tolerance mechanism using transcriptome profiling and de novo assembly of wild tomato Solanum chilense. Sci Rep 10(1):15835

Kaya C, Ashraf M, Dikilitas M, Tuna AL (2013) Alleviation of salt stress induced adverse effects on maize plants by exogenous application of indoleacetic acid (IAA) and inorganic nutrients-a field trial. Aust J Crop Sci 7:249–254

Khadhri A, Neffati M, Smiti S, Nogueira JMM, Araujo MEM (2011) Influence of salt stress on essential oil yield and composition of lemon grass (*Cymbopogon schoenanthus* L. Spreng. ssp. Laniger (Hook) Maire et Weil). Nat Prod Res 25:108–117

Khaje-Hosseini M, Powell AA, Bingham IJ (2003) The interaction between salinity stress and seed vigour during germination of soybean seeds. Seed Sci Technol 31:715–725

Kosová K, Vítámvás P, Urban MO, Prášil IT, Renaut J (2018) Plant abiotic stress proteomics: the major factors determining alterations in cellular proteome. Front Plant Sci 9:122

Lai Y, Zhang D, Wang J, Wang J, Ren P, Yao L, Si E, Kong Y, Wang H (2020) Integrative transcriptomic and proteomic analyses of molecular mechanism responding to salt stress during seed germination in Hulless Barley. Int J Mol Sci 21(1):359

Lei P, Liu Z, Hu Y, Kim H, Liu S, Liu J, Xu L, Li J, Zhao Y, Yu Z, Qu Y, Huang F, Meng F (2021) Transcriptome analysis of salt stress responsiveness in the seedlings of wild and cultivated Ricinus communis L. J Biotechnol 327:106–116

Li J, Essemine J, Shang C, Zhang H, Zhu X, Yu J, Chen G, Qu M, Sun D (2020) Combined proteomics and metabolism analysis unravels prominent roles of antioxidant system in the prevention of alfalfa (Medicago sativa L.) against salt stress. Int J Mol Sci 21(3):909

Li L, Peng Z, Mao X, Wang J, Li C, Chang X, Jing R (2021) Genetic insights into natural variation underlying salt tolerance in wheat. J Exp Bot 72(4):1135–1150

Liu A, Xiao Z, Li M-W, Wong F-L, Yung W-S, Ku Y-S, Wang Q, Wang X, Xie M, Yim AK-Y, Chan T-F, Lam H-M (2019) Transcriptomic reprogramming in soybean seedlings under salt stress. Plant Cell Environ 42(1):98–114

Luo Q, Zheng Q, Hu P, Liu L, Yang G, Li H, Li B, Li Z (2021) Mapping QTL for agronomic traits under two levels of salt stress in a new constructed RIL wheat population. Theor Appl Genet 134(1):171–189

Ma Q, Shi C, Su C, Liu Y (2020) Complementary analyses of the transcriptome and iTRAQ proteome revealed mechanism of ethylene dependent salt response in bread wheat (Triticum aestivum L.). Food Chem 325:126866

Mahajan M, Sharma S, Kumar P, Pal PK (2020) Foliar application of KNO_3 modulates the biomass yield, nutrient uptake and accumulation of secondary metabolites of *Stevia rebaudiana* under saline conditions. Ind Crop Prod 145:112102

Muthu V, Abbai R, Nallathambi J, Rahman H, Ramasamy S, Kambale R, Thulasinathan T, Ayyenar B, Muthurajan R (2020) Pyramiding QTLs controlling tolerance against drought, salinity, and submergence in rice through marker assisted breeding. PLoS ONE 15(1):e0227421

Mwando E, Angessa TT, Han Y, Zhou G, Li C (2021) Quantitative trait loci mapping for vigour and survival traits of barley seedlings after germinating under salinity stress. Agronomy 11(1):103

Niron H, Barlas N, Salih B, Türet M (2020) Comparative transcriptome, metabolome, and ionome analysis of two contrasting common bean genotypes in saline conditions. Front Plant Sci 11: 599501

Pan J, Li Z, Dai S, Ding H, Wang Q, Li X, Ding G, Wang P, Guan Y, Liu W (2020) Integrative analyses of transcriptomics and metabolomics upon seed germination of foxtail millet in response to salinity. Sci Rep 10(1):13660

Panda A, Rangani J, Parida AK (2021) Unraveling salt responsive metabolites and metabolic pathways using non-targeted metabolomics approach and elucidation of salt tolerance mechanisms in the xero-halophyte Haloxylon salicornicum. Plant Physiol Biochem 158:284–296

Pitann B, Kranz T, Mühling KH (2009) The apoplastic pH and its significance in adaptation to salinity in corn (*Zea mays* L.): comparison of fluorescence microscopy and pH-sensitive microelectrodes. Plant Sci 176:497–504

Pundir P, Devi A, Krishnamurthy SL, Sharma PC, Vinaykumar NM (2021) QTLs in salt rice variety CSR10 reveals salinity tolerance at reproductive stage. Acta Physiol Plant 43(2):35

Rasouli F, Kiani-Pouya A, Li L, Zhang H, Chen Z, Hedrich R, Wilson R, Shabala S (2020) Sugar beet (Beta vulgaris) guard cells responses to salinity stress: a proteomic analysis. Int J Mol Sci 21(7):2331

Raza A (2022) Metabolomics: a systems biology approach for enhancing heat stress tolerance in plants. Plant Cell Rep 41(3):741–763

Razzaq A, Sadia B, Raza A, Khalid Hameed M, Saleem F (2019) Metabolomics: a way forward for crop improvement. Metabolites 9(12):303

Rich-Griffin C, Stechemesser A, Finch J, Lucas E, Ott S, Schäfer P (2020) Single-cell transcriptomics: a high-resolution avenue for plant functional genomics. Trends Plant Sci 25(2):186–197

Said-Al Ahl HAH, Mahmoud AA (2010) Effect of zinc and/or iron foliar application on growth and essential oil of sweet basil (*Ocimum basilicum* L.) under salt stress. Ozean J Appl Sci 3(1):56–65

Satir O, Berberoglu S (2016) Crop yield prediction under soil salinity using satellite derived vegetation indices. Field Crop Res 192:134–143

Shalan MN, Abdel-Latif TAT, Ghadban EAE (2006) Effect of water salinity and some nutritional compounds on the growth and production of sweet marjoram plants (*Majorana hortensis* L.). Egypt J Agric Res 84(3):959

Shereen A, Mumtaz S, Raza S, Khan MA, Solangi S (2005) Salinity effects on seedling growth and yield components of different inbred rice lines. Pak J Bot 37:131–139

Singh M, Nara U, Kumar A, Choudhary A, Singh H, Thapa S (2021a) Salinity tolerance mechanisms and their breeding implications. J Genet Eng Biotechnol 19(1):173

Singh RK, Kota S, Flowers TJ (2021b) Salt tolerance in rice: seedling and reproductive stage QTL mapping come of age. Theor Appl Genet 134(11):3495–3533

Soren KR, Madugula P, Kumar N, Barmukh R, Sengar MS, Bharadwaj C, Sharma PC, Singh S, Bhandari A, Singh J, Sanwal SK, Pal M, Mann A, Sagurthi SR, Ps S, Siddique KHM, Singh NP, Roorkiwal M, Varshney RK (2020) Genetic dissection and identification of candidate genes for salinity tolerance using Axiom®CicerSNP array in chickpea. Int J Mol Sci 21(14):5058

Tang H, Zhang X, Gong B, Yan Y, Shi Q (2020) Proteomics and metabolomics analysis of tomato fruit at different maturity stages and under salt treatment. Food Chem 311:126009

Tavakkoli E, Fatehi F, Coventry S, Rengasamy P, McDonald GK (2011) Additive effects of Na^+ and Cl^- ions on barley growth under salinity stress. J Exp Bot 62(6):2189–2203

Varshney RK, Sinha P, Singh VK, Kumar A, Zhang Q, Bennetzen JL (2020) 5Gs for crop genetic improvement. Curr Opin Plant Biol 56:190–196

Varshney RK, Bohra A, Yu J, Graner A, Zhang Q, Sorrells ME (2021) Designing future crops: genomics-assisted breeding comes of age. Trends Plant Sci 26(6):631–649

Wang R, Wang X, Liu K, Zhang X-J, Zhang L-Y, Fan S-J (2020) Comparative transcriptome analysis of Halophyte Zoysia macrostachya in response to salinity stress. Plants 9(4):458

Xu Z, Chen X, Lu X, Zhao B, Yang Y, Liu J (2021) Integrative analysis of transcriptome and metabolome reveal mechanism of tolerance to salt stress in oat (*Avena sativa* L.). Plant Physiol Biochem 160:315–328. https://doi.org/10.1016/j.plaphy.2021.01.027

Yadav AK, Kumar A, Grover N, Ellur RK, Bollinedi H, Krishnan SG, Bhowmick PK, Vinod KK, Nagarajan M, Singh AK (2021) Genome-wide association study reveals marker–trait associations for early vegetative stage salinity tolerance in rice. Plants 10(3):559

Yousefirad S, Soltanloo H, Ramezanpour SS, Nezhad KZ, Shariati V (2020) The RNA-seq transcriptomic analysis reveals genes mediating salt tolerance through rapid triggering of ion transporters in a mutant barley. PLoS ONE 15(3):e0229513

Yue JY, Wang LH, Dou XT, Wang YJ, Wang HZ (2020) Comparative metabolomic profiling in the roots of salt-tolerant and salt-intolerant maize cultivars treated with NaCl stress. Biol Plant 64(1):569–577

Zhang R, Wang Y, Hussain S, Yang S, Li R, Liu S, Chen Y, Wei H, Dai Q, Hou H (2022) Study on the effect of salt stress on yield and grain quality among different rice varieties. Front Plant Sci 13:918460

Zhu J, Fan Y, Shabala S, Li C, Lv C, Guo B, Xu R, Zhou M (2020) Understanding mechanisms of salinity tolerance in barley by proteomic and biochemical analysis of near-isogenic lines. Int J Mol Sci 21(4):1516

Zhu D, Luo F, Zou R, Liu J, Yan Y (2021) Integrated physiological and chloroplast proteome analysis of wheat seedling leaves under salt and osmotic stresses. J Proteome 234:104097

Current Status and Future Perspectives of Epigenetic Gene Regulation for Salt Tolerance in Wheat

17

Avni Dahiya, Ashwani Kumar, Arvind Kumar, Satish Kumar Sanwal, Anita Mann, and Neeraj Kulshreshtha

Abstract

Wheat (*Triticum aestivum*), one of the major cereal crops being grown around the world, supports the dietary needs of a significant chunk of the global population. Crops like wheat grow in dynamic field environments where they are continuously exposed to the vagaries of nature, greatly distorting their growth, yield and reproductive success. Among the various abiotic stresses, salt stress has emerged as a major problem impacting both the crop yield and quality. Sensing the environmental cues and initiating a timely response is, therefore, critical for plants to ensure their survival. Plants have evolved a number of sophisticated adaptations and defence mechanisms to counter abiotic stresses, of which epigenetics has gained a lot of prominence in the recent times. Epigenetic regulation of gene expression is primarily achieved through mechanisms like histone modifications, DNA methylation and action of non-coding small RNAs and does not involve any direct alteration of the DNA base sequence. Epigenetic modifications have also been implicated in the development of stress memory in plants which prepares the pre-exposed plants to combat similar stress during future exposures. This review presents the current status of epigenetic mediated gene regulation in wheat and other crops under salt stress.

Keywords

Wheat · Salt stress · Plant tolerance · Epigenetics · DNA methylation · Histone modification · Small RNAs · Stress memory

A. Dahiya (✉) · A. Kumar · A. Kumar · S. K. Sanwal · A. Mann · N. Kulshreshtha
ICAR-Central Soil Salinity Research Institute, Karnal, Haryana, India

A. Kumar et al. (eds.), *Salinity and Drought Tolerance in Plants*,
https://doi.org/10.1007/978-981-99-4669-3_17

17.1 Introduction

Wheat and other field crops have to deal with environmental extremities such as temperature variations, salinity, nutrient and water stress along with bio stresses on a constant basis. Such stresses are the major causes behind significant annual crop losses and yield drops observed around the world. The concern over dwindling yield grows manifold for a crop like wheat as it supplies a fifth of the protein and food calories needed by the world population (Erenstein et al. 2022). Out of all the environmental challenges, most of the cereals are quiet sensitive to salinity. Salinity largely impacts the water and ion transport along with nutrient homeostasis thereby causing osmotic distress and nutrient imbalance in plants (Rashid et al. 2022).

Over the course of evolution, plants have developed plethora of strategies encompassing structural adaptations along with physiological and molecular mechanisms to tide over the environmental conditions unfavourable for their survival. Recent advancements in genome research technologies like deep sequencing analysis have improved our understanding about these stress responsive processes. There has been explosion of studies and findings that are reshaping our views by introducing newer and novel mechanisms to the already existing arsenal of stress responsive processes in plants.

Since the beginning of this millennium, epigenetics as a field of study has generated a lot of interest in the scientific community worldwide. The science of epigenetics in simple terms is defined as the mechanism of regulating gene expression by modifying DNA and its associated chromatin proteins without altering the DNA base sequences (Grant-Downton and Dickinson 2005).

Molecular mechanisms underpinning epigenetic events primarily includes Histone modifications, DNA methylation, production and action of a number of small RNAs together with ATP-Dependent Chromatin Remodelling (Grativol et al. 2012). The former three mechanisms are often referred to as "the three pillars of epigenetics" (Fig. 17.1) (Grant-Downton and Dickinson 2005). Epigenetic changes

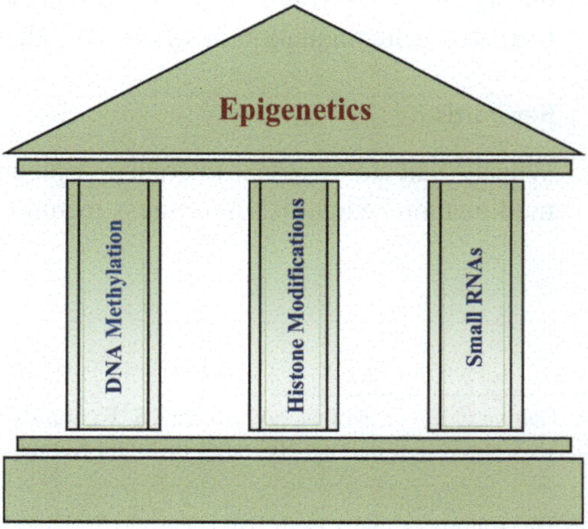

Fig. 17.1 The key players involved in epigenetic regulation include DNA Methylation, Histone Modifications and Small noncoding RNA based mechanisms, often referred to as the "three pillars of epigenetics". These players (in addition to some other molecular mechanisms) interact among themselves and coordinate the intricate process of epigenetic regulation of gene expression in an organism

may activate or repress the target genes and more often than not are reversible in nature. Environmental signals like salt stress have been shown to induce epigenetic mechanisms like DNA methylation and histone modifications ultimately leading to plant adaptation (Chang et al. 2020). The plants' ability to modify epigenetic status reversibly and rapidly imparts a flexibility in their response to the environmental stresses. Stress-induced epigenetic modifications may also get inherited as "stress memory," thereby improving the stress tolerance of plants in the subsequent generations (Sharma et al. 2022). The significance of epigenetics in stress responses also lays in the fact that, in nature the rate of formation of new gene combinations or alleles is too slow when compared to the occurrence of different environmental stresses. Therefore, for its survival the plant would largely depend upon the regulation of different stress responsive genes through epigenetic mechanisms (Peng and Zhang 2009). Understanding various components and mechanisms of epigenetic regulations during salt stress are essential to deepen our understanding of development of stress tolerance in plants. This will enable us to deal with the salinity and other abiotic stresses in a more effective manner. Such an understanding might further minimize the need for excessive genetic modification of crop plants which would be a revolutionary step forward for the breeding programmes worldwide (Yaish 2013). Though a lot of work on similar lines has already been done but the bulk of it is still concentrated with the model plants like *A. thaliana*, owing to their small genome size, short life cycle, etc. The complex and enormous genome of bread wheat (BBAADD genome) poses a vital obstruction in understanding the molecular mechanism underlying epigenetic stress responses. With the advent of next-generation sequencing and other sophisticated molecular approaches things are slowly but steadily looking up for major food crops including wheat. Here, we have discussed the different epigenetic pathways/variations employed by crops like wheat and other plants to regulate the salt stress responsiveness, which are often perceived as novel resources for developing tolerant crops.

17.2 DNA Methylation

DNA methylation, involving covalent addition of a methyl group ($-CH_3$) at the fifth carbon in the cytosine ring of the DNA molecule, is an important epigenetic mark associated with regulation and maintenance of gene expression patterns in plants (Fig. 17.2). Enzyme DNA methyltransferase, in the presence of *S*-adenosyl-L-methionine (SAM), is primarily involved in the de novo methylation and its maintenance at both the symmetric (CpG and CpNpG) and asymmetric cytosine (CpNpN) sites, where "N" stands for A, C, or T (Rashid et al. 2022). When present in the promoter region, DNA methylation typically brings about gene silencing by altering the chromatin structure and limiting the access to transcriptional activators or by facilitating the binding of transcription repressors (Zhang et al. 2018). An alternate RNA-directed DNA methylation pathway (RdDM pathway) is also known to maintain the plant's transposable elements and other repetitive DNA sequences in a repressed state, desirable for maintaining the genome stability (Matzke and Mosher

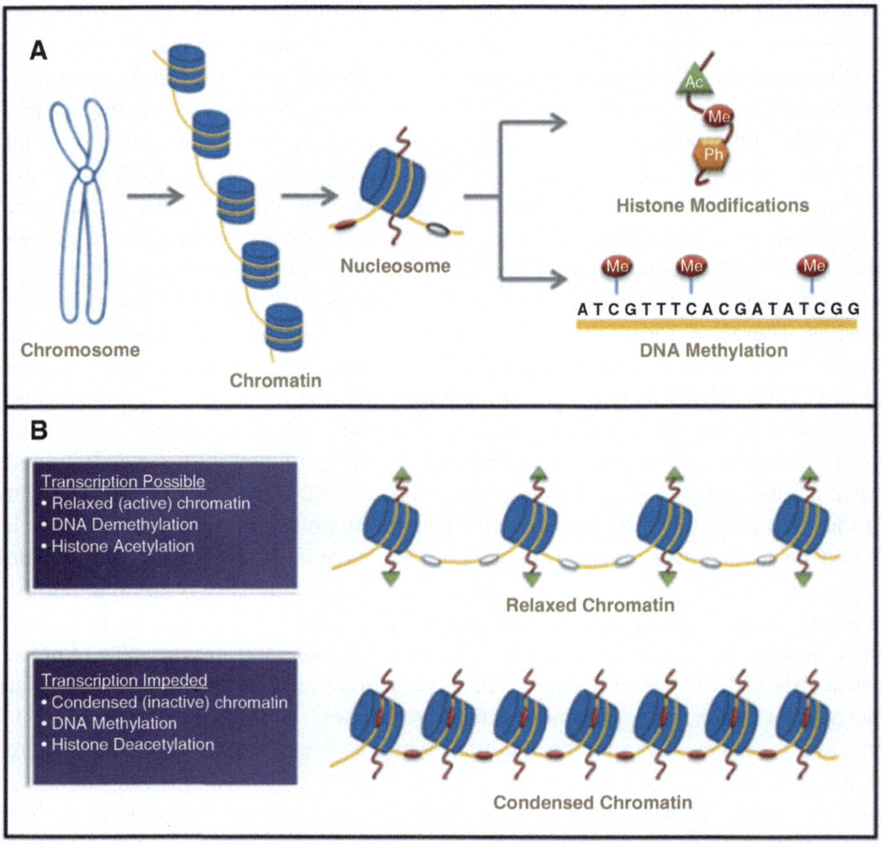

Fig. 17.2 Epigenetic regulation of gene expression. (**a**) Variations in the DNA methylation patterns and Histone modifications are involved in chromatin remodelling and transcriptional regulation. While, DNA methylation involves the covalent modification of cytosine residues in CpG dinucleotides within the gene sequences, the N-terminal tails of histones undergo a number of post-translational covalent modifications, including methylation, acetylation, phosphorylation. (**b**) DNA methylation is often associated with transcriptional silencing. However, depending upon which residue gets modified and the type of modification, histone modifications can lead to either activation or repression of gene transcription

2014). Such non-CG methylation pathway is proposed to have evolved as an epigenetic tag committed to transposon control besides mediating the plants responses to various biotic/abiotic stresses. Genomic methylation patterns also get influenced by the process of demethylation which is classified as passive or active. Passive demethylation is the loss of DNA methylation marks in the absence of active maintenance machinery. While in case of later, certain enzymes, like a family of glycosylases in plants, actively remove 5-methylcytosine from DNA (Zhang and Zhu 2012).

Abiotic stresses like salinity, drought, cold or heat induce changes in the gene expression levels in plants by modulating the cytosine methylation and demethylation either in a locus specific or pan-genomic manner (Miryeganeh 2021). A number of studies have outlined the crosstalk between chromatin methylation patterns and gene expression levels under salinity stress in wheat and other crops. For instance, Zhong et al. (2009) performed a study to assess the impact of salt stress on the DNA methylation patterns in two wheat cultivars exhibiting differential salt tolerance using methylation-sensitive amplified polymorphism (MSAP) approach. A significant, genome-wide hypomethylation at the CCGG sites was observed for both Dekang-961 (salt tolerant) and Lumai-15 (salt sensitive) cultivars. On a comparative scale though, the CCGG sequences of Dekang-961 control plants were found to be hypermethylated than that of Lumai-15. Salinity stress was also reported to elevate the 5mC levels at CHG and CHH sites in the shoots of a salt sensitive wheat cultivar, while opposite was true for the tolerant wheat genotype SR3 (Wang et al. 2014). Likewise, genome-wide bisulphite and RNA sequencing performed in three rice cultivars, namely, IR-64 (drought stress-sensitive), Nagina22 (droughttolerant) and Pokkali (salt tolerant), showed differential DNA methylation associated with varying expression of abiotic stress responsive genes in a genotype dependent manner (Garg et al. 2015). Epigenetic DNA methylation regulates the plants' response to salinity in a genotype and tissue specific manner which was further validated in rice and wheat where varying levels of DNA methylation was recorded in their shoot and root systems (Chang et al. 2020). Kumar et al. (2017) attributed the better salt tolerance ability of a wheat cultivar Kharchia-65 to elevated cytosine methylation levels and consequent repression of high-affinity potassium transporters TaHKT2,1 and TaHKT2,3 in the roots and shoot system. In a similar manner, methylation of salt responsive genes like flavonol synthase gene TaWRS15 and TaFLS1 has also been shown imparting salt tolerance in barley and wheat (Kong et al. 2020). Besides DNA methylation, DNA demethylation also contributes significantly in regulating the plant stress responses. In a 2015 study performed on transgenic tobacco, Bharti et al. provided evidence of salt stress-induced expression of AtROS1 demethylase and other genes associated with antioxidant and flavonoid biosynthesis pathways. While a number of DNA demethylase (DNA-dMTase) genes have already been identified and functionally validated in various crops, it's only recently that a group of 12 DNA-dMTase encoding genes belonging to subfamilies; DEMETER-Like (DML) and repressor of silencing 1 (ROS1) have been identified in wheat (Gahlaut et al. 2022). It is quite evident from the discussion that alteration in the DNA methylation pattern is one of the key epigenetic mechanisms regulating plants response to biotic or abiotic stresses.

17.3 Histone Modifications

Post-translational covalent modifications (PTMs) at N-terminal regions of nucleosomal histone proteins are other key mechanisms of regulating the expression levels of various stress responsive genes. Histones are vital for the packaging of DNA molecule. DNA molecule folds around the histone octamer (H2A, H2B, H3 and H4) to form a structure known as nucleosome, which is the basic unit of chromatin. Some of the common N-terminal histone tail modifications include acetylation, methylation, phosphorylation, ubiquitination and sumoylation. These modifications interfere with the integrity of the nucleosome structure and influence the chromatin organization, thus determining the transcriptional state and expression level of the associated genes (Fig. 17.2) (Berger 2007). In general, modifications such as acetylation, phosphorylation and ubiquitination are responsible for enhancing the gene expression levels, while the others like biotinylation and sumoylation have been reported to repress the gene transcription. Histone methylation, however, can lead to either activation or repression of gene transcription, depending upon the histone residues being modified. For instance, trimethylation of H3K4 (lysine 4 on the histone H3 protein subunit) seems to activate gene transcription, while dimethylation of H3K9 (lysine 9 on the histone H3 protein subunit) and H3K27 (lysine 27 on the histone H3 protein subunit) causes its repression (Yaish 2013). A combination of these site-specific post-translational modifications on different residues of histone tail constitutes a "histone code". These codes signify a particular chromatin state and regulate transcriptional activity in combination with different external and internal signals. Numerous research studies have documented the role of histone modifications brought about by histone modifying enzymes in plants responses to environmental perturbations. The histone acetyltransferases (HATs) and histone deacetylases (HDACs) have been shown regulating plant growth and gene expression by facilitating histone acetylation and deacetylation, respectively. A histone acetyltransferase (HAT), namely GCN5 (GENERAL CONTROL NONDEREPRESSIBLE 5) is a versatile epigenetic regulator and was found to improve the salt stress tolerance in *A. thaliana* (Zheng et al. 2019). In wheat, TaGNC5, an ortholog of *A. thaliana* histone acetyltransferases AtGCN5 has been attributed with regulation of responses to high temperature and salinity. This TaGCN5 was also found targeting genes overseeing the production of hydrogen peroxide, crucial for salinity stress adaptation in wheat (Zheng et al. 2019). Similarly, histone deacetylases (HDACs) such as HDA6 and HDA19 were shown to positively influence salt tolerance in Arabidopsis (Chen and Wu 2010). The overexpression of one of the rice histone deacetylases HDT701 improved the chlorophyll content and the survival rate of the seedlings under salt stress (Zhao et al. 2015). On identical lines, TaHDA4 was reported to interact with TaHOS15 protein and influence the acetylation of various defence related genes via histone deacetylation in wheat (Liu et al. 2019). Further efforts were done to gain insights into the working of these histone modifying enzymes by using chemical inhibitors. The addition of a HDAC inhibitor, Ky-2, enhanced the H4ac (acetylated H4) levels in *A. thaliana*, which was accompanied by higher SOS1 and SOS3 expression, lower accumulation of Na^+ and thus better salt tolerance (Sako et al. 2016).

17.4 Chromatin Remodelling

Multi-subunit protein complexes, referred to as chromatin remodelling factors (CRMs), have an ATPase activity and can alter the chromatin structure by modifying the DNA-histone interactions. CRMs basically regulate the assembly and displacement of nucleosomes using the energy derived from hydrolysis of ATP. These complexes can be grouped into three main classes, namely the SWI/SNF ATPases, the imitation switch (ISWI) ATPases and the chromodomain and helicase-like domain (CHD) ATPases. Out of these, SWItch/SUCROSE NON-FERMENTABLE (SWI/SNF) complex was the first one to be identified in yeast (Alvarez et al. 2010) and has been extensively studied ever since.

In plant genomes, the SWI2/SNF2 subgroup chromatin remodelling ATPases are further characterized into three types, i.e., BRAHMA (BRM), SPLAYED (SYD) and MINUSCULE (MINU) (Sang et al. 2012). These SWI2/SNF2 ATPases uses energy derived from ATP hydrolysis to alter DNA-histone interactions, thereby participating in the activation or repression of various gene expression including those responsive to abiotic stress. Antagonistic relation between SWI3B, an *Arabidopsis* homolog of the yeast SWI3 subunit of SWI/SNF chromatin remodelling complex, with the components of ABA-mediated signalling during an abiotic stress response was reported by Saez et al. (2008). The role of SWI3B as a potent negative regulator of ABA-mediated stress tolerance was further confirmed in the *swi3b* mutants of *Arabidopsis* which exhibited reduced sensitivity towards ABA-mediated inhibition of seed germination and growth along with down regulation of ABA-responsive genes like *Rab18* and *Rd29B* (Yuan et al. 2013). Further, a group of chromatin re-modellers like CLSY1 (CLASSY1), DDM1 (DEFECTIVE IN RNA-DIRECTED DNA METHYLATION1) BRM (BRAHMA), SPS (SPLAYED), all belonging to SWI2/SNF2 subfamily, were found to have a specific role in *Arabidopsis* RdDM pathway (Sang et al. 2012). Some of these re-modellers have also been found to show salt stress responsive expressions. For example, the *brm* loss of function mutant of *Arabidopsis thaliana* was more responsive to ABA levels and exhibited improved tolerance to desiccation stress (Han et al. 2012). In addition to this, a *brm-3 A. thaliana* mutant showed heightened expression of ABI5 (Abscisic Acid-INSENSITIVE 5) on sensing ABA, leading to arrested growth and better tolerance to water stress (Peirats-Llobet et al. 2015).

17.5 Small RNAs

Micro-RNA (miRNA) and small interfering RNA (siRNA) are two important classes of small RNAs (sRNAs) involved in post-transcriptional and translational regulation of gene expression. These sRNAs are usually made of small nucleotide sequences (20–30 nt long) and often bring about silencing of the genes either by repressing the transcriptional process or by targeting mRNAs to degradation (Duempelmann et al. 2020). They do so by binding to RNA induced silencing complexes (RISC) and regulating the gene expression through either chromatin remodelling, DNA methylation or mRNA degradation in a site-specific manner. Such a suppressions of target mRNA is often referred to as RNA interference and the process is now being exploited to establish the involvement of small RNAs abiotic stress responses in several crop plants (Dutta et al. 2020).

Over the course of past decade or so, the importance and involvement of sRNAs (especially miRNAs) in regulating gene expression as a response to range of abiotic stresses like salt stress have been increasingly documented. In 2008, Liu et al. while working on the model plant *A. thaliana* reported 13 miRNA transcripts (miR156, miR158, miR159, miR165, miR167, miR168, miR169, miR171, miR319, miR393, miR394, miR396 and miR397) being upregulated and one miRNA transcript (miR398) being downregulated under salinity stress. Furthermore, attempts were made to understand the interplay of abiotic stresses and regulatory miRNAs in wheat genotype C-306. Results obtained from the expression profiling revealed that out of the selected panel of conserved miRNAs (i.e., miR159, miR164, miR168, miR172, miR393, miR397, miR529 and miR1029), miR172 was upregulated and two others (miR168, miR397) were downregulated under all types of stress. A novel salt stress-induced miRNA; Ta-miR855, was also identified via a computational protocol indicating the presence of diverse set of miRNAs responsive to salt, cold and osmotic stress in wheat (Gupta et al. 2014).

A high-throughput small RNA and degradome sequencing depicted a total of 98 known and 219 novel miRNA transcripts to be associated with the superior adaptability of wheat cultivar SR3 to salt stress. Virus-Induced Gene Silencing (VIGS) was used to explore the function of these salt stress responsive miRNAs and they were found to be involved in number of stress responsive pathways like JA and auxin signalling, carbohydrate metabolism and epigenetic modifications (Han et al. 2018). Most recently, an Illumina high-throughput sequencing and in-silico study was undertaken to gain insight into root specific salinity stress response between two contrasting wheat cultivars Suntop (Salt Tolerant) and Sunmate (Salt Sensitive). A total of 191 miRNAs belonging to 23 conserved and 12 unique families were identified in both the cultivars. Also, saline conditions induced 43 and 75 miRNAs in Suntop and Sunmate cultivars, respectively (Zeeshan et al. 2021).

A lot of studies done with model plant *A. thaliana* have demonstrated that miRNAs also participate in plant salt stress responses by targeting the hormone signalling pathways. Consequent to an exposure to salt stress, miR156, miR172, miR393, miR394 and miR399 were found regulating the ABA metabolic pathway while miR319 is involved in ethylene metabolic pathway (Liu et al. 2019). Here, it is pertinent to note that the type and expression patterns of miRNAs responding to salt stress varies from one plant to other. In general, miR156, miR159, miR160, miR164, miR166, miR167, miR168, miR169, miR172, miR319, miR394, miR395, miR396, miR397, miR399 and miR408 have been reported to be associated with salt stress response in most plants (Kumar et al. 2018). On similar lines, several siRNAs have also been linked with the stress responsive epigenetic pathways. From being first identified in Arabidopsis mutants and linked with RdDM pathway, a critical pathway for gene silencing or RNAi, these sRNAs have been shown altering the methylation patterns of number of sites in plant genome (Xie and Yu 2015). Continuous and rigorous efforts to discover new regulatory sRNAs and uncover their putative role in imparting salt tolerance can be instrumental for achieving genetically improved and stress responsive wheat cultivars.

17.6 Epigenetic Stress Memory, an Important Feature of Epigenetic Inheritance

Abiotic stresses are generally recurring in nature and may even increase in their severity over time. In such situations, it is desirable on the plant's part to be able to remember past incidents and use this stored knowledge to adapt to new challenges. Plants acquire stress memory on being exposed to prior mild stressful conditions, thereby enabling them to adapt better during subsequent encounters, a process often termed as priming or hardening (Fig. 17.3) (Crisp et al. 2016). Several studies have reported development of stress memory in response to salinity, cold, drought, ABA (abscisic acid), oxidative stress in plants (Ding et al. 2012). In some cases, this priming effect gets transferred to the next generations, inducing adaptive transgenerational plasticity and conferring acclimation and adaptive benefits to the future generations as well (Sharma et al. 2022).

A number of priming strategies have been identified among plant species ranging from accumulation of intermediate compounds in the cellular compartments to

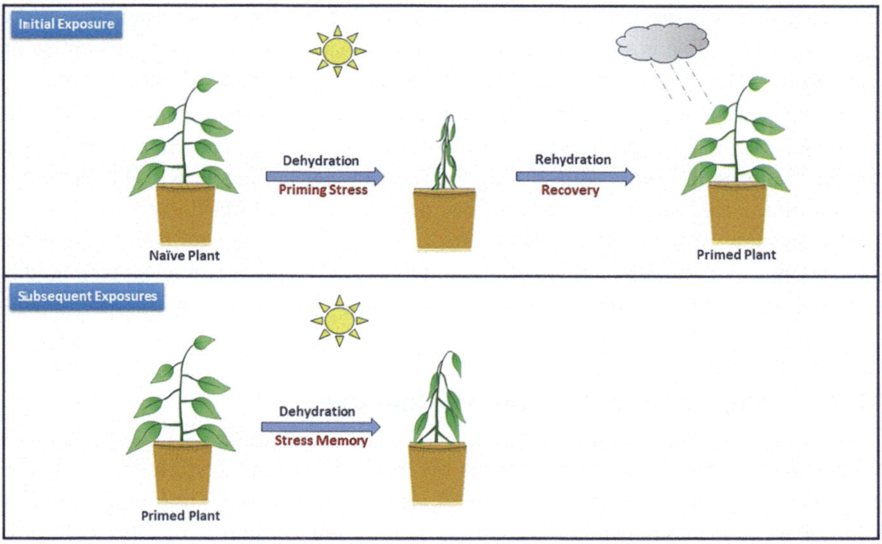

Fig. 17.3 Plants have elaborate mechanisms for stress memory. A naïve plant on experiencing a mild stress, say dehydration, for the first time undergoes wilting and recovers after rehydration. During this process it acquires the memory of that stress and gets primed. Upon subsequent exposure to stress conditions this primed plant, owing to its acquired stress memory, responds quickly and strongly which greatly improves its chances of survival

modification of key regulatory proteins to those involving epigenetic regulations (Hilker et al. 2016). There is growing evidence in support of different epigenetic mechanisms such as modification of chromatin structure, DNA methylation, chromatin remodelling, histone modifications and miRNAs, taking part in stress memory development in plants and providing means to adapt rapidly to changing local conditions (Hilker et al. 2016). Ding et al. (2012) reported a progressive accumulation of RAB18 and RD29B transcripts together with elevated H3K4me3 and Ser5P levels in Arabidopsis RNA polymerase II after multiple exposures to drought stress. Although the transcription levels of the mentioned genes fell back to their basal levels with the removal of stress, the pattern of H3K4me3 and Ser5P modifications in RNA polymerase II remained unchanged, which might be attributing to the stress memory. Such a memory has often been found to be cell specific, mitotically stable with extent of epigenetic repression being directly proportional to the duration of initial exposure to stress conditions (Crisp et al. 2016).

In fact, stress priming is now being hailed as a promising approach for improving plant stress tolerance. Scientists are evoking the epigenetic stress memory by exposing the crop plants to either stress or an elicitor treatment (Turgut-Kara et al. 2020). Enhanced salt tolerance following priming treatments is also being studied and tested by several groups and has become one of the key focus points for agricultural research and crop improvement strategies. In an *Arabidopsis* study, neighbouring plants were shown to acquire salt tolerance after receiving air-borne signals from the salt stressed plants (Lee and Seo 2014). Van Oosten et al. (2018) observed that inoculation of tomato plant roots with *Azotobacter chroococcum* 76A enhanced their tolerance to salt stress as well as their nitrogen use efficiency. With regard to wheat, Fan et al. (2018) assessed the impact of heat priming during stem elongation. The heat primed wheat plants exhibited elevated sucrose levels, improved activity of antioxidant enzymes and photosynthetic machinery and had improved tolerance to post-anthesis heat stress. Certain other research works have shown plants with compromised epigenetic machinery, say the RdDM pathway, having diminished ability for priming and transgenerational plasticity (Boyko et al. 2010). In view of these observations, stress priming followed by development of memory brought about by a targeted manipulation of the epigenome can be a potent approach for improving stress tolerance in the crop plants.

17.7 Conclusion and Future Perspectives

Wheat holds a prominent position among the food crops of the world both in terms of its grain acreage and production volume. As it is a staple food for millions, a need to improve its yield and nutritional value is constantly felt by the researchers and policy makers alike. However, the rapidly degrading environmental conditions especially the increasing problem of salinity in the cultivated lands round the world has put roadblocks in the path to achieve this goal. It is therefore crucial to speed up the breeding programmes and develop new climate and abiotic stress resilient varieties.

In the recent times, epigenetics is being seen as an important research field for crop improvement. Epigenetic control over plant responses to environmental stresses is a complex process involving not only the regulation of plant genes, but also induction of stress memories in the exposed plants and at times over couple of generations. Till now, our understanding of the various ways these stress responses get regulated by epigenetic processes is quite limited. A comprehensive understanding of these transcriptional modulations is thus needed to modify epigenetic cascades and improve agricultural productivity by improving stress responsiveness of the crop plants. It may also be added that being a self-built component of the plant, targeting major players in such epigenetic pathways should be more feasible method of crop improvement rather than introducing foreign genes to develop transgenics. Recent progress in developing powerful and versatile tools for understanding and studying the epigenetic phenomenon like DNA methylation/demethylation, histone modifications, small RNAs has lifted the prospects of harnessing these modulations and stress memory for crop management and improvement to a great extent. One way would be using the knowledge of the epigenetic adaptations to select targets for modifications by genome editing techniques like CRISPR-CAS. Though the potential is huge, most of the information available comes from the work done on the model plants. More efforts are needed to devise strategies for tapping this knowledge and transferring to crop plants before these epigenetic processes, epialleles and their inheritance can be exploited for our benefit in a directed and more convincing way.

References

Alvarez ME, Nota F, Cambiagno DA (2010) Epigenetic control of plant immunity. Mol Plant Pathol 11(4):563–576

Berger SL (2007) The complex language of chromatin regulation during transcription. Nature 447(7143):407–412

Bharti P, Mahajan M, Vishwakarma AK, Bhardwaj J, Yadav SK (2015) AtROS1 overexpression provides evidence for epigenetic regulation of genes encoding enzymes of flavonoid biosynthesis and antioxidant pathways during salt stress in transgenic tobacco. J Exp Bot 66:5959–5969

Boyko A, Blevins T, Yao Y et al (2010) Transgenerational adaptation of Arabidopsis to stress requires DNA methylation and the function of dicer-like proteins. PLoS One 5:e9514

Chang YN, Zhu C, Jiang J, Zhang H, Zhu JK, Duan CG (2020) Epigenetic regulation in plant abiotic stress responses. J Integr Plant Biol 62:563–580

Chen LT, Wu K (2010) Role of histone deacetylases HDA6 and HDA19 in ABA and abiotic stress response. Plant Signal Behav 5(10):1318–1320

Crisp P, Ganguly D, Eichten SR, Borevitz J, Pogson B (2016) Reconsidering plant memory: intersection between stress recovery, RNA turnover, and epigenetics. Sci Adv 2:e1501340

Ding Y, Fromm M, Avramova Z (2012) Multiple exposure to drought 'train' transcriptional responses in *Arabidopsis*. Nat Commun 3:740

Duempelmann L, Skribbe M, Bühler M (2020) Small RNAs in the transgenerational inheritance of epigenetic information. Trends Genet 36(3):203–214

Dutta T, Neelapua NRR, Wanib SH, Surekha C (2020) Salt stress tolerance and small RNA. In: Plant small RNA. Academic Press, London. https://doi.org/10.1016/B978-0-12-817112-7.00010-9

Erenstein O, Jaleta M, Mottaleb KA, Sonder K, Donovan J, Braun HJ (2022) Global trends in wheat production, consumption and trade. Springer, Cham. https://doi.org/10.1007/978-3-030-90673-3

Fan Y, Ma C, Huang Z et al (2018) Heat priming during early reproductive stages enhances thermo-tolerance to post-anthesis heat stress via improving photosynthesis and plant productivity in winter wheat (*Triticum aestivum* L.). Front Plant Sci 9:805

Gahlaut V, Samtani H, Gautam T, Khurana P (2022) Identification and characterization of DNA demethylase genes and their association with thermal stress in wheat (*Triticum aestivum* L.). Front Genet 13:1–14

Garg R, Chevala VVSN, Shankar R, Jain M (2015) Divergent DNA methylation patterns associated with gene expression in rice cultivars with contrasting drought and salinity stress response. Sci Rep 5:14922

Grant-Downton RT, Dickinson HG (2005) Epigenetics and its implications for plant biology. The epigenetic network in plants. Ann Bot 96:1143–1164

Grativol C, Hemerly AS, Ferreira PC (2012) Genetic and epigenetic regulation of stress responses in natural plant populations. Biochim Biophys Acta 1819:176–185

Gupta OP, Meena NL, Sharma I, Sharma P (2014) Differential regulation of microRNAs in response to osmotic, salt and cold stresses in wheat. Mol Biol Rep 41:4623–4629

Han SK, Sang Y, Rodrigues A, Wu MF, Rodriguez PL, Wagner D, Biol F (2012) The SWI2/SNF2 chromatin remodeling ATPase BRAHMA represses abscisic acid responses in the absence of the stress stimulus in Arabidopsis. Plant Cell 24:4892–4906

Han H, Wang QL, Wei Y, Liang JD, Xia G, Liu S (2018) Small RNA and degradome sequencing used to elucidate the basis of tolerance to salinity and alkalinity in wheat. BMC Plant Biol 18(1): 195

Hilker M, Schwachtje J, Baier M et al (2016) Priming and memory of stress responses in organisms lacking a nervous system. Biol Rev 91:1118–1133

Kong L, Liu Y, Wang X, Chang C (2020) Insight into the role of epigenetic processes in abiotic and biotic stress response in wheat and barley. Int J Mol Sci 21(4):1480

Kumar S, Beena AS, Awana M, Singh A (2017) Salt-induced tissue-specific cytosine methylation downregulates expression of HKT genes in contrasting wheat (*Triticum aestivum* L.) genotypes. DNA Cell Biol 36:283–294

Kumar V, Khare T, Shriram V, Wani SH (2018) Plant small RNAs: the essential epigenetic regulators of gene expression for salt-stress responses and tolerance. Plant Cell Rep 37(1):61–75

Lee K, Seo PJ (2014) Airborne signals from salt-stressed Arabidopsis plants trigger salinity tolerance in neighbouring plants. Plant Signal Behav 9:e28392

Liu HH, Tian X, Li YJ, Wu CA, Zheng CC (2008) Microarray-based analysis of stress regulated microRNAs in *Arabidopsis thaliana*. RNA 14(5):836–838

Liu J, Zhi P, Wang X, Fan Q, Chang C (2019) Wheat WD40- repeat protein TaHOS15 functions in a histone deacetylase complex to fine-tune defence responses to *Blumeriagraminis* f. sp. *tritici*. J Exp Bot 70(1):255–268

Matzke MA, Mosher RA (2014) RNA-directed DNA methylation: an epigenetic pathway of increasing complexity. Nat Rev Genet 15(6):394–408

Miryeganeh M (2021) Plants' epigenetic mechanisms and abiotic stress. Genes 12:1106

Peirats-Llobet M, Han SK, Gonzalez-Guzman M, Jeong CW, Rodriguez L, Belda-Palazon B, Wagner D, Rodriguez PL (2015) A direct link between abscisic acid sensing and the chromatin remodeling ATPase BRAHMA via core ABA signaling pathway components. Mol Plant 9:136–147

Peng H, Zhang J (2009) Plant genomic DNA methylation in response to stresses: potential applications and challenges in plant breeding. Prog Nat Sci 19(9):1037–1045

Rashid MM, Vaishnav A, Verma RK, Sharma P, Suprasanna P, Gaur RK (2022) Epigenetic regulation of salinity stress responses in cereals. Mol Biol Rep 49:761–772

Saez A, Rodrigues A, Santiago J, Rubio S, Rodriguez PL (2008) HAB1–SWI3B interaction reveals a link between abscisic acid signaling and putative SWI/SNF chromatin-remodeling complexes in *Arabidopsis*. Plant Cell 20:2972–2988

Sako K, Kim JM, Matsui A, Nakamura K, Tanaka M, Kobayashi M et al (2016) Ky-2, a histone deacetylase inhibitor, enhances high salinity stress tolerance in *Arabidopsis thaliana*. Plant Cell Physiol 57(4):776–783

Sang Y, Silva-Ortega CO, Wu S, Yamaguchi N, Wu MF, Pfuger J, Gillmor CS, Gallahar KL, Wagner D (2012) Mutations in two non-canonical Arabidopsis SWI2/SNF2 chromatin remodelling ATPases cause embryogenesis and stem cell maintenance defects. Plant J 72(6): 1000–1014

Sharma M, Kumar P, Verma V, Sharma R, Bhargava B, Irfan M (2022) Understanding plant stress memory response for abiotic stress resilience: molecular insights and prospects. Plant Physiol Biochem 179:10–24

Turgut-Kara N, Arikan B, Celik H (2020) Epigenetic memory and priming in plants. Genetica 148: 47–54

Van Oosten MJ, Di Stasio E, Cirillo V et al (2018) Root inoculation with *Azotobacter chroococcum* 76A enhances tomato plants adaptation to salt stress under low N conditions. BMC Plant Biol 18:205

Wang M, Qin L, Xie C, Li W, Yuan KL, Yu W, Xia G, Liu S (2014) Induced and constitutive DNA methylation in a salinity-tolerant wheat introgression line. Plant Cell Physiol 55(7):1354–1365

Xie M, Yu B (2015) siRNA-directed DNA methylation in plants. Curr Genom 16(1):23–31

Yaish MW (2013) DNA methylation-associated epigenetic changes in stress tolerance of plants. In: Rout GR et al (eds) Molecular stress physiology of plants. Springer, New Delhi

Yuan L, Liu X, Luo M, Yang S, Wu K (2013) Involvement of histone modifications in plant abiotic stress responses. J Integr Plant Biol 55:892–901

Zeeshan M, Qiu CW, Naz S, Cao F, Wu F (2021) Genome-wide discovery of miRNAs with differential expression patterns in responses to salinity in the two contrasting wheat cultivars. Int J Mol Sci 22:2556

Zhang H, Zhu JK (2012) Active DNA demethylation in plants and animals. Cold Spring Harb Symp Quant Biol 7:161–173

Zhang H, Lang Z, Zhu JK (2018) Dynamics and function of DNA methylation in plants. Nat Rev Mol Cell Biol 19(8):489–506

Zhao J, Zhang J, Zhang W, Wu K, Zheng F, Tian L, Liu X, Duan J (2015) Expression and functional analysis of the plant-specific histone deacetylase HDT701 in rice. Front Plant Sci 5:764

Zheng M, Liu XB, Lin JC, Liu XY, Wang ZY, Xin MM et al (2019) Histone acetyltransferase GCN5 contributes to cell wall integrity and salt stress tolerance by altering the expression of cellulose synthesis genes. Plant J 97(3):587–602

Zhong L, Xu Y, Wang J (2009) DNA-methylation changes induced by salt stress in wheat *Triticum aestivum*. Afr J Biotechnol 8(22):6201–6207

Halophytic Plants: A Potential Resource That Reduces Water Crisis in Future

18

Ajay Kumar, Neha Yadav, Heena, Naveen Kumar, Mamta, Monika, Sarita Devi, Sunil Kumar, and Sunder Singh Arya

Abstract

In the past few years, global climate change has caused severe drought, soil salinization, irregular rainfall, and other changes to the environment all over the world. These changes have decreased crop yield and upset the balance of ecosystems. Water shortages have been getting worse, and the problem is worse in dry areas. This leads to poverty and other problems in society and the economy. Therefore, for long-term development, an integrated approach is a must. One solution to the water problem could be to employ salty resources to cultivate a variety of commercially relevant halophytes in agronomic field trials, such as vegetable, fodder, and oilseed crops. Halophytes have also been used to bioremediate salt-contaminated soils, and their products may even have pharmaceutical uses. These halophytes have unique physical, biological, and anatomical characteristics that help them live in salty environments. Halophytes can do an excellent job of fixing saline soil because they have many ways to adapt. Ion compartmentalization, osmotic adjustment, ion transport and uptake, succulence, an antioxidant system, regulating redox status, and salt excretion are examples of these. Some halophytes, such as *Spartina townsendii*, *Aster tripolium*, and *Beta vulgaris* sp., exhibit low transpiration, low stomatal resistance, and low CO_2 content inside the plant at their threshold salinity tolerance. These halophytes also

A. Kumar · N. Yadav · Heena · N. Kumar · Mamta · S. S. Arya (✉)
Department of Botany, Maharshi Dayanand University, Rohtak, India

Monika
Department of Environmental Science, Maharshi Dayanand University, Rohtak, India

S. Devi
Department of Botany & Plant Physiology, CCS Haryana Agricultural University, Hisar, India

S. Kumar
Department of Botany, CRM Jat College, Hisar, India

A. Kumar et al. (eds.), *Salinity and Drought Tolerance in Plants*,
https://doi.org/10.1007/978-981-99-4669-3_18

have limited (but positive) net photosynthesis. Studies have helped us learn more about how halophytes deal with stress and how they might be used in farming and environmental services. Sodium extrusion, the SOS pathway, which helps keep the balance of ions in the cytoplasm, and vacuolar compartmentalization are all ways that tolerance is achieved.

Keywords

Halophytes · Salt-tolerant · Salinity · Water conservation · Climate change

18.1 Introduction

Sodic soils account for 434 million hectares unfavorable, whereas saline soils account for 397 million hectares of the total 831 million hectares of salt-affected soils (FAO 2000). Due to population pressure, unfavorable environmental conditions, escalating natural disasters, global climate change, agricultural land is rapidly shrinking (Hasanuzzaman et al. 2013a, b). 20% of all irrigated land, or more than 45 million hectares, is affected by salt, and each year, 1.5 million hectares of land is removed from use due to high saline levels (Pitman and Läuchli 2002; Munns and Tester 2008). If this trend keeps up, by the middle of the twenty-first century, half of the world's cropland will be destroyed (Mahajan and Tuteja 2005). Water scarcity is defined as a situation in which water demand exceeds available supply. Water shortage occurs when there is inadequate water to meet both human and environmental water needs at the same time (White 2014). There is a growing global concern over sustainable development, and water is at the center of try more than any other natural resource. Many serious environmental problems with global repercussions are tied to water issues (Gleick 1994, 2000). One-sixth of the world's population resides in arid and semi-arid areas, and these areas also have the largest population densities (World Bank 1999). There is a growing problem in many emerging countries, such as Tanzania, Sudan, Egypt, and Mexico, where a rapidly expanding population is putting a strain on arable land and other resources and pushing people to the periphery, where they must farm on marginal land in dry and semi-arid regions (Ericson et al. 2001; Darkoh 1982; Bilsborrow and Delargy 1990; Findlay 1996). Problems with water scarcity and land degradation develop when more people move to arid and semi-arid regions, leading to even more serious issues with population health, social instability, and poverty (Moench 2002). Consequently, 13–26 million people around the world become environmental refugees every year, with the majority coming from dry and semi-arid regions (Bates 2002). It is become apparent that the issues of poverty, social insecurity, and environmental refugees are all, connected challenges of population increase, water scarcity, and land degradation in arid and semi-arid countries. Now the global issue of salinity and water scarcity can be resolved by multidisciplinary approaches such as development of resistant crops, restoration of degraded land, phytoremediation, and cultivating the wild plants. Halophytes are a type of flowering plants that are capable of

Fig. 18.1 List of major halophytes discussed in this chapter. (**a**) Mesembryanthemum crystallinum, (**b**) Chenopodium album, (**c**) Bruguiera gymnorrhiza, (**d**) Salicornia europaea, (**e**) Crambe maritima, (**f**) Rhizophora mucronata (Hasanuzzaman et al. 2014)

surviving in environments with high salt; hence domesticating them is a realistic alternative. Salinity can be induced by natural (weathering of parent material, flooding of coastal land by tidal water) as well as anthropogenic activities (increase in water table brought on by extensive subsurface water irrigation, poor drainage, etc.) (Hasanuzzaman et al. 2013a, b; Munns 2005; Manchanda and Garg 2008). Halophytes are plants which complete their life cycle in salty conditions (Stuart et al. 2012). Some of the major halophytes are listed in the following (Fig. 18.1).

These halophytes, which flourish in salty conditions, exhibit distinctive morphological, anatomical, and physiological features. In order to successfully alleviate saline soil, halophytes employ a wide range of adaptation processes, including as ion compartmentalization, osmotic adjustment, ion transport and absorption, succulence, an antioxidant system, the control of redox status, and salt excretion (Lokhande and Suprasanna 2012). Halophytes can be divided into three groups based on their ecological characteristics: obligatory, facultative, and habitat-indifferent halophytes. Obligate grow only in salty habitats. Under highly salinized circumstances, they exhibit adequate growth and development. This category includes a large number of plant species from the Chenopodiaceae family. Facultative halophytes can grow in salty soils, although their best growth occurs in low- or no-salt environments. Some plant species belonging to the family Cyperaceae, Poaceae, and Brassicaceae come under this category. Plants indifferent toward their habitat, are able to survive in salty soils. But they usually grow in soils with less salt. They may live in salty soils and compete with species that are salt-sensitive. Plants' germination, growth, and reproductive capacities are all stunted by high salinity, and other physiological processes, such as cellular homeostasis, metabolism, photosynthesis, respiration, transpiration, membrane characteristics, nutritional balance, enzyme activity, and hormone control are all adversely impacted. In addition, increased salinity causes the production of reactive oxygen species (ROS), which can ultimately lead to the plant's death in extreme stress situations (Mahajan and Tuteja 2005; Hasanuzzaman et al. 2012). These halophytes have evolved unique morphological, anatomical, and physiological processes that allow them to thrive in salty conditions. Because of their varied adaptation mechanisms, such as osmotic adjustment, ion compartmentalization, ion transport, absorption, antioxidative systems, redox status maintenance, succulence and salt inclusion or excretion, halophytes are able to efficiently remediate saline soil (Lokhande and Suprasanna 2012). Coastal saline soil, mangrove forest soil, wetland, marshy land, lands of arid and semi-arid regions, and agricultural fields are a few of the places where some halophyte species thrive. These plants can be grown in high salinity soil and water, can replace more traditional crops, and provide useful products like food, fuel, fodder, fiber, medicinal oils, and fiber for textiles (Lokhande and Suprasanna 2012). It is also possible to use halophytes as a major plant species with the potential for desalination, regeneration of saline soils, and phytoremediation.

18.2 Factors Responsible for Water Crisis

1. Large quantities of potable water are primarily and readily available in the countries that are located in the northern hemisphere, whereas developing countries, which are home to approximately 40% of the world's population, have a much more limited supply.
2. The usage of water has increased by more than 600% over the course of the last century. This indicates that the rate of growth is twice as rapid as the rate of population increase. As a result, by the year 2025, it is anticipated that half of all people on earth will face difficulties associated with insufficient supplies of fresh water.

3. As a result of changes in the global climate, countries in the Mediterranean region might anticipate seeing significant setbacks by the end of this century, while other regions of the world can anticipate experiencing increasing rainfall.
4. Irrigation uses up approximately 70% of the world's available water. Since 1966, when there were just 1.5 million km^2 of irrigated land, that number has more than doubled to reach 2.70 million km^2 now. This suggests that approximately 20% of arable areas were being irrigated at the end of the previous century. Forty percent of all crops have been produced on this territory.
5. The expansion of the irrigated area leads to an increase in the amount of water that is used by irrigation systems, and this increase is progressive. This demand for additional irrigation water is a result of an expansion in arable land, in an area where blue water is already scarce or is no longer available at all. The lack of available water and the spread of desertification pose new dangers to human health, natural ecosystems, and the economies of a number of nations, significantly undermining the viability of sustainable development initiatives. As a result, an integrated strategy to problem solving is necessary, one that incorporates prospects for economically, socially, and eco-friendly growth (Duda and El-Ashry 2000).

18.3 Salinity and Freshwater Scarcity

In light of the steadily decreasing availability of freshwater resources and the salinization of the soil, one important goal is to assess the capacity of indigenous halophytes for widespread commercial usage in arid regions. One of the main goals of the research is to find and choose plant species that can handle salt stress. This is achieved by selecting and employing biomarkers to characterize halophytes, figuring out how to use water that isn't typical, like seawater, and choosing halophytes and salt-tolerant glycophytes that might be important for human or animal nutrition. There are a variety of natural salty habitats, some of which are located directly next to saltwater bodies (like a coastal salt marsh) and others that are located further inland (saline lakes, lowlands of dryland and desert terrain, and high evaporation basins). One major drawback of this idea is that even at low quantities in soil water, Na^+ (a cation) and Cl^- (an anion) are harmful to humans, plants, and most animals. That is why most estimates for managing water supply ignore it. Although we are resistant to using seawater, we must immediately find a way to the salinization. Salinization of the soil surface is increased by irrigation water use in both dry and semi-dry zones. Solutes from irrigation water can collect in soils in arid places of the world and eventually reach levels that are harmful to plant growth.

18.4 Sustainable Use of Salt-Affected Soils

Many crop species have lost systems for dealing with salt stress and other abiotic stresses as a result of being domesticated (Munns 1993; Serrano 1996). As saline levels rise, the majority of crop plants are unable to completely express their genetic potential for growth, development, and production, and their economic worth decreases (Maas 1990; Läuchli and Epstein 1990). Therefore, increasing crop plant's salt tolerance is crucial for agricultural research. Wild populations of halophytes are a powerful genetic source for crop plant improvements in salt tolerance (Glenn et al. 1999; Serrano et al. 1999; Khan et al. 2009).

Advanced salt-tolerant crop species can be developed domestically, or genetic resources can be kept in gene bank for usage in crop species development program through traditional or modern molecular breeding techniques. Alternating your crop rotation might help you save water in dry areas. In contrast to established crops like maize, soybeans, rice, etc., some of these plants now produce less. They are not acceptable in regions, where crops with good yields can be cultivated. However, some cases of halophytes being used in industry, ecology, or agriculture are well documented. In agronomic studies, halophytes have been evaluated for their potential as a vegetable, fodder, and oilseed crops due to their adaptability. Growing halophytes in seawater necessitate the use of a leaching percent to maintain optimal salinity in soil, although, at less salinity, these plants outmatch their non-halophyte counterparts in terms of both yield and water efficiency. Different plant species are employed for wastewater treatment in various nations. Different plants have been shown to have varying degrees of success at either detoxifying themselves or precipitating contaminants. Under addition, there are several coastal plants (see halophytes or xerohalophytes) that have been exploited as crops or bred to increase output thanks to their ability to thrive in saltwater irrigation. The medicinal properties of the products of several halophytes and their potential use in the bioremediation of salt-contaminated soils are discussed. It has been proven that ruminants can benefit from eating silage made from these plants.

18.5 Classification of Halophytes

Typical environments for halophytes include saline semideserts, mangrove swamps, marshes, sloughs, and seashores, where the soil and water are salty and the plant cannot develop or sustain itself without human intervention. A halophyte's adaptation to a salty environment might involve either salt tolerance (see halotolerant) or salt aversion. Plants that are not obligate halophytes but still thrive in a salty environment (by, for example, completing their reproductive life cycle during the rainy season) are sometimes called facultative halophytes (e.g., *Aster tripolium*, *Atriplex* sp., *Plantago* sp., *Chenopodium quinoa*). Plants called "obligate halophytes" (halophytes found in arid environments are known as xerohalophytes.) require saltier than 0.5% NaCl water to thrive (*Arthrocnemum* sp., *Frenkenia* sp., *Kochia* sp., *Prosopis* sp.) (Koyro and Lieth 1998) (Fig. 18.2). There are lot of closely

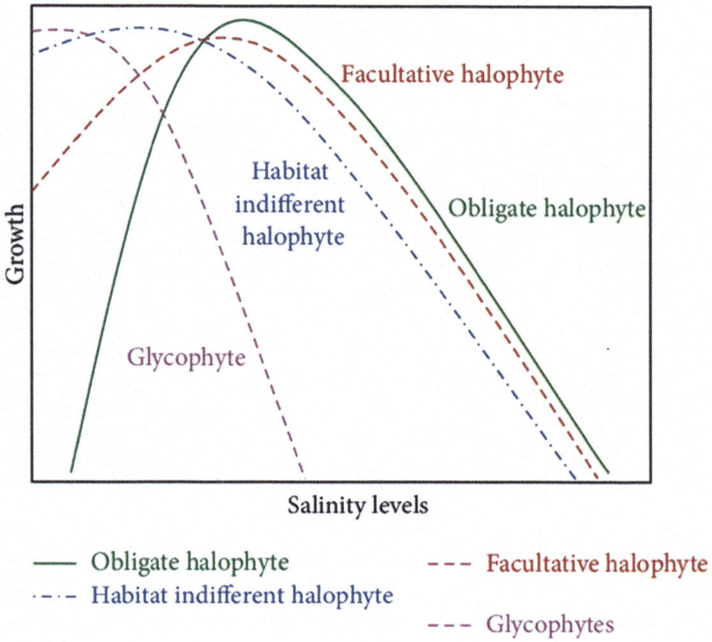

Fig. 18.2 Growth pattern of halophyte under saline condition (Hasanuzzaman et al. 2014)

related plant families, but only a handful of them have independently evolved the structural, phenological, physiological, and biochemical processes for salt tolerance. Almost all plant species are glycophytes, and thus are extremely sensitive to high salinity (e.g., most agricultural crops). Habitat-indifferent are defined as growth preferentially on salt-free soils, but in salt soils has better growth compared to glycophytes (*Sasola* sp., *Festuca rubra*, *Agrostis stolonifera*, *Juncus bufonius*).

18.6 Methods for Enhancing Generalized Salt Resistance

Why does an organism react when it is exposed to salt? The level of salt resistance is proportional to the amount of stress applied. Both salt tolerance and salt aversion are forms of resistance. There are a few common ways to categorize halophytes, including secretor vs. succulents and excluders vs. includers. Physiological and biochemical modifications are required for salt tolerance, as electrolyte accumulation threatens the survival of protoplasm. It is possible for plants to avoid salt by making structural and physiological adjustments to either reduce intracellular salt concentrations or to physically exclude salt from the roots. In theory, salt exclusion or salt inclusion both can lead to salt tolerance (Fig. 18.3).

Research into the physiology and biochemistry of halophytes has revealed a wide variety of adaptations involved in salt resistance. These include salt excretion, genetic regulation, ion compartmentalization, osmolyte synthesis, germination

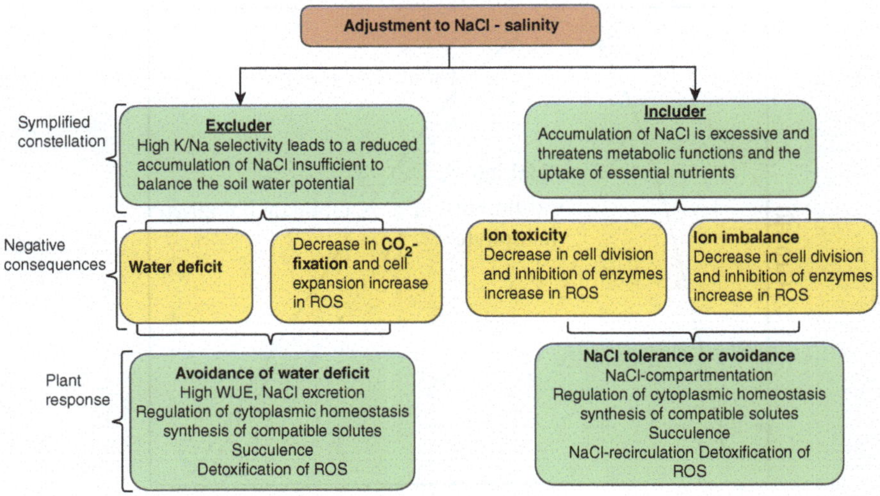

Fig. 18.3 Flow chart depicting the potential methods by which vascular halophytes respond to high external NaCl salt. (Modified after Marschner 1995)

responses, osmotic adaptation, succulence, selective ion transport, and absorption. There are a number of significant plant responses to excessive NaCl salinity that have been described in the literature that help plants escape salt harm and safeguard the symplast (Koyro and Huchzermeyer 2004; Marschner 1995; Mengel and Kirkby 2001; Munns 2002)

1. A modification in the water potential, a reduction in the matric and osmotic potentials, and an increase in the rate of organic solute synthesis are all included. To prevent protein aggregation, often known as "salting out," the majority of halophilic species and all halotolerant organisms expend energy to keep salt out of their cytoplasm. In order to be able to live in environments with high salinities, halophiles have evolved two distinct mechanisms that stop the osmotic transport of water out of their cytoplasm and keep them from drying out. Both approaches are successful because they raise the osmolarity levels found on the inside of the cell. The first step involves the accumulation of certain organic compounds with low molecular weight in cytoplasm; these substances are referred to as suitable solutes. These can either be synthesized once more or gathered from the surrounding environment. Some of the common forms of compatible (suitable) solutes are amino acids, sugars, polyols, and betaines as well as derivatives of some of these compounds. These solutes can be classified as neutral or zwitterions.
2. Controlling the ratio of water lost to carbon dioxide taken up during photosynthesis; increasing the efficiency with which water is used, or switching to a CAM photosynthesis system.
3. Ion selectivity is used to keep homeostasis, especially in the cytoplasm of important organs like salt glands. This is done through selective uptake or

exclusion, compartmentation of Na^+ and Cl^- in the vacuole, translocation in the phloem, growing parts, storage organs, selective ion transport in the shoot and flowering parts of plants.

4. High capacity for the storage of NaCl in the vacuoles of a whole plant organ, typically in older and wilting sections (for example, in leaves that are intended to be shed later) or in specialized structures like hairs. Excessive NaCl concentration can be diluted by increasing the water content in tissue. This is accomplished by expanding the volume of the tissue.

18.7 Morphological Adaptment to Salinity

In many situations, halophytes benefit from different processes and specific morphological traits that help reduce the amount of salt in their tissues and seeds, especially those that are used for photosynthesis, storage, or reproduction. Changes in halophytes' shapes include a way for them to get rid of salt and a more sticky texture. According to Marschner 1995, excreting halophytes have glands that can secrete extra salts from plant parts. Salt glands can be found in a wide variety of plant families that are not related to each other, as well as in some grasses. Alkali grass (*Puccinellia phryganodes*), saltgrass (*Distichlis spicata*), cordgrass (*Spartina alterniflora*, *S. patens*), and shoregrass have all evolved a basic structure that comprises two-celled trichomes that act as salt-collecting chambers (Monanthochloe littoralis). Additionally, *Tamarix* (Tamaricaceae), *Frankenia* (Frankeniaceae), and several common mangroves have complex salt glands. The leaf surfaces of *Atriplex* (saltbush) have vesiculated trichomes (hairs). Extra electrolytes are stored in the bladder cells, which, in the event of a bladder cell rupture, return the salt back to the environment. Additionally, the presence of trichome layer gives the leaves of *Atriplex* a silvery reflectivity, which was found to reduce the production of ROS by preventing some UV radiation from reaching the leaf tissues (ROS). Succulence is seen in many halophyte taxa that live in saline settings. To reduce salt toxicity, succulents employ increased water content within big vacuoles. By storing salt ions in vacuoles, toxins are separated from the cytoplasm and organelles of the cell. When the leaf on a stem segment sheds, salts are expelled from the plant. *Chenopodium*, *Arthrocnemum*, *Batis*, *Allenrolfea*, *Suaeda*, *Nitrophila*, *Halimione*, *Zygophyllum*, and *Salicornia all exhibit* succulence.

18.8 Cash Crop Halophyte Screening

It is not necessary for halophytic organisms to exhibit any specialized morphological adaptations in order to endure the high salt of seawater. It is necessary for plants, whether or not they have salt glands, to focus on three interrelated components of their activity in order to develop salt tolerance. Damage must be avoided at all costs, homeostatic conditions must be restored, and growth must pick up where it left off. The capacity of vascular plants to adapt to conditions of high sodium concentrations and low water potential is essential to their growth and survival in settings with high levels of salinity. This is because too much salt in the outside solution of plant cells can hurt them in many ways. It is very rare for a single trait to make a big difference in how well someone can live in an environment with a lot of NaCl. In order to obtain a survey about the processes constitution ultimately results to the salinity tolerance of specific species, it is necessary to conduct an exhaustive investigation using a QCS, with the examination of at least one or more of multiple factors (Koyro 2003). These mechanisms are associated to the primary limitations that are placed on plant development when it occurs on salty substrates. Some of these problems are a lack of water, a limit on how much CO_2 can be absorbed, nutritional imbalance and ion toxicity. Salt extrusion reduces ion toxicity while hastening the plant's loss of water and thereby lowering its capacity to take in carbon dioxide. The uptake of salt (inclusion) makes osmotic correction easier, but it also increases the risk of toxicity and nutritional imbalance (Fig. 18.2). According to Mengel and Kirkby (2001), the presence of soluble salts can influence development in a number of different ways. To begin, plants may be subjected to a stressful level of water deficit. Second, it is possible that certain ions, in high enough quantities, could be toxic and cause disorders of the body's physiological systems. Finally, high salt concentrations might lead to intracellular abnormalities.

18.9 CO_2 vs. Water Loss

In saline environments, the soil solution and atmosphere typically have low water potentials, which can make it difficult for terrestrial plants to thrive. It is essential under these conditions to reduce water loss (due to transpiration) that is greater than the rate at which water is being taken in. This is only conceivable if the plant has a lower water potential than the soil it is growing in. It was realized how important the time period was, which give rise to the concept of a "two-phase growth response" to salinity (Munns 1993, 2002; Mengel and Kirkby 2001). Initial stage of the growth reduction process can be thought of as an osmotic or water-stress phase. Even when subjected to high salinity treatment, the plants *Spartina townsendii*, *Aster tripolium*, *Sesuvium portulacastrum*, and *Beta vulgaris* sp. *maritima* have an appropriate adjustment mechanism. The entire turgescence of the leaves can be attributed to the fact that the osmotic potentials of all four halophytes, as well as many others, were sufficiently low at all salinity levels. Plants capacity to maintain high photosynthesis rate despite low water loss rates is the primary factor that determines the amount of biomass produced, it is imperative that plant water loss be kept to a minimum on soils with low water potentials. In this state of tension, the generation of biomass by a plant is dependent on both its rate of energy consumption and its rate of carbon buildup (CO_2 net photosynthesis). If the plant's rate of CO_2 fixation goes below its rate of CO_2 production (the compensation point), the plant will have reached a critical point. Because of this, the analysis of growth decrease and net photosynthesis is an essential component of the screening technique. This is especially important near the limit of salt tolerance. Koyro and Huchzermeyer state that many plants exhibit a combination of low (but positive) net photosynthesis, low transpiration, high stomatal resistance, and low internal CO_2 concentration at their threshold salinity tolerance (e.g., *Aster tripolium*, *Spartina townsendii*) (Koyro and Huchzermeyer 2004). Nevertheless, there is a broad spectrum of diversity among halophytes. Two extremely succulent halophytes, *Sesuvium portulacastrum* and *Avicennia marina*, can endure if the water balance is still positive and does not interfere with photosynthesis. Net photosynthesis and water usage efficiency (WUE) both increase in Sesuvium, but stomatal resistance decreases. These results highlight the critical need of studying the control of gas-exchange mechanisms at high salinity in connection to other factors (such as water relations). At high salinity, a water shortage is one of the most significant constraints, and it can lead to a reduction in CO_2 uptake. One such basis for evaluating their capability for use is the balance between the amount of water lost and the amount of CO_2 that is taken in.

18.10 Ion Imbalance vs. Ion Toxicity

It is important for halophytes' salt tolerance that they overcome ion toxicity and nutritional imbalance, two of the most significant growth inhibitors in saline environments. The rate at which salt-tolerant plants transfer Na^+ and Cl^- to their leaves is often slower than that of salt-sensitive plants (Munns 2002). However, for maximal growth and to reach low solute potentials, some halophytes even require an overabundance of salts (Flowers et al. 1977; Greenway and Munns 1980). Alternately, by filtering away the majority of the salt, excessive concentrations can be avoided. These halophytes, which are also known as salt excluders, adapt to salinity by excluding ions, which requires the endogenous synthesis of osmotically active solutes in order to satisfy turgor pressure requirements (Mengel and Kirkby 2001). Even animals with salt bladders may benefit from this adaptive trait. However, the literature mostly discusses NaCl salinity as though a shared reaction between Na^+ and Cl^- (both ions) is what causes a salt damage. Salt tolerance was not linked to Na^+ exclusion in maize by Schubert and Läuchli (1986). Separating the two ions is essential for comprehending the various tolerance to salt mechanisms. As opposed to that, an example of a Na-excluder that has high levels of Cl in its leaves is the *Laguncularia racemose* plant, which contains salt glands (Koyro et al. 1997). Leaves of plants such as *Atriplex papula*, *Atriplex vesicaria*, *Atriplex nummularia*, *Suaeda occidentalis*, *Suaeda brevifolia*, *Salicornia utahensis*, and amongst others, are able to store sodium and chloride despite the presence of a salty environment (salt-includers). In this case, the leaves are sticky, which is a typical adaptation for halophytes (Kinzel 1982; Mengel and Kirkby 2001). This is done to lower the concentration of toxic ions. Producing sugar alcohols (*Laguncularia racemose*, Mannitol present in leaves) or soluble carbohydrates (*Beta vulgaris* ssp. *Maritima*, Sucrose present in tap roots) or organic acids (like amino acids) or lowering the matrical potential allows these species to maintain the water potential low and the charge balanced. Creating organic solutes, on the other hand, takes a lot of energy the creation of these solutes results in a reduction in the energy condition of the plant.

The result is, a compromise in Na and Cl-excluding species is required for plant survival, and this is a slowdown in plant growth. That has nothing to do with toxicity or a lack of necessary nutrients.

18.11 Salinity Tolerance Mechanism

One common way to figure out how plants deal with salt stress at the molecular level is to look how salt stress changes the activity or expression of cellular processes and genes (Hasegawa et al. 2000a, b), which has helped us learn more about how complicated it is for plants to handle salt. Studies have provided a better understanding of how the tolerance mechanism work in halophytes (Flowers and Colmer 2008; Colmer and Flowers 2008) as well as their possible use in farming and environmental services (Rozema et al. 2013). There are three ways that halophytes work that need to be studied.

- Vacuolar compartmentalization.
- Sodium extrusion and the SOS mechanism, which permits cytoplasmic ion homeostasis.
- A process of taking and recycling sodium.

1. Since Na toxicity is the primary stressor in saline soils, much effort has been put into identifying the ion transporters and regulatory mechanisms responsible for Na^+ homeostasis and the maintenance of a high cytoplasmic Na^+/K^+ ratio. Overexpressing the vacuolar-type sodium and proton antiporter (Na^+/H^+ antiporter) gene from the halophytic plant *Atriplex gmelini* (Ag NHX1 = vacuolar Na^+/H^+ exchanger) resulted in a salt-sensitive rice cultivar with significantly greater salt tolerance than the wild type rice (Ohta et al. 2002). If large-scale genetic transfer can improve crop plants' salt tolerance, then more work like this could be beneficial in the near future. It has become clear that the Salt Overly Sensitive (SOS) signaling pathway, which is made up of the SOS1, SOS2, and SOS3 proteins, is important for both the recognition of and resistance to salt stress. A calcium signal induced by salt stress is thought to activate a protein kinase complex consisting of SOS2 and SOS3 (Zhu 2003). The Na^+/H^+ antiporter SOS1, located in the plasma membrane, is responsible for Na^+ excretion into the apoplast and is phosphorylated and activated by this protein kinase complex.
2. To a lesser extent, SOS3-like calcium-binding proteins may potentially control Na^+ transport in tonoplast. Excretion across the plasmalemma (SOS1) is an efficient route for keeping the cytosolic Na^+ content low, but vacuolar compartmentalization of sodium (Na^+) ions is also important. A Na^+/H^+ antiporter mediates the uptake of Na^+ into the vacuole. In order to generate the proton-motive force necessary for this transport, an ATPase and an H^+-pyrophosphatase work together. Na^+ sequestration into vacuoles was improved and salt tolerance

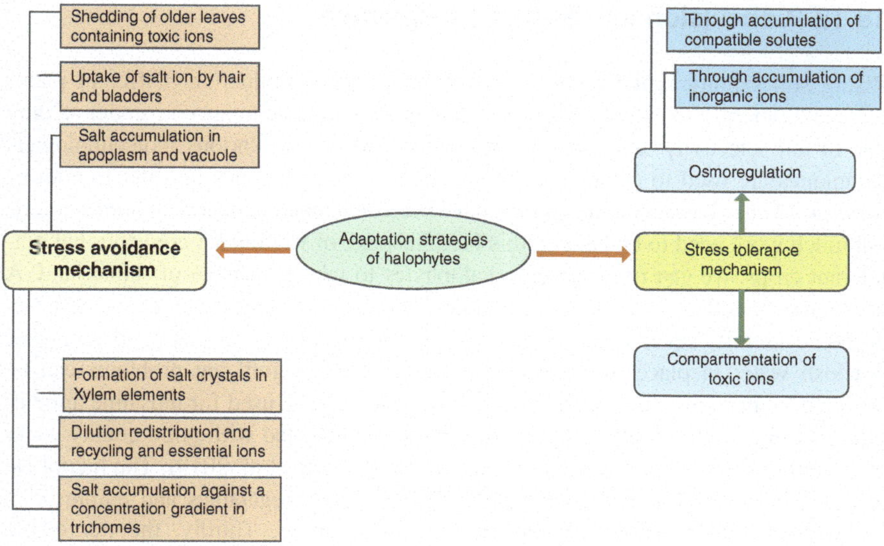

Fig. 18.4 The physiological basis for halophytes salt tolerance and resistance. (Adapted from Devi et al. 2019)

was increased when the tonoplast Na^+/H^+ antiporter and vacuolar H^+-pyrophosphatase (AVP1) both were overexpressed (Gaxiola et al. 2001).

3. It is believed that a calcium signal brought on by salt stress activates a protein kinase complex made up of SOS3 and SOS2. Phloem tissue is the only organ in which the HKT1 gene is expressed, and it plays a crucial role in regulating sodium concentration in phloem sap (Fig. 18.4).

HKT1 probably helps Na^+ ion transport from the leaves to the stem and roots by letting Na^+ ions to get into the phloem sap in the leaves and get out of it in the roots. Uozumi et al. 2000 observed that keeping K^+ uptake sustained, under salt stress requires the Na^+-K^+ symporter HKT1, a high-affinity carrier. Na^+ overload suppressed HKT1-mediated K^+ inflow and decreased HKT1 expression in transgenic wheat increased salt tolerance and decreased Na^+ absorption (Laurie et al. 2002). According to Katiyar-Agarwal et al. (2005), HKT1 being a crucial factor in tolerance of salt. Although, conventional breeding programs have had relatively little success in increasing crop salt tolerance because the trait is so complicated, both genetically and physiologically. These molecular investigations, in conjunction with rapid salt resistance testing techniques, could lead to new insights into how to boost crop yields in salty environments.

18.12 Conclusion and Future Prospective

Rising soil salinity is the main danger to agricultural production. Different plants behave differently to salinity. However, halophytes are able to survive under salinity due to ion selectivity and vacuolar compartmentalization. Various biotechnological techniques are used to create tolerant and high-yield crop types in order to address these problems. Transgenic food crops have been developed to deal with food scarcity, but much work need to be done in this field. Instead, of waiting for conventional salt-tolerant crops, we can use cash crop halophytes to take advantage of saline land. A new concept called "Biosaline Agriculture" has been developing over the last few decades. In this, several halophytes are grown to employ irrigation of saline or brackish water in place of normal crop plants. Three significant problems can be resolved in this way: First, saline or brackish water will be used for biosaline agriculture, and high-yielding glycophytic crops might be irrigated with good quality water that would otherwise be diverted for human use (Nikalje et al. 2019). The halophyte can also be utilized as attractive plants, biofuel, food (vegetables and edible oils), forage and fodder, medicinal, biofuel, and landscaping. Thirdly, the non-edible halophytes have been used in ecological balancing, phytoremediation of hazardous metals, textile dyes, etc., phytodesalination of soils impacted by salt, and other environmental clean-up projects. Halophytes can thrive in challenging circumstances without suffering a yield penalty because they have a built-in tolerance to various biotic and abiotic stresses. These multifunctional plants must be examined for their output and possible application in environmental remediation. It may be prioritized to domesticate these plants with many applications so they can replace traditional crops. Biosaline agriculture will undoubtedly offer a lot of potential as a complement to sustainable agricultural practices in the years to come (Fig. 18.5).

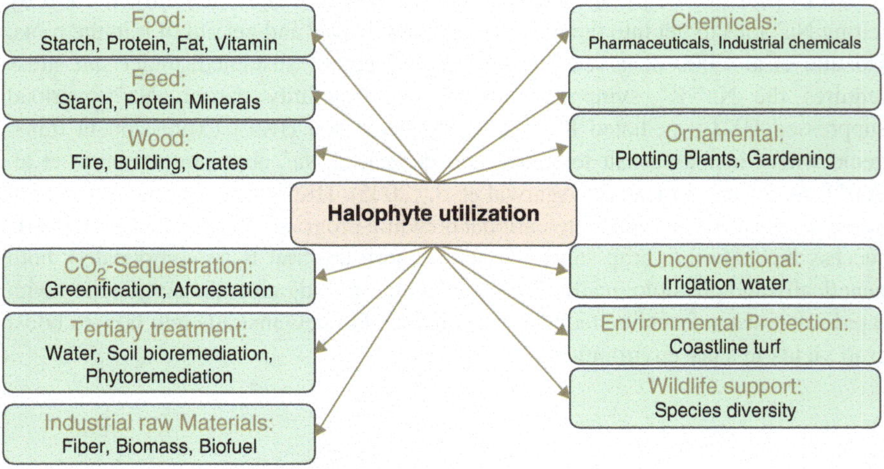

Fig. 18.5 Utilization of halophytes for the purpose of achieving economic benefits and cleaning up the environment. (Adapted from Nikalje et al. 2019)

References

Bates DC (2002) Environmental refugees? Classifying human migrations caused by environmental change. Popul Environ 23(5):465–477

Bilsborrow R, Delargy P (1990) Land use, migration and natural resource deterioration: the experience of Guatemala and the Sudan. Pop Dev Rev 16:125–147

Colmer TD, Flowers TJ (2008) Flooding tolerance in halophytes. New Phytol 179:964–974

Darkoh MBK (1982) Population expansion and desertification in Tanzania. Desertif Contr 6:26–33

Devi S, Kumar A, Mann A, Arya SS, Chand G, Kumar N, Kumari A, Pooja RB, Kumar A (2019) Intra-habitat variability of halophytic flora of north-West India. In: Hasanuzzaman M, Shabala S, Fujita M (eds) Halophytes and climate change: adaptive mechanisms and potential uses. CABI, Oxfordshire, pp 38–54

Duda AM, El-Ashry MT (2000) Addressing the global water and environment crises through integrated approaches to the management of land, water and ecological resources. Water Int 25(1):115–126

Ericson JA, Freudenberger MS, Boege E (2001) Population dynamics, migration, and the future of the Calakmul Biosphere Reserve. In: Biological diversity. CRC Press, Boca Raton, FL, pp 261–292

FAO (2000) Global network on integrated soil management for sustain-able use of salt- affected soils. FAO, Rome. http://www.fao.org/ag/agl/agll/spush

Findlay AM (1996) Population and environment in arid regions, policy and research paper No. 10. IUSSP Scientific Committee on Population and Environment, Aubervilliers

Flowers TJ, Colmer TD (2008) Salinity tolerance in halophytes. Wiley, New York, NY

Flowers TJ, Troke PF, Yeo AR (1977) The mechanism of salt tolerance in halophytes. Annu Rev Plant Physiol 28(1):89–121

Gaxiola RA, Li J, Undurraga S, Dang LM, Allen GJ, Alper SL, Fink GR (2001) Drought-and salt-tolerant plants result from overexpression of the AVP1 H+ pump. Proc Natl Acad Sci 98(20): 11444–11449

Gleick PH (1994) Reducing the risks of conflict over fresh water resources in the Middle East. In: Studies in environmental science, vol 58. Elsevier, Amsterdam, pp 41–54

Gleick PH (2000) The world's water 2000-2001. The Biennial Report on Freshwater Resources. Island Press, Washington, DC

Glenn EP, Brown JJ, Blumwald E (1999) Salt tolerance and crop potential of halophytes. Crit Rev in Plant Sci 18(2):227–255

Greenway H, Munns R (1980) Mechanisms of salt tolerance in nonhalophytes. Annu Rev Plant Physiol 31(1):149–190

Hasanuzzaman M, Hossain MA, Teixeira da Silva JA, Fujita M (2012) Plant responses and tolerance to abiotic oxidative stress: antioxidant defense is a key factor. In: Bandi V, Shanker AK, Shanker C, Mandapaka M (eds) Crop stress and its management: perspectives and strategies. Springer, Berlin, pp 261–316

Hasanuzzaman M, Nahar K, Fujita M (2013a) Plant response to salt stress and role of exogenous protectants to mitigate salt-induced damages. In: Ahmad P, Azooz MM, Prasad MNV (eds) Ecophysiology and responses of plants under salt stress. Springer, New York, NY, pp 25–87

Hasanuzzaman M, Nahar K, Fujita M et al (2013b) Enhancing plant productivity under salt stress—relevance of poly-omics. In: Ahmad P, Azooz M, Prasad MNV (eds) Salt stress in plants: omics, signaling and responses. Springer, Berlin, pp 113–156

Hasanuzzaman M, Nahar K, Mahabub Alam M, Bhowmik PC, Hossain MA, Rahman MM, Vara Prasad MN, Qzturk M, Fujita M (2014) Potential use of halophytes to remediate saline soils. Biomed Res Int 2014:589341

Hasegawa M, Bressan R, Pardo JM (2000a) The dawn of plant salt tolerance genetics. Trends Plant Sci 5(8):317–319

Hasegawa PM, Bressan RA, Zhu JK, Bohnert HJ (2000b) Plant cellular and molecular responses to high salinity. Annu Rev Plant Biol 51(1):463–499

Katiyar-Agarwal S, Verslues P, Zhu J (2005) Mechanisms of salt tolerance in plants. Plant Nutr Food Secur Hum Health Environ Protect 23:44–45

Khan MA, Ansari R, Ali H, Gul B, Nielsen BL (2009) Panicum turgidum, a potentially sustainable cattle feed alternative to maize for saline areas. Agric Ecosyst Environ 129:542–546

Kinzel H (1982) Pflanzenökologie und Mineralstoffwechsel. Verlag Eugen Ulmer, Stuttgart, p 534

Koyro HW (2003) Study of potential cash crop halophytes by a quick check system: determination of the threshold of salinity tolerance and the ecophysiological demands. In: Lieth H et al (eds) Cash crop halophytes: recent studies. Springer, Dordrecht, pp 5–17

Koyro HW, Huchzermeyer B (2004) Ecophysiological needs of the potential biomass crop *Spartina townsendii* Grov. Trop Ecol 45(1):123–139

Koyro HW, Lieth H (1998) Salinity conversion table. 2 enlarged ed.

Koyro H-W, Wegmann L, Lehmann H, Lieth H (1997) Physiological mechanisms and morphological adaptation of *Laguncularia racemosa* to high salinity. In: Lieth H, Hamdy A, Koyro H-W (eds) Water management, salinity and pollution control towards sustainable irrigation in the Mediterranean region: salinity problems and halophyte use. Technomack Publication, Bari, pp 51–78

Läuchli A, Epstein E (1990) Plant responses to saline and sodic conditions. Agric Salin Assess Manag 71:113–137

Laurie S, Feeney KA, Maathuis FJ, Heard PJ, Brown SJ, Leigh RA (2002) A role for HKT1 in sodium uptake by wheat roots. Plant J 32(2):139–149

Lokhande VH, Suprasanna P (2012) Prospects of halophytes in understanding and managing abiotic stress tolerance. In: Ahmad P, Prasad MNV (eds) Environmental adaptations and stress tolerance of plants in the era of climate change. Springer, New York, NY, pp 29–56

Maas EV (1990) Crop salt tolerance. In: Agricultural salinity assessment and management. ASCE, New York, NY, pp 262–304

Mahajan S, Tuteja N (2005) Cold, salinity and drought stresses: an overview. Arch Biochem Biophys 444(2):139–158

Manchanda G, Garg N (2008) Salinity and its effects on the functional biology of legumes. Acta Physiol Plant 30(5):595–618

Marschner H (1995) Mineral nutrition of higher plants, 2nd edn. Institute of Plant Nutrition University of Hohenheim, Stuttgart

Mengel K, Kirkby EA (2001) Principles of plant nutrition. Kluwer Academic Publishers, Dordrecht; Boston, MA; London

Moench M (2002) Water and the potential for social instability: livelihoods, migration and the building of society. Nat Res Forum 26(3):195–204

Munns R (1993) Physiological processes limiting plant growth in saline soils: some dogmas and hypotheses. Plant Cell Environ 16(1):15–24

Munns R (2002) Comparative physiology of salt and water stress. Plant Cell Environ 25(2): 239–250

Munns R (2005) Genes and salt tolerance: bringing them together. New Phytol 167(3):645–663

Munns R, Tester M (2008) Mechanisms of salinity tolerance. Annu Rev Plant Biol 59:651–681

Nikalje GC, Bhaskar SD, Yadav K, Penna S (2019) Halophytes: prospective plants for future. New Phytol 179:945–963

Ohta M, Hayashi Y, Nakashima A, Hamada A, Tanaka A, Nakamura T, Hayakawa T (2002) Introduction of a Na+/H+ antiporter gene from Atriplex gmelini confers salt tolerance to rice. FEBS Lett 532(3):279–282

Pitman MG, Läuchli A (2002) Global impact of salinity and agricultural ecosystem. In: Läuchli A, Lüttge U (eds) Salinity: environment—plants—molecules. Kluwer Academic, Dordrecht, pp 3–20

Rozema J, Muscolo A, Flowers T (2013) Sustainable cultivation and exploitation of halophyte crops in a salinising world. Environ Exp Bot 92:1–3

Schubert S, Läuchli A (1986) Na+ exclusion, H+ release, and growth of two different maize cultivars under NaCl salinity. J Plant Physiol 126(2–3):145–154

Serrano R (1996) Salt tolerance in plants and microorganisms: toxicity targets and defense responses. Int Rev Cytol 165:1–52
Serrano R, Mulet JM, Rios G, Marquez JA, de Larrinoa IF, Leube MP, Mendizabal I et al (1999) A glimpse of the mechanisms of ion homeostasis during salt stress. J Exp Bot 50:1023–1036
Stuart JR, Tester M, Gaxiola RA, Flowers TJ (2012) Plants of saline environments. Access Sci. http://www.accessscience.com
Uozumi N, Kim EJ, Rubio F, Yamaguchi T, Muto S, Tsuboi A, Bakker EP, Nakamura T, Schroeder JI (2000) Plant Physiol 122:1249–1259
White C (2014) 28. Understanding water scarcity: definitions and measurements. In: Global water: issues and insights, p 161
World Bank (1999) World development report. World Bank, Washington, DC
Zhu JK (2003) Regulation of ion homeostasis under salt stress. Curr Opin Plant Biol 6(5):441–445

Arbuscular Mycorrhizal Fungi–Plant Interaction for Salinity and Drought Stress Alleviation

19

Priyanka Chandra, Sujata Yadav, Ashwani Kumar, Parvender Sheoran, and Anita Mann

Abstract

Abiotic stresses mainly salinity and drought have become major threats to the global agricultural production. Arbuscular mycorrhizal fungi (AMF) inhabit the rhizosphere and develop a symbiotic relationship with the roots of most plant species. AMF benefits plants through mobilization as well as uptake of nutrients and improved soil structure, while the fungus receives photosynthetic carbon and nutrients from the host plant. AMF also possess the potential to alleviate abiotic stresses in crop plants through several mechanisms. Hence, this chapter will cover the importance AMF–plant interaction in ecosystem and will highlight the functions of AM fungi in alleviating the stresses in crops and plant growth promotion. Increased nutrient supply through ionic balance, biosynthesis of osmoprotectants, accumulation of metabolites, modification in plant physiology and root structure, accumulation of signal molecules are the major mechanisms of AM fungus which support crop plants to overcome such abiotic stress conditions.

Keywords

Abiotic stresses · Salinity · Drought · Arbuscular mycorrhizal fungi · Plant interaction

P. Chandra (✉) · S. Yadav · A. Kumar · A. Mann
ICAR – Central Soil Salinity Research Institute, Karnal, Haryana, India

P. Sheoran
ICAR – Central Soil Salinity Research Institute, Karnal, Haryana, India

ICAR – Agricultural Technology Application Research Institute, Ludhiana, Punjab, India

A. Kumar et al. (eds.), *Salinity and Drought Tolerance in Plants*,
https://doi.org/10.1007/978-981-99-4669-3_19

19.1 Introduction

Plant growth, development, productivity, and tolerance to climate obstacles are now hot topics in agriculture and plant-based biotechnologies. Due to biotic (phytopathogens) and abiotic (drought, salinity, flood, storm, and extreme temperatures) conditions, agricultural yield and productivity suffers large losses (Ma et al. 2020). A plant's ability to recognize such stresses and adaptation to it could be favorable for its survival (Naylor and Coleman-Derr 2018). The co-occurrence of drought and saline stress environment restrict plant growth and its survival in certain areas, perhaps making agricultural production more challenging (Dugasa et al. 2019). The combined effects of salt and drought stress on the yields of some crops, such as wheat, potatoes, and barley, are more detrimental than either stress state alone (Levy et al. 2013; Yousfi et al. 2010, 2012). Plants respond differently to combinatorial stress conditions (Li et al. 2021). As a result, plants respond to mixed stresses in a complex and flexible manner, employing a variety of transcripts, metabolites, and proteins that are specific to the combination of stresses as well as the total of individual-to-individual challenges (Guo et al. 2021a, b; Hosseini et al. 2021; Osthoff et al. 2019; Prasch and Sonnewald 2013; Zandalinas et al. 2021). Agriculture sustainability is threatened by two major stressors: drought and soil salinity, which have a negative impact on crop development and productivity globally (Arif et al. 2020; Gupta et al. 2020, 2022). Bacterial biomass is reduced as a result of the effects of drought stress on both the plant host and the soil around its roots (Sadeghi et al. 2012). The symbiotic relationship between soil microorganisms and plants is well known to be an effective and environmentally friendly defense against harmful environmental stresses (Zhu et al. 2016a, b, c). Mycorrhizas are symbiotic relationships between higher plants and fungi that occur in the plant's root apparatus and extend into the rhizosphere and surrounding soil (Giovannetti and Mosse 1980; Giovannetti et al. 2010; Ganugi et al. 2019). Over 80% of plant species can form symbiotic relationships with arbuscular mycorrhizal fungus (AMF), which is a type of beneficial and ancient soil microorganism belonging to the monophyletic phylum Glomeromycota (Smith and Read 2010).

Symbiotic AMF has been found to have a number of benefits for maintaining the functioning of agro-ecosystems, including improving the plant's ability to absorb water and nutrients as well as promoting growth (Gianinazzi et al. 2010). Plant yield is affected by AMF because they induce a number of morphological, physiological, biochemical, and molecular changes (Zhu et al. 2016a, b, c). Ectomycorrhiza, in which the mycelium does not penetrate the root cell wall specifically, and endomycorrhiza, in which the mycelium penetrates the cell membrane, are two commonly distinguished types of mycoses. Arbuscular, ericoid, arbutoid, monotropoid, and orchid mycorrhizae are other classifications of endomycorrhizae, which are different types of mycorrhizae. Arbuscular mycorrhizal fungi are ubiquitous in soil and symbiotically associate with the roots of angiosperms and other plants to induce beneficial effects (Sadhana 2014; Gerdemann 1968).

19.2 Impact of Salinity and Drought Stress on Crops Physiology

Salinity and drought are known to limit agricultural production worldwide (Verma et al. 2022). The effects of drought and salinity on physiological, biochemical, and molecular changes in plants are very similar (Sairam and Tyagi 2004; Wang et al. 2022). Collectively and individually, these stressors reduce soil water potential around the roots and in soil. Plant roots are naturally capable of sensing soil water deficits as well as other signals influenced by subterranean and/or above-ground processes. Soil dryness and salinity deprives plant roots of adequate water absorption to support normal growth (Cruz de Carvalho 2008; Mickelbart et al. 2015; Ribba et al. 2020; Uddin et al. 2016). Different signaling pathways are activated under the effects of drought and salinity, leading to the changes in the cellular redox state of plants (Ahanger et al. 2017; Dat et al. 2000; Mittler 2002; Ribba et al. 2020). Plant defense responses to individual drought or salinization stress may alter from those displayed under simultaneous two stresses (Verma et al. 2022).

Water stress reduces stomatal opening and leaf size, leaf water potential, inhibits root growth, slows flowering, limits fruit set, and limits growth as well as yield (Osakabe et al. 2014). Thus, plants have evolved a variety of mechanisms to minimize water resource use or manage their growth until they encounter adverse conditions (Athar and Ashraf 2009; Gupta et al. 2021b, c). Stress induced by drought and salinity first leads to intracellular ionic toxicity, followed by disruption of osmotic equilibrium under prolonged stress. This combination of osmotic and ionic shocks alters plant growth and development. Ionic homeostasis and intracellular osmolarity must be rapidly regulated to maintain salt tolerance. Both can harm plants through osmosis, leading to significant accumulation of reactive oxygen species (ROS) such as superoxide anions and hydrogen peroxide in plants. When accumulation exceeds tolerance, these stressors can damage cell structures such as DNA, proteins, and cell membranes, and interfere with normal cellular metabolism (Patel et al. 2020; You and Chan 2015; Wang et al. 2022).

19.3 Arbuscular Mycorrhizal Fungi–Plant Interaction

AMF and plants are thought to coexist 400 million years ago (Selosse et al. 2015). These associations are generated by a number of biological processes that have several beneficial effects on the biotic population of agro- and natural ecosystems (van der Heijden et al. 2015). A well-known example of a mutual interaction that can regulate plant growth and development is the symbiotic association of AMF. AMFs have a network of mycelium under the plant's roots that enhances the absorption of nutrients that cannot be obtained. Although derived from different plant species, the mycelium invades the roots of both to create a common mycorrhizal network (CMN). According to Pringle et al. (2009), CMN is recognized as an important component of terrestrial ecosystems, due to its significant impact on various plant communities, especially invasive species, and fungal transport of phosphate (P) and nitrogen (N) to plants (Smith and Read 2010). Furthermore, among other benefits, shared nutrients also flow from fungi to plants, which may be why AMF increases plants' tolerance to biotic and abiotic factors (Plassard and Dell 2010). They can improve soil quality and thus promote plant growth under low and high stress situations (Navarro et al. 2014; Alqarawi et al. 2014a, b). The morphological-physiological characteristics of the plants are modified by AMF invasion to make them more resistant to more severe stressors (Alqarawi et al. 2014a, b; Hashem et al. 2015).

Most terrestrial flora are thought to have AMF as a natural growth regulator. Researchers are promoting the use of AMF as an important biofertilizer for sustainable crop yields and productivity (Barrow 2012). In addition, compared with soils not treated with AMF, AMF inoculated soils are more stable mass and significantly stronger extraradical hyphal mycelium (Syamsiyah and Herawati 2018) which improves soil health. According to Wu et al. (2014), glomalin-associated soil protein (GRSP) is believed to maintain soil water content in the face of various abiotic challenges. This protein then controls the frequency of water between the soil and the plant, thus automatically initiating plant development and growth process. Glomalin is composed of 30–40% of related chemicals, which prevent the soil from drying out by increasing the water holding capacity of the soil (Sharma et al. 2017). AMF inoculation impacts growth-related processes including stomatal conductance, leaf water potential, relative water content (RWC), PSII, and CO_2 efficiency (He et al. 2017a, b; Chandrasekaran et al. 2019). By altering the physiology of aerial organs and tissues, AMF also improves water stress tolerance (Bárzana et al. 2012). In addition, AMF inoculation increases dry matter accumulation and moisture absorption from water, thereby enhancing the plant's tolerance to environmental challenges including salinity and drought. Organic farming to promote growth and maximize yield can significantly benefit from the use of AMF for plant growth in various biological habitats. According to Bieleski (1973), AMF can increase the material uptake capacity of the surface host roots up to ten times. Ions such as P, Zn, and Cu do not easily diffuse through the soil. The roots deplete these immobile nutrients from the soil of the area immediately surrounding the roots due to this poor diffusion. Increased nutrient uptake by mycelium in soil is often associated with increased plant growth due to AM symbiosis. It is well known that a network of mycelium attached to the roots of a living tree has the potential to infect the roots of other nearby plants (Chiariello et al. 1982; Francis and Read 1984, 1995; Newman 1988). Several reports are being found, which stated the amelioration effect of AMF in several crops and are described in Table 19.1.

19.4 Effect of AMF on Crops Under Salinity Stress

Soil salinization is a well-known environmental problem that poses a serious threat to the security of global food supplies. It is well established that salinity stress inhibits plant growth by reducing vegetative growth rates and net assimilation rates, leading to lower yields (Ahanger et al. 2017). Furthermore, it encourages overproduction of reactive oxygen species (Ahanger and Agarwal 2017a, b;

Table 19.1 Effect of AMF in abiotic stress mitigation

S no.	AMF	Crop	Mechanism	Abiotic stress	References
1.	*Rhizophagus intraradices; Funneliformis mosseae; F. geosporum*	*Triticum aestivum*	Restored maximum phytochemistry	Drought	Mathur et al. (2018)
2.	*Rhizophagus irregularies*	*Triticum aestivum*	Enhanced NUE by altering plant C assimilation and N uptake	Salinity	Zhu et al. (2016a, b, c)
3.	*Funneliformis mosseae*	*Triticum durum Triticum aestivum*	Significantly improved the plants' biomass production	Drought	Bernardo et al. (2019)
4.	*Glomus claroideum*	*Triticum aestivum*	Significant increase in total dry weight, relative water content and leaf chlorophyll content	Drought	Beltrano and Ronco (2008)
5.	*Glomus mosseae; Glomus etunicatum*	*Triticum aestivum*	Improved growth, nutrient uptake and yield	Drought	Al-karaki et al. (2004)
6.	*Rhizophagus intraradices, Funneliformis mosseae, F. geosporum*	*Triticum durum*	Exhibited an increase in RWC for both leaf and soil	Alkaline soils drought	Al-Karaki and Al-Omoush (2002)
7.	*Glomus* spp.	*Triticum aestivium*	Higher gas exchange capacity, higher osmolyte concentration, reduced oxidative damage by increasing antioxidant activity	Salinity	Talaat and Shawky (2014)
8.	*Funneliformis mosseae*	*Triticum aestivium*	Reducing Na uptake; increasing P, N, K, and Mg contents and stimulating photosynthetic pigments and some metabolic contents	Salinity	Abdel-Fattah and Asrar (2012)

(continued)

Table 19.1 (continued)

S no.	AMF	Crop	Mechanism	Abiotic stress	References
9.	*Rhizophagus intraradices, Funneliformis mosseae, Funneliformis geosporum*	*Triticum aestivum*	Mitigation of the metabolic inhibition of photosynthesis through the maintenance of PSI and PSII integrity and stability	Salinity	Mathur et al. (2018, 2019)
10.	*Glomus* sp., *Acaulospora* sp., *Gigaspora* sp., *and Scutellospora* sp.	*Maccadamia integrifolia*	Increased soluble sugars and proline for osmoregulation defense response, improves water relation in plant tissues	Drought	Yooyongwech et al. (2013)
11.	*Glomus versiforme*	*Zea mays*	Increased growth and photosynthesis, compatible solutes, up-regulation of antioxidant system	Drought	Begum et al. (2019)
12.	*Rhizophagus intraradices*	*Euonymus maackii*	Increasing photosynthesis capacity, accelerating nutrient absorption and activating antioxidant enzyme activities	Salt stress	Li et al. (2019)
13.	*Rhizophagus intraradices*	*Pisum sativum*	Higher nutrient uptake, accumulation of compatible osmolytes, and lower cellular leakage of electrolyte	Salt stress	Parihar et al. (2020)
14.	*Azospirillum* sp.	*Lectuca sativa*	Encourage aerial biomass, vitamin C content, proline levels and reduce stomatal conductance	Drought	Fasciglione et al. (2015)

(continued)

Table 19.1 (continued)

S no.	AMF	Crop	Mechanism	Abiotic stress	References
15.	*Bacillus subtilis Rhizo*	*Solanum lycopersicum*	ACC deaminase activity; protects against oxidative damage and enhances plant growth	Drought	Gowtham et al. (2020)
16.	*Bacillus amyloliquefaciens*	*Oryza sativa*	Induces metabolic and physiological parameters via different enzymes	Salinity	Bisht et al. (2020)
17.	*Bacillus* sp. *RhStr_71, RhStr_223, and RhStr_JH5*	*Pisum sativum*	Increased IAA production, phosphate solubilization, ammonia production, ACC deaminase activity, and siderophore production	Salinity	Gupta et al. (2021a)
18.	*Bacillus* sp. *(CHR JH 203 and BST YS1_42)*	*Pisum sativum*	ACC deaminase production; enhanced plant biomass, improved root system architecture, increased photosynthetic capacity and modulate antioxidant enzyme production	Salinity	Gupta et al. (2021a)
19.	*Pseudomonas putida and Bacillus amyloliquefaciens*	*Cicer arietinum* L.	Promote ACC deaminase activity, mineral solubilization, hormones production, biofilm formation, and siderophore activity	Drought	Kumar et al. (2016)
20.	*Unneliformis mosseae, Hizophagus diaphanum, Glomus versiforme*	*Poncirus trifoliata*	Higher plant growth and biomass, acid and total phosphatase activity, leaf and root P contents	Drought	Wu et al. (2011)

(continued)

Table 19.1 (continued)

S no.	AMF	Crop	Mechanism	Abiotic stress	References
			particularly in *F. mosseae*		
21.	*Unneliformis mosseae*	*Helianthus annuus* L.	Inoculated plants produced more dry matter, heavier seeds and greater seed and oil yields	Drought	Gholamhoseini et al. (2013)
22.	*Funneliformis mosseae*, *Paraglomus occultum*	*Poncirus trifoliata*	Increased hyphal length, hyphal water absorption rate, and leaf water potential	Drought	Zhang et al. (2018a, b)
23.	*Glomus etunicatum*, *Glomus intraradices*, *Glomus mosseae*	*Cucumis sativus* L.	Increased biomass, photosynthetic pigment synthesis, and enhanced antioxidant enzymes	Salinity	Hashem et al. (2018)
24.	*Rhizophagus irregularis*	*Solanum lycopersicum* L.	Enhanced shoot FW, leaf area, leaf number, root FW, and levels of growth hormones	Salinity	Khalloufi et al. (2017)
25.	*Rhizophagus irregularis*	*Digitaria eriantha*	Increased shoot dry matter, stomatal conductance, lipid peroxidation, H_2O_2 in shoot and root	Drought	Pedranzani et al. (2016)

Ahanger et al. 2018). Efforts are underway to investigate potential strategies to improve agricultural productivity on saline soils. One such promising method is the use of AMF to reduce the harmful effects of salinity on plants (Santander et al. 2019). Many studies have documented the effectiveness of AMF in promoting plant growth and yield under salinity stress (Fig. 19.1) (Talaat and Shawky 2014). According to El-Nashar (2017), AMF improves the growth rate, leaf water potential, and water use efficiency of *Antirrhinum majus* plants. The positive effects of AMF symbiosis on physiological variables such as photosynthesis rate, stomatal conductance, and leaf water relations under salinity were recently documented by Ait-El-Mokhtar et al. (2019). According to Sheng et al. (2011), AMF significantly reduced the negative effects of salt stress on photosynthesis. Under saline conditions, root microbial culture significantly increased photosynthesis rate, as well as other gas

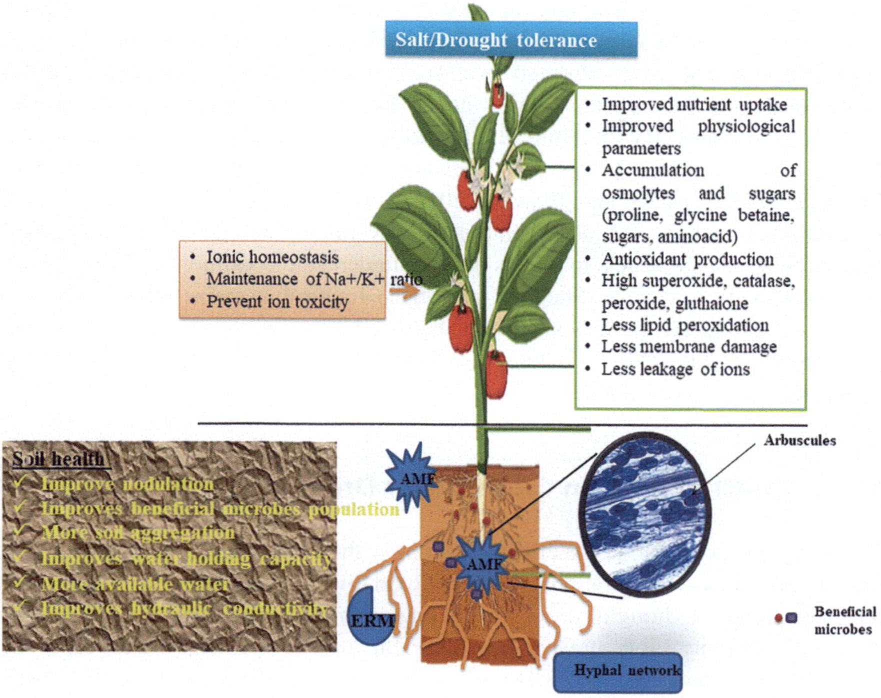

Fig. 19.1 Mechanism of AMF colonization in alleviation of salt and drought stress

exchange properties, chlorophyll content, and water use efficiency in *Ocimum basilicum* L. (Elhindi et al. 2017).

Under saline conditions, AMF-inoculated *Allium sativum* plants exhibited superior growth characteristics such as leaf area index and fresh and dry biomass (Borde et al. 2010). Wang et al. (2018) recently showed that mycorrhizal inoculation under moderate salinity conditions significantly improved shoot and root fresh and dry weights and N concentrations. In addition, plants containing AMF have enhanced synthesis of salicylic acid, jasmonic acid, and many important mineral nutrients. For example, AMF-treated *Cucumis sativus* plants had increased amounts of total P, Ca^{2+}, N, Mg^{2+}, and K^{+} under salt stress conditions compared to untreated plants (Hashem et al. 2018).

Under saline conditions, mycorrhizal inoculation of hot pepper increased chlorophyll levels, Mg^{2+} and N uptake, and decreased Na^{+} transport (Çekiç et al. 2012). In addition, Santander et al. (2019) using lettuce demonstrated that stressed mycorrhizal plants produce more biomass, increase proline synthesis, take up more nitrogen, and have significant changes in ionic relationships, especially reduced Na^{+} accumulation. The concentrations of key growth regulators can be efficiently controlled by AMF inoculation. Under salt stress, several studies reported that AMF-mediated increases in cytokinin concentrations lead to significant photosynthetic translocation

(Hameed et al. 2014; Talaat and Shawky 2014). In addition, an altered polyamine pool was found to be responsible for AMF-mediated growth promotion under salt stress (Kapoor et al. 2013). Furthermore, Aroca et al. (2013) showed how the effects of salt on lettuce plants were significantly reduced by increasing strigolactones in AMF-treated plants. By inhibiting lipid membrane peroxidation in response to salt stress, AMF-colonized plants can alleviate oxidative stress (Abdel Latef and Miransari 2014; Talaat and Shawky 2014). Furthermore, we found that AMF inoculation increased the accumulation of several organic acids and enhanced osmoregulatory processes in plants grown under salt conditions (Zhang et al. 2017, 2020). For instance, Sheng et al. (2011) found that maize plants grown in saline soil produced more betaine and increased specific organic acid synthesis, suggesting that his AMF in osmoregulation of plants under salt stress indirectly confirmed their involvement.

19.5 Effect of AMF on Crops Under Drought Stress

Many factors contribute to the effects of drought stress on plants. For example, root dehydration slows transpiration and causes oxidative stress (Impa et al. 2012). The adverse effects of drought stress on plant growth are caused by the changes in enzymatic activity, ion uptake, and nutrient uptake (Ahanger and Agarwal 2017a, b; Ahanger et al. 2017). However, there is compelling evidence that AMF reduces the effects of drought stress through several mechanisms (Fig. 19.1) in various crops such as wheat, barley, maize, soybean, strawberry, and onion (Mena-Violante et al. 2006; Yooyongwech et al. 2016; Moradtalab et al. 2019). Plant resistance to drought can be attributed primarily to roots that have access to many soils and fungi with extra-root mycelium (Gianinazzi et al. 2010; Gutjahr and Paszkowski 2013). Various physiological and biochemical processes in plants are associated with increased osmotic adaptation (Kubikova et al. 2001), stomatal regulation by regulation of ABA metabolism (Duan et al. 1996), proline accumulation (Ruiz-Sánchez et al. 2010; Yooyongwech et al. 2013) or increased glutathione (Rani 2016). Finally, under impending drought conditions, the symbiotic relationship between different plants and AMF may increase root size and efficiency, leaf area index, and biomass (Al-Karaki et al. 2004; Gholamhoseini et al. 2013). In addition, AMF and host plants interact and help protect plants from harsh environmental conditions (Ruiz-Lozano 2003). In addition, AMF symbiosis improves stomatal conductivity, leaf water status, gas exchange, and transpiration rate (Morte et al. 2000; Mena-Violante et al. 2006). AMF can support ABA responses that regulate stomatal conductance and other related physiological processes. By upregulating the antioxidant system, Li et al. (2019) recently showed that her AMF-mediated increase in growth and photosynthesis occurs in plant species of C3 (*Leymus chinensis*) and C4 (*Hemarthria altissima*).

19.6 Role of AM Fungi in Plant Growth Promotion

Arbuscular mycorrhizal fungi (AMF) are a group of soil microorganisms that plays an important role in maintaining plant diversity and above-ground productivity in ecosystems (Collins and Foster 2009). AMF is often viewed as a traditional mutualist that promotes plant growth (van der Heijden et al. 1998, 2006). AMF provides soil nutrients to plants in exchange for photosynthesis (Johnson 2010). The costs and benefits of AMF-plant symbiosis can be altered by nutrient supplementation, leading to antagonism (Johnson et al. 1997; Grman 2012; Grman and Robinson 2013). However, the effects of nutritional supplementation on plant-mycorrhizal interactions are not yet fully understood (Yang et al. 2014). The interaction of plants and AMF in this context is defined by plant growth responses. Adverse plant growth is associated with antagonism, whereas positive plant growth is associated with mutualism in his AMF (Johnson 2010; Grman 2012). The term "parasitism" is synonymous with the term "antagonism" (Yang et al. 2014). If the net costs to plants exceed the net benefits to fungi, nutrient additions can have antagonistic effects (Grman 2012).

19.7 Mechanisms of AMF

19.7.1 Increased Nutrient Supply

AMF have tremendous power to change nutritional elements from an unavailable to an available form via biological processes (Ma et al. 2019a, b). A variety of environmental factors, such as soil type and characteristics, metal pollution, and abiotic stress, have a significant impact on their performance (e.g., drought and salinity). Mycorrhizal association with drought-stressed plants can increase nutrient accumulation by promoting the formation of large hyphal networks and glomalin secretion, both of which promote water and nutrient uptake (Bahadur et al. 2019). Salinity is known to limit plant growth by causing plants to have higher Na^+ concentrations and lower K^+/Na^+ ratios. Over a quarter of a million words and a half of a million pounds (Bharti et al. 2014; Tewari and Arora 2014). Kohler et al. (2009) investigated the effects of inoculation of plant growth-promoting rhizobacterium *Pseudomonas mendocina* on the growth, nutrient accumulation, and physiological parameters of *Lactuca sativa* exposed to salt stress using AMF *Glomus intraradices* or *Glomus mosseae* alone or in combination. In comparison to the control treatments, *L. sativa* inoculated with *P. mendocina* reduced plant Na^+ uptake while increasing K^+ uptake, resulting in a higher K^+/Na^+ ratio and higher shoot biomass. This finding suggests that AMF could be used to alleviate salinity stress in salt-sensitive plants. Increased K^+ transporter (HKT1) transcriptional expression was downregulated in *Arabidopsis* roots by volatile organic chemicals produced by these beneficial microorganisms, but increased in shoots, resulting in decreased Na^+ uptake by roots and increased Na^+ expulsion from shoots (Zhang et al. 2008).

19.7.2 Biosynthesis of Osmoprotectants

AMF can produce suitable osmolytes in response to drought or salinity stress, which work in tandem with osmolytes released by plants (such as proline, trehalose, and polyamines) to promote plant growth and development (Wu et al. 2013). Plant stress tolerance has been strongly linked to the ability to accumulate proline under stress (Nxele et al. 2017). By directly scavenging ROS and stabilizing proteins and membranes, proline has a tremendous ability to reduce lipid peroxidation and modify cytosolic pH under salinity and drought stress (Gill and Tuteja 2010). Numerous studies have found that AMF-inoculated plants have higher proline content, which improves their ability to withstand salinity and drought stress (Bharti et al. 2014; Shintu and Jayaram 2015). However, it is unclear whether it is derived from rhizosphere soils or as a result of the proline biosynthesis pathway being activated (Herbinger et al. 2002). A number of studies, on the other hand, discovered that AMF inoculation reduced proline concentration in plants under salinity and drought stress (Ma et al. 2017, 2019a; Singh and Jha 2017). This is due to the possibility that AMF can mitigate the negative effects of salinity and drought by regulating osmotic balance and preserving the bioenergetics of plant cells. Trehalose also has a significant impact on reducing the damage that salinity and drought do to plant cells because it is a high-level glucoside. This could be explained by trehalose's ability to keep proteins and membranes stable. It was recently reported that salt-stressed mycorrhizal plants accumulate more trehalose than non-mycorrhizal plants. This may be because trehalose-6-phosphate synthase and phosphatase, two enzymes involved in the production of trehalose, have increased activity when AMF is present, but the activity of trehalose-degrading enzymes, e.g., trehalase, has decreased (Garg and Pandey 2016). Almost all organisms produce polyamines, which are biogenic amines with an aliphatic nitrogen structure and are involved in a variety of plant growth and development processes, including cell division and differentiation, root growth, flower development, fruit ripening, senescence, programmed cell death, DNA replication, and transcription Translation (Alcázar et al. 2011). According to Cassan et al. (2009) inoculation of Oryza seedlings under osmotic stress with the cadaverine (polyamine)-producing *A. brasilense* Az39 significantly accelerated root growth (Cassan et al. 2009). Zhang et al. (2019) recently performed a pot experiment to assess the function of AMF in *Poncirus trifoliata's* root polyamine homeostasis, activity, and gene expression of polyamine-related synthesizing and degrading enzymes. The findings demonstrate that mycorrhizas can enhance plant drought tolerance through modulation of polyamine metabolism by increasing putrescine and cadaverine levels as well as polyamine catabolic enzyme and putrescine synthase activity under drought stress (Ma et al. 2020).

19.7.3 Synthesis of antioxidant enzymes

Abiotic stresses such as high salinity or drought leads to overproduction of ROS, leading to altered cellular redox homeostasis (Hare et al. 1998). The increased ROS level causes inactivation of membrane-bound proteins, reduced membrane fluidity, DNA damage, inhibition of protein synthesis and enzymatic activities. There is considerable evidence that plants inoculated with AMF can survive under abiotic stress-induced oxidative stress through manipulation of antioxidant enzymes. Pedranzani et al. (2016) found that application of AMF *Rhizophagus irregularis* improved the physiological performance of *Digitaria eriantha* under drought, salinity, and cold stress by upregulating antioxidant enzyme activity (catalase (CAT) and ascorbate peroxidase (APX)) and jasmonate synthesis. Recently reports proved that application of the AMF stimulated the activity of CAT and superoxide dismutase (SOD) under different stress conditions (e.g., drought, salinity). Although plants inoculated with AMF have been shown to mitigate oxidative damage, the underlying mechanisms behind AMF-induced changes in antioxidant enzyme activities are poorly understood. Many factors, such as host plant species, AMF type and strain, as well as the type, degree and duration of abiotic stress, can be responsible for such changes in enzyme levels.

19.7.4 Modification in Plant Physiology and Root Structure

Drought is the greatest constraint on crop growth and the spread of natural plant communities worldwide, and several articles address the physics and physiology of water uptake by roots. Bi et al. (2019) suggested that AMF can alleviate root damage stress caused by cracking, but the underlying mechanisms are unclear. While, under root damage stress, the hormone levels in root and leaf of AMF-inoculated plants differed from those of uninoculated controls. Mycorrhizal plants alleviates changes in root and leaf hormone levels caused by root damage stress and promoted root development, particularly by increasing fine root fraction and leaf area and chlorophyll content. Zheng et al. (2022) suggested that AMF inoculation may alleviate stress from root damage by altering hormone levels and leaf and root response to crack-induced root damage. The experiment was developed by Zheng et al. (2022) concluded that the proportion of fine roots at 20 cm depth differed significantly between different treatments of mycorrhiza. However, the percentage of fine roots at 40 cm depth did not differ significantly between the different treatments, probably due to an inadequate root system. Fine roots can respond to environmental stress by self-regulating and changing growth mode (Pregitzer et al. 2002), and these are the main nutrient-absorbing roots. A higher proportion of fine roots have a positive effect on nutrient uptake and resistance to abiotic stress indicating that AMF can affect plant root development. For instance, AMF has been found to significantly increase root hair density of orange trees (Wu et al. 2016) and improve growth of wheat under drought stress by regulating endogenous IAA levels and promoting lateral root formation (Arzanesh et al. 2011). Mycorrhizal-induced changes in root hair growth are closely related to upregulated expression of auxin-related genes and concomitant accumulation of IAA in roots (Liu et al. 2018). Thus, the results show that inoculation with AMF significantly increased root IAA levels, 1 week after soil crevice interference (Zheng et al. 2022). Some studies show that while high concentrations of IAA inhibit primary root elongation, it stimulates lateral and adventitious root growth (Stoeckle et al. 2018). In horticulture, the stimulating effects of IAA on adventitious and lateral roots are also often used for vegetative propagation. The proportion of fine roots was significantly positively correlated with root IAA. Higher IAA levels in mycorrhizal plants can stimulate lateral root growth, leading to a significant increase in the proportion of fine roots. After soil splitting, AMF significantly increased the number of root tips, total root length, surface area and volume. However, AMF did not significantly affect the morphology of the root system of some plants (Zheng et al. 2022; Lenoir et al. 2016).

19.8 Production of phytohormones

AMF produces IAA, GA, ABA, and other phytohormones that promote plant growth and help plants to adapt and tolerate environmental challenges (Ma et al. 2019a, b). IAA and GA synthesis from AMF improved plant health under drought and salinity stress by increasing plant growth and nutrient uptake (e.g., root length, root area, and number of root tips) (Egamberdieva and Kucharova 2009; Ma et al. 2016, 2017). According to the several studies, AMF can promote the growth of several plant species which inludes *S. lycopersicum, Solanum pimpinellifolium, Capsicum annuum, B. napus, Helianthus annuus, P. vulgaris, T. aestivum,* and *L. sativa* under drought and saline conditions (Yildirim and Taylor 2005; Barassi et al. 2006; Ma et al. 2016, 2017, 2019a; Khan et al. 2017). Abscisic acid produced by AMF can also help plants to withstand salinity or drought. In addition, inoculation of AMF significantly improved the growth and proline content of *Glycine max* under salt stress. Similarly, the ability of *O. sativa* to tolerate salt stress was significantly improved by inoculation with AMF. This was achieved by altering the endogenous hormone and incorporating essential amino acids (e.g., glutamic acid, aspartic acid, phenylalanine, proline, and cysteine) (Shahzad et al. 2017).

19.9 Conclusion

Arbuscular mycorrhizal fungi play a variety of roles in the development of various plant species. AMF influence nutrient availability and absorption, increase photosynthetic rate, enhance antioxidant processes, and increase resilience to environmental stress. In addition to improving primary and secondary metabolites, AMF also increases growth, yield, and productivity. Although AM fungi are currently used on a small scale in greenhouses and agricultural systems, they have much broader potential for application hence further evaluation needs to be carried out for the better usage of such valuable eco-friendly biofertilizer.

References

Abdel Latef AAH, Miransari M (2014) The role of arbuscular mycorrhizal fungi in alleviation of salt stress. In: Use of microbes for the alleviation of soil stresses. Springer, New York, pp 23–38

Abdel-Fattah GM, Asrar AWA (2012) Arbuscular mycorrhizal fungal application to improve growth and tolerance of wheat (*Triticum aestivum* L.) plants grown in saline soil. Acta Physiol Plant 34(1):267–277

Ahanger MA, Agarwal RM (2017a) Potassium up-regulates antioxidant metabolism and alleviates growth inhibition under water and osmotic stress in wheat (*Triticum aestivum* L). Protoplasma 254(4):1471–1486

Ahanger MA, Agarwal RM (2017b) Salinity stress induced alterations in antioxidant metabolism and nitrogen assimilation in wheat (*Triticum aestivum* L) as influenced by potassium supplementation. Plant Physiol Biochem 115:449–460

Ahanger MA, Tomar NS, Tittal M, Argal S, Agarwal R (2017) Plant growth under water/salt stress: ROS production; antioxidants and significance of added potassium under such conditions. Physiol Mol Biol Plants 23(4):731–744

Ahanger MA, Alyemeni MN, Wijaya L, Alamri SA, Alam P, Ashraf M, Ahmad P (2018) Potential of exogenously sourced kinetin in protecting *Solanum lycopersicum* from NaCl-induced oxidative stress through up-regulation of the antioxidant system, ascorbate-glutathione cycle and glyoxalase system. PLoS One 13(9):e0202175

Ait-El-Mokhtar M, Laouane RB, Anli M, Boutasknit A, Wahbi S, Meddich A (2019) Use of mycorrhizal fungi in improving tolerance of the date palm (*Phoenix dactylifera* L.) seedlings to salt stress. Sci Hortic 253:429–438

Alcázar R, Bitrián M, Bartels D, Koncz C, Altabella T, Tiburcio AF (2011) Polyamine metabolic canalization in response to drought stress in Arabidopsis and the resurrection plant *Craterostigma plantagineum*. Plant Signal Behav 6(2):243–250

Al-Karaki GN, Al-Omoush M (2002) Wheat response to phosphogypsum and mycorrhizal fungi in alkaline soil. J Plant Nutr 25:873

Al-Karaki G, McMichael B, Zak J (2004) Field response of wheat to arbuscular mycorrhizal fungi and drought stress. Mycorrhiza 14(4):263–269

Alqarawi AA, Abd Allah EF, Hashem A (2014a) Alleviation of salt-induced adverse impact via mycorrhizal fungi in *Ephedra aphylla* Forssk. J Plant Interact 9(1):802–810

Alqarawi AA, Hashem A, Abd Allah EF, Alshahrani TS, Huqail AA (2014b) Effect of salinity on moisture content, pigment system, and lipid composition in *Ephedra alata* Decne. Acta Biol Hung 65(1):61–71

Arif Y, Singh P, Siddiqui H, Bajguz A, Hayat S (2020) Salinity induced physiological and biochemical changes in plants: an omic approach towards salt stress tolerance. Plant Physiol Biochem 156:64–77

Aroca R, Ruiz-Lozano JM, Zamarreño ÁM, Paz JA, García-Mina JM, Pozo MJ, López-Ráez JA (2013) Arbuscular mycorrhizal symbiosis influences strigolactone production under salinity and alleviates salt stress in lettuce plants. J Plant Physiol 170(1):47–55

Arzanesh MH, Alikhani HA, Khavazi K, Rahimian HA, Miransari M (2011) Wheat (*Triticum aestivum* L.) growth enhancement by *Azospirillum* sp. under drought stress. World J Microbiol Biotechnol 27(2):197–205

Athar HR, Ashraf M (2009) Strategies for crop improvement against salinity and drought stress: an overview. In: Salinity and water stress. Springer, Dordrecht, pp 1–16

Bahadur A, Batool A, Nasir F, Jiang S, Mingsen Q, Zhang Q, Pan J, Liu Y, Feng H (2019) Mechanistic insights into arbuscular mycorrhizal fungi-mediated drought stress tolerance in plants. Int J Mol Sci 20(17):4199

Barassi CA, Ayrault G, Creus CM, Sueldo RJ, Sobrero MT (2006) Seed inoculation with *Azospirillum* mitigates NaCl effects on lettuce. Sci Hortic 109(1):8–14

Barrow CJ (2012) Biochar: potential for countering land degradation and for improving agriculture. Appl Geogr 34:21–28

Bárzana G, Aroca R, Paz JA, Chaumont F, Martinez-Ballesta MC, Carvajal M, Ruiz-Lozano JM (2012) Arbuscular mycorrhizal symbiosis increases relative apoplastic water flow in roots of the host plant under both well-watered and drought stress conditions. Ann Bot 109(5):1009–1017

Begum N, Qin C, Ahanger MA, Raza S, Khan MI, Ashraf M, Zhang L (2019) Role of arbuscular mycorrhizal fungi in plant growth regulation: implications in abiotic stress tolerance. Front Plant Sci 10:1068

Beltrano J, Ronco MG (2008) Improved tolerance of wheat plants (*Triticum aestivum* L.) to drought stress and rewatering by the arbuscular mycorrhizal fungus *Glomus claroideum*: effect on growth and cell membrane stability. Braz J Plant Physiol 20:29–37

Bernardo L, Carletti P, Badeck FW, Rizza F, Morcia C, Ghizzoni R, Lucini L (2019) Metabolomic responses triggered by arbuscular mycorrhiza enhance tolerance to water stress in wheat cultivars. Plant Physiol Biochem 137:203–212

Bharti N, Barnawal D, Awasthi A, Yadav A, Kalra A (2014) Plant growth promoting rhizobacteria alleviate salinity induced negative effects on growth, oil content and physiological status in Mentha arvensis. Acta Physiol Plant 36(1):45–60

Bi Y, Zhang J, Song Z, Wang Z, Qiu L, Hu J, Gong Y (2019) Arbuscular mycorrhizal fungi alleviate root damage stress induced by simulated coal mining subsidence ground fissures. Sci Total Environ 652:398–405

Bieleski RL (1973) Phosphate pools, phosphate transport, and phosphate availability. Annu Rev Plant Physiol 24:225–252

Bisht N, Mishra SK, Chauhan PS (2020) *Bacillus amyloliquefaciens* inoculation alters physiology of rice (*Oryza sativa* L. var. IR-36) through modulating carbohydrate metabolism to mitigate stress induced by nutrient starvation. Int J Biol Macromol 143:937–951

Borde M, Dudhane M, Jite PK (2010) AM fungi influences the photosynthetic activity, growth and antioxidant enzymes in *Allium sativum* L. under salinity condition. Notulae Scientia Biologicae 2(4):64–71

Cassan F, Maiale S, Masciarelli O, Vidal A, Luna V, Ruiz O (2009) Cadaverine production by *Azospirillum brasilense* and its possible role in plant growth promotion and osmotic stress mitigation. Eur J Soil Biol 45(1):12–19

Çekiç FÖ, Ünyayar S, Ortaş İ (2012) Effects of arbuscular mycorrhizal inoculation on biochemical parameters in *Capsicum annuum* grown under long term salt stress. Turk J Bot 36(1):63–72

Chandrasekaran M, Chanratana M, Kim K, Seshadri S, Sa T (2019) Impact of arbuscular mycorrhizal fungi on photosynthesis, water status, and gas exchange of plants under salt stress—a meta-analysis. Front Plant Sci 10:457

Chiariello N, Hickman JC, Mooney HA (1982) Endomycorrhizal role for interspecific transfer of phosphorus in a community of annual plants. Science 217(4563):941–943

Collins CD, Foster BL (2009) Community-level consequences of mycorrhizae depend on phosphorus availability. Ecology 90(9):2567–2576

Cruz de Carvalho MH (2008) Drought stress and reactive oxygen species: production, scavenging and signaling. Plant Signal Behav 3(3):156–165

Dat J, Vandenabeele S, Vranova EVMM, Van Montagu M, Van Breusegem F (2000) Dual action of the active oxygen species during plant stress responses. Cell Mol Life Sci 57(5):779–795

Duan X, Neuman DS, Reiber JM, Green CD, Saxton AM, Augé RM (1996) Mycorrhizal influence on hydraulic and hormonal factors implicated in the control of stomatal conductance during drought. J Exp Bot 47(10):1541–1550

Dugasa MT, Cao F, Ibrahim W, Wu F (2019) Differences in physiological and biochemical characteristics in response to single and combined drought and salinity stresses between wheat genotypes differing in salt tolerance. Physiol Plant 165(2):134–143

Egamberdieva D, Kucharova Z (2009) Selection for root colonising bacteria stimulating wheat growth in saline soils. Biol Fertil Soils 45(6):563–571

Elhindi KM, El-Din AS, Elgorban AM (2017) The impact of arbuscular mycorrhizal fungi in mitigating salt-induced adverse effects in sweet basil (*Ocimum basilicum* L.). Saudi J Biol Sci 24:170–179

El-Nashar YI (2017) Response of snapdragon (*Antirrhinum majus* L.) to blended water irrigation and arbuscular mycorrhizal fungi inoculation: uptake of minerals and leaf water relations. Photosynthetica 55(2):201–209

Fasciglione G, Casanovas EM, Quillehauquy V, Yommi AK, Goñi MG, Roura SI, Barassi CA (2015) *Azospirillum* inoculation effects on growth, product quality and storage life of lettuce plants grown under salt stress. Sci Hortic 195:154–162

Francis R, Read DJ (1984) Direct transfer of carbon between plants connected by vesicular–arbuscular mycorrhizal mycelium. Nature 307(5946):53–56

Francis RM, Read DJ (1995) Mutualism and antagonism in the mycorrhizal symbiosis, with special reference to impacts on plant community structure. Can J Bot 73(S1):1301–1309

Ganugi P, Masoni A, Pietramellara G, Benedettelli S (2019) A review of studies from the last twenty years on plant–arbuscular mycorrhizal fungi associations and their uses for wheat crops. Agronomy 9(12):840

Garg N, Pandey R (2016) High effectiveness of exotic arbuscular mycorrhizal fungi is reflected in improved rhizobial symbiosis and trehalose turnover in *Cajanus cajan* genotypes grown under salinity stress. Fungal Ecol 21:57–67

Gerdemann JW (1968) Vesicular-arbuscular mycorrhiza and plant growth. Annu Rev Phytopathol 6(1):397–418

Gholamhoseini M, Ghalavand A, Dolatabadian A, Jamshidi E, Khodaei-Joghan A (2013) Effects of arbuscular mycorrhizal inoculation on growth, yield, nutrient uptake and irrigation water productivity of sunflowers grown under drought stress. Agric Water Manag 117:106–114

Gianinazzi S, Gollotte A, Binet MN, van Tuinen D, Redecker D, Wipf D (2010) Agroecology: the key role of arbuscular mycorrhizas in ecosystem services. Mycorrhiza 20(8):519–530

Gill SS, Tuteja N (2010) Reactive oxygen species and antioxidant machinery in abiotic stress tolerance in crop plants. Plant Physiol Biochem 48(12):909–930

Giovannetti M, Mosse B (1980) An evaluation of techniques for measuring vesicular arbuscular mycorrhizal infection in roots (cowpeas, maize, beans). New Phytol 84:489

Giovannetti M, Avio L, Sbrana C (2010) Fungal spore germination and pre-symbiotic mycelial growth–physiological and genetic aspects. In: Arbuscular mycorrhizas: physiology and function. Springer, Berlin, pp 3–32

Gowtham HG, Singh B, Murali M, Shilpa N, Prasad M, Aiyaz M, Niranjana SR (2020) Induction of drought tolerance in tomato upon the application of ACC deaminase producing plant growth promoting rhizobacterium *Bacillus subtilis* Rhizo SF 48. Microbiol Res 234:126422

Grman E (2012) Plant species differ in their ability to reduce allocation to non-beneficial arbuscular mycorrhizal fungi. Ecology 93(4):711–718

Grman E, Robinson TM (2013) Resource availability and imbalance affect plant–mycorrhizal interactions: a field test of three hypotheses. Ecology 94(1):62–71

Guo J, Sun B, He H, Zhang Y, Tian H, Wang B (2021a) Current understanding of bHLH transcription factors in plant abiotic stress tolerance. Int J Mol Sci 22(9):4921

Guo Q, Li X, Niu L, Jameson PE, Zhou W (2021b) Transcription-associated metabolomic adjustments in maize occur during combined drought and cold stress. Plant Physiol 186(1): 677–695

Gupta A, Rico-Medina A, Caño-Delgado AI (2020) The physiology of plant responses to drought. Science 368(6488):266–269

Gupta A, Bano A, Rai S, Dubey P, Khan F, Pathak N, Sharma S (2021a) Plant growth promoting rhizobacteria (PGPR): a sustainable agriculture to rescue the vegetation from the effect of biotic stress: a review. Lett Appl NanoBiosci 10:2459–2465

Gupta A, Bano A, Rai S, Kumar M, Ali J, Sharma S, Pathak N (2021b) ACC deaminase producing plant growth promoting rhizobacteria enhance salinity stress tolerance in Pisum sativum. 3 Biotech 11(12):1–17

Gupta A, Rai S, Bano A, Khanam A, Sharma S, Pathak N (2021c) Comparative evaluation of different salt-tolerant plant growth-promoting bacterial isolates in mitigating the induced adverse effect of salinity in *Pisum sativum*. Biointerface Res Appl Chem 11(5):13141–13154

Gupta A, Bano A, Rai S, Mishra R, Singh M, Sharma S, Pathak N (2022) Mechanistic insights of plant-microbe interaction towards drought and salinity stress in plants for enhancing the agriculture productivity. Plant Stress 4:100073

Gutjahr C, Paszkowski U (2013) Multiple control levels of root system remodeling in arbuscular mycorrhizal symbiosis. Front Plant Sci 4:204

Hameed A, Dilfuza E, Abd-Allah EF, Hashem A, Kumar A, Ahmad P (2014) Salinity stress and arbuscular mycorrhizal symbiosis in plants. In: Use of microbes for the alleviation of soil stresses. Springer, Berlin, pp 139–159

Hare PD, Cress WA, Van Staden J (1998) Dissecting the roles of osmolyte accumulation during stress. Plant Cell Environ 21(6):535–553

Hashem A, Abd Allah EF, Alqarawi AA, Aldubise A, Egamberdieva D (2015) Arbuscular mycorrhizal fungi enhances salinity tolerance of *Panicum turgidum* Forssk by altering photosynthetic and antioxidant pathways. J Plant Interact 10(1):230–242

Hashem A, Alqarawi AA, Radhakrishnan R, Al-Arjani ABF, Aldehaish HA, Egamberdieva D, Abd Allah EF (2018) Arbuscular mycorrhizal fungi regulate the oxidative system, hormones and ionic equilibrium to trigger salt stress tolerance in *Cucumis sativus* L. Saudi J Biol Sci 25(6): 1102–1114

He F, Sheng M, Tang M (2017a) Effects of *Rhizophagus irregularis* on photosynthesis and antioxidative enzymatic system in *Robinia pseudoacacia* L. under drought stress. Front Plant Sci 8:183

He L, Li C, Liu R (2017b) Indirect interactions between *arbuscular mycorrhizal* fungi and *Spodoptera exigua* alter photosynthesis and plant endogenous hormones. Mycorrhiza 27(6): 525–535

Herbinger K, Tausz M, Wonisch A, Soja G, Sorger A, Grill D (2002) Complex interactive effects of drought and ozone stress on the antioxidant defence systems of two wheat cultivars. Plant Physiol Biochem 40(6–8):691–696

Hosseini SZ, Ismaili A, Nazarian-Firouzabadi F, Fallahi H, Nejad AR, Sohrabi SS (2021) Dissecting the molecular responses of lentil to individual and combined drought and heat stresses by comparative transcriptomic analysis. Genomics 113(2):693–705

Impa SM, Nadaradjan S, Jagadish SVK (2012) Drought stress induced reactive oxygen species and anti-oxidants in plants. In: Abiotic stress responses in plants. Springer, Berlin, pp 131–147

Johnson NC (2010) Resource stoichiometry elucidates the structure and function of arbuscular mycorrhizas across scales. New Phytol 185(3):631–647

Johnson NC, Graham JH, Smith FA (1997) Functioning of mycorrhizal associations along the mutualism–parasitism continuum. New Phytol 135(4):575–585

Kapoor R, Evelin H, Mathur P, Giri B (2013) Arbuscular mycorrhiza: approaches for abiotic stress tolerance in crop plants for sustainable agriculture. In: Plant acclimation to environmental stress. Springer, Berlin, pp 359–401

Khalloufi M, Martínez-Andújar C, Lachaâl M, Karray-Bouraoui N, Pérez-Alfocea F, Albacete A (2017) The interaction between foliar GA3 application and arbuscular mycorrhizal fungi inoculation improves growth in salinized tomato (*Solanum lycopersicum* L.) plants by modifying the hormonal balance. J Plant Physiol 214:134–144

Khan AL, Waqas M, Asaf S, Kamran M, Shahzad R, Bilal S, Lee IJ (2017) Plant growth-promoting endophyte *Sphingomonas* sp. LK11 alleviates salinity stress in *Solanum pimpinellifolium*. Environ Exp Bot 133:58–69

Kohler J, Hernández JA, Caravaca F, Roldán A (2009) Induction of antioxidant enzymes is involved in the greater effectiveness of a PGPR versus AM fungi with respect to increasing the tolerance of lettuce to severe salt stress. Environ Exp Bot 65(2–3):245–252

Kubikova E, Jennifer LM, Bonnie HO, Michael DM, Augé MR (2001) Mycorrhizal impact on osmotic adjustment in *Ocimum basilicum* during a lethal drying episode. J Plant Physiol 158(9): 1227–1230

Kumar M, Mishra S, Dixit V, Kumar M, Agarwal L, Chauhan PS, Nautiyal CS (2016) Synergistic effect of *Pseudomonas putida* and *Bacillus amyloliquefaciens* ameliorates drought stress in chickpea (*Cicer arietinum* L.). Plant Signal Behav 11(1):e1071004

Lenoir I, Fontaine J, Sahraoui ALH (2016) Arbuscular mycorrhizal fungal responses to abiotic stresses: a review. Phytochemistry 123:4–15

Levy D, Coleman WK, Veilleux RE (2013) Adaptation of potato to water shortage: irrigation management and enhancement of tolerance to drought and salinity. Am J Potato Res 90(2): 186–206

Li X, Li M, Zhou B, Yang Y, Wei Q, Zhang J (2019) Transcriptome analysis provides insights into the stress response crosstalk in apple (*Malus domestica*) subjected to drought, cold and high salinity. Sci Rep 9(1):1–10

Li PC, Yang XY, Wang HM, Ting PAN, Yang JY, Wang YY, Xu CW (2021) Metabolic responses to combined water deficit and salt stress in maize primary roots. J Integr Agric 20(1):109–119

Liu C, Ravnskov S, Liu F, Rubæk GH, Andersen MN (2018) Arbuscular mycorrhizal fungi alleviate abiotic stresses in potato plants caused by low phosphorus and deficit irrigation/partial root-zone drying. J Agric Sci 156(1):46–58

Ma T, Zeng W, Li Q, Wu J, Huang J (2016) Effects of water, salt and nitrogen stress on sunflower (*Helianthus annuus* L.) at different growth stages. J Soil Sci Plant Nutr 16(4):1024–1037

Ma Y, Rajkumar M, Moreno A, Zhang C, Freitas H (2017) Serpentine endophytic bacterium *Pseudomonas azotoformans* ASS1 accelerates phytoremediation of soil metals under drought stress. Chemosphere 185:75–85

Ma Y, Rajkumar M, Oliveira RS, Zhang C, Freitas H (2019a) Potential of plant beneficial bacteria and arbuscular mycorrhizal fungi in phytoremediation of metal-contaminated saline soils. J Hazard Mater 379:120813

Ma Y, Vosátka M, Freitas H (2019b) Beneficial microbes alleviate climatic stresses in plants. Front Plant Sci 10:595

Ma Y, Dias MC, Freitas H (2020) Drought and salinity stress responses and microbe-induced tolerance in plants. Front Plant Sci 11:591911

Mathur S, Sharma MP, Jajoo A (2018) Improved photosynthetic efficacy of maize (*Zea mays*) plants with arbuscular mycorrhizal fungi (AMF) under high temperature stress. J Photochem Photobiol B Biol 180:149–154

Mathur S, Tomar RS, Jajoo A (2019) Arbuscular mycorrhizal fungi (AMF) protects photosynthetic apparatus of wheat under drought stress. Photosynth Res 139(1):227–238

Mena-Violante HG, Ocampo-Jiménez O, Dendooven L, Martínez-Soto G, González-Castañeda J, Davies FT, Olalde-Portugal V (2006) Arbuscular mycorrhizal fungi enhance fruit growth and quality of Chile ancho (*Capsicum annuum* L. cv San Luis) plants exposed to drought. Mycorrhiza 16(4):261–267

Mickelbart MV, Hasegawa PM, Bailey-Serres J (2015) Genetic mechanisms of abiotic stress tolerance that translate to crop yield stability. Nat Rev Genet 16(4):237–251

Mittler R (2002) Oxidative stress, antioxidants and stress tolerance. Trends Plant Sci 7(9):405–410

Moradtalab N, Hajiboland R, Aliasgharzad N, Hartmann TE, Neumann G (2019) Silicon and the association with an arbuscular-mycorrhizal fungus (*Rhizophagus clarus*) mitigate the adverse effects of drought stress on strawberry. Agronomy 9(1):41

Morte A, Lovisolo C, Schubert A (2000) Effect of drought stress on growth and water relations of the mycorrhizal association *Helianthemum almeriense*-Terfezia claveryi. Mycorrhiza 10(3): 115–119

Navarro JM, Pérez-Tornero O, Morte A (2014) Alleviation of salt stress in citrus seedlings inoculated with arbuscular mycorrhizal fungi depends on the rootstock salt tolerance. J Plant Physiol 171(1):76–85

Naylor D, Coleman-Derr D (2018) Drought stress and root-associated bacterial communities. Front Plant Sci 8:2223

Newman EI (1988) Mycorrhizal links between plants: their functioning and ecological significance. In: Advances in ecological research, vol 18. Academic, New York, pp 243–270

Nxele X, Klein A, Ndimba BK (2017) Drought and salinity stress alters ROS accumulation, water retention, and osmolyte content in sorghum plants. S Afr J Bot 108:261–266
Osakabe Y, Osakabe K, Shinozaki K, Tran LSP (2014) Response of plants to water stress. Front Plant Sci 5:86
Osthoff A, Donà dalle Rose P, Baldauf JA, Piepho HP, Hochholdinger F (2019) Transcriptomic reprogramming of barley seminal roots by combined water deficit and salt stress. BMC Genomics 20(1):1–14
Parihar M, Rakshit A, Rana K, Tiwari G, Jatav SS (2020) The effect of arbuscular mycorrhizal fungi inoculation in mitigating salt stress of pea (*Pisum Sativum* L.). Commun Soil Sci Plant Anal 51(11):1545–1559
Patel MK, Kumar M, Li W, Luo Y, Burritt DJ, Alkan N, Tran LSP (2020) Enhancing salt tolerance of plants: from metabolic reprogramming to exogenous chemical treatments and molecular approaches. Cell 9(11):2492
Pedranzani H, Rodríguez-Rivera M, Gutiérrez M, Porcel R, Hause B, Ruiz-Lozano JM (2016) Arbuscular mycorrhizal symbiosis regulates physiology and performance of *Digitaria eriantha* plants subjected to abiotic stresses by modulating antioxidant and jasmonate levels. Mycorrhiza 26(2):141–152
Plassard C, Dell B (2010) Phosphorus nutrition of mycorrhizal trees. Tree Physiol 30(9):1129–1139
Prasch CM, Sonnewald U (2013) Simultaneous application of heat, drought, and virus to Arabidopsis plants reveals significant shifts in signaling networks. Plant Physiol 162(4): 1849–1866
Pregitzer KS, DeForest JL, Burton AJ, Allen MF, Ruess RW, Hendrick RL (2002) Fine root architecture of nine north American trees. Ecol Monogr 72:293–309
Pringle A, Bever JD, Gardes M, Parrent JL, Rillig MC, Kilronomos JN (2009) Mycorrhizal symbioses and plant invasions. Annu Rev Ecol Evol Syst 40:699–715
Rani B (2016) Effect of arbuscular mycorrhiza fungi on biochemical parameters in wheat *Triticum aestivum* L. under drought conditions. Doctoral dissertation, CCSHAU, Hisar
Ribba T, Garrido-Vargas F, O'Brien JA (2020) Auxin-mediated responses under salt stress: from developmental regulation to biotechnological applications. J Exp Bot 71(13):3843–3853
Ruiz-Lozano JM (2003) Arbuscular mycorrhizal symbiosis and alleviation of osmotic stress. New perspectives for molecular studies. Mycorrhiza 13(6):309–317
Ruiz-Sánchez M, Aroca R, Muñoz Y, Polón R, Ruiz-Lozano JM (2010) The arbuscular mycorrhizal symbiosis enhances the photosynthetic efficiency and the antioxidative response of rice plants subjected to drought stress. J Plant Physiol 167(11):862–869
Sadeghi A, Karimi E, Dahaji PA, Javid MG, Dalvand Y, Askari H (2012) Plant growth promoting activity of an auxin and siderophore producing isolate of Streptomyces under saline soil conditions. World J Microbiol Biotechnol 28(4):1503–1509
Sadhana B (2014) Arbuscular mycorrhizal fungi (AMF) as a biofertilizer—a review. Int J Curr Microbiol App Sci 3(4):384–400
Sairam RK, Tyagi A (2004) Physiology and molecular biology of salinity stress tolerance in plants. Curr Sci 86:407–421
Santander C, Sanhueza M, Olave J, Borie F, Valentine A, Cornejo P (2019) Arbuscular mycorrhizal colonization promotes the tolerance to salt stress in lettuce plants through an efficient modification of ionic balance. J Soil Sci Plant Nutr 19(2):321–331
Selosse MA, Strullu-Derrien C, Martin FM, Kamoun S, Kenrick P (2015) Plants, fungi and oomycetes: a 400-million year affair that shapes the biosphere. New Phytol 206(2):501–506
Shahzad R, Khan AL, Bilal S, Waqas M, Kang SM, Lee IJ (2017) Inoculation of abscisic acid-producing endophytic bacteria enhances salinity stress tolerance in *Oryza sativa*. Environ Exp Bot 136:68–77
Sharma S, Prasad R, Varma A, Sharma AK (2017) Glycoprotein associated with *Funneliformis coronatum*, *Gigaspora margarita* and *Acaulospora scrobiculata* suppress the plant pathogens in vitro. Asian J Plant Pathol 11(4):199–202

Sheng M, Tang M, Zhang F, Huang Y (2011) Influence of arbuscular mycorrhiza on organic solutes in maize leaves under salt stress. Mycorrhiza 21(5):423–430

Shintu PV, Jayaram KM (2015) Phosphate solubilising bacteria (*Bacillus polymyxa*)—an effective approach to mitigate drought in tomato (*Lycopersicon esculentum* Mill.). Trop Plant Res 2:17–22

Singh RP, Jha PN (2017) The PGPR *Stenotrophomonas maltophilia* SBP-9 augments resistance against biotic and abiotic stress in wheat plants. Front Microbiol 8:1945

Smith SE, Read DJ (2010) Mycorrhizal symbiosis. Academic, New York

Stoeckle D, Thellmann M, Vermeer JE (2018) Breakout-lateral root emergence in *Arabidopsis thaliana*. Curr Opin Plant Biol 41:67–72

Syamsiyah J, Herawati A (2018) The potential of arbuscular mycorrhizal fungi application on aggregrate stability in alfisol soil. IOP Conf Ser Earth Environ Sci 142(1):012045

Talaat NB, Shawky BT (2014) Protective effects of arbuscular mycorrhizal fungi on wheat (*Triticum aestivum* L.) plants exposed to salinity. Environ Exp Bot 98:20–31

Tewari S, Arora K (2014) Talc based exopolysaccharides formulation enhancing growth and production of *Hellianthus annuus* under saline conditions. Cell Mol Biol 60(5):73–81

Uddin MN, Hossain MA, Burritt DJ (2016) Salinity and drought stress: similarities and differences in oxidative responses and cellular redox regulation. In: Water stress and crop plants: a sustainable approach, vol 1. Wiley, New York, pp 86–101

Van Der Heijden MG, Klironomos JN, Ursic M, Moutoglis P, Streitwolf-Engel R, Boller T, Wiemken A, Sanders IR (1998) Mycorrhizal fungal diversity determines plant biodiversity, ecosystem variability and productivity. Nature 396(6706):69–72

Van Der Heijden MG, Streitwolf-Engel R, Riedl R, Siegrist S, Neudecker A, Ineichen K, Boller T, Wiemken A, Sanders IR (2006) The mycorrhizal contribution to plant productivity, plant nutrition and soil structure in experimental grassland. New Phytol 172(4):739–752

van der Heijden MG, Martin FM, Selosse MA, Sanders IR (2015) Mycorrhizal ecology and evolution: the past, the present, and the future. New Phytol 205(4):1406–1423

Verma S, Negi NP, Pareek S, Mudgal G, Kumar D (2022) Auxin response factors in plant adaptation to drought and salinity stress. Physiol Plant 174:e13714

Wang Y, Wang M, Li Y, Wu A, Huang J (2018) Effects of arbuscular mycorrhizal fungi on growth and nitrogen uptake of *Chrysanthemum morifolium* under salt stress. PLoS One 13(4):e0196408

Wang Z, Wei Y, Zhao Y, Wang Y, Zou F, Huang S, Hu H (2022) Physiological and transcriptional evaluation of sweet sorghum seedlings in response to single and combined drought and salinity stress. S Afr J Bot 146:459–471

Wu QS, Zou YN, He XH, Luo P (2011) Arbuscular mycorrhizal fungi can alter some root characters and physiological status in trifoliate orange (*Poncirus trifoliata* L. Raf.) seedlings. Plant Growth Regul 65(2):273–278

Wu QS, Srivastava AK, Zou YN (2013) AMF-induced tolerance to drought stress in citrus: a review. Sci Hortic 164:77–87

Wu Y, Wang Y, Xie X (2014) Spatial occurrence and geochemistry of soil salinity in Datong basin, northern China. J Soils Sediments 14(8):1445–1455

Wu N, Li Z, Wu F, Tang M (2016) Comparative photochemistry activity and antioxidant responses in male and female *Populus cathayana* cuttings inoculated with arbuscular mycorrhizal fungi under salt. Sci Rep 6(1):1–15

Yang Y, Tang M, Sulpice R, Chen H, Tian S, Ban Y (2014) Arbuscular mycorrhizal fungi alter fractal dimension characteristics of *Robinia pseudoacacia* L. seedlings through regulating plant growth, leaf water status, photosynthesis, and nutrient concentration under drought stress. J Plant Growth Regul 33(3):612–625

Yildirim E, Taylor AG (2005) Effect of biological treatments on growth of bean plants under salt stress. Science 123(1):1

Yooyongwech S, Phaukinsang N, Cha-um S, Supaibulwatana K (2013) Arbuscular mycorrhiza improved growth performance in *Macadamia tetraphylla* L. grown under water deficit stress involves soluble sugar and proline accumulation. Plant Growth Regul 69(3):285–293

Yooyongwech S, Samphumphuang T, Tisarum R, Theerawitaya C, Cha-um S (2016) Arbuscular mycorrhizal fungi (AMF) improved water deficit tolerance in two different sweet potato genotypes involves osmotic adjustments via soluble sugar and free proline. Sci Hortic 198: 107–117

You J, Chan Z (2015) ROS regulation during abiotic stress responses in crop plants. Front Plant Sci 6:1092

Yousfi S, Serret MD, Araus JL (2010) A comparative effect of salinity and drought on growth, ion concentration and δ13C and δ15N in barley. Asp Appl Biol 105:73–81

Yousfi S, Serret MD, Márquez AJ, Voltas J, Araus JL (2012) Combined use of δ13C, δ18O and δ15N tracks nitrogen metabolism and genotypic adaptation of durum wheat to salinity and water deficit. New Phytol 194(1):230–244

Zandalinas SI, Sengupta S, Fritschi FB, Azad RK, Nechushtai R, Mittler R (2021) The impact of multifactorial stress combination on plant growth and survival. New Phytol 230(3):1034–1048

Zhang H, Xie X, Kim MS, Kornyeyev DA, Holaday S, Paré PW (2008) Soil bacteria augment Arabidopsis photosynthesis by decreasing glucose sensing and abscisic acid levels in plants. Plant J 56(2):264–273

Zhang X, Wang L, Ma F, Yang J, Su M (2017) Effects of arbuscular mycorrhizal fungi inoculation on carbon and nitrogen distribution and grain yield and nutritional quality in rice (*Oryza sativa* L.). J Sci Food Agric 97(9):2919–2925

Zhang A, Zhang T, Kang X, Guo C (2018a) Changes of spectral characteristics of plant leaves before and after dust-retention under hyperspectral imaging. Trans Chin Soc Agric Eng 34(19): 170–176

Zhang G, Sun Y, Sheng H, Li H, Liu X (2018b) Effects of the inoculations using bacteria producing ACC deaminase on ethylene metabolism and growth of wheat grown under different soil water contents. Plant Physiol Biochem 125:178–184

Zhang YC, Zou YN, Liu LP, Wu QS (2019) Common mycorrhizal networks activate salicylic acid defense responses of trifoliate orange (*Poncirus trifoliata*). J Integr Plant Biol 61(10): 1099–1111

Zhang F, Zou YN, Wu QS, Kuča K (2020) Arbuscular mycorrhizas modulate root polyamine metabolism to enhance drought tolerance of trifoliate orange. Environ Exp Bot 171:103926

Zheng J, Cui M, Wang C, Wang J, Wang S, Sun Z, Han S (2022) Elevated CO_2, warming, N addition, and increased precipitation affect different aspects of the arbuscular mycorrhizal fungal community. Sci Total Environ 806:150522

Zhu C, Ling N, Guo J, Wang M, Guo S, Shen Q (2016a) Impacts of fertilization regimes on arbuscular mycorrhizal fungal (AMF) community composition were correlated with organic matter composition in maize rhizosphere soil. Front Microbiol 7:1840

Zhu X, Song F, Liu S, Liu F (2016b) Arbuscular mycorrhiza improve growth, nitrogen uptake, and nitrogen use efficiency in wheat grown under elevated CO_2. Mycorrhiza 26(2):133–140

Zhu X, Song F, Liu S, Liu F (2016c) Role of arbuscular mycorrhiza in alleviating salinity stress in wheat (*Triticum aestivum* L.) grown under ambient and elevated CO_2. J Agron Crop Sci 202(6): 486–496

An Overview of Phytohormones Mediated Drought and Salinity Tolerance in Plants 20

Anita Kumari, Pooja Ahlawat, Kiran, Babita Rani, Alisha Goyal, Pooja, Adhini S. Pazhany, Ashwani Kumar, Sarita Devi, Nisha Kumari, Sukham Madaan, and Minakshi Jattan

Abstract

Abiotic stresses including drought and salinity due to the changing environment are continuous threat to crop production in terms of reduced yield and quality. Preventing global crop losses and increasing food production is crucial to meet the rising demands of the growing human population. Identification of physiological, biochemical, and molecular mechanisms developed by plants to counteract abiotic stresses and maintain their growth and survival under adverse condition is need of time. Phytohormones are in a prominent position, playing important regulatory roles in plant physiology affecting both developmental processes and responses to a wide range of abiotic and biotic stresses. They play central roles in the ability of plants to adapt the changing environments, by mediating growth, development, nutrient allocation, and source/sink transitions. They are involved in cellular regulation via various signaling pathways. Auxins, gibberellins, cytokinins, brassinosteroids, ethylenes, abscisic acid are major phytohormones along with growth regulators like jasmonic acid and salicylic

A. Kumari (✉) · P. Ahlawat · Kiran · B. Rani · S. Devi · N. Kumari · S. Madaan · M. Jattan
CCS Haryana Agricultural University, Hisar, India

A. Goyal
CCS Haryana Agricultural University, Hisar, India

ICAR – Central Soil Salinity Research Institute, Karnal, Haryana, India

Pooja
ICAR – Sugarcane Breeding Institute, Regional Centre, Karnal, Haryana, India

A. S. Pazhany
ICAR – Sugarcane Breeding Institute, Coimbatore, India

A. Kumar
ICAR – Central Soil Salinity Research Institute, Karnal, Haryana, India

A. Kumar et al. (eds.), *Salinity and Drought Tolerance in Plants*,
https://doi.org/10.1007/978-981-99-4669-3_20

acid involved in plant stress tolerance. Crosstalk between the different plant hormones results in synergetic or antagonic interactions that play crucial roles in response of plants to stress adaptation. Here, we review recent advances made in understanding the role of these hormones in modulating plant defense responses against drought and salt stress.

Keywords

Phytohormones · Abiotic stress · Salinity · Drought stress · Plant growth regulators

20.1 Introduction

Climate change poses a significant threat to human civilization. To feed increasing world population the crop productivity must be increased by 70% up to 2050 (Tilman et al. 2011). Plants are vulnerable to a variety of abiotic and biotic challenges as a result of climate change; over the past few decades, research on plants under abiotic stress has yielded crucial insights. According to Zandalinas et al. (2016), a significant progressive research work has been done in understanding the physiological, biochemical, and molecular mechanism of various plant species under abiotic stress conditions. Plants that are subjected to two or more stress conditions also employ particular mechanisms to adapt and complete the life cycle. Two significant abiotic factors that have a negative impact on a third-world population's quality of life include salinity and drought stress. These stresses also have an impact on agricultural output.

Drought stress is exacerbated by the reduction in water availability to plants roots and a rise in the atmospheric vapor pressure deficit, causing detrimental effect on plant growth and agricultural yield (Ahanger et al. 2013). Productivity losses differed by plant species, growth stage, drought longevity, and severity, which determines response of plant toward stress. Under water deficit conditions plant cell membranes gets affected due to disturbance in ion exchange which ultimately disturbs the structure of cell membranes. In response to drought stress, plants employ a variety of adaptive mechanisms to reduce the negative effects on plant growth. For example, morphological adaptation includes reduction in leaf area, stem elongation, root proliferation (Krasensky and Jonak 2012), while physiological mechanisms include increasing water-use efficiency, stomata closing, reduced turgor pressure.

High concentrations of salts present in the soil are another threat to human lives, similar to water scarcity. Salinity is a long term problem although the degradation of arable land has been caused by a gradual rise in salinity over many centuries, the degradation of cultivated land may take less than a century. It has now become a highly major issue for agricultural production, especially in arid and semi-arid countries, as its extent is currently steadily expanding throughout the world (Munns and Tester 2008). Soil salinity can be divided into primary and secondary salinization depending on the type of source from which it was salinized. Primary salinization arises from weathering of minerals and soil obtained from saline parent rocks. Secondary salinization is brought on by human intervention such as irrigation, deforestation, overgrazing, or intense farming. According to Abobatta (2018), salinity stress restricts plant development by adversely affecting various metabolic processes in plants. The dangers of salinity depend on the stage of growth, the severity of the stress, and how long it lasts.

In environments with high temperatures, both water stress and salt stress naturally occur. More than 10% of fertile land was damaged by both salt and water stress, which hastened the desertification and salinization of the entire planet. Both abiotic stresses retard the growth and development of plant causing metabolic changes, nutritional imbalance, reduction in leaf area, overall biomass disturbance, osmotic imbalance, Na^+-Cl^- ions toxicity (Isayenkov and Maathuis 2019; Petretto et al. 2019). There is need for the identification of promising traits that leads to development of stress tolerant crop varieties through conventional breeding or genetic engineering. Certain types of salt-tolerant plants may have evolved specific complex systems, according to comparisons of adaptive responses among different species. Plants have distinct and more complex mechanisms for responding and tolerating environmental stress than do animals (Qin et al. 2011). How plants react to different environmental conditions is one of the most important things to understand for plant biologists. Reason behind the development of adaptive mechanism for abiotic stress tolerance in plant includes combination of various physiological, metabolic, and molecular responses. For stress tolerance, the genes which control ion and water homeostasis are crucial (Bartels and Sunkar 2005; Munns and Tester 2008).

Phytohormone signaling, ROS signaling, cellular osmotic adjustment to maintain water potential are some of the processes that regulate the decreased plant growth and output under drought and salt stress (Khan et al. 2015; Tardieu et al. 2014). Numerous tactics have been used to lessen the negative effects of salinity and drought stress, and attempts are being undertaken to investigate the mechanisms of stress tolerance. However, phytohormones have a wide range of mechanisms for tolerating drought and salinity stress and established as essential for acclimatization of plants in response to biotic and abiotic stresses (Ullah et al. 2017). A number of physiological functions and biochemical systems have been found to be modulated by phytohormones, which in turn affect stress tolerance (Fatma et al. 2013). Drought stress causes production of phytohormones and these phytohormones are further involved in regulation of drought tolerance in plants. There has been research on the role of phytohormones in stress tolerance, including ethylene (Iqbal et al. 2012), gibberellins (Iqbal et al. 2011), abscisic acid (ABA) (Gurmani et al. 2013), cytokinins (CK) (Wu et al. 2014), salicylic acid (SA) (Khan and Khan 2013). Phytohormones like abscisic acid, ethylene, jasmonic acid, and salicylic acid participate in activities connected to drought tolerance via osmotic adjustment (Vishwakarma et al. 2017). Phytohormones act as chemical messengers that activate a variety of physiological processes in plants, such as osmolyte accumulation, closing of stomata, and more root growth to reduce water loss (Ullah et al. 2018). In this chapter, we will discuss on involvement of different phytohormones in salinity and drought stress tolerance in various crop plants.

20.2 The Impact of Drought and Salinity Stress on Plants

20.2.1 Impact of Drought on Plants

Numerous severe drought occurrences have happened in the last few decades, making farming extremely difficult in a number of nations. According to studies, drought stress during plant reproduction has an impact on a number of biological functions in leaves, including protein levels, ROS balance, and energy metabolism (Wang et al. 2019a, b). Under drought conditions the distribution and breakdown of phytohormones disturbed, there are reports on auxin (indole acetic acid), cytokinin, as well as the starch content of developed grains, early senescence, impaired seed germination, and seedling establishment (Chen et al. 2018; Liu et al. 2018). Drought stress reduced chlorophyll content, dry weight of root and shoot, plant height, spike length, and yield production (Fig. 20.1). Additionally, it leads to oxidative stress, hormonal instability, and decreased antioxidant enzyme activity (Abbas et al. 2018), slowed down the pace of photosynthesis, metabolic processes, and nutrient absorption causing eventual reduction in plant growth and productivity (Hasanuzzaman et al. 2018).

Avoidance, escape, tolerance, and recovery are the four main types of drought stress response mechanisms that plants typically exhibit to cope with stress conditions. The two main defenses of plants against water deficit stress are drought tolerance and drought avoidance. The ability of a plant to withstand a dry environment through a variety of physiological processes, such as osmotic adjustment using osmoprotectants, is known as drought tolerance. The continuation of physiological functions during a water deficit, including as stomata regulation, root system development, and other processes, is known as drought avoidance. The ability to alter their life cycle is how drought resistance works (short life cycle to avoid drought stress). The capacity of a plant to resume growth after being exposed to intense drought stress is known as drought recovery (Manavalan et al. 2009). Drought stress causes the stomata closure which leads to reduction in intercellular CO_2 concentration and increased photorespiration. Plant cell water potential reduced under drought conditions causes decline in the activity of RuBISCo enzyme and rate of photosynthesis. Drought stress enhanced the ROS production, down regulate the non-cyclic transport of e- and decline in nutrient uptake which finally leads to oxidative damage in plants, obstruction of ATP synthesis and enhance the susceptibility to diseases incidence.

20.2.2 Impact of Salinity on Plants

Salinity is another most important abiotic stress that reduces plant productivity globally (FAO 2009). Salinity stress lowers the capacity of the roots to absorb water, which primarily inhibits plant growth. Major effects of salinity stress to plants are due to ionic and osmotic effects of salts (Fig. 20.2).

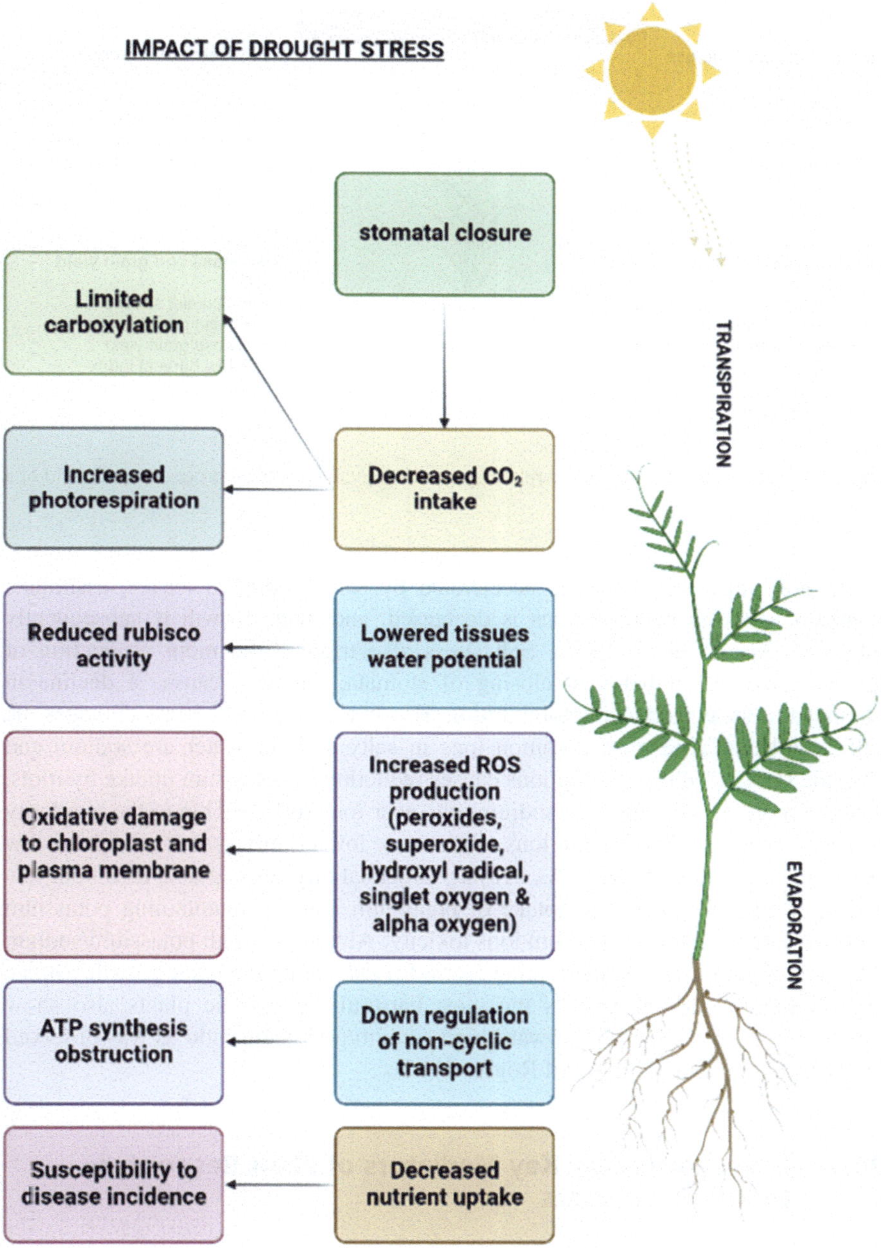

Fig. 20.1 Impact of drought stress on plants

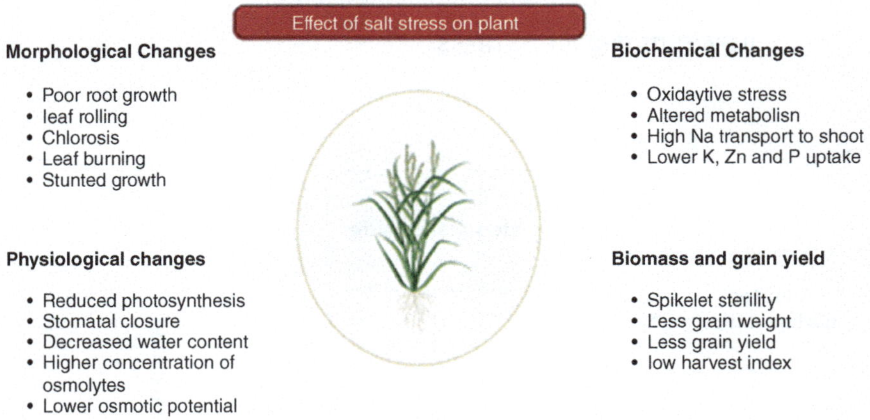

Fig. 20.2 Effects of salt stress on morphology, physiology, biochemical processes, and yield of a plant

As premature senescence is accelerated by salt buildup in leaves, assimilate transport to the growing regions is decreased, and plant growth is subsequently inhibited (Munns et al. 1995). Salt stress also triggers the more production of abscisic acid which helps in closing of stomata, where it causes a decline in photosynthesis as well as photoinhibition. However, severe salt stress damages the cell completely. The most common ions in salty soils or water are sodium and chloride (Levitt 1980). Sodium ions causes reduction in potassium uptake by roots, leads to more accumulation of sodium ions near root surface. Due to the similarity between sodium and potassium ions, plants have low-affinity systems that have low potassium/sodium selectivity. As a result, under salinity stress plants uses both low and high affinity system for uptake of potassium ions for maintaining potassium nutrition and to lessen the sodium ions toxicity. Although a high potassium/sodium ratio is thought to be the determining factor for enhancing the plant growth while a high sodium/potassium ratio is the most harmful. Halophytic plants also show cytosolic enzymes sensitivity toward high sodium/potassium ratio as was observed in glycophyte plants (Noble and Rogers 1992).

20.3 Phytohormones: Key Mediators of Plant Responses to Abiotic Stresses

In order to react to diverse internal and external stimuli, plants must control their growth and development (Wolters and Jürgens 2009). These reactions are mediated by phytohormones, a broad class of signaling chemicals that are present in minute amounts in cells. It is generally known that they play crucial role in helping plants adapt to constantly changing surroundings by mediating growth, development, source/sink transitions, and nutrient allocation (Fahad et al. 2015).

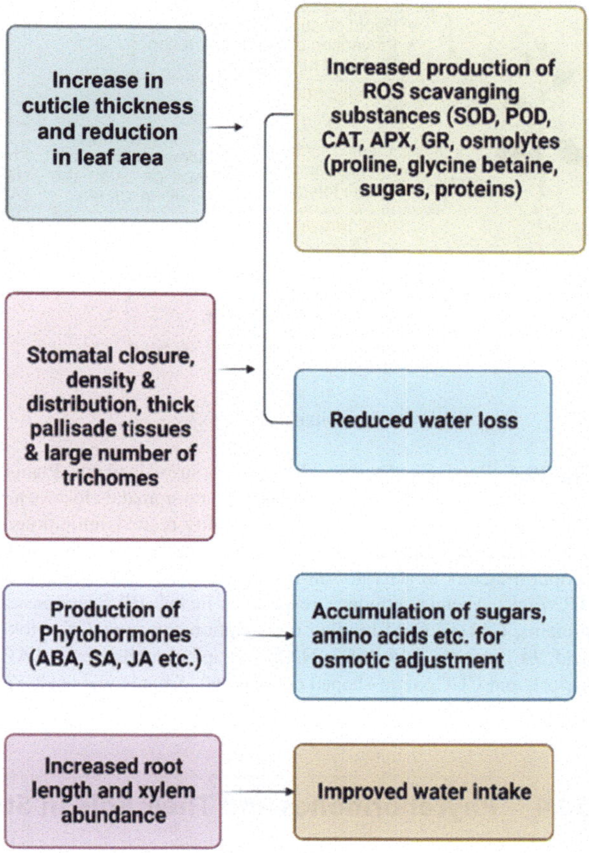

Fig. 20.3 Plant response toward drought tolerance:. The thickness of cuticle increased and leaf area decreased; closing of stomata, reduced stomatal density and distribution; thickned palisade tissues and large number of trichomes under drought stress leads to increase in production of ROS Scavenging substances like CAT, POX, SOD etc. and reduce the water loss from plants. Phytohormones production (ABA, SA, JA etc) causes osmotic adjustment by accumulating sugars and amino acids. The abundance of xylem and increase of root length improve the water intake under drought stress

Endogenous phytohormones are critically required by plant for proper growth and development. Under stress conditions phytohormone helps in modulating plant's metabolism; however, plant response to abiotic stresses depends on a variety of factors (Fahad et al. 2015). Phytohormones can exert their effects either where they were synthesized or where they were transported (Peleg and Blumwald 2011). The growth of plastic and the development of plants both heavily rely on phytohormones. They include auxin (IAA), cytokinins (CKs), abscisic acid (ABA), ethylene (ET), gibberellins (GAs), salicylic acid (SA), brassinosteroids (BRs), and jasmonates (JAs). The strigolactones (SL) are relatively new phytohormones. Figures 20.3 and 20.4 show various responses and involvements of various phytohormones under drought and salinity stress. Detailed account on role of phytohormones under stress conditions will be discussed in the coming sections.

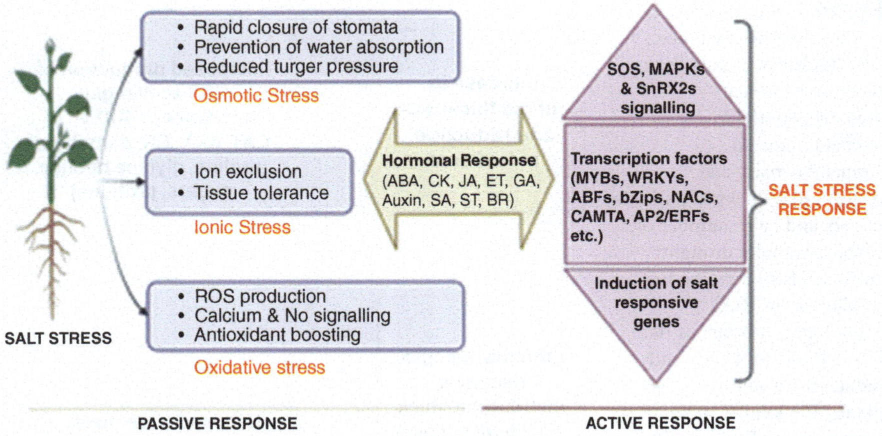

Fig. 20.4 Passive and active salt responses stress in plants. Plants are subject to oxidative, osmotic, and ionic stress due to salinity. Plants regulate stomatal closure, ion exclusion, ROS generation, and Ca^{2+} signaling in response to these salinity-related limitations. In addition to regulating these processes, hormonal pathways influence downstream signaling via the activation of several transcription factors to activate salt-responsive genes. *ABA* bscisic acid, *ABF* ABRE binding factors, *AP2/ERFs* Apetala2/ethylene responsive factor, *BR* Brassinosteroids, *bZIP* basic leucine zipper proteins, *CAMTA* CaM-binding transcription activator, *CK* cytokinin, *ET* ethylene, *GA* gibberellic acid, *JA* jasmonic acid, *NACs* NAM (no apical meristem), ATAF (Arabidopsis thaliana activating factor), and CUC2 (cup-shaped cotyledons), *SA* salicylic acid, *ST* strigolactones

20.4 Phytohormones and Their Role in Stress Management

The exposure of plants to several stresses can hamper overall processes, instigating several adversities and increased production of reactive oxygen species (ROS) which further leads to degradation of integral macromolecules like carbohydrates, proteins, amino acids, nucleic acids, lipid peroxidation and generate oxidative injuries to cell membrane (Yadav et al. 2011; Fahad et al. 2015; Isayenkov and Maathuis 2019). The activation of antioxidant defense mechanism after exposure to stress leads to production of both enzymatic and non-enzymatic antioxidants and metabolites including SOD, CAT, POX, AsA, Glu etc. that play a major role in maintaining proper metabolism of plants (Sharma et al. 2012). There is positive relationship between antioxidant enzyme action and various stress-induced genes (*RAB18*, *RD22*, *RD29A*, *RD29B*, *DREB2A*, and *DREB2B*) (Bielach et al. 2017). Different phytohormones alter different physiological factors discussed below individually.

20.4.1 AUXIN

Among various phytohormones, auxin plays vital role in growth, development, and response to various stresses in plants including water deficit, salinity, and pathogen

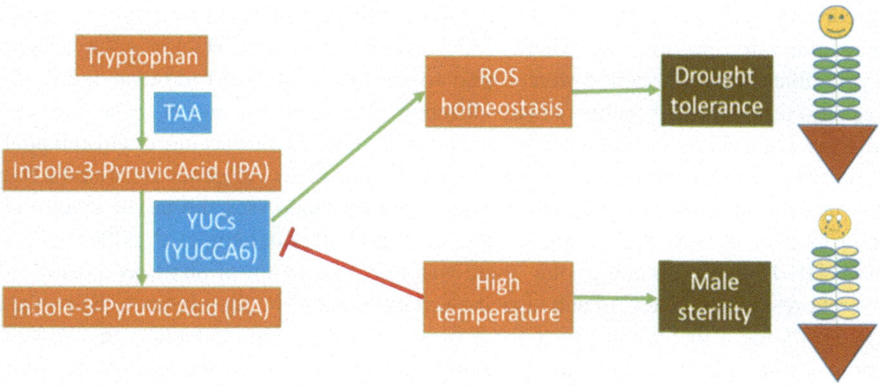

Fig. 20.5 Representation of model proposed for the function of YUCCA6 in drought stress. YUCCA6 protein is involved in auxin biosynthesis as a rate-limiting step converting IPA to IAA. *TAA* tryptophan aminotransferase of Arabidopsis, *IAA* indole acetic acid, *ROS* reactive oxygen species

attack (Ghanashyam and Jain 2009; Lee et al. 2012; Singh et al. 2017). Kaya et al. (2010) demonstrated that young growing apical meristematic tissues mainly produce IAA which further induces and stimulates cell division, cell elongation, flower development, gametogenesis, embryogenesis, vascular differentiation, and root initiation (Khadr et al. 2020). Other than developmental processes, auxin plays multifunctional role in combating the stress conditions in plants (Abdel Latef et al. 2021). Different environmental stresses including light, temperature, pathogens, metal toxicity induce auxin synthesis (Zhao 2018). Root plasticity is controlled by auxin transit from limited storage regions or by oxidation, biosynthesis, and conjugation under diverse stress conditions (Korver et al. 2018). Naser and Shani (2016) suggested that osmotic imbalance also influenced auxin biosynthesis and transport through YUC and PIN proteins, respectively (Fig. 20.5). Auxin is transported in *Arabidopsis thaliana* through many routes, including ATP-binding cassette subfamily B (ABCB), influx channels or carriers named AUX1 and Like-AUX1 (LAX), PIN proteins (PIN), and PIN-LIKES (PILS) proteins (Smith et al. 2017). Wang et al. (2019a, b) reported that during the water scarcity in tobacco plants, different PIN proteins, viz. *PIN1*, *PIN3*, *PIN3b*, *PIN4*, and *PIN9* were upregulated and stimulation of root development was analyzed. During water deficit conditions in Arabidopsis (*Arabidopsis thaliana*), ROS production causes stomata to close by increased level of glucosinolates (GLS) via *DREB2A/B* which further induces *IAA5/6/19 and* upgrades its resistance toward drought (Salehin et al. 2019). To date, however, some interconnected pathways for the production of auxin in plants have been postulated, including four tryptophan-dependent and a tryptophan-independent pathway (Lekshmy et al. 2017). Indole 3 pyruvate (IPA) is formed from tryptophan pathway which ultimately forms IAA by flavin monooxygenases of *YUC* genes. Current evidence also supports auxin biosynthesis modifies the environmental changes through YUCCA (YUC) pathway (Kim et al. 2013; Lee et al. 2012).

According to Zhao (2010) also during IAA synthesis via TAM (tryptamine) pathway, the rate-limiting step where TAM is converted into HTAM (*N*-hydroxyl tryptamine), *YUCCA* genes were found to be associated by following indole-3-acetaldoxime (IAOX) pathway with different intermediates, and HTAM progress to form IAA. Thus different pathways are associated with varying environmental conditions and developmental phases. *YUCCA2* and *YUCCA6* genes are found to be suppressed under high temperature which in turn suppresses the auxin level and induces male sterility in *Arabidopsis thaliana* and barley (Cheng et al. 2006; Sakata et al. 2010). In potato and poplar, there was overproduction of auxin as a result of overexpression of *YUCCA6* and also showed increased drought tolerance (Kim et al. 2013). Plants also remain green for prolong time but with reduction in yield as compared to wild type. Therefore, in potato, biosynthesis of auxin through *YUCCA* genes can be explored under different environmental conditions with respect to the plant responses. There was amplified ROS production and lesser drought tolerance in the phenotypes which were triple mutants *yuc1 yuc2 yuc6 auxin genes* (Shi et al. 2014).

Drought and salinity stress also activate auxin response factors (ARFs). They facilitate plants to cope up different stress conditions by adjusting compatible solutes, chlorophyll development, and root promotion via small RNA-mediated post-transcriptional regulation (Verma et al. 2022). IAA8 auxin gene enhanced the formation of lateral roots under stress conditions (Zhang et al. 2020). Khadr et al. (2020) reported that foliar spray of 30 mg/L of naphthalene acetic acid (NAA) showed positive effects on physiological and morphological characteristics of wheat. Similarly, treating salt-stressed faba bean plants (*Vicia faba* L.) with exogenous IAA resulted in an enhancement of growth and also reinforced the antioxidative defense mechanisms. There is upregulation of auxin genes like *AtRD26* and *AtDREB2* which are induced under drought conditions and also play a vital role in growth and development of plant. In addition, auxin-responsive gene *TaSAUR75* activates genes that respond to stress, such as *AtRD26* and *AtDREB2*, which are crucial for plant growth and development in the presence of water deficits (Guo et al. 2018). Thus according to different studies it is concluded that auxin plays a crucial and central role in governing and regulating the plant growth, development and metabolic activities with respect to environmental signals. Several IAA related genes alter and regulate under abiotic stress conditions intensifying antioxidant mechanism via conserving ionic balance.

20.4.2 Gibberellins

The gibberellins (GAs) are a wide group of tetracyclic diterpenoid carboxylic acids, but only a few of them, GA_1 and GA_3, operate as growth hormones in higher plants (Sponsel and Hedden 2010). Majorly derived by geranylgeranyl diphosphate via terpenes pathway, they possess an ent-gibberellin tetracyclic skeleton as its essential structure (Olszewski et al. 2002). The biosynthetic process is split into three phases based on the enzymes involved. Solvent-producing enzymes in plastids will first create ent-kaurene in first stage. Then, in second stage, microsomal MONOOXYGENASES oxidize the end product of the first step, and in the third step, 2-OXOGLUTARATE-DEPENDENT DIOXYGENASES catalyze the reaction. The GAs are favorable for germination of seeds, expansion of leaf, elongation of stem, initiation of flower and trichome, flower and fruit development as well (Yamaguchi 2008). Due to their capacity to stimulate development, they are essential for plant's whole life cycles. All other phytohormones interact with GAs in a variety of developmental and stimulus-response mechanisms (Munteanu et al. 2014). Both positive and detrimental mutual regulation can occur in the relationships between GA and ET, depend upon that tissue and signaling system (Munteanu et al. 2014). It functions as a growth regulator, promoting blooming, stem elongation, fruit enlargement, and seed germination in particular (Camara et al. 2018). GA was shown to control the vegetative and reproductive growth of *Hibiscus cannabinus* L. and increase the plant's prebiotic fiber (Muniandi et al. 2018). Without altering the fruit's shape, exogenous GA treatment of the rabbit eye blueberry (*Vaccinium ashei*) improves the number of viable seeds, SOD activity, individual fruit weight, and delayed fruit ripening time (Zang et al. 2016). Additionally, this phytohormone was found to raise the amount of sucrose and dry matter in sugarcane (Rai et al. 2017). GA treatment encourages early sprouting and an abundance of sprouted buds to form in potato tubers (Alexopoulos et al. 2017).

The evidence for their critical role in abiotic response to stress and adaptation is intriguingly expanding (Colebrook et al. 2014). Studies (Skirycz et al. 2011; Claeys et al. 2012) have investigated the function of GAs in the osmotic stress response of Arabidopsis thaliana seedlings. When rice is under salt stress, exogenous GA treatment boosts lipid synthesis (Liu et al. 2018). The use of GA boosts the production of H_2S, which acts as an antioxidant to combat oxidative stress, promote plant development, and lower boron levels (Kaya et al. 2020; Iftikhar et al. 2019). The cell wall and plastid stability are altered, which increases *Solanum lycopersicum* L. resilience to temperature stress (Gamel et al. 2017). By increasing the levels of macro-elements (K, Ca, Mg, and Fe) and osmoprotectants (proline and soluble protein), 0.1 mM GA and 0.1 mM ascorbic acid reduce the detrimental effects of saline stress on okra. This improves antioxidant enzymes and lowers H_2O_2 content, lipid peroxidation, and electrolyte leakage (Wang et al. 2019a). Due to its role in maintaining the source-sink relationship and ability to explain why sink enzyme activity declines under salinity, GA signaling is crucial for plants to survive salt (Iqbal et al. 2011). It has been shown that giving plants GA enhances their levels of nitrogen and magnesium while they are under salt stress. GA aids plants in preserving their normal development and growth when exposed to salt stress. Additionally, 10^{-6} M GA treatment of tomato enhanced the amount of proline and sugar in the cytoplasm, which serves as an osmoprotectant and improves a variety of developmental indices under 100 mM salt stress (Ben Rhouma et al. 2020). The *AtGAMT1* gene is overexpressed and generates an enzyme that enhances the methylation of active GA, resulting in the production of inactive GA. Reduced GA concentration in transgenic plants causes them to have smaller leaves with higher stomatal conductance and reduced stomatal intensity, which slows down transpiration (Nir et al.

2014). Genes involved in GA biosynthesis are also downregulated by *SlDREB* overexpression. A reduced GA concentration enhances tomatoes' drought tolerance mechanism by decreasing internode elongation and foliage development (Li et al. 2012). In addition, *PtGA2ox1* overexpression lowers the GA level in the roots, stems, and leaves of tobacco plants, enhancing their ability to withstand drought (Zhong et al. 2014). Due to the upregulation of the *A. thaliana* gene, transgenic tomatoes have decreased levels of bioactive GA. Due to reduced stomata and pores, which resulted in less whole-plant transpiration, the *GA METHYL TRANSFERASE 1 (AtGAMT1)* gene displayed drought tolerance (Nir et al. 2014). In small grains, GA deficiency confers both lodging resistance and drought tolerance. We explore that GA negatively regulated plants' response to drought stress, in comparison to other phytohormones. Therefore, GA downregulation may be the first focus in order to develop plants that can tolerate drought.

20.4.3 Cytokinins

Cytokinins (CKs) are a group of multifaceted plant hormone that have a significant impact on a variety of processes which take place during growth and development (Nishiyama et al. 2011). These are adenine derivatives with isoprenoid side chains flanking the adenine ring at the N6 position. It has been classified by a substituted base into three groups: zeatin (Z), dihydrozeatin (diHZ), and N6-(Δ-isopentenyl) adenine (ZIP). In many species, the zeatin-type cytokinins are the most active and most prevalent forms of cytokinin (Kieber and Schaller 2018). In the biosynthetic process, isopentenyl transferase catalyzes the transfer of an isopentenyl group from dimethylallyl diphosphate to AMP, ADP, or ATP which is a crucial step for the production of CK. These are group of compounds that stimulate water uptake, increase cell division, promote organ development, and lead to regeneration and proliferation of shoot as in *Brassica juncea* L., CK signaling encourages callus regeneration (Lu et al. 2020). In transgenic and mutant cells, the changed activity of metabolic enzymes is critical for a number of crop attributes such as productivity and enhanced stress tolerance (Zalabak et al. 2013). In stressful conditions, the endogenous level of CKs was increased through absorption and increased production. Abiotic stresses such as drought (Kang et al. 2012) and salinity (Nishiyama et al. 2011) indicate the changes in the endogenous levels of CKs in response to stress (O'Brien and Benková 2013). CK response accelerates leaf senescence in Arabidopsis which is regulated by Type-B Arabidopsis response (Raines et al. 2016; Zubo et al. 2017). The Ck signaling pathway in *Arabidopsis* is essential for LZR216 which promote root structural alterations and plant growth and development (Naulin et al. 2020). Additionally, CK has a big impact on how plants react to stress caused by nutrient, salt, osmotic, temperature, and drought (Cortleven et al. 2019). CK induces seeds to come out of dormancy, which is the opposite of how ABA suppress seed germination (Fahad et al. 2015). It is usually regarded as ABA antagonists (Pospíšilová 2003). The ABA/CK ratio rises when CK content decreases and ABA accumulation increases in plants under water stress condition. Adaptation to drought stress is aided by the increased apical dominance, which is enhanced by the lower CK levels and combined with ABA modulation of stomatal aperture (O'Brien and Benková 2013). Cytokinin is a potential second messenger for auxin signaling in controlling bud activity (Müller and Leyser 2011). In MAX2 signaling pathway, Arabidopsis experiences growth and callus formation as a result of interaction between the CK and F-box protein (Li et al. 2019). *Nicotiana tobaccum* is protected from the Chilli veinal mottle virus by CK (Zou et al. 2020). Moreover, it also activates proteins that have detrimental effects on growth, enhancing plants ability to tolerate osmotic stress (Karunadasa et al. 2020). The Arabidopsis root-derived CK, i.e. trans-*zeatin* guards the plants against circadian stress (Frank et al. 2020). Osmotic stress induces the production of CK, which prevents leaf senescence, inhibits ABA signaling and ABA-mediated reactions, reduces lipid peroxidation and ROS damage, increases the plant ability to withstand osmotic stress, and promotes growth and development (Gujjar and Supaibulwatana 2019). It has also been suggested that CK can be up- and downregulated to improve drought resistance. The negative effects of stress on photosynthesis are mitigated when CK is elevated during drought stress (Prerostova et al. 2018).

The primary method of CK elevation is the overexpression of the biosynthetic gene, i.e. isopentenyl transferase, which enhances root development and, in turn, it raises the antioxidant activity. Thus, it improves tolerance for drought stress (Xu et al. 2016). Plants accumulate CK in root tissues during drought stress due to decrease in cytokinin dehydrogenase/oxidase activity (Havlova et al. 2008). Overexpressions of these enzymes increase the content of protective compounds which ultimately increase tolerance in tobacco, Arabidopsis and barley under drought stress (Nishiyama et al. 2011; Macková et al. 2013). This has been the main method for the enzymes to downregulate CK. It also promotes signaling in a variety of ways in response to drought stress. The increase in CK levels under stress along with the activation of stress-responsive genes may provide us some indications about a target that can help plants become more resistant to drought stress. Wu et al. (2014) proposed that the increase in proline content caused by exogenous CK increased salt resistance of *Solanum melongena*.

20.4.4 Abscisic Acid

Abscisic acid (ABA) commonly known as a "stress hormone" is derived from its function in the abscission of plant leaves. It is also one of the phytohormone that has received the most research due to its particular role in plant adaptation to different abiotic stresses. Abscisic acid, a sesquiterpenoid molecule with 15 carbon atoms that is generated in the plastids (Xiong and Zhu 2003). It is widely distributed throughout many kingdoms including cyanobacteria, sponges, algae, lichens, mosses, and mammals (Valona et al. 2008; Cutler et al. 2010; Mehrotra et al. 2014). Higher plants produce ABA via a two-step process that converts the xanthoxin to ABA via ABA-aldehyde after cleaving a C40 carotenoid precursor (Xiong and Zhu 2003). ABA regulates a vast range of physiological, molecular, and cellular functions in plants at various phases of development. Additionally, it is necessary for the synthesis of lipids and storage proteins, as well as for stomatal opening, embryo morphogenesis, seedling growth, plant development, seed maturation (Sreenivasulu et al. 2012) and also influences the differentiation of xylem fibers. The function of stress tolerance in ABA has attracted a lot of attention since it is thought to be a crucial messenger in plants' adaptive response to abiotic stresses. ABA serves as an intracellular signal that enables plants to withstand unfavorable climatic circumstances (Keskin et al. 2010). Intrinsic ABA levels rise quickly in response to abiotic stresses which modified the gene expression by activating particular signaling pathways (O'Brien and Benková 2013). When plants were subjected to salt stress, Zhang et al. (2006) observed a significant increase in ABA content.

ABA is created in the roots and transferred to the leaves when a plant is exposed to water-deficient conditions. Here, it activates drought-resilience mechanisms in the plants that restrict growth and cause stomata to close (Qi et al. 2018). ABA regulates the expression of genes that respond to stress as well as the production of protective proteins like LEA and dehydrins to protect plants from water deprivation (Sreenivasulu et al. 2012). The type-A ARR5 is phosphorylated by ABA-activated SnRK2s, increasing its stability and amplifying the stress response that is mediated by ABA. In addition, type-A ARR5, which inhibits CK signaling in Arabidopsis through a negative feedback loop, limits plant development. In transgenic Arabidopsis, activation of IbARF5 upregulates the biosynthetic genes such as IbZEP, IbABA2, and IbNCED, leads to the development of tolerance under drought stress condition (Kang et al. 2018). Furthermore, SAPK2 is a crucial molecule of ABA-dependent pathway which enhances the expression of number of stress-responsive genes, such as OsOREB1, OsLEA3, OsRab16b, OsRab21, and OsbZIP23 under water deficit condition (Lou et al. 2017). All these stress-responsive genes that are affected by drought may be suitable targets for the gene engineering of plants to make them drought tolerant plant. Furthermore, REL1 overexpression modulates drought tolerance in rice plants (Liang et al. 2018). The local ABA reactions in guard cells are similarly stimulated by AtABCG25 overexpression, which promotes plant water usage efficiency in water-scarce circumstances (Kuromori et al. 2016). Exogenous ABA application has been observed to increase the activity of antioxidant enzymes, ASR1 expression, interacellular ABA level however it reduce oxidative damage to minimize the detrimental effects of drought on maize seedlings (Yao et al. 2019). According to research on the Arabidopsis, a model plant, ABA is a prominent drought-responsive plant growth regulator that is significantly involved in the management of drought stress. Nemhauser et al. (2006) reported that up to 10% of protein-encoding genes are transcriptionally controlled by ABA. Desiccation tolerance is mediated by osmoprotectants and antioxidant enzymes, both of which are synthesized by mechanisms that are upregulated by ABA (Chaves et al. 2003). According to reports, the Camellia sinensis bud activity-dormancy transition depends on genes responsible for the metabolism and signaling of ABA and GA (Yue et al. 2018). ABA controls the expression of genes involved to cell wall alterations and anthocyanin production in the ripening fruits of *Vaccinium myrtillus* L. (Karppinen et al. 2018). Plants acquire ABA in response to abiotic stresses, which causes a reaction to the harsh environment. Guajardo et al. (2016) reported that Pyropia orbicularis, a species of seaweed, exhibited the levels of ABA that were 4–7 times higher during the oxidative stress at the time of water deficit condition. In water-stressed leaves, mesophyll cells are identified to be the principal location of ABA production (McAdam and Brodribb 2018). It also promotes the expression of AGO2 and AGO3 to promote resistance to the bamboo mosaic virus (Alazem et al. 2017). ABA regulates stomatal closure to prevent water loss; hence, the genes associated in stomatal closure could be a key target to alter in order to improve drought resistance in plants.

20.4.5 Ethylene

Ethylene (ET) is a gaseous phytohormone which is significant in mitigating the negative effects of abiotic stress conditions. Along with being a crucial stress response regulator, it also has a major impact on petal and leaf abscission, floral senescence, and ripening of fruit (Groen and Whiteman 2014; Pech et al. 2012; Wang et al. 2020). This phytohormone is a basic 2-C compound that is biosynthesized from methionine using ACC (1-aminocyclopropane-1-carboxylic acid) and *S*-AdoMet (*S*-adenosyl-L-methionine). ACC oxidase turns ACC into ethylene, whereas ACC synthase turns *S*-AdoMet into ACC (Vandenbussche and Van Der Straeten 2012). Low temperature and salinity are abiotic stresses that affect plant endogenous ET levels. Higher ET concentrations were consequently effective in improving tolerance of plants (Shi et al. 2012). It also has a significant impact on how plants respond defensively to heat stress (Larkindale et al. 2005). ET usually works in conjugation with another plant growth regulators including jasmonic acid and salicylic acid. These are the key factors which control the plant defense against pathogens and pests (Klay et al. 2014). A series of signaling pathways involved in plant defense are triggered by the biosynthesis, transport, and storage of the hormones (Matilla-Vázquez and Matilla 2014). According to Yin et al. (2015), ethylene and abscisic acid tend to have synergistic and antagonistic effects on growth and development of plants. It plays an important part in the starting of autophagy and inducing ROS amelioration, which supports survival during re-oxygenation stress, low level of oxygen and flooding. Furthermore, the plant growth-promoting fungus, *Penicillium Viridicatum* also plays a role in the ET signaling pathway which contributes to systemic resistance in Arabidopsis (Hossain et al. 2017). It has been found that during abiotic and biotic stresses, ET is essential for nodule development and signaling in legumes (Khalid et al. 2017). It also encourages nodal root emergence, which slows down root development and eventually has a detrimental impact on Zea mays ability to resist root lodging (Shi et al. 2019). ET also controls the regions where adventitious roots start forming in Arabidopsis hypocotyls (Rasmussen et al. 2017). Gómez-Soto et al. (2019) reported that the exogenous administration of plant hormones like ET and ABA causes the expression of AtNIP1 and AtNIP5, which improves the boron uptake in *Arabidopsis* plant. In tobacco plant, drought tolerance is improved by overexpressing GmERF3, which increases proline content, soluble sugar content, and reduces the accumulation of MDA content (Zhai et al. 2017). Furthermore, transgenic tomato plants overexpressing SlERF5 also showed high drought resistance (Pan et al. 2012). Cotton was also found to contain the AP2/EREBP gene, which is adapted to water stress (Liu and Zhang 2017). ET controls a number of signaling pathways to mitigate drought just like other phytohormones. Therefore, the information provided above regarding the stress-responsive gene mediated by ET may be useful for the future research.

20.4.6 Brassinosteroids

A new class of polyhydroxy steroidal phytohormones with significant growth and development promoting abilities is known as brassinosteroids (BRs). They were initially identified and isolated from the pollen of rape plants (Brassica napus). More than 70 BRs have been synthesized from plant sources. The three most bioactive BRs are brassinolide, 24-epibrassinolide, and 28-homobrassinolide. These BRs are widely used in experimental and physiological studies (Vardhini et al. 2006). They are also found in basically every part of the plant, including the vascular cambium, roots, leaves, flower, stems, pollen, buds, fruits, and seeds (Bajguz and Hayat 2009). They are essential for a number of developmental activities, such as floral initiation, root and stem development, and the formation of flowers and fruits (Bajguz and Hayat 2009). However, new research suggests that BRs and related chemicals have stress-mitigating effects in plants exposed to diverse abiotic stressors. These abiotic stressors include high temperature (Janeczko et al. 2011), low temperature (Wang et al. 2014), soil salinity (Abbas et al. 2013), light (Kurepin et al. 2012), drought (Mahesh et al. 2013), metals/metalloids (Bajguz and Hayat 2009), and organic contaminants (Ahammed et al. 2020). Vardhini and Anjum (2015) explored the potential of BRs and related compounds to modify the antioxidant defense system components after exposure to abiotic stress to avoid the stress-induced oxidative burst.

There is still much to learn about the sources, pathways, and enzymology of their biosynthesis, source-sink relationships, stress physiology, interactions with microbes, fungi, and mammals, and the realization of their powerful uses (Fahad et al. 2015). At high temperature, BZR1, i.e. BRs-activated transcription factor attaches to the PIF4 promoter which activates a number of genes that promote growth and development (Ibanez et al. 2018). In *Oryza sativa*, overexpression of OsMIR396d gene affects BRs and GA signaling pathways which control plant morphology and potential of crop yield (Tang et al. 2018). Important BRs signaling regulator, i.e. *Oryza sativa* mediator subunit 25 interacts with OsBZR1 gene to control rice plant structure and BRs signaling (Ren et al. 2020). Root hydrotropism in Arabidopsis is dependent on BRs-associated H^+ efflux (Miao et al. 2018). Brassinosteroids are also exploited as priming agents and are known to control plant growth, development, and abiotic stress tolerance (Ahammed et al. 2020). It has been shown that brassinolide (5 M/L) seed priming of lucerne (*Medicago sativa*

L.) increased seed germination and seedling development under salt stress (Zhang et al. 2007). Similarly, peanut seed priming with 0.15 ppm brassinosteroids enhanced drought resistance and boosted peanut output (Huang et al. 2020). During water stress, BRs encourage Rubisco and efficient leaf water use to improve CO_2 assimilation and leaf water economy (Farooq et al. 2009). Numerous studies have also demonstrated that wheat, *Arabidopsis*, and *Brassica napus* have been beneficial under water stress condition. Extracellular 24-epibrassinolide treatment enhances brassinosteroids content leading to decreased ABA and ROS levels and hence aiding in stomatal expansion during drought tolerance (Tanveer et al. 2019). Three WRKY transcription factors, WRKY46, WRKY54, and WRKY70 have been recognized as crucial signaling elements that, in turn, either favorably or unfavorably influence BRs controlled growth and drought responses (Chen et al. 2017). Similar to this, Brassica napus increased drought resistance by overexpressing the Arabidopsis BRs biosynthesis gene AtDWF4 (Sahni et al. 2016). Furthermore, ABA signaling is negatively regulated by ABI1 and ABI2, which has been seen to engage with BIN2 and affect BRs signaling, which leads to Arabidopsis stress responses. The localized phytohormone BRs interact with other important stress hormones such as ABA that plays a significant function in drought stress.

20.4.7 Jasmonates

The term "jasmonates" refers to the ubiquitous cyclopentanone plant hormones produced by the metabolism of membrane fatty acids, particularly methyl jasmonate (MeJA) and its free acid jasmonic acid (JA). These multifunctional chemicals are engaged in vital plant growth and survival activities, such as reproduction, blooming, fruiting, secondary metabolism, senescence, and direct and indirect defensive responses (Seo et al. 2001; Fahad et al. 2015). JA increases plant defensive responses to pathogens attack and environmental stresses, including salt, low temperature, and drought (Pauwels et al. 2009), along with the developmental activities in plants. JAs are essential signaling molecules triggered by a variety of environmental stresses such as drought (Du et al. 2013), salt (Pauwels et al. 2009), and exposure of UV (Demkura et al. 2010). They are equipped to combat various environmental difficulties (Dar et al. 2015). Exogenous MeJA treatment significantly reduced salinity stress responses in soybean seedlings (Yoon et al. 2009). Unexpectedly, endogenous levels of JA increased under salt stress in rice roots and were shown to reduce the negative effects of salt stress (Wang et al. 2001). There has been a widespread usage of jasmonic acid derivatives as priming agents to mitigate abiotic stressors. It has been shown that rice seed priming with MeJA at concentrations of 2.5 and 5 mM improved Chl content and photochemical efficiency in response to PEG stress (Sheteiwy et al. 2018). Similarly, MeJA priming enhanced the development of broccoli sprouts under salt stress (Hassini et al. 2017). Exogenous administration of JA is hypothesized to enhance antioxidant activity under drought stress by inducing ascorbate peroxidase (APX), glutathione reductase (GR) monodehydroascorbate reductase (MDHAR), and dehydroascorbate reductase

(DHAR) (Shan et al. 2015). Additionally, it enhances plant water uptake by controlling root hydraulic conductivity under the conditions of low moisture content (Sanchez-Romera et al. 2014). Methyl JA increases the rate of photosynthetic activity, grain yield, and resistance to drought in a number of plant species, including maize, banana, soybean, and *Cistus albidus* (Yu et al. 2018). Furthermore, methyl JA promotes the accumulation of appropriate solutes and osmoprotectants, which enhances antioxidant activity, gas exchange parameters and chlorophyll content, hence inducing stomatal closure and improving water usage efficiency and water status. In addition, methyl JA alters endogenous phytohormones and polyamine to minimize the negative effects of drought stress (Xiong et al. 2020). It is claimed that exogenous administration of 0.5 mM methyl JA maintains wheat development and grain yield under water deficit condition (Anjum et al. 2016). Furthermore, 10 M of methyl JA reduces the negative effects of the extreme drought stress on sugar beets (Fugate et al. 2018) then up to 20 M of methyl JA affects a variety of physiological processes, including the production of more solutes and secondary metabolites like flavonoids and phenolic compounds, hence promoting plant growth in drought-stressed environment (Mohamed and Latif 2017). Endogenous JA greatly reduces the impact of drought stress on oat through fatty acids and lipids. MeJA might modulate drought tolerance in plants by accelerating antioxidant and osmotic adjustment molecules (Xiong et al. 2020). We infer that JA signaling is linked to the regulation of secondary metabolites for drought control in plants.

20.4.8 Salicylic Acid

A naturally present phenolic compound called salicylic acid (SA) controls the expression of proteins which is involved in pathogenesis (Miura and Tada 2014). In addition to controlling defensive responses, it also controls abiotic stress responses, ripening, and plant growth and development (Rivas-San Vicente and Plasencia 2011; Hara et al. 2012). Two mechanisms are involved in the production of SA: the isochorismate (IC) route and the phenylalanine ammonia lyase (PAL) pathway. The IC pathway is the most crucial pathway in tomato and *Nicotiana benthamiana* (Uppalapati et al. 2007). During low SA concentration, antioxidant activity of plants is increased, whereas high SA concentrations may produce cell death or make plants more sensitive to abiotic stresses (Jumali et al. 2011). This is an intriguing and notable common notion. The majority of genes responsible for reduced sensitivity to SA treatment are associated with stress and signaling pathways that ultimately lead to cell death. SA includes genes that make heat shock proteins, antioxidants, chaperones, and secondary metabolite-producing enzymes such as cytochrome P450, sinapyl alcohol dehydrogenase, and cinnamyl alcohol dehydrogenase (Jumali et al. 2011). SA has a role in how plants react to abiotic stresses such as heat (Fayez and Bazaid 2014), cold (Munne-Bosch and Penuelas 2003), salt (Fahad and Bano 2012), and drought (Miura et al. 2013). Both SA and ABA are involved in drought response control (Miura and Tada 2014). Phillyrea angustifolia's endogenous concentrations of SA increased fivefold in response to drought stress

(Munne-Bosch and Penuelas 2003). The SA content of barley roots almost doubled as a result of water deprivation (Bandurska 2005). The pathogenesis-related genes PR1 and PR2 are induced by drought stress (Miura et al. 2013). However, the precise molecular mechanisms underpinning SA's functions in abiotic stress tolerance remain mostly unknown, necessitating additional study in this field. Increasing levels of SA in plants suggest that it is essential for maintaining plant disease resistance, adaptability to abiotic stress, thermogenesis, seedling growth, fruit development, DNA damage/repair, and other functions (Dempsey and Klessig 2017). Exogenous SA is administered to improve enzyme activity, alter cell redox status, protect plants from oxidative stress, and reduce the negative effects of salt stress on mustard plants (Husen et al. 2018). SA is also involved in cell growth and division. When SA is applied externally to canola, more pods and seeds are produced (Keshavarz and Sanavy 2016) and when applied to marigold under water deficit, it enhances various physiological mechanisms, increases bioproductivity, and reduces the negative effects of water stress (Abbas et al. 2019).

Additionally, SA boosts antioxidant enzymes and proline content, both of which are necessary for mung bean seedlings to endure less aluminum stress (Ali 2017). Also, the development and output of commercially important crops like maize, strawberries, and other plants under salt stress are improved by the exogenous administration of SA (Tahjib-Ul-Arif et al. 2018). According to Csiszár et al. (2018), SA maintains glutathione levels and redox equilibrium in situations of salt stress (Csiszár et al. 2018). It has also been noted that the application of SA enhances the root and shoot's osmolyte and proline content in order to preserve the cell's turgor pressure without compromising with other metabolic functions. During drought stress, its application stimulates non-enzymatic defensive systems, such as sugar accumulation for energy saving, osmoregulation, and decreases the levels of free radicals and malondialdehyde in a variety of agricultural plants, such as safflower, wheat, and Brassica rapa (Chavoushi et al. 2019). SA treatment improves the tolerance of agricultural plants to drought stress through redox balance and proline synthesis (Chavoushi et al. 2019). A mechanism for drought tolerance in Arabidopsis loss-of-function lines cpr5 and acd6 was shown to be mediated by endogenously accumulating SA (Miura et al. 2013). When the pathogen-induced pepper CAPIP2 gene was transferred into Arabidopsis, it increased the plant's tolerance to a variety of biotic and abiotic conditions, such as drought stress (Lee et al. 2006). Additionally, it has been proven that SIZ1-mediated accumulation of internal SA increases drought resistance and encourages stomatal closure in Arabidopsis (Miura et al. 2013). External SA treatment was evaluated for its potential to improve drought tolerance by activating a variety of defensive mechanisms, such as the antioxidant system and increasing osmolyte content in the vegetative phase of safflower, maize, and barley, respectively (Abdelaal et al. 2020). Thus, a potential transgenic target for developing drought-resistant plants involves genes engaged in signaling in response to exogenous administration of SA.

20.5 Crosstalk Between Phytohormone

The perception of abiotic stressors initiates signal transduction cascades that interact with the baseline pathways transduced by phytohormones (Harrison 2012) in order to respond to environmental challenges. Changes in stress-responsive hormone levels influence cellular dynamics and are crucial for coordinating the growth responses to stress (Kohli et al. 2013). A signaling network is created by the convergence points between hormone signal transduction cascades, which are known as crosstalk (Harrison 2012). Insights into the main phytohormones ABA, IAA, BRs, GAs, JA, and ET's biosynthetic and fundamental signaling components have been disclosed in recent decades (Singh and Jwa 2013). In order to fine-tune the defense against environmental threats, various plant hormones collaborate to form signal defense networking. Among these, SA, JA, and ABA are of particular note and are thought to be important participants in the control of signaling pathways. Understanding the interconnections between phytohormones and how they work together to combat abiotic stressors has received more attention recently. Incorporating various input signals, regulating growth, and developing stress tolerance in plants all depend on interactions between phytohormones, according to unambiguous experimental results (Kohli et al. 2013). The extensive and repetitive signaling intermediates of each hormone offer insights into their presumed roles in this crosstalk. Understanding the interactions between defense signaling pathways and phytohormonal signaling pathways is crucial because it could point to new potential targets for the creation of phytohormones and host resistance mechanisms (Grant and Jones 2009). Expression of possible genes implicated in drought tolerance is regulated by ERF1 (ETHYLENE RESPONSE FACTOR1). By activating many transcription factors (TFs), including those whose expression is regulated by the ABA-responsive component SnRK2, increased drought tolerance. The phytohormones jasmonic acid and abscisic acid are encouraged to be produced, as well as the ROS detoxification process, which closes stomata and ultimately leads to stress tolerance. Cytokinin, brassinosteroid, and auxin also promote the synthesis of ethylene, which activates the DELLA and causes root growth. Photosynthesis is improved by JA (Fig. 20.6).

20.6 Conclusion and Prospective

Abiotic stresses intensively reduce the yield of major crops by affecting the growth and development of plants. Significant studies showed that phytohormone helps in abiotic stress tolerance. In this chapter we focused on role of different phytohormone under stress conditions. Plant hormones as growth regulators regulate various physiological processes under salt and drought stress. Phytohormones regulate the metabolic processes under stress condition by up- or downregulation of phytohormone-associated genes. To deal with stress conditions, genes which are involved in phytohormones biosynthesis can be identified with the help of recent techniques. Exogenous application of phytohormones may be used to maintain

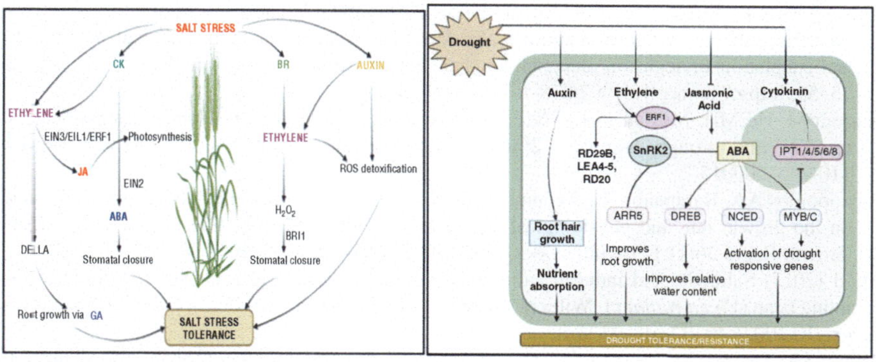

Fig. 20.6 Interactions phytohormone under salinity and drought conditions

growth and productivity of plants under stress conditions; however, more research work could be done under salt and drought stress. Possible outcome in this field is development of salt and drought tolerant varieties for major crops with the help of genetic engineering. Soil microbes may also be helpful in increasing stress resistance by altering the levels of osmoprotectant, phytohomones, secondary metabolites. Phytohormones could enhance drought and salinity resistance by employing various physiological, biochemical, and molecular mechanisms.

References

Abbas S, Latif HH, Elsherbiny EA (2013) Effect of 24-epibrassinolide on the physiological and genetic changes on two varieties of pepper under salt stress conditions. Pak J Bot 45(4): 1273–1284

Abbas T, Rizwan M, Ali S, Adrees M, Mahmood A, Zia-ur-Rehman M, Ibrahim M, Arshad M, Qayyum MF (2018) Biochar application increased the growth and yield and reduced cadmium in drought stressed wheat grown in an aged contaminated soil. Ecotoxicol Environ Saf 148:825–833

Abbas SM, Ahmad R, Waraich EA, Qasim M (2019) Exogenous application of salicylic acid at different plant growth stages improves physiological processes in marigold (*Tagetes erecta* L.). Pak J Agric Sci 56(3):541–548. https://doi.org/10.21162/pakjas/197276

Abdel Latef AAH, Tahjib-Ul-Arif M, Rhaman MS (2021) Exogenous auxin-mediated salt stress alleviation in faba bean (*Viciaf aba* L). Agronomy 11:547. https://doi.org/10.3390/agronomy11030547

Abdelaal KA, Attia KA, Alamery SF, El-Afry MM, Ghazy AI, Tantawy DS, Al-Doss AA, El-Shawy ESE, Abu-Elsaoud AM, Hafez YM (2020) Exogenous application of proline and salicylic acid can mitigate the injurious impacts of drought stress on barley plants associated with physiological and histological characters. Sustainability 12(5):1736. https://doi.org/10.3390/su12051736

Abobatta WF (2018) Some physiological mechanisms of salt tolerance in the glycophytes plant: overview. Acta Sci Agric 2(12):154–156

Ahammed GJ, Li X, Liu A, Chen S (2020) Brassinosteroids in plant tolerance to abiotic stress. J Plant Growth Regul 39(4):1451–1464

Ahanger MA, Tyagi SR, Wani MR, Ahmad P (2013) Drought tolerance: role of organic osmolytes, growth regulators, and mineral nutrients. In: Ahmad P, Wani M (eds) Physiological mechanisms and adaptation strategies in plants under changing environment. Springer, New York, NY, pp 25–55. https://doi.org/10.1007/978-1-4614-8591-9_2

Alazem M, He MH, Moffett P, Lin NS (2017) Abscisic acid induces resistance against bamboo mosaic virus through Argonaute 2 and 3. Plant Physiol 174(1):339–355. https://doi.org/10.1104/pp1600015

Alexopoulos AA, Karapanos IC, Akoumianakis KA, Passam HC (2017) Effect of gibberellic acid on the growth rate and physiological age of tubers cultivated from true potato seed. J Plant Growth Regul 36(1):1–10. https://doi.org/10.1007/s00344-016-9616-z

Ali B (2017) Salicylic acid induced antioxidant system enhances the tolerance to aluminium in mung bean (*Vigna radiata* L Wilczek) plants. Indian J Plant Physiol 22(2):178–189. https://doi.org/10.1007/s40502-017-0292-1

Anjum SA, Tanveer M, Hussain S, Tung SA, Samad RA, Wang L, Khan I, Shah AN, Shahzad B (2016) Exogenously applied methyl jasmonate improves the drought tolerance in wheat imposed at early and late developmental stages. Acta Physiol Plant 38(1):1–11. https://doi.org/10.1007/s11738-015-2047-9

Bajguz A, Hayat S (2009) Effects of brassinosteroids on the plant responses to environmental stresses. Plant Physiol Biochem 47(1):1–8

Bandurska H (2005) The effect of salicylic acid on barley response to water deficit. Acta Physiol Plant 27(3):379–386

Bartels D, Sunkar R (2005) Drought and salt tolerance in plants. Crit Rev Plant Sci 24(1):23–58

Ben Rhouma M, Kriaa M, Ben Nasr Y, Mellouli L, Kammoun R (2020) A new endophytic Fusarium oxysporum gibberellic acid: optimization of production using combined strategies of experimental designs and potency on tomato growth under stress condition. Biomed Res Int 2020:4587148. https://doi.org/10.1155/2020/4587148

Bielach A, Hrtyan M, Tognetti VB (2017) Plants under stress: involvement of auxin and cytokinin. Int J Mol Sci 18(7):1427. https://doi.org/10.3390/ijms18071427

Camara MC, Vandenberghe LP, Rodrigues C, de Oliveira J, Faulds C, Bertrand E, Soccol CR (2018) Current advances in gibberellic acid (GA_3) production, patented technologies and potential applications. Planta 248(5):1049–1062. https://doi.org/10.1007/s00425-018-2959-x

Chaves MM, Maroco JP, Pereira JS (2003) Understanding plant responses to drought—from genes to the whole plant. Funct Plant Biol 30(3):239–264

Chavoushi M, Najafi F, Salimi A, Angaji SA (2019) Improvement in drought stress tolerance of safflower during vegetative growth by exogenous application of salicylic acid and sodium nitroprusside. Ind Crop Prod 134:168–176

Chen J, Nolan TM, Ye H, Zhang M, Tong H, Xin P, Chu J, Chu C, Li Z, Yin Y (2017) Arabidopsis WRKY46, WRKY54, and WRKY70 transcription factors are involved in brassinosteroid-regulated plant growth and drought responses. Plant Cell 29(6):1425–1439. https://doi.org/10.1105/tpc1700364

Chen T, Feng B, Fu W, Zhang C, Tao L, Fu G (2018) Nodes protect against drought stress in rice (*Oryza sativa*) by mediating hydraulic conductance. Environ Exp Bot 155:411–419

Cheng Y, Dai X, Zhao Y (2006) Auxin biosynthesis by the YUCCA flavin monooxygenases controls the formation of floral organs and vascular tissues in Arabidopsis. Genes Dev 20(13):1790–1799. https://doi.org/10.1101/gad1415106

Claeys H, Skirycz A, Maleux K, Inzé D (2012) DELLA signaling mediates stress-induced cell differentiation in Arabidopsis leaves through modulation of anaphase-promoting complex/cyclosome activity. Plant Physiol 159(2):739–747

Colebrook EH, Thomas SG, Phillips AL, Hedden P (2014) The role of gibberellin signalling in plant responses to abiotic stress. J Exp Biol 217(1):67–75

Cortleven A, Leuendorf JE, Frank M, Pezzetta D, Bolt S, Schmülling T (2019) Cytokinin action in response to abiotic and biotic stresses in plants. Plant Cell Environ 42(3):998–1018. https://doi.org/10.1111/pce13494

Csiszár J, Brunner S, Horváth E, Bela K, Ködmön P, Riyazuddin R, Gallé Á, Hurton Á, Papdi C, Szabados L, Tari I (2018) Exogenously applied salicylic acid maintains redox homeostasis in salt-stressed Arabidopsis gr1 mutants expressing cytosolic roGFP1. Plant Growth Regul 86(2): 181–194. https://doi.org/10.1007/s10725-018-0420-6

Cutler SR, Rodriguez PL, Finkelstein RR, Abrams SR (2010) Abscisic acid: emergence of a core signaling network. Annu Rev Plant Biol 61:651–679

Dar TA, Uddin M, Khan MMA, Hakeem KR, Jaleel H (2015) Jasmonates counter plant stress: a review. Environ Exp Bot 115:49–57

Demkura PV, Abdala G, Baldwin IT, Ballare CL (2010) Jasmonate-dependent and-independent pathways mediate specific effects of solar ultraviolet B radiation on leaf phenolics and antiherbivore defense. Plant Physiol 152(2):1084–1095

Dempsey D, Klessig DF (2017) How does the multifaceted plant hormone salicylic acid combat disease in plants and are similar mechanisms utilized in humans? BMC Biol 15(1):1–11. https://doi.org/10.1186/s12915-017-0364-8

Du H, Liu H, Xiong L (2013) Endogenous auxin and jasmonic acid levels are differentially modulated by abiotic stresses in rice. Front Plant Sci 4:397. https://doi.org/10.3389/fpls201300397

Fahad S, Bano A (2012) Effect of salicylic acid on physiological and biochemical characterization of maize grown in saline area. Pak J Bot 44(4):1433–1438

Fahad S, Nie L, Chen Y, Wu C, Xiong D, Saud S, Hongyan L, Cui K, Huang J (2015) Crop plant hormones and environmental stress. Sustain Agric Rev 15:371–400

FAO (2009) How to feed the world in 2050. FAO, Rome. https://www.fao.org/fileadmin/templates/wsfs/docs/expert_paper/How_to_Feed_the_World_in_2050.pdf

Farooq M, Wahid A, Basra SMA (2009) Improving water relations and gas exchange with brassinosteroids in rice under drought stress. J Agron Crop Sci 195(4):262–269

Fatma M, Khan MIR, Masood A, Khan NA (2013) Coordinate changes in assimilatory sulfate reduction are correlated to salt tolerance: involvement of phytohormones. Ann Rev Res Biol 3(3):267–295

Fayez KA, Bazaid SA (2014) Improving drought and salinity tolerance in barley by application of salicylic acid and potassium nitrate. J Saudi Soc Agric Sci 13(1):45–55

Frank M, Cortleven A, Novák O, Schmülling T (2020) Root-derived trans-zeatin cytokinin protects Arabidopsis plants against photoperiod stress. Plant Cell Environ 43(11):2637–2649

Fugate KK, Lafta AM, Eide JD, Li G, Lulai EC, Olson LL, Deckard EL, Khan MF, Finger FL (2018) Methyl jasmonate alleviates drought stress in young sugar beet (*Beta vulgaris* L) plants. J Agron Crop Sci 204(6):566–576. https://doi.org/10.1111/jac12286

Gamel RE, Elsayed A, Bashasha J, Haroun S (2017) Priming tomato cultivars in β-sitosterol or gibberellic acid improves tolerance for temperature stress. Int J Bot 13(1):1–14. https://doi.org/10.3923/ijb2017114

Ghanashyam C, Jain M (2009) Role of auxin-responsive genes in biotic stress responses. Plant Signal Behav 4(9):846–848. https://doi.org/10.4161/psb499376. PMID:19847104

Gómez-Soto D, Galván S, Rosales E, Bienert P, Abreu I, Bonilla I, Bolaños L, Reguera M (2019) Insights into the role of phytohormones regulating pAtNIP5; 1 activity and boron transport in *Arabidopsis thaliana*. Plant Sci 287:110198

Grant MR, Jones JD (2009) Hormone (dis)harmony moulds plant health and disease. Science 324: 750–752

Green SC, Whiteman NK (2014) The evolution of ethylene signaling in plant chemical ecology. J Chem Ecol 40(7):700–716

Guajardo E, Correa JA, Contreras-Porcia L (2016) Role of abscisic acid (ABA) in activating antioxidant tolerance responses to desiccation stress in intertidal seaweed species. Planta 243(3):767–781. https://doi.org/10.1007/s00425-015-2438-6

Gujjar RS, Supaibulwatana K (2019) The mode of cytokinin functions assisting plant adaptations to osmotic stresses. Plants 8(12):542. https://doi.org/10.3390/plants8120542

Guo Y, Jiang Q, Hu Z, Sun X, Fan S, Zhang H (2018) Function of the auxin-responsive gene TaSAUR75 under salt and drought stress. Crop J 6:181–190

Gurmani AR, Bano A, Ullah N, Khan H, Jahangir M, Flowers TJ (2013) Exogenous abscisic acid (ABA) and silicon (Si) promote salinity tolerance by reducing sodium (Na+) transport and bypass flow in rice (*Oryza sativa indica*). Aust J Crop Sci 7(9):1219–1226

Hara M, Furukawa J, Sato A, Mizoguchi T, Miura K (2012) Abiotic stress and role of salicylic acid in plants. In: Ahmad P, Prasad M (eds) Abiotic stress responses in plants. Springer, New York, NY. https://doi.org/10.1007/978-1-4614-0634-1_13

Harrison MA (2012) Cross-talk between phytohormone signaling pathways under both optimal and stressful environmental conditions. In: Khan NA, Nazar R, Iqbal N, Anjum NA (eds) Phytohormones and abiotic stress tolerance in plants. Springer, Berlin, pp 49–76

Hasanuzzaman M, Mahmud JA, Anee TI, Nahar K, Islam MT (2018) Drought stress tolerance in wheat: omics approaches in understanding and enhancing antioxidant defense. In: Zargar S, Zargar M (eds) Abiotic Stress-mediated sensing and signaling in plants: an omics perspective. Springer, Singapore

Hassini I, Martinez-Ballesta MC, Boughanmi N, Moreno DA, Carvajal M (2017) Improvement of broccoli sprouts (*Brassica oleracea* L. var. italica) growth and quality by KCl seed priming and methyl jasmonate under salinity stress. Sci Hort 226:141–151

Havlova M, Dobrev PI, Motyka V, Štorchová H, Libus J, Dobra J, Malbeck J, Gaudinov AA, Vankov AR (2008) The role of cytokinins in responses to water deficit in tobacco plants over-expressing trans-zeatin O-glucosyltransferase gene under *35S* or *SAG12* promoters. Plant Cell Environ 31(3):341–353

Hossain MM, Sultana F, Hyakumachi M (2017) Role of ethylene signalling in growth and systemic resistance induction by the plant growth-promoting fungus *Penicillium viridicatum* in *Arabidopsis*. J Phytopathol 165(7–8):432–441

Huang L, Zhang L, Zeng R, Wang X, Zhang H, Wang L, Liu S, Wang X, Chen T (2020) Brassinosteroid priming improves peanut drought tolerance via eliminating inhibition on genes in photosynthesis and hormone signaling. Genes 11(8):919

Husen A, Iqbal M, Sohrab SS, Ansari MKA (2018) Salicylic acid alleviates salinity-caused damage to foliar functions, plant growth and antioxidant system in Ethiopian mustard (*Brassica carinata* A. Br.). Agric Food Secur 7(1):1–14

Ibanez C, Delker C, Martinez C, Burstenbinder K, Janitza P, Lippmann R, Ludwig W, Sun H, James GV, Klecker M, Grossjohann A, Schneeberger K, Prat S, Quint M (2018) Brassinosteroids dominate hormonal regulation of plant thermomorphogenesis via BZR1. Curr Biol 28:303–310

Iftikhar A, Ali S, Yasmeen T, Arif MS, Zubair M, Rizwan M, Alhaithloul HAS, Alayafi AA, Soliman MH (2019) Effect of gibberellic acid on growth, photosynthesis and antioxidant defense system of wheat under zinc oxide nanoparticle stress. Environ Pollut 254:113109

Iqbal N, Nazar R, Khan MIR, Masood A, Khan NA (2011) Role of gibberellins in regulation of source–sink relations under optimal and limiting environmental conditions. Curr Sci 100(7): 998–1007

Iqbal N, Masood A, Khan NA (2012) Hytohormones in salinity tolerance: ethylene and gibberellins cross talk. In: Khan N, Nazar R, Iqbal N, Anjum N (eds) Phytohormones and abiotic stress tolerance in plants. Springer, Berlin. https://doi.org/10.1007/978-3-642-25829-9_3

Isayenkov SV, Maathuis FJ (2019) Plant salinity stress: many unanswered questions remain. Front Plant Sci 10:80

Janeczko A, Oklešťková J, Pociecha E, Kościelniak J, Mirek M (2011) Physiological effects and transport of 24-epibrassinolide in heat-stressed barley. Acta Physiol Plant 33(4):1249–1259

Jumali SS, Said IM, Ismail I, Zainal Z (2011) Genes induced by high concentration of salicylic acid in Mitragyna speciosa. Aust J Crop Sci 5(3):296–303

Kang NY, Cho C, Kim NY, Kim J (2012) Cytokinin receptor-dependent and receptor-independent path ways in the dehydration response of *Arabidopsis thaliana*. J Plant Physiol 169:1382–1391

Kang C, He S, Zhai H, Zhao N, Liu Q (2018) A sweet potato auxin response factor gene (*IbARF5*) is involved in carotenoid biosynthesis and salt and drought tolerance in transgenic *Arabidopsis*. Front Plant Sci 9:1307. https://doi.org/10.3389/fpls201801307

Karppinen K, Tegelberg P, Häggman H, Jaakola L (2018) Abscisic acid regulates anthocyanin biosynthesis and gene expression associated with cell wall modification in ripening bilberry (*Vaccinium myrtillus* L.) fruits. Front Plant Sci 9:259

Karunadasa SS, Kurepa J, Shull TE, Smalle JA (2020) Cytokinin-induced protein synthesis suppresses growth and osmotic stress tolerance. New Phytol 227(1):50–64

Kaya C, Tuna AL, Okant AM (2010) Effect of foliar applied kinetin and indole acetic acid on maize plants grown under saline conditions. Turk J Agric For 34(6):529–538

Kaya C, Sarıoğlu A, Ashraf M, Alyemeni MN, Ahmad P (2020) Gibberellic acid-induced generation of hydrogen sulfide alleviates boron toxicity in tomato (*Solanum lycopersicum* L) plants. Plant Physiol Biochem 153:53–63

Keshavarz H, Sanavy SAMM (2016) How salicylic acid modulate photosynthetic pigments, growth and yield and yield components of canola. J Genet Resour 2:1–9

Keskin BC, Yuksel B, Memon AR, Topal-Sarıkaya A (2010) Abscisic acid regulated gene expression in bread wheat (*Triticum aestivum L*). Aust J Crop Sci 4(8):617–625

Khadr A, Wang GL, Wang YH, Zhang RR, Wang XR, Xu ZS, Tian YS, Xiong AS (2020) Effects of auxin (indole-3-butyric acid) on growth characteristics, lignification, and expression profiles of genes involved in lignin biosynthesis in carrot taproot. Peer J 8:e10492. https://doi.org/10.7717/peerj.10492

Khalid A, Ahmad Z, Mahmood S, Mahmood T, Imran M (2017) Role of ethylene and bacterial ACC deaminase in nodulation of legumes. In: Khan MS, Musarrat J, Zaidi A (eds) Microbes for legume improvement. Springer, Vienna

Khan MIR, Khan NA (2013) Salicylic acid and jasmonates: approaches in abiotic stress tolerance. J Plant Biochem Physiol 1(4):e113

Khan MIR, Asgher M, Fatma M, Per TS, Khan NA (2015) Drought stress vis a vis plant functions in the era of climate change. Clim Change Environ Sustain 3(1):13–25

Kieber JJ, Schaller GE (2018) Cytokinin signaling in plant development. Development 145(4): dev149344

Kim J, Baek D, Park HC, Chun HJ, Oh DH, Lee MK, Cha JY, Kim WY, Kim MC, Chung WS, Bohnert HJ (2013) Overexpression of Arabidopsis YUCCA6 in potato results in high-auxin developmental phenotypes and enhanced resistance to water deficit. Mol Plant 6(2):337–349

Klay I, Pirrello J, Riahi L, Bernadac A, Cherif A, Bouzayen M, Bouzid S (2014) Ethylene response factor Sl-ERF.B.3 is responsive to abiotic stresses and mediates salt and cold stress response regulation in tomato. Sci World J 2014:167681. https://doi.org/10.1155/2014/167681. Epub 2014 Aug 6. PMID: 25215313; PMCID: PMC4142182

Kohli A, Sreenivasulu N, Lakshmanan P, Kumar PP (2013) The phytohormone crosstalk paradigm takes center stage in understanding how plants respond to abiotic stresses. Plant Cell Rep 32: 945–957

Korver RA, Koevoets IT, Testerink C (2018) Out of shape during stress: a key role for auxin. Trends Plant Sci 23(9):783–793. https://doi.org/10.1016/jtplants201805011

Krasensky J, Jonak C (2012) Drought salt and temperature stress-induced metabolic rearrangements and regulatory networks. J Exp Bot 63(4):1593–1608. https://doi.org/10.1093/jxb/err460

Kurepin LV, Joo SH, Kim SK, Pharis RP, Back TG (2012) Interaction of brassinosteroids with light quality and plant hormones in regulating shoot growth of young sunflower and Arabidopsis seedlings. J Plant Growth Regul 31(2):156–164

Kuromori T, Fujita M, Urano K, Tanabata T, Sugimoto E, Shinozaki K (2016) Overexpression of AtABCG25 enhances the abscisic acid signal in guard cells and improves plant water use efficiency. Plant Sci 251:75–81. https://doi.org/10.1016/jplantsci201602019

Larkindale J, Hall JD, Knight MR, Vierling E (2005) Heat stress phenotypes of Arabidopsis mutants implicate multiple signaling pathways in the acquisition of thermotolerance. Plant Physiol 138(2):882–897

Lee SC, Kim SH, An SH, Yi SY, Hwang BK (2006) Identification and functional expression of the pepper pathogen-induced gene CAPIP2 involved in disease resistance and drought and salt stress tolerance. Plant Mol Biol 62(1):151–164

Lee M, Jung JH, Han DY, Seo PJ, Park WJ, Park CM (2012) Activation of a flavin monooxygenase gene YUCCA7 enhances drought resistance in Arabidopsis. Planta 235(5):923–938. https://doi.org/10.1007/s00425-011-1552-3

Lekshmy S, Krishna GK, Jha SK, Sairam RK (2017) Mechanism of auxin mediated stress signaling in plants. Mech Plant Horm Signal Under Stress 1:37–52

Levitt J (1980) Responses of plants to environmental stresses: water radiation salt and other stresses. 2 Academic Press, New York, NY, 365

Li J, Sima W, Ouyang B, Wang T, Ziaf K, Luo Z, Liu L, Li H, Chen M, Huang Y, Feng Y (2012) Tomato SlDREB gene restricts leaf expansion and internode elongation by down regulating key genes for gibberellin biosynthesis. J Exp Bot 63(18):6407–6420

Li W, Nguyen KH, Van Ha C, Watanabe Y, Tran LSP (2019) Crosstalk between the cytokinin and MAX2 signaling pathways in growth and callus formation of *Arabidopsis thaliana*. Biochem Biophys Res Commun 511(2):300–306

Liang J, Guo S, Sun B, Liu Q, Chen X, Peng H, Zhang Z, Xie Q (2018) Constitutive expression of REL1 confers the rice response to drought stress and abscisic acid. Rice 11(1):1–11

Liu C, Zhang T (2017) Expansion and stress responses of the AP2/EREBP superfamily in cotton. BMC Genomics 18(1):1–16

Liu L, Xia W, Li H, Zeng H, Wei B, Han S, Yin C (2018) Salinity inhibits rice seed germination by reducing α-amylase activity via decreased bioactive gibberellin content. Front Plant Sci 9:275. https://doi.org/10.3389/fpls201800275

Lou D, Wang H, Liang G, Yu D (2017) OsSAPK2 confers abscisic acid sensitivity and tolerance to drought stress in rice. Front Plant Sci 8:993

Lu H, Xu P, Hu K, Xiao Q, Wen J, Yi B, Ma C, Tu J, Fu T, Shen J (2020) Transcriptome profiling reveals cytokinin promoted callus regeneration in Brassica juncea. Plant Cell Tissue Organ Cult 141(1):191–206

Macková H, Hronková M, Dobrá J, Turečková V, Novák O, Lubovská Z, Motyka V, Haisel D, Hájek T, Prášil IT, Gaudinová A (2013) Enhanced drought and heat stress tolerance of tobacco plants with ectopically enhanced cytokinin oxidase/dehydrogenase gene expression. J Exp Bot 64(10):2805–2815

Mahesh K, Balaraju P, Ramakrishna B, Rao SSR (2013) Effect of brassinosteroids on germination and seedling growth of radish (*Raphanus sativus* L) under PEG-6000 induced water stress. Am J Plant Sci 4(12):2305–2313

Manavalan LP, Guttikonda SK, Phan Tran LS, Nguyen HT (2009) Physiological and Mol approaches to improve drought resistance in soybean. Plant Cell Physiol 50(7):1260–1276

Matilla-Vázquez MA, Matilla AJ (2014) Ethylene: role in plants under environmental stress. In: Ahmad P, Wani M (eds) Physiological mechanisms and adaptation strategies in plants under changing environment. Springer, New York, NY

McAdam SA, Brodribb TJ (2018) Mesophyll cells are the main site of abscisic acid biosynthesis in water-stressed leaves. Plant Physiol 177(3):911–917

Mehrotra R, Bhalothia P, Bansal P, Basantani MK, Bharti V, Mehrotra S (2014) Abscisic acid and abiotic stress tolerance–different tiers of regulation. J Plant Physiol 171(7):486–496

Miao R, Wang M, Yuan W, Ren Y, Li Y, Zhang N, Zhang J, Kronzucker HJ, Xu W (2018) Comparative analysis of Arabidopsis ecotypes reveals a role for brassinosteroids in root hydrotropism. Plant Physiol 176(4):2720–2736

Miura K, Tada Y (2014) Regulation of water salinity and cold stress responses by salicylic acid. Front Plant Sci 5:4

Miura K, Okamoto H, Okuma E, Shiba H, Kamada H, Hasegawa PM, Murata Y (2013) SIZ1 deficiency causes reduced stomatal aperture and enhanced drought tolerance via controlling salicylic acid-induced accumulation of reactive oxygen species in Arabidopsis. Plant J 73(1): 91–104

Mohamed HI, Latif HH (2017) Improvement of drought tolerance of soybean plants by using methyl jasmonate. Physiol Mol Biol Plants 23(3):545–556

Müller D, Leyser O (2011) Auxin cytokinin and the control of shoot branching. Ann Bot 107(7): 1203–1212

Muniandi SKM, Hossain MA, Abdullah MP, AbShukor NA (2018) Gibberellic acid (GA_3) affects growth and development of some selected kenaf (*Hibiscus cannabinus* L.) cultivars. Ind Crop Prod 118:180–187

Munne-Bosch S, Penuelas J (2003) Photo-and antioxidative protection and a role for salicylic acid during drought and recovery in field-grown *Phillyrea angustifolia* plants. Planta 217(5): 758–766

Munns R, Tester M (2008) Mechanisms of salinity tolerance. Annu Rev Plant Biol 59:651

Munns R, Schachtman DP, Condon AG (1995) The significance of a two-phase growth response to salinity in wheat and barley. Funct Plant Biol 22(4):561–569

Munteanu V, Gordeev V, Martea R, Duca M (2014) Effect of gibberellin cross talk with other phytohormones on cellular growth and mitosis to endoreduplication transition. Int J Adv Res Biol Sci 1(6):136–153

Naser V, Shani E (2016) Auxin response under osmotic stress. Plant Mol Biol 91(6): 661–672. PMID: **27052306**

Naulin PA, Armijo GI, Vega AS, Tamayo KP, Gras DE, de la Cruz J, Gutiérrez RA (2020) Nitrate induction of primary root growth requires cytokinin signaling in Arabidopsis thaliana. Plant Cell Physiol 61(2):342–352

Nemhauser JL, Hong F, Chory J (2006) Different plant hormones regulate similar processes through largely nonoverlapping transcriptional responses. Cell 126(3):467–475

Nir IDO, Moshelion M, Weiss D (2014) The Arabidopsis gibberellin methyl transferase 1 suppresses gibberellin activity reduces whole-plant transpiration and promotes drought tolerance in transgenic tomato. Plant Cell Environ 37(1):113–123

Nishiyama R, Watanabe Y, Fujita Y, Le DT, Kojima M, Werner T, Vankova R, Yamaguchi-Shinozaki K, Shinozaki K, Kakimoto T, Sakakibara H (2011) Analysis of cytokinin mutants and regulation of cytokinin metabolic genes reveals important regulatory roles of cytokinins in drought salt and abscisic acid responses and abscisic acid biosynthesis. Plant Cell 23(6): 2169–2183

Noble CL, Rogers ME (1992) Arguments for the use of physiological criteria for improving the salt tolerance in crops. Plant Soil 146(1):99–107

O'Brien JA, Benková E (2013) Cytokinin cross-talking during biotic and abiotic stress responses. Front Plant Sci 4:451

Olszewski N, Sun TP, Gubler F (2002) Gibberellin signaling: biosynthesis catabolism and response pathways. Plant Cell 14:S61–S80. https://doi.org/10.1105/tpc010476

Pan Y, Seymour GB, Lu C, Hu Z, Chen X (2012) Chen G (2012) An ethylene response factor (ERF5) promoting adaptation to drought salt tolerance in tomato. Plant Cell Rep 31(2):349–360

Pauwels L, Inzé D, Goossens A (2009) Jasmonate-inducible gene: what does it mean? Trends Plant Sci 14(2):87–91

Pech JC, Purgatto E, Bouzayen M, Latché A (2012) Ethylene and fruit ripening. Annu Plant Rev 44: 275–304

Peleg Z, Blumwald E (2011) Hormone balance and abiotic stress tolerance in crop plants. Curr Opin Plant Biol 14(3):290–295

Petretto GL, Urgeghe PP, Massa D, Melito S (2019) Effect of salinity (NaCl) on plant growth nutrient content and glucosinolate hydrolysis products trends in rocket genotypes. Plant Physiol Biochem 141:30–39

Pospíšilová J (2003) Participation of phytohormones in the stomatal regulation of gas exchange during water stress. Biol Plant 46(4):491–506

Prerostova S, Dobrev PI, Gaudinova A, Knirsch V, Körber N, Pieruschka R, Fiorani F, Brzobohatý B, Černý M, Spichal L, Humplik J (2018) Cytokinins: their impact on molecular and growth responses to drought stress and recovery in *Arabidopsis*. Front Plant Sci 9:655

Qi J, Song CP, Wang B, Zhou J, Kangasjärvi J, Zhu JK, Gong Z (2018) Reactive oxygen species signaling and stomatal movement in Plant responses to drought stress and pathogen attack. J Integr Plant Biol 60(9):805–826

Qin F, Shinozaki K, Yamaguchi-Shinozaki K (2011) Achievements and challenges in understanding plant abiotic stress responses and tolerance. Plant Cell Physiol 52(9):1569–1582

Rai RK, Tripathi N, Gautam D, Singh P (2017) Exogenous application of ethrel and gibberellic acid stimulates physiological growth of late planted sugarcane with short growth period in sub-tropical India. J Plant Growth Regul 36(2):472–486

Raines T, Shanks C, Cheng CY, McPherson D, Argueso CT, Kim HJ, Franco-Zorrilla JM, López-Vidriero I, Solano R, Vaňková R, Schaller GE (2016) The cytokinin response factors modulate root and shoot growth and promote leaf senescence in *Arabidopsis*. Plant J 85(1):134–147

Rasmussen A, Hu Y, Depaepe T, Vandenbussche F, Boyer FD, Van Der Straeten D, Geelen D (2017) Ethylene controls adventitious root initiation sites in Arabidopsis hypocotyls independently of strigolactones. J Plant Growth Regul 36(4):897–911

Ren Y, Tian X, Li S, Mei E, He M, Tang J, Xu M, Li X, Wang Z, Li C, Bu Q (2020) Oryza sativa mediator subunit OsMED25 interacts with OsBZR1 to regulate brassinosteroid signaling and plant architecture in rice. J Integr Plant Biol 62(6):793–811

Rivas-San Vicente M, Plasencia J (2011) Salicylic acid beyond defence: its role in plant growth and development. J Exp Bot 62(10):3321–3338

Sahni S, Prasad BD, Liu Q, Grbic V, Sharpe A, Singh SP, Krishna P (2016) Overexpression of the brassinosteroid biosynthetic gene DWF4 in *Brassica napus* simultaneously increases seed yield and stress tolerance. Sci Rep 6(1):1–14

Sakata T, Oshino T, Miura S, Tomabechi M, Tsunaga Y, Higashitani N, Miyazawa Y, Takahashi H, Watanabe M, Higashitani A (2010) Auxins reverse plant male sterility caused by high temperatures. Proc Natl Acad Sci 107(19):8569–8574

Salehin M, Li B, Tang M, Katz E, Song L, Ecker JR, Kliebenstein DJ, Estelle M (2019) Auxin-sensitive Aux/IAA proteins mediate drought tolerance in Arabidopsis by regulating glucosinolate levels. Nat Commun 10:4021

Sanchez-Romera B, Ruiz-Lozano JM, Li G, Luu DT, Martinez-Ballesta Mdel C, Carvajal M, Zamarreno AM, Garcia-Mina JM, Maurel C, Aroca R (2014) Enhancement of root hydraulic conductivity by methyl jasmonate and the role of calcium and abscisic acid in this process. Plant Cell Environ 37(4):995–1008

Seo HS, Song JT, Cheong JJ, Lee YH, Lee YW, Hwang I, Lee JS, Choi YD (2001) Jasmonic acid carboxyl methyltransferase: a key enzyme for jasmonate-regulated plant responses. Proc Natl Acad Sci 98(8):4788–4793

Shan C, Zhou Y, Liu M (2015) Nitric oxide participates in the regulation of the ascorbate-glutathione cycle by exogenous jasmonic acid in the leaves of wheat seedlings under drought stress. Protoplasma 252(5):1397–1405

Sharma P, Jha AB, Dubey RS, Pessarakli M (2012) Reactive oxygen species oxidative damage and antioxidative defense mechanism in plants under stressful conditions. J Bot 2012:217037

Sheteiwy MS, Gong D, Gao Y, Pan R, Hu J, Guan Y (2018) Priming with methyl jasmonate alleviates polyethylene glycol-induced osmotic stress in rice seeds by regulating the seed metabolic profile. Environ Exp Bot 153:236–248

Shi Y, Tian S, Hou L, Huang X, Zhang X, Guo H, Yang S (2012) Ethylene signaling negatively regulates freezing tolerance by repressing expression of CBF and type-A ARR genes in Arabidopsis. Plant Cell 24(6):2578–2595

Shi H, Chen L, Ye T, Liu X, Ding K, Chan Z (2014) Modulation of auxin content in *Arabidopsis* confers improved drought stress resistance. Plant Physiol Biochem 82:209–217

Shi J, Drummond BJ, Habben JE, Brugire N, Weers BP, Hakimi SM, Lafitte HR, Schussler JR, Mo H, Beatty M, Zastrow-Hayes G (2019) Ectopic expression of ARGOS 8 reveals a role for ethylene in root-lodging resistance in maize. Plant J 97(2):378–390

Singh R, Jwa NS (2013) Understanding the responses of rice to environmental stress using proteomics. J Proteome Res 12:4652–4669

Singh M, Gupta A, Laxmi A (2017) Striking the right chord: signaling enigma during root gravitropism. Front Plant Sci 8:1304

Skirycz A, Claeys H, De Bodt S, Oikawa A, Shinoda S, Andriankaja M, Maleux K, Eloy NB, Coppens F, Yoo SD, Saito K, Inze D (2011) Pause-and-stop: the effects of osmotic stress on cell proliferation during early leaf development in Arabidopsis and a role for ethylene signaling in cell cycle arrest. Plant Cell 23(5):1876–1888

Smith SM, Li C, Li J (2017) Hormone function in plants. In: Li J, Li C, Smith SM (eds) Hormone metabolism and signaling in plants. Academic Press, London, pp 1–38. ISBN 9780128115626

Sponsel VM, Hedden P (2010) Gibberellin biosynthesis and inactivation. In: Davies PJ (ed) Plant hormones. Springer, Dordrecht

Sreenivasulu N, Harshavardhan VT, Govind G, Seiler C, Kohli A (2012) Contrapuntal role of ABA: does it mediate stress tolerance or plant growth retardation under long-term drought stress? Gene 506(2):265–273

Tahjib-Ul-Arif M, Siddiqui M, Sohag AAM, Sakil M, Rahman M, Polash MAS, Mostofa MG, Tran LSP (2018) Salicylic acid-mediated enhancement of photosynthesis attributes and antioxidant capacity contributes to yield improvement of maize plants under salt stress. J Plant Growth Regul 37(4):1318–1330

Tang Y, Liu H, Guo S, Wang B, Li Z, Chong K, Xu Y (2018) OsmiR396d affects gibberellin and brassinosteroid signaling to regulate plant architecture in rice. Plant Physiol 176(1):946–959

Tanveer M, Shahzad B, Sharma A, Khan EA (2019) 24-Epibrassinolide application in plants: an implication for improving drought stress tolerance in plants. Plant Physiol Biochem 135:295–303

Tardieu F, Parent B, Caldeira CF, Welcker C (2014) Genetic and physiological controls of growth under water deficit. Plant Physiol 164(4):1628–1635

Tilman D, Balzer C, Hill J, Befort BL (2011) Global food demand and the sustainable intensification of agriculture. Proc Natl Acad Sci 108(50):20260–20264

Ullah A, Sun H, Yang X, Zhang X (2017) Drought coping strategies in cotton: increased crop per drop. Plant Biotechnol J 15(3):271–284

Ullah A, Manghwar H, Shaban M, Khan AH, Akbar A, Ali U, Ali E, Fahad S (2018) Phytohormones enhanced drought tolerance in plants: a coping strategy. Environ Sci Pollut Res 25(33):33103–33118

Uppalapati SR, Ishiga Y, Wangdi T, Kunkel BN, Anand A, Mysore KS, Bender CL (2007) The phytotoxin coronatine contributes to pathogen fitness and is required for suppression of salicylic acid accumulation in tomato inoculated with Pseudomonas syringae pv tomato DC3000. Mol Plant-Microbe Interact 20(8):955–965

Valona C, Freya NF, Leunga J (2008) An update on abscisic acid signaling in plants and more. Mol Plant 1:198–217

Vandenbussche F, Van Der Straeten D (2012) The role of ethylene in plant growth and development. Annu Plant Rev 44:222

Vardhini BV, Anjum NA (2015) Brassinosteroids make plant life easier under abiotic stresses mainly by modulating major components of antioxidant defense system. Front Environ Sci 2:67

Vardhini BV, Anuradha S, Rao SSR (2006) Brassinosteroids-new class of plant hormone with potential to improve crop productivity. Indian J Plant Physiol 11(1):1

Verma S, Negi NP, Pareek S, Mudgal G, Kumar D (2022) Auxin response factors in plant adaptation to drought and salinity stress. Physiol Plant 174(3):e13714

Vishwakarma K, Upadhyay N, Kumar N, Yadav G, Singh J, Mishra RK, Kumar V, Verma R, Upadhyay RG, Pandey M, Sharma S (2017) Abscisic acid signaling and abiotic stress tolerance in plants: a review on current knowledge and future prospects. Front Plant Sci 8:161

Wang B, Lüttge U, Ratajczak R (2001) Effects of salt treatment and osmotic stress on V-ATPase and V-PPase in leaves of the halophyte *Suaeda salsa*. J Exp Bot 52(365):2355–2365

Wang X, Shu C, Li H, Hu X, Wang Y (2014) Effects of 001% brassinolide solution application on yield of rice and its resistance to autumn low-temperature damage. Acta Agric Jiangxi 26(5): 36–38

Wang Y, Cao JJ, Wang KX, Xia XJ, Shi K, Zhou YH, Yu JQ, Zhou J (2019a) BZR1 mediates brassinosteroid-induced autophagy and nitrogen starvation in tomato. Plant Physiol 179(2): 671–685

Wang YH, Zhang G, Chen Y, Gao J, Sun YR, Sun MF, Chen JP (2019b) Exogenous application of gibberellic acid and ascorbic acid improved tolerance of okra seedlings to NaCl stress. Acta Physiol Plant 41(6):1–10

Wang S, Zhou Q, Zhou X, Zhang F, Ji S (2020) Ethylene plays an important role in the softening and sucrose metabolism of blueberries postharvest. Food Chem 310:125965

Wolters H, Jürgens G (2009) Survival of the flexible: hormonal growth control and adaptation in plant development. Nat Rev Genet 10(5):305–317

Wu X, He J, Chen J, Yang S, Zha D (2014) Alleviation of exogenous 6-benzyladenine on two genotypes of eggplant (Solanum melongena Mill) growth under salt stress. Protoplasma 251(1): 169–176

Xiong L, Zhu JK (2003) Regulation of abscisic acid biosynthesis. Plant Physiol 133(1):29–36

Xiong B, Wang Y, Zhang Y, Ma M, Gao Y, Zhou Z, Wang B, Wang T, Lv X, Wang X, Wang J, Deng H, Wang Z (2020) Alleviation of drought stress and the physiological mechanisms in Citrus cultivar (*Huangguogan*) treated with methyl jasmonate. Biosci Biotechnol Biochem 84(9):1958–1965

Xu Y, Burgess P, Zhang X, Huang B (2016) Enhancing cytokinin synthesis by overexpressing ipt alleviated drought inhibition of root growth through activating ROS-scavenging systems in *Agrostis stolonifera*. J Exp Bot 67(6):1979–1992

Yadav S, Irfan M, Ahmad A, Hayat S (2011) Causes of salinity and plant manifestations to salt stress: a review. J Environ Biol 32(5):667

Yamaguchi S (2008) Gibberellin metabolism and its regulation. Annu Rev Plant Biol 59:225–251

Yao C, Zhang F, Sun X, Shang D, He F, Li X, Zhang J, Jiang X (2019) Effects of S-Abscisic acid (S-ABA) on seed germination seedling growth and Asr1 gene expression under drought stress in maize. J Plant Growth Regul 38(4):1300–1313

Yin CC, Ma B, Collinge DP, Pogson BJ, He SJ, Xiong Q, Duan KX, Chen H, Yang C, Lu X, Wang YQ (2015) Ethylene responses in rice roots and coleoptiles are differentially regulated by a carotenoid isomerase-mediated abscisic acid pathway. Plant Cell 27(4):1061–1081

Yoon JY, Hamayun M, Lee SK, Lee IJ (2009) Methyl jasmonate alleviated salinity stress in soybean. J Crop Sci Biotechnol 12(2):63–68

Yu X, Zhang W, Zhang Y, Zhang X, Lang D, Zhang X (2018) The roles of methyl jasmonate to stress in plants. Funct Plant Biol 46(3):197–212. https://doi.org/10.1071/FP18106

Yue C, Cao H, Hao X, Zeng J, Qian W, Guo Y, Ye N, Yang Y, Wang X (2018) Differential expression of gibberellin-and abscisic acid-related genes implies their roles in the bud activity-dormancy transition of tea plants. Plant Cell Rep 37(3):425–441

Zalabak D, Pospisilová H, Smehilova M, Mrizova K, Frebort I, Galuszka P (2013) Genetic engineering of cytokinin metabolism: prospective way to improve agricultural traits of crop plants. Biotechnol Adv 31(1):97–117

Zandalinas SI, Balfagón D, Arbona V, Gomez-Cadenas A, Inupakutika MA, Mittler R (2016) ABA is required for the accumulation of APX1 and MBF1c during a combination of water deficit and heat stress. J Exp Bot 67(18):5381–5390

Zang YX, Chun IJ, Zhang LL, Hong SB, Zheng WW, Xu K (2016) Effect of gibberellic acid application on plant growth attributes return bloom and fruit quality of rabbit eye blueberry. Sci Hort 200:13–18

Zhai Y, Shao S, Sha W, Zhao Y, Zhang J, Ren W, Zhang C (2017) Over expression of soybean GmERF9 enhances the tolerance to drought and cold in the transgenic tobacco. Plant Cell Tissue Organ Cult 128(3):607–618

Zhang J, Jia W, Yang J, Ismail AM (2006) Role of ABA in integrating plant responses to drought and salt stresses. Field Crop Res 97(1):111–119

Zhang S, Hu J, Zhang Y, Xie XJ, Knapp A (2007) Seed priming with brassinolide improves lucerne (*Medicago sativa* L) seed germination and seedling growth in relation to physiological changes under salinity stress. Aust J Agric Res 58(8):811–815

Zhang Y, Li Y, Hassan MJ, Li Z, Peng Y (2020) Indole-3-acetic acid improves drought tolerance of white clover via activating auxin, abscisic acid and jasmonic acid related genes and inhibiting senescence genes. BMC Plant Biol 20:150. https://doi.org/10.1186/s12870-020-02354-y

Zhao Y (2010) Auxin biosynthesis and its role in plant development. Annu Rev Plant Biol 61:49–64

Zhao Y (2018) Essential roles of local auxin biosynthesis in plant development and in adaptation to environmental changes. Annu Rev Plant Biol 69:417–435

Zhong T, Zhang L, Sun S, Zeng H, Han L (2014) Effect of localized reduction of gibberellins in different tobacco organs on drought stress tolerance and recovery. Plant Biotech Rep 8(5): 399–408

Zou W, Chen L, Zou J, Han H, Fei C, Lin H, Xi D (2020) Cytokinin receptor CRE1 is required for the defense response of *Nicotiana tabacum* to Chilli veinal mottle virus. Plant Growth Regul 90(3):545–555

Zubo YO, Blakley IC, Yamburenko MV, Worthen JM, Street IH, Franco-Zorrilla JM, Zhang W, Hill K, Raines T, Solano R, Kieber JJ (2017) Cytokinin induces genome-wide binding of the type-B response regulator ARR_{10} to regulate growth and development in Arabidopsis. Proc Natl Acad Sci 114(29):E5995–E6004

Selenium Mediated Amelioration of Drought in Crop Plants: A Review

21

Sapna, Sinky, Babita Rani, Pooja, Adhini S. Pazhany, K. D. Sharma, and Ashwani Kumar

Abstract

Drought is among the most dangerous consequences of climate change. It poses a substantial threat to agriculture, which reduces crop yields and impedes the realization of land's full potential around the globe. This review article is divided into two sections: the effect of water stress on the physiological and biochemical features of crop plants and the amelioration of this effect by selenium treatment. In the first section, physiological and biochemical drought responses of crop plants are discussed. Drought stress has a detrimental effect on plant growth characteristics such as plant height, number of leaves and tillers per plant, and dry weight of various plant parts. This is as a result of decreased relative water content (RWC), which closes stomata and consequently decreases transpiration rate, stomatal conductance, and photosynthesis. In the second section, physiological and biochemical alterations generated by selenium (Se) in plants under drought stress are reviewed. Selenium's antioxidant properties enable it to mitigate the negative effects of drought-induced stress on plants. The accumulation of metabolites is accompanied by the activation of catalase, superoxide dismutase, ascorbate peroxidase, and glutathione peroxidase. By promoting the accumulation of osmolytes, Se plays a crucial function in the maintenance of water relations such as leaf water potential and RWC. Osmotic adjustment increases

Sapna · Sinky · B. Rani · K. D. Sharma
CCS Haryana Agricultural University, Hisar, India

Pooja
ICAR – Sugarcane Breeding Institute, Regional Centre, Karnal, Haryana, India

A. S. Pazhany
ICAR – Sugarcane Breeding Institute, Coimbatore, India

A. Kumar (✉)
ICAR – Central Soil Salinity Research Institute, Karnal, Haryana, India

A. Kumar et al. (eds.), *Salinity and Drought Tolerance in Plants*,
https://doi.org/10.1007/978-981-99-4669-3_21

drought tolerance by allowing cell expansion, plant development, and stomata to remain partially open, as well as by maintaining stomatal conductivity, photosynthesis, and transpiration. Therefore, Se improves the quality of grains by promoting the absorption and translocation of many micro- and macroelements whose uptake was reduced under drought.

Keywords

Biofortification · Drought · Physiochemical traits · Selenium · Yield

21.1 Introduction

Drought is one of the most threatening aspects of climate change that poses serious challenges for the survival of crop plants, thus reducing fitness and potential yield of the plants. The world's dry and semi-arid regions, particularly those in developing nations, are in grave danger due to severe water scarcity. For instance, it impacts at least 60 million hectares of wheat-growing land in industrialized countries and 32% of the 99 million hectares of land that are used for wheat cultivation in developing nations (Shamsi and Kobraee 2011). The primary causes of these losses are lower net photosynthetic rates brought on by metabolic constraints, including stomatal closure, chloroplast oxidative damage, and inadequate grain setting (Farooq et al. 2014). Understanding the intricate reactions of plants to a lack of water depends heavily on physiological and biochemical approaches. The use of treatment with substances that reduce abiotic stress could be a more practical, less expensive method of improving plant drought tolerance.

The action of Se on plants has been studied for almost 70 years and its essentiality has not yet been confirmed (Lyons et al. 2008). Currently, numerous studies have been made with trace elements to improve the response of plants subjected to drought, as in the case of selenium (Se) and silicon (Si) (Feng et al. 2013). Studies have shown that in adequate concentrations Se can be beneficial to plants. Physiological and antioxidant properties of Se play beneficial roles in plants exposed to various abiotic stresses (El-Ramady et al. 2016). For example, selenium plays an important role in mitigating salinity (Abul-Soud and Abd-Elrahman 2016), heavy metals (Li et al. 2016), and high temperature stress (Balal et al. 2016) by increasing activity of GPX and antioxidants with a concurrent decrease in lipid peroxidase activity. Selenium improves plant tolerance to drought stress (Nawaz et al. 2015a) by regulating water status (Yao et al. 2009), increasing chlorophyll in plant leaves (Dong et al. 2013), and stimulating the activity of the antioxidant enzymes like CAT and SOD (Proietti et al. 2013). Figure 21.1 displays the list of physiological and biochemical roles of selenium.

Selenium has been reported to improve the yield of multiple food crops like wheat (Ducsay et al. 2016), barley (Ducsay et al. 2009), rice (Wang et al. 2013; Zhang et al. 2014), and maize (Chilimba et al. 2012) in multiple studies. Furthermore, selenium (Se) is an essential element for humans and animals, being a component of nearly

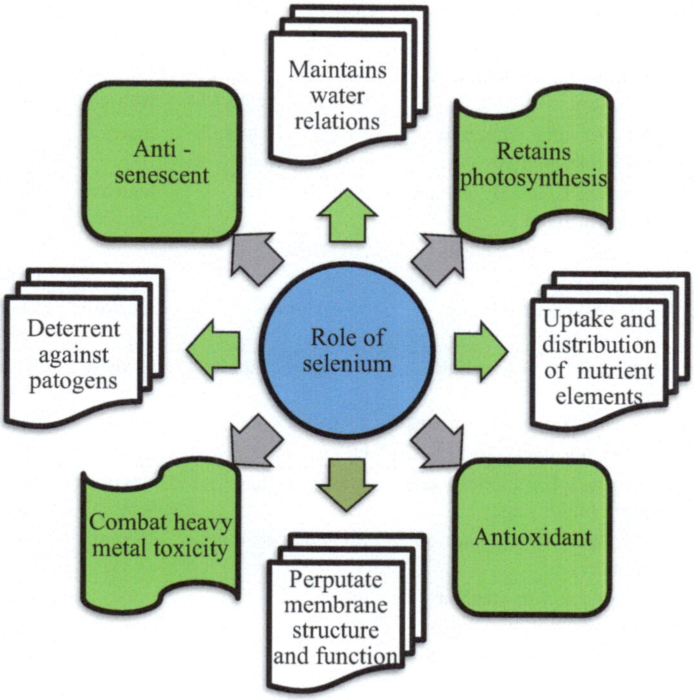

Fig. 21.1 Brief outline of positive effects of selenium in plants

30 selenoproteins or selenoenzymes (Rayman 2012) having anti-cancer, antioxidant, and anti-stress properties (Carey et al. 2012). Selenium deficiency causes many diseases. Worldwide, it is estimated that over one billion people ingest Se below the recommended dose of 55 μg/day (Mora et al. 2015; Banuelos et al. 2017). Crop biofortification represents a great strategy (Galinha et al. 2014) which speaks for the need of supplementation of this nutrient via fertilizers for increasing its level in edible parts as represented in Table 21.1. A comprehensive understanding of the impact of Se on crop exposed to drought is critical for improving drought resistance. Although there are several excellent reviews on plant drought stress available, there are relatively few thorough reviews on selenium mediated biofortification under drought stress In this review, the responses of crops to Se application during water deficit are discussed, including common morphological, physiological, and biochemical responses.

Table 21.1 Effect of selenium on yield and nutritional quality of cereal crop grown under drought stress

Crop	Base supplementation of selenium	Effect on grain yield	Nutrients	References
Winter wheat	60 g/ha foliar from 6 to 18 weeks after sowing	7.62 ± 0.72 to 8.08 ± 0.75 g/plant	Increase	Hajiboland et al. (2015)
Wheat	0.50 mg/L Se fertigation at flag leaf stage	2.73–4.72 t/ha	Increase	Nawaz et al. (2015b)
Buckwheat	1 g Se/L foliar spray at 55 DAS	No effect	–	Tadina et al. (2007)
Rice	0.03 mM seed priming	15–21 g/plant	Increase	Emam et al. (2014)
Maize	285 g Se/ha at tasseling	No significant effect	Increase	Wang et al. (2013)
Wheat	36 mg/L foliar spray at 50% flowering stage	12,880–13,690 kg/ha	–	Teimouri et al. (2014)
Barley	20 g/ha foliar application at grain filling stage	4.4–6.3 kg/ha	– (not seen)	Sajedi et al. (2011)
Rice	21 g/ha at 15 day interval till harvest	2.1–2.7 g/plant	Increase	Wang et al. (2013)
Wheat	18 g/ha at fifth node on stem	9%	–	Sajedi (2017)
Barley	18 g/ha at 75% florescence spikelets	6%	–	Sajedi (2017)

21.2 Effect of Drought on Crop Growth

Growth is one of the most important markers for evaluating the responses of plants to environmental stress, such as drought, salinity, heat stress, etc. Water stress may hinder internode elongation and leaf expansion by limiting cell division (Namich 2007). Drought leads to the loss of water content, which reduces turgor pressure in the cell, preventing cell enlargement and division, resulting in a decrease in plant growth and accumulation of dry material (Delfine et al. 2002).

Wheat growth characteristics, including plant height, number of leaves per plant, leaf area per plant, flag leaf area, number of tillers, shoot dry weight, and root dry weight per plant, were severely impacted by water stress as compared to a normally irrigated control plants (Hammad and Ali 2014). Similarly significant reduction in fresh weight of shoot and root was noticed along with shoot/root ratio, total biomass, and chlorophyll content in 24 days old wheat seedlings exposed to drought conditions (Yao et al. 2012). Crop phenology, dry matter, and grain yield responses to water stress varied on plant growth stage, crop cultivar, and drought severity (Abid et al. 2018).

21.3 Effect of Drought on Crop Physiological Characteristics

Water inaccessibility directly or indirectly disturbs almost every physiological process in plants. The maintenance of favorable water status in cells is considered a prime defense mechanism in water-stressed plants (Kaldenhoff et al. 2008). Plant water relations were significantly influenced by drought stress in wheat and this effect was more pronounced at anthesis causing 21% and 14.5% more reduction in leaf water potential (Ψw) and in leaf RWC, respectively, than at the tillering stage (Nawaz et al. 2017). Maize plants grown under 60% field capacity showed markedly reduced water status and more negative Ψw (Hussain et al. 2016). The relative water content (RWC) is a crucial drought stress measure that indicates the soil water availability to the plant. It is specified that stomata remain closed when RWC is low; hence, reduced stomatal conductance ultimately leads to decrease in photosynthesis (Cornic 2000). In studies performed on four cultivars of bread wheat by Siddique et al. (2000), RWC reduced to 43% (from 88% to 45%) by moisture stress.

Photosynthetic processes in plants are affected to a great extent by drought. Due to the closure of stomata under low water availability, the intake of carbon dioxide (CO_2) is blocked and plants are unable to perform photosynthesis (Yang et al. 2006). Drought-induced reduction in gas exchange characteristics has been well documented in many crops such as wheat (Ahmad et al. 2018), chickpea (Rahbarian et al. 2011), cowpea (Singh and Reddy 2011), and mungbean genotypes (Baroowa et al. 2015).The value of Fv/Fm reflects the capacity to trap electrons by the photosystem II (PSII), and the decrease in Fv/Fm is attributed to damage of D1 reaction-center protein of photosystem II (Zhang et al. 2014). Water stress caused degradation of PS II oxygen evolving complex and the PS II reaction centers (Murata et al. 2007). The quantum yield of photosystem II (Fv/Fm) was significantly reduced under drought stress in both HC-1 and RSG 931 chickpea genotypes and it was found to be positively correlated with RWC (Kumar et al. 2018).

The role of the intact cell membrane remains to be more critical for plant adaptation in drought stress conditions. Changes in cell membranes under water deficit are often associated with the increase in the cell permeability and electrolyte leakage (Ahmad et al. 2018). According to a study conducted by Beltrano et al. (2008), wheat seedlings with a soil water potential of nearly −1.2 MPa from watery ripe, milk stage, and soft drought stage to harvesting showed a significant increase in electrolyte leakage compared to well-watered plants at each of the aforementioned stages. Water stress diminishes the availability, uptake, transfer, and metabolism of mineral nutrients in plants (Farooq et al. 2009). The availability of N, P, K, Ca, Mn, Mo, Fe, B, and Zn uptake to the plant decreases with decreasing soil water content, due to the decreasing mobility of these elements under water stress (Hu and Schmidhalter 2005). But under moist condition, the availability of Mn and Fe is increased due to its conversion to reduced and more soluble forms (Havlin et al. 1999). In wheat water stress caused reduction of 24%, 42%, 14%, and 9% in P, Zn, Fe, and Mg contents, respectively, than that of normal irrigated plants, whereas K accumulation in grain was not affected by water deficit conditions (Nawaz et al. 2015b). Contrary to this, in an experiment conducted by Ashraf et al. (1998) to determine the effect of water stress on nutritional changes in wheat, Ca, Mg, and P concentration decreased, K increased and Na content was not affected with water stress in all the wheat genotypes under examination.

21.4 Effect of Drought on Crop Biochemical Traits

Chlorophyll is an important photosynthetic pigment which is directly linked to the yield of the plant. Wheat requires highest amount of chlorophyll at the flowering stage for proper development of grains (Lehari et al. 2019). Decline in chlorophyll *a*, chlorophyll *b*, and total carotenoids in wheat plant grown under water stress have been noticed by Ibrahim (2016). Analysis of MDA, which is a final product of lipid peroxidation and, a marker of stress in plants reflects the effects of stress on the membranes (Anjum et al. 2017). Yao et al. (2012) noticed that drought treatment significantly increased the rate of ROS production and MDA content in wheat. Reactive oxygen species (ROS) perform many vital roles in inter- and intracellular signaling to control plant growth and development (Breusegem et al. 2001). Water stress agitates the equilibrium between antioxidant defense and the amount of ROS (Gill and Tuteja 2010). CAT is one of the most important enzymatic antioxidants that protect cells against stress by converting the detrimental H_2O_2 directly to H_2O and O_2 (Akbulut et al. 2018). Wheat at 40% FC had 66.29% higher CAT activities as compared to irrigated control (Sattar et al. 2019). This enhanced activity of antioxidant enzyme might be due to osmotic stress and ROS production (Sheoran et al. 2015; Caverzan et al. 2016). Superoxide dismutase (SOD) is considered to be first line of defense to safeguard plant against antioxidant stress as it regulates the concentration superoxide in the cell. Mohammadi et al. (2015) documented enhancement in SOD activity with an increase in stress intensity in all the ten tested wheat genotypes. This also indicated high and significant correlations of SOD and GPX activities with grain yield pointing out that these parameters can be used to estimate potential field performance of wheat under different irrigation conditions. APX participates in the ascorbate-glutathione pathway to combat enhanced generation of ROS. During abiotic stress, this enzyme becomes active to protect plant cells (Ghosh and Biswas 2017). Shabbir et al. (2016) studied APX action in wheat leaves underneath drought and concluded that drought stress upsurges APX activity up to 55% compared to normal irrigated wheat plants. Contrary to this, APX activity of PEG treated wheat seedlings showed descending trends upon increased levels of drought stress (Lan et al. 2020).

Ascorbic acid (AsA) plays an important role in scavenging the reactive oxygen species (ROS) under abiotic stress. Drought stress significantly increased ascorbic acid contents in wheat crop grown under different moisture regimes (Shahzadi et al. 2017). Contrary to this, a decreasing trend of ascorbate content was observed in 29 wheat genotypes under water deficit conditions compared to irrigated environments, and genotypes with higher reduced ascorbate under irrigated conditions had a lower drought susceptibility index than genotypes with low endogenous reduced ascorbate (Roy et al. 2017). Glutathione (GSH) is a low molecular

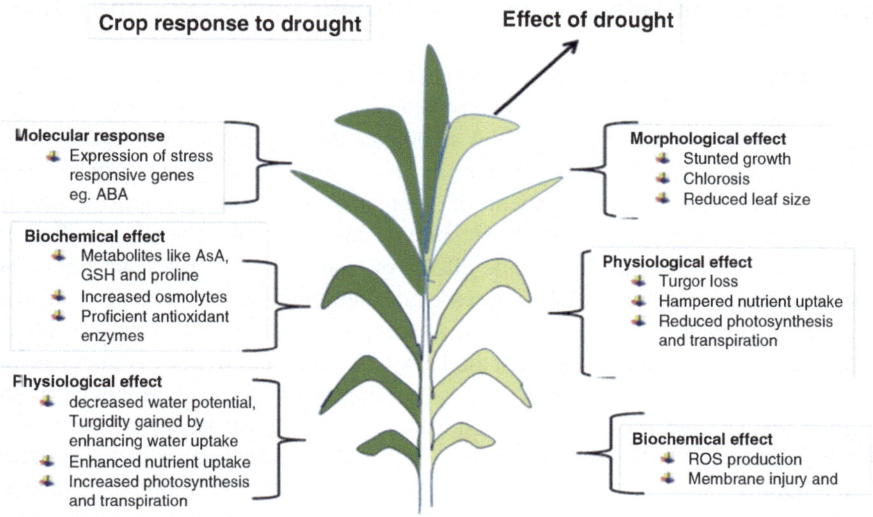

Fig. 21.2 Drought stress mediated changes in physiological and biochemical traits

weight thiol and the chemical reactivity of the thiol group makes it particularly suitable to serve a broad range of biochemical functions in all organisms. This reactivity along with the relative stability and high water solubility of GSH makes it an ideal biochemical to protect plants against various stresses (Ghosh and Biswas 2017). Furthermore, AsA and GSH serve as electron donors to APX and GPX (Batth et al. 2017). Hydroxymethylglutathione (hmGSH) and glutathione (GSH) were present in greater quantities after PEG treatment in the two wheat tolerant genotypes than in the sensitive ones (Kocsy et al. 2004).

High accumulation of proline is a very common but an important adaptive mechanism to drought stress (Alexieva et al. 2001). Keyvan (2010) reported that drought stress reduces total chlorophyll content in wheat but increases the amount of proline and soluble carbohydrates. Proline acts as an osmolyte that reduces the osmotic potential to minimize the water loss and also contributes in stabilizing membranes and scavenging hydroxyl radical under stress conditions (Hayat et al. 2012). Drought stress significantly increased proline content in wheat (Simova-Stoilova et al. 2008), maize (Bocchini et al. 2018), and rice (Mostajeran and Rahimi-Eichi 2009). Total soluble sugars (TSS) are also important compatible solutes that have main function in water regulation. They impart tolerance to the plant exposed to drought stress when accumulated in higher amounts (Kelaleche et al. 2018). An experiment conducted by Ibrahim (2016) indicated that the limited irrigations in wheat significantly increased TSS up to 24% compared to control. Fructose and TSS both increased more under severe drought at the field capacity of 35–40% than moderate stress of 55–60% FC during tillering stage (Abid et al. 2018). Drought stress mediated changes in physiological and biochemical traits are presented in Fig. 21.2.

21.5 Effect of Selenium on Crop Growth Under Drought Stress

Selenium promotes the development of healthier crops by helping in development of more vigorous radicles with extensive root hairs, which escalates root activity under drought conditions (Yao et al. 2009). A significant increase in number of productive tillers by exogenous Se supply indicates its vital role in seedling establishment and early growth through development of vegetative structure and production of assimilates to fill an economically important sink (Boldrin et al. 2013). Proper mineral nutrition can ease the hostile effects of water scarcity. Selenium regulates the uptake and distribution of some essential nutrients such as Zn, Fe, Cu, and Mn. These elements are involved in reactivation of associated antioxidants such as SOD and CAT, thus helps to quench the ROS levels and improve stress tolerance in plants (Feng et al. 2013; Yao et al. 2013). The supplementation of wheat plants with sodium selenate escalates total nitrogen up to 20% than control (Lara et al. 2019). The regulation of the N metabolism was associated with increased nitrate reductase (NR) activity in wheat (Hajiboland and Sadeghzadeh 2014). Foliar and drench application of selenium significantly reduced the electrolyte leakage% (Moussa and Hassen 2017). Further selenite-treatment delayed leaf senescence in rice by regulating the activity of enzymatic antioxidants and also increasing chlorophyll content which sustained growth and yield formation (Duan et al. 2019).

Supplemental Se was effective in improving the yield and yield components and enhanced grain weight by 26%, biomass yield by 12%, and grain yield by 24%, which attributed to the influence of Se on water relations, gas exchange characteristics, and enhanced activity of osmo-protectants and antioxidant machinery (Nawaz et al. 2015a). In an experiment by Bocchini et al. (2018) Se-biofortified (CSe) maize plants showed improved water stress tolerance by increasing aerial biomass by 13% as compared to untreated (-Se) plants grown under water limiting conditions. Response of exogenous Se application on plants varies based on its concentrations applied as at lower rates, Se stimulated growth of ryegrass seedlings in pot experiments, while at high doses it acted as pro-oxidant reducing yields and induced metabolic disturbances (Hartikainen et al. 2000). At present, the effect of Se on wheat yield under drought stress is still controversial mainly due to different concentrations and methods used as presented in Table 21.1.

21.6 Effect of Selenium on Crop Physiology Under Drought Stress

Plant cell reduces the detrimental effect of drought by actively accumulating solutes, in response to low soil water potential known as osmotic adjustment (Hammad and Ali 2014). Further, disturbance in the osmotic balance reduces turgidity and cell elongation (this has been repeated multiple times in the paper) (Shabbir et al. 2016). The supplemented Se significantly reduced solute potential which contributed to maintenance of turgor and resulted in the higher pressure potential in water-stressed wheat (Nawaz et al. 2015a). Selenium foliar spray at 40 mg L^{-1} resulted in 41% less

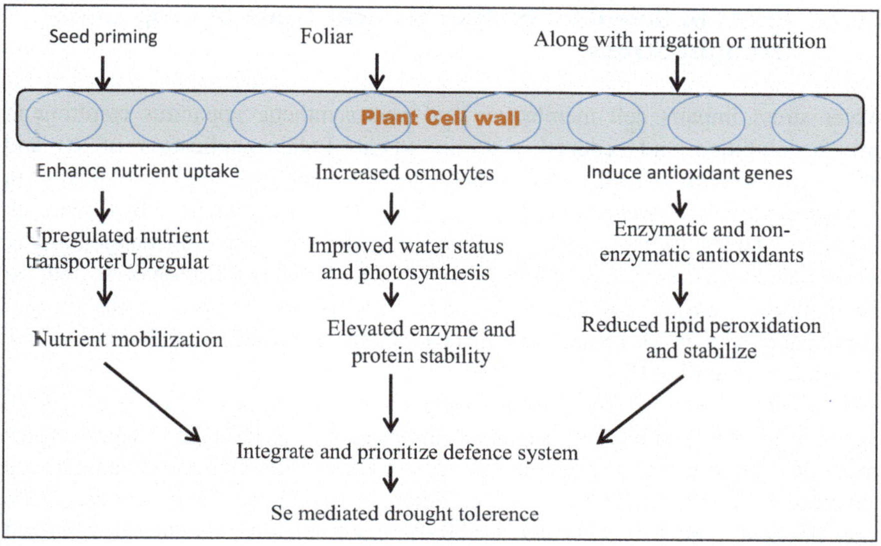

Fig. 21.3 Simplified pattern indicating potential mechanism involved in Se mediated drought tolerance in crop plants

negative leaf water potential in maize (*Zea mays* L.) under drought stress (Nawaz et al. 2016). Similar effect of Se was also reported on potato (Germ et al. 2007) and maize (Sajedi et al. 2011). Selenium-induced gains in water potential might be due to its positive role in osmotic balance and ion homeostasis to increase water uptake (Kuznetsov et al. 2003) and reduce transpiration under water deficit conditions (Yao et al. 2009). Selenium treatments significantly increase leaf RWC as described by Wang (2011) in water-stressed *Triticum aestivum, Trifolium repens*. Aissa et al. (2018) demonstrated the ability of Se to regulate the water status in *Sorghum* under drought stress.

It is found that water insufficiency leads to degradation in photosynthetic pigments and cell membranes; however, the application of Se minimizes the damage to chloroplast structure and PSII reaction centers (Germ 2008). Hajiboland et al. (2015) conducted an experiment and concluded that drought stress affected gas exchange parameters and photosynthetic efficiency (Fv/Fm, F′v/F′m, and ETR) and this quantum yield (Fv/Fm) was completely returned back to its control levels by Se application in "Homa" genotype of wheat. In this study Se treatment boosted stomatal conductance, transpiration rate," and particularly net assimilation rates in both the normal and water deficit conditions. Germ et al. (2007) noted increase in transpiration rate due to Se under well watered conditions, while non-significant differences were observed under drought stress. Suitable amount of Se is beneficial for mitigating the drought stress in crops as it modulates various physio-biochemical traits (Fig. 21.3).

21.7 Effect of Selenium on Biochemical Traits of Crop Under Drought Stress

Water stress impairs cell membranes and photosynthetic apparatus resulting in growth inhibition and reduced yield in wheat. Foliar application of 20 and 40 mg L^{-1} sodium selenite reduced the degree of membrane lipid peroxidation in two wheat varieties: "*Shunmai-1718*" and "*Jintai-102*" as documented by Chen et al. (2020). Contrary to this, MDA content in barley remained unchanged under Se-supplemented drought conditions (Habibi 2013). Foliar application of Se reduced the damage to the chloroplasts (Nawaz et al. 2016) and helped to sustain leaf anthocyanin content of wheat under drought stress (Shahzadi et al. 2017) and low temperature condition (Chu et al. 2010).

Selenium is essential component of glutathione peroxidase (Rios et al. 2009). Application of Se significantly enhanced ascorbate and glutathione in wheat grown under drought stress as documented by Ibrahim (2014). Selenium implementation enhanced CAT (Wu et al. 2020), APX (Lara et al. 2019), and GPX activities (Chu et al. 2013) in wheat leaf. This increased level of antioxidative enzymes is attributed to antagonistic effects of Se in response to ROS production (Yao et al. 2011; Malik et al. 2012). There are reports that Se treatment did not significantly affect CAT activity in wheat under different moisture regimes (Shahzadi et al. 2017). Se induced effects were varied under different situations due to its concentration (Hu et al. 2013), stress environment and crop type (Iqbal et al. 2015). The higher activity of the APX enzyme and the constant values of MDA and H_2O_2 contents in Se-treated plants can be favorable regarding phyto-mass production and grain yield (Lara et al. 2019) because they reduce the damage caused by the oxidative stress and maintain cellular homeostasis, which is of fundamental importance for preserving cell membrane permeability (Ahmad et al. 2016). Djanaguiraman et al. (2005) and Xue et al. (2001) showed increased activity of SOD following the treatment with Se in soybean and lettuce, respectively, while Saidi et al. (2014) reported a reduction in SOD activity of sunflower plants whose seeds were primed with Se (Table 21.2).

The synthesis of total sugars and proline are distinctive features of wheat plants under drought stress. When amassed in huge amount they maintain potential integrity in cells by avoiding cellular dehydration and act as membrane osmo-protectants to avoid the disintegration of proteins. Wheat cultivars (Pasban-90 and Kohistan-97) seedlings were primed with 75 μM Se-solution resulted in 75% and 64% higher soluble sugars, respectively under drought in comparison to seedlings without Se priming (Nawaz et al. 2013). In another study, Se foliar spray of 40 ppm increased the accumulation of TSS by 33% in wheat (Nawaz et al. 2015a). In an experiment conducted by Khan et al. (2015) plants treated with Se showed higher proline accumulation as a result of decreased activity of proline oxidase. The increase in proline content, K concentrations, and nitrogen metabolism in aerial parts of drought stressed maize plants might be responsible for Se-biofortification and increased plant's resistance to water deficit conditions (Bocchini et al. 2018).

Table 21.2 Selenium mediated changes in antioxidative enzymes in different crops under drought

Se treatment	Crop	CAT	SOD	POX	APX	GPX	GR	GST	References
40 mg Se/L at tasseling stage	Maize	↑	↑	↑	↑	–	–	–	Nawaz et al. (2016)
5 μM sol of sodium selenite in hydroponics	Cucumber	↑	↑	↑	↑	–	–	–	Jóźwiak and Politycka (2019)
0.05 mg Se/kg before sowing	Wheat	↑	–	↑	–	–	–	–	Yao et al. (2009)
16 ppm for 3 days at 10 DAS in hydroponic	Barley	↑	NAS		↑		↑	↑	Akbulut and Çakır (2010)
10–40 mg Se/L at 7 DAS	Wheat	↑	↑	↑	↑			↑	Chen et al. (2020)
30 g Se/ha at heading stage	Barley	↑	↑	↑	↓	↑	–	–	Habibi (2013)
40 mM Se at 27 DAS	Wheat	↑	↑	↑	–	–	–	–	Sattar et al. (2019)
10–20 mg/L soaking of grains pre sowing	Wheat	↑	↑	↓	–	–	–	–	Ibrahim (2014)
2.0 mg/kg soil treatment	Rice	NAS	↓	–	NAS	–	–	–	Andrade et al. (2018)
0.5 mg/kg soil treatment		NAS	↑	–	NAS	–	–	–	

↑ = increase, ↑ = decrease, NAS = not affected significantly, – = not seen in mentioned study

21.8 Conclusions

Drought is a significant threat to global wheat production. It disrupts the growth, physiological, and biochemical properties of wheat, ultimately resulting in significant output losses. In recent years, there has been an increasing interest in selenium (Se) as an essential element for not only animals and people but also plants. Optimal Se treatment not only protects wheat from ROS-induced oxidative damage by activating antioxidative systems but also increases Se content in edible plant portions. Additionally, it improves yield-related factors in an effective manner. Role of Se under water constraints is indispensable in the future of wheat production.

References

Abid M, Ali S, Qi LK, Zahoor R, Tian Z, Jiang D, Snider JL, Dai T (2018) Physiological and biochemical changes during drought and recovery periods at tillering and jointing stages in wheat (Triticum aestivum L.). Sci Rep 8:1–15

Abul-Soud MA, Abd-Elrahman SH (2016) Foliar selenium application to improve the tolerance of eggplant grown under salt stress conditions. Int J Plant Soil Sci 9:1–10

Ahmad R, Waraich EA, Nawaz F, Ashraf MY, Khalid M (2016) Selenium (Se) improves drought tolerance in crop plants – a myth or fact? J Sci Food Agric 96:372–380

Ahmad Z, Waraich EA, Akhtar S, Anjum S, Ahmad T, Mahboob W, Hafeez OBA, Tapera T, Labuschagne M, Rizwan M (2018) Physiological responses of wheat to drought stress and its mitigation approaches. Acta Physiol Plant 40(80):1–13

Aissa N, Malagoli M, Radhouane L (2018) An approach to alleviate the impact of drought stress with selenium amendment. Iran J Sci Technol Trans A Sci 2:1–6

Akbulut M, Çakır S (2010) The effects of Se phytotoxicity on the antioxidant systems of leaf tissues in barley (Hordeum vulgare L.) seedlings. Plant Physiol Biochem 48:160–166

Akbulut GB, Yigit E, Kaya A, Aktas A (2018) Effects of salicylic acid and organic selenium on wheat (Triticum aestivum L.) exposed to fenoxaprop-p-ethyl. Ecotoxicol Environ Saf 148:901–909

Alexieva V, Sergiev I, Mapelli S, Karanov E (2001) The effect of drought and ultraviolet radiation on growth and stress markers in pea and wheat. Plant Cell Environ 24:1337–1344

Andrade FR, Silva GN, Guimarães KC, Barreto HBF, Souza KRD, Guilherme LRG, Faquin V, Reis AR (2018) Selenium protects rice plants from water deficit stress. Ecotoxic Environ Safety 164: 562–570

Anjum SA, Ashraf U, Tanveer M, Khan I, Hussain S, Shahzad B (2017) Drought induced changes in growth, osmolyte accumulation and antioxidant metabolism of three maize hybrids. Front Plant Sci 8(69):1–13

Ashraf MY, Ala SA, Bhatti AS (1998) Nutritional imbalance in wheat (Triticum aestivum L.) genotypes grown at soil water stress. Acta Physiol Plant 20(3):307–310

Balal RM, Shahid MA, Javaid MM, Iqbal Z, Anjum MA, Garcia- Sanchez F et al (2016) The role of selenium in amelioration of heat-induced oxidative damage in cucumber under high temperature stress. Acta Physiol Plant 38:1–14

Banuelos GS, Lin ZQ, Broadley M (2017) Selenium biofortification. In: Selenium in plants: molecular, physiological, ecological and evolutionary aspects, 1st edn. Springer, New York, pp 231–255

Baroowa B, Gogoi N, Paul S, Baruah KK (2015) Response of leaf water status, stomatal characteristics, photosynthesis and yield in black gram and green gram genotypes to soil water deficit. Funct Plant Biol 42:1010–1018

Batth R, Singh K, Kumari S, Mustafiz A (2017) Transcript profiling reveals the presence of abiotic stress and developmental stage specific ascorbate oxidase genes in plants. Front Plant Sci 8(198):1–15

Beltrano J, Ronco MG, Arango MC (2008) Soil drying and rewatering applied at three grain development stages affect differential growth and grain protein deposition in wheat (*Triticum aestivum* L.). Braz J Plant Physiol 18:341–350

Bocchini M, Amato RD, Ciancaleoni S, Fontanella MC, Palmerini CA, Beone GM, Onofri A, Negri V, Marconi G (2018) Soil selenium (Se) biofortification changes the physiological, biochemical and epigenetic responses to water stress in Zea mays L. by inducing a higher drought tolerance. Front Plant Sci 9(389):1–14

Boldrin PF, Faquin V, Ramos SJ, Boldrin KVF, Avila FW, Guilherme LRG (2013) Soil and foliar application of selenium in rice biofortification. J Food Compos Anal 31:238–244

Breusegem FV, Vranova E, Dat JF, Inze D (2001) The role of active oxygen species in plant signal transduction. Plant Sci 161:405–414

Carey AM, Scheckel KG, Lombi E, Newville M, Choi YS, Norton GJ, Price AH, Meharg AA (2012) Grain accumulation of selenium species in rice (Oryza sativa L.). Environ Sci Technol 46:5557–5564

Caverzan A, Casassola A, Brammer SP (2016) Antioxidant responses of wheat plants under stress. Genet Mol Biol 39:1–6

Chen RJ, Wang LX, Zhang XJ, Wang SL, Li H, Gao SJ (2020) Physiological and molecular basis of the effects of exogenous selenium application on wheat seedling performance under drought stress. Appl Ecol Environ Res 18(1):1227–1236

Chilimba ADC, Young SD, Black CR, Meacham MC, Lammel J, Broadley MR (2012) Agronomic biofortification of maize with selenium (Se) in Malawi. Field Crop Res 125:118–128. https://doi.org/10.1016/j.fcr.2011.08.014

Chu J, Yao X, Zhang Z (2010) Responses of wheat seedlings to exogenous selenium supply under cold stress. Biol Trace Elem Res 136:355–363

Chu J, Yao X, Yue Z, Li J, Zhao J (2013) The effects of selenium on physiological traits, grain selenium content and yield of winter wheat at different development stages. Biol Trace Elem Res 151:434–440

Cornic G (2000) Drought stress inhibits photosynthesis by decreasing stomatal aperture- not by affecting ATP synthesis. Trends Plant Sci 5:187–198

Delfine S, Tognettir R, Loreto F, Alvino A (2002) Physiological and growth responses to water stress in field grown bell pepper (Capsicum annuum, L.). J Hortic Sci Biotechnol 77(6):697–704

Djanaguiraman M, Devi DD, Shanker AK, Sheeba JA, Bangarusamy U (2005) Selenium—an antioxidative protectant in soybean during senescence. Plant Soil 272:77–86

Dong JZ, Wang Y, Wang SH, Yin LP, Xu GJ, Zheng C (2013) Selenium increases chlorogenic acid, chlorophyll and carotenoids of Lycium chinense leaves. J Sci Food Agric 93:310–315

Duan MY, Cheng SR, Lu RH, Lai RF, Zheng AX, Ashraf U, Fan PS, Du B, Luo HW, Tang XR (2019) Effect of foliar sodium selenate on leaf senescence of fragrant rice in South China. Appl Ecol Environ Res 17:3343–3351

Ducsay L, Lozek O, Varga L (2009) Effect of selenium foliar application on its content in spring barley. Agrochemia 12:3–6

Ducsay L, Lozek O, Marcek M, Varenyiova M, Hozlar P, Losak T (2016) Possibility of selenium biofortification of winter wheat grain. Plant Soil Environ 62:379–383

El-Ramady H, Abdalla N, Taha HS, Alshaal T, El-Henawy A, Faizy SEDA, Shams MS, Youssef SM, Shalaby T, Bayoumi Y, Elhawat N, Shehata S, Sztrik A, Prokisch J, Fári M, Domokos-Szabolcsy É, Pilon-Smits EA, Selmar D, Haneklaus S, Schnug E (2016) Selenium and nano-selenium in plant nutrition. Environ Chem Lett 14:123–147

Eman MM, Khattab HE, Helal NM, Deraz AE (2014) Effect of selenium and silicon on yield quality of rice plant grown under drought stress. Aust J Crop Sci 8(4):596–605

Farooq M, Wahid A, Kobayashi N, Fujita D, Basra SMA (2009) Plant drought stress: effects, mechanisms and management. Agron Sustain Dev 29:185–212

Farooq M, Hussain M, Siddique KHM (2014) Drought stress in wheat during flowering and grain-filling periods. Crit Rev Plant Sci 33:331–349. https://doi.org/10.3389/fpls.2017.01950

Feng R, Wei C, Tu S (2013) The roles of selenium in protecting plants against abiotic stresses. Environ Exp Bot 87:58–68

Galinha C, Sánchez-Martínez M, Pacheco AMG, Freitas MC, Coutinho J, Maçãs B, Almeida AS, Pérez-Corona MT, Madrid Y, Wolterbeek HT (2014) Characterization of selenium-enriched wheat by agronomic biofortification. J Food Sci Technol 52(7):4236–4245

Germ M (2008) The response of two potato cultivars on combined effects of selenium and drought. Acta Agric Slov 91(1):121–137

Germ M, Stibilj V, Osvald J, Kreft I (2007) Effect of selenium foliar application on chicory (*Cichorium intybus* L.). J Agric Food Chem 55:795–798

Ghosh S, Biswas AK (2017) Selenium modulates growth and thiol metabolism in wheat (*Triticum aestivum* L.) during arsenic stress. Am J Plant Sci 8:363–389

Gill SS, Tuteja N (2010) Reactive oxygen species and antioxidant machinery in abiotic stress tolerance in crop plants. Plant Physiol Biochem 48:909–930

Habibi G (2013) Effect of drought stress and selenium spraying on photosynthesis and antioxidant activity of spring barley. Acta Agric Slov 101(1):31–39

Hajiboland R, Sadeghzadeh N (2014) Effect of selenium supplementation on CO_2 and NO^{3-} assimilation under low and adequate N supply in wheat (*Triticum aestivum* L.) plants. Photosynthetica 52(4):501–510

Hajiboland R, Sadeghzadeh N, Ebrahimi N, Sadeghzadeh B (2015) Influence of selenium in drought-stressed wheat plants under greenhouse and field conditions. Acta Agric Slov 105(2): 175–191

Hammad SAR, Ali OAM (2014) Physiological and biochemical studies on drought tolerance of wheat plants by application of amino acids and yeast extract. Ann Agric Sci 59(1):133–145

Hartikainen H, Xue T, Piironen V (2000) Selenium as an anti-oxidant and pro-oxidant in ryegrass. Plant Soil 225:193–200

Havlin JL, Beaton JD, Tisdale SL, Nelson WL (1999) Soil fertility and fertilizers: an introduction to nutrient management. Prentice-Hall, London, pp 406–425

Hayat S, Hayat Q, Alyemeni MN, Wani AS, Pichtel J, Ahmad A (2012) Role of proline under changing environments: a review. Plant Signal Behav 7:1456–1466

Hu Y, Schmidhalter U (2005) Drought and salinity: a comparison of their effects on mineral nutrition of plants. J Plant Nutr Soil Sci 168:541–549

Hu KL, Zhang JW, You Y (2013) Influence of selenium on growth, lipid peroxidation and antioxidative enzyme activity in melon (*Cucumis melo* L.) seedlings under salt stress. Acta Soc Bot Pol 82(3):193–197

Hussain S, Khan F, Cao W, Wu L, Geng M (2016) Seed priming alters the production and detoxification of reactive oxygen intermediates in rice seedlings grown under sub-optimal temperature and nutrient supply. Front Plant Sci 7:439

Ibrahim HM (2014) Selenium pretreatment regulates the antioxidant defense system and reduces oxidative stress on drought-stressed wheat (*Triticum aestivum* L.) plants. Asian J Plant Sci 54: 120–128

Ibrahim HM (2016) Selenium supplementation alters biochemical constitutes and improves yield of wheat plants subjected to drought stress. Int J Agric Innov Res 4(4):649–654

Iqbal M, Hussain I, Liaqat H, Ashraf MA, Rasheed R, Rehman AU (2015) Exogenously applied selenium reduces oxidative stress and induces heat tolerance in spring wheat. Plant Physiol Biochem 94:95–103. https://doi.org/10.1016/j.plaphy.2015.05.012

Jóźwiak W, Politycka B (2019) Effect of selenium on alleviating oxidative stress caused by a water deficit in cucumber roots. Plants (Basel) 8(7):217. https://doi.org/10.3390/plants8070217. Epub 2019 Jul 11. PMID: 31336811; PMCID: PMC6681189

Kaldenhoff R, Ribas-Carbo M, Sans JF, Lovisolo C, Heckwolf M, Uehlein N (2008) Aquaporins and plant water balance. Plant Cell Environ 31:658–666

Kelaleche H, Guendouz A, Hafsi M (2018) The effect of water stress on some physiological and biochemical traits in five durum wheat (Triticum durum Desf.) genotypes. Int J Biosci 12(1): 90–97

Keyvan S (2010) The effects of drought stress on yield, relative water content, proline, soluble carbohydrates and chlorophyll of bread wheat cultivars. J Anim Sci 8:1051–1060

Khan MIR, Nazir F, Asgher M, Per TS, Khan NA (2015) Selenium and sulfur influence ethylene formation and alleviate cadmium-induced oxidative stress by improving proline and glutathione production in wheat. J Plant Physiol 173:9–18

Kocsy G, Szalai G, Galiba G (2004) Effect of osmotic stress on glutathione and hydroxymethyl glutathione accumulation in wheat. J Plant Physiol 161:785–794

Kumar P, Boora KS, Kumar N, Batra R, Goyal M, Sharma KD, Yadav RC (2018) Traits of significance for screening of chickpea (*Cicer arietinum* L.) genotypes under terminal drought stress. J Agrometeorol 20(1):40–45

Kuznetsov VV, Kholodova VP, Kuznetsov VV, Yagodin BA (2003) Selenium regulates the water status of plants exposed to drought. Dokl Biol Sci 390:266–268

Lan C, Lin K, Huang W, Chen C, Huang W, Chen C (2020) Comparisons of chlorophyll fluorescence and physiological characteristics of wheat seedlings influenced by iso-osmotic stresses from polyethylene glycol and sodium chloride. Agronomy 10(325):4–14

Lara TS, Lessa JHL, Souza KRD, Corguinha APB, Martins FAD, Lopes G, Guilherme LRG (2019) Selenium biofortification of wheat grain *via* foliar application and its effect on plant metabolism. J Food Compos Anal 81:10–18

Lehari K, Kumar M, Burman V, Kumar V, Chand P, Singh R (2019) Morphological, physiological and biochemical analysis of wheat genotypes under drought stress. J Pharmacogn Phytochem 2: 1026–1030

Li MQ, Hasan M, Li CX, Ahammed GJ, Xia XJ, Shi K (2016) Melatonin mediates selenium-induced tolerance to cadmium stress in tomato plants. J Pineal Res 61:291–302

Lyons GH, Genc Y, Soole K, Stangoulis JCR, Liu F, Graham RD (2008) Selenium increases seed production in brassica. Plant Soil 318:73–80

Malik JA, Goel S, Kaur N, Sharma S, Singh I, Nayyar H (2012) Selenium antagonises the toxic effects of arsenic on mungbean (Phaseolus aureus Roxb.) plants by restricting its uptake and enhancing the antioxidative and detoxification mechanisms. Environ Exp Bot 77:242–248

Mohammadi S, Rezaie M, Mahootchi AH, Hamze H, Kalilagdam N (2015) Effect of water stress on oxidative damage and antioxidant enzyme activity of bread wheat genotypes. WALIA J 31:163–169

Mora ML, Durán P, Acuña AJ, Cartes P, Demanet R, Gianfreda L (2015) Improving selenium status in plant nutrition and quality. J Soil Sci Plant Nutr 15:486–503

Mostajeran A, Rahimi-Eichi V (2009) Effects of drought stress on growth and yield of rice (*Oryza sativa* L.) cultivars and accumulation of proline and soluble sugars in sheath and blades of their different ages leaves. Am Eurasian J Agric Environ Sci 5:264–272

Moussa HR, Hassen AM (2017) Selenium affects physiological responses of Phaseolus vulgaris in response to salt level. Int J Veg Sci 38:1–18

Murata N, Takahashi S, Nishiyama Y, Allakhverdiev SI (2007) Photoinhibition of photosystem II under environmental stress. Biochim Biophys Acta 1767:414–421

Namich AAM (2007) Response of cotton cultivar Giza80 to application of glycine betaine under drought conditions. Minufiya J Agric Res 32(6):1637–1651

Nawaz F, Ashraf MY, Ahmad R, Waraich EA (2013) Selenium (Se) seed priming induced growth and biochemical changes in wheat under water deficit conditions. Biol Trace Elem Res 51:284–293. https://doi.org/10.1007/s12011-012-9556-9

Nawaz F, Ahmad R, Ashraf MY, Waraich EA, Khan SZ (2015a) Effect of selenium foliar spray on physiological and biochemical processes and chemical constituents of wheat under drought stress. Ecotoxicol Environ Saf 113:191–200

Nawaz F, Ashraf MY, Ahmad R, Waraich EA, Shabbir RN, Bukhari MA (2015b) Supplemental selenium improves wheat grain yield and quality through alterations in biochemical processes under normal and water deficit conditions. Food Chem 175:350–357

Nawaz F, Naeem M, Ashraf MY, Tahir MN, Zulfiqar B, Salahuddin M, Shabbir RN, Aslam M (2016) Selenium supplementation affects physiological and biochemical processes to improve fodder yield and quality of maize (*Zea mays* L.) under water deficit conditions. Front Plant Sci 7(1438):1–13

Nawaz F, Ashraf MY, Ahmad R, Waraich EA, Shabbir RN (2017) Selenium supply methods and time of application influence spring wheat (*Triticum aestivum* L.) yield under water deficit conditions. J Agric Sci 155:643–656

Proietti P, Nasini L, Del Buono D, D'Amato R, Tedeschini E, Businelli D et al (2013) Selenium protects olive (*Olea europaea* L.) from drought stress. Sci Hortic 164:165–171

Rahbarian R, Khavari-Nejad R, Ganjeali A, Bagheri A, Najafi F (2011) Drought stress effects on photosynthesis, chlorophyll fluorescence and water relations in tolerant and susceptible chick-pea genotypes. Acta Biol Crac Ser Bot 53:47–56

Rayman MP (2012) Selenium and human health. Lancet 379(9822):1256–1126

Rios JJ, Blasco B, Cervilla LM (2009) Production and detoxification of H_2O_2 in lettuce plants exposed to selenium. Ann Appl Biol 154:107–116

Roy S, Arora A, Chinnusamy V, Singh VP (2017) Endogenous reduced ascorbate: an indicator of plant water deficit stress in wheat. Indian J Plant Physiol 22:365–368

Saidi I, Chtourou Y, Djebali W (2014) Selenium alleviates cadmium toxicity by preventing oxidative stress in sunflower (Helianthus annuus) seedlings. J Plant Physiol 171:85–91

Sajedi NA (2017) Evaluation of selenium and salicylic acid effect on physiological and qualitative characteristics of dry-land wheat cultivars. Iran Agr Res 36(2):91–100

Sajedi NA, Ardakani MR, Madani H, Naderi A, Miransari M (2011) The effects of selenium and other micronutrients on the antioxidant activities and yield of corn (*Zea mays* L.) under drought stress. Physiol Mol Biol Plants 17:215–222

Sattar A, Akhtar M, Ahmad C, Muhammad S, Sami I, Allah U, Nawaz A, Abbas T, Ali Q (2019) Physiological and biochemical attributes of bread wheat (*Triticum aestivum* L.) seedlings are influenced by foliar application of silicon and selenium under water deficit. Acta Physiol Plant 41(146):1–11

Shabbir RN, Waraich EA, Ali H, Nawaz F, Ashraf MY, Ahmad R (2016) Supplemental exogenous NPK application alters biochemical processes to improve yield and drought tolerance in wheat (*Triticum aestivum* L.). Environ Sci Pollut Res 23:2651–2662

Shahzadi I, Iqbal M, Rasheed R et al (2017) Foliar application of selenium increases fertility and grain yield in bread wheat under contrasting water availability regimes. Acta Physiol Plant 39: 173. https://doi.org/10.1007/s11738-017-2477-7

Shamsi K, Kobraee S (2011) Bread wheat production under drought stress conditions. Ann Biol Res 2:352–358

Sheoran S, Thakur V, Narwal S, Turan R, Mamrutha HM, Singh V, Tiwari V, Sharma I (2015) Differential activity and expression profile of antioxidant enzymes and physiological changes in wheat (*Triticum aestivum* L.) under drought. Appl Biochem Biotechnol 177:1282–1298

Siddique MRB, Hamid A, Islam MS (2000) Drought stress effects on water relations of wheat. Bot Bull Acad Sin 41:35–39

Simova-Stoilova L, Demirevska K, Petrova T, Tsenov N, Feller U (2008) Antioxidative protection in wheat varieties under severe recoverable drought at seedling stage. Plant Soil Environ 54: 529–536

Singh SK, Reddy KR (2011) Regulation of photosynthesis, fluorescence, stomatal conductance and water-use efficiency of cowpea (*Vigna unguiculata* [L.] Walp.) under drought stress. J Photochem Photobiol B 105:40–50

Tadina N, Germ M, Kreft I, Breznik B, Gaberscلk A (2007) Effects of water deficit and selenium on common buckwheat (Fagopyrum esculentum Moench.) plants. Photosynthetica 45:472–476

Teimouri S, Hasanpour J, Tajali AA (2014) Effect of selenium spraying on yield and growth indices of wheat (*Triticum aestivum* L.) under drought stress condition. Int J Adv Biol Biomed Res 2(6): 2091–2103

Wang CQ (2011) Water-stress mitigation by selenium in Trifolium repens L. J Plant Nutr Soil Sci 174:276–282

Wang YD, Wang X, Wong YS (2013) Generation of selenium-enriched rice with enhanced grain yield, selenium content and bioavailability through fertilisation with selenite. Food Chem 141: 2385–2393. https://doi.org/10.1016/j.foodchem.2013.05.095

Wu C, Dun Y, Zhang Z, Li M, Wu G (2020) Foliar application of selenium and zinc to alleviate wheat (Triticum aestivum L.) cadmium toxicity and uptake from cadmium-contaminated soil. Ecotoxicol Environ Saf 90:1–9. https://doi.org/10.1016/j.ecoenv.2019.110091

Xue TL, Hartikainen H, Piironen V (2001) Antioxidative and growth-promoting effects of selenium on senescing lettuce. Plant Soil 273:55–61

Yang X, Chen X, Ge Q, Li B, Tong Y, Zhang A, Li Z, Kuang T, Lu C (2006) Tolerance of photosynthesis to photo-inhibition, high temperature and drought stress in flag leaves of wheat: a comparison between a hybridization line and its parents grown under field conditions. Plant Sci 171:389–397. https://doi.org/10.1016/j.plantsci.2006.04.010

Yao X, Chu J, Wang G (2009) Effects of selenium on wheat seedlings under drought stress. Biol Trace Elem Res 130:283–290

Yao X, Chu J, He X, Ba C (2011) Protective role of selenium in wheat seedlings subjected to enhanced UV-B radiation. Russ J Plant Physiol 58:283–289

Yao X, Chu J, Liang L, Geng W, Li J, Hou G (2012) Selenium improves recovery of wheat seedlings at rewatering after drought. Russ J Plant Physiol 59(6):701–707. https://doi.org/10.1134/s1021443712060192

Yao X, Jianzhou C, Xueli H, Binbin L, Jingmin L, Zhaowei Y (2013) Effect of selenium on agronomical characters of winter wheat exposed to enhanced ultraviolet-B. Ecotox Environ Saf 92:320–326

Zhang M, Tang SH, Huang X, Zhang FB, Pang YW, Huang QY, Yi Q (2014) Selenium uptake, dynamic changes in selenium content and its influence on photosynthesis and chlorophyll fluorescence in rice (Oryza sativa L.). Environ Exp Bot 107:39–45

22 Multifaceted Mechanisms of Silicon in Alleviation of Drought Stress in Plants

Pooja Singh, Vikram Kumar, Jyoti Sharma, Sakshi Saini, Priyanka Sharma, Sandeep Kumar, and Asha Sharma

Abstract

Constant climate changes, high fluctuations in weather conditions, extremely low rainfall, and enormous drought conditions are attracting significant consideration in today's world. Drought stress is one of the challenging abiotic stresses to agricultural fields and the environment. It affects plant physiology by delaying cell division, decreasing CO_2 assimilation, water potential in leaves, plant water status, photosynthetic pigment, closing stomata, and other biochemical processes. To antagonize the drought condition, plants develop different mechanisms through different physiological and biochemical processes. Silicon (Si), as a multi-talented element, shows great performance in alleviating drought conditions in plants. Despite being a non-essential element, its role in stress tolerance is recommendable. Multiple effects of Si in mitigating water-deficit conditions through different mechanisms of action were presented in the chapter. Moreover, this chapter comprehends on different physiological, morphological, biochemical, and molecular aspects of plants dealing with drought stress. Application of Si has an important and notable effect on increasing plant tolerance to drought conditions by regulating homeostasis, osmoprotectant, antioxidant enzymes related to a defense mechanism, reducing transpiration, enhancing compatible solutes, increasing uptake of water by roots, nutrient balancing, and many more. In addition, modulation in gene expression, phytohormone synthesis, activation of different defense-related mechanisms by acting as a signaling molecule, improving the water status of the plants under drought-stress situations,

P. Singh · V. Kumar · J. Sharma · S. Saini · P. Sharma · A. Sharma (✉)
Department of Botany, Maharshi Dayanand University, Rohtak, India
e-mail: drasha.botany@mdurohtak.ac.in

S. Kumar
Department of Botany, Baba Mastnath University, Rohtak, India

A. Kumar et al. (eds.), *Salinity and Drought Tolerance in Plants*,
https://doi.org/10.1007/978-981-99-4669-3_22

and increasing the production and growth of the plants are some other facts of Si. Si also enhances plant resistance to particular stress conditions by altering the expression of stress-related genes in plants. So, this chapter helps in understanding the multidimensional role of Si in drought-stressed plants via different alleviation mechanisms.

Keywords

Antioxidant enzymes · Drought stress · Silicon · Stress tolerance · Water uptake

22.1 Introduction

Drought stress is one of the most devastating abiotic stresses among all environmental stresses, which severely affects agricultural production and has various adverse effects on plant morphology, physiology, and yield production. Due to the constant problem of drought stress, sustainable agriculture production is decreasing day by day globally (Azeem et al. 2022; Ma et al. 2019). A decrease in agronomic yield leads to a reduction in income for local farmers. Low production affects the maintenance and economy of the farmers to a great extent (Boudiar et al. 2020). Plants mainly consist of about 90% water, which contributes greatly to various physiological, morphological, biochemical, and molecular aspects of plant development and growth (Brodersen et al. 2019; Abbasi and Abbasi 2010). Hence, drought stress is considered adverse abiotic stress in areas where plants face water scarcity conditions (Anjum et al. 2011; Diatta et al. 2020). Drought becoming a major serious issue in dealing with world food demands due to persistent poor crop production (Okorie et al. 2019). Plants can grow under drought conditions, but their growth and production are halted due to water-deficit conditions. In addition, growth parameters like seed germination; physiological parameters like plant water status and membrane stress injury; biochemical parameters like chlorophyll pigments, carotenoids, and photosynthetic rate are also affected under water-stressed conditions (Seleiman et al. 2021; McDowell et al. 2022; Pepe et al. 2022). Along with this, it decreases plant height by disturbing the internal system of plants (Lei et al. 2021). It was suggested that this reduction in plant height is due to reduced growth of plant cells, early senescence, and retarded cell division rate (Liang et al. 2018; Elnaggar et al. 2018). Plants decreased the transpiration rate to slow down the water loss, reduced the number of leaves, and promoted wilting of leaves under drought conditions (Alam et al. 2021a; Wu et al. 2022). In addition, drought induces oxidative stress in plants by increasing reactive oxygen species (ROS) content and inhibiting plant's growth and physiology (Kapoor et al. 2020). During high water scarcity, water transport via xylem becomes interrupted, impairs cell growth in higher plants. The cellular damage, structural instability, wilting and abscission of leaves, leaf rolling, decrease in leaf area, leaf number, and closure of stomata are some important alterations that occur during low to high water scarcity conditions and inhibit the plant growth (Correia et al. 1997; Kim et al. 2017; Ahmad et al. 2018). According to

Ahmad et al. (2018), drought affects the various development stages of plants, such as cell enlargement, cell division, gene expression, morpho-physiological, and homeostasis conditions. These various aspects are affected by the low water condition of the plant, which directly retards the overall plant growth and production (Wahab et al. 2022).

Silicon (Si) is the richest element found on the earth, which plays a valuable role in plant growth and production, especially during stressful conditions (Shahzad et al. 2022; Pooja et al. 2022). Its emerging role under stress conditions is incredible. In many studies, it was observed that Si defends plants from the negative impact of drought stress in many crops. In addition, Si is one of the macronutrients in many plant species, particularly plants belonging to the Poaceae family (Epstein 1999). Plants that can accumulate Si in their plant parts, such as rice, wheat, and sugarcane, are Si-accumulator plants and take maximum benefits by increasing their growth in stressful conditions. Those plants that are low Si-accumulating plants also defend themselves in stressed conditions with Si application. Si is a multifaceted element present in the soil, which shows multiple effects on plants by improving plant growth, germination rate, photosynthetic efficiency, plant height, biomass, and lastly, yield production under all environmental stresses (Ali et al. 2020; Mir et al. 2020; Mundada et al. 2021). Si application helps in the regulation of different morpho-physiological aspects of plants which are disturbed during drought stress conditions. Furthermore, it has been stated that even a high concentration of Si does not affect the physiology of plants (Bhardwaj and Kapoor 2021). Si supplementation enhances plant growth, antioxidant enzyme activities, water use efficiency, photosynthetic pigments, and ionic balance via regulating the ROS level in plant cells further decreasing the oxidative stress under water stress (Rizwan et al. 2015). The major mechanism of Si through which it improves the growth of plant includes regulation of aquaporin activity, water use efficiency, biosynthesis of compatible solutes like phenol, proline, and sugar, biosynthesis of phytohormones, enhancement in antioxidant defense system by regulating enzyme activities, and downregulating the gene expressions of different stress-related proteins under drought conditions (Wang et al. 2021; Pooja et al. 2022). In tomato plants, it was found that Si enhanced the root hydraulic conductance and membrane stability and suppressed ROS generation via improving antioxidant enzyme activities under water-deficit environments (Shi et al. 2016).

In this chapter, we will highlight the different mechanisms of Si under drought-stress conditions involving morphological adaptations, photosynthetic rate, stomatal conductance, transpiration, ion homeostasis, regulation of aquaporin activity, water use efficiency, and biosynthesis of compatible solutes. Moreover, we will also discuss the different antioxidant enzyme activities in defense system, reduction of oxidative stress by suppressing the level of ROS production, regulation of different phytohormones, and regulation of gene expressions in plants under drought conditions (Fig. 22.1).

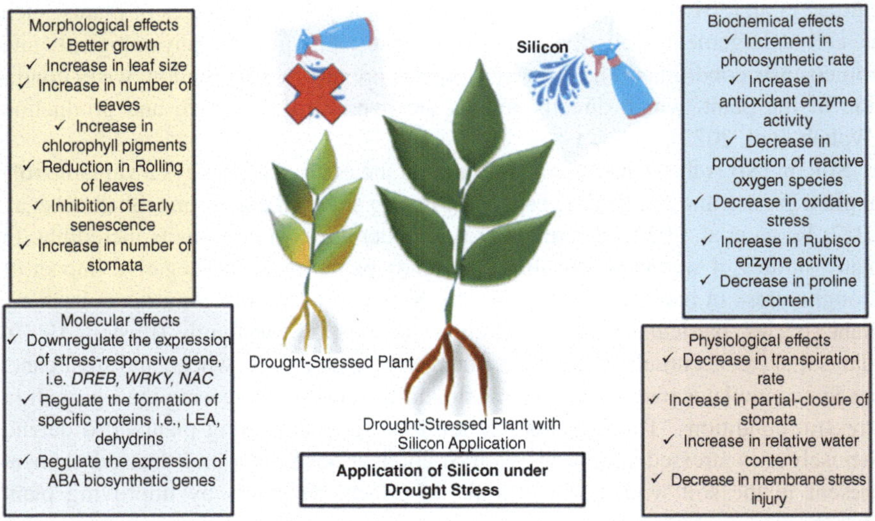

Fig. 22.1 Si-induced different biochemical, morphological, physiological, and molecular changes in plants during drought-stress conditions

22.2 Si as a Fertilizer

Generally, fertilizers are used to produce high crop production and yield worldwide. Local farmers use different commercial chemical fertilizers in order to deal with the decreased production of crops. These chemical fertilizers are cost-effective and provide instant results on crops but lower the fertility of soil (Purbajanti et al. 2019). Si application emerging as a new approach to agriculture in different environmental stress conditions enhances soil fertility and nutrients and improves plant growth. Si is recognized as a biofertilizer after knowing its role in stress alleviation, increasing soil fertility, and improving the performance of many plants under stressful conditions (Pooja et al. 2022). In soil, Si supplementation improves the properties of soil and increases the NPK (nitrogen, phosphorous, and potassium) value in soil with different macro- and micronutrients (Tables 22.1 and 22.2).

Moreover, applying Si as a fertilizer improves soil fertility at a greater level. Besides this, Si acts as a multifunctional element by enhancing plant growth under different biotic and abiotic stresses and providing stress tolerance to plants against many stresses like drought or salinity. Si provides nutrition to the plants through soil by converting the nutrients into a plant-available form (Tayade et al. 2022). According to a report on rice plants, it was found that Si as a fertilizer improved plant weight and yield production. In this report, they suggested that to increase the uptake of Si in plants, we must apply Si in different fertilizer forms (Kovács et al. 2022).

Table 22.1 Role of different types of Si fertilizers in improving agricultural production in different crops under stressed conditions

Sr. no.	Type of Si fertilizer	Type of stress and plant species	Application in agricultural field	Mechanisms	References
1.	Sodium silicate	Biotic stress; *Cucumis melo*	Promoting wound healing increases the hardness and reduces the flexibility of healing tissue	Accumulation of suberin polyphenolic and Si at wounds	Xue et al. (2022)
2.	Potassium silicate	Salinity stress	Removal of sodium from red mud for sustainable agriculture	Hydrothermal leaching	Chao et al. (2022)
3.	Potassium silicate	Drought stress; *Avena sativa*	Increases the physiological parameters like relative water content, photosynthetic pigments	Enhances water use efficiency	Kutasy et al. (2021)
4.	Monosilicic acid	Drought stress; *Cucumis melo*	Enhances the fruit quality and fruit yield	Increases flesh thickness of fruit	Alam et al. (2021b)
5.	Foliar Si	Drought stress; *Triticum aestivum; Beta vulgaris; Coriandrum sativum*	Improves growth, morpho-physiological parameters	Accumulation of Si in plant parts	Aurangzaib et al. (2022), Artyszak et al. (2021), Afshari et al. (2021)
6.	Aerosil 200 (SiO_2)	Salt stress, drought stress; *Cucurbita pepo*	Improves the growth parameters, seed germination, photosynthetic efficiency	Improved antioxidant enzyme activities	Siddiqui et al. (2014), Tayade et al. (2022)
7.	Agribooster™	Drought stress; *Allium cepa; Glycine max*	Improves the photosynthetic pigments; increases the water availability	Enhances antioxidant enzyme activity	Rangwala et al. (2018, 2019), Tayade et al. (2022)
8.	Absorbent silicates	Drought stress	Improves soil fertility, water holding capacity of soil	Enhances soil moisture retention	Kovács et al. (2022)

(continued)

Table 22.1 (continued)

Sr. no.	Type of Si fertilizer	Type of stress and plant species	Application in agricultural field	Mechanisms	References
9.	Silicate rock powder	Environmental stress	Increases the rate of Si uptake in plants	Enhances weathering rate of silicate in the soil by releasing organic acids	Harley and Gilkes (2000), Kovács et al. (2022)
10.	Slag silicate fertilizer	Environmental stress; *Oryza sativa*	Enhances atmospheric carbon dioxide sequestration and Si content in rice plant	Phytolith production in field soil and trap of carbon in phytoliths	Makabe-Sasaki et al. (2014), Kovács et al. (2022)
11.	Si-rich rice husk ash and rice husk biochar	Abiotic and biotic stress; *Solanum melongena; Oryza sativa*	Decreases insect infection; reduces arsenic level	Deposition of Si in plants	Bakhat et al. (2021), Leksungnoen et al. (2019)
12.	Sodium silicate	Drought stress; *Zea mays*	Increases photosynthetic pigments in plants	Increases the uptake of potassium and calcium	Bijanzadeh et al. (2022)
13.	Nano-Si	Drought stress; *Triticum aestivum*	Improves yield production	Increases water use efficiency	Ahmadian et al. (2021)

22.3 Different Drought Tolerance Mechanisms Driven by Si in Plants

Bukhari et al. (2020) reported the study on Si under different environmental stresses and found that it increased crop quality and production under stressful conditions by reducing the negative effect of various stresses. Under drought conditions, the exogenous Si application improves the water status of the plants by increasing the relative water content (RWC) and water potential of the plants (Salim 2014). In addition, it reduces the leaf rolling and increases the rigidity and stiffness of the leaves, leaf area, number of stomata, and water status of the plants under drought conditions (Saud et al. 2014). Furthermore, Si alleviates drought stress by reducing oxidative stress and enhancing the antioxidant enzyme activities, and carboxylase activities in plants (Gong and Chen 2012). In a report, it was also found that addition of Si improves the photosynthetic pigments under water-deficit conditions and enhances the photosynthetic rate (Pilon et al. 2014). Its ameliorative role has also been seen in wheat plant defense system by regulation of different antioxidant

Table 22.2 Si-mediated different mechanisms of action in various plant species under drought conditions

Sr. no.	Plant species	Drought stress	Form and dose of Si applied	Mechanism of action	References
1.	*Fragaria ananassa*	0.6 soil moisture depletion	K_2SiO_3 10 mmol L^{-1}	Improved plant growth by enhancing cell wall extensibility to promote cell elongation; increased photosynthetic rate and chlorophyll content	Dehghanipoodeh et al. (2018)
2.	*Solanum lycopersicum* L.	1.0% PEG	Na_2SiO_3 1.2 mM	Improved cell wall resistance by regulating synthesis of lignin; protects plant from oxidative damage by ROS scavenging	Cao et al. (2020)
3.	*Cucumis melo var. Cantalupensis*	75% field capacity	H_4SiO_4 200 kg ha^{-1}	Improved fruit quality by increasing total soluble solid content, increased length, and yield of fruit	Alam et al. (2021b)
4.	*Triticum aestivum*	2-level irrigation frequency	K_2SiO_3 12 kg ha^{-1}	Improved water potential in plant by preventing excessive transpiration; increased grain yield, number of spikelet and straw yield by improving photosynthetic rates	Ahmad et al. (2016a)
5.	*Hordeum vulgare*	80% water holding capacity	Nanosilica ($nSiO_2$) 100 ppm	Enhanced photosynthetic rates by increasing synthesis of photosynthetic pigments; improved yield attributes, i.e. number of tillers, plant height.	Hellala et al. (2020)

(continued)

Table 22.2 (continued)

Sr. no.	Plant species	Drought stress	Form and dose of Si applied	Mechanism of action	References
6.	*Arachis hypogea*	15% PEG	Na_2SiO_3 2 mM	Helps in osmotic adjustment by improving uptake of Ca^+, K^+, and Mg^{2+} ions; enhanced photosynthetic rate by inhibiting chlorophyll degradation; increased the activity of antioxidative enzymes	Patel et al. (2021)
7.	*Zea mays* L.	10% (w/v) PEG	Na_2SiO_3 1 mM	Improved photosynthetic rate by enhancing chlorophyll and carotenoid content; increased membrane integrity; decreased electrolyte leakage from mitochondrial electron transport chain	Bijanzadeh et al. (2022)
8.	*Avena sativa*	6 mm rainfall	Si (silicon) 3.0 L ha^{-1}	Improved water use efficiency by adjusting stomatal conductance; increased photosynthetic rate, air-leaf temperature difference, and improved yield	Kutasy et al. (2021)
9.	*Solanum lycopersicum* L.	75% field capacity	H_4SiO_4 (silicic acid) 0.25 mM	Enhanced cell wall integrity and improved hydraulic conductivity by increasing root water uptake; improved photosynthetic parameters by decreasing level of ROS in chloroplast;	Chakma et al. (2021)

(continued)

Table 22.2 (continued)

Sr. no.	Plant species	Drought stress	Form and dose of Si applied	Mechanism of action	References
				increased overall growth and yield of plant	
10.	*Saccharum officinarum* L.	55% field capacity	$Ca_3MgO_8Si_2$ 600 kg ha^{-1}	Alleviated oxidative stress damage by increasing proline content and improving antioxidative enzymes activity	Bezerra et al. (2019)
11.	*Mangifera indica*	10 m^3 irrigation water with − 0.77 bars	K_2SiO_3 (potassium silicate) 1.5 mM	Enhanced photosynthetic efficiency by increasing pigments—chlorophyll, carotenoids, and flavonoids; improved yield by increasing synthesis of metabolites	Helaly et al. (2017)
12.	*Brassica napus var. napus* L.	30% field water capacity	$Si(OC_2H_5)_4$ 3.4 mM	Maintains leaf water potential by improving water uptake by roots	Saja-Garbarz et al. (2021)

enzymes like peroxidase (POX), superoxide dismutase (SOD), and catalase (CAT) under high drought stress (Bukhari et al.2020).

22.3.1 Si-Induced Morphological Adaptations for Drought Tolerance

According to a morphological study, it has been observed that leaf is one of the major aspects of assessing drought stress in any plant. Leaf rolling, abscission, and wilting of leaf are the main symptoms of drought stress in plants. Many studies showed that due to rolling of leaf under drought stress, leaf size and leaf area are also reduced, and by this, plants cannot receive sufficient sunlight to process their different reactions like photosynthetic activities, which further affect the plant growth and metabolism (Li-feng et al. 2012). Leaf rolling also reduces the transpiration rate and water loss under water-deficit conditions which is a vital feature of drought tolerance in plants (Zhang et al. 2021). However, Si supplementation recovers the leaf rolling in the plants, which ultimately enhances the photosynthetic pigments due to sufficient light. This helps increase number of leaves and leaf area and improves transpiration so that plant can easily conduct their metabolic processes (Pei et al. 2010). Furthermore, Si application improves the RWC of the plant and enhances the rigidity of the leaf by reducing the rate of leaf rolling under water-stressed situations (Emam et al. 2012; Saud et al. 2014). Si improves the RWC and water capacity of plant leaves by Si accumulation in the leaves and increases the thickness of leaves under drought conditions (Gong et al. 2003). This deposition of Si in plant leaves also decreases the transpirational rate by preventing the leakage of water molecules between the polymerized silica in the leaves (Ahmed et al. 2014; Keller et al. 2015). Ning et al. (2020) found an improvement in the photosynthetic pigments, number of leaves, leaf area, and in antioxidant enzyme activities of maize plants with the Si treatment under water-deficit conditions. In addition, Si also helps in the improvement of root growth and increases root hydraulic conductance and water absorption through roots under high drought stress. The enhancement in root hydraulic conductance is due to reduced oxidative stress in the plant cells and an improved antioxidant defense system with Si supplementation (Wang et al. 2021). Furthermore, Si helps in the development of root growth by accumulating in the root endodermis and forming a dense silica layer with suberin (Fleck et al. 2011). This formation of silica layer in the root cells increases the water holding capacity of roots and helps in mitigation of the adverse effect of drought. Si improves the root/shoot ratios in plants under water-deficit conditions. Casparian band, suberin, and silica formation in the root endodermis is the major development in the root, which further improves the cell wall thickness under drought conditions (Wang et al. 2021).

22.3.2 Si-Induced Decrease in Transpiration and Increase in Stomatal Conductance for Drought Tolerance

The benefits of Si on plant health somewhere link to the regulation of the transpiration process. The deposition of Si on the leaf epidermal cells of the plant helps reduce water loss through reduced transpiration rate in drought conditions (Pooja et al. 2022). This reduced transpiration rate by Si-mediated deposition is an important drought tolerance mechanism under water-deficit conditions (Zhu and Gong 2014). Moreover, accumulation of Si around stomatal cells was also reported in rice plants. It has been found that Si decreased the transpiration rate by up to 30% with the deposition in stomatal cells and cuticles (Rea et al. 2022). Si interacts with the cell wall components like hemicellulose and maintains structural and mechanical stability in plants (Ma et al. 2015). After cross-linking with hemicellulose, Si also helps in holding a high quantity of water within them. In this way, plants can survive the adverse conditions of drought stress with Si treatment (Souri et al. 2021). Furthermore, the application of nano-Si is also emerging in drought-stress conditions. Nano-Si treatment was reported to improve the sugar beet plants' performance under drought conditions by increasing the photosynthetic pigments, plant biomass, and stomatal conductance and reducing the transpiration rate (Namjoyan et al. 2020). Similarly, in tomatoes, Si lessened the negative impact of drought stress by increasing stomatal conductance, CO_2 assimilation rate, and photosynthetic rate (Cao et al. 2020).

22.3.3 Si-Facilitated Photosynthetic Pigments and Photosynthetic Rate for Drought Tolerance

Photosynthesis is directly associated with different biochemical, metabolic, and physiological processes in plants. Drought stress decreases the photosynthetic rate and pigments like chlorophyll and carotenoids in plants halt the overall growth of the plants. This reduction in photosynthesis was due to the reduced movement of CO_2 through stomata and mesophyll cells (Bhardwaj and Kapoor 2021). The lower movement of CO_2 in plant leaves resulted in a decreased rate of chloroplast CO_2, which further damaged both photosystems, finally decreasing the photosynthetic rate (Singh and Thakur 2018). Chlorophylls are also the best indicator of any type of stress in plants. It plays a major part in maintaining plant health and biomass as it directly links to photosynthetic rate (Wang et al. 2002). Si application improves the photosynthetic pigments by enhancing the activity of Rubisco enzyme, which contributes to increasing the photosynthetic rate under drought stress (Seal et al. 2018). Along with Rubisco, GADPH is also enhanced by Si application, which declines oxidative stress by decreasing ROS production and increasing different antioxidant enzyme activities. Increment of these two enzymes under drought conditions by Si treatment provides drought tolerance to plants (Zhang et al. 2020). It was found that in chestnut plants, Si treatment enhances the stomatal conductance, transpiration rate, CO_2 assimilation, and gaseous exchange, which ultimately increases the photosynthetic pigments in plants during water-deficit conditions (Zhang et al. 2013). Along with it, Si increases the drought tolerance in plants by maintaining the photosynthetic rate, which further improves the overall processes in plants under water-stressed conditions (Bhardwaj and Kapoor 2021). According to a study on oat plants, it was found that Si increased the rate of photosynthesis up to double and improved the water uptake by maintaining transpiration rate and photosynthetic pigments in plants under water-stressed conditions. Si also lessens the adverse effect of drought and increases the oat plants' yield (Kutasy et al. 2021). Similarly, under water-deficit conditions, Si application proved beneficial by showing better growth and improved photosynthetic rate (Ahmad et al. 2016a) (Fig. 22.2).

This figure illustrates that (a) application of Si improves root/shoot ratio which increases the water absorption capacity of roots and improves aquaporin activity and osmotic driving force, thereby increasing root hydraulic conductance, thus providing tolerance to water-stressed plants; (b) Si ameliorates ROS-mediated inhibition of the activity of aquaporins, thus enhancing the activity of aquaporins to transport water from root cells to the xylem by upregulating plasma membrane intrinsic protein (PIP) aquaporin gene expression; (c) Si activates the expression of SKOR gene (Stelar K^+ Outward Rectifier) for transport of K^+ ions from root cells to xylem sap. The accumulation of K^+, soluble sugars, and amino acids inside xylem sap leads to

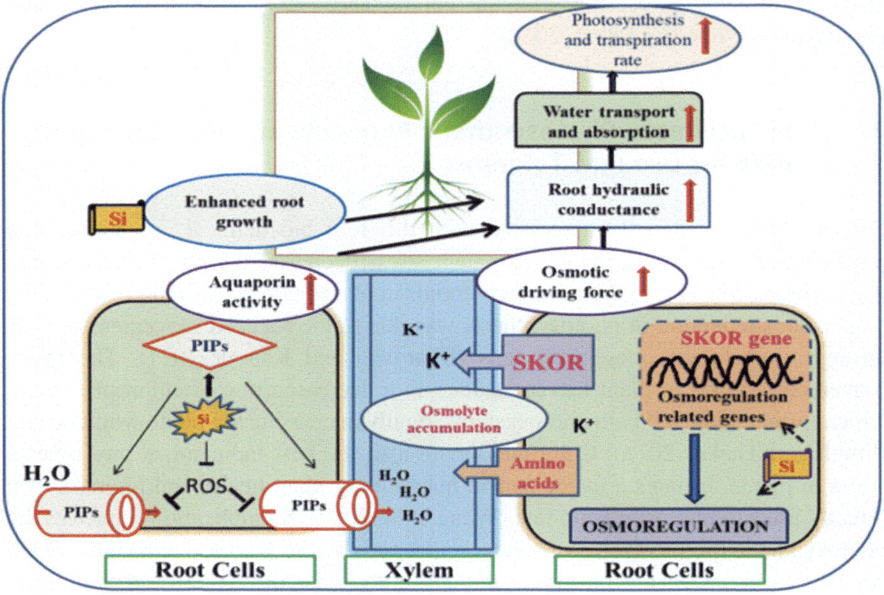

Fig. 22.2 A brief overview of mechanisms of Si to maintain osmotic balance in plants under drought stress (Chen et al. 2018)

osmoregulation which involves the accumulation of osmolytes to maintain osmotic potential, which further increases the osmotic driving force.

22.3.4 Si-Induced Regulation of Aquaporins in Plants for Drought Tolerance

Aquaporins is the member of the major intrinsic protein (MIP) family, which helps regulate water levels in plants and small solutes through membranes. The regulation of aquaporins is an essential feature in plants, especially during water-deficit conditions. Its activity maintains plant development and growth under adverse stress. The inflow and outflow of water molecules are channelized by aquaporins' action, which links it with drought tolerance in plants under drought-stress situations (Ahmed et al. 2021). The aquaporins family consists of different transporters, proteins, and channels that activate water channels in plants along with varying solutes through cell membranes. Si influx transporters (*Lsi*1) also belong to the aquaporins family, which helps in the uptake of Si through the soil and enhances the absorption of water in plants (Ma et al. 2006). After the Si uptake in plants by Si influx transporters, it is transported upward through the xylem by efflux transporters (*Lsi*2) to other parts of the plants. Si uptake and transport into the plants accumulate it into polymerized silica or amorphous silica in different plant parts. Some plants show the accumulation of Si in phytolith formation in plant parts (Mandlik et al.

2020). It has also been found that Si upregulates the transcription of different root aquaporin [*SbPIP1;3/1;4(2)*, *SbPIP1;6*, *SbPIP2;2*, and *SbPIP2;6*] gene expression under water-stressed conditions and increases the hydraulic conductance of root (Liu et al. 2014). In addition, Si treatment under drought-stressed plants activated the gene expression of different aquaporins, which resulted in improved water status of the plants with improved plant growth (Manivannan and Ahn 2017). Si recovers the ionic balance in plants with improved water uptake under drought conditions. Si-facilitated upregulation of different aquaporin genes activity under water-deficit conditions provides drought tolerance to plants and helps in the alleviation of negative effects of drought. Apart from this, Si also helps in activation of root aquaporins which directly recovers the root growth and development in stressed subjected plants (Liu et al. 2014). A study on sorghum plants showed that Si enhanced the upregulation of different genes of *SbPIP* aquaporins and improved the transpiration rate with water loss by reducing the H_2O_2 level in plants (Liu et al. 2015).

22.3.5 Si-Induced Water Use Efficiency (WUE) for Drought Tolerance

Water use efficiency (WUE) is a very important parameter in plants and a determining factor of crop production and yield under water-stressed conditions (Bhardwaj and Kapoor 2021). The application of Si maintains WUE in many ways, transpiration is one of them. The accumulation of Si on the leaf epidermal cells and below the cuticular layer reduces the transpiration process and water loss and increases stomatal conductance (Thorne et al. 2020). It was found that Si improved the plant health, production, and yield with quality parameters in maize plants by increasing the WUE under water-deficit conditions (Gomaa et al. 2021). Si-mediated increment in WUE under water-stressed conditions provides drought tolerance to the plants and helps them by decreasing water loss (Rea et al. 2022). Water retention is also found to be increased in many plants, such as broad bean, tomato, and sugarcane, with Si treatment under drought stress (Verma et al. 2020; Desoky et al. 2020; Khan et al. 2020).

22.3.6 Si Modulates the Biosynthesis of Compatible Solutes for Drought Tolerance

Plants accumulate different kinds of compatible solutes along with upregulation of antioxidant enzyme activities to lessen the adverse effects under a stress environment. Major compatible solutes include proline, phenol, soluble sugar, polyols, glycine betaine, sorbitol, mannitol, etc., which is also called osmoprotectant in plants (Ahammed et al. 2020). They also contribute to improving water retention in various plants under stress conditions. Recent studies showed that *Glycyrrhiza uralensis* and *Salvadora persica* also accumulate amino acids, glycine betaine, and sugars to protect themselves under drought conditions (Zhang et al. 2018; Rangani et al.

2020). Si applications have been reported to enhance drought tolerance efficiency via increasing the accumulation of different compatible solutes and osmolytes. Those plants which accumulate Si in their plant parts are found to be more protective under drought-stress conditions by Si-mediated regulation of different osmolytes. Similar results were found in rice and sorghum plants exposed to stress conditions (Ming et al. 2012; Sonobe et al. 2010).

22.3.6.1 Soluble Sugars

Soluble sugars are the most significant compatible solute or osmoprotectant, which perform a vital function in providing drought tolerance in plants after accumulation. This is a kind of adaptation by which plants protect themselves from any stress (Patel et al. 2021). Soluble sugar accumulation in plants activates different gene expressions related to photosynthetic rate, sugar metabolism, and structural stability (Rangani et al. 2020). Sonobe et al. (2010) stated that Si treatment improves the concentration of soluble sugars in sorghum plants under water-deficit conditions. Similarly, in peanut plants, it was found that the addition of Si increased the rate of sugar content such as fructose, mannose, sucrose, and trehalose under water-stressed conditions (Patel et al. 2021). This improvement in sugar content in the plants during water-deficit situations is somewhere associated with the improved antioxidant defense system with the application of Si. Ning et al. (2020) also found a similar trend of sugar accumulation under stress conditions with Si treatment in maize plants.

22.3.6.2 Polyphenols

Polyphenols are a member of a large family of secondary metabolites, which are formed in plants by phenylpropanoid, shikimic acid, and pentose phosphate pathway (Patel et al. 2021). Polyphenols are low molecular weight compounds that also act as osmoprotectants and effective antioxidants. They alleviate the negative effect of adverse stress up to a great extent and lower ROS production in plants. Rangani et al. (2018) also reported the role of polyphenol in improving drought tolerance in water-deficit situations with its accumulation in plants. In many studies, its accumulation is linked with Si application in plants. It was observed that supplementation of Si enhances the rate of polyphenols accumulation under water stress (Hajiboland et al. 2018).

22.3.6.3 Proline

Proline is measured as a significant indicator of stress tolerance in plants to alleviate different types of abiotic stresses like drought, heavy metal stress, and salinity. It acts as an osmoregulatory amino acid that protects plants from the detrimental result of stress. It was found that proline accumulation under high-stress conditions amended the plant biomass, water status, and photosynthetic pigments by partially alleviating the negative impact of drought (Abdelaal et al. 2020). However, in maize, it was observed that Si reduced the level of high proline content by reducing ROS formation and oxidative stress (Parveen et al. 2019). Si treatments help in the regulation of different types of osmolytes and balance the redox equilibrium in plants exposed to water-deficit conditions. Si regulates the expression of different genes such as *DREB2A* and *NAC5* under water-deficit situations and maintains the proline and other osmolytes concentrations from overaccumulation in plants and increases drought tolerance (Khattab et al. 2014). Moreover, in sorghum, Si supplementation reduced the proline concentrations and increased the sugar concentrations in drought-stressed plants (Yin et al. 2014). The overaccumulation of proline content is harmful to plants and damages plants by disturbing the redox equilibrium. In that case, Si addition in plants during drought conditions lowers the proline concentration and improves plant health (Kaya et al. 2006).

22.3.7 Si-Induced Regulation of Ion Homeostasis in Plants for Drought Tolerance

Minerals are vital nutrients for plant growth and development. The ionic homeostasis mechanism determines the fate of a plant by controlling the downstream events (Tripathi et al. 2020). Water deficits also affect plant growth and production by hindering the absorption and transport of nutrients by the roots and shoots. Because of this reason, nutrient supply is limited, which further disturbs the overall status of plants (Ratnakumar et al. 2016; Xu et al. 2017). In plants exposed to drought stress, Si application enhances the absorption, distribution, and transportation of mineral ions, thus, reducing stress levels in plants (Etesami and Jeong 2018). Si foliar spraying has been reported to increase wheat plant growth and development by increasing the content of phosphorus and potassium (Ratnakumar et al. 2016). Supplementation of Si enhances the development of wheat plants in water-restricted conditions by optimizing nutrient absorption (Xu et al. 2017). It was observed that Si regulates the water potential by improving the K^+ level in shoots and grains (Ahmad et al. 2016a, b). It has also been shown that Si application enhances root growth by stimulating root elongation and increasing cell wall stability in the root growing region (Etesami and Jeong 2018; Hattori et al. 2003). The addition of Si improves ionic balance in plants by enhancing root growth and improving water absorption during water-deficit conditions (Malik et al. 2021).

Moreover, Si-facilitated different mechanisms of drought tolerance in plants like (a) enhancing water potential and water status with increased absorption of essential nutrients from the soil (Liu et al. 2014; Chen et al. 2018); (b) increasing the capacity of roots for different ions (Hernandez-Apaolaza et al. 2014); (c) activating ion transporter proteins (Pavlovic et al. 2013); (d) regulating aquaporin proteins in plants under water stress (Feng Shao et al. 2017). In short, the addition of Si helps in the improvement of ions absorption through roots, balances the ions, and increases the water potential, and growth under water-stressed situations (Wang et al. 2021).

22.3.8 Si Suppresses ROS-Induced Oxidative Damage for Drought Tolerance

Different environmental stress generates oxidative stress in the plants and disturbs the various functions of plant tissues and their constituents. Additionally, these stresses also distort the antioxidant defense machinery through the high production of ROS and alter various biological functions and metabolic processes (Bhardwaj et al. 2021). Due to drought stress, ROS accumulates excessively, which damages cellular components. Along with this, drought stress also causes oxidative damage in plant tissues via altering the cellular metabolism and disturbing the cell membrane and other biomolecules like lipid, protein, and nucleic acid (Kapoor et al. 2020). Si enhances the drought tolerance efficiency of plants via regulating the antioxidant defense machinery to alleviate oxidative stress (Tayyab et al. 2018). Si plays a positive role in enhancing antioxidative enzyme activity as well as the elimination of O^{2-} and H_2O_2 by triggering a decrease in an oxidative burst through the reduction of ion toxicity and accumulating nucleo-proteins and providing defense to plants from stress (Abdelaal et al. 2020). As a result of Si supplementation in drought-stressed maize plants, antioxidant enzyme activity like POD, CAT, and SOD was increased, thereby declining the MDA content (Ning et al. 2020; Parveen et al. 2019). Nano-Si application in sugar beet increased the antioxidant enzymes activities like SOD, GPX, and CAT while diminishing the level of MDA and H_2O_2 (Namjoyan et al. 2020). Biju et al. (2017) also reported the decline in lipid peroxidation and H_2O_2content through the application of Si in lentil leaves subjected to drought stress.

22.3.9 Si-Induced Regulation of Antioxidant Defense Mechanism for Drought Tolerance

Overproduction of ROS is often associated with plant adaptation to drought stress (Patel et al. 2021). Thus, plants must have effective antioxidant defense machinery to regulate the ROS content. Plants have enzymatic (GR, GPX, APX, CAT, SOD) and non-enzymatic antioxidant defense systems to protect themselves from environmental damage (Ahammed et al. 2021). Along with the direct detoxification of ROS, these enzymes may also generate non-enzymatic antioxidants (Patel et al. 2021). Several studies documented to show the positive impact of Si in mitigating the harmful effect of water deficits on different plant species. Si application improves the activities of SOD, APX, CAT, and POD in wheat and peanuts under water stress conditions (Ahmad and Haddad 2011; Patel et al. 2021). In stress conditions, SOD catalyzes the generation of H_2O_2 from O^{2-}, which is catalyzed by CAT, APX, and GPX to produce water (Hasanuzzaman et al. 2020).

22.3.10 Si-Induced Regulation of Phytohormone Synthesis in Plants for Drought Tolerance

Phytohormones like jasmonic acid (JA), salicylic acid (SA), abscisic acid (ABA), etc. play a vital role in increasing plant tolerance to different stresses by regulating the various growth as well as physiological responses (Ahmad et al. 2016a, b; Khan et al. 2020). Under drought conditions, exogenous applications of SA maintain the optimal water status by enhancing RWC in plants (Sohag et al. 2020). In plants, the application of Si differentially regulates endogenous levels of phytohormone, including JA, ABA, and indole acetic acid (IAA) (Xu et al. 2017; Jang et al. 2018). By increasing phytohormone levels via Si application, plants became more tolerant to drought conditions (Yin et al. 2016). It was found that Si supplementation reduced the levels of JA in plant shoots and increased levels of salicylic acid in plants under drought stress (Hamayun et al. 2010). There is a possibility that Si alleviates the downregulation of JA-synthesis enzymes like lipoxygenase, allene oxide synthase 2, and 12-oxophytodienoate reductase 3, allene oxide synthase 1, which account for the reduced levels of JA (Kim et al. 2014a).

A similar observation was reported by Pei et al. (2010). It was observed that Si regulates the level of ABA phytohormone in the leaves under water-deficit situations. This regulation of the ABA level may be due to the enhancement of *OsZEP* gene expression by Si (Kim et al. 2014b). Therefore, ABA synthesis enhances the ABA-mediated gene expression and activates various ABA-regulated functions like stomatal closure, which lessens water loss through transpiration and ultimately restricts plant development (Yoshida et al. 2019). Si-treated leaves under drought stress accumulate ABA, which promotes partial closure of stomata, thereby maintaining the plant's water relations as shown in Fig. 22.3 (Zhang et al. 2006),

Fig. 22.3 Foliar application of Si in inducing ABA-mediated partial closure of stomata. This figure illustrates that drought stress induces complete stomatal closure in plants which lead to inhibiting plant growth; however, the addition of Si under drought-stressed plants regulates the gene expression of *OsZEP* which further maintains the concentration of ABA phytohormone (Kim et al. 2014b). This regulation of ABA by Si treatment induces partial stomatal closure in the plant leaves so that plants can save themselves from high water loss and function normally under stress situations

thus improving the plant's drought tolerance. In drought-stressed plants, Si application enhanced cytokinin (CK) gibberellic acid (GA) and IAA but declined ABA level (Helaly et al. 2017). In addition, Si application enhances the levels of GA, IAA, and ABA in maize (Merhij et al. 2019) and sugarcane (Verma et al. 2019) under drought stress. As a result of these findings, Si may boost drought tolerance by regulating the synthesis of phytohormones. Additionally, Si supply increased the expression of polyamine biosynthesis genes in sorghum under drought conditions (*SPDS, SAMDC06, SAMDC04, ADC, CAP, ODC1, -2, -3*) and decreased the ACC synthesis genes expression (*ACS1* and *ACS2*) (Yin et al. 2014). In plants, polyamines influence many physiological and biochemical functions that contribute to stress mitigation (Milon et al. 2020).

22.3.11 Si-Induced Gene Expression in Plants for Drought Tolerance

The first microarray study of Si-induced gene expression was conducted in the rice plant. In this plant, it was found that Si increases the expression of zinc finger proteins (*ZFPs*) and reduces the expression of chlorophyll a/b binding protein and metallothionein-like protein (Bhardwaj and Kapoor 2021). Under drought-stress conditions, Si upregulates the gene expression of some transcription factors, such as *ZFPs*, which is a stress-responsive gene expression and improves the plant conditions by its enhanced expression and providing drought tolerance to plants. *ZFPs* help in improving the stress defensive genes in plants under drought stress with the application of Si (Watanabe et al. 2004; Manivannan and Ahn 2017). In addition to *ZFPs*, no apical meristem (*NAC*), cup-shaped cotyledon (*CUC*), and C-repeat/dehydration-responsive element binding factors (*CBF/DREB*) are the most common stress mediator genes, whose expressions are regulated by Si application in plants under drought-stress situations. These gene expressions under stress conditions increase drought-stress tolerance in plants at the molecular level (Cong et al. 2008; Kapoor et al. 2020). Furthermore, Si also increases the expressions of some other drought-specific genes such as *OsRDCP1, OsCMO* (coding rice choline monooxygenase), and dehydrin *OsRAB16b* by stimulating the *DREB2A* (dehydration-responsive element binding protein 2A) and *NAC5* under water-stressed conditions. The increase in different gene expressions leads to the formation of different osmolytes in plants, such as soluble sugars, betaine, and proline, which improve plant stress tolerance (Bhardwaj and Kapoor 2021). Bae et al. (2011) observed the importance of *OsRDCP1* drought-specific gene in different morpho-physiological processes of plants to alleviate drought stress. In addition, *OsRAB16b* is also an important stress-related gene that increases the tolerance capacity of plants under water-stressed conditions (Lenka et al. 2011). Furthermore, Si also upregulates the gene expressions of four ASH-GSH cycle genes such as *TaDHAR, TaGR, TaGS*, and *TaMDHAR* in drought-stressed plants by reducing the H_2O_2 and MDA content (Kim et al. 2017; Ma et al. 2016) (Table 22.3).

Table 22.3 Effect of Si on different gene expressions and functions induced during drought stress

S. no	Plant species	Genes	Genes/protein function	References
1.	*Oryza sativa*	*OsDREB2A; OsRDCP1*	Activates *DREB2A* protein by phosphorylation; inactivates and degrades water stress-related proteins	Khattab et al. (2014)
2.	*Sorghum bicolor*	*SAMDC*	Encodes enzyme which initiates the synthesis of ethylene	Yin et al. (2014)
3.	*Brassica napus var. napus* L.	*BnTIP;1*	Improves aquaporins activity and increases the activity of antioxidants	Saja-Garbarz et al. (2022)
4.	*Triticum aestivum*	*TaSOD*, *TaCAT*, and *TaAPX*	Production and activation of antioxidant enzymes	Ma et al. (2016)
5.	*Arabidopsis thaliana*	*At5g22460*	Encodes esterase lipase thioesterase family protein transporter gene, which increases drought-stress tolerance	Li et al. (2008), Naz et al. (2021)
6.	*Arabidopsis thaliana*	*At5g59030*	Encodes copper transporter protein which maintains ion homeostasis	Li et al. (2008), Carneiro et al. (2017)
7.	*Sorghum bicolor*	*SbPIP1;6*	Activates plasma membrane intrinsic protein *PIP1;6* aquaporins and regulates the activities of aquaporins	Liu et al. (2015)
8.	*Oryza sativa*	*OsCMO*	Encodes *Oryza sativa* choline monooxygenase, increasing the accumulation of different osmolytes for drought tolerance	Rehman et al. (2021)

22.4 Conclusion and Future Outlook

The issue of a water shortage is extremely prevalent in the agricultural sectors, where plants must deal with water-stressed circumstances that cause a sharp decline in production. The alleviation of drought stress by Si application is emerging as a great opportunity for farmers. From this chapter, we can conclude that Si has multiple stress alleviation mechanisms by which it improves plant growth and production. Supplementation of Si provides drought-stress tolerance to plants under water-deficit conditions by regulating ion homeostasis, aquaporins activity, osmolytes accumulation, antioxidant enzyme activity, and many more. Its role in the biosynthesis of phytohormones under stress conditions is also noticeable. Apart from this, Si plays a great role in regulating the different gene expressions which are induced during drought-stress situations. Si upregulates many stress-related genes, which help in alleviating stress from plants. It was observed that Si-accumulator plants such as wheat, rice, and sugarcane get more benefits from their Si-accumulating capacity in their plant parts which provide stress tolerance and mechanical support to plants. So, in near future, we can try to genetically manipulate the plant's ability to uptake Si from soil and becomes stress-resistant to drought stress. Also, we may encourage farmers to use Si as a biofertilizer in areas vulnerable to drought.

References

Abbasi T, Abbasi SA (2010) Biomass energy and the environmental impacts associated with its production and utilization. Renew Sust Energ Rev 14(3):919–937

Abdelaal KA, Mazrou YS, Hafez YM (2020) Silicon foliar application mitigates salt stress in sweet pepper plants by enhancing water status, photosynthesis, antioxidant enzyme activity and fruit yield. Plants 9(6):733

Afshari M, Pazoki A, Sadeghipour O (2021) Foliar-applied silicon and its nanoparticles stimulate physio-chemical changes to improve growth, yield and active constituents of coriander (*Coriandrum sativum* L.) essential oil under different irrigation regimes. Silicon 13(11): 4177–4188

Ahammed GJ, Li X, Liu A, Chen S (2020) Brassinosteroids in plant tolerance to abiotic stress. J Plant Growth Regul 39:1451–1464

Ahammed GJ, Li CX, Li X, Liu A, Chen S, Zhou J (2021) Overexpression of tomato RING E3 ubiquitin ligase gene SlRING1 confers cadmium tolerance by attenuating cadmium accumulation and oxidative stress. Physiol Plant 173(1):449–459

Ahmad ST, Haddad R (2011) Study of silicon effects on antioxidant enzyme activities and osmotic adjustment of wheat under drought stress. Czech J Genet Plant Breed 47(1):17–27

Ahmad M, El-Saeid MH, Akram MA, Ahmad HR, Haroon H, Hussain A (2016a) Silicon fertilization—a tool to boost up drought tolerance in wheat (*Triticum aestivum* L.) crop for better yield. J Plant Nutr 39(9):1283–1291

Ahmad P, Rasool S, Gul A, Sheikh SA, Akram NA, Ashraf M, Kazi AM, Gucel S (2016b) Jasmonates: multifunctional roles in stress tolerance. Front Plant Sci 7:813

Ahmad Z, Waraich EA, Akhtar S, Anjum S, Ahmad T, Mahboob W, Hafeez OBA, Tapera T, Labuschagne M, Rizwan M (2018) Physiological responses of wheat to drought stress and its mitigation approaches. Acta Physiol Plant 40(4):1–13

Ahmadian K, Jalilian J, Pirzad A (2021) Nano-fertilizers improved drought tolerance in wheat under deficit irrigation. Agric Water Manag 244:106544

Ahmed M, Asif M, Hassan FU (2014) Augmenting drought tolerance in sorghum by silicon nutrition. Acta Physiol Plant 36(2):473–483

Ahmed S, Kouser S, Asgher M, Gandhi SG (2021) Plant aquaporins: a frontward to make crop plants drought resistant. Physiol Plant 172(2):1089–1105

Alam H, Khattak JZ, Ksiksi TS, Saleem MH, Fahad S, Sohail H, Ali Q, Zamin M, El-Esawi MA, Saud S, Jiang X (2021a) Negative impact of long-term exposure of salinity and drought stress on native *Tetraena mandavillei* L. Physiol Plant 172(2):1336–1351

Alam A, Hariyanto B, Ullah H, Salin KR, Datta A (2021b) Effects of silicon on growth, yield and fruit quality of cantaloupe under drought stress. Silicon 13(9):3153–3162

Ali N, Réthoré E, Yvin JC, Hosseini SA (2020) The regulatory role of silicon in mitigating plant nutritional stresses. Plants 9(12):1779

Anjum SA, Xie XY, Wang LC, Saleem MF, Man C, Lei W (2011) Morphological, physiological and biochemical responses of plants to drought stress. Afr J Agric Res 6:2026–2032

Artyszak A, Gozdowski D, Siuda A (2021) Effect of the application date of fertilizer containing silicon and potassium on the yield and technological quality of sugar beet roots. Plants 10(2): 370

Aurangzaib M, Ahmad Z, Jalil MI, Nawaz F, Shaheen MR, Ahmad M, Hussain A, Ejaz MK, Tabassum MA (2022) Foliar spray of silicon confers drought tolerance in wheat (*Triticum aestivum* L.) by enhancing morpho-physiological and antioxidant potential. Silicon 14(9): 4793–4807

Azeem M, Haider MZ, Javed S, Saleem MH, Alatawi A (2022) Drought stress amelioration in maize (*Zea mays* L.) by inoculation of Bacillus spp. strains under sterile soil conditions. Agriculture 12(1):50

Bae H, Kim SK, Cho SK, Kang BG, Kim WT (2011) Overexpression of OsRDCP1, a rice RING domain-containing E3 ubiquitin ligase, increased tolerance to drought stress in rice (*Oryza sativa* L.). Plant Sci 180(6):775–782

Bakhat HF, Bibi N, Fahad S, Hammad HM, Abbas S, Shah GM, Zakir A, Murtaza B, Ashraf MR (2021) Rice husk bio-char improves brinjal growth, decreases insect infestation by enhancing silicon uptake. Silicon 13(10):3351–3360

Bezerra BKL, Lima GPP, dos Reis AR, Silva MDA, de Camargo MS (2019) Physiological and biochemical impacts of silicon against water deficit in sugarcane. Acta Physiol Plant 41(12): 1–12

Bhardwaj S, Kapoor D (2021) Fascinating regulatory mechanism of silicon for alleviating drought stress in plants. Plant Physiol Biochem 166:1044–1053

Bhardwaj S, Kapoor D, Singh S, Gautam V, Dhanjal DS, Jan S, Ramamurthy PC, Prasad R, Singh J (2021) Nitric oxide: a ubiquitous signal molecule for enhancing plant tolerance to salinity stress and their molecular mechanisms. J Plant Growth Regul 40:1–13

Bijanzadeh E, Barati V, Egan TP (2022) Foliar application of sodium silicate mitigates drought stressed leaf structure in corn (*Zea mays* L.). S Afr J Bot 147:8–17

Biju S, Fuentes S, Gupta D (2017) Silicon improves seed germination and alleviates drought stress in lentil crops by regulating osmolytes, hydrolytic enzymes and antioxidant defense system. Plant Physiol Biochem 119:250–264

Boudiar R, Casas AM, Gioia T, Fiorani F, Nagel KA, Igartua E (2020) Effects of low water availability on root placement and shoot development in landraces and modern barley cultivars. Agronomy 10(1):134

Brodersen CR, Roddy AB, Wason JW, McElrone AJ (2019) Functional status of xylem through time. Annu Rev Plant Biol 70(1):407–433

Bukhari MA, Sharif MS, Ahmad Z, Barutçular C, Afzal M, Hossain A, Sabagh AE (2020) Silicon mitigates the adverse effect of drought in canola (*Brassica napus* L.) through promoting the physiological and antioxidants activity. Silicon 13(11):3817–3826

Cao BL, Ma Q, Xu K (2020) Silicon restrains drought-induced ROS accumulation by promoting energy dissipation in leaves of tomato. Protoplasma 257(2):537–547

Carneiro JM, Chacón-Madrid K, Galazzi RM, Campos BK, Arruda SC, Azevedo RA, Arruda MA (2017) Evaluation of silicon influence on the mitigation of cadmium-stress in the development of Arabidopsis thaliana through total metal content, proteomic and enzymatic approaches. J Trace Elem Med Biol 44:50–58

Chakma R, Saekong P, Biswas A, Ullah H, Datta A (2021) Growth, fruit yield, quality, and water productivity of grape tomato as affected by seed priming and soil application of silicon under drought stress. Agric Water Manag 256:107055

Chao X, Zhang TA, Lyu G, Liang Z, Chen Y (2022) Sustainable application of sodium removal from red mud: cleaner production of silicon-potassium compound fertilizer. J Clean Prod 352: 131601

Chen D, Wang S, Yin L, Deng X (2018) How does silicon mediate plant water uptake and loss under water deficiency? Front Plant Sci 9:281

Cong L, Chai TY, Zhang YX (2008) Characterization of the novel gene BjDREB1B encoding a DRE-binding transcription factor from *Brassica juncea* L. Biochem Biophys Res Commun 371(4):702–706

Correia MJ, Rodrigues ML, Ferreira MI, Pereira JS (1997) Diurnal change in the relationship between stomatal conductance and abscisic acid in the xylem sap of field-grown peach trees. J Exp Bot 48(9):1727–1736

Dehghanipoodeh S, Ghobadi C, Baninasab B, Gheysari M, Shiranibidabadi S (2018) Effect of silicon on growth and development of strawberry under water deficit conditions. Hortic Plant J 4(6):226–232

Desoky ESM, Mansour E, Yasin MA, El Sobky ESE, Rady MM (2020) Improvement of drought tolerance in five different cultivars of *Vicia faba* with foliar application of ascorbic acid or silicon. Span J Agric Res 18(2):16

Diatta AA, Fike JH, Battaglia ML, Galbraith JM, Baig MB (2020) Effects of biochar on soil fertility and crop productivity in arid regions: a review. Arab J Geosci 13(14):1–17

Elnaggar A, El-Keblawy A, Mosa KA, Soliman S (2018) Drought tolerance during germination depends on light and temperature of incubation in Salsola imbricata, a desert shrub of Arabian deserts. Flora 249:156–163

Emam MM, Khattab HI, Helal NM (2012) Effects of silicon or selenium on photosynthetic apparatus and antioxidant capacity of rice plant grown under drought condition. Egypt J Exp Biol 8(2):271–283

Epstein E (1999) Silicon. Annu Rev Plant Biol 50:641

Etesami H, Jeong BR (2018) Silicon (Si): review and future prospects on the action mechanisms in alleviating biotic and abiotic stresses in plants. Ecotoxicol Environ Saf 147:881–896

Feng Shao J, Che J, Yamaji N, Fang Shen R, Feng Ma J (2017) Silicon reduces cadmium accumulation by suppressing expression of transporter genes involved in cadmium uptake and translocation in rice. J Exp Bot 68(20):5641–5651

Fleck AT, Nye T, Repenning C, Stahl F, Zahn M, Schenk MK (2011) Silicon enhances suberization and lignification in roots of rice (*Oryza sativa*). J Exp Bot 62(6):2001–2011

Gomaa MA, Kandil EE, El-Dein AA, Abou-Donia ME, Ali HM, Abdelsalam NR (2021) Increase maize productivity and water use efficiency through application of potassium silicate under water stress. Sci Rep 11(1):1–8

Gong H, Chen K (2012) The regulatory role of silicon on water relations, photosynthetic gas exchange, and carboxylation activities of wheat leaves in field drought conditions. Acta Physiol Plant 34(4):1589–1594

Gong HJ, Chen KM, Chen GC, Wang SM, Zhang CL (2003) Effects of silicon on growth of wheat under drought. J Plant Nutr 26(5):1055–1063

Hajiboland R, Moradtalab N, Eshaghi Z, Feizy J (2018) Effect of silicon supplementation on growth and metabolism of strawberry plants at three developmental stages. N Z J Crop Hortic Sci 46(2): 144–161

Hamayun M, Sohn EY, Khan SA, Shinwari ZK, Khan AL, Lee IJ (2010) Silicon alleviates the adverse effects of salinity and drought stress on growth and endogenous plant growth hormones of soybean (*Glycine max* L.). Pak J Bot 42(3):1713–1722

Harley AD, Gilkes RJ (2000) Factors influencing the release of plant nutrient elements from silicate rock powders: a geochemical overview. Nutr Cycl Agroecosyst 56(1):11–36

Hasanuzzaman M, Bhuyan MHM, Zulfiqar F, Raza A, Mohsin SM, Mahmud JA, Fujita M, Fotopoulos V (2020) Reactive oxygen species and antioxidant defense in plants under abiotic stress: revisiting the crucial role of a universal defense regulator. Antioxidants 9(8):681

Hattori T, Inanaga S, Tanimoto E, Lux A, Luxová M, Sugimoto Y (2003) Silicon-induced changes in viscoelastic properties of sorghum root cell walls. Plant Cell Physiol 44(7):743–749

Helaly MN, El-Hoseiny H, El-Sheery NI, Rastogi A, Kalaji HM (2017) Regulation and physiological role of silicon in alleviating drought stress of mango. Plant Physiol Biochem 118:31–44

Hellala F, Amerb AK, El-Sayeda S, El-Azab K (2020) Mitigation the negative effect of water stress on barley by nano silica application. Plant Arch 20:3224–3231

Hernandez-Apaolaza L (2014) Can silicon partially alleviate micronutrient deficiency in plants? A review. Planta 240(3):447–458

Jang SW, Kim Y, Khan AL, Na CI, Lee IJ (2018) Exogenous short-term silicon application regulates macro-nutrients, endogenous phytohormones, and protein expression in *Oryza sativa* L. BMC Plant Biol 18(1):1–12

Kapoor D, Bhardwaj S, Landi M, Sharma A, Ramakrishnan M, Sharma A (2020) The impact of drought in plant metabolism: how to exploit tolerance mechanisms to increase crop production. Appl Sci 10(16):5692

Kaya C, Tuna L, Higgs D (2006) Effect of silicon on plant growth and mineral nutrition of maize grown under water-stress conditions. J Plant Nutr 29(8):1469–1480

Keller C, Rizwan M, Davidian JC, Pokrovsky OS, Bovet N, Chaurand P, Meunier JD (2015) Effect of silicon on wheat seedlings (*Triticum turgidum* L.) grown in hydroponics and exposed to 0 to 30 μM Cu. Planta 241(4):847–860

Khan A, Khan AL, Imran M, Asaf S, Kim YH, Bilal S, Numan M, Al-Harrasi A, Al-Rawahi A, Lee IJ (2020) Silicon-induced thermotolerance in *Solanum lycopersicum* L. via activation of antioxidant system, heat shock proteins, and endogenous phytohormones. BMC Plant Biol 20(1):1–18

Khattab HI, Emam MA, Emam MM, Helal NM, Mohamed MR (2014) Effect of selenium and silicon on transcription factors *NAC5* and *DREB2A* involved in drought-responsive gene expression in rice. Biol Plant 58(2):265–273

Kim YH, Khan AL, Waqas M, Jeong HJ, Kim DH, Shin JS, Kim JG, Yeon MH, Lee IJ (2014a) Regulation of jasmonic acid biosynthesis by silicon application during physical injury to *Oryza sativa* L. J Plant Res 127(4):525–532

Kim YH, Khan AL, Waqas M, Shim JK, Kim DH, Lee KY, Lee IJ (2014b) Silicon application to rice root zone influenced the phytohormonal and antioxidant responses under salinity stress. J Plant Growth Regul 33(2):137–149

Kim YH, Khan AL, Waqas M, Lee IJ (2017) Silicon regulates antioxidant activities of crop plants under abiotic-induced oxidative stress: a review. Front Plant Sci 8:510

Kovács S, Kutasy E, Csajbók J (2022) The multiple role of silicon nutrition in alleviating environmental stresses in sustainable crop production. Plants 11(9):1223

Kutasy E, Buday-Bódi E, Virág IC, Forgács F, Melash AA, Zsombik L, Nagy A, Csajbók J (2021) Mitigating the negative effect of drought stress in oat (*Avena sativa* L.) with silicon and sulphur foliar fertilization. Plants 11(1):30

Lei C, Bagavathiannan M, Wang H, Sharpe SM, Meng W, Yu J (2021) Osmopriming with polyethylene glycol (PEG) for abiotic stress tolerance in germinating crop seeds: a review. Agronomy 11(11):2194

Leksungnoen P, Wisawapipat W, Ketrot D, Aramrak S, Nookabkaew S, Rangkadilok N, Satayavivad J (2019) Biochar and ash derived from silicon-rich rice husk decrease inorganic arsenic species in rice grain. Sci Total Environ 684:360–370

Lenka SK, Katiyar A, Chinnusamy V, Bansal KC (2011) Comparative analysis of drought-responsive transcriptome in Indica rice genotypes with contrasting drought tolerance. Plant Biotechnol J 9(3):315–327

Li J, Leisner SM, Frantz J (2008) Alleviation of copper toxicity in Arabidopsis thaliana by silicon addition to hydroponic solutions. J Am Soc Hortic Sci 133(5):670–677

Liang B, Gao T, Zhao Q, Ma C, Chen Q, Wei Z, Li C, Li C, Ma F (2018) Effects of exogenous dopamine on the uptake, transport, and resorption of apple ionome under moderate drought. Front Plant Sci 9:755

Li-feng W, Hao F, Yun-He J (2012) Photosynthetic characterization of a rolled leaf mutant of rice (*Oryza sativa* L.). Afr J Biotechnol 11(26):6839–6846

Liu P, Yin L, Deng X, Wang S, Tanaka K, Zhang S (2014) Aquaporin-mediated increase in root hydraulic conductance is involved in silicon-induced improved root water uptake under osmotic stress in *Sorghum bicolor* L. J Exp Bot 65(17):4747–4756

Liu SC, Yao MZ, Ma CL, Jin JQ, Ma JQ, Li CF, Chen L (2015) Physiological changes and differential gene expression of tea plant under dehydration and rehydration conditions. Sci Hortic 184:129–141

Ma JF, Tamai K, Yamaji N, Mitani N, Konishi S, Katsuhara M, Ishiguro M, Murata Y, Yano M (2006) A silicon transporter in rice. Nature 440(7084):688–691

Ma J, Cai H, He C, Zhang W, Wang L (2015) A hemicellulose-bound form of silicon inhibits cadmium ion uptake in rice (*Oryza sativa*) cells. New Phytol 206(3):1063–1074

Ma D, Sun D, Wang C, Qin H, Ding H, Li Y, Guo T (2016) Silicon application alleviates drought stress in wheat through transcriptional regulation of multiple antioxidant defense pathways. J Plant Growth Regul 35(1):1–10

Ma Y, Vosátka M, Freitas H (2019) Beneficial microbes alleviate climatic stresses in plants. Front Plant Sci 10:595

Makabe-Sasaki S, Kakuda KI, Sasaki Y, Ando H (2014) Effects of slag silicate fertilizer on silicon content of rice plants grown in paddy fields on the Shonai Plain, Yamagata, Japan. Soil Sci Plant Nutr 60(5):708–721

Malik MA, Wani AH, Mir SH, Rehman IU, Tahir I, Ahmad P, Rashid I (2021) Elucidating the role of silicon in drought stress tolerance in plants. Plant Physiol Biochem 165:187–195

Mandlik R, Thakral V, Raturi G, Shinde S, Nikolić M, Tripathi DK, Sonah H, Deshmukh R (2020) Significance of silicon uptake, transport, and deposition in plants. J Exp Bot 71(21):6703–6718

Manivannan A, Ahn YK (2017) Silicon regulates potential genes involved in major physiological processes in plants to combat stress. Front Plant Sci 8:1346

McDowell NG, Sapes G, Pivovaroff A, Adams HD, Allen CD, Anderegg WR, Arend M, Breshears DD, Brodribb T, Choat B, Cochard H (2022) Mechanisms of woody-plant mortality under rising drought, CO_2 and vapour pressure deficit. Nat Rev Earth Environ 3(5):294–308

Merhij IE, Al-Timmen WM, Jasim AH (2019) The effect of silicon, tillage and the interaction between them on some antioxidants and phytohormones during drought stress of maize (*Zea mays* L.) plants. Plant Arch 19:67–74

Milon MAA, Lutfa Khatun M, Islam MA (2020) Polyamines—a positive modulator against biotic and abiotic stresses on plants: review of current knowledge. Preprints.org 2020100639. https://doi.org/10.20944/preprints202010.0639.v1

Ming DF, Pei ZF, Naeem MS, Gong HJ, Zhou WJ (2012) Silicon alleviates PEG-induced water-deficit stress in upland rice seedlings by enhancing osmotic adjustment. J Agron Crop Sci 198(1):14–26

Mir RA, Bhat KA, Shah AA, Zargar SM (2020) Role of silicon in abiotic stress tolerance of plants. In: Improving abiotic stress tolerance in plants. CRC Press, Boca Raton, pp 271–290

Mundada PS, Barvkar VT, Umdale SD, Kumar SA, Nikam TD, Ahire ML (2021) An insight into the role of silicon on retaliation to osmotic stress in finger millet (*Eleusine coracana* (L.) Gaertn). J Hazard Mater 403:124078

Namjoyan S, Sorooshzadeh A, Rajabi A, Aghaalikhani M (2020) Nano-silicon protects sugar beet plants against water deficit stress by improving the antioxidant systems and compatible solutes. Acta Physiol Plant 42(10):1–16

Naz R, Batool S, Shahid M, Keyani R, Yasmin H, Nosheen A, Hassan MN, Mumtaz S, Siddiqui MH (2021) Exogenous silicon and hydrogen sulfide alleviates the simultaneously occurring drought stress and leaf rust infection in wheat. Plant Physiol Biochem 166:558–571

Ning D, Qin A, Liu Z, Duan A, Xiao J, Zhang J, Liu Z, Zhao B, Liu Z (2020) Silicon-mediated physiological and agronomic responses of maize to drought stress imposed at the vegetative and reproductive stages. Agronomy 10(8):1136

Okorie VO, Mphambukeli TN, Amusan SO (2019) Exploring the political economy of water and food security nexus in BRICS. Afr Insight 48(4):21–38

Parveen A, Liu W, Hussain S, Asghar J, Perveen S, Xiong Y (2019) Silicon priming regulates morpho-physiological growth and oxidative metabolism in maize under drought stress. Plants 8(10):431

Patel M, Fatnani D, Parida AK (2021) Silicon-induced mitigation of drought stress in peanut genotypes (*Arachis hypogaea* L.) through ion homeostasis, modulations of antioxidative defense system, and metabolic regulations. Plant Physiol Biochem 166:290–313

Pavlovic J, Samardzic J, Maksimović V, Timotijevic G, Stevic N, Laursen KH, Hansen TH, Husted S, Schjoerring JK, Liang Y, Nikolic M (2013) Silicon alleviates iron deficiency in cucumber by promoting mobilization of iron in the root apoplast. New Phytol 198(4):1096–1107

Pei ZF, Ming DF, Liu D, Wan GL, Geng XX, Gong HJ, Zhou WJ (2010) Silicon improves the tolerance to water-deficit stress induced by polyethylene glycol in wheat (*Triticum aestivum* L.) seedlings. J Plant Growth Regul 29(1):106–115

Pepe M, Crescente MF, Varone L (2022) Effect of water stress on physiological and morphological leaf traits: a comparison among the three widely-spread invasive alien species *Ailanthus altissima, Phytolacca americana*, and *Robinia pseudoacacia*. Plants 11(7):899

Pilon C, Soratto RP, Broetto F, Fernandes AM (2014) Foliar or soil applications of silicon alleviate water-deficit stress of potato plants. Agron J 106(6):2325–2334

Pooja V, Sharma J, Verma S, Sharma A (2022) Importance of silicon in combating a variety of stresses in plants: a review. J Appl Nat Sci 14(2):607–630

Purbajanti ED, Slamet W, Fuskhah E (2019) Effects of organic and inorganic fertilizers on growth, activity of nitrate reductase and chlorophyll contents of peanuts (*Arachis hypogaea* L.). In: IOP conference series: earth and environmental science, vol 250, no. 1, p 012048). IOP Publishing, Bristol

Rangani J, Panda A, Patel M, Parida AK (2018) Regulation of ROS through proficient modulations of antioxidative defense system maintains the structural and functional integrity of photosynthetic apparatus and confers drought tolerance in the facultative halophyte *Salvadora persica* L. J Photochem Photobiol B Biol 189:214–233

Rangani J, Panda A, Parida AK (2020) Metabolomic study reveals key metabolic adjustments in the xerohalophyte *Salvadora persica* L. during adaptation to water deficit and subsequent recovery conditions. Plant Physiol Biochem 150:180–195

Rangwala T, Bafna A, Vyas N, Gupta R (2018) Role of soluble silica in alleviating oxidative stress in soybean crop. Indian J Agric Res 52(1):9–15

Rangwala T, Bafna A, Vyas N, Gupta R (2019) Beneficial role of soluble silica in enhancing chlorophyll content in onion leaves. Curr Agric Res J 7(3):358

Ratnakumar P, Deokate PP, Rane J, Jain N, Kumar V, Berghe DV, Minhas PS (2016) Effect of ortho-silicic acid exogenous application on wheat (*Triticum aestivum* L.) under drought. J Funct Environ Bot 6(1):34–42

Rea RS, Islam MR, Rahman MM, Nath B, Mix K (2022) Growth, nutrient accumulation, and drought tolerance in crop plants with silicon application: a review. Sustainability 14(8):4525

Rehman MU, Ilahi H, Adnan M, Wahid F, Rehman FU, Ullah A, Ullah A, Zia A, Raza MA (2021) Application of silicon: a useful way to mitigate drought stress: an overview. Curr Res Agric Farming 2:9–17

Rizwan M, Ali S, Ibrahim M, Farid M, Adrees M, Bharwana SA, Zia-ur-Rehman M, Qayyum MF, Abbas F (2015) Mechanisms of silicon-mediated alleviation of drought and salt stress in plants: a review. Environ Sci Pollut Res 22(20):15416–15431

Saja-Garbarz D, Ostrowska A, Kaczanowska K, Janowiak F (2021) Accumulation of silicon and changes in water balance under drought stress in *Brassica napus* var. *napus* L. Plants 10(2):280

Saja-Garbarz D, Libik-Konieczny M, Fellner M, Jurczyk B, Janowiak F (2022) Silicon-induced alterations in the expression of aquaporins and antioxidant system activity in well-watered and drought-stressed oilseed rape. Plant Physiol Biochem 174:73–86

Salim BB (2014) Effect of boron and silicon on alleviating salt stress in maize. Middle East J Agric Res 3(4):1196–1204

Saud S, Li X, Chen Y, Zhang L, Fahad S, Hussain S, Sadiq A, Chen Y (2014) Silicon application increases drought tolerance of Kentucky bluegrass by improving plant water relations and morphophysiological functions. Sci World J 2014:1

Seal P, Das P, Biswas AK (2018) Versatile potentiality of silicon in mitigation of biotic and abiotic stresses in plants: a review. Am J Plant Sci 9(7):1433

Seleiman MF, Al-Suhaibani N, Ali N, Akmal M, Alotaibi M, Refay Y, Dindaroglu T, Abdul-Wajid HH, Battaglia ML (2021) Drought stress impacts on plants and different approaches to alleviate its adverse effects. Plants 10(2):259

Shahzad S, Ali S, Ahmad R, Ercisli S, Anjum MA (2022) Foliar application of silicon enhances growth, flower yield, quality and postharvest life of tuberose (*Polianthes tuberosa* L.) under saline conditions by improving antioxidant defense mechanism. Silicon 14(4):1511–1518

Shi Y, Zhang Y, Han W, Feng R, Hu Y, Guo J, Gong H (2016) Silicon enhances water stress tolerance by improving root hydraulic conductance in *Solanum lycopersicum* L. Front Plant Sci 7:196

Siddiqui MH, Al-Whaibi MH, Faisal M, Al Sahli AA (2014) Nano-silicon dioxide mitigates the adverse effects of salt stress on *Cucurbita pepo* L. Environ Toxicol Chem 33(11):2429–2437

Singh J, Thakur JK (2018) Photosynthesis and abiotic stress in plants. In: Biotic and abiotic stress tolerance in plants. Springer, Singapore, pp 27–46

Sohag AAM, Tahjib-Ul-Arif M, Brestic M, Afrin S, Sakil MA, Hossain MT, Hossain MA, Hossain MA (2020) Exogenous salicylic acid and hydrogen peroxide attenuate drought stress in rice. Plant Soil Environ 66(1):7–13

Sonobe K, Hattori T, An P, Tsuji W, Eneji AE, Kobayashi S, Kawamura Y, Tanaka K, Inanaga S (2010) Effect of silicon application on sorghum root responses to water stress. J Plant Nutr 34(1):71–82

Souri Z, Khanna K, Karimi N, Ahmad P (2021) Silicon and plants: current knowledge and future prospects. J Plant Growth Regul 40(3):906–925

Tayade R, Ghimire A, Khan W, Lay L, Attipoe JQ, Kim Y (2022) Silicon as a smart fertilizer for sustainability and crop improvement. Biomol Ther 12(8):1027

Tayyab M, Islam W, Zhang H (2018) Promising role of silicon to enhance drought resistance in wheat. Commun Soil Sci Plant Anal 49(22):2932–2941

Thorne SJ, Hartley SE, Maathuis FJ (2020) Is silicon a panacea for alleviating drought and salt stress in crops? Front Plant Sci 11:1221

Tripathi DK, Vishwakarma K, Singh VP, Prakash V, Sharma S, Muneer S, Nikolic M, Deshmukh R, Vaculík M, Corpas FJ (2020) Silicon crosstalk with reactive oxygen species, phytohormones and other signaling molecules. J Hazard Mater 408:124820

Verma KK, Singh RK, Song QQ, Singh P, Zhang BQ, Song XP, Chen GL, Li YR (2019) Silicon alleviates drought stress of sugarcane plants by improving antioxidant responses. Biomed J Sci Techn Res 17:002957

Verma KK, Anas M, Chen Z, Rajput VD, Malviya MK, Verma CL, Singh RK, Singh P, Song XP, Li YR (2020) Silicon supply improves leaf gas exchange, antioxidant defense system and growth in *Saccharum officinarum* responsive to water limitation. Plants 9(8):1032

Wahab A, Abdi G, Saleem MH, Ali B, Ullah S, Shah W, Mumtaz S, Yasin G, Muresan CC, Marc RA (2022) Plants' physio-biochemical and phyto-hormonal responses to alleviate the adverse effects of drought stress: a comprehensive review. Plants 11(13):1620

Wang DONG, Poss JA, Donovan TJ, Shannon MC, Lesch SM (2002) Biophysical properties and biomass production of elephant grass under saline conditions. J Arid Environ 52(4):447–456

Wang M, Wang R, Mur LAJ, Ruan J, Shen Q, Guo S (2021) Functions of silicon in plant drought stress responses. Hortic Res 8:254

Watanabe S, Shimoi E, Ohkama N, Hayashi H, Yoneyama T, Yazaki J, Fujii F, Shinbo K, Yamamoto K, Sakata K, Sasaki T (2004) Identification of several rice genes regulated by Si nutrition. Soil Sci Plant Nutr 50(8):1273–1276

Wu J, Wang J, Hui W, Zhao F, Wang P, Su C, Gong W (2022) Physiology of plant responses to water stress and related genes: a review. Forests 13(2):324

Xu L, Islam F, Ali B, Pei Z, Li J, Ghani MA, Zhou W (2017) Silicon and water-deficit stress differentially modulate physiology and ultrastructure in wheat (*Triticum aestivum* L.). 3 Biotech 7(4):1–13

Xue S, Li Z, Li B, Ackah S, Wang B, Zheng X, Wang Y, Bi Y, Prusky D (2022) Sodium silicate treatment promotes suberin poly phenolic and silicon deposition, and enhances hardness and brittleness at muskmelon wounds. Sci Hortic 300:111087

Yin L, Wang S, Liu P, Wang W, Cao D, Deng X, Zhang S (2014) Silicon-mediated changes in polyamine and 1-aminocyclopropane-1-carboxylic acid are involved in silicon-induced drought resistance in *Sorghum bicolor* L. Plant Physiol Biochem 80:268–277

Yin L, Wang S, Tanaka K, Fujihara S, Itai A, Den X, Zhang S (2016) Silicon-mediated changes in polyamines participate in silicon-induced salt tolerance in Sorghum bicolor L. Plant Cell Environ 39(2):245–258

Yoshida T, Obata T, Feil R, Lunn JE, Fujita Y, Yamaguchi-Shinozaki K, Fernie AR (2019) The role of abscisic acid signaling in maintaining the metabolic balance required for Arabidopsis growth under nonstress conditions. Plant Cell 31(1):84–105

Zhang T, Liu Y, Yang T, Zhang L, Xu S, Xue L, An L (2006) Diverse signals converge at MAPK cascades in plant. Plant Physiol Biochem 44(5–6):274–283

Zhang C, Moutinho-Pereira JM, Correia C, Coutinho J, Gonçalves A, Guedes A, Gomes-Laranjo J (2013) Foliar application of Sili-K® increases chestnut (*Castanea* spp.) growth and photosynthesis, simultaneously increasing susceptibility to water deficit. Plant Soil 365(1):211–225

Zhang W, Yu X, Li M, Lang D, Zhang X, Xie Z (2018) Silicon promotes growth and root yield of *Glycyrrhiza uralensis* under salt and drought stresses through enhancing osmotic adjustment and regulating antioxidant metabolism. Crop Prot 107:1–11

Zhang L, Lei D, Deng X, Li F, Ji H, Yang S (2020) Retracted: cytosolic glyceraldehyde-3-phosphate dehydrogenase 2/5/6 increase drought tolerance via stomatal movement and reactive oxygen species scavenging in wheat. Plant Cell Environ 43(4):836–853

Zhang G, Hou X, Wang L, Xu J, Chen J, Fu X, Shen N, Nian J, Jiang Z, Hu J, Zhu L (2021) PHOTO-SENSITIVE LEAF ROLLING 1 encodes a polygalacturonase that modifies cell wall structure and drought tolerance in rice. New Phytol 229(2):890–901

Zhu Y, Gong H (2014) Beneficial effects of silicon on salt and drought tolerance in plants. Agron Sustain Dev 34(2):455–472

Crop Plants and Grasses as Potential Phytoremediators: Physiological Perspectives and Efficient Mitigating Strategies

23

Seva Nayak Dheeravathu, S. Srividhya, R. Swarna, S. Avinash, G. Rajesha, K. Venkatesh, Reetu, Tejveer Singh, Nilamani Dikshit, Thulasi Bai Vadithe, T. N. Usha, D. Venkateshwarlu, Anita Mann, and Ashwani Kumar

Abstract

Rapid industrialization and urbanization have led to increased contamination in the environment with heavy metal and toxic element. The contamination of agricultural soils and irrigation water due to these toxic elements is a serious threat to ecosystem and human health. Using plants to scavenge, extract, or reduce the bioavailability of toxic elements in the environment, including soil, water, and air, is called phytoremediation. It is a cost-effective and sustainable solution to manage heavy metal contamination from soil water, prevent soil pollution and air pollution. Some of the cereals, grasses, and crop plants are

S. N. Dheeravathu (✉)
ICAR-Indian Institute of Millets Research, Hyderabad, India

ICAR-Indian Grassland and Fodder Research Institute, Jhansi, UP, India

S. Srividhya · R. Swarna · S. Avinash · G. Rajesha · K. Venkatesh
ICAR-Indian Institute of Millets Research, Hyderabad, India

Reetu · T. Singh · N. Dikshit
ICAR-Indian Grassland and Fodder Research Institute, Jhansi, UP, India

T. B. Vadithe
Department of Microbiology, Acharya Nagarjuna University, Guntur, AP, India

T. N. Usha
Zonal Agricultural and Horticultural Research Station, Keladi Shivappa Nayaka University of Agricultural and Horticultural Sciences, Shivamogga, Karnataka, India

D. Venkateshwarlu
CSIR-Central Food Technological Research Institute, Mysore, Karnataka, India

A. Mann · A. Kumar
ICAR – Central Soil Salinity Research Institute, Karnal, Haryana, India

A. Kumar et al. (eds.), *Salinity and Drought Tolerance in Plants*,
https://doi.org/10.1007/978-981-99-4669-3_23

found useful for phytoremediation. They have heavy metal and toxic element carriers along with favorable morpho-physiological characteristics which make them suitable for phytoremediation. The ideal crop plants and grasses–mean they have anatomical traits viz. long and profuse roots, high growth rate etc and physiological mechanism for phytoremediation. By the use of phytoremediators, soil erosion and leaching of contaminants can be prevented. Phytoremediation mechanisms known in plant species are phytofiltration, extraction, stabilization, transformation/degradation, volatilization, removal of aerial contaminants and desalination. The proof of concept of phytoremediation was long established. However large scale utility is limited worldwide primarily because it is a slow process. This chapter highlights the potential crop plants and grass species for phytoremediation strategies. Further, the physiological, biochemical, and molecular approaches in revealing the uptake, transport, and localization mechanisms must be studied in detail for employing these crops as potential candidates for phytoremediation. There is scope to bridge the gap for cost-effective, efficient, and environment-friendly remediation technologies through advanced biotechnological tools and strategies.

Keywords

Cadmium · Heavy metals · Lead · Phytoremediation · Plant growth promoting bacteria · Grasses

23.1 Introduction

Globally pollution levels have increased ever since industrial revolution. Soil, water, and air pollution has caused long term health concern to human and environmental hazards. Measures are taken to reduce pollution globally like use of lead free automotive fuel, restricting use of aerosol pesticides, water treatment plants, etc. However, the major concern is the rising concentration of heavy and toxic metals in the environment (FAO and UNEP 2021), Table 23.1. The anthropogenic activities such as untreated sewage disposal, mining, gas emission, and municipal waste production, smelting, and the use of pesticides and fertilizers in agriculture have aggravated the conditions severely (Kabata-Pendias and Pendias 1989). Heavy metals are generally defined as elements with metallic properties and an atomic number more than 20 or these are a group of elements with a high density, i.e., above 5 g/cm^3 (Appenroth 2010). In the context of environmental pollution, the most important trace elements are arsenic (As), chromium (Cr), cadmium (Cd), copper (Cu), lead (Pb), mercury (Hg), nickel (Ni), zinc (Zn), cobalt (CO), selenium (Se), manganese (Mn), and uranium (U) (He et al. 2015). These trace elements can be subdivided into essential elements (Cu, Ni, and Zn) and nonessential elements (As, Cd, Cr, Hg, and Pb) for plants, but both nonessential and trace elements can be toxic even at low prevailing concentrations (Clemens 2006). These elements may be biogenic, i.e., essential for proper plant functioning, since they regulate the processes

Table 23.1 Types of heavy metals, source, toxicity, and permissible limit of heavy metals and threshold for hyperaccumulator plant

Pollutants/heavy metals	Major sources[a]	Effect on human health/ toxic effect[a]	Permissible levels[a] (drinking water mg/L)	Average range in soils[b] ($\mu g\ g^{-1}$ dry weight)	Average range in plant tissues[c] ($\mu g\ g^{-1}$ dry weight)	Threshold for hyperaccumulators[d,e] ($\mu g\ g^{-1}$ dry weight)
Arsenic (As)	Metal processing, mining sites, timber storage, coke ovens emission, smelters, pesticides, fungicides, and fertilizer industries	Bronchitis, dermatitis, and poisoning, accumulate inside the cell and carcinogen	0.02–0.05			
Cadmium (Cd)	Welding, electroplating, pesticides, phosphate fertilizer, planting industries	Nephrotoxicity, carcinogenicity, affect gastrointestinal and pulmonary tract/renal dysfunction, lung disease, lung cancer, bone defects, kidney damage, bone marrow	0.005	1–2	0.03–0.5	100
Cr (chromium)	Fertilizer, fossil, fuel burning, oil drilling sites, metal tanneries and planting industries/mine, mineral sources	Damage to nervous system, irritability, and carcinogen	0.05	5–1000	0.2–1	1000
Cu (copper)	Electroplating industries/ mining, pesticide production, chemical industry	Anemia, liver and kidney damage, stomach irritation	1.3	2–60	2–20	1000
Iron (Fe)	Metal refining, engine parts	Seizures or coma	0.3			

(continued)

Table 23.1 (continued)

Pollutants/heavy metals	Major sources[a]	Effect on human health/ toxic effect[a]	Permissible levels[a] (drinking water mg/L)	Average range in soils[b] ($\mu g\ g^{-1}$ dry weight)	Average range in plant tissues[c] ($\mu g\ g^{-1}$ dry weight)	Threshold for hyperaccumulators[d,e] ($\mu g\ g^{-1}$ dry weight)
Lead (Pb)	Paint, pesticides, smoking, automobile emission, mining, burning of coal, batteries, and use of lead products	Neurodegeneration, crosses the blood–brain barrier, mental retardation in children, development delay, fatal infant encephalopathy, chronic damage to the nervous system, liver, kidney damage	0.05	10–150	0.1–5	1000
Mercury (Hg)	Pesticides, batteries industry, paper industry, caustic soda, gold and cement production	Chromosome breakage, bronchitis asthma, hunter Russell syndrome/ tremors, gingivitis, protoplasm poisoning, damage to nervous system, spontaneous abortion	0.002	≤0.1	0.005–0.2	1000
Manganese (Mn)				100–4000	1–7000	10,000
Nickle (Ni)	Electroplating industries	Eczematous reaction	0.1	2–200	0.4–4	1000
Selenium (Se)	Mining and agricultural irrigation	Bronchitis and gastrointestinal disturbances	0.05	1–2	0.01–0.2	1000

Zinc (Zn)	Refineries, brass manufacture, metal plating	Damage to nervous system, dermatitis	15	25–200	15–150	10,000
Volatile organic compounds (VOCs), e.g. secondary organic aerosols (SOAs), benzene, toluene	Biogenic and anthrophonic	Respiratory diseases, dizziness, eye and lung irritation and nausea				

[a] Modified after the USEPA-United States Environmental Protection Agency
[b] Lepp (1981)
[c] Markert (1996)
[d] McIntyre (2003)
[e] Baker and Brooks (1989)

of photosynthesis, respiration, nitrogen metabolism (Fe, Zn, Cu, and Co, among others), or toxic that cause diseases and disrupt many plant processes (Szarek-Łukaszewska 2014; Tibbett et al. 2021). Some volatile organic compounds (VOCs) such as benzene and toluene are monocyclic aromatic hydrocarbons that are emitted into the atmosphere, from biogenic and anthrophonic sources. Volatile organic compounds promote the production of secondary organic aerosols (SOAs), which affect human health and climate change (Shrivastava et al. 2017). Toxic heavy metals such as Pb, Co, Cd can be differentiated from other pollutants, since they do not biodegrade, rather accumulate in living organisms, soil, surface and underground water, thus causing various diseases and disorders even in relatively lower concentrations (Pehlivan et al. 2009). Soil contamination with an excessive amount of metals and toxic elements (Na^+ and Cl^-) can result in reduced soil microbial activity, soil fertility, soil quality, and eventually significant yield reduction (McGrath et al. 1995). The entry of toxic materials into the food chain (Haan and Lubbers 1983) has serious consequences of biomagnification. In past few decades agriculture land deteriorated tremendously due to the accumulation of heavy and toxic metals which has caused impaired soil and reduced soil fertility. There is reduced soil biomome which is important to keep soil healthy and fertile. Contamination of food with the heavy metals is a serious health concern for humans and animals. The ecosystems are polluted which is endangering all life forms on earth (Sidhu 2016). The lack of quality irrigation water and use of sewage water for growing vegetables have led to increased heavy metal accumulation such as Cu, Fe, Ni, and Zn (Rattan et al. 2005). The places in vicinity of industry are more prone to heavy metal contamination. A pilot study in the industrial area of Surat region of India reveals high concentration of toxic elements than the permissible limits (Krishna and Govil 2007). Similarly, it is estimated that elemental toxicity and heavy metal contamination (Hg, Pb, Cr and Cd) are serious threat to 66 million people globally (Rahman and Singh 2020). Furthermore, water contamination alone has affected more than 150 million people globally (Ravenscroft et al. 2011). Chemical methods for heavy metals (HMs) remediation such as heat treatment, chemical extraction, ion exchange, membrane separation, electro remediation, soil replacement, precipitation, and chemical leaching are costly (Selvi et al. 2019). Plant and microbial-based technologies are generally called phytoremediation and bioremediation, respectively. These techniques are eco-friendly, non-invasive, energy-efficient, and cost-effective to reclaim sites with low to moderate concentrations of trace elements are proposed as alternative approaches (Vangronsveld et al. 2009). Phytoremediation is an integrated multidisciplinary approach to clean up contaminated soils and irrigation water, which combines the disciplines of plant physiology, soil chemistry, microbiology, environmental science, plant breeding, and biotechnology. Phytoremediation is a green strategy that uses hyperaccumulator plants/halophytes and their rhizospheric microorganisms to stabilize, degrade, transfer, and volatilize pollutants in soil, irrigation water, and the environment (Liu et al. 2020). According to the soil conditions, pollutants, and the species of plants used, seven types of phytoremediation are phytofiltration, extraction, stabilization, transformation/degradation, volatilization, removal of aerial contaminates, and

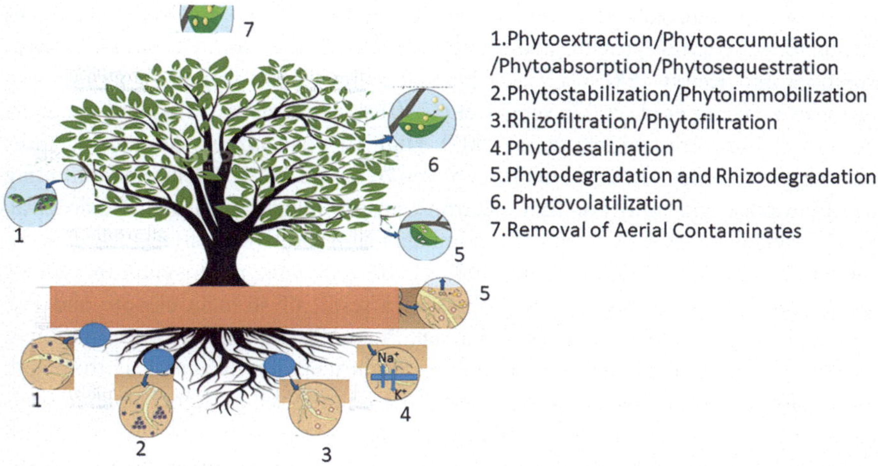

Fig. 23.1 Different types of phytoremediation techniques (phytoextraction, phytofiltration, phytovolatilization, phytostabilization, and phytotransformation)

desalination (Fig. 23.1). Understanding phytoremediation through crop plants especially grasses necessitates to investigate the molecular synthesis, enzyme induction, and membrane transport. Some crop species especially grasses that are reported as potential candidates for phytoremediation. Grasses have elaborate root structure and may have specialized propagating underground structures. Below ground structures like roots, rhizomes and bulbs bind the soil and to prevent erosion and leaching of contaminant. An ideal phytoremediation plant species is expected to grow profusely even in less fertile soils (Zurek et al. 2013; Gołda and Korzeniowska 2016). In addition, they provide multiple benefits like improvement in soil structure and quality, reduction in erosion, and increase in biodiversity. Growing perennial grasses on degraded lands are much more sustainable than regular agricultural practices. To prevent toxic buildup of mineral ions in the cytosol, many plants may localize/sequester them in the vacuole (Stewart and Ahmad 1983). This chapter highlights the mechanisms which crop plants and grasses exhibit the detoxifying mechanism. Research efforts are under way to bestow heavy metal and toxic elements tolerance species using both classical/conventional plant breeding and molecular biology (Hasegawa et al. 2000).

23.2 Heavy Metals and Salt Ions Toxicity in Plants and Grasses

The plants that grow in heavy metal or toxic elements or organic/inorganic contaminated soil, water, or air alter the anatomical, morpho-physiological, and biochemical changes which affect normal growth and development of plant. Germination is severely affected by Hg which inhibits amylase and protease enzyme in the seed (Islam et al. 2007; Sengar et al. 2010). In general high concentration of heavy metals in the environment adversely affects plants. Cu and Cd toxicity reduces germination, seedling growth, and number of lateral roots in *Solanum melongena* (Neelima and Reddy 2003). Cr and Pb reduce root fresh weight in *Pennisetum purpureum* (Islam et al. 2007; Sengar et al. 2010; Zhang et al. 2014) and Cd alters the root architecture (Sridhar et al. 2007). High Cr concentration in soil increases reactive oxygen species (ROS) in plants by removing Fe from proteins and inhibiting the chloroplast and mitochondrial electron transport chains in plants (Cosio et al. 2004; Scandalios 2005). Cd causes ion imbalance in root which affects Mg absorption and metabolism (Sridhar et al. 2007). Cd decreases the photosynthesis process of the plant and photorespiration process as a result of stomata closure and low uptake of CO_2 into plant tissues (Aravind and Prasad 2015). Higher concentration of Na^+ and Cl^- inhibits uptake of essential macronutrients like K^+ and Ca^+ from soil. Sodium ion in higher concentration breaks down chlorophyll in Bajra Napier hybrids leaves (Dheeravathu et al. 2021). Many reports reveal that relative water content (RWC) and membrane stability index (MSI) are reduced under heavy metals, salinity, and drought (Naseem et al. 2021). Dheeravathu et al. (2021) reported that salt-tolerant BN-hybrids maintain high MSI and RWC under stress conditions. Naseem et al. (2021) reported that MSI and RWC contents decrease as the Se level increases in maize leaves and Xia et al. (2012) observed that RWC content decreases at high levels of Se in *Hordeum vulgare* plants.

23.3 Mechanism of Phytoremediation

23.3.1 Phytoextraction/Phytoaccumulation/Phytoabsorption/ Phytosequestration

Phytoextraction involves the absorption of toxic metals by roots followed by translocation and accumulation in the aerial parts (deposition at cell wall, cell membrane, and vacuole). Mechanism of heavy metal extraction/accumulations include absorption of metal cations followed by metal–ligand complex formation inside the plant cell or metal-phytochelatin complex (Lajayer et al. 2019) (Fig. 23.1 and Tables 23.2 and 23.3).

23.3.2 Phytostabilization/Phytoimmobilization

Phytoimmobilization/Phytostabilization can be achieved through immobilization or inactivation of heavy metal toxicants or pollutants within the roots or in the rhizosphere. Toxicants can be immobilized or stabilized with the help of root exudates or trapped in the root surface with the help of sequestered or transport protein in the vacuoles of root cell through cellular process (ITRC 2009). The main objective is to avoid mobilization of contaminates and limit their diffusion in the soil (Fig. 23.1 and Tables 23.2 and 23.3).

Table 23.2 Biochemical composition of root exudates

Classification	Composition	Functions	References
Carbohydrates	Arabinose, fructose, fucose, galactose, glucose, lactose, mannose, raffinose, rhamnose, ribose, sucrose, xylose	Nutrient and energy sources, anchoring of bacteria to plant surfaces	Slaveykova et al. (2010), Venkatesh and Vedaraman (2012)
Organic acids	Acetic, isocitric, lactic, aconitic, adipic, butyric, citric, cyclic, fumaric, gluconic, glutaric, glycolic, glyoxylic, hydroxybutyric, indole-3-acetic, maleic, malic, malonic, oxalic, piscidic, propionic, pyruvic, succinic, tartaric, valeric	Chemotaxis signals to microbes for chelation and solubilization of mineral nutrients, act as acidifiers of soil	Mucha et al. (2005), Magdziak et al. (2011), Ramachandran et al. (2011), Vranova et al. (2013)
Amino acids	All the amino acids	They serve as nutrient and energy sources to microbes for mediating the phytoremediation process	Ma et al. (2009a, b, 2011), Ahemad and Kibret (2014)
Flavonols	Naringenin, kaempferol, quercetin, myricetin, rutin, genistein, strigolactone isoflavonoids, neoflavonoids, etc.	Activation of secondary metabolism defense again pathogens	Hofmann (2013), Dakora and Phillips (2002), Zhao et al. (2005)
Phenols	Caffeic acid, ferulic acid, styrene, *N*-hexanoyl-D,L-homoserine-lactone, 7-hydroxy-6-methoxy, pyrocatechol	Inductors of resistance against phytopathogens	Steinkellner and Mammerler (2007), Steinkellner et al. (2007), Von Rad et al. (2008)
Enzymes	Amylase, DNase, protease, phosphatase, RNase, xylanase, sucrase, urease, polygalacturonase,	Release of phosphorus from organic molecules, organic matter transformations in soil	Wu et al. (2014)
Coumarins	Umbelliferone	Activation of secondary metabolism defense again pathogens	Ahemad and Kibret (2014)

23.3.3 Rhizofiltration/Phytofiltration

In the rhizofiltration/phytofiltration technique, plant roots either absorb or adsorb the heavy metals or toxic contaminations from surface water and waste streams. The

Table 23.3 Bioremediation of heavy metals by the PGPB

Sl. no.	Name of PGPB	Phytoremediation method	% of removal of heavy metals	References
1.	*Pseudomonas fluorescens* and *P. putida*	Adsorb	Uptake of 38% of Cr	Hussein et al. (2004)
2.	*Rhizobium. Leguminosarum* (M5) + *P. fluorescens* (K23) + *Luteibacter* sp. + *Variovorax* sp.	Phytoextraction (*Lathyrus sativus*)	Uptake of 35% of Pb	Abdelkrim et al. (2018)
3	*Bacillus cereus*	Phytoextraction (maize)	92% of Ni	Khan and Bano (2016)
4	*Bacillus* sp. CIK-516	Phytoextraction (radish)	Increase the uptake of Ni by 10 folds	Akhtar et al. (2018)
5	*Streptomyces pactum*	Phytoextraction (wheat)	Uptake of 121% of Cd, Cu, and Zn in the roots and shoots and Pb only in the roots	Ali et al. (2021)
6	*Bacillus*	Phytoextraction (groundnut)	66% of Pb uptake	Banik et al. (2018)
7	*Brucella* sp. K12	Phytostabilization (okra)	Reduced the 43% uptake of Cr in okra plant	Maqbool et al. (2015)
8	*B. cereus* TCR17	Phytostabilization (*Sorghum bicolor*)	Reduced the 60% uptake of Cr in okra plant	Bruno et al. (2020)
9	*B. megaterium* H3	Phytostabilization (vegetables)	Reduced the 41–80% uptake of Cd and Pb in vegetables	Wang et al. (2018)
10	*Pseudomonas* sp. K32	Phytostabilization (rice)	Reduced the 90% uptake of Cd in rice	Pramanik et al. (2021)
11	*Proteus mirabilis*	Phytostabilization (maize)	Reduced the 78% uptake of Cr in rice	Vishnupradeep et al. (2022)

rhizofiltration of heavy metals contaminated water sites leads to root-mediated fixation and precipitation inside the root in an insoluble form (Fig. 23.1 and Tables 23.2 and 23.3).

23.3.4 Phytodegradation and Rhizodegradation (Phytotransformation)

Plant roots can metabolize heavy metals or toxic pollutants/or organic contaminants within the root tissue through the process known as rhizodegradation. While in leaf and shoot tissues the process is known as phytodegradation. In phytodegradation, the organic contaminants are metabolized/or mineralized inside the plant cells by specific enzymes such as dehalogenases and nitroreductases which ultimately reduce the toxicity (Fig. 23.1 and Tables 23.2 and 23.3).

23.3.5 Phytovolatilization

Phytovolatilization is the release of the contaminant or a modified form of the contaminant to the atmosphere from the plant. The contaminants is uptaken by the plant then metabolized, and released through transpiration. In phytovolatilization, contaminants are converted into a volatiles and released in the air through stomata viz Cd, Pb, Cr is converted into Hg by plants. However, this mechanism merely transfers contaminants from one environmental compartment to another, which may somehow return back to the original source (soil) by precipitation and hence could be less effective than phytoextraction and phytofiltration phytoremediation (Sarma 2011; Nikolić and Stevović 2015). It is commonly employed when treating groups of highly volatile metals like Hg and As. Phytovolatilization of As involves the conversion of elemental As to selenoamino acids, such as selenomethione, which is modified by methylation to a volatile and less toxic form, dimethyl selenide (Wang et al. 2012) (Fig. 23.1 and Tables 23.2 and 23.3).

23.3.6 Removal of Aerial Contaminates

Some studies reported that some species of volatile organic compounds particularly oxygenated VOCs are taken by plant leaves (Dela Cruz et al. 2014; Niinemets et al. 2014) (Fig. 23.1 and Tables 23.2 and 23.3). The process of VOCs uptake by the plant leaves is analogous to CO_2 diffusion and assimilation.

23.3.7 Phytodesalination

Phytodesalination refers to the use of halophytic plants for removal of excess amounts of salts from soil and irrigation water. Removal of excessive salt from saline soil and saline water to increase the crop productivity by use of halophytes is called phytodesalination. Phytodesalination is a recently reported emerging technique. Remediation can be achieved through extraction of excess amount of toxic salt ions (Na^+ and Cl^-) by halophyte plants (*Suaeda maritima* and *Sesuvium portulacastrum*) from the soil and store the excess amount of toxic salt ions (Na^+, Cl) in root and shoot vacuoles (Fig. 23.1 and Tables 23.2 and 23.3). Ravindran et al. (2007) reported that *Suaeda maritime* and *Sesuvium portulacastrum* can remove 504 and 474 kg of Na and Cl, respectively, from 1 ha of saline soil in a period of 4 months.

23.4 Heavy Metals and Salt Ions Transport and Metabolism

23.4.1 Uptake and Translocation of Heavy Metals and Salts in Plant Species

A few plant species are known to release root exudates that mobilize the non-bioavailable heavy metals ions in the rhizosphere by altering pH. The soluble form of heavy metal is sorbed on the root surface. Entry into the cellular atmosphere of plant could be symplastic or apoplastic. Unlike apoplastic pathway symplastic pathway is energy consuming process and works against the ionic gradient. After entering into the root cells, the heavy metal ions form complex with the chelators and organic acids present in the root cells. The high-affinity binding sites and plasma membrane-localized transporters render metal ion entry across plasma membrane.

Generally, uptake of metals/salt ions occurs through secondary transporters such as channels proteins, H^+ coupled carrier proteins, and low- and high-affinity transporters. Various studies suggest that the heavy and toxic metals are uptaken using the essential element transporters in the plants. Some of the illustrations are given following: PO_4^{3-} transporters, aquaglyceroporin facilitates transport for As^{2+}, zinc-iron permease (ZIP) gene homolog proteins for Cd^{2+}, Zn^{2+}, and cation Fe^{2+}. Some of the plant species have inherent resistance-associated macrophages proteins (Nramps) gene encoding the metal transporter for Fe and also toxic metal Cd. Nickle/cobalt transporter (NiCOT) Ni^{2+} permease transport Pb and Co. H^+-ATPase and H^+-pyrophosphatase enzymes transport toxic Na^+ ions from the cell wall into the cytosol. ATPase/H^+-pyrophosphatase generates a proton electrochemical gradient that facilitates Ni^{2+}, Cd^{2+}, and Na^+ to cross the plasma membrane. These ions enter the vacuoles through cation diffusion facilitators (CDFs) ABC type ATPase and Cd^{2+}/H^+ antiporter and Na^+/H^+ antiporter and ATPase and pyrophosphatase (Fig. 23.2). Shruti and Dubey (2006), Sunkar et al. (2000) reported the uptake and regulation of Pb through NtCBP4 and AtCNGC1 protein expression (Fig. 2).

23.4.2 Phytoremediation in Plants, Mode of Action

A good phytoremediator must be efficient in metal uptake and translocation to shoots (have the ability to accumulate and tolerate high concentration levels of metals and show rapid and profuse growth with deep root system) (DalCorso et al. 2019). The involvement of components of membrane transporters, ion channels, and aquaporins in uptake and sequestration of heavy metals into plant parts has been shown (Manara 2012). For example, arsenic is a phosphate analogue which is taken up in plants via phosphate transporter. In plants arsenic is detoxified by reduction and subsequently forming complexes with thiol-reactive peptides such as γ-glutamylcysteine (γ-EC), glutathione (GSH), and phytochelatins (Vatamaniuk et al. 2002). Further these thiol complexes are routed to vacuoles of root and shoot through involvement of glutathione conjugating pumps (Wang et al. 2002).

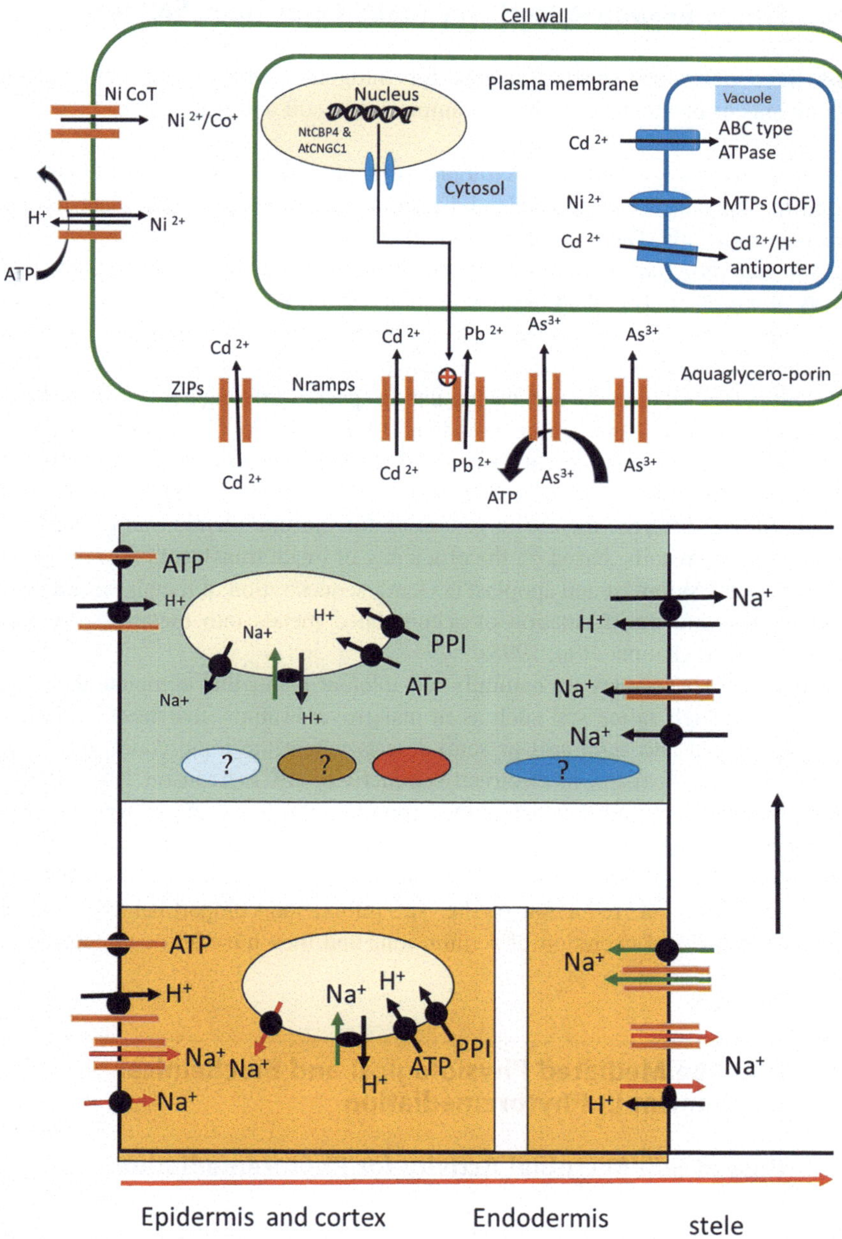

Fig. 23.2 Uptake and translocation of heavy metals and salts in plant species

23.5 Plants Response to Heavy Metals and Toxic Salt Ions

Plants have developed four strategies for mitigating heavy metal or organic contaminants or elemental toxicity contamination in soil and water.

1. *Metal excluders/or avoidance*: Avoidance strategy refers to the ability of plants to limit the uptake of heavy metals and restrict their movement into plant tissues through root cells (Dalvi and Bhalerao 2013).
2. *Metal indicator plants*: Metal indicator plants accumulate metals in their arial parts of the plant. The plant tissue reflects the level of metal in the soil.
3. *Accumulator plant (hyperaccumulators)*: Hyperaccumulators plant species can concentrate metals in their above-ground tissues to levels far exceeding those present in the soil or in the non-accumulating species growing nearby. It has been proposed that a plant containing more than 0.1% of Ni, Co, Cu, Cr or Pb or 1% of Zn in its leaves on a dry weight basis is called a hyperaccumulator, irrespective of the metal concentration in the soil (Baker and Walker 1990). Hyperaccumulator plants utilize different metabolic processes for the mobilization and uptake of metal ions from soils, based on the efficiency of metal translocation to the plant shoots via the symplast and apoplast (xylem), sequestration of metals within cells and tissues, and transformation of accumulated metals into metabolically less harmful forms (Kumar et al. 1995).
4. *Halophytes:* Halophytes are naturally salt-tolerant plants that complete their life cycle under high saline soil such as in mangrove swamps, marshes, seashores, and saline arid and semi-arid or semi-deserts where the ionic concentration is equal or above 8–10 dS m electrical conductivity (EC) (Aronson 1985, 1989). Some authors reported that halophytes complete the life cycle in water with a rather high salinity level of about 11.70 g NaCl/L (Flowers and Colmer 2008). Halophytes have different salinity-tolerant adaptation mechanisms that include osmotic adjustment, reduction of the Na^+ influx, ion compartmentalization in their vacuoles, and excretion of sodium ions and they have less competition in saline environments.

23.6 Microbe Mediated Physiological and Biochemical Mechanism in Phytoremediation

23.6.1 Role of Soil Microbial Activity for Phytoremediation

Microorganisms are ubiquitous and are capable to evolve rapidly to the changing environmental conditions. When they are continuously exposed to pollutants, they become tolerant and exhibit exceptional levels of capability to transform pollutants as their source of energy and raw material (Nanda et al. 2019). Microbes coexist with plants and their interaction plays a major role in uptake and detoxification of harmful heavy metal or elements in the soil which through the process of acidification, precipitation, chelation, complexation, and redox reactions include root exudate acids, synthesis of phytohormones, activation of defense mechanism, solubilization of mineral nutrients, secretion of specific enzymes assisted phytoremediation (Ma et al. 2016). Root exudates from the phytoremediators solubilize the nutrients and minerals from the soil for uptake by the processes like (a) acidification which is due to proton (H^+) release, (b) by forming complexes of metals/minerals with organic acids, (c) intracellular binding compounds (e.g., phytochelatins, organic acids, and amino acids), (d) redox reactions at the rhizosphere where the electron transfer takes place, (e) production of siderophores, phytohormones, biochelators (Rajkumar et al. 2012; Pérez-Montaño et al. 2013; Sessitsch et al. 2013). The benefial microbes interact via chemotaxis to the roots of the plant for release of exudates like organic acids, phytochelatins, amino acids, and enzymes that bring about the alteration in the availability, mobility, and uptake of the metals in the soil.

The process of acidification causes displacement of heavy metal cations adsorbed on soil particles by the action of hydrogen ions exuded from the plant roots. The root exudates help in mobility of element by lowering the rhizosphere pH (Sheoran et al. 2011) which enhances bioavailability of elements fixed in the soil (Alford et al. 2010; Kim et al. 2010), Table 23.2. Also, protonation by soil microbes acidifies their environment by exporting protons to replace heavy metal cations at sorption sites (Rajkumar et al. 2012). Chelation is a process by which organic compounds released from both plants and microorganisms chelate/scavenge metal ions from sorption sites and heavy metal-bearing soils. Chelating compounds are generally known as metal-binding compounds, organic acid anions, siderophores, biosurfactants, and metallophores (Sessitsch et al. 2013) and involved in various processes including nutrient acquisition, mineral weathering, heavy metal detoxification, and mobilization/solubilization in soil (Rajkumar et al. 2012). Precipitation is a process where certain rhizosphere associated microorganisms have the ability to stimulate the enzymatically catalyzed precipitation of radionuclides (e.g., U, Tc) and toxic metals (e.g., Cr, Se) by microbial reduction processes, which is one of the important mechanisms for phytoremediation (Payne and DiChristina 2006), Table 23.2. The alkalinizing effect which is induced by AMF and certain bacteria leads to release of OH^-, can result in an active nitrate uptake by microbes and a reduction in metal phytoavailability in the rhizosphere by secreting glomalin (Giasson et al. 2008). These bacteria act as metal sinks to reduce the mobile and available metal cations in soil, thereby creating a more suitable environment for plants. Another important biochemical mechanism in the process of phytoremediation is transformation of metal contaminants where microbes execute a biogeochemical redox process, which leads to microbial oxidation of heavy metals, such as As, Cr, Hg, Fe, Mn, and Se reducing their phytotoxicity levels (Majumder et al. 2013; Kashefi and Lovley 2000), Table 23.2. One or more of these physiological, biochemical, and molecular changes helps in achieving the phytoremediation by plants that are mediated through the microbes. The strategies used in developing phytoremediation plants are (a) Screening of hyperaccumulation plants, (b) plant breeding, (c) development of improved hyperaccumulation plants using advance genetic tools and techniques.

23.7 Mitigating Strategies to Heavy Metal Contamination

23.7.1 Phytohormone Assisted Mitigating Strategy

Number of studies have demonstrated that these phytohormones favorably affect the amount of heavy metal (HM) accumulation and enhance plant development and HM tolerance. Ji et al. (2015) reported that exogenous application of 10,100 and 1000 mgL of gibberellic acid 3 (GA3) can significantly increase the biomass and phytoremediation efficiency of *Solanum nigrum*. Exogenous application of 10 and 100 mM IAA in nutritional solutions reduces the negative effect of *Trigonella foenumgraecum* of Cd by preventing Cd absorption and controlling the ascorbate glutathione cycle (Bashri and Prasad 2016).

23.7.2 Nanoparticle-Assisted Mitigating Strategy

Use of nanoparticle (NP) is novel technology for efficient removal of heavy metals (Zhu et al. 2019). These particles boost the capacity of phytoremediation through a variety of mechanism, such as: (a) contact with HMs through adsorption/redox processes, (b) promotion of plant growth, or (c) facilitation of HMs phytoremediation (Song et al. 2019). The result of the chemical interaction demonstrated that plants use nanoparticles to stabilize heavy metals by electrostatic adsorption. Rhizospheric bacteria and fungi can promote plant development through NPs. Numerous studies have shown the beneficial effects of nanoparticles in phytoremediation. In this regard, Khan and Bano (2016) observed that the inclusion of Ag nanoparticles (AgNPs) with plant growth promoting rhizobacteria regulated the growth and phytoextraction potential of maize plants (PGPR). Cd contaminated soil was treated with nano-titanium dioxide (nano-TiO) particles to enhance the potential of soybean plants to remove Cd (*Glycine max*) (Singh and Lee 2016).

23.7.3 Nutrients and Beneficial Elements Mitigating Strategy

Nitrogen, sulfur, and silicon are among the essential elements for plant growth and development. Nitrogen can reduce the toxicity of trace elements by the mechanism of increasing the chlorophyll contents and enhancing the photosynthetic activity, anti-oxidative enzymes, and production of N-containing metabolites such as amino acids and derivatives, GSH, and PCs (Sharma and Dietz 2006; Sarwar et al. 2010; Singh et al. 2016). Sulfur reduces trace element toxicity by encouraging the manufacture of S-containing metabolites like GSH and PCs, increasing the activity of the AsA-GSH cycle, and controlling ethylene signaling, among other things (Sarwar et al. 2010; Singh et al. 2016). An adequate supply of sulfur allows increased Cd translocation from the roots to the stems in *Panicum maximum* (Rabêlo et al. 2018b). Sulfur enhances dGSH and PCs synthesis, AsA-GSH cycle enzyme activities are crucial in grasses. These enzymes help in increased Cd phytoextraction efficiency

(Rabêlo et al. 2017a, b, 2018a). In general, the trace element toxicity alleviation mechanism includes the regulation of trace element uptake and root-to-shoot translocation, modulation of the cation binding capacity of the cell walls, increase of enzymatic (e.g., SOD, APX, and DHAR) and non-enzymatic (e.g., AsA and GSH) antioxidants, and the complexation or co-precipitation of trace element ions with Si in the cytoplasm, followed by sequestration of the trace elements in the vacuoles (Filon-Smits et al. 2009). This is especially significant for monocots with substantial, Si accumulations (10–15%) (Hodson et al. 2005). In *P. maximum*, adding Si reduced the amount of Cu that moved from roots to the shoots, reducing Cu toxicity in the leaves and increased biomass production (Vieira Filho and Monteiro 2020).

23.7.4 Plant and Microorganism Interactions Mitigating Strategy

The plant growth promoting bacteria (PGPB) are beneficial group of bacteria that develop symbiotic relationship with plants and promote various plant growth promotion (PGP) activities in adverse conditions (Hashem et al. 2019). These PGPB may act as free-living or rhizosphere bacteria that form specific symbiotic relationships with roots, endophytic bacteria that can colonize plant interior tissues like *Rhizobia* spp. and cyanobacteria (Glick 2012). During this association with the plants, the PGPR are also involved in alleviating the toxicity of heavy metal present in soil through various mechanism, viz. metal biosorption, bioaccumulation, redox reaction, mobilization, precipitation, transformation, and translocation activities (Ma et al. 2016; Wang et al. 2022). PGPB increase the tolerance level of plants against metal stress by improving detoxification rates of plants, enzymes secreted by plant roots, and modifying soil pH (Guo et al. 2020). In addition to this, certain metal-resistant PGPB alter metal mobility and bioavailability and consequently plant usage rate by releasing chelating agents, acidification, and redox changes (Verma and Kuila 2019). Bacteria may also enhance nutrient uptake, increasing plant growth and defense by diminish heavy metals intake and their toxic effects (Chaudhary and Khan 2022). Therefore, use of PGPB strains having the capacity of heavy metal resistant can be a suitable candidate for phytoremediation of heavy metals to minimize the adverse effect on production and productivity. Chen et al. (2017) reported that during Zn and Cd stress, *Pseudomonas fluorescens* induced the growth promotion, chlorophyll content, and enzyme activity of *Sedum alfredii* by producing IAA, and improve plant Cd absorption by regulating the expression and transport genes of Cd.

The microbial-induced carbonate precipitation (MICP) is a viable bioremediation approach for metal immobilization where carbonates can bind to the metals (e.g., Pb^{2+} and Cu^{2+}). These metal elements change from soluble forms to insoluble forms, thus reducing their toxicity (Tamayo-Figueroa et al. 2019). The MICP caused by *Bacillus pasteurii* (ATCC 11859) maintained the microbial growth while reducing the available Pb content in the soil, resulting in a decrease in Pb extraction and available Pb content by 76.34% and 41.65%, respectively (Chen et al. 2021). It has been reported that a group of metal-resistant PGPB, such as *Pseudomonas*,

Arthrobacter, *Acinetobacter* sp., *Agrobacterium*, *Bacillus*, *Azoarcus*, *Azospirillum*, *Azotobacter*, *Burkholderia*, *Klebsiella*, *Alcaligenes*, *Serratia*, *Rhizobium*, and *Enterobacter* species have great potential to promote the growth of various plants in the metal-contaminated environments (Enebe and Babalola 2018; Ke et al. 2021; Ma et al. 2020). The level of removal of heavy metal by the PGPB is mentioned in Table 23.3.

23.8 Improving Phytoremediation Ability in Plants

Traditional breeding (plant hybridization) or genetic engineering (creation of transgenic plants) are employed to either improve growth rate and biomass of hyperaccumulator or introduce hyperaccumulation traits to fast growth, high biomass plants (DalCorso et al. 2019).

23.9 Traditional Plant Breeding Mitigating Strategy

Phytoremediation efficient plant species tolerant to toxic chemicals can be selected from contaminated habitats (Laxman et al. 2015) followed by hybridization and progeny selection (Dushenkov et al. 2002). Brewer et al. (1999) used electrofusion to fuse protoplasts isolated from the Zn hyperaccumulator *T. caerulescens* and *Brassica napus*. The selected hybrids (somatic hybrid) showed enhanced hyperaccumulation capability and tolerance derived from *T. caerulescens* and higher biomass production derived from *B. napus*. The hybrid derivatives showed the ability to accumulate high levels of Zn and Cd. The somaclonal variation in *C. dactylon* enhanced the tolerance and accumulation of Pb. The occurrence of somaclonal variation via somatic embryogenesis and organogenesis of *C. dactylon* cultures was as high as 33% (Taghizadeh et al. 2015; Nehnevajova et al. 2007) used chemical mutagen ethyl methanesulfonate (EMS) to treat sunflowers and obtained sunflower "giant mutant," which exhibited significantly enhanced heavy metal extraction ability with 7.5 times accumulation for Cd, 9.2 times for Zn, and 8.2 times for Pb compared to control plants.

23.10 Biotechnological Mitigating Strategy

Despite successful efforts in development of phytoremediation efficient plants through traditional breeding and genetic approaches, the evolution of molecular biology, genetic engineering, and availability of genome sequence of different plant species have paved the way for development of transgenic and genome edited plant species. Transgenic technology can be exploited to develop heavy metal tolerant and phytoremediation efficient plants through transfer of genes controlling metal uptake, transport, accumulation, and detoxification (Fasani et al. 2018). The best example for this approach is improvement of Eastern cottonwood trees (*Populus deltoides* Bartr. ex Marsh.) for decontamination of Hg^{2+} contaminated soil (Lyyra et al. 2007). Agarwal and Rani (2022) have published a comprehensive review of transgenic plants developed for enhancing the phytoremediation potential. Czako et al. (2005) assessed the transfer of gene encoding enzymes for the breaking down of organomercurials [organomercurial lyase (MerB, EC 4.99.1.2)] and the reduction of Hg^{2+} to HgO [mercuric reductase (MerA) (EC 1.16.1.1)] into the *S. alterniflora*. Song et al. (2019) showed that the expression of ZIP and detoxification related genes increase the effect of supplementary ABA on Zn in *Vitis vinifera*.

23.11 Gene Editing Mitigating Strategy

The other recent development in the form of genome editing through CRISPR-Cas9 technology has proven to be a promising approach in developing heavy metal tolerant plant species. CRISPR-Cas9 technology might prove handy in tweaking the expression of genes coding for these pumps and transporters to promote enhanced uptake and sequestration into vacuoles (Guo et al. 2008). Few of the reports demonstrating use of CRISPR-Cas9 in enhancing phytoremediation potential of crop plants have become available. Tang et al. (2017) showed in that knockout for the metal transporter gene *OsNRAMP5* led to low Cd accumulation in shoots, roots, and in grains of rice grown on Cd contaminated soil. There are other reports listing the successful application of CRISPR-Cas9 mediated gene editing in enhancing phytoremediation potential of plants (Miglani 2017; Rai et al. 2021).

23.12 Conclusion

Phytoremediation is a low cost and sustainable technology to decontaminate soils polluted with heavy metal and toxic elements. Few crop plants and grasses are identified as potential remediators. Their utility in phytoremediation can be enhanced through various approaches like genetic engineering, gene editing, nano-technology, and nano-biotechnology with the use phytohormones. However, much elaborative studies employing physiological and molecular approaches are required to better understand the processes involved in the uptake, transport, accumulation, localization, and detoxification of toxic and trace elements. Several types of crop plants and grass have been investigated, but only a limited number of transporters of As, Cd, Cu, and Zn, Na^+ and Cl^- have been identified in few species of crops and grasses. The phytoremediation is an emerging technology for reclamation of toxic elements in the soil; however, information on how plants uptake and assimilate heavy metals or volatiles from air into the plant system through leaf and areal plant parts is limited. Hence, understanding the role of each trace element transporters, genomics, and proteomics in crop plants and grasses is essential to optimize the phytoremediation of trace elements.

References

Abdelkrim S, Jebara SH, Jebara M (2018) Antioxidant systems responses and the compatible solutes as contributing factors to lead accumulation and tolerance in Lathyrus sativus inoculated by plant growth promoting rhizobacteria. Ecotoxicol Environ Saf 166:427–436

Agarwal P, Rani R (2022) Strategic management of contaminated water bodies: omics, genome-editing and other recent advances in phytoremediation. Environ Technol Innov 27:102463

Ahemad M, Kibret M (2014) Mechanisms and applications of plant growth promoting rhizobacteria: current perspective. J King Saud Univ 26:1–20

Akhtar MJ, Ullah S, Ahmad I, Rauf A, Nadeem SM, Khan MY et al (2018) Nickel phytoextraction through bacterial inoculation in *Raphanus sativus*. Chemosphere 190:234–242

Alford EA, Pilon-Smits EAH, Paschke M (2010) Metallophytes - a view from the rhizosphere. Plant Soil 337:33–50

Ali A, Guo D, Li YM, Shaheen SM, Wahid F, Antoniadis V et al (2021) Streptomyces pactum addition to contaminated soils improved soil quality and plant growth and enhanced metals phytoextraction by wheat in a green remediation trial. Chemosphere 273:129692

Appenroth KJ (2010) What are "heavy metals" in plant sciences? Acta Physiol Plant 32:615–619

Aravind P, Prasad MNV (2015) Cadmium-zinc interactions in a hydroponic system using ceratophyllum demersum L.: adaptive ecophysiology, biochemistry and molecular toxicology. Braz J Plant Physiol 17:3–20

Aronson J (1985) Economic halophytes—a global review. In: Wickens GE, Goodin JR, Field DV (eds) Plants for arid lands. George Allen & Unwin, London, pp 177–188

Aronson J (1989) Salt-tolerant plants of the world. University of Arizona, Tucson

Baker AJM, Brooks RR (1989) Terrestrial higher plants which hyperaccumulate metallic elements - a review of their distribution. Ecol Phytochem Biorecov 1:81

Baker AJM, Walker PL (1990) Ecophysiology of metal uptake by tolerant plants, heavy metal tolerance in plants. In: Shaw AJ (ed) Evolutionary aspects. CRC Press, Boca Raton, pp 155–177

Banik A, Pandya P, Patel B, Rathod C, Dangar M (2018) Characterization of halotolerant, pigmented, plant growth promoting bacteria of groundnut rhizosphere and its in-vitro evaluation of plant-microbe protocooperation to withstand salinity and metal stress. Sci Total Environ 630: 231–242

Bashri G, Prasad SM (2016) Exogenous IAA differentially affects growth, oxidative stress and antioxidants system in Cd stressed Trigonella foenum-graecum L. seedlings: toxicity alleviation by up-regulation of ascorbate-glutathione cycle. Ecotoxicol Environ Saf 132:329–338

Brewer EP, Saunders JA, Angle JS, Chaney RL, McIntosh MS (1999) Somatic hybridization between the zinc accumulator Thlaspi caerulescens and Brassica napus. Theor Appl Genet 99: 761–771

Bruno LB, Karthik C, Ma Y, Kadirvelu K, Freitas H, Rajkumar M (2020) Amelioration of chromium and heat stresses in Sorghum bicolor by Cr6+ reducing-thermotolerant plant growth promoting bacteria. Chemosphere 244:125521

Chaudhary K, Khan S (2022) Role of plant growth promoting bacteria (PGPB) for bioremediation of heavy metals: an overview. In: I. Management Association (ed) Research anthology on emerging techniques in environmental remediation. IGI Global, Hershey, pp 663–680

Chen C, Xin K, Liu H, Cheng J, Shen X, Wang Y et al (2017) Pantoea alhagi, a novel endophytic bacterium with ability to improve growth and drought tolerance in wheat. Sci Rep 7:41564

Chen MJ, Li YF, Jiang XR, Zhao DR, Liu XF, Zhou JL et al (2021) Study on soil physical structure after the bioremediation of Pb pollution using microbial-induced carbonate precipitation methodology. J Hazard Mater 411:125103

Clemens S (2006) Toxic metal accumulation, responses to exposure and mechanisms of tolerance in plants. Biochimie 88:1707–1719

Cosio C, Martinoia E, Keller C (2004) Hyperaccumulation of cadmium and zinc in Thlaspi caerulescens and Arabidopsis halleri at the leaf cellular level. Plant Physiol 134:716–725

Czako M, Feng X, He Y, Liang D, Marton L (2005) Genetic modification of wetland grasses for phytoremediation. Z Naturforsch 60c:285–291
Dakora FD, Phillips DA (2002) Root exudates as mediators of mineral acquisition in low-nutrient environments. Plant Soil 245:35–47
DalCorso G, Fasani E, Manara A, Visioli G, Furini A (2019) Heavy metal pollutions: state of the art and innovation in phytoremediation. Int J Mol 20(14):3412
Dalvi AA, Bhalerao SA (2013) Response of plants towards heavy metal toxicity: an overview of avoidance, tolerance and uptake mechanism. Ann Plant Sci 2:362–368
Dela Cruz M, Christensen JH, Thomsen JD, Müller R (2014) Can ornamental potted plants remove volatile organic compounds from indoor air?—A review. Environ Sci Pollut Res 21:13909–13928
Dheeravathu SN, Singh K, Ramteke PW, Reetu, Dikshit N, Prasad M, Deb D, Vadithe TB (2021) Physiological responses of Bajra-Napierhybrids and a tri-specific hybrid to salinity stress. Tropical Grasslands-Forrajes Tropicales 9:337–347
Dushenkov S, Skarzhinskaya M, Glimelius K, Gleba D, Raskin I (2002) Bioengineering of a phytoremediation plant by means of somatic hybridization. Int J Phytoremediation 4:117–126
Enebe MC, Babalola OO (2018) The influence of plant growth-promoting rhizobacteria in plant tolerance to abiotic stress: a survival strategy. Appl Microbiol Biotechnol 102:7821–7835
Fasani E, Manara A, Martini F, Furini A, DalCorso G (2018) The potential of genetic engineering of plants for the remediation of soils contaminated with heavy metals: transgenic plants for phytoremediation. Plant Cell Environ 41:1201–1232
Flowers TJ, Colmer TD (2008) Salinity tolerance in halophytes. New Phytol 179:945–963
Food and Agriculture Organization of the United Nations and the United Nations Environment Programme (2021) Global assessment of soil pollution-summary for policy makers. FAO, Rome
Giasson P, Karam A, Jaouich A (2008) Arbuscular mycorrhizae and alleviation of soil stresses on plant growth. In: Siddiqui ZA, Akhtar MS, Futai K (eds) Mycorrhizae: sustainable agriculture and forestry. Springer, Dordrecht, pp 99–134
Glick BR (2012) Plant growth-promoting bacteria: mechanisms and applications. Scientifica 2012: 963401
Gołda S, Korzeniowska J (2016) Comparison of phytoremediation potential of three grass species in soil contaminated with cadmium. Environ Prot Nat Resource 27:8–14
Guo J, Dai X, Xu W, Ma M (2008) Overexpressing GSH1 and AsPCS1 simultaneously increases the tolerance and accumulation of cadmium and arsenic in Arabidopsis thaliana. Chemosphere 72:1020–1026
Guo JK, Muhammad H, Lv X, Wei T, Ren XH, Jia HL, Atif S, Hua L (2020) Prospects and applications of plant growth promoting rhizobacteria to mitigate soil metal contamination: a review. Chemosphere 246:125823
Haan SD, Lubbers J (1983) Microelements in potatoes under normal conditions, and as affected by micro-elements in municipal waste compost, sewage sludge and degraded materials from harbours. Rapport Institute Voor Bodemvruchtbaarheld 83:22
Hasegawa PM, Bressan RA, Zhu J-K, Bohnert HJ (2000) Plant cellular and molecular responses to high salinity. Annu Physiol Mol Biol Plants 51:463–499
Hashem A, Tabassum B, Abd Allah EF (2019) Bacillus subtilis: a plant-growth promoting rhizobacterium that also impacts biotic stress. Saudi J Biol Sci 26:1291–1297
He Z, Shentu J, Yang X, Baligar VC, Zhang T, Stoffella PJ (2015) Heavy metal contamination of soils: sources, indicators, and assessment. Environ Sustain Indicators 9:17–18
Hodson MJ, White PJ, Mead A, Broadley MR (2005) Phylogenetic variation in the silicon composition of plants. Ann Bot 96:1027–1046
Hofmann NR (2013) Volatile organic compounds: a bacterial contribution to plant sulfur nutrition. Plant Cell 25:2381
Hussein H, Ibrahim SF, Kandeel K, Moawad H (2004) Biosorption of heavy metals from waste water using Pseudomonas sp. Electron J Biotechnol 7(1):30–37

Interstate Technology & Regulatory Council (2009) Phytotechnology technical and regulatory guidance and decision trees (revised). Interstate Technology & Regulatory Council, Washington, DC, pp 1–131

Islam E, Yang X, Li T, Liu D, Jin X, Meng F (2007) Effect of Pb toxicity on root morphology, physiology and ultrastructure in the two ecotypes of Elsholtzia argyi. J Hazard Mater 147:806–816

Ji P, Tang X, Jiang Y et al (2015) Potential of gibberellic acid 3 (GA3) for enhancing the phytoremediation efficiency of Solanum nigrum L. Bull Environ Contam Toxicol 95:810–814

Kabata-Pendias A, Pendias H (1989) Trace elements in the soil and plants. CRC Press, Boca Raton, pp 83–34

Kashefi K, Lovley DR (2000) Reduction of Fe(III), Mn(IV), and toxic metals at 100 degrees C by Pyrobaculum islandicum. Appl Environ Microbiol 66:1050–1056

Ke T, Guo GY, Liu JR, Zhang C, Tao Y, Wang PP et al (2021) Improvement of the Cu and Cd phytostabilization efficiency of perennial ryegrass through the inoculation of three metal-resistant PGPR strains. Environ Pollut 271:116314

Khan N, Bano A (2016) Modulation of phytoremediation and plant growth by the treatment with PGPR, Ag nanoparticle and untreated municipal waste water. Int J Phytoremediation 18:1258–1269

Kim S, Lim H, Lee I (2010) Enhanced heavy metal phytoextraction by Echinochloa crus-galli using root exudates. J Biosci Bioeng 109:47–50

Krishna AK, Govil PK (2007) Soil contamination due to heavy metals from an industrial area of Surat, Gujarat, Western India. Environ Monitor Assess 124:263–275

Kumar PBAN, Dushenkov V, Motto H, Raskin I (1995) Phytoextraction the use of plants to remove heavy metals from soils. Environ Sci Technol 29:1232–1238

Lajayer BA, Moghadam NK, Maghsoodi MR, Ghorbanpour M, Kariman K (2019) Phytoextraction of heavy metals from contaminated soil, water and atmosphere using ornamental plants: mechanisms and efficiency improvement strategies. Environ Sci Pollut Res 26:8468–8484

Laxman SK, Sudhakar G, Swaminathan SK, Muralidhar RC (2015) Identification of elite native plants species for phytoaccumulation and remediation of major contaminants in uranium tailing ponds and its affected area. Environ Dev Sustain 17:57–81

Lepp NW (1981) Effect of heavy metal pollution on plants. Metals in the environment. Applied Science Publishers, London

Liu S, Yang B, Liang Y, Xiao Y, Fang J (2020) Prospect of phytoremediation combined with other approaches for remediation of heavy metal-polluted soils. Environ Sci Pollut Res 27:16069–16085

Lyyra S, Meagher RB, Kim T, Heaton A, Montello P, Balish RS, Merkle SA (2007) Coupling two mercury resistance genes in Eastern cottonwood enhances the processing of organomercury. Plant Biotechnol J 5:254–262

Ma Y, Rajkumar M, Freitas H (2009a) Improvement of plant growth and nickel uptake by nickel resistant-plant growth promoting bacteria. J Hazard Mater 166:1154–1161

Ma Y, Rajkumar M, Freitas H (2009b) Inoculation of plant growth promoting bacterium Achromobacter xylosoxidans strain Ax10 for the improvement of copper phytoextraction by Brassica juncea. J Environ Manag 90:831–837

Ma Y, Prasad MNV, Rajkumar M, Freitas H (2011) Plant growth promoting rhizobacteria and endophytes accelerate phytoremediation of metalliferous soils. Biotechnol Adv 29:248–258. https://doi.org/10.1016/j.biotechadv.2010.12.001

Ma Y, Oliveira RS, Freitas H, Zhang C (2016) Biochemical and molecular mechanisms of plant-microbe-metal interactions: relevance for phytoremediation. Front Plant Sci 7:918

Ma H, Wei MY, Wang ZR, Hou SY, Li XD, Xu H (2020) Bioremediation of cadmium polluted soil using a novel cadmium immobilizing plant growth promotion strain Bacillus sp. TZ5 loaded on biochar. J Hazard Mater 388:122065

Magdziak Z, Kozlowska M, Kaczmarek Z, Mleczek M, Chadzinikolau T, Drzewiecka K et al (2011) Influence of Ca/Mg ratio on phytoextraction properties of Salix viminalis. II. Secretion of low molecular weight organic acids to the rhizosphere. Ecotoxicol Environ Saf 74:33–40

Majumder A, Bhattacharyya K, Bhattacharyya S, Kole SC (2013) Arsenic-tolerant, arsenite-oxidising bacterial strains in the contaminated soils of West Bengal, India. Sci Total Environ 46:1006–1014

Manara A (2012) Plant responses to heavy metal toxicity. In: Furini A (ed) Plants and heavy metals. Springer, Dordrecht, pp 27–53

Maqbool Z, Asghar HN, Shahzad T, Hussain S, Riaz M, Ali S, Arif MS, Maqsood M (2015) Isolating, screening and applying chromium reducing bacteria to promote growth and yield of okra (*Hibiscus esculentus* L.) in chromium contaminated soils. Ecotoxicol Environ Saf 114: 343–349

Markert B (1996) Instrumental element and multi-element analysis of plant samples. John Wiley & Sons, Chichester

McGrath SP, Chaudri AM, Giller KE (1995) Long-term effects of metals in sewage sludge on soils, microorganisms and plants. J Ind Microbiol 14:94–104

McIntyre TC (2003) Databases and protocol for plant and microorganism selection: hydrocarbons and metals. In: McCutcheon SC, Schnoor JL (eds) Phytoremediation. Transformation and control of contaminants. John Wiley & Sons, Hoboken, pp 887–904

Miglani GS (2017) Genome editing in crop improvement: present scenario and future prospects. J Crop Improv 31:453–559

Mucha AP, Marisa C, Almeida R, Bordalo AA, Teresa M, Vasconcelos SD (2005) Exudation of organic acids by a marsh plant and implications on trace metal availability in the rhizosphere of estuarine sediments. Estuar Coast Shelf Sci 65:191–198

Nanda M, Kumar V, Singh DK (2019) Multimetal tolerance mechanisms in bacteria: the resistance strategies acquired by bacteria that can be exploited to 'clean-up' heavy metal contaminants from water. Aquat Toxicol 2019(212):1–10

Naseem M, Anwar-ul-Haq M, Wang X, Farooq N, Awais M, Sattar H, Malik HA, Mustafa A, Ahmad J, El-Esawi MA (2021) Influence of selenium on growth, physiology, and antioxidant responses in maize varies in a dose-dependent manner. J Food Qual 2021:6642018

Neelima P, Reddy KJ (2003) Differential effect of cadmium and mercury on growth and metabolism of Solanum melongena L. seedlings. J Environ Biol 24:453–460

Nehnevajova E, Herzig R, Federer G, Erismann K-H, Schwitzguébel, J.- P. (2007) Chemical mutagenesis—a promising technique to increase metal concentration and extraction in sunflowers. Int J Phytoremediation 9:149–165

Niinemets Ü, Fares S, Harley P, Jardine KJ (2014) Bidirectional exchange of biogenic volatiles with vegetation: emission sources, reactions, breakdown and deposition. Plant Cell Environ 37: 1790–1809

Nikolić M, Stevović S (2015) Family Asteraceae as a sustainable planning tool in phytoremediation and its relevance in urban areas. Urban For Urban Green 14:782–789

Payne AN, DiChristina TJ (2006) A rapid mutant screening technique for detection of technetium [Tc(VII)] reduction-deficient mutants of Shewanella oneidensis MR-1. FEMS Microbiol Lett 259:282–287

Pehlivan E, Ozkan AM, Dinç S, Parlayici S (2009) Adsorption of Cu^{2+} and Pb^{2+} ion on dolomite powder. J Hazard Mater 167:1044–1049

Pérez-Montaño F, Jiménez-Guerrero I, Sánchez-Matamoros RC, LópezBaena FJ, Ollero FJ, Rodríguez-Carvajal MA et al (2013) Rice and bean AHL-mimic quorum-sensing signals specifically interfere with the capacity to form biofilms by plant-associated bacteria. Res Microbiol 164:749–760

Pilon-Smits EAH, Quinn CF, Tapken W, Malagoli M, Schiavon M (2009) Physiological functions of beneficial elements. Curr Opin Plant Biol 12:267–274

Pramanik K, Mandal S, Banerjee S, Ghosh A, Maiti TK, Mandal NC (2021) Unraveling the heavy metal resistance and biocontrol potential of Pseudomonas sp. K32 strain facilitating rice seedling growth under Cd stress. Chemosphere 274:129819

Rabêlo FHS, Azevedo RA, Monteiro FA (2017a) Proper supply of S increases GSH synthesis in the establishment and reduces tiller mortality during the regrowth of Tanzania Guinea grass used for Cd phytoextraction. J Soils Sediments 17:1427–1436

Rabêlo FHS, Azevedo RA, Monteiro FA (2017b) The proper supply of S increases amino acid synthesis and antioxidant enzyme activity in Tanzania Guinea grass used for Cd phytoextraction. Water Air Soil Pollut 228:394

Rabêlo FHS, Fernie AR, Navazas A, Borgo L, Keunen E, Silva BKA et al (2018a) A glimpse into the effect of sulfur supply on metabolite profiling, glutathione and phytochelatins in Panicum maximum cv. Massai exposed to cadmium. Environ Exp Bot 151:76–88

Rabêlo FHS, Lux A, Rossi ML, Martinelli AP, Cuypers A, Lavres J (2018b) Adequate S supply reduces the damage of high Cd exposure in roots and increases N, S and Mn uptake by Massai grass grown in hydroponics. Environ Exp Bot 148:35–46

Rahman Z, Singh VP (2020) Bioremediation of toxic heavy metals (THMs) contaminated sites: concepts, applications and challenges. Environ Sci Pollut Res 27:27563–27581

Rai KK, Pandey N, Meena RP, Rai SP (2021) Biotechnological strategies for enhancing heavy metal tolerance in neglected and underutilized legume crops: a comprehensive review. Ecotoxicol Environ Saf 208:111750

Rajkumar M, Sandhya S, Prasad MNV, Freitas H (2012) Perspectives of plant-associated microbes in heavy metal phytoremediation. Biotechnol Adv 30:1562–1574

Ramachandran VK, East AK, Karunakaran R, Downie JA, Poole PS (2011) Adaptation of Rhizobium leguminosarum to pea, alfalfa and sugar beet rhizospheres investigated by comparative transcriptomics. Genome Biol 12:106

Rattan RK, Datta SP, Chhonkar PK, Suribabu K, Singh AK (2005) Long-term impact of irrigation with sewage effluents on heavy metal contents in soils, crops and ground water – a case study. Agric Ecosyst Environ 109:210–322

Ravenscroft P, Brammer H, Richards K (2011) Arsenic pollution: a global synthesis. John Wiley &Sons, Hoboken, p 616. ISBN: 9781444355468

Ravindran KC, Venkatesan K, Balakrishnan V, Chellappan KP, Balasubramanian T (2007) Restoration of saline land by halophytes for Indian soils. Soil Biol Biochem 39:2661–2664

Sarma H (2011) Metal hyperaccumulation in plants: a review focusing on phytoremediation technology. J Environ Sci Technol 4:118–113

Sarwar N, Saifullah, Malhi SS, Zia MH, Naeem A, Bibi S et al (2010) Role of mineral nutrition in minimizing cadmium accumulation by plants. J Sci Food Agric 90:925–937

Scandalios JG (2005) Oxidative stress: molecular perception and transduction of signals triggering antioxidant gene defenses. Braz J Med Biol Res 38:995–1014

Selvi A, Rajasekar A, Jayaraman T, Ananthaselvam A, Sathishkumar K, Madhavan J, Pattanathu K, Rahman SM (2019) Integrated remediation processes toward heavy metal removal/ recovery from various environments—a review. Front Environ Sci 7:66

Sengar RS, Gautam M, Sengar K, Chaudhary R, Garg S (2010) Physiological and metabolic effect of mercury accumulation in higher plants system. Toxicol Environ Chem 92:1265–1281

Sessitsch A, Kuffner M, Kidd P, Vangronsveld J, Wenzel WW, Fallmann K et al (2013) The role of plant-associated bacteria in the mobilization and phytoextraction of trace elements in contaminated soils. Soil Biol Biochem 60:182–194

Sharma SS, Dietz K-J (2006) The significance of amino acids and amino acid-derived molecules in plant responses and adaptation to heavy metal stress. J Exp Bot 57:711–726

Sheoran V, Sheoran AS, Poonia P (2011) Role of hyperaccumulators in phytoextraction of metals from contaminated mining sites, a review. Crit Rev Environ Sci Technol 41:168–214

Shrivastava M, Cappa CD, Fan J, Liu AH, Guenther AB, Jimenez JL, Kuang C, Laskin A, Martin ST, Ng NL, Petaja T, Pierce JR, Rasch PJ, Roldin P, Seinfeld JH, Shilling J, Smith JN, Thornton JA, Volkamer R, Wang J, Worsnop DR, Zaveri RA, Zelenyuk A, Zhang Q (2017) Recent

advances in understanding secondary organic aerosol: implications for global climate forcing. Rev Geophys 55:509–559
Shruti M, Dubey RS (2006) Heavy metal uptake and detoxification mechanisms in plants. Int J Agric Res 1:122–141
Sidhu GPS (2016) Heavy metal toxicity in soils: sources, remediation technologies and challenges. Adv Plants Agric Res 5:445–446
Singh J, Lee BK (2016) Influence of nano-TiO2 particles on the bioaccumulation of Cd in soybean plants (*Glycine max*): a possible mechanism for the removal of Cd from the contaminated soil. J Environ Manag 170:88–96
Singh S, Parihar P, Singh R, Singh VP, Prasad SM (2016) Heavy metal tolerance in plants: role of transcriptomics, proteomics, metabolomics, and ionomics. Front Plant Sci 6:1143
Slaveykova VI, Parthasarathy N, Dedieu K, Toescher D (2010) Role of extracellular compounds in Cd-sequestration relative to Cd uptake by bacterium Sinorhizobium meliloti. Environ Pollut 158:2561–2565
Song B, Xu P, Chen M, Tang W, Zeng G, Gong J, Zhang P, Ye S (2019) Using nanomaterials to facilitate the phytoremediation of contaminated soil. Crit Rev Environ Sci Technol 49:791–824
Sridhar BBM, Han FX, Diehl SV et al (2007) Effects of Zn and cd accumulation on structural and physiological characteristics of barley plants. Braz J Plant Physiol 19:15–22
Steinkellner S, Mammerler R (2007) Effect of flavonoids on the development of Fusarium oxysporum f. sp. Lycopersici. J Plant Interact 2:17–23
Steinkellner S, Lendzemo V, Langer I, Schweiger P, Khaosaad T, Toussaint JP et al (2007) Flavonoids and strigolactones in root exudates as signals in symbiotic and pathogenic plant–fungus interactions. Molecules 12:1290–1306
Stewart GR, Ahmad I (1983) Adaptation to salinity in angiosperm halophytes. In: Robb DA, Pierpoint WS (eds) Metals and micronutrients: uptake and utilization by plants. Academic, New York, pp 33–50
Sunkar R, Kaplan B, Bouché N, Arazi T, Dolev D et al (2000) Expression of a truncated tobacco NtCBP4 channel in transgenic plants and disruption of the homologous Arabidopsis CNGC1 gene confer Pb2+ tolerance. Plant J 24:533–542
Szarek-Łukaszewska G (2014) Heavy metals hyper accumulating plants. Kosmos 63:443–453
Taghizadeh M, Kaf M, Fattahi Moghadam MR (2015) Breeding by in vitro culture to improve tolerance and accumulation of lead in cynodon dactylon L. J Agric Sci Technol 17:1851–1860
Tamayo-Figueroa DP, Castillo E, Brandão PFB (2019) Metal and metalloid immobilization by microbiologically induced carbonates precipitation. World J Microbiol Biotechnol 35:35–58
Tang L, Mao B, Li Y, Lv Q, Zhang L, Chen C, He H, Wang W, Zeng X, Shao Y et al (2017) Knockout of OsNramp5 using the CRISPR/Cas9 system produces low Cd-accumulating indica rice without compromising yield. Sci Rep 7:1–12
Tibbett M, Green I, Rate A, De Oliveira VH, Whitaker J (2021) The transfer of trace metals in the soil-plant-arthropod system. Sci Total Environ 779:146260
Vangronsveld J, Herzig R, Weyens N, Boulet J, Adriaensen K, Ruttens A et al (2009) Phytoremediation of contaminated soils and groundwater: lessons from the field. Environ Sci Pollut Res 16:765–794
Vatamaniuk OK, Bucher EA, Ward JT, Rea PA (2002) Worms take the 'phyto' out of 'phytochelatins'. Trends Biotechnol 20:61–64
Venkatesh NM, Vedaraman N (2012) Remediation of soil contaminated with copper using rhamnolipids produced from Pseudomonas aeruginosa MTCC 2297 using waste frying rice bran oil. Ann Microbiol 62:85–91
Verma S, Kuila A (2019) Bioremediation of heavy metals by microbial process. Environ Technol Innov 14:100369
Vieira Filho LO, Monteiro FA (2020) Silicon modulates copper absorption and increases yield of Tanzania Guinea grass under copper toxicity. Environ Sci Pollut Res 27:31221–31232
Vishnupradeep R, Bruno LB, Taj Z, Rajkumar M, Challabathula D, Karthik C et al (2022) Plant growth promoting bacteria improve growth and phytostabilization potential of *Zea mays* under

chromium and drought stress by altering photosynthetic and antioxidant responses. Environ Technol Innov 25:102154

Von Rad U, Klein I, Dobrev PI, Kottova J, Zazimalova E, Fekete A et al (2008) Response of Arabidopsis thaliana to N-hexanoyl-DL-homoserine lactone, a bacterial quorum sensing molecule produced in the rhizosphere. Planta 229:73–85

Vranova V, Rejsek K, Formanek P (2013) Aliphatic, cyclic and aromatic organic acids, vitamins and carbohydrates in soil: a review. Sci World J 2013:524239

Wang J, Zhao FJ, Meharg AA, Raab A, Feldmann J, McGrath SP (2002) Mechanisms of arsenic hyperaccumulation in *Pteris vittata*. Uptake kinetics, interactions with phosphate, and arsenic speciation. Plant Physiol 130:1552–1561

Wang J, Feng X, Anderson CW, Xing Y, Shang L (2012) Remediation of mercury contaminated sites—a review. J Hazard Mater 221:1–18

Wang Q, Zhang WJ, He LY, Sheng XF (2018) Increased biomass and quality and reduced heavy metal accumulation of edible tissues of vegetables in the presence of Cd-tolerant and immobilizing Bacillus megaterium H3. Ecotoxicol Environ Saf 148:269–274

Wang Y, Narayanan M, Shi X, Chen X, Li Z, Natarajan D, Ma Y (2022) Plant growth promoting bacteria in metal-contaminated soil: current perspectives on remediation mechanisms. Front Microbiol 13:966226

Wu ZP, McGrouther K, Huang JD, Wu PB, Wu WD, Wang HL (2014) Decomposition and the contribution of glomalin-related soil protein (GRSP) in heavy metal sequestration: field experiment. Soil Biol Biochem 68:283–290

Xia YX, Liu SQ, Li H, Chen XW (2012) Effects of selenium on physiological characteristics, selenium content and quality of garlic. Plant Nutr Fertilizer Sci 18:733–774

Zhang X, Zhang X, Gao B, Li Z, Xia H, Li H et al (2014) Effect of cadmium on growth, photosynthesis, mineral nutrition and metal accumulation of an energy crop, king grass (Pennisetum americanum × P. purpureum). Biomass Bioenergy 67:179–187

Zhao J, Davis LT, Verpoort R (2005) Elicitor signal transduction leading to production of plant secondary metabolites. Biotechnol Adv 23:283–333

Zhu Y, Xu F, Liu Q, Chen M, Liu X, Wang Y, Sun Y, Zhang L (2019) Nanomaterials and plants: positive effects, toxicity and the remediation of metal and metalloid pollution in soil. Sci Total Environ 662:414–421

Zurek G, Pogrzeba M, Rybka K, Prokopiuk K (2013) Suitability of grass species for phytoremediation of soils polluted with heavy-metals. In: Breeding strategies for sustainable forage and Turfgrass improvement. Springer, Dordrecht, p 245

24 Challenges and Strategies to Improve Drought Tolerance in Plants Through Agronomic Managements

Aradhana Bali, Sandeep Rawal, and Kapil Singla

Abstract

Agriculture production and productivity are prone to abiotic and biotic stresses. Biotic stresses comprise insect–pest incidences, while abiotic stresses come from environmental factors that include drought, floods (both due to irregular pattern of rainfall), temperature extremes (heat, cold chilling/frost), chemical stress (excess of soluble salts, low pH/acid), nutritional deficiencies and imbalances, physical factors (susceptibility to erosion, steep slopes, surface crusting and sealing, low water-holding capacity, impeded drainage, low structural stability, root-restricting layer, high swell/shrink potential) (Minhas et al. 2017 Abiotic stress management for resilient agriculture. Springer, Berlin). These factors coupled with inherent productivity potential of cultivated/selected varieties define the crop productivity in an ecosystem. The intensity and impact of any stress amplify under limited natural resources availability. Indian agriculture being rainfed is subjected to extremities of rainfall pattern. Drought is the recurring phenomenon in rainfed areas although from past few years' floods have also become more frequent owing to changing climatic conditions. The impact of drought is not just confined to agriculture production, as it severely affects livestock also and brings in misery in the affected areas with people struggling to survive and threatening national food and nutritional security. To mitigate drought stress, initiatives have to be taken at microlevel and management includes agronomic (in situ water conservation, integrated farming system, crop diversification, contingency plan, watershed management, mulching, etc.), engineering (grading, bunding, terracing, land capability classification, runoff management, etc.), and physiological interventions (role of plant growth regulators, enzymatic activities, use of

A. Bali (✉) · S. Rawal · K. Singla
Krishi Vigyan Kendra, CCS Haryana Agricultural University, Dambla, Yamunanagar, India
e-mail: a.bali@hau.ac.in

A. Kumar et al. (eds.), *Salinity and Drought Tolerance in Plants*,
https://doi.org/10.1007/978-981-99-4669-3_24

reflectants/protectants, developing drought tolerance, and mitigating mechanisms in plants, etc.).

Keywords

Stress · Drought · Tolerance

24.1 Introduction

World's population is increasing at a faster pace, while the land under cultivation is shrinking due to demand from other sectors. This is posing a major threat to environmental sustainability and food security (Yazdani et al. 2007). Shrinking resources creates additional pressure on the limited available resources and leads to overexploitation of natural resources especially water along with excessive use of fertilizers and pesticides which has deteriorated the soil quality further (Bhat et al. 2009). Abido and Zsombik (2018) had observed that 25% of the world's agricultural lands are now affected by water stress and by 2025, half of the world's population will be living in water-stressed areas (WHO report).

Agriculture is an input driven enterprise and water is the most critical input for successful crop production. Any deviation from normal can offset the expected output. Owing to the burgeoning population and scarcity of natural resources, the stress on already limited water resources in India is increasing every year (Table 24.1). Per capita availability of water in India has reduced from 1816 m^3 to 1140 m^3 with simultaneous population rise from 1029 to 1640 million in 50 years (pib.gov.in).

Along with it, water demand for different sectors has increased from 813 to 1447 billion cubic meters in 40 years (Basin Planning Directorate, CWC 2019) (Table 24.2). It is inferred from the data discussed that water demand exceeds water availability and gap is increasing with each year. To minimize the gap, sector-wise water distribution and consumption have to be regulated and critical measures are needed to be taken. Agriculture sector has the major share in water demand and consumption among all sectors, so agronomic interventions involving less water for irrigation with high productivity have to be promoted at large scale.

Crop productivity is determined by the crops genetic potential and its interaction with the environment which can be modified with best agronomic practices. In a meeting at Stanford University, a group of experts—including crop scientists from

Table 24.1 Average annual per capita availability of water in India

Year	Population (million)	Per capita water availability (m^3/year)
2001	1029	1816
2011	1210	1545
2025	1394	1340
2050	1640	1140

Source: pib.gov.in (2015)

Table 24.2 Estimated water demand in India for different sectors

Sector	Water demand (billion cubic meters)		
	2010	2025	2050
Irrigation	688	910	1072
Industry	12	23	63
Energy	5	15	130
Drinking water	56	73	102
Others	52	72	80
Total	813	1093	1447

Source: Basin Planning Directorate, CWC (2019)

seed companies—concluded as part of their recommendations that "particularly for managing moisture stress in rainfed systems, agronomy may well offer even greater potential benefits than improved crop varieties" (Lobell 2009). Different crops/varieties have different genetic potential depending upon their adaption abilities. Stresses in plants lead to crop failures and crops adapt different defense mechanisms to cope up with the stress imposed on it. Biotic stresses like insect-pest attack can be managed through integrated approach and timely intervention can help in reducing yield loss to a great extent. Abiotic stresses like low or high water stress, temperature stress, salt stress, etc. are difficult to predict and manage as these are generally weather driven with limited scope of modification, thus leading to major yield losses. Although plants have their own adaptive mechanism against any adverse environment condition the degree of tolerance and adaptability to abiotic stresses varies among different crops and within crop, different varieties. Crop physiology undergoes significant changes throughout its life cycle as plants grow from seed to its vegetative phase and ultimately culminating into reproductive phase to complete its life cycle. But under stress, the physiological changes within the plants vary depending upon the degree of stress imposed on plants and plant's ability to withstand that stress. Water and temperature stress are the two major abiotic constraints that occur frequently and in tandem with each other in Indian conditions. The response time to deal with these stresses is very limited and chances of crop failure become high. It requires mid-season corrections through agronomic management but the abrupt physiological changes within plants lead to losses and damage to plant growth. Low water stress or drought inhibits crop growth and restricts its life cycle.

Drought Irrigation or water is an important critical input in agricultural production. Any deviation from normal can offset the expected output. Drought in acute terms can be explained as shortage of available water to crops during major portion of cropping season which may be due to substantial deviation of rainfall from normal or uneven and untimely temporal/spatial distribution pattern leading to severely impaired crop production (DA&FW 2016). It can be further classified into meteorological, agricultural, hydrological, and socioeconomic drought depending upon the impact on ground and surface water resources, cropping pattern and choices adopted, agro-climatic features, socioeconomic vulnerability of local population. It is considered as the major factor which can limit the agricultural production to a large extent and the major drought years have also coincided with substantial production decline and increased commodity prices. Drought can manifest itself into three major impacts on agriculture:

1. Impact on farm production and farmer: Drought induced stress are directly related with reduced farm production and increased farmer distress. However in low production years, higher commodity prices can slightly offset this impact.
2. Market level impact: Agricultural output is known to drive market prices. Any reduction is going to increase the retail prices of agricultural commodities. This impacts the sectors which depend upon agriculture inputs the most. For example, reduced maize production can spiral upwards the price of feed used in poultry which can further eat away the buying power of purchasers.
3. Ecological impact: Reduced production during drought years put extra burden on environment through disturbed food chain and other cascading effects like soil erosion, dust bowls, reduced acreage, and reduced biodiversity.

24.2 Concept of Drought

In arid regions, drought is a common phenomenon and the terms, arid and drought, are sometimes used interchangeably. Arid climate is a permanent feature in an area, while drought is a temporary event. Drought is related to the timing (i.e., principal season of occurrence, delays in the start of the rainy season, occurrence of rains in relation to crop growth stages) and the effectiveness (i.e., rainfall intensity, number of rainfall events) of the rains. It should not be viewed as a physical or natural event only as its impact is multifaceted. It can bring economic, social, and environmental hardship, which increases vulnerability of societies limiting their capacity to overcome its stress and impacts cultural, health, and welfare of the affected area. So in order to plan strategies for drought management it is better to understand its concept and its manifestation. In conceptual terms, drought can be simplified as a period of deficient precipitation, resulting in extensive damage to crop growth and development leading to loss of crop yield. In operational definition, drought can be categorized on the basis of its effect on different sectors or its relevance to users and on the basis of the degree of severity or period of its incidence:

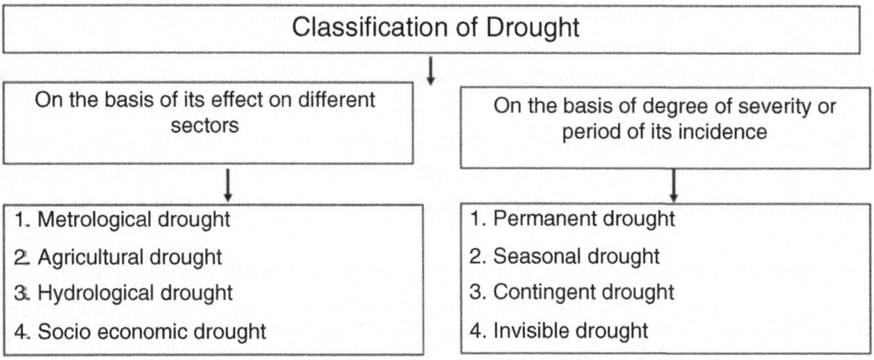

Classification of drought on the basis of its effect on different sectors or its relevance to users:

1. *Meteorological drought*: Meteorological drought is commonly based on precipitation amount and time. Its significant departure from normal average over a certain period of time leads to meteorological drought and is region-specific.
2. *Agricultural drought*: Agricultural drought is more concerned with the impact of drought on crop production. It is said to occur when there is not enough soil moisture to meet a crop water requirement at a particular time. This leads to significant physiological changes and negatively impacts crop productivity of that area. Impact of agricultural drought can be modified to some extent through agronomic management options.
3. *Hydrological drought*: Hydrological drought refers to deficiencies in surface and subsurface water supplies. This drought can directly be measured as stream/river flow and as lake, reservoir, and groundwater levels. Because there is a time lag between the time of rain falling and its appearance in streams, rivers, lakes, and reservoirs, hydrological measurements are not the earliest indicators of drought. Engineering measures are more suited to modify hydrological drought.
4. *Socioeconomic drought*: Socioeconomic drought occurs when physical water shortage starts to affect people, individually and collectively. Most socioeconomic definitions of drought associate it with the supply and demand. Socioeconomic drought occurs when the demand exceeds the supply as a result of a weather-related shortfall in water supply.

Classification of drought on the basis of the degree of severity or period of its incidence and its impact on agriculture:

1. *Permanent drought*: It refers to recurrent and continuous phenomenon of limited water availability or drought conditions owing to arid agroecological conditions. These areas are categorized as deserts which have sparse vegetation cover but its vegetation is well adapted to limited moisture conditions. Agriculture

opportunities are limited under these situations and need major land configuration modification to increase area under cultivation.

2. *Seasonal drought*: The areas where precipitation offers some scope for crop cultivation and where annual rainfall is defined and period of dryness is already known. Mono cropping can be adopted in these areas. Moisture conservation through agronomic modification like planting date, choice of crop and variety offers major scope to cope with drought stress and in achieving better germination and good crop stand.
3. *Contingent drought*: Sudden and unexpected deviation in precipitation amount and time leads to contingent drought where normally intensive agriculture is practiced. Sub-humid to humid climate comes under this category with ample rainfall to support at least two crops in an year. Measures like residue retention or mulching, removal of older leaves, use of nanoparticles, soil absorbents, antitranspirants offer opportunities for reduced evaporative and transpiration losses leading to better adaptation during contingent drought.
4. *Invisible drought*: Drought which occurs in humid climate is known as invisible drought. It occurs when evapotranspiration losses are higher than soil moisture supplying capacity even though there is ample rainfall in that area. Higher temperature, salt content in soil may the reason for invisible drought. This leads to changes in crop physiology and decline in crop yields.

24.3 Effect of Drought on Plants

1. *Impact on plant germination and growth*: Drought brings in abrupt changes in physiology and bio-chemical behavior of the plants like photosynthesis, nutrient metabolism, translocation, and absorption (Hussain et al. 2018). Under water stress conditions, water required for seed imbibition is limited (Molina et al. 2018) which affects several metabolic processes (Farooq et al. 2009; Fahad et al. 2017) like synthesis of hydrolytic enzymes. These enzymes are important for hydrolysis of reserve food into simple available form for embryo uptake (Ali and Elozeiri 2017) and hence germination. Therefore drought during initial growth stages is known to drastically impact germination of crops. For plant growth and development, cell expands in size due to turgor pressure which is generated by water absorption, but under limited moisture availability, plants tend to maintain its turgidity by reducing cell size (Weijde et al. 2017) which leads to poor growth, reduced photosynthesis, and crop yield (Christophe et al. 2011).
2. *Effect on photosynthesis and water absorption*: Plant physiology is determined by enzymatic activities being influenced by abiotic factors especially temperature and water stress (Fahad et al. 2017). Plant-water relation is affected by temperature, water availability that directly influences leaf water potential, canopy temperature, and transpiration and respiration activities. Trabelsi et al. (2019) reported decreased leaf water potential and relative water content under water stress conditions which reduces the activities of photosynthetic pigments (Fathi and Tari 2016). Due to high temperature or reduced availability of moisture,

plants tend to close their stomata to minimize transpiration losses. But this abnormal closure of stomata hampers photosynthesis (Shareef et al. 2018) which affects biomass partitioning and yield under drought conditions.

3. *Impact on nutrient uptake and yield*: Plants absorb nutrients through transpirational pull and drought stress in plants reduces plant nutrient uptake by reducing nutrient diffusion and mass flow in the soil (Silva et al. 2011). Under water deficit, root activity gets reduced and it slows down ion diffusion and finally nutrient uptake by the plants through roots (Christophe et al. 2011). Yield of a crop is the assimilate of its partitioning into reproductive form in cereals (Pandey et al. 2017). Mid-season moisture stress brings abrupt physiological changes in the plants by accelerating its phenology (Marjani et al. 2016). High temperature at grain ripening phase hastens leaf senescence, reduces grain filling duration, and increases grain filling rate (Bali and Pannu 2017). The yield obtained under such conditions has shrinked grains with less individual grain weight and size. So, drought brings a drastic change in phenology and physiology of the crops.

Drought Management Strategies Drought poses a significant threat to our social and economic life by reducing crop productivity. Its impact spans over many sectors of the economy as water is integral part for our survival and it is an important component to produce goods and provide services. So, drought management strategies are to be formulated with a contingency plan for real-time implementation and important interventions in this regard are:

- Tolerant or adaptive varieties
- Short-duration crops
- Land preparation and tillage
- Irrigation management
- ET loss management through different methods
- Rainwater management

The process of formulating the implementation strategies in drought management involves preparedness and real-time contingency measures in the fields. Preparedness involves anticipation of timing of drought occurrence which can be gauged from observing past events/meteorological data or from utilizing the local indigenous knowledge. In areas with limited access to early weather warning system and lack of formal education, indigenous knowledge comes in handy in preparing to cope better with abiotic stresses. In a study conducted by Muyambo et al. (2017), several early signs of imminent drought have been documented which help farmers in coping better to drought related stress by adopting suitable agronomic management practices. Along with it land and soil management options like broad bed furrows, compartmental bunding, slope within the fields, trenches etc. also offer advantages in drought years (Table 24.3).

Table 24.3 Indigenous knowledge to mitigate drought risk

Species	Behavior	Description in relation to drought or rain
Snakes	Movement of snakes in one direction	Drought expected
Bees	When bees fly in a certain direction	Drought expected
Frogs	Noise by frogs in afternoon	Drought expected
Horse	Horse jumping playfully	Precipitation expected
Butterflies	Butterflies flying together	Drought with a good farming season

Source: Muyambo et al. (2017)

Table24.4 Recommended cropping pattern depending upon precipitation availability

Annual rainfall (mm)	Cropping pattern to be adopted
350–625	Single crop in kharif
650–750	Intercropping
780–900	Sequential cropping
900 and above	Sequential cropping

Real-time contingency measures include mulching, water conservation techniques, antitranspirants' spray, etc. to avoid ET losses and to conserve moisture for critical stage in crops (Srinivasarao et al. 2013).

24.4 Various Agronomic Interventions for Managing Drought Stress/Risk for Sustainable Production Are

1. *Selection of appropriate cropping system, drought resistant crops and cultivars*: In arid climates, drought preparedness is the only way to survive and sustain. Selection of drought tolerant crops and varieties are important pre-requisites for sustainable crop production in drought-prone areas (Singh et al. 2014). Short duration and less water requiring crops are to be encouraged. Moreover, crops with waxy leaves, awns, narrow leaf structures, etc. should be preferred as they tend to save moisture loss from the plants. Selection of crops like pearl millet, gram, mustard, cotton, sunflower etc. should be encouraged in drought-prone areas to achieve good crop production. Depending upon the amount of rainfall in area, cropping pattern and varieties are to be selected (Tables 24.4, 24.5, and 24.6).

2. *Tillage modification to manage drought stress*: Tillage is mechanical manipulation of soil to prepare land for cultivation of crops. Number of tillage operation, timing, and depth plays an important role from the viewpoint of water conservation. Conventional tillage which is also known as intensive tillage includes multiple tillage operations to incorporate residue into the soil and preparing fine seed bed for fast and better seed establishment, fertilizer incorporation, and per emergence herbicide application. Conversely, conservation tillage or minimum

Table 24.5 Efficient crops in drought-prone areas

Location	Traditional crop	Yield (kg ha^{-1})	Efficient crops	Yield (kg ha^{-1})
Bellary	Cotton	200	Sorghum	2670
Varanasi	Wheat	860	Chickpea	2850
Ranchi	Upland rice	2880	Corn	3360
Indore	Green gram, wheat	1180, 1120	Soybean, safflower	3330, 2420
Agra	Wheat	1030	Rapeseed-mustard	2040
Hissar	Wheat	320	Sativa	1610
Udaipur	Corn	1800	Sorghum	2900
Rewa	Soybean	400	Soybean	1200

Adapted from DA&FW (2016)

Table 24.6 Drought tolerant cultivars released in India

Crop	Drought tolerant cultivars
Rice	Anjali, Vandana, Sahabhagi Dhan, DRR Dhan 42 (IR64 Drt 1), Satyabhama, Birsa Vikas Dhan 203, DRR Dhan 43, Rajendra Bhagwati, Birsa Vikas Dhan 111, JaldiDhan 6
Maize	HM 4, Pusa Hybrid Makka 1, DHM 121, Pusa Hybrid Makka 5, Buland
Wheat	HI 1531, PBW 527, HI 8627, NIAW 1415, K 8962, HD 2888, PBW 644, HD 2987, WH 1080, HD 3043, PBW 396, K 9465, MP 3288, HPW 349, HD 4672
Pearl millet	HHB-226, HHB 67 improved, Dhanshakti, GHB 757, GHB 719, Pusa Composite 443, HHB 234, Mandor Bajra Composite 2, RHB-177
Sorghum	CSH 19 R, CSV 18, CSH 15R
Chickpea	Vijay, RSG 14, Vikas, ICCV 10, RSG 888, Pusa 362, Vijay
Barley	RD 2660, K603
Groundnut	TAG-24, Ajaya, ICGV 91114, Girnar 1, Kadiri 6
Sugarcane	Co 86032, Co 98014 (Karan-1), Co 0238, Co 0403, Co 0239, Co 0118, Co 06927
Soybean	JS 95-60, NRC 7
Cotton	Veena, Raj DH 7, HD 324, CICR-1,Surabhi, Pratap Kapi, Suraj, AK 235

Source: PIB (2015)

tillage involves minimum disturbance to soil and retention of residues on the soil surface. The selection of tillage operation and residue management option has profound effect on movement of water (runoff/infiltration/retention/evaporation). In drought stress management, water conservation is the key; therefore, selection of appropriate tillage operation is very crucial to mitigate drought damage. Retention of crop residues on the topsoil surface can improve water infiltration rate due to surface roughness which further decreases the surface runoff and erosion chances (Moore 2015). It has been observed that zero tillage combined with residue retention has potential to trap 70% more water as compared to conventional tillage, which is very crucial in water limited conditions observed during drought (Al-Kaisi 2020). This observation is also reflected in adoption of no till by farmers during drought years where extreme dry conditions favor increased adoption rate of no tillage by farmers whereas wet conditions (floods) do not reflect change in tillage adoption pattern (Ding

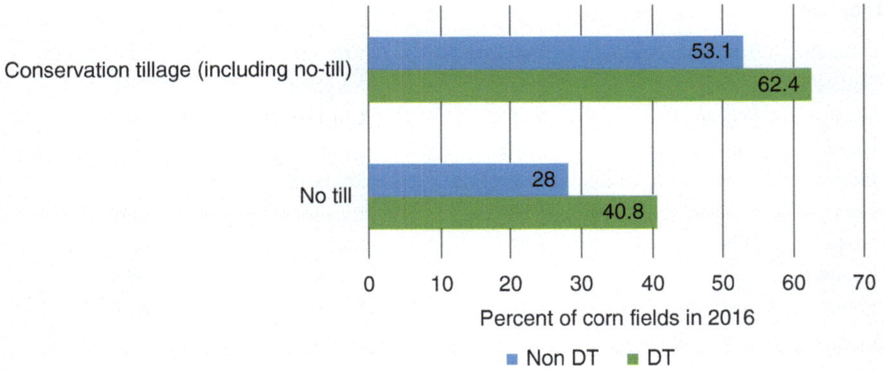

Fig. 24.1 Share of US drought tolerant (DT) and non-DT fields that used conservation tillage and no-tillage in 2016. Source: USDA (2016)

et al. 2009). No till or conservation tillage also compliment where other drought management options are adopted, for example, farmers in USA who planted drought tolerant (DT) maize mostly adopted conservation tillage or no tillage at their farms (Fig. 24.1). Reduced evaporation losses coupled with better water retention may be the reasons for adoption of conservation tillage which improves the chances of survival and better productivity of drought tolerant maize.

3. *Improved irrigation efficiency to manage drought*: India is a dynamic country with many agroecological zones falling into its geography. The increasing population depending on limited natural resources put great pressure on its ecology. Rainfall plays a critical role in Indian agriculture. It is a gamble of monsoon and a good amount of rainfall which ensures high groundwater recharge with proper management. Both surface water and groundwater are used for irrigating the fields. In rainfed areas, it is very important to utilize the water very efficiently. In drought-prone areas, irrigation with utmost precision and high efficiency system is the priority. The efficiency of canal system is less than 40%, while groundwater irrigation efficiency is approximately 50%. Still there is huge water loss that needs to be tapped in drought-prone areas. Hence, high water efficiency systems are to be understood and implemented. Strategies should be made to bring every single drop of water to the plants directly without getting wasted midway. Micro-irrigations systems have very high potential in improving the water efficiency to more than 90% and can provide the critical irrigation to crop in drought scenario. Micro-irrigation system has immense scope for water conservation and attaining higher crop productivity. Crop water requirement can be fulfilled by just providing adequate moisture to the roots. Conversely, excess water applied to the soil works as an early shock to the crop plants due to anoxia and subsequently, when water subsides, then roots are able to absorb the

moisture through capillaries and provide it to whole plant. Through micro-irrigation system, that initial shock to the plants can be avoided and rather drop wise irrigation ensures direct absorption by crop roots. This helps in improving water use efficiency and thus helps in attaining optimum crop productivity (Ashoka et al. 2015) in water limited environments. Extreme climate events like drought, floods, cyclones, heat waves, frost, etc. have a direct impact on agriculture followed by adverse impact on human and animal lives. Loss in crop productivity hampers our food security which has larger impact on the society and economy. With extremities in temperature and rainfall pattern in India, it is very important to implement strategies with extensive planning which give a wider scope of adaption to the farmers.

4. *Increase in soil organic matter*: A large amount of scientific evidence shows that organic matter is the most important trait in making soils more resistant to drought and able to cope better with less and more erratic rainfall (Bot and Benites 2005; Lal 2008; Pan et al. 2009; Riley et al. 2008). To minimize the impact of drought, soil needs to capture the rainwater that falls on it, store as much of that water as possible for future plant use, and allow for plant roots to penetrate and proliferate. Problems with or constraints on one or several of these conditions cause soil moisture to be one of the main limiting factors for crop growth. The capacity of soil to retain and release water depends on a broad range of factors such as soil texture, soil depth, soil architecture (physical structure including pores), organic matter content, and biological activity. However, appropriate soil management can improve this capacity. Practices that increase soil moisture content can be categorized into three groups: (1) those that increase water infiltration; (2) those that manage soil evaporation; and (3) those that increase soil moisture storage capacities. All three are related to soil organic matter. And this organic matter can be efficiently increased through crop residue mulching. Let us look how crop residue mulching can perform all these three practices. The crop residue amount increasing on the soil surface reduces the evaporation rate (Gill and Jalota 1996; Prihar et al. 1996). Consequently, the application of crop residue is the best practice to add organic amendment in soil and cover it surface. For obtaining sustainable development, crop residue properly manages to simultaneously increase soil organic carbon, soil nutrients, water availability and productivity requirement as well as livestock fodder. The availability of plant water content was significantly lower in conventional tillage as compared to zero tillage under rice wheat cropping system as reported by Bhattacharyya et al. (2006, 2008). Increased levels of organic matter and associated soil fauna lead to greater pore space with the immediate result that water infiltrates more readily and can be held in the soil (Roth 1985). The improved pore space is a consequence of the bioturbating activities of earthworms and other macro-organisms and channels left in the soil by decayed plant roots. Organic matter contributes to the stability of soil aggregates and pores through the bonding or adhesion properties of organic materials, such as bacterial waste products, organic gels, fungal hyphae, and worm secretions and casts. Moreover, organic matter intimately mixed with mineral soil materials has

a considerable influence in increasing moisture holding capacity. Especially in the topsoil, where the organic matter content is greater, more water can be stored. Hudson (1994) showed that for each 1-percent increase in soil organic matter, the available water-holding capacity in the soil increased by 3.7%. Soil water is held by adhesive and cohesive forces within the soil and an increase in pore space will lead to an increase in water-holding capacity of the soil. As a consequence, less irrigation water is needed to irrigate the same crop. In general, a 1% increase in SOM content increased AWHC, on average, up to 1.5% times its weight, depending on soil texture and clay mineralogy. These values were consistent with the theoretical calculations that showed that the potential AWHC increase (on a volumetric basis) from a unit increase in SOM (% weight) is about 1.5–1.7% for the 0–8% SOM range. This equates to 10,800 L of water for each additional 1% increase in SOM (up to 8% SOM) for a layer thickness of 15 cm covering 0.4 ha area (an acre furrow slice).

5. *Residue retension/mulching*: Evapotranspiration losses are major water losses from crop lands (Morison et al. 2008). So, under drought, it becomes very important to minimize the losses so that all the water should be diverted to crop plants. Land without plantation is exposed to high radiation and thus evaporation losses are high in such areas while if temperature is high and relative humidity is low then transpiration losses from the plants will be high. To avoid evaporation losses, mulching is very important mitigation strategy. It improves water regulation in the plant system, reduces runoff, reduces weed infestation, and regulates the soil temperature (Waraich et al. 2011; Kazemia and Safaria 2018). Soil cover protects the soil against the impact of raindrops, prevents the loss of water from the soil through evaporation, and also protects the soil from the heating effect of the sun. Soil temperature influences the absorption of water and nutrients by plants, seed germination and root development, as well as soil microbial activity and crusting and hardening of the soil. Mulching with crop residues or cover crops regulates soil temperature. The soil cover reflects a large part of solar energy back into the atmosphere and thus reduces the temperature of the soil surface. This results in a lower maximum soil temperature in mulched compared with unmulched soil.
6. *Biodiversity*: Under present and future scenarios of a changing climate, farmers' reliance on crop diversity is particularly important in drought-prone areas where irrigation is not available. Diversity allows the agroecosystem to remain productive over a wider range of conditions, conferring potential resistance to drought (Naeem et al. 1994). In the dry-hot habitats of the Middle East, some wild wheat cultivars have an extraordinary capacity to survive drought and make highly efficient use of water, performing especially well under fluctuating climates (Peleg et al. 2009). Researching the diversity and drought coping traits of wild cultivars provides scientists with new tools to breed crops better adapted to less rainfall. In Italy, a high level of genetic diversity within wheat fields on nonirrigated farms reduces the risk of crop failure during dry conditions. In a modeling scenario, where rainfall declines by 20%, the wheat yield would fall sharply, but when diversity is increased by 2% not only is this decline reversed,

above average yields can also be achieved (Di Falco and Chavas 2006, 2008). In semi-arid Ethiopia, growing a mix of maize cultivars in the same field acts like an insurance against dry years. Fields with mixed maize cultivars yielded about 30% more than pure stands under normal rainfall years but outperformed with 60% more yield than monocultures in dry years (Tilahun 1995). Crop genetic diversity is the richness of different genes within a crop species. This term includes diversity that can be found among the different varieties of the same crop plant (such as the thousands of traditional rice varieties in India), as well as the genetic variation found within a single crop field (potentially very high within a traditional variety, very low in a genetically engineered or hybrid rice field). Cropping diversity at the farm level is the equivalent to the natural species richness within a prairie, for example. Richness arises from planting different crops at the same time (intercropping a legume with maize, for example), or from having trees and hedges on the farm (agroforestry). Farm-level cropping diversity can also include diversity created over time, such as with the use of crop rotations that ensure the same crop is not grown constantly in the same field. Farm diversity at the regional level is the richness at landscape level, arising from diversified farms within a region. It is high when farmers in a region grow different crops in small farms as opposed to large farms growing the same cash crop (for example, large soya monocultures in Argentina).

7. *On-farm rainwater harvesting (RWH) pond*: There are numerous research works which show that drought/water scarcity is very harmful for crop production. But nowadays climate change has worsen the situation even in the areas where water scarcity was not so much. Climate change has significantly affected the livelihood and income of farmers across the globe (IPCC 2014). Rainwater harvesting (RWH) in ponds can be a promising way to include resilience in the system against water scarcity and climate change. RWH in this context can be described as a method of inducing, collecting, storing, and conserving local surface runoff for agriculture production (Ibraimo 2007). This harvested water can be easily used without significant treatment (Nolan and Lartigue 2017). Farm pond technology has the potential to increase availability of water for supplemental irrigation, increase in cropped area and productivity leading to increase in net returns from crops (Rao et al. 2017). Nearly 25–30% of crop productivity may enhance through farm pond intervention as harvested rainwater available for providing one or two protective irrigations to crops at critical growth stages during dry spells and droughts (Dupdal et al. 2020). Rainwater buffer tank significantly reduces the runoff peak flow hence had the capacity to protect against the adverse effects of flood such as damage to properties and loss of life (Qin et al. 2019).

References

Abido WAE, Zsombik L (2018) Effect of water stress on germination of some Hungarian wheat landraces varieties. Acta Ecol Sin 38:422–428

Ali A, Elozeiri AA (2017) Metabolic processes during seed germination. Intech Open, London, p 141

Al-Kaisi M (2020) Residue management consideration for this fall [WWW Document]. Iowa State Univ. Ext. Outreach. https://crops.extension.iastate.edu/blog/mahdi-al-kaisi/residue-management-consideration-fall

Ashoka P, Kadasiddappa MM, Sanjey MT (2015) Enhancing water productivity through microirrigation technologies in Indian agriculture. Ann Plant Soil Res 17:601–605

Bali A, Pannu RK (2017) Quality parameters and yield of wheat as influenced bysowing dates and irrigation levels under semi-arid conditions of Hisar. Ann Agric Res 39(1):20–25

Basin Planning Directorate, CWC (2019) XI Plan Document. Report of the standing sub-committee on "Assessment of availability and requirement of water for diverse uses-2000". Accessed 23 Mar 2019

Bhat NR, Suleiman MK, Abdal M (2009) Selection of crops for sustainable utilization of land and water resources in Kuwait. World J Agric Sci 5:201–206

Bhattacharyya R, Singh RD, Chandra S, Kundu S, Gupta HS (2006) Effect of tillage and irrigation on yield and soil properties under rice (Oryza sativa)-wheat (Triticum aestivum) system on a sandy clay loam soil of Uttaranchal. Indian J Agric Sci 76(7):405–409

Bhattacharyya R, Kundu S, Pandey SC, Singh KP, Gupta HS (2008) Tillage and irrigation effects on crop yields and soil properties under the rice–wheat system in the Indian Himalayas. Agric Water Manag 95(9):993–1002

Bot A, Benites J (2005) Drought-resistant soils. Optimization of soil moisture for sustainable plant production. Food and Agriculture Organization of the United Nations. Land and Water Development Division

Christophe S, Jean-Christophe A, Annabelle L, Alain O, Marion P, Anne-Sophie V (2011) Plant N fluxes and modulation by nitrogen, heat and water stresses: a review based on comparison of legumes and non-legume plants. In: Shanker A, Venkateswarlu B (eds) Abiotic stress in plants-mechanisms and adaptations. IntechOpen, Rijeka, pp 79–118

DA&FW (2016) Manual for drought management by Department of Agriculture & Farmer Welfare

Di Falco S, Chavas J-P (2006) Crop genetic diversity, farm productivity and the management of environmental risk in rainfed agriculture. Eur Rev Agric Econ 33:289–314

Di Falco S, Chavas J-P (2008) Rainfall shocks, resilience, and the effects of crop biodiversity on agroecosystem productivity. Land Econ 84:83–96

Ding Y, Schoengold K, Tadesse T (2009) The impact of weather extremes on agricultural production methods: does drought increase adoption of conservation tillage practices? J Agric Resour Econ 34:395–411

Dupdal R, Patil SL, Naik BS, Ramesha MN (2020) Role of farm pond in improving productivity and farm income in dryland area. Life Sci Leaflets 128(2020):09–14

Fahad S, Bajwa AA, Nazir U, Anjum SA, Farooq A, Zohaib A, Sadia S, Nasim W, Adkins S, Saud S, Muhammad Ihsan Z, Alharby H, Wu C, Wang D, Huang J (2017) Crop production underdrought and heat stress: plant responses and management options. Front Plant Sci 8(1147): 1–16

Farooq M, Wahid A, Kobayashi N, Fujita D, Basra SMA (2009) Plant drought stress: effects, mechanisms and management. Agron Sustain Dev 29:185–212

Fathi A, Tari DB (2016) Effect of drought stress and its mechanism in plants. Int J Life Sci 10(1): 1–6

Gill BS, Jalota SK (1996) Evaporation from soil in relation to residue rate, mixing depth, soil texture and evaporativity. Soil Technol 8(4):293–301

Hudson BD (1994) Soil organic matter and available water capacity. J Soil Water Conserv 49(2): 189–194

Hussain HA, Hussain S, Khaliq A, Ashraf U, Anjum SA, Men S, Wang L (2018) Chilling and drought stresses in crop plants: implications, cross talk, and potential management opportunities. Front Plant Sci 9(393):1–21

Ibraimo NM (2007) Rainwater harvesting technologies for small scale rainfed agriculture in arid and semi-arid areas. Int J Educ Res 1(6):1–3

IPCC (2014) Climate change impacts, adaptation, and vulnerability. Part B: regional aspects. Contribution of Working Group II to the Fifth Assessment Report of the Intergovernmental Panel on Climate Change. Cambridge University Press, Cambridge, UK and New York, NY, USA, pp 688. Sweetbriar Brook, Ampthill, Storbritannien, p 66

Kazemia F, Safaria N (2018) Effect of mulches on some characteristics of a drought tolerant flowering plant for urban landscapes. Desert 23(1):75–84

Lal R (2008) Managing soil water to improve rainfed agriculture in India. J Sustain Agric 32:51–75

Lobell DB (2009) Climate extremes and crop adaptation. Summary statement from a meeting at the program on Food Security and Environment, Stanford, CA, held on June 16-18, 2009

Marjani A, Farsi M, Rahimizadeh M (2016) Response of chickpea (Cicer arietinum L.) genotypes to drought stress at different growth stages. Int J Agric Biosyst Eng 10(9):1–5

Minhas PS et al (2017) Abiotic stress management for resilient agriculture. Springer, Berlin. https://doi.org/10.1007/978-981-10-5744-1_1

Molina R, Lopez-Santos C, Gomez-Ramirez A, Vilchez A, Espinos JP, Gonzalez-Elipe AR (2018) Influence of irrigation conditions in the germination of plasma treated nasturtium seeds. Sci Rep 8:16442

Moore LE (2015) Tillage, residue management and their effect on soil moisture. Drought Manage Factsheet 2

Morison JIL, Baker NR, Mullineaux PM, Davies WJ (2008) Improving water use in crop production. Philos Trans R Soc B 363:639–658

Muyambo F, Bahta YT, Jordaan AJ (2017) The role of indigenous knowledge in drought risk reduction: a case of communal farmers in South Africa. Jamba 9:420. https://doi.org/10.4102/jamba.v9i1.420

Naeem S, Thompson LJ, Lawler SP, Lawton JH, Woodfin RM (1994) Declining biodiversity can alter the performance of ecosystems. Nature 368:734–737

Nolan, Lartigue (2017) Evaluation of the performance of rainwater harvesting systems for domestic use in Tlalpan, Mexico City. Civil Eng J. https://doi.org/10.28991/cej-2017-00000080

Pan G, Smith P, Pan W (2009) The role of soil organic matter in maintaining the productivity and yield stability of cereals in China. Agric Ecosyst Environ 129:344–348

Pandey G, Yadav L, Tiwari A, Khatri HB, Basnet S, Bhattarai K, Gyawali B, Rawal N, Khatri N (2017) Analysis of yield attributing characters of different genotypes of wheat in Rupandehi, Nepal. Int J Environ Agric Biotechnol 2(5):2374–2379

Peleg Z, Fahima T, Krugman T, Abbo S, Yakir D, Korol AB, Saranga Y (2009) Genomic dissection of drought resistance in durum wheat x wild emmer wheat recombinant inbreed line population. Plant Cell Environ 32:758–779

PIB (2015) Press information bureau [WWW Document]. Press release by PIB, Gov. India, Minist. Agric. Farmers Welf. https://pib.gov.in/newsite/PrintRelease.aspx?relid=123999. Accessed 11 August 2022

Prihar SS, Jalota SK, Teiner. (1996) Residue management for reducing evaporation in relation to soil type and evaporativity. Soil Use Manag 12:150–157

Qin Y, Huang Z, Yu Z, Liu Z, Wang L (2019) A novel buffer tank to attenuate the peak flow of runoff. Civil Eng J 5(12):2525–2532

Rao S, Rejani R, Rama Rao CA, Rao KV, Osman M, Srinivasa Reddy K, Kumar M, Kumar P (2017) Farm ponds for climate-resilient rainfed agriculture. Curr Sci 112(3):471–477

Riley H, Pommeresche R, Eltun R, Hansen S, Korsaeth A (2008) Soil structure, organic matter and earthworm activity in a comparison of cropping systems with contrasting tillage, rotations, fertilizer levels and manure use. Agric Ecosyst Environ 124:275

Roth CH (1985) Infiltrabilität von Latosolo-Roxo-Böden in Nordparaná, Brasilien, in Feldversuchen zur Erosionskontrolle mit verschiedenen Bodenbearbeitungs-systemen und Rotationen. Gött. Bod. Ber. 83:1–104

Shareef M, Gui D, Zeng F, Ahmed Z, Waqas M, Zhang B, Iqbal H, Fiaz M (2018) Impact of drought on assimilates partitioning associated fruiting physiognomies and yield quality attributes of desert grown cotton. Acta Physiol Plant 40:71

Silva EC, Nogueira RJMC, Silva MAM, Albuquerque B (2011) Drought stress and plant nutrition. Plant Stress 5(1):32–41

Singh L, Beg MKA, Akhter S, Qayoom S, Lone BA, Singh P, Singh P (2014) Efficient techniques to increase water use efficiency under rainfed eco-systems. J Agric Search 1(4):193–200

Srinivasarao C, Chary R, Mishra G, Nagarjuna PK, Maruthi S, Venkateswarlu B, Sikka AK (2013) Real time contingency planning: initial experiences from AICRPDA. All India CRIDA, ICAR, Hyderabad

Tilahun A (1995) Yield gain and risk minimization in maize (Zea mays) through cultivar mixtures in semi-arid zones of the Rift Valley in Ethiopia. Exp Agric 31:161–168

Trabelsi L, Gargouria K, Hassena AB, Mbadra C, Ghrab M, Ncube B, Staden JV, Gargour R (2019) Impact of drought and salinity on olive water status and physiological performance in an arid climate. Agric Water Manag 213:749–759

USDA (2016). No-till and conservation tillage practices are more common on fields planted with drought-tolerant corn [WWW Document]. USDA Natl. Stat. Serv. https://www.ers.usda.gov/data-products/chart-gallery/gallery/chart-detail/?chartId=93264. Accessed 11 November 2022

Waraich EA, Ahmad R, Ashraf MY, Saifullah AM (2011) Improving agricultural wateruse efficiency by nutrient management in crop plants. Acta Agric Scand B Soil Plant Sci 61(4): 291–304

Weijde TVD, Huxley LM, Hawkins S, Sembiring EH, Farrar K, Dolstra O, Visser RGF, Trindade LM (2017) Impact of drought stress on growth and quality of miscanthus for biofuel production. Glob Change Biol Bioenergy 9:770–782

Yazdani F, Allahdadi I, Abas AG (2007) Impact of superabsorbent polymer on yield and growth analysis of soybean (Glycine max L.) under drought stress condition. Pak J Biol Sci 10:4190–4196

Nanoweapons to Fight with Salt and Drought Stress

25

Prinka Goyal and Norah Johal

Abstract

Plants face various abiotic and biotic stresses throughout their life cycle which adversely affect plant growth development and ultimately yield. Drought, salinity, and heat stress are most prevalent abiotic stresses, threatening the global food security. Plants fight with these stresses by altering their physiological, molecular, and biochemical pathways but stress sensitive plants are unable to cope with stresses. In this regard, exploration of some novel strategies and their exploitation are need of the hour to mitigate stress and improve yield. Nanoparticles emerged as magic bullets for agriculturists, farmers, and scientists to improve plant performance under stress conditions. Several studies have depicted the use of nanoparticles in mitigation of abiotic stresses to enhance crop productivity. The size of these particles ranges from 1 to 100 nm, and are available in the form of plant growth promoters, herbicides, pesticides as well as fertilizers. Several reports showed that application of inorganic and/or organic nanoparticles confer tolerance in plants against stresses. These particles enhance plants tolerance by modulating their physiological, biochemical and molecular routes as well as their gene expression. Nanoparticles minimize oxidative stress by enhancing the radical scavenging potential and antioxidant enzymatic and non-enzymatic activities of plants. These particles crosstalk with various plant hormones to make plants thrive under stress. Thus, supplementation of nanoparticles emerged as novel strategy to improve plant tolerance.

P. Goyal (✉)
Department of Botany, Punjab Agricultural University, Ludhiana, India
e-mail: prinkagoyal2015@pau.edu

N. Johal
Department of Agronomy, Punjab Agricultural University, Ludhiana, India

A. Kumar et al. (eds.), *Salinity and Drought Tolerance in Plants*,
https://doi.org/10.1007/978-981-99-4669-3_25

Keywords

Abiotic stress · Salinity · Drought · Nanoparticles

25.1 Introduction

Plants encounter with various abiotic (natural or anthropogenic stress) and biotic stresses throughout their course of life. Among abiotic stresses, temperature stress (heat and cold stress), water stress (drought stress and flooding), and salt stress are major threats for sustainable crop production (Fig. 25.1) (Zhang et al. 2020). These stresses modulate plants at morphological, physiological, anatomical, biochemical, and molecular level temporarily or permanently which hamper plant growth and development and ultimately reduce crop yield. Plants exposed to salt stress experience osmotic stress (water deficit) during initial stage of stress followed by hyperionic stress (high concentration of ions Na^+ and Cl^- in cytosol) as well as oxidative stress (production of reactive oxygen species) at later stage of salt stress. The adverse effect of osmotic stress (physiologically dry soil) due to salinity on plants is similar to drought stress (physical dry soil). Oxidative stress occurs due to production of reactive oxygen species (superoxide radicals ($O_2^{\cdot -}$), singlet oxygen ($^1O_2$), hydrogen peroxide (H_2O_2), hydroxyl radical ($OH^{\cdot -}$), hydroperoxyl radical ($HO_2^{\cdot}$), alkoxy radical ($RO^{\cdot}$), peroxyl radicals ($ROO^{\cdot}$), and excited carbonyl (RO)) which leads to disruption of cell membrane and ultimately cell death (Hasanuzzaman et al. 2013). Ionic stress leads to accumulation of sodium ions (Na^+) in cytosol, which disturbs ion homeostasis by disturbing uptake of potassium (K^+) and calcium ions (Ca^+) which are essential macroelements for plant growth and development. It also leads to premature leaf fall and affects crop yield. Drought stress (low moisture content in soil) causes drastic loss in crop yield in arid regions. It negatively affects plant growth and development, one of them is reduction in leaf expansion (decrease in photosynthetic area) due to decrease in water content of leaf. Other effects include production of reactive oxygen species, decrease in leaf area, and in severe conditions wilting and ultimately plant death. Plants either fight with these stresses or avoid it to nullify the negative effects of stresses on themselves. The plants which fight with stresses are called stress tolerant plants, and the ability of plants to fight with stresses is known as their tolerance. The plants having low tolerance for stresses are known

Fig. 25.1 Major stresses faced by plants during its life

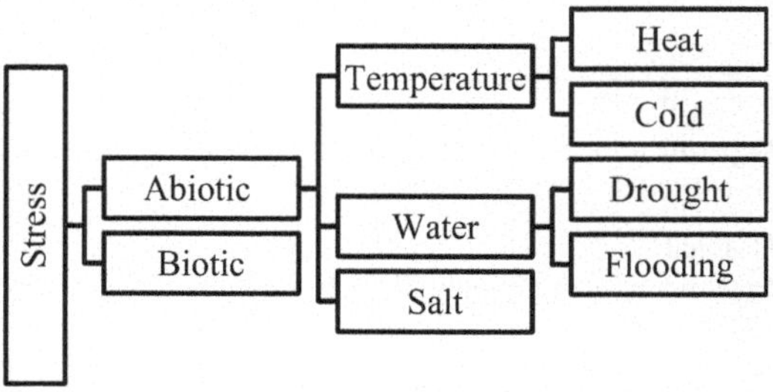

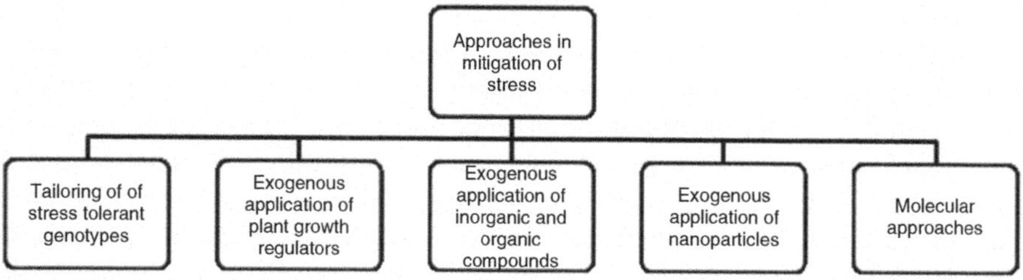

Fig. 25.2 Approaches in mitigation of drought and salt stress

as stress sensitive plants. The plants which avoid stresses are called escapers. Halophytes have in-built mechanisms to fight with salt stress (tolerance mechanism). These are called salt stress tolerant plants. Ephemeral plants complete their life cycle before the onset of dry period (avoidance mechanism). These are known as drought escapers.

The various strategies are sought and adopted to cope with stresses or to improve plant tolerance against stresses (Fig. 25.2). One of them is tailoring of stress tolerant genotypes with high grain yield followed by their screening under natural stress conditions in fields. The development of tolerant varieties is time consuming and labour-intensive with limited success. The screening of these genotypes for their tolerance against stress is extra burden for scientists and researchers. The degree of tolerance varies among different genotypes. So alternative approaches are sought, and applied for mitigation of these stresses, and maximize yield potential of crops. The most prevalent approach is upgradation of plant tolerance mechanism against stresses by exogenous application of plant growth regulators, osmoprotectants, biofertilizers, inorganic and organic compounds as well as nanoparticles. The utilization of nanotechnology in agriculture to cope with stress as well as for sustainable crop production drew attention of scientists, researchers, agriculturists, and farmers (Heidarabadi 2022).

25.2 What Are Nanoparticles?

The term nanotechnology was first earned by scientist Norio Taniguichi, a professor at Tokyo University of Science, in 1974, and this technology came into limelight during twentieth century (Khan and Rizvi 2014) and attained prominent position in field of science, engineering, medical, agriculture, and industries in the twenty-first century. The nanomaterials (NMs) are proved as Kohinoor gems in the field of agriculture. These are used in the form of plant growth promoters or stimulants, pesticides, herbicides, fertilizers, pesticides carriers, and plant growth regulators. Plant growth stimulants nanoparticles are employed for alleviation of negative effects of stress by modulation of their tolerance mechanisms at physiological, biochemical as well as at molecular level. The dimensions of NMs range from 1 nm to 100 nm with large surface-to-volume ratio, and these are available in one dimension (surface films), two dimensions (strands and films), and three dimensions (nanoparticles (NPs)). On the basis of mode of synthesis, the NPs are of three types: natural, incidental, and synthetic (engineered). The synthetic NPs are synthesized either from bulk chemicals (top-down approach) or from atoms or ions (bottom-up approach) by physical, chemical, and biological methods (using microorganisms and plants (green factories)). The NPs prepared using plant extracts are called green nanoparticles, and this is cost-effective and eco-friendly approach for synthesis of NPs. The synthesis of nanoparticles using pant extract is called green synthesis or phytosynthesis (not confused with photosynthesis), e.g. synthesis of selenium NPs using stem extract of *Leucas lavandulifolia* (Kirupagaran et al. 2016). On the basis of their nature, NPs are of two types: (a) inorganic NPs and (b) organic NPs. a) Inorganic NPs are further divided into two groups:

1. Metallic and metallic oxides NPs (Au NPs, Ag NPs, CuO, ZnO, Fe_2O_3)
2. Metalloid and metalloid oxides NPs (SiO_2 NPs)

b) Carbon nanoparticles (quantum dots, carbon nanotubes, Fullerenol NPs ($C_6O(OH)_{24}$)) are involved in the category of organic NPs (Taran et al. 2017).

25.3 Inorganic Nanoparticles

25.3.1 Silver NPs (Ag NPs)

Hojjat (2016) observed positive effect of AgNPs in lentils exposed to drought stress. The application of Ag NPs enhanced germination rate, germination percentage, root length, root fresh, and dry biomass of lentils under drought stress.

25.3.2 Gold NPs (Au NPs)

The gold nanoparticles are included in the category of metal nanoparticles. The potential role of Au NPs in inducing tolerance against salinity was observed in wheat by Wahid et al. (2022). These NPs minimized oxidative stress by modulating activities of antioxidant enzymes. The improvement in plant growth attributes under saline conditions with application of Au NPs was also reported (Wahid et al. 2022).

25.3.3 Zinc Oxide NPs (ZnO NPs)

Zinc is included in the category of essential microelement. In tomato, the role of these NPs to cope with salt stress was observed by Faizan et al. (2021). The tomato seedlings were subjected to salt stress (150 mM), and foliar application of ZnO NPs at concentration of 10, 50, and 100 mg/L was given at 25 DAS (days after sowing).

The application improved tomato plant growth in terms of shoot length, root length as well as fresh and dry biomass and leaf area under salt stress. The physiological attributes (photosynthesis rate, chlorophyll content, and carotenoid content) as well as biochemical parameters (antioxidant enzymes and protein content) also improved with foliar spray of them. Semida et al. (2021) also studied the role of ZnO NPs in stress (water deficit) tolerance in eggplant. The foliar application of NPs alleviated the negative effects of drought stress by improving plant growth, membrane stability, photosynthetic rate, water productivity as well as yield. Application of 50 and 100 ppm ZnO NPs improved fruit yield by 12.2% and 22.6%, respectively of plants exposed to water stress, in comparison with plants grown in fully irrigated soil.

25.3.4 Copper Oxide NPs (CuO NPs)

Copper is one of the microelements in plants. CuO NPs are included in the category of inorganic metal oxide nanoparticles. These NPs amplified drought tolerance in maize when 12-day old seedlings were treated with it. The grain yield reduced significantly under water deficit conditions but application of CuO NPs increased grain yield as well as seed number. It ameliorated the negative effects of water deficient stress on maize yield by enhancing chlorophyll and carotenoid content, which directly increased the photosynthesis rate and grain yield. Significant increase in level of antioxidant enzymes under drought stress with application of CuONPs was also reported. These enzymes fight with reactive oxygen species and minimize the effects of drought stress on plants (Van Nguyen et al. 2022).

25.3.5 Titanium Oxide NPs (TiO_2 NPs)

Titanium oxide is one of the inorganic nanoparticles. Various studies reported the application of TiO_2 NPs in mitigation of salt and drought stress by improving their growth, and enhanced yield. Dawood et al. (2019) fertigated the water deficit soil with these NPs and assessed the performance of four wheat cultivars (Sohag 3, Benisuif 5, Sakha 93, and Sed 12). Under control conditions (water deficit soil), wheat growth in terms of leaf traits is negatively affected. TiO_2 NPs promoted photosynthetic rate, improved leaf health, increased leaf chlorophyll content, LAI (leaf area index) as well as leaf growth (thicker and heavier leaves) and decreased leaf aging.

In other study, the positive effect of foliar application of these NPs under water deficit condition was reported in Dragon head (*Dracocephalum moldavica* L.) (Mohammadi et al. 2016). Water deficit conditions damage cell membrane and lead to oxidative stress. TiO_2 NPs application at different concentrations (10 and 40 ppm) ameliorated effect of water stress by increasing proline content and reducing hydrogen peroxide and MDA content (MDA content is indicator of degree of membrane damage). The stabilization of membrane was also reported by Sompornpailin and Chayaprasert (2020) in *Nicotiana tabacum*. The increase in level of enzymatic and nonenzymatic antioxidants in *Vicia faba* was reported (Khan et al. 2020). Shariatzadeh Bami et al. (2021) reported that application of these NPs at different concentration 10, 20, and 30 ppm modulated molecular pathway in *Artemisia absinthium* L. (a herb), grown under salt stress conditions. The degree of expression of two genes ADS and DBR2 (key gene in biosynthesis pathway of artemisinin) was noted under salt stress with application of NPs. The expression of ADS gene recorded maximum in plants, sprayed with 30 ppm NPs followed by 20 and 10 ppm with salinity stress 50 ppm. While expression of DBR gene showed reverse trend. The maximum expression was observed in plants grown under control conditions (no application of NPs) followed by plants with NPs application. The alternation in biochemical pathways for salt stress tolerance with NPs was reported in *Artemisia absinthium* L. With application of TiO_2 NPs, a significant increase in antioxidant enzymes catalase, peroxidase, superoxide dismutase, polyphenol oxidase, and guaiacol as well as protein concentration was reported as compared to control. These enzymes engulf reactive oxygen species synthesized during later phase of salt stress (Shariatzadeh Bami et al. 2021).

25.3.6 Iron Oxide NPs (Fe_2O_3 NPs)

Iron is one of the micronutrients in plants. It acts as cofactors for various enzymes. It exists in two forms: Fe^{2+} and Fe^{3+} form. The maghemite (yttrium doping-stabilized γ-Fe_2O_3NPs) are used as fertilizers and improve plant growth and development, but their role in drought stress tolerance using maghemite as nanozyme was studied by Palmqvist et al. (2017) in rapeseed. The increase in catalase activity (decrease in H_2O_2 content) as well as membrane stability (decrease in lipid peroxidation) with application of NPs contributed for stress tolerance in these plants. The role of Fe_2O_3 NPs as nanozyme against salt stress tolerance by improving their growth parameters as well as modulating their gene activity in *Eucalyptus tereticornis* was studied by Singh et al. (2022).

25.3.7 Cerium Oxide NPs (Ce_2O_3 NPs)

Cerium oxide NPs act as weapon to fight with stress. The application of these NPs in soil at 500 mg/kg soil, equipped the roots of *Brassica napus* with large apoplastic barriers which led to reduction in transport of Na^+ ions in shoot as well as their accumulation. This provided physiological tolerance to plants against salt stress (Rossi et al. 2017).

25.4 Organic Nanoparticles

25.4.1 Chitosan Nanoparticles (CSNPs)

Chitosan polysaccharide is deacetylated form of chitin, hydrophilic in nature. The use of chitosan NPs in soil, in hydroponics and through foliar spray equipped the plants with protective mechanisms against stress. In periwinkle (*Catharanthus roseus*), 1 g/L chitosan NPs improvised plant tolerance against drought stress (50% FC) (Ali et al. 2021). The drought mediated oxidative stress was also minimized by increasing the activities of enzymes catalase and ascorbate peroxidase. Membrane disruption during salt stress and drought stress leads to leakage of ion. This leakage is indication of membrane stability. The malondialdehyde (MDA) content also indicates membrane stability. In chitosan NPs treated plants, MDA content decreased which is indication of membrane integrity under stress.

Hassan et al. (2021) exposed the same plant to salt stress (NaCl 150 mM), and applied 1% CSNPs as foliar spray. The application of NPs helped the plants to cope with salt stress by modulating their antioxidant enzyme activities (catalase, glutathione reductase, and ascorbate peroxidase) as well as gene expression of MAPK3 (mitogen-activated protein kinases), GS (geissoschizine synthase), and ORCA3 (octadecanoid-derivative responsive AP2-domain). These genes are related to biosynthesis of alkaloids which improve plant tolerance against stress.

25.4.2 Nanoparticles in Alleviation of Salt and Drought Stress

NPs are used as weapons to fight with stress, and these nullify or minimize the deteriorative effects of stresses on plants. These are employed either in soil or through foliar spray. From soil, these particles enter inside the plants through root and from leaves these enter through stomata.

In the plants, NPs alter their morphological, physiological, and biochemical as well as molecular states positively and negatively. Most of the studies reported positive modulation in their tolerance mechanisms against drought and salt stress and improved plant growth and development and ultimately production (Ali et al. 2021; Zulfiqar and Ashraf 2021).

25.4.3 Nanoparticles in Alleviation of Salt Stress

The protective role of ZnO NPs in tomato against salt stress was reported by Hosseinpour et al. (2020). In other study, salt stress negatively affected growth parameters (plant height, number of leaves, fresh and dry biomass of root and shoot) of *Moringa peregrina* but application of Hoagland solution containing ZnO and Fe_3O_4 NPs improved growth attributes as well as biochemical parameters of plants under normal and saline conditions (Soliman et al. 2015). Various reports on the exogenous application of NPs to mitigate salt stress are depicted in Table 25.1.

Table 25.1 List of nanoparticles used in amelioration of salt stress in plants

Sr. No.	Type of nanoparticles used	Plant/crop	Dose of nanoparticles used and mode of application	Level of salinity	References
1	Ag NPs	Wheat	300 ppm, foliar application	100 mM NaCl	Wahid et al. (2022)
2	ZnO NPs	Rice	50 mg/L with hydroponic technique	60, 80, and 100 mM NaCl	Singh et al. (2022)
3	Si NPs	Tomato	0.5, 1, 2, and 3 mM, seed treatment	150 and 200 mM NaCl	Almutairi (2016)
4	Fe_2O_3 and ZnO NPs (alone and in combination)	Wheat	2 g/L, foliar spray	0, 75, and 150 mM NaCl	Fathi et al. (2017)
5	Cu NPs absorbed on 1 g of Cs-PVA hydrogel	Tomato	10 mg application of hydrogel (having Cu NPs) in soil	100 mM NaCl	Hernández-Hernández et al. (2018)
6	Chitosan NPs	Bean	0.1%, 0.2%, and 0.3% dissolved in 0.1% HCl, seed treatment	0, 50, 100, and 150 mM NaCl	Zayed et al. (2017)
7	Poly(acrylic acid) coated cerium oxide nanoparticles (PNC) and DiI-PNC	Cotton	0.9 mM, foliar delivery	200 mM NaCl	Liu et al. (2021)
8	TiO_2 NPs	Moldavian balm	0, 50, 100, and 200 mg/L with hydroponic technique	0, 50, and 100 mM NaCl	Gohari et al. (2020)
9	K_2SO_4 NPs	Alfa alfa	1/10, 1/8, and 1/4 of the full K rate in Hoagland solution with hydroponic technique	0 and 6 dS m^{-1} using $CaCl_2$ $2H_2O$: NaCl (2:1)	El-Sharkawy et al. (2017)
10	Se-NPs	Wheat	50, 75, and 100 mg/L, seed treatment	50, 100, and 150 mM NaCl	Ghazi et al. (2022)
11	TiO_2 NPs	Maize	40,60, and 80 ppm, seed treatment	200 mM NaCl	Shah et al. (2021)
12	TiO_2 NPs	*Artemisia absinthium* L.	0, 10, 20, 30 mg/L, foliar spray	0, 50, 100, and 150 mM NaCl	Bami et al. (2022)

(continued)

Table 25.1 (continued)

Sr. No.	Type of nanoparticles used	Plant/crop	Dose of nanoparticles used and mode of application	Level of salinity	References
13	ZnO NPs	Okra	10 mg/L, foliar application	0, 10, 25, 50, 75, and 100% sea water	Alabdallah and Hasan (2021)
14	Se-NPs and ZnO NPs	Rapeseed	150 µmol/L and 100 mg/L seed priming	150 mM NaCl	El-Badri et al. (2021)
15	Ag NPs	Wheat	1 mg/L, seed priming	25 and 100 mM NaCl	Abou-Zeid and Ismail (2018)

25.4.4 Nanoparticles in Alleviation of Drought Stress

Yang et al. (2017), Borišev et al. (2016), and Mohammadi et al. (2016) also reported the protective role of CuO and ZnO NPs in *Triticum aestivum*, Fullerenol nanoparticles in *Beta vulgaris* L., and TiO_2 NPs in *Dracocephalum moldavica* L. against drought stress. In wheat, ZnO and CuO NPs interacted with microorganisms in rhizosphere and improved plant growth and development under drought stress. Various reports on the exogenous application of NPs to mitigate drought stress are depicted in Table 25.2.

Table 25.2 List of nanoparticles used in amelioration of drought stress in plants

Sr. No.	Type of nanoparticles used	Plant/crop	Dose of nanoparticles used and mode of application	Level of stress	References
1	Fe NPs	Wheat	25, 50, and 100 mg/kg, soil application	Drought stress (DC, 35% of soil water holding capacity)	Adrees et al. (2020)
2	Ag NPs and Cu NPs	Wheat	10, 20, and 30 mg/L of Ag NPs and 3, 5, and 7 mg/L of Cu NPs, hydroponic	Osmotic stress of -6, -8, and -10 bars using PEG-6000 (polyethylene glycol)	Ahmed et al. (2021)
3	Ag NPs	Eggplant	0.1, 0.2, 0.5 μmol, foliar application	80% Field Capacity (FC) as control, 50% FC, 35% FC, and 20% FC after the plant establishment	Alabdallah and Hasan (2021)
4	Chitosan NPs	Barley	30, 60, and 90 ppm, soil and foliar application	Withholding of irrigation for 15 days after pollination to maturity (late season stress)	Behboudi et al. (2018)
5	CeO_2 NPs (nanoceria)	Sorghum	10 mg/L, foliar spray	Control: Plants maintained at 100% pot capacity moisture and drought stress: withholding water for 21 days	Djanaguiraman et al. (2018)
6	Chitosan nanoparticles	Salvia abrotanoides	30, 60, and 90 ppm, foliar application	30% FC (severe DS), 50% FC (medium DS), and 100% FC (well-watered, No DS)	Dawood et al. (2019)
7	Se NPs	Wheat	10, 20, 30, and 40 mg/L, foliar application	35% FC (DS) and 100% FC (control)	Ikram et al. (2020)

(continued)

Table 25.2 (continued)

Sr. No.	Type of nanoparticles used	Plant/crop	Dose of nanoparticles used and mode of application	Level of stress	References
8	Fe, ZnO, Cu and Co NPs	Soyabean	50 mg/L of Fe, ZnO and Cu NPs, 0.05 mg/L of Co NPs, seed treatment	65–70% soil moisture content (SMC, control) and 30–40% SMC (DS)	Linh et al. (2020)
9	ZnO NPs	Rice	5, 10, 15, 25, and 50 ppm, seed priming	Water holding capacity (WHC, control) at 70% and WHC at 35%	Waqas Mazhar et al. (2022)
10	Yttrium doping-stabilized γ-Fe_2O_3 NPs	Rapeseed	0.5, 0.8, 1, or 2 mg/ml, soil application	–	Palmqvist et al. (2017)
11	ZnO NPs	Sunflower	100 ppm, foliar spray (three times during life cycle)	Irrigation at 3, 6, and 9 days of intervals	Al-Dhalimi and Al-Ajeel (2020)
12	ZnO NPs	Wheat	0.5 and 1.0 mg/L, foliar spray	35%, 60%, and 85% FC	Sadati et al. (2022)
13	Fe_3O_4 NPs (magnetite)	*Setaria italica*	5, 10, 15, 20, 50, 90, and 120 mg/L, seed treatment	Using 10% PEG	Sreelakshmi et al. (2021)
14	Zn and Cu NPs	Wheat	Seed treatment	70% total moisture capacity (TMC, control), 30% TMC	Taran et al. (2017)
15	SiO_2-NPs and Se-NPs alone and in combinations	Strawberry	Se-NPs (25 mg/L), SiO_2-NPs (125 mg/L), and Se/SiO_2-NPs (50 and 100 mg/L) foliar spray (three times during life cycle)	100% FC (control), 60% FC (moderate stress) and 25% FC (severe stress)	Zahedi et al. (2020)

References

Abou-Zeid H, Ismail G (2018) The role of priming with biosynthesized silver nanoparticles in the response of Triticum aestivum L. to salt stress. Egypt J Bot 58:73–85

Adrees M, Khan ZS, Ali S, Hafeez M, Khalid S, Ur Rehman MZ, Hussain A, Hussain K, Chatha SA, Rizwan M (2020) Simultaneous mitigation of cadmium and drought stress in wheat by soil application of iron nanoparticles. Chemosphere 238:124681

Ahmed F, Javed B, Razzaq A, Mashwani ZR (2021) Applications of copper and silver nanoparticles on wheat plants to induce drought tolerance and increase yield. IET Nanobiotechnol 15(1): 68–78

Alabdallah NM, Hasan MM (2021) Plant-based green synthesis of silver nanoparticles and its effective role in abiotic stress tolerance in crop plants. Saudi J Biol Sci 28(10):5631–5639

Al-Dhalimi AM, Al-Ajeel SAH (2020) Effect of plant regulators, zinc nanoparticles and irrigation intervals on leaf content of endogenous hormones and nutrients in sunflower (Helianthus annuus l.). Plant Arch 20(1):2720–2725

Ali EF, El-Shehawi AM, Ibrahim OHM, Abdul-Hafeez EY, Moussa MM, Hassan FAS (2021) A vital role of chitosan nanoparticles in improvisation the drought stress tolerance in Catharanthus roseus (L.) through biochemical and gene expression modulation. Plant Physiol Biochem 161: 166–175

Almutairi ZM (2016) Effect of nano-silicon application on the expression of salt tolerance genes in germinating tomato (Solanum lycopersicum L.) seedlings under salt stress. Plant Omics 9(1): 106–114

Bami SS, Khavari-Nejada RA, Ahadib AM, Rezayatmand Z (2022) TiO2 nanoparticles and salinity stress in relation to artemisinin production and ADS and DBR2 expression in Artemisia absinthium L. Brazilian J of Biol 82:e237214. https://doi.org/10.1590/1519-6984.237214

Behboudi F, Tahmasebi Sarvestani Z, Kassaee MZ, ModaresSanavi SAM, Sorooshzadeh A, Ahmadi SB (2018) Evaluation of chitosan nanoparticles effects on yield and yield components of barley (Hordeum vulgare L.) under late season drought stress. J Water Environ Nanotechnol 3(1):22–39

Borišev M, Borišev I, Župunski M, Arsenov D, Pajević S, Ćurčić Ž, Djordjevic A (2016) Drought impact is alleviated in sugar beets (Beta vulgaris L.) by foliar application of fullerenol nanoparticles. PLoS One 11(11):e0166248

Dawood MF, Abeed AH, Aldaby EE (2019) Titanium dioxide nanoparticles model growth kinetic traits of some wheat cultivars under different water regimes. Plant Physiol Rep 24(1):129–140

Djanaguiraman M, Nair R, Giraldo JP, Prasad PVV (2018) Cerium oxide nanoparticles decrease drought-induced oxidative damage in sorghum leading to higher photosynthesis and grain yield. ACS Omega 3(10):14406–14416

El-Badri AM, Batool M, Wang C, Hashem AM, Tabl KM, Nishawy E, Wang B (2021) Selenium and zinc oxide nanoparticles modulate the molecular and morpho-physiological processes during seed germination of Brassica napus under salt stress. Ecotoxicol Environ Saf 225:112695

El-Sharkawy MS, El-Beshsbeshy TR, Mahmoud EK, Abdelkader NI, Al-Shal RM, MissaouiA M (2017) Response of alfalfa under salt stress to the application of potassium sulfate nanoparticles. Am J Plant Sci 8(8):1751–1773

Faizan M, Bhat JA, Chen C, Alyemeni MN, Wijaya L, Ahmad P, Fangyuan Y (2021) Zinc oxide nanoparticles (ZnO-NPs) induce salt tolerance by improving the antioxidant system and photosynthetic machinery in tomato. Plant Physiol Biochem 161:122–130

Fathi A, Zahedi M, Torabian S, Khoshgoftar A (2017) Response of wheat genotypes to foliar spray of ZnO and Fe_2O_3 nanoparticles under salt stress. J Plant Nutr 40(10):1376–1385

Ghazi AA, El-Nahrawy S, El-Ramady H, Ling W (2022) Biosynthesis of nano-selenium and its impact on germination of wheat under salt stress for sustainable production. Sustainability 14(3):1784

Gohari G, Mohammadi A, Akbari A, Panahirad S, Dadpour MR, Fotopoulos V, Kimura S (2020) Titanium dioxide nanoparticles (TiO_2 NPs) promote growth and ameliorate salinity stress effects on essential oil profile and biochemical attributes of Dracocephalum moldavica. Sci Rep 10(1):1–14

Hasanuzzaman M, Nahar K, Alam MM, Roychowdhury R, Fujita M (2013) Physiological, biochemical and molecular mechanisms of heat stress tolerance in plants. Int J Mol Sci 14:9643–9684

Hassan FAS, Ali E, Gaber A, Fetouh MI, Mazrou R (2021) Chitosan nanoparticles effectively combat salinity stress by enhancing antioxidant activity and alkaloid biosynthesis in Catharanthus roseus (L.) G. Don. Plant Physiol Biochem 162:291–300

Heidarabadi MD (2022) Metal nanoparticles and abiotic stress tolerance. In: Advances in plant defense mechanisms. IntechOpen, London

Hernández-Hernández H, González-Morales S, Benavides-Mendoza A, Ortega-Ortiz H, Cadenas-Pliego G, Juárez-Maldonado A (2018) Effects of chitosan–PVA and Cu nanoparticles on the growth and antioxidant capacity of tomato under saline stress. Molecules 23(1):178

Hojjat SS (2016) The effect of silver nanoparticle on lentil seed germination under drought stress. Int J Farming Allied Sci 5:208–212

Hosseinpour A, Haliloglu K, TolgaCinisli K, Ozkan G, Ozturk HI, Pour-Aboughadareh A, Poczai P (2020) Application of zinc oxide nanoparticles and plant growth promoting bacteria reduces genetic impairment under salt stress in tomato (Solanum lycopersicum L. 'Linda'). Agriculture 10(11):521

Ikram M, Raja NI, Javed B, Hussain M, Hussain M, Ehsan M, Rafique N, Malik K, Sultana T, Akram A (2020) Foliar applications of bio-fabricated selenium nanoparticles to improve the growth of wheat plants under drought stress. Green Process Synth 9(1):706–714

Khan MR, Rizvi TF (2014) Nanotechnology: scope and application in plant disease management. Plant Pathol J 13:214–231

Khan MN, AlSolami MA, Basahi RA, Siddiqui MH, Al-Huqail AA, Abbas ZK, Siddiqui ZH, Ali HM, Khan F (2020) Nitric oxide is involved in nano-titanium dioxide-induced activation of antioxidant defense system and accumulation of osmolytes under water-deficit stress in Viciafaba L. Ecotoxicol Environ Saf 190:110152. https://doi.org/10.1016/j.ecoenv.2019.110152

Kirupagaran R, Saritha A, Bhuvaneswari S (2016) Green synthesis of selenium nanoparticles from leaf and stem extract of Leucas lavandulifolia Sm. and their application. J Nanosci Technol 2:224–226

Linh TM, Mai NC, Hoe PT, Lien LQ, Ban NK, Hien LTT, Chau NH, Van NT (2020) Metal-based nanoparticles enhance drought tolerance in soybean. J Nanomater 2020:4056563

Liu J, Li G, Chen L, Gu J, Wu H, Li Z (2021) Cerium oxide nanoparticles improve cotton salt tolerance by enabling better ability to maintain cytosolic K^+/Na^+ ratio. J Nanobiotechnol 19(1):1–16

Mohammadi H, Esmailpour M, Gheranpaye A (2016) Effects of TiO2 nanoparticles and waterdeficit stress on morpho-physiological characteristics of dragonhead (Dracocephalum moldavica L.) plants. Acta Agric Slovenica 107(2):385–396

Palmqvist NG, Seisenbaeva GA, Svedlindh P, Kessler VG (2017) Maghemite nanoparticles acts as nanozymes, improving growth and abiotic stress tolerance in Brassica napus. Nanoscale Res Lett 12(1):1–9

Rossi L, Zhang W, Ma X (2017) Cerium oxide nanoparticles alter the salt stress tolerance of Brassica napus L. by modifying the formation of root apoplastic barriers. Environ Pollut 229:132–138

Sadati SYR, Godehkahriz SJ, Ebadi A, Sedghi M (2022) Zinc oxide nanoparticles enhance drought tolerance in wheat via Physio-Biochemical changes and stress genes Expression. Iran J Biotech 20(1):e3027

Semida WM, Abdelkhalik A, Mohamed GF, Abd El-Mageed TA, Abd El-Mageed SA, Rady MA, Ali EF (2021) Foliar application of zinc oxide nanoparticles promotes drought stress tolerance in eggplant (Solanum melongena L.). Plan Theory 10:421. https://doi.org/10.3390/plants10020421

Shah T, Latif S, Saeed F, Ali I, Ullah S, Alsahli AA, Ahmad P (2021) Seed priming with titanium dioxide nanoparticles enhances seed vigor, leaf water status, and antioxidant enzyme activities in maize (Zea mays L.) under salinity stress. J King Saud Univ Sci 33(1):101207

Shariatzadeh Bami S, Khavari-Nejad RA, Ahadi AM et al (2021) TiO2 nanoparticles effects on morphology and physiology of Artemisia absinthium L. under salinity stress. Iran J Sci Technol Trans Sci 45:27–40. https://doi.org/10.1007/s40995-020-00999-w

Singh A, Sengar RS, Rajput VD, Minkina T, Singh RK (2022) Zinc oxide nanoparticles improve salt tolerance in rice seedlings by improving physiological and biochemical indices. Agriculture 12(7):1014

Soliman AS, El-feky SA, Darwish E (2015) Alleviation of salt stress on Moringa peregrine using foliar application of nanofertilizers. J Hortic For 7:36–47

Sompornpailin K, Chayaprasert W (2020) Plant physiological impacts and flavonoid metabolic responses to uptake TiO_2 nanoparticles. Austr J Crop Sci 14:1995. https://doi.org/10.21475/ajcs.20.14.04

Sreelakshmi B, Induja S, Adarsh PP, Rahul HL, Arya SM, Aswana S, Haripriya R, Aswathy BR, Manoj PK, Vishnudasan D (2021) Drought stress amelioration in plants using green synthesised iron oxide nanoparticles. Mater Today Proc 41:723–727

Taran N, Storozhenko V, Svietlova N, Batsmanova L, Shvartau V, Kovalenko M (2017) Effect of zinc and copper nanoparticles on drought resistance of wheat seedlings. Nanoscale Res Lett 12(1):1–6

Van Nguyen D, Nguyen HM, Le NT et al (2022) Copper nanoparticle application enhances plant growth and grain yield in maize under drought stress conditions. J Plant Growth Regul 41:364–375. https://doi.org/10.1007/s00344-021-10301-w

Wahid I, Rani P, Kumari S, Ahmad R, Hussain SJ, Alamri S, Nirmalya Tripathy M, Khan IR (2022) Biosynthesized gold nanoparticles maintained nitrogen metabolism, nitric oxide synthesis, ions balance, and stabilizes the defense systems to improve salt stress tolerance in wheat. Chemosphere 287(2):132–142

Waqas Mazhar M, Ishtiaq M, Hussain I, Parveen A, Hayat Bhatti K, Azeem M, Thind S, Ajaib M, Maqbool M, Sardar T, Muzammil K (2022) Seed nano-priming with zinc oxide nanoparticles in rice mitigates drought and enhances agronomic profile. PLoS One 17(3):e0264967

Yang KY, Doxey S, McLean JE, Britt D, Watson A, Al Qassy D, Anderson AJ (2017) Remodeling of root morphology by CuO and ZnO nanoparticles: effects on drought tolerance for plants colonized by a beneficial pseudomonad. Botany 96(3):175–186

Zahedi SM, Moharrami F, Sarikhani S, Padervand M (2020) Selenium and silica nanostructure-based recovery of strawberry plants subjected to drought stress. Sci Rep 10(1):1–18

Zayed MM, Elkafafi SH, Zedan AM, Dawoud SF (2017) Effect of nano chitosan on growth, physiological and biochemical parameters of Phaseolus vulgaris under salt stress. J Plant Prod 8(5):577–585

Zhang H, Zhao Y, Zhu JK (2020) Thriving under stress: how plants balance growth and the stress response. Dev Cell 55(5):529–543

Zulfiqar F, Ashraf M (2021) Nanoparticles potentially mediate salt stress tolerance in plants. Plant Physiol Biochem 160:257–268

Potential Use of Nanofertilizers in Alleviating Stresses in Plants

26

Sunita Devi, Naresh Kumar, Ashwani Kumar, Pooja Dhansu, Adhini S. Pazhany, Anita Mann, Ajay Kumar Bhardwaj, and Parvender Sheoran

Abstract

Due to climate change scenario, the world's natural resources are depleting at a considerably faster rate than they are being replenished, and the agriculture sector is no exception to this trend. In such diverse environmental circumstances, it is critical to identify the possible solution to overcome technological problems such as the yield barrier, resource efficiency, and the creation of ecologically acceptable technologies. Nowadays, nutrient use efficiency is one of the most pressing and difficult scientific challenges for evaluating crop production systems, which can be highly influenced by various factors such as fertilizer management and soil and plant–water relationships. Nanotechnology has made it possible to use nanoscale or nano-structured materials as fertilizer carriers or controlled release carriers in the development of "smart fertilizers" that cut environmental protection expenditures. Nanofertilizers are well-known for their huge surface areas and nanoscale size, both of which contribute to their high potential for reducing plant abiotic stress. Photosynthetic rate, nutrient absorption efficiency, phytohormone

S. Devi (✉) · A. Kumar · A. Mann · A. K. Bhardwaj
ICAR – Central Soil Salinity Research Institute, Karnal, Haryana, India

N. Kumar
Eternal University, Baru Sahib, Himachal Pradesh, India

P. Dhansu
ICAR – Sugarcane Breeding Institute, Regional Centre, Karnal, Haryana, India

A. S. Pazhany
ICAR – Sugarcane Breeding Institute, Coimbatore, India

P. Sheoran
ICAR – Central Soil Salinity Research Institute, Karnal, Haryana, India

ICAR – Agricultural Technology Application Research Institute, Ludhiana, Punjab, India

A. Kumar et al. (eds.), *Salinity and Drought Tolerance in Plants*,
https://doi.org/10.1007/978-981-99-4669-3_26

regulation, and plant defense system are some of the morphological, biochemical, and physiological markers that it helps with. Various studies show that nanofertilizers reduce oxidative stress, enhance proline, chlorophyll, and relative water content and so increase plant stress tolerance to salt toxicity. Antioxidant enzymes including catalase, peroxidase, and superoxide dismutase have also been found to benefit from nanoparticles. So, keeping in mind the importance of nanotechnology, the present chapter deals with the role of nanofertilizers in alleviating the stress induced changes in plants.

Keywords

Nanofertilizers · Abiotic stress · Antioxidant enzymes · Plant defense system

26.1 Introduction

Indian agriculture is highly dependent on environment conditions so alteration in weather and global changes directly influences the crop productivity. Although technology advancement is improving the status of the agriculture but still it is unable to achieve the sustainable agriculture scenarios. Our global food demand is anticipated to climb by approximately 70% in 2050 as a result of increase in population (Verma et al. 2022). According to previous data analysis it was predicted that world population is continuously growing with 1.24% per year and estimated to reach 8.5 billion in 2030. So to feed such large population agriculture output should be increased by 70% (Alexandratos and Bruinsma 2012). More stress will be on agriculture so more chemical fertilizers are used to fulfill the necessities. Such excessive use of chemical adversely affects soil health, nutrient imbalance which consequently reduced productivity instead of increasing. A recent emerging technology known as nanotechnology has the potential to transform agriculture, health, medicine, environment challenge, and defense system. The term "nanotechnology" was first coined by Norio Taniguichi, a professor at Tokyo University of Science, in 1974 (Khan and Rizvi 2014). Its widespread application in agriculture has been anticipated for twentieth century. There are several uses for nanomaterials in medical, environmental science, agriculture, and food processing as a result of recent advances in their manufacturing. Agriculture has always profited from these developments throughout history. Its beneficial impact furnishes a sustainable use of resources which makes future perspective for next green revolution. Additionally, since agriculture encounters various and novel difficulties, such as decreased crop yield brought on by biotic and abiotic pressures, such as nutrient insufficiency and environmental pollution, the development of nanotechnology has provided exciting opportunities for precision agriculture.

The promising solution offered by nanotechnology is considerate to the tremendously increasing population's feed necessity. Nanotechnology is associated with target specific delivery of nanofertilizers that might help in minimizing the nutrient losses along with reducing environmental pollution that could resulted into higher yield. Under current scenario, Nanotechnology will meets the demands of efficient fertilizers use for food production. In India, Former President, late Dr. APJ Abdul Kalam also emphasized nanotechnology in agriculture by saying "We have to launch vertical missions under an umbrella organization with the public-private investment in at least 10 nanotechnology products in water, energy, agriculture, healthcare, space, defense sectors." Nanofertilizers are well known for controlled release of nutrients. Nanofertilizers are used in smaller quantities as compared to bulky chemical fertilizers that would reduce the nutrient run-off and environmental pollution. So intervention of this smarter technology or nanotechnology is the prime request for present problems with added advantages. Still now there is lack of practice for risk assessment and risk management of agriculture nanoproducts with a permissible dose (time × concentration) which is major reason for delayed commercial development of these products.

Nanomaterials in agriculture aim in particular to reduce the amount of sprayed chemical products by smart delivery of active ingredients, minimize nutrient losses in fertilization, and increase yields through optimized water and nutrient management. Nanotechnology derived devices are also being explored in the field of plant breeding and genetic transformation. Additionally, agriculture could be a source of bio-nano composites with enhanced physical mechanical properties based on traditionally harvested materials for bio-industrial purposes. Agricultural institutions are not well versed in concepts and potential of nanobiotechnology to integrate them in crop production programs. Large companies are investigating the potential that nanotech solutions offer in the agricultural field. However, multinationals do not demonstrate a sufficiently high economic interest. Further, nanobiotechnology requires a multidisciplinary research approach. According to definition nanofertilizers are nanomaterials which are either micro- or macronutrients or acting as carriers/additives for nutrients, encapsulated nutrients inside the nanomaterials (Kah et al. 2018; DeRosa et al. 2010). Nanotechnology was precisily described by European Union (EU) as a key enabling technology which is "any intentiolly produced materials that has one or more dimensions of the order of 100 nm or less or that is composed of discrete functional parts, either internally or at the surface, many of which have one or more dimensions of the order of 100 nm or less, including structures, agglomerates or aggregates, which may have a size above the order of 100nm but retain properties that are characteristic of the Nanoscale" (EU 2011). The superiority of nanomaterial containing nutrients is that such fertilizer responds to various biochemical or physical stimuli such as ethylene production in roots and acidification of rhizosphere (in response to *P*/*K* deficiency in soil) (DeRosa et al. 2010). Nanofertilizers are efficient source of balanced crop nutrition with lowest dose as comparative to chemical fertilizers. Such fertilizers with lower dose and higher availability are the prerequisite for sustainable agriculture development (Shang et al. 2019). Nano-fertilizers are macro and micronutrient encapsulated by nanomaterial which further coated with thin protective polymeric films or derived from nano-emulsions/nanoparticles. Such formulation is more potent than conventional fertilizer that is associated with various adverse pollutions. Such formulation has higher surface tension, target specific delivery, moderate release, and required in

Table 26.1 Enlisted various nanomaterials used for stress mitigation

Nanofertilizers/ nanoparticles	Plant species	Abiotic stress	Impact on yield and other attributes	References
Silica NPs	Cucumber	Salinity and drought	Increased photosynthetic activity, antioxidant enzymes and biomass	Alsaeedi et al. (2019)
Selenium NPs	Wheat	Drought	Increased plant growth and development	Ikram et al. (2020)
Iron oxide NPs	Wheat	Salinity and heavy metal	Enhances photosynthetic rate and restricts cadmium uptake	Manzoor et al. (2021)
Cerium oxide NPs	Cotton	Salinity	Promotes plant growth through balanced cytosolic K^+/Na^+	Liu et al. (2021)
Silver NPs	Summer savory	Salinity	Enhance s plant growth and germination	Nejatzadeh (2021)
Zinc oxide NPs	Safflower	Salinity	Improves plant germination, growth, and development	Yasmin et al. (2021)
Silicon NPs	Sweet orange	Salinity	Oxidative stress tolerance improved	Mahmoud et al. (2022)
Gold NPs	Wheat	Salinity	Improves plant defense	Wahid et al. (2022)
Calcium NPs	Barley	Heavy metal	Increased photosynthetic efficiency and antioxidant enzymes	Nazir et al. (2022)
Iron oxide NPs	Rice	Cadmium and drought stress	Increased biomass, antioxidant enzymes, and photosynthesis efficiency	Ahmed et al. (2021)
Nano-se, nano-Si NPs	Barley	Abiotic stress	Activates defense mechanism through antioxidant activity	Shalaby et al. (2021)
CeO_2, N, P, K, Zn, Fe nanofertilizers	Cabbage	Control	Increases nutrient use efficiency and plant growth	Abdulhameed et al. (2021)
Nano-B, nano-si, nano-Zn	Wheat	Drought	Increased protein percentage in wheat grain, minimizes drought damage	Ahanger et al. (2021)
Nanoscale zero-valent iron	Rice	No-stress	Priming rice with nZVI (10–80 mg L^{-1}) enhances yield; promoted the distribution of nutrients in grains and their contents	Guha et al. (2021)
NPK nanofertilizers	Potato	Control	NPK nanofertilizers significantly improved potato yield and its parameters compared to NPK chemical fertilizers	Abd El-Azeim et al. (2020)

(continued)

Table 26.1 (continued)

Nanofertilizers/ nanoparticles	Plant species	Abiotic stress	Impact on yield and other attributes	References
Nano-Cu NPs	Maize (*Zea mays*)	Drought	NPs regulated protective mechanism of maize	Van Nguyen et al. (2022)
Glycine nano-Si NPs	Feverfew	Drought	Mitigator of the adverse effects of drought	Esmaili et al. (2020)
Nano-Si	Sugar beet	Water stress	Protects plants by enhancing GB, antioxidants, and flavonols	Namjoyan et al. (2020)
Nano-Fe and -Zn	Rosemary	Salinity	Increased total phenolic and total flavonoid contents; growth and salt tolerance	Hassanpouraghdam et al. (2020)
ZnO NPs	Maize	Drought	Promoted the synthesis of melatonin and activated enzymatic antioxidants, which alleviated damage in chloroplast due to drought	Sun et al. (2020)
SiO_2 nanoparticle + Zn	Maize	Nutrients stress	Increases grain yield of maize by 37% and increases thelinoleic acid, compared to control	Asadpour et al. (2020)

lesser quantity comparative to chemical one, e.g. nutrient-augmented zeolites (Malekian et al. 2011; Perrin et al. 1998; Zwingmann et al. 2011) (Table 26.1).

Nanoparticles are designed in such a way that didn't interfere with the residing molecule but increases the activity of some enzymes and photosynthetic efficiency e.g. TiO_2-NPs (Titanium dioxide), carbon nanotubes (CNTs). These particles promoted seed germination and plant growth. Titanium dioxide nanoparticles induced active oxygen like superoxide and hydroxide anions which initiate capsule penetration of seed, increase seed stress resistance which further facilitates intake of water and oxygen for fast germination of seed. Carbon nanotubes also penetrate seed and promotes seed germination (Khodakovskaya et al. 2009, 2013).

Various materials are used for formulation of nanofertilizers like clay minerals, hydroxyapatite, chitosan, polyacrylic acid, and zeolite. Such formulation is very effective for soil and foliar application. These materials with large surface area and strong binding with urea facilitate the slow release of bound fertilizers up to 60 days as compared to conventional fertilizers (Kottegoda et al. 2011). Both biotic and abiotic stress tremendously affect the crop productivity. The major abiotic stresses are heat stress, cold stress, salinity, drought, chemical or heavy metals accumulations and flooding, whereas virus, fungi, and bacteria are the key biotic stresses. The biomagnifications of heavy metals are increasing in agricultural soils as a result of widespread chemical fertilizer use, air settling, sewage waste, and quick industrial

Fig. 26.1 General effects of biotic and abiotic stresses on plants

expansion. Toxic HMs have consistently impacted crop yield in recent decades (Ahmed et al. 2021a; Chen et al. 2022). The determinate effect of heavy metals inhibits plant growth by interfering with the intake of nutrients, antioxidant enzymes, photosynthetic machinery, and by raising reactive oxygen species. Other stress like soil salinity negatively affects at a physiological level, which disturbs ionic and water homeostasis (Abdel Latef et al. 2018). Additionally, heat stress results in the production of ROS, stomata opening, leaf size, leaf water potential, root growth, and seed number, size, and tolerance, which can impede flowering and fruiting and diminish crop yield (Xu et al. 2021). The approaches such as conventional breeding, marker-assisted breeding, and transgenic crop engineering are moderate, tedious methods but new approaches nano-based application have recently emerged as a promising tool to control impacts on stress imposed plant. Therefore, the main goal is to facilitate plants' rapid adaptability to environmental challenges without endangering already-vulnerable ecosystems (Vermeulen et al. 2012).

Plants either adapt itself to stress but some artificial antistressors such as nitric acids, rhizobia, pyraclostrobin, plant growth promoting rhizobacteria (PGPR), melatonin, strigolactones, phytohormones, and nutrients such as selenium and silicon are used to mitigate the stress in plants. This task necessitates a multifaceted approach, including the activation of the plant's enzymatic system, hormonal control, stress gene expression, toxic metal absorption regulation, and avoiding water deficiency stress or flash floods by reducing the plant's life cycle. Researchers have undertaken a variety of attempts to create technology and methods that will lead to sustainable agricultural systems while minimizing negative effects on the environment (Dubey and Mailapalli 2016; Pretty 2008). Nowadays stress mitigation is the major thematic area for the research in agriculture sector. Potential advancement of technology combined with agriculture necessitates the stress adaptation of plant. Recently stress tolerance in plant due to effect of nanomaterials is the keen interest among researchers targeting toward agriculture sustainability. Heat, salinity, cold

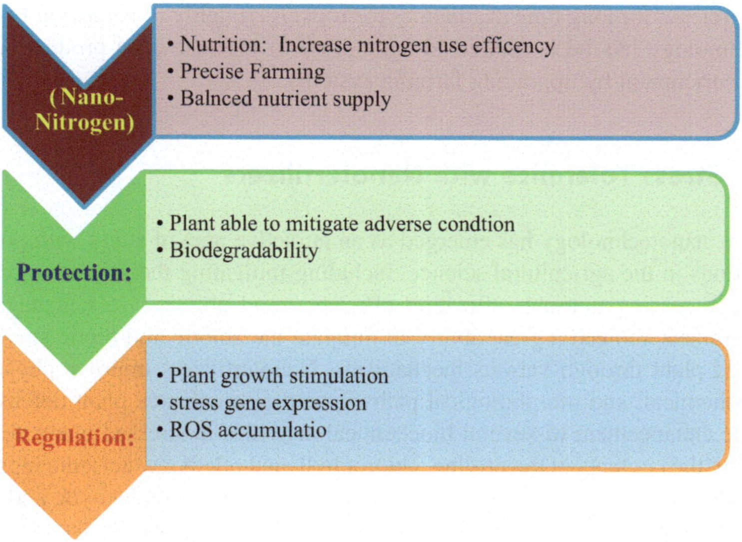

Fig. 26.2 Detailed effect of nanofertilizers on plant physiological and biochemical process

drought, flooding/submersion (anoxia), chemical toxicities, and excessive light are all major abiotic factors that influence plants (Fig. 26.1). In such diverse environmental circumstances, it is critical to identify a study topic to overcome technological problems such as the yield barrier, resource efficiency, and the creation of ecologically acceptable technologies (Bhardwaj et al. 2019).

Nanotechnology which deals with the production of nanomaterial having nanometer diameter is recently on boom due to its wider application. When compared to their bulk counterparts, these nano-sized entities or materials display exceptional qualities in terms of reactivity and other physico-chemical properties. They serve as the transitional state between individual atoms or molecules. Such particles possess distinct property with high utility (Bhardwaj et al. 2019). Nanoparticles are widely used in every sector like medical, agriculture, defense, and education. The characteristics of nanoparticles (NPs) are attributable to their large surface area to volume ratio, which improves both their chemical and physical responses. Thus, a smaller amount of a nanomaterial can perform the same or even better than its bulk version. The characteristics of nanoparticles (NPs) are due to their high surface area to volume ratio, which boosts their chemical reactivity as well as physical responses. This implies that a smaller amount of nanomaterials can perform the same or even higher activity than their bulk forms (Zulfiqar et al. 2018). Unlike conventional fertilizers, which require heavy application, nanofertilizers can be used sparingly to attain the desired production target. Various studies showed toxicity symptoms despite of several advantages of nanofertilizers, various nanotoxicity and genotoxicity observed with some formulation (Khalifa and Hasaneen 2018). Nanofertilizers wider utility and stability dominate its effect in plants under stress condition. Nanomaterials are small in size as compared with bulky ions (100–1000 times lager) and have unique properties that assists lower run-off, less leaching, slow nutrient release for long time availability to crops. As roughly illustrated in Fig. 26.2 nanotechnology has been applied to agriculture to increase crop production with quality enrichment by upgrading farming systems.

26.2 Stress Tolerance with Nanofertilizers

Recently, nanotechnology has emerged as an attractive area of study with potential applications in the agricultural science, including mitigating the impacts of climate change, increasing nutrient utilization efficiency and abiotic stress management. Nanofertilizers showed a great ability to mitigate the abiotic and biotic stresses on cultivated plant through various mechanisms. Nanofertilizers improve physiological, biochemical, and morphological pathway and also enhance plant defense system. The enhancement of several biochemical activities in stressed plants (e.g., by increasing the contents of the proline, chlorophyll, and relative water); the regulation of the salt toxicity; a reduction in the accumulation of malondialdehyde and H_2O_2; and the maintenance of the ionic equilibrium, depending on the type of stress. Nanofertilizers helps in mitigating the stress induced effects such as reduced generation of ROS, increased antioxidative enzyme activities (CAT, POX, SOD) reduce oxidative stress (H_2O_2 and MDA), activating specific genes, enhance photosynthetic pigments, promote water and nutrients uptake, regulation of plant hormones, decreased plasma membrane damage and chlorophyll degradation. Various researches intimate the mechanisms of nanofertilizers on crop productivity under stress condition, nanofertilizers helps in reduction of oxidative stress, which leads to an increase in plant stress tolerance, the enhancement of several biochemical activities in stressed plants (e.g., by increasing the contents of proline, chlorophyll, and relative water) the regulation of salt toxicity; a reduction in the accumulation of malondialdehyde and H_2O_2. Guha et al. (2021) discussed briefly the following aspects of nanofertilizers: (1) foliar and soil treatments offer the proper nutrients for increasing plant growth, (2) they are low-cost and sustainable sources of plant nutrients, (3) they have a high fertilization efficiency, and (4) they play a major role in pollution prevention. Nanofertilizers also help to clean up water contamination and could be considered new fertilizer alternatives. Instead of abiotic, biotic stress influenced the crop production. Farmers use excessive pesticides which are not ecologically suitable. Recent research on nanomaterials also proved that nanomaterials could successfully reduce the risk of disease due to biotic factor. Like AgNPs nanoparticles are synthesized using cotton plants and possess antibacterial activity against bacterial pathogens of *Malvaceae* and *Brassicaceae* family of crops (Vanti et al. 2019). Despite of its advantages, researchers also studied the risk associated with nanofertilizers and found that nanotoxicity to soil, water, air and its reactivity to human body through inhalation, ingestion and surface contact. Compared to conventional fertilizers nanofertilizers have many distinguished attributes such as nutrient uptake efficiency, controlled release modes, effective duration of the nutrients release, and a reduced loss rate of in the production of higher crop yields of high quality.

26.3 Role of Nanofertilizers to Combat the Abiotic Stress

About 23% of the world's cultivated lands experience severe salinity stress, which severely affects crop yield (Onaga and Kerstin 2016). Due to their large surface areas and nanoscale size, nanofertilizers have a remarkable power to reduce the abiotic/biotic stressors on plants through a variety of ways. Nanofertilizers can enhance the morphological, biochemical, and physiological indicators of cultivated plants such as efficiency of photosynthetic activity, nutrient uptake, the regulation of phytohormones, and the improvement of the plant defense system. Due to its higher surface area and reactivity, nanofertilizers help in mitigating the affects of stresses by providing proper nutrition at proper time. Nanofertilizers have the capacity to decrease the effect of oxidative stress, increased biochemical activity (increase the content of proline, chlorophyll), relative water content, regulated salt toxicity, decrease in the accumulation of melondialdehyde, hydrogen peroxide and maintenance of ionic equilibrium in plant under stress conditions (Fig. 26.3). So use of nanofertilizers for stress resistance has been gaining immense attention in today's scenario. Conversely, it has been documented that the use of nano-SiO_2 enhances seed germination, increases plant fresh weight, dry weight, and chlorophyll content with proline accumulation in tomato and squash plants under NaCl stress (Siddiqui and Al-Whaibi 2014). Similarly, Torabian et al. (2017) showed that foliar application of iron sulfate ($FeSO_4$) nanoparticles increased the tolerance of sunflower cultivars to salinity stress. They found that applying nano $FeSO_4$ increased leaf area, shoot dry weight, sub-stomatal CO_2 concentration (Ci), chlorophyll content,

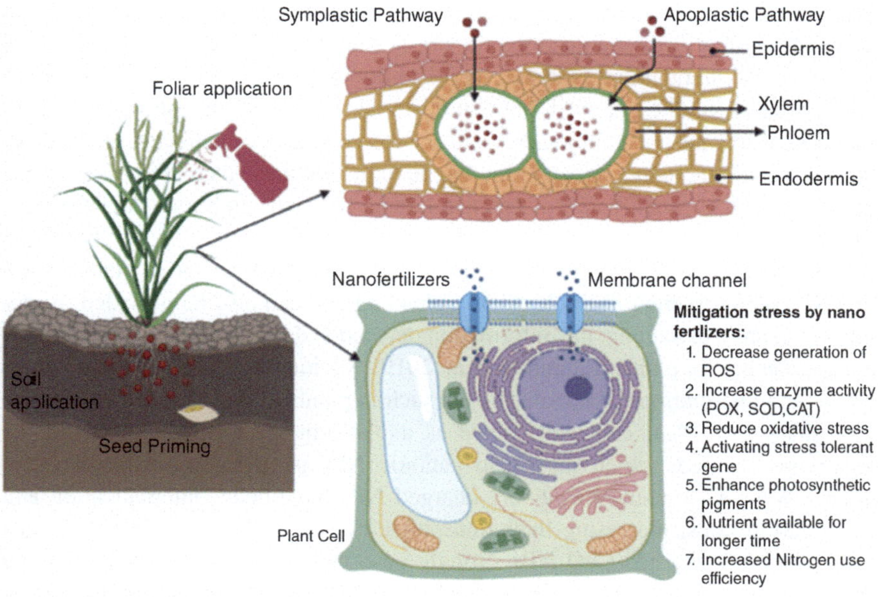

Fig. 26.3 Mechanism of action of nanofertilizers in plants

maximum photochemical efficiency of photosystem II (Fv/Fm), and iron (Fe) content, but also significantly lowered sodium (Na) content in leaves. Stresses hastens the photosynthesis in plants, an important metabolic process due to alterations in activities of PS II, Rubisco and ATP synthase (Adisa et al. 2019). Rajput et al. (2021) noted that SiO_2 nanoparticles enhanced the activity of carbonic anhydrase, photosynthetic and water use efficiency of pumpkin plants. Nanofertilizers has potential to accelerate plant growth and productivity which facilitate to be useful tools in agricultural practices, particularly in drought-prone or even flash flood-prone areas where the early maturity of crops is a crucial component for sustainable crop production. Additionally, the detoxification or remediation of dangerous pollutants such heavy metals has been demonstrated to be successful with the use of nanoparticles. For instance, Wang et al. (2014) demonstrate that the foliar application of nano-Si at a dosage of 2.5 mM considerably increases the ability of rice plants to withstand Cd stress by controlling Cd accumulation. The same group demonstrated in a different investigation that nano-Si is similarly efficient when combined with Cd against Pb, Cu, and Zn. It suggests that using nano-Si fertilizers may be better than using conventional fertilizers at preventing the build-up of heavy metals. Abiotic stress is primarily characterized by the production of reactive oxygen species (ROS) by cell organelles. When plants are exposed to oxidative stress from their surroundings, they have the enzymatic tools to handle it. However, if the defense system fails, plants will experience the consequences. Stress is reduced by nanomaterials, as demonstrated by their ability to activate particular genes, collect osmolytes, and provide free nutrients and amino acids. As plant faces abiotic stress major process of plant metabolism is affected like photosynthesis and gas change. Various researches focused on impact of nano-fertilizer on photosynthetic leaf gas exchange capacity to study major preventive effect of nanofertilizers against abiotic stress. The physiological and biochemical indicators of agricultural plants significantly improved with the usage of nanofertilizers. Studies showed that nanomaterials enhanced the antioxidant, accumulation of free proline, amino acids and activity of superoxide dismutase, catalase, peroxidase, nitrate reductase, and glutathione reductase these increase physiological parameters and eventually improve plant tolerance to harsh conditions (Shalaby et al. 2021).

It is critical to evaluate how NFs build up and move from the soil to plants. Since NFs/NPs prefer to enter through xylem uploading and can also be provided exogenously via phloem loading, this information may aid in determining the best methods and approaches for plants (Pitambara et al. 2019). Various factors affect the efficacy, uptake and distributions of nanofertilizers such in plants as particle size, concentration, shape etc. Plant cell walls serve as a platform for ion exchange and are negatively charged, which may help cationic NPs penetrate more readily than anionic NPs. As a result, negatively charged NPs have better internalization and translocation rates, increasing their transport efficiency (Zhu et al. 2012). In this case, positively charged CeO NPs strongly adsorbed onto the negatively charged root surfaces, whereas negatively charged CeO_2 NPs showed limited root accumulation but improved shoot internalization, mostly by overcoming electrostatic resistance (Liu et al. 2019). Nanomaterials can be used wisely to boost crop output without endangering the environment. Due to their high efficacy and environmentally friendly makeup, research on the application of nanocomposites in the field of plant protection has significantly increased in recent years (Gomathi et al. 2019). Microarray analysis revealed that a number of genes were up-regulated or down-regulated by the application of AgNPs in Arabidopsis (Banerjee and Kole 2016). Such nanomaterial-induced responses directly contribute to plant stress defense. However, depending on the plant species, stage of growth, and type of nanomaterials utilized, different plants respond differently to nanofertilizers (Abdel-Aziz et al. 2016). Before the technology reaches the farm gate, more research is required to determine the signaling cascades and the genes activated by particular nanomaterials in various plant species

26.4 Future Perspectives

Still practices are going on to make nanotechnology more advantageous for agriculture sector. However concrete contributions are still uncertain (Parisi et al. 2015). Several factors are considered before commercial application of nanotechnology like nanofertilizers, nanopesticides etc. in agriculture. Precise use of nanofertilizers will counterbalance the high initial production investment. Currently, lots of nanotechnology based products are available in the market for which public perception is positive and demand for more future sensitive application. Such responses facilitate the promising application nanotechnology for improvement agriculture and generate benefits to human being. Massive agriculture waste (wheat straw and soy hulls) deposition is a big problem but incorporation of nanotechnology to use this waste as a source of bionanocomposites will be speculative solution.

26.5 Conclusion

Till now, nanotechnology is widely explored in the field of medicines, plant breeding, genetic transformation so to make more safe and beneficial different regulations and regulatory bodies are involved in its safety assessment. Wide use of nanotechnology is also concerned with phytotoxicity, bioaccumulation, and bioavailability of nanoparticle in plant. It is estimated that some health and environmental hazards may emerge from use of nanotechnology in agriculture. The introduction of engineered nanomaterials and their applications in the context of sustainable agriculture have fundamentally altered the world's agricultural landscape by their novelty, rapid expansion, and size in order to satisfy projected increases in global food consumption. The preservation of the environment from contamination is a key objective for commerce in sustainable agriculture, and nanomaterials guarantee improved input management and conservation. With lowered farming hazards, the potential of nanomaterials inspires a new green revolution. However, there are still a lot of unknowns regarding the ecotoxicity, acceptable limit, and absorption capacity of various nanomaterials. To understand the behavior and outcome of changed agriculture inputs and their interactions with bio-macromolecules found in biological systems, more research is urgently needed. The introduction of engineered nanomaterials and their applications in the context of sustainable agriculture have fundamentally altered the world's agricultural landscape by their novelty, rapid expansion, and size in order to satisfy projected increases in global food consumption. The preservation of the environment from contamination is a key objective for commerce in sustainable agriculture, and nanomaterials guarantee improved input management and conservation. With lowered farming hazards, the potential of nanomaterials inspires a new green revolution. However, there are still a lot of unknowns regarding the ecotoxicity, acceptable limit, and absorption capacity of various nanomaterials. To understand the behavior and outcome of changed agriculture inputs and their interactions with bio-macromolecules found in biological systems, more research is urgently needed. Remarkable progess has been made in the field of nanotechnology based product, but due to lack of legislative framework and health hazards delayed their commercialization. Proper nano-based material size should be addressed with up to date about global acceptance and rejection of agricultural nanoproducts. Precautionary guidelines related to formulation, application, and storage should be mentioned properly. The proper testing, validation, and certification of nonmaterial and nanoproducts in agriculture are also necessary for commercialization of technology.

References

Abd El-Azeim MM, Sherif MA, Hussien MS, Tantawy IAA, Bashandy SO (2020) Impacts of nano- and non-nanofertilizers on potato quality and productivity. Acta Ecol Sin 40:388–397

Abdel Latef AAH, Srivastava AK, El-sadek MSA, Kordrostami M, Tran LSP (2018) Titanium dioxide nanoparticles improve growth and enhance tolerance of broad bean plants under saline soil conditions. Degrad Dev 29:1065–1073

Abdel-Aziz HMM, Hasaneen MNA, Omer AM (2016) Nano chitosan-NPK fertilizer enhances the growth and productivity of wheat plants grown in sandy soil. Span J Agric Res 14:17

Abdulhameed MF, Taha AA, Ismail RA (2021) Improvement of cabbage growth and yield by nanofertilizers and nanoparticles. Environ Nanotechnol Monit Manage 15:100437

Adisa IO, Pullagurala VLR, Peralta-Videa JR, Dimkpa CO, Elmer WH, Gardea-Torresdey JL, White JC (2019) Recent advances in nano-enabled fertilizers and pesticides: a critical review of mechanisms of action. Environ Sci Nano 6:2002–2030

Ahanger MA, Qi M, Huang Z, Xu X, Begum N, Qin C, Zhang C, Ahmad N, Mustafa NS, Ashraf M (2021) Improving growth and photosynthetic performance of drought stressed tomato by application of nano-organic fertilizer involves up-regulation of nitrogen, antioxidant and osmolyte metabolism. Ecotoxicol Environ Saf 216:112195

Ahmed T, Noman M, Manzoor N, Shahid M, Abdullah M, Ali L (2021) Nanoparticle-based amelioration of drought stress and cadmium toxicity in rice via triggering the stress responsive genetic mechanisms and nutrient acquisition. Ecotoxicol Environ Saf 209:111829

Ahmed T, Noman M, Ijaz M, Ali S, Rizwan M, Ijaz U (2021a) Current trends and future prospective in Nano remediation of heavy metals contaminated soils: a way forward towards sustainable agriculture. Ecotoxicol Environ Saf 227:112888. https://doi.org/10.1016/j.ecoenv.2021.112888

Alexandratos N, Bruinsma J (2012) World agriculture towards 2030/2050: the 2012 Revision. ESA Working paper No. 12-03. Food and Agriculture organization of United Nations, Rome

Alsaeedi A, El-Ramady H, Alshaal T, El-Garawany M, Elhawat N, Al-Otaibi A (2019) Silica nanoparticles boost growth and productivity of cucumber under water deficit and salinity stresses by balancing nutrients uptake. Plant Physiol Biochem 139:1–10

Asadpour S, Madani H, Mohammadi GN, Heravan IM, Abad HHS (2020) Improving maize yield with advancing planting time and nano-silicon foliar spray alone or combined with zinc. SILICON 14:201–209

Banerjee J, Kole C (2016) Plant nanotechnology: an overview on concepts, strategies and tools. In: Kole C, Kumar D, Khodakovskaya M (eds) Plant nanotechnology. Springer, Cham, pp 1–14

Bhardwaj AK, Hamed LM, Sharma N, Rajwar D, Meti S, Nagaraja MS (2019) Engineered polymeric and nano-materials for taming salty soils and water used for crop production. In: Research developments in saline agriculture. Springer, Singapore, pp 391–405

Chen F, Bashir A, Rehman MZ, Adrees M, Qayyum MF, Ma J (2022) Combined effects of green manure and zinc oxide nanoparticles on cadmium uptake by wheat (Triticum aestivum L.). Chemosphere 298:134348

DeRosa MC, Monreal C, Schnitzer M, Walsh R, Sultan Y (2010) Nanotechnology in fertilizers. Nat Nanotechnol 5:91

Dubey A, Mailapalli DR (2016) Nanofertilisers, nanopesticides, nanosensors of pest and nanotoxicity in agriculture. In: Lichtfouse E (ed) Sustainable agriculture reviews. Springer, Cham, pp 307–330

Esmaili S, Tavallali V, Amiri B (2020) Nano-silicon complexes enhance growth, yield, water relations and mineral composition in Tanacetum parthenium under water deficit stress. SILICON 13:2493–2508

European Commission (2011) Commission recommendation of 18 October 2011 on the definition of nanomaterial (2011/696/EU)

Gomathi T, Rajeshwari K, Kanchana V, Sudha PN, Parthasarathy K (2019) Impact of nanoparticle shape, size, and properties of the sustainable nanocomposites. In: Inamuddin A, Thomas S, Kumar Mishra R, Asiri AM (eds) Sustainable polymer composites and nanocomposites. Springer, Cham, pp 313–336

Guha T, Mukherjee A, Kundu R (2021) Nano-scale zero valent iron (nZVI) priming enhances yield, alters mineral distribution and grain nutrient content of Oryzasativa L. cv. gobindobhog: a field study. J Plant Growth Regul 41:710–733

Hassanpouraghdam MB, Mehrabani LV, Tzortzakis N (2020) Foliar application of nano-zinc and iron affects physiological attributes of Rosmarinus officinalis and Quietens NaCl salinity depression. J Soil Sci Plant Nutr 20:335–345

Ikram M, Raja NI, Javed B, Mashwani ZUR, Hussain M, Hussain M (2020) Foliar applications of bio-fabricated selenium nanoparticles to improve the growth of wheat plants under drought stress. Green Process Synth 9:706–714

Kah M, Kookana RS, Gogos A, Bucheli TD (2018) A critical evaluation of nanopesticides and nanofertilizers against their conventional analogues. Nat Nanotechnol 13:677–684

Khalifa NS, Hasaneen MN (2018) The effect of chitosan nanofertilizer on pisumsativum plants. 3 Biotech 8(4):1–12

Khan MR, Rizvi TF (2014) Nanotechnology: Scope and application in plant disease management. Plant Pathol J 13:214–231

Khodakovskaya M, Dervishi E, Mahmood M, Xu Y, Li Z, Watanabe F, Biris AS (2009) Carbon nanotubes are able to penetrate plant seed coat and dramatically affect seed germination and plant growth. ACS Nano 3:3221–3227

Khodakovskaya MV, Kim BS, Kim JN, Alimohammadi M, Dervishi E, Mustafa T, Cernigla CE (2013) Carbon nanotubes as plant growth regulators: effects on tomato growth, reproductive system, and soil microbial community. Small 9:115–123

Kottegoda N, Munaweera I, Madusanka N, Karunaratne VA (2011) Green slow-release fertilizer composition based on urea-modified hydroxyapatite nanoparticles encapsulated wood. Curr Sci 101:73–78

Liu M, Feng S, Ma Y, Xie C, He X, Ding Y, Zhang J, Luo W, Zheng L, Chen D (2019) Influence of surface charge on the phytotoxicity, transformation, and translocation of CeO2 nanoparticles in cucumber plants. ACS Appl Mater Interfaces 2019(11):16905–16913

Liu J, Li G, Chen L, Gu J, Wu H, Li Z (2021) Cerium oxide nanoparticles improve cotton salt tolerance by enabling better ability to maintain cytosolic K+/na+ ratio. J Nanobiotechnol 19:153

Mahmoud LM, Shalan AM, El-Boray MS, Vincent CI, El-Kady ME, Grosser JW (2022) Application of silicon nanoparticles enhances oxidative stress tolerance in salt stressed 'Valencia' sweet orange plants. Sci Hortic 295:110856

Malekian R, Abedi-Koupai J, Eslamian SS (2011) Influences of clinoptilolite and surfactant-modified clinoptilolite zeolite on nitrate leaching and plant growth. J Hazard Mater 185:970–976

Manzoor N, Ahmed T, Noman M, Shahid M, Nazir MM, Ali L (2021) Iron oxide nanoparticles ameliorated the cadmium and salinity stresses in wheat plants, facilitating photosynthetic pigments and restricting cadmium uptake. Sci Total Environ 769:145221

Namjoyan S, Sorooshzadeh A, Rajabi A, Aghaalikhani M (2020) Nano-silicon protects sugar beet plants against water deficit stress by improving the antioxidant systems and compatible solutes. Acta Physiol Plant 42:157

Nazir MM, Li Q, Noman M, Ulhassan Z, Ali S, Ahmed T (2022) Calcium oxide nanoparticles have the role of alleviating arsenic toxicity of barley. Front Plant Sci 13:843795

Nejatzadeh F (2021) Effect of silver nanoparticles on salt tolerance of Satureja hortensis l. during in vitro and in vivo germination tests. Heliyon 7:e05981

Onaga G, Kerstin W (2016) Advances in plant tolerance to biotic stresses. In: Abdurakhmon IY (ed) Plant genomics. IntechOpen, Rijeka, pp 167–228

Parisi C, Vigani M, Rdríguez-Cerezo E (2015) Agricultural nanotechnologies: what are the current possibilities. Nano Today 10:124–127

Perrin TS, Drost DT, Boettinger JL, Norton JM (1998) Ammonium-loaded clinoptilolite: a slow-release nitrogen fertilizer for sweet corn. J Plant Nutr 21:515–530

Pitambara A, Archana A, Shukla YM (2019) Nanofertilizers: a recent approach in crop production. Nanotechnol Agric Crop Prod Prot 2019:25–58

Pretty J (2008) Agricultural sustainability: concepts, principles and evidence. Philos Trans R Soc Lond Ser B Biol Sci 363:447–465

Rajput VD, Minkina T, Feizi M, Kumari A, Khan M, Mandzhieva S, Sushkova S, El-Ramady H, Verma KK, Singh A (2021) Effects of silicon and silicon-based nanoparticles on rhizosphere microbiome, plant stress and growth. Biology 10:791

Shalaby TA, Abd-Alkarim E, El-Aidy F, Hamed E, Sharaf-Eldin S, Taha N, El-Ramady H, Bayoumi Y, Dos Reis AR (2021) Nano-selenium, silicon and H2O2 boost growth and productivity of cucumber under combined salinity and heat stress. Ecotoxicol Environ Saf 212:111962

Shang Y, Hasan MK, Ahammed GJ, Li M, Yin H, Zhou J (2019) Applications of nanotechnology in plant growth and crop protection: a review. Molecules 24:2558

Siddiqui MH, Al-Whaibi MH (2014) Role of nano-SiO2 in germination of tomato (Lycopersicum esculentum seeds Mill.). Saudi J Biol Sci 21:13–17

Sun L, Song F, Guo J, Zhu X, Liu S, Liu F, Li X (2020) Nano-ZnO-induced drought tolerance is associated with melatonin synthesis and metabolism in maize. Int J Mol Sci 21:782

Torabian S, Zahedi M, Khoshgoftar AH (2017) Effects of foliar spray of nano-particles of FeSO4 on the growth and ion content of sunflower under saline condition. J Plant Nutr 40:615–623

Van Nguyen D, Nguyen HM, Le NT, Nguyen KH, Nguyen HT, Le HM, Nguyen AT, Dinh NTT, Hoang SA, Van Ha C (2022) Copper nanoparticle application enhances plant growth and grain yield in maize under drought stress conditions. J Plant Growth Regul 41:364–375

Vanti GL, Nargund VB, Basavesha KN, Vanarchi R, Kurjogi M, Mulla SI, Tubaki S, Patil RR (2019) Synthesis of Gossypium hirsutum-derived silver nanoparticles and their antibacterial efficacy against plant pathogens. Appl Organomet Chem 33:e4630

Verma KK, Song X-P, Joshi A, Tian DD, Rajput VD, Singh M, Arora J, Minkina T, Li YR (2022) Recent trends in nanofertilizers for sustainable agriculture under climate change for global food security. Nanomaterials 12:173

Vermeulen SJ, Aggarwal PK, Ainslie A, Angelone C, Campbell BM, Challinor AJ, Hansen JW, Ingram JSI, Jarvis A, Kristjanson P (2012) Options for support to agriculture and food security under climate change. Environ Sci Policy 15:136–144

Wahid I, Rani P, Kumari S, Ahmad R, Hussain SJ, Alamri S (2022) Biosynthesized gold nanoparticles maintained nitrogen metabolism, nitric oxide synthesis, ions balance, and stabilizes the defence systems to improve salt stress tolerance in wheat. Chemosphere 287: 132142

Wang F, Haftka JJH, Sinnige TL, Hermens JLM, Chen W (2014) Adsorption of polar, nonpolar, and substituted aromatics to colloidal graphene oxide nanoparticles. Environ Pollut 186:226–233

Xu DM, Fu RB, Liu HQ, Guo XP (2021) Current knowledge from heavy metal pollution in Chinese smelter contaminated soils, health risk implications and associated remediation progress in recent decades: acritical review. J Clean Prod 286:124989

Yasmin H, Mazher J, Azmat A, Nosheen A, Naz R, Hassan MN (2021) Combined application of zinc oxide nanoparticles and biofertilizer to induce salt resistance in safflower by regulating ion homeostasis and antioxidant defence responses. Ecotoxicol Environ Saf 218:112262

Zhu Z-J, Wang H, Yan B, Zheng H, Jiang Y, Miranda OR, Rotello VM, Xing B, Vachet RW (2012) Effect of surface charge on the uptake and distribution of gold nanoparticles in four plant species. Environ Sci Technol 46:12391–12398

Zulfiqar F, Navarro M, Ashraf M, Akram NA, Munné-Bosch S (2018) Nanofertilizers use for sustainable agriculture: advantages and limitations. Plant Sci 289:1–11

Zwingmann N, Mackinnon IDR, Gilkes RJ (2011) Use of a zeolite synthesised from alkali treated kaolin as a K fertiliser: glasshouse experiments on leaching and uptake of K by wheat plants in sandy soil. Appl Clay Sci 53:684–690

Cut-Soiler Constructed Drainage: A Prospective Technique for Improving Crop Physiology and Yield Under Salt Stress Conditions

27

Rajender Kumar Yadav, Neha, Gajender Yadav, Ashwani Kumar, Govind Prasad, Manisha, and Junya Onishi

Abstract

Salts are the primary source of salinity in soil and water. Around one billion ha of global land area is affected by different kinds and concentrations of salt and associated threats. Soil salinity influences the performance of plants and determines their establishment and distribution. Salinity is regarded as one of the major limiting factors to crop production in arid and semiarid regions. Secondary salinization of irrigated drylands is often caused by either over irrigation causing rise in groundwater level and/or use of saline groundwater without adequate drainage. Cut-soiler constructed residue filled preferential shallow subsurface drainage (PSSD) can be a possible solution to prevent salt accumulation and facilitate salt removal by providing preferential outflow of the salt laden water from soil profiles. The cut-soiler PSSD is easy to construct without involvement of heavy machinery and cost. It is easy to adopt and can be applied on small scale at individual farm. This chapter describes the challenges of soil salinity for agricultural production. It also covers the use of cut-soiler constructed PSSD for salinity management, placement of amendments at desired subsurface depth for subsurface sodicity management, and role of cut-soiler in residue management. The information provided is based on research and feasibility trials on this technique in India under ICAR-CSSRI-JIRCAS project and elsewhere. Cut-soiler PSSD has shown effectiveness in reducing ~50% soil salinity during initial 3 years of its construction. Subsurface placement of gypsum + rice straw residue at 40 cm depth also tended to reduce soil sodicity (in terms of ESP) up to 24%,

R. K. Yadav (✉) · Neha · G. Yadav (✉) · A. Kumar · G. Prasad · Manisha
ICAR – Central Soil Salinity Research Institute, Karnal, Haryana, India
e-mail: RK.Yadav@icar.gov.in

J. Onishi
Japan International Research Center for Agricultural Sciences (JIRCAS), Tsukuba, Ibaraki, Japan

A. Kumar et al. (eds.), *Salinity and Drought Tolerance in Plants*,
https://doi.org/10.1007/978-981-99-4669-3_27

15%, and 6% at lateral distance of 0.30, 0.60, and 1.25 m, respectively. The resultant improvement in pearl millet and mustards yields under reduced soil salinity was ~23.5% and 31%, respectively. Similarly reduction in subsurface sodicity increased rice and wheat up to 15% and 16%, respectively. The yield enhancement in these crops is mainly due to improved plant water relations and physiological functioning under cut-soiler assisted salinity/sodicity management. These results suggest that cut-soiler constructed PSSD is a potential technique for the sustainable agricultural production under both saline and sodic salt stress environments. Further pilot scale studies are needed on its feasibility, cost-effectiveness, and adoptability by farming community to make it a viable option for the management of salty lands.

Keywords

Cut-soiler preferential shallow subsurface drainage (PSSD) · Salinity · Subsurface sodicity · Drainage · Crop physiology

27.1 Introduction

Currently, the food security of the world is severely challenged by ever-increasing population, deterioration in land and water resources, and climate vagaries. Salt-affected soils (SAS) occur in more than 100 countries of the world and the total global salt-affected soils area extends over 835 million ha, comprising sodic (438 million ha) and saline (397 million ha) soils, respectively (FAO/ UNESCO Soil Map of the World; FAO 2008). Additionally, approximately 76 million ha area is affected by human-induced salinization and sodification (Hossain 2019; Wicke et al. 2011) (Fig. 27.1).

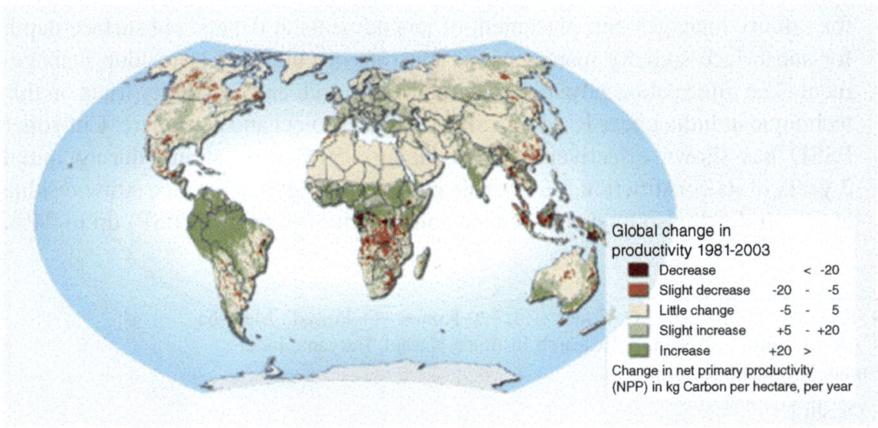

Fig. 27.1 Countries map affected by soil salinity/sodicity problem (https://www.researchg ate.net/publication/262495450)

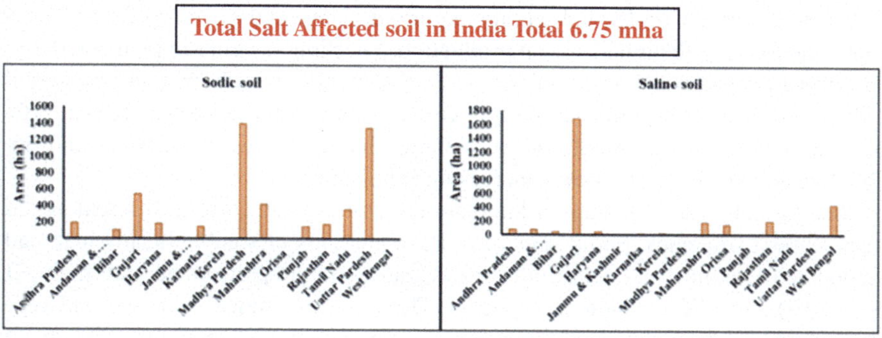

Fig. 27.2 Total salt-affected soil in India

At present, India has 6.73 million ha SAS area distributed in 16 states, of which 2.95 million ha are saline (Mandal et al. 2010). Moreover, another 16.2 M ha are likely to be salt-affected by 2050 (CSSRI 2015) in the predicted climate change scenario. Groundwater surveys indicate that poor quality waters being utilized for irrigation in different states are 32–84% of the total groundwater development (Minhas and Gupta 1992). Many more areas with good quality aquifers are endangered with contamination as a consequence of excessive withdrawals of groundwater (Fig. 27.2).

Soil salinization is a major environmental issue leading to acceleration of land degradation processes such as loss of biodiversity, hydrological resources, nutrient recycling that restrict valuable ecosystem services (Montanarella et al. 2018). It limits the agricultural and biomass production (Li et al. 2014) and at high levels poses a conspicuous threat to sustainable agricultural development. SAS are an important ecological entity in the landscape of any arid and semiarid region. Countries that reside in arid and semiarid regions do suffer from the build-up of salt balance that suppresses crop growth, nutrition, and subsequently in productivity failure, which result in lower food production and consequently insecurity for food and livelihood. Increased irrigation supply and its poor management lead to overirrigation and rise in groundwater. The rising groundwater in irrigated areas without proper drainage and use of saline groundwater for irrigation (Raheja et al. 2019) cause salt accumulation in dryland regions and thus bring more areas under salinization. As such indiscriminate use of poor quality waters in the absence of proper soil-water-crop management practices poses grave risks to soil health and environment. Development of salinity, sodicity, acidity, waterlogging, and toxicity problems in soils not only deteriorates the quality and quantity of produce and limits the choice of cultivable crops, many times the effects manifest so severe that lands eventually go out of cultivation. Yield losses are particularly detrimental at a local scale because SAS are not uniformly distributed and threaten the continued existence of agriculture in some regions and countries (Eynard et al. 2005). According to an estimate, the losses due to salt induced land degradation are about US $27.3 billion per year.

The management of dryland salinity requires drainage and leaching of salts below root zone to bring them under crop production. Leaching is being implemented as a measure to remove the salt, and because of that, we can observe the increased effluent water in its volume and the salt concentration in the drainage channel. The existing salinity management options using drainage are also insufficient and not able to catch up the pace of ever-increasing salinization.

Besides salinity, 3.78 million ha sodic soils in India require a different set of management strategies. Sodic soils have large amounts of soluble bicarbonate and carbonate of sodium and very high exchangeable sodium percentage with high pH (9.5–10.5) and ESP (as high as 80–95%). These become much dispersed and very impermeable to both water and air. Soil and water quality beckon major concerns of some sodic soil tracts mainly because of fine textured subsurface soil, high residual sodium carbonate (RSC), and high SAR (sodium adsorption ratio) water which restrict the natural drainage and a rationale to development of surface and subsurface sodicity. Amelioration of these soils needs a source of calcium (Ca^{2+}) that can replace the excess exchangeable sodium (Na^{+}). In addition to supplying readily available Ca^{2+} and SO_4^{2-} ions, gypsum is an excellent soil amendment, used as a soil conditioner to improve physical and chemical properties by promoting better aggregation, increasing water infiltration and movement through the profile consequently reclaim sodic soils and supply Ca and S for plant nutrition (Qadir et al. 2001). In addition to mitigation of salt stress, cut-soiler operation can help to manage the surplus amount out of estimated 84 and 141 Mt year^{-1}crop residues available in India. Of the 82 Mt of surplus crop residues nearly 70 Mts (44.5 Mt rice straw and 24.5 Mt wheat straw) are burned annually (Singh and Sidhu 2014). The working mechanism of the cut-soiler machine can place the amendment and straw both at desired depth and thus serve the purpose to reclaim subsurface sodicity and leaching out the excessive salts through in situ semi-natural lateral drainage system. Therefore, research and development for standardization of the supplemental/alternative techniques like construction of residue/residue + gypsum filled cut-soiler drainage are required for sustainable management of SAS.

27.2 Processes of Salinization

The USSL (1954) characterization of SAS in three categories, with slight modifications later by Soil Science Society of America (1987), is the most accepted classification. Likewise, poor quality water has also been characterized based on its suitability for irrigation purpose. Saline soils refer to soils that contain excess of neutral soluble salts to impair their productivity. The dominant soluble salts are chloride and sulfate of sodium along with appreciable quantities of chlorides and sulfates of calcium and magnesium. Sodic or alkali soils are influenced by dominance of exchangeable sodium. Sodic soils contain sodium salts capable of alkaline hydrolysis, mainly CO_3^{2-} and HCO_3^{-} of Na^{+}. These soils are also termed as 'alkali' soils. While the saline-alkali soils contain both excess of soluble salts and exchangeable sodium. For the management point of view, the SAS are broadly categorized

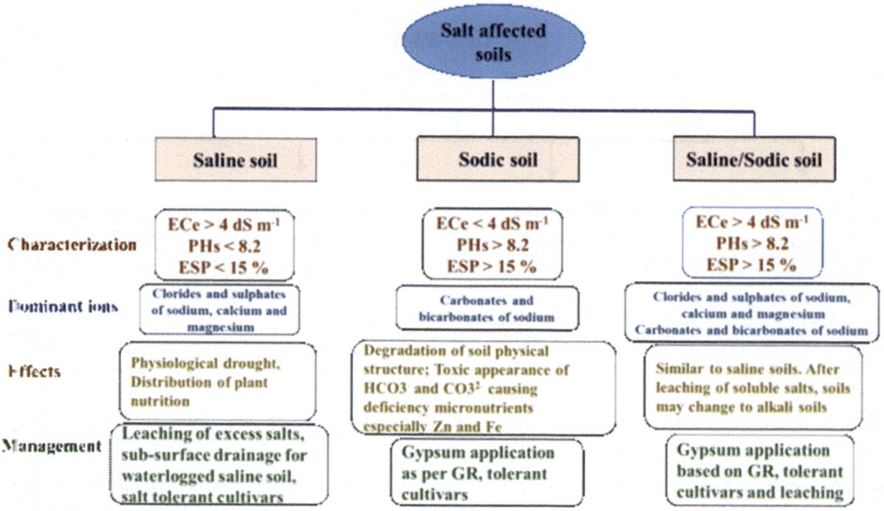

Fig. 27.3 Characteristics, dominant ions, effects, and management of salt-affected soils

into saline and sodic. The characteristics, dominant ions, effects, and management of salt-affected soils are shown in Fig. 27.3.

27.3 On Farm Management of SAS and Role of Cut-Soiler

The term reclamation of saline soils refers to the methods used to remove soluble salts from the root zone. To improve crop growth in saline soils the excess salts must be removed from the root zone. Leaching is the most effective procedure for removing salts from the root zone. In surface water stagnation and saline water irrigated saline soil areas proper land shaping and provision of drainage are needed to solve the problems. The cut-soiler PSSD is a residue filled shallow subsurface drainage that helps in salt removal and to manage the seasonal surface waterlogging to good extent. Whereas for the reclamation of sodic soil, amendment materials like gypsum is required that directly supply Ca^+ for the replacement of exchangeable Na^+. Despite of reclamation of vast sodic areas in India, the reclamation process is confined to the surface layer only. It is because of involvement of huge investment and manpower or machinery to place the amendments into deeper soil layers. Cut-soiler machine can place the amendment along with residue/straw at 40–60 cm depth and thus serves the purpose to reclaim subsurface sodicity up to desired depth. Further, cut-soiler PSSD construction using the surface scattered residue as a filling material tends to improve soil hydraulic properties. For construction of these drains the cut-soiler has been found suitable to manage a large quantity of residue and a potential option for in situ residue management. The detailed description of cut-soiler, its specifications, mode of operation, and advisory guidelines are provided in the subsequent sections.

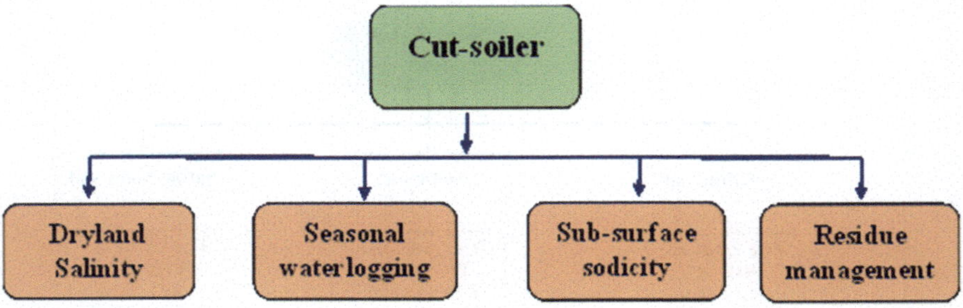

Fig. 27.4 Potential scope of cut-soiler for the management of SAS

27.4 Cut-Soiler

Cut-soiler is a tractor operated machine that uses soil surface scattered materials like straw, residue, or remaining stems for construction of PSSD channels while running on the field. It could be useful to manage the twin problems of surface water stagnation and salinity. At the same time, it is a very helpful machine for subsurface sodicity and residue management.

Cut-soiler has been developed jointly by National Agriculture and Food Research Organization (NARO) and Hokkai Koki Corporation, Japan (patented in 2017). It was introduced in India under Japan International Research Centre for Agricultural Sciences (JIRCAS)-ICAR-Central Soil Salinity Research Institute (ICAR-CSSRI) collaborative research project. Cut-soiler constructed PSSD has been successfully applied in the multipurpose paddy fields of Japan for discharging saline water (Okuda et al. 2020). Previous studies (Chiba et al. 2012; Kaneko et al. 2002) also reported the desalinization effect of cut-soiler for bringing problematic salt-affected soils under sustainable crop production (Fig. 27.4).

27.4.1 Cut-Soiler Specifications

The specifications and tractor power requirements of cut-soiler are given below (Fig. 27.5):

27.4.2 Cut-Soiler

27.4.2.1 Specifications

- Dimensions ($L \times W \times H$): 2.0 m × 1.5 m × 1.65 m.
- Weight: 800 kg.

Fig. 27.5 Cut-soiler machine

27.4.2.2 Recommendations

- Tractor capacity range: 60–120 HP.
- Operating speed: 2–3 km/h.
- Straw material treatment: maximum 10 cm length.
- Construction depth: 40–60 cm.
- Machine maker company: Hokkai Koki Corporation.

27.4.3 Operation Mode

Cut-soiler machine cuts the soil and opens V-shape furrow by cutting and lifting the soil, followed by simultaneous placing of the surface scattered straw and residue at bottom with overlaying the lifted soil (Fig. 27.6). Such cut-soiler construction lines serve as drainage channels and help to manage surface waterlogging and soil salinity. In subsurface drainage, pipes are laid at 1.5 m depth to drain out excess water from the fields, while the cut-soiler machine makes drains at 40–60 cm depth. Crop residue is filled in the drains and that helps to drain out water.

27.4.4 Advantages of Cut-Soiler Based Shallow Subsurface Drainage Technology

1. Cut-soiler is a tractor mounted machine and its operation is cost-effective and easy to adopt by individual farmer.
2. It uses surface lying residue and straw as a filling material, thus likely to help in surface residue management and reducing residue burning problem.
3. Construction line being shallow, it helps in reducing drainage effluent volume.
4. It can also help in managing subsurface sodicity.
5. Putting residue at bottom of cut-soiler opening will help in carbon sequestration and improvement in soil health.

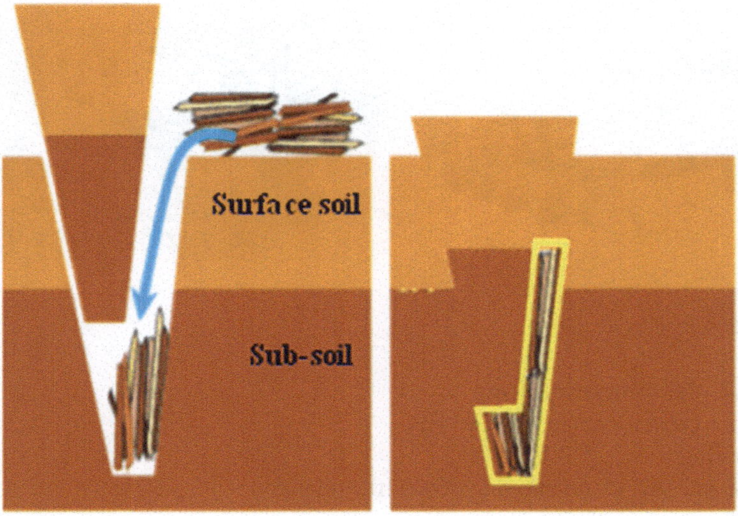

Fig. 27.6 Method of cut-soiler operation for shallow subsurface preferential drainage construction

27.4.5 Advisory

1. The tractor running should be straight in the direction of natural slope, so as to maintain the proper interval and to improve water flow.
2. To maintain the stable uniform depth of cut-soiler construction lines, the field should be laser levelled followed by operation with slow speed along with constant rotation of PTO (power take-off).
3. The filling material, such as crop residues, should be preferably uniform in size (8–10 cm) and spread on the ground. Otherwise, it may choke and hamper the uniformity of cut-soiler operation and also reduce tractor pulling ability.
4. Life of cut-soiler construction line depends on the quality of straw/residues used. However, life of channels, prepared using rice straw residue, remains effective for ~3 years. The farmers themselves can check by digging through soil in the cut-soiler line.
5. The mulcher is used to cut the rice straw in finer length up to 10 cm and gypsum with agricultural grade having ≥70% purity per cent is broadcasted on the basis of 50% GR (gypsum requirement) in the field.
6. The depth of cut-soiler wheel should be adjusted before running in the field depending upon subsurface sodic layer.
7. The maximum spacing between two consecutive cut-soiler drainage lines should be ≤2.5 m for better results.
8. All the cut-soiler lateral drains should be connected with main drain line to lay off drained water from field.

27.5 Soil Salinity Management

To quantify the salt removal effect of cut-soiler and its subsequent effect on crop growth and physiology, the experiments were laid out in semicontrolled lysimeters under ICAR-JIRCAS collaborative project. The cut-soiler operation simulated drains were constructed by filling with rice residue at 60 cm below the surface. The pearl millet and mustard were taken as test crops for 2 subsequent years (2019–2020 and 2020–2021) to study the effect of cut-soiler PSSD on growth and physiology of both crops. The PSSD were connected to an outlet provided at the same depth (60 cm) in the lysimeter set up for this purpose.

The cut-soiler assisted PSSD helped to reduce soil salinity during 3 years as compared to the initial values. The overall salinity reduction from the construction of simulated cut-soiler drain from July 2018 to October 2020 was 59.92% (Neha et al. 2022a). Figure 27.7 shows the reduction in soil salinity in cut-soiler (PSSD) treated plots, during 2019 and 2020. There was a 23.28% and 41.46% reduction in soil salinity under cut-soiler PSSD in April 2019 and April 2020, respectively. A similar reduction in soil salinity in cut-soiler treated plots was also recorded during October in both years. The lower soil salinity observed in residue filled simulated cut-soiler PSSD plots throughout the experimental period shows the positive effect of preferential outflow of water and salts through cut-soiler PSSD (Neha et al. 2022a). The enhanced outflow of water in the cut-soiler plots also resulted in lower soil moisture (%) in these plots. The lower soil moisture content (%) in cut-soiler treatments implies the higher outflow of water through the cut-soiler preferential drains that resulted in larger removal of salts by the PSSD system. The temporal decrease in soil salinity was due to continuous removal of salts by cut-soiler PSSD.

The total salt removal through cut-soiler PSSD was estimated by Neha et al. (2022b) (Table 27.1). The total volume of drained water and thus amount of salt removal were directly related to the total water input available for leaching mainly through rainfall as the same amount of irrigation water applied in all studied durations. The amount of salt removal was also related to the salinity of drained

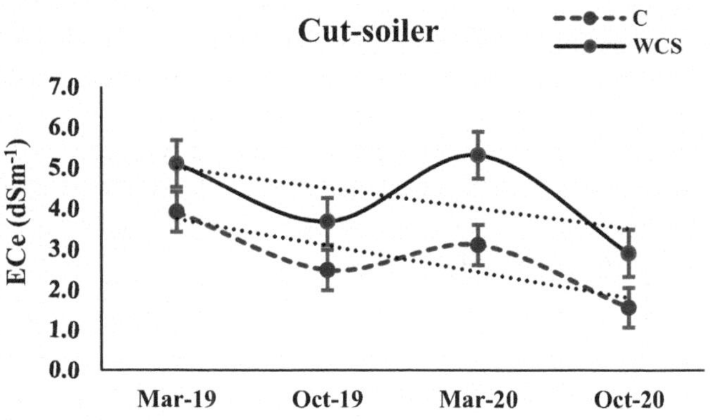

Fig. 27.7 The changes in soil salinity (ECe, dS m^{-1}) under cut-soiler

Table 27.1 Soil water and salt balance components at different times

Period	Rainfall (mm)	Irrigation (mm)	Evaporation (mm)	Drain water (mm)	Drain EC[a] (dS m^{-1})	Salt removed (kg/lys[a])
June 2018 to October 2018	984.7	3 × 50	524.1	188.4 (753 L)	3.50	1.7
November 2018 to April 2019	79.6	3 × 50	491.9	–	–	–
May 2019 to October 2019	400.3	3 × 50	984.8	42.7 (170 L)	2.93	0.32
November 2019 to April 2020	322.3	3 × 50	444.4	14.6 (58.4 L)	3.14	0.12
May 2020 to October 2020	995.3	3 × 50	873.0	137.5 (550 L)	2.16	0.77
November 2020 to April 2021	113.0	3 × 50	542.1	–	–	–

Source: Neha et al. (2022b)
[a] It is calculated individually for each duration

water along with water input. The water and salt drainage were consistently higher during the rainy seasons. Maximum water (188.4 mm) and salt (1.7 kg) were drained out from June 2018 to October 2018 followed by May 2020 to October 2020 (137.5 mm; 0.77 kg) when maximum rainfall received. Low water and salt drainage during May 2019 to October 2019 were mainly due to lesser rainfall in that year. However, there was very low (November 2019 to April 2020) or no drainage in winter season.

The reduction in soil salinity by executing preferential drainage was also reported earlier (Technical Manual. Okuda et al. 2017) with the cut-drains being created under optimum moisture in saline fields in Uzbekistan. Here the vertical voids were created by the blade of the drain-drilling machine. Such drainage by cut-drains was found effective in reducing soil salinity up to 44% near to cut-drains in comparison to control after 1 year of leaching salts in salt-affected regions of Uzbekistan (Okuda et al. 2020). In this study the constructed cut-drain provided supplemental drain when used as a connection to the main drain and facilitated the leaching effect. They found the higher reduction in soil salinity under cut-drain and near to the drainpipe (Fig. 27.8).

The average soil salinity (EC_e) of 0–100 cm soil layer before and after leaching is shown in Fig. 27.8. Percentage reductions of soil salinity after leaching and the relative soil salinity of the control field, cut-drain field, and near the drainpipe were 75%, 73%, and 56%, respectively.

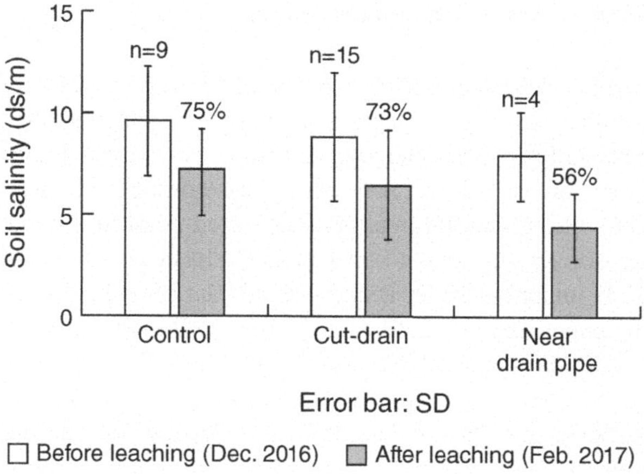

Fig. 27.8 Rate of soil salinity change (0–100 cm)

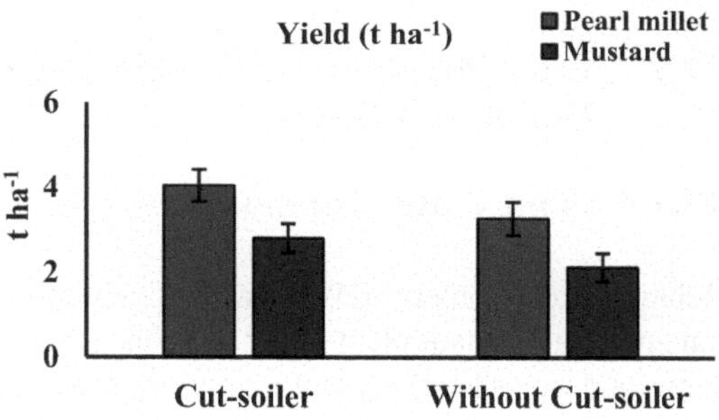

Fig. 27.9 Effect of cut-soiler on grain yield of pearl millet and mustard crops

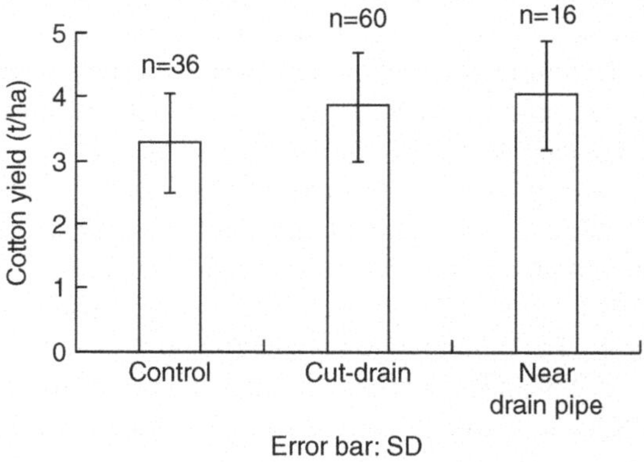

Fig. 27.10 Cotton yield (t ha^{-1}) in 2017

27.6 Yield Improvements

Increased yield effect of cut-soiler PSSD was the result of consistent reduction in soil salinity along with higher leaching of salts that created better soil conditions, especially in cut-soiler operated plots. Improved drainage conditions of cut-soiler plots that led to lower soil salinity also contributed toward increased crop yields. The cut-soiler operation treatment increased pearl millet grain yield (~23.54%) in comparison to without cut-soiler plots (Neha et al. 2022a). The increase in mustard seed yield under cut-soiler PSSD was ~31% (2.80 t ha^{-1}) yield in comparison to without the cut-soiler, i.e. (2.13 t ha^{-1}), Fig. 27.9 (Neha et al. 2022b).

Okuda et al. (2018) reported 20% increase in cotton yield (Fig. 27.10) after 1 year of construction of preferential drainage by cut-drains in salt-affected fields of Uzbekistan. They concluded that shallow subsurface drainage was effective in reducing soil salinity and accordingly improving crop yields.

27.7 Crop Physiological Responses of Cut-Soiler Drainage Under Salt Stress

27.7.1 Plant Water Relations

Relative water content (RWC), water potential, osmotic potential, and turgor potential are the important plant water relations parameters used for quantifying the water status and stress level in plants. Neha et al. (2022a) studied these traits in pearl millet and showed significant decline under increased level of salinity stress (Table 27.2) in without cut-soiler plots. Relatively higher rhizosphere salinity under without cut-soiler (control) resulted in decreasing the soil solution water potential and thus reduction of water uptake by plant roots and ultimately caused decrease in relative water content (RWC), water potential, osmotic potential, and turgor potential (Nandwal et al. 2007).

The physiological traits also showed decreasing trend under salinity stress (Table 27.2) as usually observed under different abiotic stresses. Cut-soiler enhanced the photosynthetic rate, transpiration rate, and chlorophyll fluorescence by 9.35%, 7.14%, and 7.57%, respectively, in comparison to without cut-soiler in pearl millet crop (Table 27.2), Neha et al. (2022a).

All plant water relations parameters were also maintained higher under cut-soiler PSSD in mustard crop at both vegetative and reproductive stages (Neha et al. 2022b). RWC under cut-soiler was recorded higher at vegetative (91.26%) and reproductive (82.75%) than without cut-soiler (81.48; 74.17%) (Table 27.3). The cut-soiler caused 28.72% and 22.44% improvement in water potential; 15.15% and 18.69% in osmotic potential at vegetative and reproductive stages. Turgor potential under cut-soiler was recorded higher at vegetative (0.430 MPa) stage only.

Table 27.2 Effect of cut-soiler on plant water relation and physiological traits of pearl millet

Treatments/traits	Relative water content (%)	Water potential (Mpa)	Osmotic potential (Mpa)	Turgor pressure	Photosynthetic rate (μmol m^{-2} s^{-1})	Transpiration rate (μmol m^{-2} s^{-1})	Chlorophyll fluorescence (Fv/Fm)
Cut-soiler							
Cut-soiler	72.95^A	1.58^B	1.72^B	0.14	25.25^A	10.66^A	0.71^A
Without cut-soiler	70.01^B	1.96^A	2.14^A	0.16	23.09^B	9.59^B	0.66^B

Source: Neha et al. (2022a)

Symbols A and B indicate the significance of difference based on LSD (Least Significance Difference) at 5% level of significance using ANOVA

Table 27.3 Effect of cut-soiler on plant water relation of mustard crop

	Relative water content (%)		Water potential (MPa)		Osmotic potential (MPa)		Turgor pressure	
Treatments/traits	Veg	Repro	Veg	Repro	Veg	Repro	Veg	Repro
Cut-soiler	91.26^{A}	82.75^{A}	-1.88^{B}	-1.47^{B}	-2.31^{B}	-2.14^{B}	0.430^{A}	0.670
Without cut-soiler	81.48^{B}	74.17^{B}	-2.42^{A}	-1.80^{A}	-2.66^{A}	-2.54^{A}	0.232^{B}	0.742

Symbols A and B indicate the significance of difference based on LSD (Least Significance Difference) at 5% level of significance using ANOVA

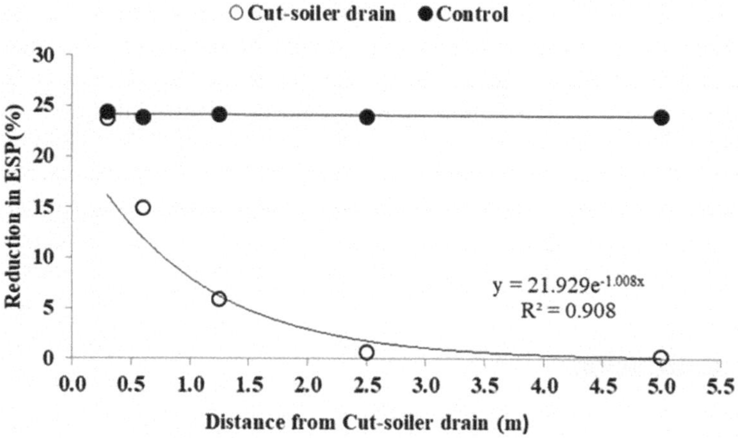

Fig. 27.11 Reduction (%) in soil subsurface (40 cm depth) ESP at different lateral distance intervals from cut-soiler construction line

Table 27.4 Changes in subsurface (40 cm below surface) sodicity (ESP%) at variable lateral distance from cut-soiler drain and increase in grain yield (t ha^{-1}) of wheat (2019–2020) and rice (2020)

	Grain yield (t ha^{-1})		Soil ESP (%) distance cut-soiler line		
Cut-soiler spacing	Wheat (2019–2020)	Rice (2020)	1/8 middle of quarter	1/4 middle of half	1/ 2 Middle
Control	4.08	2.08	24.4	23.8	24.2
2.5	4.83	2.45	18.6	20.7	23.4
5	4.52	2.24	20.3	21.9	24.1
10	4.12	2.12	22.8	23.6	23.9
CD (0.05)	0.58	0.38	–	–	–

Source: Yadav et al. (2022)

27.8 Subsurface Sodicity Reclamation and Residue Management Using Cut-Soiler

Yadav et al. (2022) placed the gypsum (50% GR) along with rice residue at 40 cm below the surface in a field experiment using cut-soiler. Gypsum, rice residue, and gypsum + rice residue were applied in subsurface at 2.5, 5.0, and 10.0 m of lateral spacing. Subsurface sodicity (in terms of ESP) was reduced by up to 23.77, 14.92, and 5.95 per cent at lateral distance of 0.30, 0.60, and 1.25 m, respectively, by the placement of gypsum + rice straw residue at 40 cm depth (Fig. 27.11).

Rice and wheat yields increased by 15% and 16% in 2.5 m spacing and 6% and 10.7% in 5.0 m spacing, respectively, over control (Table 27.4). The results showed that closer spacing (2.5 m or less) appears to be more promising for subsurface sodicity management.

Another column study shows that the application of amendment and rice straw residue has positive effect in mitigating the salt stress in rice crop. The subsurface placement of residue and gypsum along with the surface application of gypsum (50% GR) increased the RWC (relative water content), photosynthetic rate, transpiration rate, and stomatal conductance in rice plants over control. In addition to these traits, subsurface placement of residue and gypsum also reduced membrane injury, Na^+K^+ levels, and lowered water potential and osmotic potential in rice. Therefore, the cut-soiler assisted subsurface placement of these amendments found promising in reclamation of subsurface sodicity and increase rice yields in sodic soils.

27.9 Conclusion

The cut-soiler PSSD can reduce up to 50–60% soil salinity during 3 years of its construction in different soils. Cut-soiler PSSD resulted in a higher outflow of free water and salts with drainage. Subsurface placement of gypsum and rice residue by cut-soiler operation was found effective in reducing subsurface sodicity (ESP %) and also helped in in situ management of the rice residue and thus avoid residue burning. The cut-soiler PSSD could be an effective technique for management of soil salinity, subsurface sodicity, and crop residues. The reduction in soil salinity and subsurface sodicity provided congenial soil conditions for crop production. The improvement in pearl millet and mustard yield by managing dryland soil salinity and increased rice wheat yields by penetrating the sodicity reclamation process deeper in subsurface layers was evident from their positive effect on plant water relations and physiological parameters of these crops.

Therefore, cut-soiler constructed PSSD has proved to be a potential technique for the sustainable agricultural production under both saline and sodic salt stress environments. The yield enhancement in the agricultural crops is mainly due to improved soil physic-chemical properties and consequently plant water relations and physiological functioning of these crops. Further studies are needed on its feasibility, cost-effectiveness, and adoptability by farming community to make it a viable option for the management of salty lands and saline water use for irrigation. A more precise understanding of water and salt dynamics and economic viability of cut-soiler PSSD in the management of root zone salinity and crop residues will be a step towards achieving land degradation neutrality and food and livelihood security for farmers of salt-affected areas.

References

Chiba K, Kato T, Togashi C, Kanmuri H (2012) Effectiveness of desalinization by infiltration of water and desalinization of tsunami-hit farmlands in Miyagi Prefecture. Water Land Environ Eng 80(7):3–6. (In Japanese)

CSSRI (2015) Vision 2050. ICAR-Central Soil Salinity Research Institute, Karnal

Eynard A, Lal R, Wiebe K (2005) Crop response in salt-affected soils. J Sustain Agric 27(1):5–50

FAO/UNESCO Soil Map of the World (2008) https://www.fao.org/soils-portal/data-hub/soil-maps-anddatabases/faounesco-soil-map-of-the-world/en/

Hossain MS (2019) Present scenario of global salt affected soils, its management and importance of salinity research. Int Res J Biol Sci 1(1):1–3

Kaneko T, Murakawa M, Kozai N, Mitsugi K (2002) Desalting technique for illuviated salts soil at paddy field by using underdrainage. J Jpn Soc Irrig Drain Reclam Eng 70:27–30. (In Japanese)

Li J, Pu L, Zhu M, Zhang J, Li P, Dai X, Liu L (2014) Evolution of soil properties following reclamation in coastal areas: a review. Geoderma 226:130–139

Mandal T, Maity S, Dasgupta D, Datta S (2010) Advanced oxidation process and biotreatment: their roles in combined industrial wastewater treatment. Desalination 250(1):87–94

Minhas PS, Gupta RK (1992) Quality of irrigation water: assessment and management. Indian Council of Agricultural Research, New Delhi

Montanarella L, Scholes R, Brainich A (2018) The IPBES assessment report on land degradation and restoration (p. 744). Secretariat of the intergovernmental science-policy platform on biodiversity and ecosystem services. ISBN (electronic) 978-3-947851-09-6

Nandwal AS, Kukreja S, Kumar N, Sharma PK, Jain M, Mann A, Singh S (2007) Plant water status, ethylene evolution, N_2-fixing efficiency, antioxidant activity and lipid peroxidation in Cicer arietinum L. nodules as affected by short-term salinization and desalinization. J Plant Physiol 164(9):1161–1169. https://doi.org/10.1016/j.jplph.2006.05.017

Neha, Yadav G, Yadav RK, Kumar A, Rai AK, Sehwag M, Onishi J, Omori K, Sharma PC (2022a) Desalinization effect of simulated cut-soiler based residue filled preferential shallow sub surface drainage improves physiology and yield of rainfed pearl millet in saline semiarid regions. Land Degrad Dev 33(10):1758–1770. https://doi.org/10.1002/ldr.4237

Neha, Yadav G, Yadav RK, Kumar A, Rai AK, Onishi J, Omori K, Sharma PC (2022b) Salt removal through residue filled cut-soiler simulated preferential shallow subsurface drainage improves yield, quality and plant water relations of mustard (*Brassica juncea* L.). Sustainability 14(7):4146. https://doi.org/10.3390/su14074146

Okuda Y, Omori K, Onishi J (2017) Shallow sub-surface drainage for mitigating salinization, Technical Manual, Japan International Research Center for Agricultural Sciences (JIRCAS). Farmers' Council of Uzbekistan (FC):1–82

Okuda Y, Onishi J, Shirokova YI, Kitagawa I, Kitamura Y, Fujimaki H (2018) Salt removal technology by shallow subsurface drainage in combination with a cut-drain. J Arid Land Stud 28(Suppl):127–130

Okuda Y, Goto K, Kitagawa I (2020) A trial of desalinization by using a mole-drain in the Republic of Uzbekistan. Development of subsurface drainage and water-saving irrigation technology for mitigation of soil salinization in Uzbekistan, p 65

Qadir M, Ghafoor A, Murtaza G (2001) Use of saline–sodic waters through phytoremediation of calcareous saline–sodic soils. Agric Water Manag 50(3):197–210

Raheja H, Goel A, Bundela DS (2019) Reclamation of waterlogged saline lands in Haryana through subsurface drainage technology—a review. In: Sustainable engineering. Springer, Berlin, pp 167–175

Singh Y, Sidhu HS (2014) Management of cereal crop residues for sustainable rice-wheat production system in the Indo-Gangetic plains of India. Proc Indian Natl Sci Acad 80(1):95–114

Soil Science Society of America (1987) Glossary of Soil Science Terms. Soil Science Society of America, Wisconsin, USA, Technology Committee

USSL (1954) Diagnosis and improvement of saline and alkali soils. Handbook 60.United States Salinity Laboratory. United States Department of Agriculture, Washington, DC, p 160

Wicke B, Smeets E, Dornburg V, Vashev B, Gaiser T, Turkenburg W, Faaij A (2011) The global technical and economic potential of bioenergy from salt-affected soils. Energy Environ Sci 4(8): 2669–2681

Yadav RK, Yadav G, Neha, Narjary B, Sharma PC, Omori K, Onishi J, Watanabe T, Anzai T, Okamoto K (2022) Gypsum and crop residue placement by cut-soiler help to manage soil sub-surface sodicity in semi-arid Indo-Gangetic Plains. J Arid Land Stud 32-S:113–116

Tree Adaptive and Reclamation Mechanisms in Managing Waterlogged Saline Habitats: Prospectus and Impediments

28

Rakesh Banyal, Varun Saini, Neha Saini, Rajkumar, Manish Kumar, and Kuldeep

Abstract

Worldwide, the twin problem of waterlogging and salinity is upsurging with each passing day and increasing at an alarming rate in irrigation command areas, affecting agricultural productivity in 1/3rd total irrigated area. This is now a big concern if not addressed on scientific lines and will accelerate at a greater pace in the future at the behest of climate change. These soils are classified into waterlogged (0.0–2.0 m), sensitive/critical (2.0–3.0 m), protected (3.0–5.0 m) and safe (>5.0 m) classes, helping in devising the management strategies. The solution lies in drainage improvement. The limitations of the available conventional engineering approaches gave a call for effective, affordable, socially acceptable and environmental friendly alternatives. With this backdrop, the only viable option is bio-drainage to use waterlogged saline soils for productive functions. It can be used as a corrective and/or preventive measure by intercepting soil water before it arrives at the water table. The trees with inbuilt capability of salt tolerance with phreatophytic nature are suitable to plant in waterlogged saline landscapes. Such tolerant plants adapted three basic mechanisms of salt exclusion, excreting and accumulation realized through the strategies of ion compartmentalisation, osmotic adaptation, osmolyte production, succulence, selective uptake and transport of ions for establishment, growth, production and soil reclamation. Trees are grouped into very promising and promising classes based on their overall performance rating and fast, medium and slow bio-drainers on evapotranspirative potential. *Eucalyptus* emerged as a potential biodrainage

R. Banyal (✉) · Rajkumar · M. Kumar
ICAR-Central Soil Salinity Research Institute, Karnal, Haryana, India

V. Saini · N. Saini · Kuldeep
ICAR-Central Soil Salinity Research Institute, Karnal, Haryana, India

College of Agriculture, CCS Haryana Agricultural University, Hisar, Haryana, India

A. Kumar et al. (eds.), *Salinity and Drought Tolerance in Plants*,
https://doi.org/10.1007/978-981-99-4669-3_28

tree which gave the best biodrainage models in addressing the problem of inland ecosystem. Block models are best to recede the water table than boundary models in severely affected areas. However, boundary plantation model was preferred by the farmers in moderately and less affected areas in Haryana state. Six years in 1 × 1 m spacing (twin row strip plantation) on farm acre line was the recommended *Eucalyptus* based biodrainage model for achieving higher water table drawdown, biomass, carbon sequestration and crop yield. Willow clone (J799) is comparable to *Eucalyptus* in bio-drained capability and would be useful in managing waterlogged saline soils in semi-arid regions. Although the biodrainage technology is known from the last decade of the twentieth century still lot to do in addressing some of the bottlenecks with this technology. The impediments are decrease in tree transpiration rate with the advancing age, trees could not survive if the soil salinity goes beyond 12 dS/m, allelopathy of trees, social acceptability, etc. To address the impediments, the broad future research domains would be on the trees adaptation and reclamation aspects are tree water use, salt uptake and tolerance and sustainability. Biodrainage is an emerging technology for managing waterlogged saline soils and its feasibility needs more scientific backup. At present state of knowledge, biodrainage should be combined with conventional drainage options as an integrated approach for economic and ecological solutions of the waterlogged saline landscapes.

Keywords

Waterlogged · Saline soil · Soil reclamation · Adaptation · Physiology · Mechanisms

28.1 Introduction

India is bestowed with a wide range of climatic and physiographic conditions. The incessant increase in global population and decrease in average landholding results in soil degradation with each passing day is now a big challenge. In the Indian context, in the 1960s, Green Revolution not only helped in achieving self-sufficiency but made it a food surplus country. The practices done during that time included excessive use of fertilizers, groundwater irrigation, expansion of canal irrigation, growing of high yielding varieties and mechanization, etc. but now resulted in numerous problems of high cultivation costs, decreased productivity, loss of soil fertility and above all degradation of natural resources. It has been seen that the problem of salinity coupled with waterlogging is common across the globe. Globally, one billion hectares of arid and semi-arid area is salt affected and out of which 1/3rd irrigated area is waterlogged in nature. In India, about 6.73 m ha area is salt affected out of which 3.77 m ha is sodic and 2.96 m ha is saline in nature. Problem is more intense as the underlain aquifer is of poor quality (saline and RSC water) in salt affected soils. Worldwide, as per estimate, 255 m ha area experiences the risk of waterlogging which accounted for one-third of total irrigated area. Out of which

60 m ha is waterlogged and 20 m ha is salt affected (Heuperman et al. 2002). Annual loss of irrigated land at the behest of secondary salinization with waterlogging is at the rate of 1.50 m ha. The introduction of a canal network with poor drainage options and faulty irrigation management practices are the reasons for secondary salinization and associated waterlogging in inland regions of the country. This is now a big concern especially in semi-arid regions of northwestern part of India which if not addressed on scientific lines will accelerate at a greater pace in the future. Hence there is a need to drain excess water and salts to make such areas suitable for cultivation. Conventional engineering based approaches like subsurface drainage (SSD) (vertical, horizontal and mole) are in practice but their adoption is regulated by high capital, operational and maintenance cost and above all environmental concerns. The limitations and shortcomings of conventional engineering approaches call for alternatives to keep agricultural practices sustainable over time. We need alternative techniques which are effective, affordable, socially acceptable and environmental friendly. The only one option which could fulfil all the characteristics is the biodrainage technique.

Bioremediation is an environment friendly and cost-effective approach that can tackle the twin problem of waterlogging and salinity. Trees are usually referred as perennial plants with certain distinct characteristics, viz. having a stem, well defined crown, supporting branches with well-developed root structure. It plays a significant role in regulating soil and surrounding environment. Owing to their prolonged existence, trees are often entrusted upon to address numerous climatic adversities with existing biotic and abiotic stresses. Trees can easily adapt to dynamic environments with ease and also provide other related products either economical and/or ecological significance. With the increase in the extent of area under waterlogging and salinity worldwide, trees are now seen to be a promising way to address this problem. Adaptations of trees to such conditions are quite interesting with reclamation of such harsh landscapes especially salt affected lands for their productive utilization.

Excessive salt in soils can lead to affect almost every stage of growth in any plant (Dmuchowski et al. 2022). Furthermore, the salt when absorbed by the plants directly affects the cells and distorts them. Excessive salt disturbs the osmotic potential and reduces water absorption. Waterlogging is a drainage problem noticeable when the soil pores in the root zone of plants are saturated and the insufficient air circulation renders the soil barren and not suitable for agriculture. It is due to combined factors, viz. geographical location, climatic factors and anthropological causes. The twin situation of waterlogging and salinity is most common and threatening in areas where drainage is the problem and the soil stability is disturbed. It is high time now to look into this twin hazard and find out possible and viable solutions.

The severity of waterlogging is said to be less if the water table remains below 3 m for a substantial period of time during a year. This problem alleviates further and is classified into moderately affected when the water table in any area is found to be less than 1.5 m and critical when the groundwater is present up to the depths of 0–30 cm with surface ponding where the water table comes above the surface.

Similarly, presence of excess salts in soils hence lifting the osmotic pressure and inhibiting the plant water uptake results in soil salinity (Singh and Lal 2018). Waterlogging and salinity problem is prominent in all types of lands whether they are irrigated or non-irrigated. Researchers have directed toward drainage technology as a remedy for waterlogging and salinity problems yet being costly, labor intensive, and polluting (dumping of effluent), is now lacking interest. Searching for economically and ecologically viable alternatives is the need of the hour. Such alternative is biodrainage which involves extracting water through evapotranspiration done by fast growing and deep-rooted plants (Dash et al. 2005). It can be adapted by planting vegetation comprising tree species that have a phreatophytic nature. It will result in lowering of water table, moderation of soils and surrounding climate, cycling of minerals and nutrition helps in removal of salt ions. This chapter is in central stage of providing the basic concept of biodrainage through adaptive tree and reclamation mechanisms, challenges and new paths in biodrainage research with reflections of successful models from Haryana state *vis-a-vis Eucalyptus* as model tree in biodrainage measures.

28.2 Trees Adaptive Mechanism in Salt Laden Landscapes

Salinity is often related to barren lands with white crusted surface devoid of vegetation, patchy trees growing stunted and slender. Excessive salt has an inverted relation with plant mechanisms and productivity. Halophytes are trees or plants which are capable to survive in harsh salt-affected soils either by adapting or modifying certain parts or by adapting mechanisms to cope up or exclude salts. Halophytes or salt-tolerant plants have a stunning range of abating salt stress and can be grown all over the world be it in the dry lands or the salty gusts and tides of sea or ocean. Salt stress in trees results in reduced osmotic potential of plants (Liu et al. 2020) eventually causing its death commonly termed as physiological drought. Further, the excessive quantities of Na^+ and Cl^- ions in the form of ionic poisoning (Luo et al. 2017) usually disrupt the cell organelles due to secondary oxidative stress (Qi et al. 2018) and eventually hamper the tree growth and development. Similarly, increased concentration of these ions inside the plants due to excessive uptakes eventually hampers the nutrient balance inside the plants. Signal perception, integration and processing are the basic steps adopted by any tree in the form of number of traits which control cell functions and development in salt stressed environment. Looking into this aspect, the exploration of salt tolerance among trees is now gaining importance worldwide and is nowadays a hot topic for research in the forestry sector.

It can be inferred from the literature available that the salt tolerance mechanism in any plant can be achieved by adapting to more than one of the strategies either simultaneously or separately (Hanin et al. 2016). Further, the salt tolerance mechanism in any plant falls into two categories, namely avoidance and tolerance where, avoidance is the ability to avoid salt ions uptake and tolerance refers to the situation when compartmentation of the salt ions is absorbed inside the plant body (Greenway and Munns 1980). Plants can adapt salt tolerance by employing the three basic mechanisms:

1. Salt avoidance or ion selectivity.
2. Salt accumulation or ion accumulation.
3. Osmotic adjustments.

Besides these three basic mechanisms, the plants imply different adaptive strategies, viz. ion compartmentalisation, osmotic adaptation, osmolyte production, succulence, selective uptake and transport of ions (Koyro et al. 2011). Salinity is nowadays one of the leading problems in any ecosystem worldwide. It is known that some plants have the ability to grow under harsh saline conditions by adapting to soil salinity through (a) osmotic tolerance, (b) salt exclusion and (c) salt accumulation (Munns and Tester 2008). In a study pertaining to Date Palm, it was observed that it adapted to the salt accumulation mechanism by accumulating excessive sodium ions in its leaves (Alrasbi et al. 2010).

In spite of many studies elucidating the salt tolerance mechanism in trees, further focussed and high-end studies are needed to identify the pin-pointed traits, helpful in enabling the survival and growth of trees in salt affected landscapes. Mechanisms of salinity tolerance in trees are thought to be complex because of interlinking of various on-going growth and physiological processes in the plant system.

28.3 Ion Selectivity

It is the adaptive strategy implied by salt-tolerant plants to discriminate chemically similar ions by restricting their uptake. The organelles responsible for this discrimination are present throughout the plant membranes (Bliss et al. 1984; Kuiper 1968). Several species of mangroves are able to even store these ions in special organelles called sweat glands (Jacoby 1964). According to Munns and Tester (2008), the sodium:potassium ratio is the key determinant of salt tolerance in plants. The plants have moderated such that they selectively avoid absorbing the toxic sodium ion. The plasma membrane helps in transporting the excess sodium in and out of the cell in order to maintain the sodium:potassium ratio in the plants. Here, when the salt stress is sensed by the plant, the sensors signalling takes place where the Na^+/H^+ anti-porter is stimulated, Ca^{2+} pathways also get activated and the plant membrane functions are so arranged that excessive sodium is neglected selectively.

The ion avoidance works mainly on the role played by the SOS (salt overly sensitive) stress signalling pathways as described by Gupta and Huang (2014) and presented in Fig. 28.1. The SOS pathway consists of three proteins, namely SOS1, SOS2 and SOS3 as evident from the illustration in the figure. The main work is performed by the SOS1 protein which encodes the plasma membrane that consists of the NA^+/H^+ anti-porter. Upon signalling, it further helps in regulating the sodium ion efflux at cell levels. The SOS2 and SOS3 protein are complimentary to this and are activated when salt stress is sensed.

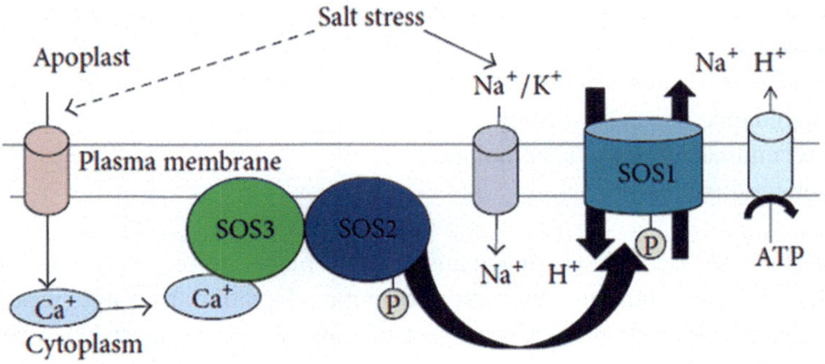

Fig. 28.1 Ion selectivity model adapted by plants to tolerate excessive sodium ion. (**Source:** Gupta and Huang 2014)

28.4 Ion Accumulation

It is the capability of plant species to store salt in the plants either in glands or excreting it through leaves. This mechanism is adapted by very few and special group of plants which develop specific and specialized structures like salt glands. The main function of such glands is to accumulate excessive salt. These structures are further classified into two types that are those which secrete excess salt on leaf surface and the other those that accumulate salts into vacuoles (Dassanayake and Larkin 2017).

28.5 Osmotic Adjustments

It is the capability of the plant to maintain the osmoregulation in such a way that the turgor potential is maintained. When excessive salt ions are present in the soils the resultant water potential in the soil reduces and the plant experiences difficulty in absorbing soil solution. This leads to the drying and dehydration of cells to maintain the osmotic potential. Several small molecular compounds, viz. glycine, betaine, proline, etc. are responsible to these compounds which get accumulated in large quantity. Hence, the potential in the plant cell starts increasing and leads to increase absorption and retention of water by the cells (Zhang et al. 2020; Tang et al. 2015; Luo et al. 2017; Flowers et al. 1977) (Fig. 28.2).

Over time, the plants have adapted to survive and grow well in saline and waterlogged habitats. Though, the performance of the plants might get affected based on the site properties and prevailing conditions. Halophytes are group of plants which have higher level of tolerance to salts by adapting to several physiological and biochemical traits like osmoregulation, ion-compartmentalization, homeostasis, etc. (Lombardi et al. 2022). Hence, halophytes are suggested and found suitable to be grown in such harsh landscapes affected by the twin problems

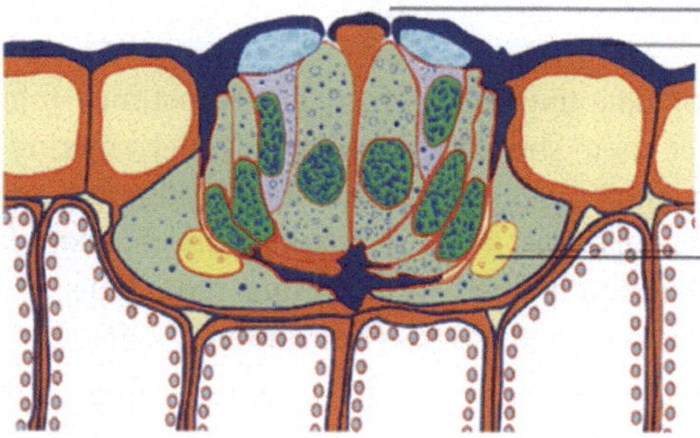

Fig. 28.2 Salt glands in plants showing salt accumulation as salt tolerance mechanism. (**Source:** Hasanuzzaman et al. 2014)

of waterlogging and salinity. Halophytes are generally adopted as one of the undermentioned adaptive mechanisms (Lombardi et al. 2022; Walter 1961) for survival and establishment in salt affected landscapes. Halophytes have the capability to absorb ions on selective basis, regulate ion accumulation and eliminate the toxicity of Na^+ ions from the cytosol (Han et al. 2014). The basic mechanism of tree tolerance to salt stress is based on three adopted mechanisms (a) salt exclusion, (b) salt excretion and (c) salt accumulation for which a pictorial is given in Fig. 28.3 (Banyal et al. 2017).

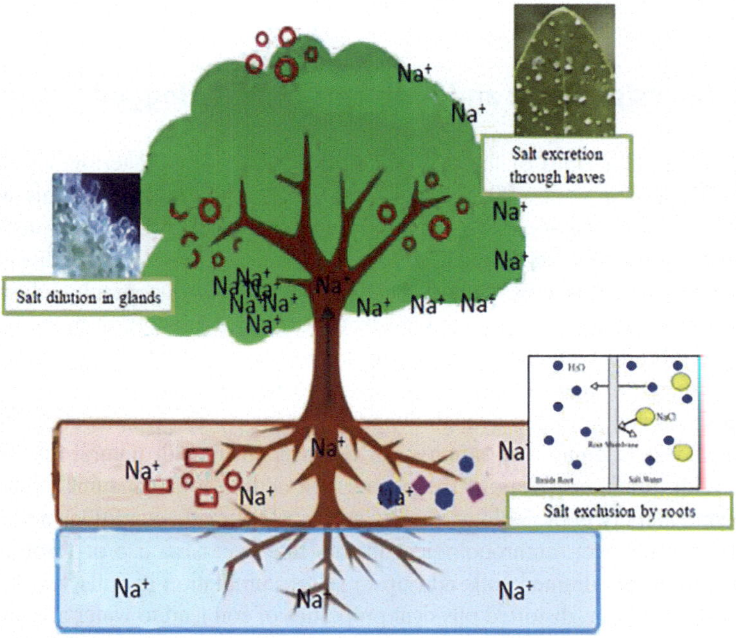

Fig. 28.3 Tree adaptive mechanisms in salt laden ecologies. (***Source:*** Banyal et al. 2017)

28.6 Salt Exclusion Tolerance Mechanism

The plants adopted this type of adaptive mechanism are sometimes referred as pseudo-halophytes. They protect their shoots by inhibiting salts entering to cell system through roots with the help of certain membranes and ionic recycling. Mangroves are an example to adopt this kind of adaptive mechanism for their survival and growth. Salt level is maintained by secreting more ions to outside through salt glands on stem and leaves of the trees (Basyuni et al. 2019).

28.7 Salt Excreting Mechanism

These plants are capable of releasing the excess salts through specialized modified structures named as salt glands. This adaptive mechanism helps in order to avoid distortion of cells. The plants under this adaptive mechanism are *Limonium* spp., *Tamarix* spp., *Acacia ampliceps* (salt wattle plant), etc.

28.8 Salt Accumulating Mechanism

Plants accumulate excessive salts inside their body vacuoles in such a way that it does not hamper the physiological functioning and growth of the plants. *Salvadora persica*, *Suaeda nudiflora,* etc. are plants capable of adopting salt accumulation mechanism to survive in such harsh landscapes.

28.9 Genesis, Extent and Problems of Waterlogged Saline Soils

Globally more than 1/3rd of the irrigated land is affected by waterlogging and soil salinity (Heuperman et al. 2002). Almost every developing country in this world is either affected or on the verge of getting affected by the twin problem of waterlogging and salinity especially in the arid and semi-arid climatic regions. In India, this problem is most prevalent and is a resultant of combined factors, viz. topography, climate and anthropological factors. The problem has its genesis due to distortion of the naturally present equilibrium between the groundwater recharge and discharge (Datta et al. 2000). Further, the main contributing factors to this problem are the faulty farm operations coupled with the availability of poor quality of groundwater. The twin problem of waterlogging and salinity can be asserted to both natural and artificial origins. With time weathering of the parent rock adding to the natural salinity. The other factors namely poor geology, floods and erratic climatic conditions adds fuel to the problem. However, anthropological factors like excessive use of fertilizers and chemicals in poorly drained soils add up to salt accumulation in soils which further seep down with water, distorted physical properties of soil lead to water accumulation with time and the salts further are crusted on surface with evaporation (Dikeogu et al. 2021). Any soil is said to be waterlogged and saline when the groundwater table reaches and remains up to the crop zone for a prolonged period of a year (Michael and

Ojha 2006). This situation when occurring in arid and semi-arid areas is accompanied by salinity, hence rendering such soils incapable of cultivation and barren with time. The solution to the problem lies in lowering the groundwater table and leaching away excess salts below the root zone. It is here where the conceptualization of biodrainage comes as a saviour.

Waterlogging may be defined as stagnation of water on the land surface or where the water table rises to an extent that soil pores in the crop root zone become saturated. This creates the condition low in oxygen and high in carbon dioxide level (Settler et al. 2009). Waterlogging condition arises when soil remains saturated with water for a significant period of time (a few weeks to months). It occurs due to poor management of water and obstruction of natural drainage systems by the disorganized embankment construction (disrupting the balance of inflow and outflow of water), excessive rainfall, drainage congestion and blocking of natural drainage, dense haphazard embankments, over-irrigation cyclones, flooding, river basin encroachment through siltation and finally human activities.

In India, the area under waterlogging is 6.00 m ha out of which 3.40 and 2.60 m ha are under subsurface and surface ponding, respectively. It is covering 1.80% of the total geographical area of the country. However, waterlogged saline soils occupy 3.00 m ha area in total of which >2.00 and 1.00 m ha in arid/semi-arid northwestern states and coastal and vertisols regions of the country, respectively. Further, as per Masilamani et al. (2020) area affected by waterlogging in the country is about 4.50 m ha among which about 48% is in canal command areas whereas the remaining 51% area lies outside the canal command areas. Another study estimated that the area severely affected by both waterlogging and salinity in India is about 5.50 m ha (Singh 2018). In India, the worst affected states due to waterlogging in the country are Andhra Pradesh (0.33 m ha), Gujarat (0.48 m ha), Haryana (0.62 m ha), Bihar (0.11 m ha), Punjab (1.09 m ha), UP (0.81 m ha) and West Bengal (1.85 m ha). Similarly, about 6.73 m ha area is salt affected area. The area which is affected by twin problem of waterlogging and salinity is either found separate or coexisting.

The causes of waterlogging and salinity are illustrated as below:

28.10 Rise in Groundwater Table

In areas of good amount of rainfall, the soil pores are then filled with water until the soil becomes fully saturated. Now, as time passes, the infiltration capacity of such soils decreases due to saturation of soil with water which results in accumulation of excess water. Seepage of water in ground leads to alleviation of groundwater table resulting in waterlogging and salinity.

28.11 Inadequate Drainage

The groundwater begins to rise with continuous irrigation because of accretion through leakage, seepage and deep infiltration. This is common cause of waterlogging and salinity in areas where topography is low lying than the adjacent

Plate 28.1 Twin problem of waterlogging and salinity

areas impounding the condition of water stagnation without natural flow of water resulting in poor drainage. The rate of infiltration to the runoff in such low lying areas is low. Drainage of excess water is affected due to elevated surface relief and it gets accumulated leading to waterlogging coupled with soil salinization due to evaporation. The fate of most irrigation projects in India leads to secondary salinization (Bowonder et al. 1986; Gupta et al. 2019). The farmers used plenty of water initially with the inception of irrigation project and the salts and water move upward with the help of capillary action, once the water table rose to 1–2 m from the surface. The water present in the upper surface gets transpired and evaporated, while the salts remain in the root zone resulting in soil salinization (Plate 28.1).

28.12 Seepage and Leakage

Unlined irrigation canals are the reasons for water seepage to the adjacent low lying areas especially in irrigation commands. The percolated water from the banks of the canals is often seen to be accumulated leading to waterlogging situation in adjoining areas.

28.13 Irrigation

The quality of irrigation water is nowadays depleting due to use of poor quality groundwater (saline and RSC) and surface contamination on a larger front. Irrigation with salts (chlorides, sulphates, carbonates and bicarbonates of sodium, magnesium and calcium) laden water adds to the salinity. These salts seep downwards and are later seen affecting the soil quality with continuous practice of use of such poor quality water especially in irrigation commands.

28.14 Natural Salt Occurrence

In the areas, where salt rich parent rocks are present, the salts are released continuously with weathering which further gets flushed and leached down to groundwater with rainfall and adding to the problem of soil salinity coupled with waterlogging. Waterlogging and salinity affected the physical, chemical and biological properties of the soils. The first and the most common problem arising due to waterlogging and salinity is difficulty in irrigation and related agronomical procedures. It is clear from the observation made by Ahmad and Singh (1991) that rise in water table along with secondary salinization has affected irrigated agriculture to a larger extent. Further, they also raised the concern that none less than 2.00–3.00 lakh ha of irrigated lands is added to the degradation annually with the twin problem of waterlogging and salinity, worldwide. Another estimate of the rate of addition in the quantum of waterlogged and saline lands globally is about 1.50 m ha that too in the irrigation commands (Brundtland and Khalid 1987).

28.15 Remedial Measures to Use Waterlogged Saline Lands

Improvement in drainage of any area has been suggested to be an ideal remedy for the problem of waterlogging and salinity (Ministry of Water Resources 1991). The most serious common factors for twin situation are drainage inadequacies, seepage from water channels, excess water availability, poor maintenance and negligence of drainage systems. In addition to these factors, change in cropping patterns, poor farm management, faulty irrigation and water management are also contributing to the expansion of waterlogged saline areas. The management of such landscapes involves draining of excess water through surface drainage, subsurface drainage and biodrainage (Fig. 28.4).

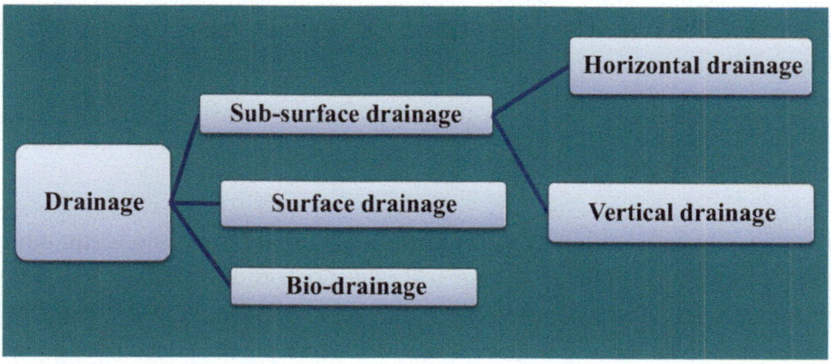

Fig. 28.4 Schematic presentation of waterlogged saline soils management

Table 28.1 Suitable tree species for saline and waterlogged situations

Performance ratings	Tree species
Very promising	*Acacia nilotica, Eucalyptus tereticornis, Eucalyptus camaldulensis, Prosopis juliflora, Pithecellobium dulce, Tamarix articulata, Tamarix troupi, Tamarix ericoides, Salix* spp., *Acacia mangium, Acacia auriculiformis, Bamboo arundinacea, Acacia ampliceps*
Promising	*Azadirachta indica, Casuarina glauca, Dalbergia sissoo, Feronia limonia, Tecomella undulata, Terminalia arjuna, Syzygium cumini, Hardwickia binata, Populus* spp., *Pongamia* spp., *Bauhinia variegata, Cassia glauca, Cassia fistula*

The site does not remain ideal for vegetation growth. The plants tolerant to salinity with inbuilt phreatophytes virtue are capable of growing in waterlogged saline lands. It is where phreatophytes (a group of plants which are capable of adapting excess water availability) come to rescue. Though these plants do not belong to any specific plant family but they are characterized based on their common characteristics of utilizing a large amount of water (Fletcher and Elmendorf 1955). These plants are highly recommended for such landscapes where high transpiring plants can aid to lower the water table and also provide several economic and ecological benefits. While advocating for biodrainage, care is taken while selecting suitable plant species based on agro-climatic conditions, topography, quality of groundwater, etc.

Plants belonging to phreatophyte group are divided into very promising and promising trees like *Tamarix gallica, Eucalyptus* spp., *Bamboo* spp., *Acacia ampliceps, Salix* spp., *Prosopis* spp., *Syzygium cumini,* etc. and detailed list is given in Table 28.1.0. As per Ellis and VanDjik (2009) block plantation with densities of 500 or 600 trees per ha and 1000–1200 trees per ha are suitable for carrying out biodrainage works in canal command areas. Similar models can be suggested for other areas based on physical conditions and social acceptability in the area. *Eucalyptus* spp. is found to be the most successful common tree adapted in such areas for better utilization of waterlogged saline landscapes (Banyal et al. 2019). The other favoured tree species are *Tamarix, Prosopis* and *Acacias* (Tomar and Minhas 1998). Accordingly, Heuperman et al. (2002) have suggested that shifting plantations can also be adapted to manage waterlogging and salinity situation.

28.16 Biodrainage Concept

Biodrainage can be used either as a corrective measure by lowering water tables or as a preventative measure by intercepting soil water before it arrives at the water table (Kapoor and Denecke 2001). Although the term biodrainage is relatively new, Indian farmers are using vegetation to dry out soil profiles for a long time. The first documented use of the term biodrainage is credited to Gafni (1994). Before that date, the term bio-pumping was used to describe the application of trees to control water table (Heuperman 1992). However in India, the biodrainage research is attributed to start at Central Soil Salinity Research Institute, Karnal, Haryana as a Lysimeter experiment (a microcosm experiment) in the 1990s and the term biodrainage was first time documented by Chhabra and Thakur (1998).

Biodrainage may be defined as.

Pumping of excess soil water into the atmosphere by deep-rooted plants using their bio-energy.

It is analogous to energy-operated water pumps and is proven technology to arrest salinity build-up in canal commands with growing of suitable tree species. The trees absorb water from the capillary fringe located above the groundwater table. The absorbed water is translocated to different parts of plant and finally more than 98% of the absorbed water is transpired into the atmosphere mainly through the stomata (Akram et al. 2008). This combined process of absorption (by roots), translocation (through xylem) and transpiration (through leaf stomata) of excess groundwater into the atmosphere by the deep-rooted vegetation describes biodrainage (Fig. 28.5) (Dagar 2014). Phreatophytes use the water from the saturated part of the soil profile. Tree roots intercept seepage and leakage when planted along the canal banks. The application of biodrainage is not only restricted to irrigated areas but also extended to rainfed areas where water and salt balances get disturbed by land use changes.

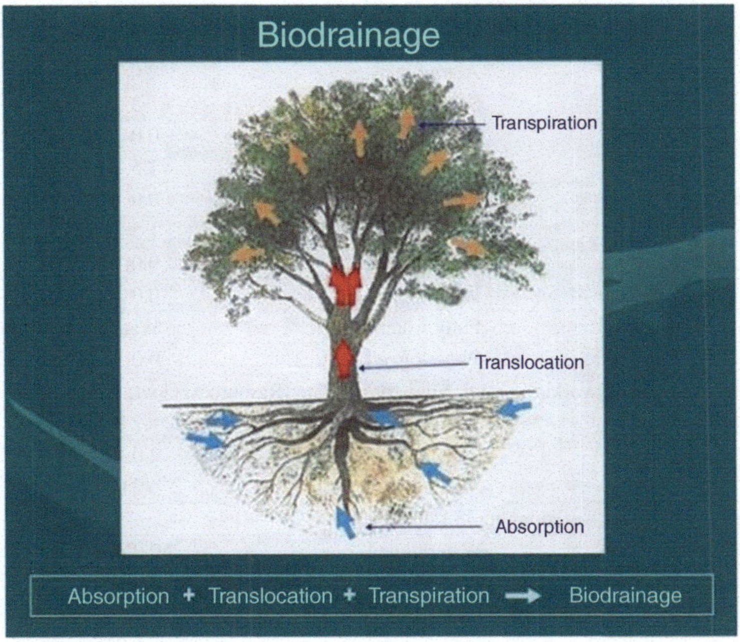

Fig. 28.5 Biodrainage concept

28.17 Possible Biodrainage Scenario

Biodrainage can be imperative in two land use context, viz. irrigated and dry land/rainfed systems. Under irrigated system, recharge control, groundwater flow interception and discharge enhancement are the points of consideration. However, in unirrigated system water table control, channel seepage and biodrainage-cum-conventional drainage systems are the possible scenarios. It is reported that if the water table depth is less than 3 m from the surface, then it has deleterious effect on the crops. The norms for classifying the waterlogged soils in general and waterlogged saline soils in particular proposed by different Indian states and Working Group of Ministry of Water Resources, Govt. of India are given in Table 28.2.

The varying norms for defining waterlogged soils are adopted across the Indian states. In Uttar Pradesh, the soils having water table up to 3 m, in Punjab up to 2.0 m, in Haryana up to 3.0 m, in Karnataka and Himachal Pradesh up to 2.0 m and in Maharashtra up to 1.2 m are termed as waterlogged soils. However, the severity of the problem accentuated further classification of waterlogged soils into fully, worst, bad, critical, alarming, moderately critical, etc. adopted in Indian states. The classification of waterlogged soils was also put forward by the Working Group of Ministry

Table 28.2 Norms and criteria for classification of waterlogged and waterlogged saline soils in India

Domain of classification	Waterlogged soils class	Water table depth (m)
(a) *General criteria adopted by some Indian states*		
Uttar Pradesh	Worst zone	<1.0
	Bad zone	1.0–2.0
	Alarming	2.0–3.0
	Safe zone	>3.0
Punjab	Very critical	0.0–1.5
	Critical	1.5–2.0
Haryana	Critical	0.0–1.5
	Moderately critical	1.5–3.0
Karnataka	Waterlogged	0.0–2.0
Himachal Pradesh	Waterlogged	0.0–2.0
Maharashtra	Fully waterlogged	Water at the surface
	Waterlogged	0.0–1.2
(b) Criteria adopted by working group, Ministry of Water Resources, Govt. of India		
	Waterlogged	<2.0
	Potentially waterlogged	2.0–3.0
	Safe	>3.0
(c) Criteria adopted for waterlogging in saline lands		
	Waterlogged	0.0–2.0
	Sensitive/critical	2.0–3.0
	Protected	3.0–5.0
	Safe	>5.0

Source: Sarkar (1997), Banyal et al. (2022)

of Water Resources, Govt. of India in which three categories were identified as waterlogged (<2.0 m), potentially waterlogged (2.0–3.0 m) and safe (>3.0 m). Separate classification of waterlogged saline soils is in place covering from waterlogged (0.0–2.0 m), sensitive/critical (2.0–3.0 m), protected (3.0–5.0 m) and safe (>5.0 m). It has been observed that the water table between 3.0 and 5.0 m in saline soils has been kept under protected category envisages that water table up to 5.0 m can give deleterious effect on the cropping pattern of the area. So, the protective strategies should be followed to keep the water table beyond 3.0–5.0 m to avoid colossal damage in future.

28.18 Significance of Trees in Amelioration and Managing Waterlogged Saline Soils

It is evident from the nature of waterlogged saline soils that normal crop cultivation is not possible in such landscapes. Hence, the scientists looked towards complex rehabilitation using mechanical measures to help salt leaching and force excessive water to drain using subsurface drainage technologies. It is a very good option when installed in a large area but somehow requires a lot of expertise, budget and furthermore looking towards the recent trends of rising population. Biodrainage seems a healthy, cost-effective and safe option to counter this twin problem of waterlogging and salinity. Further, utilizing the trees energy to drain out excess water from the ground to atmosphere makes it an environmental friendly approach, requiring less maintenance and less overall cost of the project. This method proves out to be more significant compared to other available conventional methods of engineering approach to reclaim such soils. Using trees in managing these soils does not produce any effluent or pollutant into environment, besides this the benefit received from adapting trees will increase with time as well as the tree provides for intermediate yields during the life cycle. Hence, adapting trees to reclaim such lands is a productive and preventive method of control. Utilizing trees to reclaim such landscapes will yield benefits on the aesthetics ends too. The site where plantation works will be carried out for reclamation will modify into green open spaces, serving recreational needs. The trees will also act as windbreaks and shelter belts hence helping in moderating the cold wave and hot wave prevalent in the area.

When one computes the ecological benefits from such plantations, trees on such harsh landscapes will certainly yield a lot of benefit in form of accumulating higher carbon through sequestration, resulting in increased carbon credits. Vegetation will also benefit the surroundings by moderation of temperature, alleviating the value of the site. The presence of trees will also result in mitigation of greenhouse gases which will also aid in fulfilment of the obligations of reducing global warming.

Trees transpire a lot which helps to maintain the groundwater table, being deep-rooted they are firm which provides various ecologically and economically outputs. They help in regulating the hydrological cycle, carbon cycle, stabilize the soils, increase build-up of carbon in soils, leading to amelioration and increase the productivity and value of such landscapes. Trees adapt to such habitats commonly found in arid and semi-arid climates. Hence, these can certainly provide an upper hand in waterlogged saline habitats by returning appropriate benefits. With proper and larger root biomass present these plants help in controlling soil erosion, building of nutrients in soil, help in evapotranspiration of excess water from ground, provide basic needs (food, fuel, fodder, fibre, etc.). Therefore, there is a need to look into the possibilities of adapting phreatophytes in reclaiming the waterlogged saline soils in the near future. Some of the case studies highlighting the significance of trees in using waterlogged saline soils are summarized here as under and presented in Table 28.3 including the names of trees, plantation techniques and salient points of recommendations.

Table 28.3 Biodrainage potential of different salt-tolerant trees and their field models for managing waterlogged saline soils

S. no.	Tree species	Plantation technique	Salient points
1.	*Eucalyptus tereticornis* (clonal) (Ram et al. 2008)	Parallel strip	• Groundwater table drawdown was 0.91 m during span of 3 years experimentation. • There was no increase in the net salinity in groundwater beneath the plantation. • Sinker root of 4.40 m depth was recorded and the groundwater table depth was found to be 4.70 m indicating that the plant acquired water from the groundwater table through capillary fringe.
2.	*Eucalyptus tereticornis* (clonal) (Ram et al. 2008)	Ridge planting in strips along bunds	• The average drawdown of water table was 0.85 m 3 years of *eucalyptus* plantations. • Average rate of transpiration was 438 mm annually. • Sinker roots were found up to the depth of 3.35 m indicating the tree drawing water from groundwater table.
3.	*Eucalyptus* spp., *Prosopis juliflora, Callistemon lanceolatus, Melia azedarach, Terminalia arjuna and Pongamia pinnata* (Toky et al. 2011)	–	• Based on the capability of drawdown of water table trees are classified into (a) **Fast bio-drainers:** *Eucalyptus* spp. *and Prosopis* spp. (b) **Medium bio-drainers:** *Eucalyptus*

(continued)

Table 28.3 (continued)

S. no.	Tree species	Plantation technique	Salient points
			spp., *Melia* spp. *and Callistemon* spp. **(c) Slow bio-drainers**: *Terminalia* spp. *and Pongamia* spp. • Overall reduction in groundwater table was 20 cm.
4.	*Eucalyptus* spp. *and Casuarina* spp. (Chowdhury et al. 2011)	–	• Efficiency of *Eucalyptus* was found better than *Casuarina* spp. however, both the species are suitable for use in biodrainage plantations. • Water table drawdown was 15–25 cm.
5.	*Eucalyptus camaldulensis, Dalbergia sissoo and Acacia nilotica* (Sharma 2001)	Canal side plantation along Indira Gandhi Nahar Project	• Excess seepage was checked through biodrainage plantation in 6 years of time period.
6.	*Eucalyptus* spp. (Banyal et al. 2019)	Block plantation and boundary plantation	• There was appreciable effect of *Eucalyptus* on drawdown of groundwater table in closer spaced block plantations compared to boundary plantation. • *Eucalyptus* plantation resulted in improved physico-chemical properties of soil underneath the plantations than bare fallow adjoining areas.
7.	*Eucalyptus camaldulensis, Eucalyptus fastigata, Eucalyptus rudis and Corymbia tessellaris* (Bala et al. 2014)	Raised bunds	• Planting of these trees resulted in improved vegetation cover in waterlogged saline areas. • Drawdown in water table. • *Eucalyptus rudis* emerged best among other with respect to the growth, biomass and overall biodrainage potential.
8.	*Prunus armeniaca, Populus nigra, Salix nigra, Catalpa bignonioides, Populus euphratica, Elaeagnus angustifolia, Tamarix androssowii,*	MPTs plantation	• *Elaeagnus angustifolia* (Russian Olive) ranked highest showing faster growth and higher water use. • *Populus* spp. *and Ulmus*

(continued)

Table 28.3 (continued)

S. no.	Tree species	Plantation technique	Salient points
	Ulmus pumila, Fraxinus spp. *and Morus alba* (Khamzina et al. 2006)		*pumila* emerged as ideal tree spp. in biodrainage models.
9.	*Eucalyptus tereticornis* (clonal) (Dagar et al. 2016)	Block and twin row strip (staggered/continuous) plantations, spacing 1 × 1, 1 × 2 and 1 × 3 m	• Water table drawdown was 43.0 cm in 1 × 1 m, 38.5 cm in 1 × 2 m and 31.5 cm in 1 × 3 m spacing in 4 year plantation. • Rotation of 6 years, closer spacing of 1 × 1 m under strip plantation of *Eucalyptus* in paired rows on farm acre line was the optimum for achieving higher water table drawdown and other system benefits.

28.19 *Eucalyptus* as Embryonic Biodrainage Tree Spp. in Haryana (Banyal et al. 2019)

In Haryana, out of the total 4.42 m ha of area, more than 0.05 m ha is having shallow water table less than 1.50 m deep and is under critical condition. An additional 0.38 m ha area has water table between 1.50 and 3.00 m, rendering it potentially waterlogged and saline land. Nine districts of Haryana state are suffering from waterlogging and salinity problem (Tribune 2016). Most of the existing and potential waterlogged saline areas occur in arid and semi-arid regions in central inland depression basin of the state, encompassing Rohtak, Jhajjar, Bhiwani, Hisar, Sonipat districts and some parts of Jind, Fatehabad, Sirsa and Palwal districts. Rohtak is the worst hit district in the state with 47.2% of its area falling under 'potentially waterlogged category' and 9.90% of the area covering under 'waterlogged and saline' area. This is because of the fact that alarming rise in the water table is causing threat to the highly productive agricultural land of the state. About 200 km long axis of arable land (mostly underlain by saline groundwater) comprising the districts of Jhajjar, Rohtak, Hisar, Fatehabad and Sirsa in Haryana forms an inland basin with no natural drainage. Introduction of canal irrigation has resulted in a chain of adverse social, agricultural, hydrological and environment changes in southwestern Haryana with the inland basin as its epicentre. The *Eucalyptus* plantations were done by the state forest department and farmers on large scale in waterlogged saline areas of Haryana. Two types of plantation models, i.e. block and boundary were commonly prevalent in the state. In pure blocks of *Eucalyptus,* the spacing was 1.5 × 3.0 m, 1.5 × 4.0 m and 1.5 × 6.0 m and growing of intercrops was under wider spacing of

1.5 × 4.0 and 1.5 × 6.0 m in moderately and less affected areas. In boundary model, parallel ridges were constructed in the north-south (N-S) direction along the bunds of agricultural waterlogged fields and two ridges in the field 66 m apart on *kila-lines* (bunds separating one acre field area; 66 × 60 m). In single row, the spacing was 1.50 m between plant to plant and 1.5 × 2.0 m and 1.5 × 1.5 m in parallel and staggered twin row strip plantations.

The *Eucalyptus* plantations were done in block and boundary plantation models successfully and gave significant increase in growth parameters (plant height, diameter at breast height-DBH) per cent current annual increment (CAI) in both the plantation models across the state. The plantation age on selected sites was varying from 2 to 5 years (Fig. 28.6). It is clearly seen from Fig. 28.6 that the major chunk of plant height and DBH was allocated to fifth year of plantations followed by fourth, third and second year. Plant height showed higher values compared to DBH in fifth, fourth and third year. But the trend was reverse during the second year of plantation. The increase was consistent from second to fifth year of plantation age. It is true for the obtained results as per the existing growth pattern of trees in close spacing under block model and with the ageing of plantations.

The *Eucalyptus* plantations showed direct impact in drawing down the water table in waterlogged soils. The higher drawdown was observed in block model than the boundary model. There were significant improvements in soil quality by higher values of soil organic carbon, nitrogen, phosphorus coupled with lower soil electrical conductivity and pH underneath *Eucalyptus* plantation compared with bare saline waterlogged areas. The presence of organic carbon, nitrogen and phosphorus increased with the advancing of plantation age from second to fifth year.

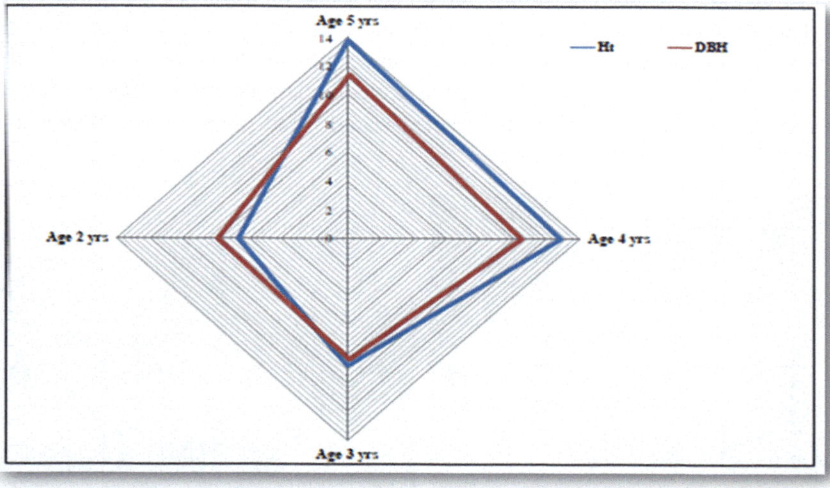

Fig. 28.6 Growth increment trends of *Eucalyptus* plantations across the age gradations in waterlogged saline soils in Haryana

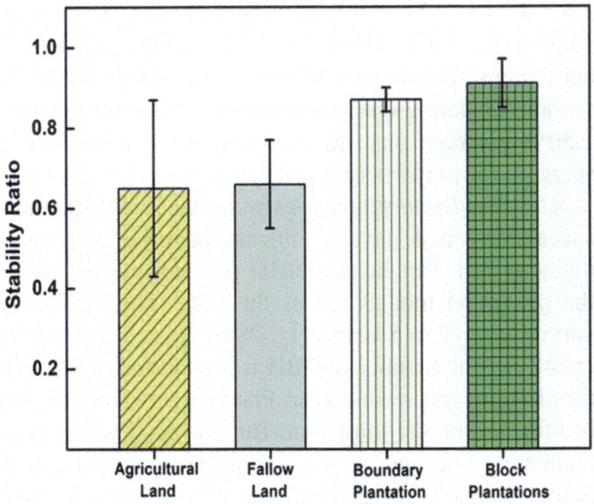

Fig. 28.7 Soil stability trends under the influence of *Eucalyptus* plantations

The soil aggregate stability was analysed through high energy moisture characteristics (HEMC) method. The stability ratio increased from 0.65 in agriculture land to 0.87 underneath boundary and finally 0.91 in block plantation models, indicating the significant increase in the stability of soil under erosive processes (Fig. 28.7). It is recommended from the present investigation on soil stability that *Eucalyptus* trees are the main source for having the stable soil structure to rehabilitate the waterlogged saline ecology in better and long-term effective way (Plate 28.2).

Social acceptance of *Eucalyptus* is an issue because of various myths at farmers' level in the state. Therefore, efforts have been made to work out the status of the social acceptance of *Eucalyptus* plantations under biodrainage technology with the help of well-structured and pre-tested interview schedule. It has been observed that farmers were in deep distress and forced to leave the farming and migrated to nearby cities to earn livelihood in severely affected waterlogged saline areas of Fatehabad,

Plate 28.2 Eucalyptus based biodrainage models in waterlogged saline landscapes

Sonipat, Rohtak and Hisar districts. Some farmers were hopeful their lands will be improved by planting *Eucalyptus*. Every farmer has one question to the survey team that after how many days, months and years, the land will become cultivable for agronomical and other crops. This question is still remained unanswered.

The provisioning of ecosystem services obtained from *Eucalyptus* plantations and technical knowledge of the plantation was also tested and the responses were categorized into moderate (40.3%) followed by low (25.4%), no (20.6%) and high (13.7%) group and in high (46.5%) followed by no (31.3%), low (16.8%) and moderate (5.61%) categories, respectively. Farmers were having good knowledge about the plantation technique in biodrainage. Overall, the provisioning of ecosystem services and farmers knowledge about biodrainage was grouped in moderate and high category as per the existing awareness of the respondents. The farmers gave mixed response about the social acceptability of *Eucalyptus* as biodrainage species. In areas where the problem was not intense then farmer showed their reluctance for *Eucalyptus* and vice-versa. Farmers also want some other high evapotranspirative tree species that may be multipurpose and/or fruit trees. However, it has been found to be boon for moderately and severely affected areas of the Haryana state for its reclamation and productive outputs. *Eucalyptus* emerged as potential option which needs to be promoted for eco-friendly remediation and in improving the livelihood security of waterlogged saline areas of the state.

28.20 *Eucalyptus* Based Biodrainage Models (Dagar et al. 2016; Ram et al. 2011)

The first systematic and successful *Eucalyptus* based biodrainage model was developed at village Puthi in Hansi block of Hisar district, Haryana. It is known worldwide as successful plantation model for lowering of water table through *Eucalyptus* plantations. Before the *Eucalyptus* plantations, the water table was 2–3 ft and lowered down to >15 ft in span of 10 years and farmers are now practising wheat-rice rotation with significant yield with additional tangible and non-tangible benefits of plantations.

- Standardized the planting spacing of *Eucalyptus* as biodrainage plantations favouring the drawdown of groundwater table.
- The average transpiration rate of *Eucalyptus* in block plantation was 40.0 L day^{-1} tree^{-1} compared to 68.0, 71.5 and 73.8 L day^{-1} tree^{-1} in 1 × 1, 1 × 2 and 1 × 3 m tree spacing in strip plantation under boundary model.
- Total amount of water transpired per annum was 1825 mm in block plantation and 745, 391, 269 mm in 1 × 1, 1 × 2 and 1 × 3 m tree spacing. Water table was lowered by 43.0 cm in 1 × 1 m, 38.5 cm in 1 × 2 m and 31.5 cm in 1 × 3 m spacing during the fourth year of plantation than adjacent fields without plantation.
- Benefit–cost ratio of strip plantations at the discounted rate of 12% of first rotation (5 years and 4 months) in boundary model was 3:1 against 1.3:1 of agricultural crops in Haryana and projected to be many-fold for next 3–4 rotations.

Plate 28.3 *Eucalyptus* in parallel strip row boundary plantation model at Puthi, Hisar

- The yield of wheat grains was 3.34 times higher than the yield in the adjacent area without biodrainage plantation.
- Block model lowered the g.w.t. effectively with substantial biomass but provides economic returns after a gap of 6 years, therefore may not be preferred by the farmers.
- Rotation of 6 years in 1 × 1 m spacing (twin row strip plantation) on farm acre line was the optimum for achieving higher water table drawdown, biomass, carbon sequestration and crop yield from waterlogged saline areas of the state and preferred by the farmers (Plate 28.3).

28.21 Potential of Willow as Biodrainage Tree Spp. in Semi-arid Region

Willow tree is capable of producing high biomass as it utilizes maximum soil water for its growth and development. It would be possible alternative to *Eucalyptus* in waterlogged saline areas especially in semi-arid regions. The experiment was done at ICAR-CSSRI to find out water bio-drained potential of four (J799, SI-64-017, 131/25 and PN731) 2 year aged identified clones in closed Lysimeter (microcosm) conditions by simulating three scenarios of waterlogging coupled with non-saline water in scenario 1.0 (ECiw <1.0 dS/m), low saline water in scenario 2.0 (ECiw 4 dS/m) and high saline water in scenario 3.0 (ECiw 10 dS/m). Results showed that all the three scenarios of waterlogging and salinity had significant effect on growth attributes and biodrainage potential of the four Willow clones. The performance of clones under scenario-1.0 was better than scenario 2.0 and 3.0 in terms of the growth, physiological parameters and water bio-drained. Clone J799 outperformed than the other three clones by registering the highest per cent increments in plant height (3.87%, 14.5% and 17.6%), collar diameter (6.37%, 14.6% and 17.1%), longest branch (10.5%, 17.2% and 22.0%) and crown spread (14.7%, 31.8% and 38.6%) in three seasons, respectively. Whereas clone PN731 registered minimum per cent increments in growth attributes (Plate 28.4).

Plate 28.4 Willow clones in lysimeter setup

Results suggested that the growth increments showed decreasing trend with the increase in the salt load. The clone J799 also showed better photosynthetic activity and higher transpiration rate in all the simulated scenarios. The total water bio-drained on season basis by Willow clones was calculated and presented in Fig. 28.8. Clone J799 displayed highest biodrainage potential in terms of water bio-drained on season basis (287, 487 and 651 mm) as well as annual basis (1425 mm), followed by clone 131/25, SI-64-017 and PN731. The total bio-drained water after 1 year was recorded highest in scemario-1.0 (1888 mm)

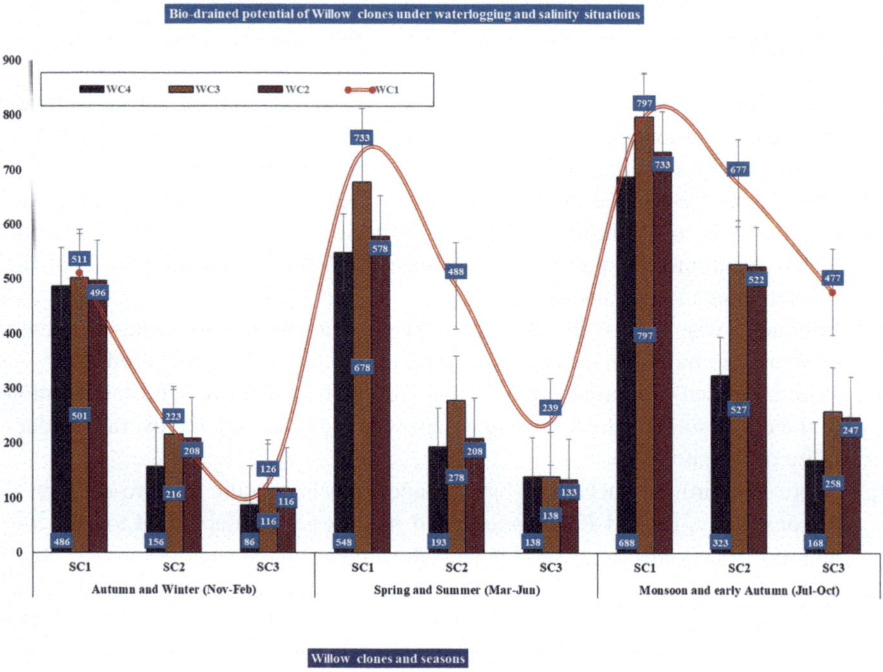

Fig. 28.8 Water-biodrained potential of Willow clones

and lowest in scenario-3 (561 mm). Overall, clone J799 showed higher amount of water bio-drained in terms of biodrainage under twin situation of waterlogging and salinity compared to rest of the three clones. Henceforth, willow tree clone J799 could be viable option for managing the waterlogged saline landscapes for achieving the ecological and environmental sustenance (Neha et al. 2022). The amount of water bio-drained is comparable to *Eucalyptus* and it can be visualized that this species has the potential to come up in waterlogged saline areas in semi-arid regions.

28.22 Impediments in Using Trees for Managing Waterlogged and Saline Soils

Biodrainage is economically viable and ecological sound technology globally. However, use of trees on waterlogged saline landscapes would certainly go with some limitations. However, when one looks into the scenario on long-term basis, then it is quite certain that the benefits will overcome the limitations. The prominent limitations which one may encounter while reclaiming the waterlogged saline soils with trees are enlisted here as under:

(a) It needs large portion under tree plantations for higher drawdown rate in groundwater table and thus not suited to areas where land is scarce and expensive.
(b) Due to prevailing harsh conditions in such landscapes, establishment of trees might be nightmare in initial stages. Hence, the early care of the trees in such cases is of utmost significance.
(c) During the initial years, it is evident that the plantation might not be so much effective in drawdown of groundwater table. Therefore, little patience is required when adapting biodrainage as an option of reclamation.
(d) When trees are planted in close proximity, then it is somewhat natural occurrence of the competition with other companion flora for sunlight, nutrients, water requirements, space which ultimately affects the overall growth of the planted trees in biodrainage models.
(e) Similarly, when the trees have attained maturity or rotation age, then their efficiency to bio-drain excess water from the ground will also get reduced.
(f) With increased evapotranspiration, there are high chances of salt accumulation in the upper soil profile, depleting the growth and hindering the overall productivity of the site.
(g) There is dearth of successful biodrainage models specific to agro-ecological regions. Only, limited *Eucalyptus* based models are available that too for low and moderately affected waterlogged saline sites.
(h) The efficiency of the biodrainage technique in managing waterlogged saline soils depends on the groundwater salinity and it is not workable when salinity goes beyond 12 dS/m due to salt accumulation underneath tree plantation strips (Kapoor and Denecke 2001).

(i) Some adverse effect of trees in the form of release of toxic chemicals (allelopathy) on companion crops may be agronomical conventional crops. However, this point was confronted by Ram et al. (2011) that the allelopathic chemicals released by Eucalyptus got neutralized and thus could not affect the crops as well as existing soil microflora.
(j) Willingness and motivation are also among one of the common limitations in success of biodrainage especially in small land holding farmers. They are not willing to spare larger area for tree plantation.

28.23 Future Research Domains in Tree Adaptive Mechanisms in Biodrainage Technology

28.23.1 Tree Water Use

(a) Wide variations are reported in tree water use values. Transpiration rates depend on climatic conditions, type and age of tree species, size of plantation, density of tree plantations (spacing), soil moisture regime, etc. So, there is need to ponder upon to accurate designs and criteria for biodrainage tree plantings as per the regional suitability.
(b) *Eucalyptus* is mainly used in biodrainage which necessitates to study that it has special features in terms of rooting and stomatal conductance besides physical and physiological adaptability. Do *Eucalyptus* and similar transpiring plants continue to transpire indefinitely or would there be gradual decrease in effectiveness of the trees in their capacity for consuming water and the response mechanism therefore?
(c) Density of tree plantings for biodrainage should aim at maximum evapotranspiration per unit area. So, still we do not have the exact number of plants maintained per unit area for specific drawdown of water table in unit time. However, some location specific models are in place which needs further validation.
(d) What other plant species could be considered for use in biodrainage besides Eucalyptus? It is still a big question to solve that which other potential tree species other than *Eucalyptus* can be boon for such soils. At ICAR-CSSRI, a work on exploring the possibility of *Salix* spp. has been initiated to find out its establishment, growth, physiological, biochemical and evapotranspiration potential in controlled environment. There is need of such experimentation to diversify the tree kitty suitable for managing these soils.
(e) It is need of the hour to work on integrated approach of biodrainage with conventional engineering and traditional methods of drainage to have longer duration functional models for using waterlogged saline soils in productive perspectives.

28.23.2 Salt Uptake and Tolerance

(a) Research is required on the mineral absorption by trees and salt-tolerant crops and the soil salt balance.
(b) Growth of trees and salt-tolerant crops in relation to increase in salt levels in biodrainage systems affecting the evapotranspirative capacity requires detailed assessment and should be explicitly related to tree species' variety.
(c) Below a bio-drain, viz. trees, salt may accumulate and redistribute in the soil profile and also in the groundwater. This process is not well known and needs more study.

28.23.3 Sustainability

(a) Issue of salt accumulation and the longevity of deep-rooted tree plantings in shallow water table conditions continue to be a topic of discussion. A mechanism for salt removal from the root zone of deep-rooted plants is critical for the long-term survival of plants growing under these conditions. To overcome this problem, a mechanism should be worked out either on conventional drainage or natural deep drainage for maintaining the salt balance in the tree growing zone.
(b) Modelling of salt dynamics under tree plantations is important to provide field practitioners with guidelines on establishment and management. Sound data sets (soil salinity, soil water content, water table levels, piezometric levels) based on proven measurement techniques are important to validate these models and to test hypothesis on sustainability. It requires long-term investigation programmes.
(c) Standardized approach is required to allow inter-site comparison. Positive environmental benefits of biodrainage systems lack recognition in general and should be paid more attention for percolating the information to wide arena of audience including planners, designers and project implementers.
(d) Effectiveness of trees as biodrainage system depends on socio-economic conditions. So, collaboration of the local rural farming families is to work out which is otherwise adamant to its adoptability because of lack in knowledge on biodrainage aspects.

References

Ahmad A, Singh PP (1991) Environmental impact assessment for sustainable development: Chittaurgarh Irrigation Project in Outer Himalayas. JSTOR: Ambio 20(7):298–302

Akram NA, Shahbaz M, Ashraf M (2008) Nutrient acquisition in differentially adapted population of *Cynodon dactylon* (L.) Pers. and *Cenchrus ciliaris* L. under drought stress. Pak J Bot 40: 1433–1440

Alrasbi SAR, Hussain N, Schmeisky H (2010) Evaluation of the growth of date palm seedlings irrigated with saline water in the Sultanate of Oman. In: Proceedings of the IV international date palm conference, Abu Dhabi, p 882

Bala N, Singh G, Bohra NK, Limba NK, Baloch SR (2014) Biodrainage for restoration of canal command waterlogged area in Indian Desert. Indian For 140:462–467

Banyal R, Rajkumar, Manish K, Yadav RK, Dagar JC (2017) Agroforestry for rehabilitation and sustenance of saline ecologies. In: Dagar et al (eds) Agroforestry anecdotal to modern science. Springer Nature Singapore Pte Ltd., Singapore, pp 413–454

Banyal R, Bhardwaj AK, Singh RK, Gajender KM, Aslam P, Jagdish C, Bhatia VK (2019) Impact of *Eucalyptus* plantations on waterlogged saline ecologies in Indo-Gangetic plains, Project Report. ICAR-CSSRI, Karnal, p 84

Banyal R, Neha S, Varun K (2022) Insights of waterlogging and salinity management through bio-drainage: reflections from Haryana state. In: Banyal et al (eds) 2022 compendium of short course on recent developments in agroforestry dimensions for managing salt affected ecologies. ICAR-CSSRI, Karnal, pp 122–128

Basyuni M, Ramayani, Hayullah A, Prayunita Hamka M, Putri LA, Baba S (2019) Growth of salt-secretor and non-salt secretor mangrove seedlings with varying salinity and their relations to habitat zonation. In: IOP conference series earth and environmental, p 236

Bliss RD, Platt-Alota KA, Thompson WW (1984) Changes in plasma lemma organization in cowpea radicel during inhibition in water and NaCl. Plant Cell Environ 7:606

Bowonder B, Ramana KV, Rajagopal R (1986) Waterlogging in irrigation projects. Sadhna 9(3): 177–190

Brundtland GM, Khalid M (1987) Our common future: report of the world commission on the environment. UNEP Governing Council, Oxford University Press, Oxford

Chhabra R, Thakur NP (1998) Lysimeter study on the use of bio-drainage to control waterlogging and secondary salinization in (canal) irrigated arid/semi-arid environment. Irrig Drain Syst 12(3):265–288

Chowdhury RS, Kumar A, Brahmanad PS, Ghosh S, Mohanty RK, Jena SK, Sahoo N, Panda GC (2011) Application of bio-drainage for reclamation of waterlogged situations in deltaic Orissa. Research Bulletin-53. Directorate of Water Management, Bhubaneswar

Dagar JC (2014) Greening salty and waterlogged lands through agroforestry systems. In: Dagar JC, Singh AK, Arunachalam A (eds) Agroforestry systems in India: livelihood security & environmental services-advances in agroforestry, vol 10. Springer Publishers, New Delhi, pp 333–344

Dagar JC, Lal K, Ram J, Kumar M, Chaudhary SK, Yadav RK, Ahamad S, Singh G, Kaur A (2016) Eucalyptus geometry in agroforestry on waterlogged saline soils influences plant and soil traits in North-West India. Agric Ecosyst Environ 233:33–42

Dash CJ, Sarangi A, Singh AK, Dahiya S (2005) Bio-drainage: an alternative drainage technique to control waterlogging and salinity. J Soil Water Conserv India 4(3–4):149–155

Dassanayake M, Larkin JC (2017) Making plants break a sweat: the structure, function and evolution of plant salt glands. Front Plant Sci 8:406. https://doi.org/10.3389/fpls.2017.00406

Datta KK, Jong CD, Singh OP (2000) Reclaiming salt affected land through drainage in Haryana, India: a financial analysis. Agric Water Manag 46:55–71

Dikeogu TC, Okeke OC, Nwachukwu HGO, Agbo CC (2021) Salinization and waterlogging in agricultural lands: causes, effects and mitigation. Int J Innov Environ Stud Res 9(1):45–57

Dmuchowski W, Bragoszewska P, Gozdowski D, Dabrowska AHB, Chojnacki T, Jozwlak A, Swlezewska E, Suwara I, Gworek B (2022) Strategies of urban trees for mitigating salt stress: a case study of eight plant species. Trees 36:899–914

Ellis T, VanDjik A (2009) Agroforestry for management of water, salt and agricultural diffuse source pollutants. In: Nuberg L, George B, Ried R (eds) Agroforestry for natural resource management. CSIRO Publishing, Collingwood

Fletcher HC, Elmendorf HB (1955) Phreatophytes—a serious problem in the west. Yearbook of Agriculture, pp 423–429

Flowers TJ, Troke PF, Yeo AR (1977) The mechanism of salt tolerance in halophytes. Annu Rev Plant Physiol 28:875–884

Gafni A (1994) Biological drainage-rehabilitation option for saline-damaged lands. Water Irrig 337: 33–36. (translation from Hebrew to English)

Greenway H, Munns R (1980) Mechanisms of salt tolerance in non-halophytes. Annu Rev Plant Physiol 31:149–190

Gupta B, Huang B (2014) Mechanism of salinity tolerance in plants: physiological, biochemical and molecular characterization. Int J Genomics 2014:701596. https://doi.org/10.1155/2014/701596

Gupta SK, Sharma PC, Chaudhari SK (2019) Handbook of saline and alkali soils: diagnosis, reclamation and management. Scientific Publishers (India), New Pali Road, Jodhpur, Rajasthan

Han GL, Wang MJ, Yuan F, Sui N, Song J, Wang BS (2014) The CCCH zinc finger protein gene AtZFP1 improves salt resistance in Arabidopsis thaliana. Plant Mol Biol 86:237–253

Hanin M, Ebel C, Ngom M, Laplaze L, Masmoudi K (2016) New insights on plant salt tolerance mechanisms and their potential use for breeding. Front Plant Sci 7:1–17

Hasanuzzaman M, Nahar K, Alam MM, Bhowmik PC, Hossain MA, Rahman MM, Prasad MNV, Ozturk M, Fujita M (2014) Potential use of halophytes to remediate saline soils. Biomed Res Int 2014:589341. https://doi.org/10.1155/2014/589341

Heuperman AF (1992) Trees in irrigation areas; the bio-pumping concept. Trees Nat Resour 34:20–25

Heuperman AF, Kapoor AS, Denecke HW (2002) Biodrainage—principles, experiences and applications, knowledge synthesis report-6, international programme for technology and research in irrigation and drainage. FAO, Rome

Jacoby B (1964) Function of bean roots and stems in sodium retention. Plant Physiol 39:445–449

Kapoor AS, Denecke HW (2001) Bio-drainage and bio-disposal: the Rajasthan experience. GRID 17:39–61

Khamzina A, Lamers JPA, Martius C, Worbes M, Vlek PLG (2006) Potential of nine multipurpose tree species in reducing saline groundwater tables in lower Amu Darya River region of Uzbekistan. Agrofor Syst 66:129–141

Koyro HW, Khan MA, Lieth H (2011) Halophytic crops: a resource for the future to reduce the water crisis. Emir J Food Agric 23(11):1–16

Kuiper PJC (1968) Lipids in grape roots in relation to chloride transport. Plant Physiol 43:1367–1371

Liu X, Chen C, Liu Y, Liu Y, Zhao Y, Chen M (2020) The presence of moderate salt can increase tolerance of *Elaeagnus angustifolia* seedlings to waterlogging stress. Plant Signal Behav 10: e1743512–e1743518. https://doi.org/10.1080/15592324.2020.1743518

Lombardi T, Bertacchi A, Pistelli L, Pardossi A, Pecchia S, Toffanin A, Sanmartin C (2022) Biological and agronomic traits of the main halophytes widespread in the Mediterranean region as potential new vegetable crops. Horticulturae 8(3):195

Luo ZJ, Sun YH, Lu N, Li Y (2017) Research advances on salt-tolerance mechanism and genetic transformation of poplar. J Nucl Agric Sci 31(3):482–492. (Acta Agriculturae Nucleatae Sinica)

Masilamani P, Arulmozhiselvan K, Alagesan A (2020) Prospects of bio-drainage to mitigate problems of waterlogging and soil salinity in context of India—a review. J Appl Nat Sci 12(2):229–243

Michael AM, Ojha TP (2006) Principles of agricultural engineering, vol II. Jain Brothers Publishing, New Delhi

Ministry of Water Resources (1991) Report of working group on problem identification in irrigated areas with suggested remedial measures. Government of India, New Delhi

Munns R, Tester M (2008) Mechanism of salinity tolerance. Annu Rev Plant Biol Rep 59:651–681

Neha S, Banyal R, Anita M, Bhardwaj AK, Dhillon RS, Jogender K, Varun S, Yadav RK, Sharma PC (2022) Determining the bio-drainage potential of willow tree clones under twin situation of waterlogging and salinity. In: Abstract of 6th national conference on salinity management for land degradation neutrality and livelihood security under changing climate of ISSS&WQ, Karnal, Haryana, India, p 217

Qi Y, Li JP, Chen CX, Li LX, Zheng XL, Liu J, Zhu TT, Pang CH, Wang BS, Chen M (2018) Adaptive growth response of exotic *Elaeagnus angustifolia* L. to indigenous saline soil and its beneficial effects on the soil system in the Yellow River Delta, China. Trees 32(6):1723–1735

Ram J, Dagar JC, Singh G, Lal K, Tanwar VS, Sheoran SS, Kaledhonkar MJ, Dar SR, Kumar M (2008) Bio-drainage: eco-friendly technique for combating waterlogging and salinity. Technical Bulletin. ICAR-CSSRI, Karnal

Ram J, Dagar JC, Lal K (2011) Bio-drainage to combat waterlogging, increase farm productivity and sequester carbon in canal command areas of northwest India. Curr Sci 100(11):1673–1680

Sarkar TK (1997) Prevention measures – its role and strategy for effective control of potential waterlogging and salt problems in irrigated command. In: Refresher course on land drainage in India, p 1–15

Settler TL, Waters I, Sharma SK, Singh SN, Kulshreshtha N, Yaduvanshi NPS, Ram PC, Singh BN, Rane JMC, Donald G, Khabz-Saberi H, Biddolph TB, Wilson R, Barcley I, Mclean R, Cakir M (2009) Review of wheat improvement for waterlogging tolerance in Australia and India: the importance of anaerobiosis and element toxicities associated with different soils. J Ann Bot 103: 221–235

Sharma KD (2001) Indira Gandhi Nahar Pariyojana-lessons learnt from past management practices in the Indian arid zones. In: Regional Management of Water Resources, IAHS Scientific Assembly, Maastricht

Singh A (2018) Alternative management options for irrigation-induced salinization and waterlogging under different climatic conditions. Ecol Indic 90:184–192

Singh G, Lal K (2018) Review and case studies on biodrainage: an alternative drainage system to manage waterlogging and salinity. Irrig Drain 67:51. https://doi.org/10.1002/ird.2252

Tang XL, Mu XM, Shao HB, Wanh HY, Brestic M (2015) Global plant responding mechanisms to salt stress: physiological and molecular levels and implications in biotechnology. Crit Rev Biotechnol 35(4):425–437

Toky OP, Angish R, Datta KS, Arora V, Rani C, Vasudevan P, Harris PJC (2011) Bio-drainage for preventing waterlogging and concomitant wood yields in arid agro-ecosystems in North-western India. J Sci Ind Res 70:639–644

Tomar OS, Minhas PS (1998) Afforestation of salt-affected soils. In: Tyagi NK, Minhas PS (eds) Agricultural salinity management in India. Central Soil Salinity Research Institute, Karnal

Tribune (2016) Nine districts hit by soil salinity, waterlogging, The Tribune (Daily Newspaper), 15

Walter H (1961) Salinity problems in the arid zones. The adaptations of plants to saline soils. Arid Zone Res 14:65–68

Zhang M, Liu Y, Han G, Zhang Y, Wang B, Chen M (2020) Salt tolerance mechanism in trees: research progress. Trees 35:717. https://doi.org/10.1007/s00468-020-02060-0

Sandalwood: A Potential High-Value Tree Species for Salinity Stress Conditions

29

Kamlesh Verma, Raj Kumar, Aarju Sharma, Sunita Devi, Prashant Sharma, Ajay Kumar Bhardwaj, and Ashwani Kumar

Abstract

Sandalwood (*Santalum album* L.) is a high-value tree species known for its fragrant wood and oil and has been widely used in traditional medicine, cosmetics, and religious ceremonies. Over the years, the demand for this species is increasing which needs for commercial plantation of this species beyond its natural habitat especially in the agroforestry systems. The sandalwood is a root hemi-parasite and is unable to absorb the water and most of nutrients (macro- and micro-nutrient) required for its growth directly from the soil due to lack of root hairs. Thus, a suitable host is required to improve the growth, survival, and yield of sandalwood as the transfer nutrients between sandalwood and host also depend on the particular host species. Simultaneously, in recent years, the soil salinity is increasing at a very fast pace leading to think about the establishment of sandalwood plantations in the saline environment. However, it is imperative to have the knowledge about tolerance of sandalwood to salinity stress with particular host, including its morphological, physiological, and biochemical responses along with haustorial anatomy, as sandalwood responds differently toward its hosts at the molecular level. Moreover, the transfer of nutrients between sandalwood and host also depends on the particular host species. Overall, the potential for sandalwood to thrive under salinity stress conditions can offer an opportunity for timber growers to diversify their crops and contribute to sustainable land use practices.

K. Verma · R. Kumar · A. Sharma · S. Devi · A. K. Bhardwaj · A. Kumar (✉)
ICAR – Central Soil Salinity Research Institute, Karnal, Haryana, India

P. Sharma
YSP University of Horticulture and Forestry, Solan, India

A. Kumar et al. (eds.), *Salinity and Drought Tolerance in Plants*,
https://doi.org/10.1007/978-981-99-4669-3_29

Keywords

Sandalwood · Salinity · Host · Parasite · Haustorium

29.1 Introduction

Sandalwood belongs to family Santalaceae of genus Santalum, is second most expensive wood in the world right after African Blackwood, and is used for the industrial manufacture of various products (Viswanath 2014; Rashkow 2014). Across the globe, 18 sandalwood species belonging to genus *Santalum* have been reported, and among them, the Indian sandalwood (*Santalum album* L.) and Australian sandalwood (*S. spicatum*) have been observed to be economically important and most prominent species. In India, sandalwood (*S. album* L.) occupies around 9600 km^2 area (Gairola et al. 2007), 90% of which occurs in Karnataka and Tamil Nadu state (Jeeva et al. 1998), and remaining in the discrete pockets of Northern India (Jain et al. 1998; MoEFCC 2012). Indian Sandalwood is considered as one of the most valuable species, because it yields a high-quality aromatic oil (up to 6%) rich in santalols (up to 90%), obtained from its heartwood, and used in the medicinal, perfumery, and aromatic industry (Burdock and Carabin 2008; Baldovini et al. 2011). The annual global demand for sandalwood is estimated to be between 6000 and 7000 metric tonnes. India roughly produces 200 tonnes of sandalwood per year, while 400 tonnes is produced in other countries, resulting in a gap of nearly 5400 tonnes (Viswanath and Chakraborty 2022). At present in India, there is an immense potential for cultivating sandalwood at large scale to meet both domestic and global demand of its products.

Sandalwood tree is a root hemi-parasite that depends on host plant to supply nutrients and water through haustorial connections, especially during the early stages of development, as it absorbs less (<30%) material from the natural environment. The term "haustorium" refers to a collection of well-defined structural and physiological links that connect the conductive systems of the xylem, phloem, or both, and facilitate parasitic plants by feeding them with water and minerals (Mower et al. 2004; Aly et al. 2011). *S. album* is an aggressive hemi-parasite with 70% of seedlings able to generate haustoria within 30 days after germination (Nagaveni and Srimathi 1985). Species lacks root hairs, and in absence of the host, the deficiency of various essential elements, such as N, P, K, S, Ca, Mg, Fe, Zn, and Cu, may occur in plant tissues (Srikantaprasad et al. 2022). In nature, as many as 300 species can act as host of sandalwood, which plays a significant role in maintaining the growth of sandalwood (Radomiljac et al. 1998; Nagaveni and Vijayalakshmi 2003). Therefore, sandalwood can only be grown successfully with a suitable host, which makes the selection of host as the most important silvicultural factor affecting the establishment of sandalwood plantations (Radomiljac 1994). Moreover, sandalwood requires different host species under contracting edapho-climatic conditions, especially under abiotic stress conditions. Since, photosynthesis and water-use efficiency of sandalwood largely depend on host plant response to environmental conditions (Radomiljac et al. 1999a, b), hence, the cultivation of sandalwood becomes more complicated than the traditional monoculture plantations (Radomiljac 1998). Moreover, the selection and management of hosts remained one of the most important silvicultural research topics since the discovery of the parasitic nature of sandalwood (*S. album*) in 1871. Evidences have also indicated that different host species create contrasting differences in the various growth attributes of sandalwood (Radomiljac 1998; Sukarna 2002; Singh et al. 2018). Beside this, number of host plants, spacing between plants, and soil type also affect the growth of sandalwood (Gardner et al. 1991; Kasim 2002; Gomes and Adnyana 2017). Overall, the selection and planting of suitable host trees for sandalwood are immensely important to increase returns from the plantation investments (Rocha and Santhoshkumar 2022).

29.2 Growth Performance, Morphology, and Biomass Production of Sandalwood

In nature, despite the preponderance of parasitic plants having a wide range of hosts, there is significant level of host choice in some parasitic plants (Surendran et al. 1998). Similarly, the semi-root parasite sandalwood depends on host plant for nutrient and water supply, as it absorbs less material from the natural environment (Ouyang et al. 2016; Rocha et al. 2017). Sandalwood seedlings get their nutrition from the seed reserve at first, but in the later stages, they rely on their host species for water and nutrient requirement for performing the various growth processes. However, sandalwood photosynthesizes at its own, but the host is essentially required for its growth and development in the later stages (Ananthapadmanabha et al. 1984; Yadav et al. 2019). If sandalwood is not provided with any suitable host, the growth of the leaves is affected, and leaves turn yellowish or even chlorosis occurs which may lead to death of plant within 1 year (Surachman 1989). Simultaneously, inadequate understanding of host-parasite relationships makes regeneration and establishment to be problematic. Therefore, the selection of an appropriate host is essential for the successful establishment of sandalwood (Glatzel and Geils 2009; Srikantaprasad et al. 2021). Moreover, the growth pattern, morphological traits, and biomass production are considered to be best indicators of performance of sandalwood with a specific host (Surendran et al. 1998).

In nature, a large number of species can be utilized as the host; however, each host species has different affinity to assist in the growth and development of the sandalwood (Table 29.1). Good host have a better capability to provide nutrient to the sandalwood leading to better growth, morphology, physiological processes, and haustorial connection. For instance, in a 24-month-old sandalwood plantation Padmanabha et al. (1988) demonstrated that among 30 different host forest species only 10 species, namely *Terminalia alata, Casuarina equisetifolia, Terminalia arjuna, Melia dubia, Wrightia tinctoria, Acacia nilotica, Pongamia pinnata, Dalbergia sissoo, Cassia siamea,* and *Bauhinia biloba* are recognized as the best host. Taide et al. (1994) recommended *Casuarina equisetifolia* followed by

Table 29.1 Suitable host for sandalwood

Suitable host	Remarks	Sources
Terminalia alata, Casuarina equisetifolia, Terminalia arjuna, Melia dubia, Wrightia tinctoria, Acacia nilotica, Pongamia pinnata, Dalbergia sissoo, Cassia siamea and *Bauhinia biloba*	6- to 24-month-old plantation; better height and biomass	Padmanabha et al. (1988)
Desmanthus virgatus, *Crotalaria juncea* L. and *Alrernanthera* cv	26-week-old seedling to 4-year-old plantation; enhanced survival and growth	Fox et al. (1996)
Casuarina equisetifolia and *Terminalia catappa*	Better growth	Taide et al. (1994)
Sesbania formosa, and *Alternanthera nana*	287 days old seedling; better survival and growth	Radomiljac et al. (1998)
Sesbania formosa	38 week seedlings; higher root-shoot dry weight, stem diameter, height, and leaf area	Radomiljac et al. (1999a, b)
Acacia acuminata	*Santalum spicatum* seedlings; higher plant survival	Brand et al. (2000), Brand (2013)
Acacia saligna, A. acuminata	*Santalum spicatum* seedlings; enhanced survival and diameter increment	Brand et al. (2003)
Cajanus cajan	Higher survival per cent (80%) and height increment	Dutt and Verma (2004)
Mimosa pudica, Alternanthera sessilis	Better growth, haustorial formations, and nutrient status	Annapurna et al. (2006)
Dalbergia sissoo	Enhanced growth	Guleria (2013)
Casuarina equisetifolia	High number of haustoria (44 live, 6 dead)	Rocha et al. (2014)
Acacia confusa, Dalbergia odorifera	Higher number of shoots, number of roots and haustorial biomass production	Lu et al. (2014)
Mimosa pudica	Better growth and morphological parameters along with large-sized haustoria	Deepa and Yusuf (2015)
Cajanus cajan	Enhanced growth, chlorophyll content, and seedling quality index	Mohapatra et al. (2018), Mohapatra and Anil (2022)
Ocimum sanctum, Cajanus cajan	Better growth and development	Das and Tah (2016)
Alternanthera sp., *Casuarina junghuhniana*	Greater number of haustorium and better growth parameters	Gomes and Adnyana (2017)
Citrus aurantium, Punica granatum	For initial 6–7 years, the survival, collar diameter, crown size, height, and clear bole of *S. album* were observed maximum with the host *P. granatum* for the long-term	Singh et al. (2018)

(continued)

Table 29.1 (continued)

Suitable host	Remarks	Sources
Acacia saligna	*Santalum acuminatum* seedlings; increased plant biomass, haustorial size and frequency, root mass ratios, δ15N and δ13C isotope signatures	Nge et al. (2019)
Crotalaria retusa, Mimosa pudica	During nursery stage; higher number of leaves, plant height, and collar diameter	Nakandalage et al. (2021)
Sesbania grandiflora, Albizia lebbeck, Casuarina junghuhniana	Enhanced growth (height and basal diameter)	Balasubramanian et al. (2021)
Cymbopogon nardus, Melaleuca cajuputi	Increased height, diameter, number of leaves, and number of haustorium in the sandalwood	Tefnai et al. (2021)
Prosopis juliflora	Eleven-year-old plantations in the northern dry zone of Karnataka; enhanced growth and increment of tree with increased chlorophyll, soluble sugar, and free proline content in leaves	Srikantaprasad et al. (2022)
Melia dubia, Dalbergia sissoo, Azadirachta indica	Suitable host	Verma et al. (2023)

Terminalia catappa as a best sandalwood host in pot condition among the selected 15 host species. Similarly, Mohapatra et al. (2018) evaluated 14 different primary host plant species and showed that *Cajanus cajan* to be best primary host to produce highest seedling height (42.23 cm), collar diameter (3.83 mm), leaf number (50.14), branch number (4.43), shoot biomass (7.48 g), root biomass (3.25 g), total biomass (10.73 g), and seedling quality index (0.80) in the sandalwood, while the sandalwood plants without any host showed poor growth performance. The haustorial connection of sandalwood seedling with the bad host can eventually lead to the death of the plant. For example, within 2 years of establishment, every sandalwood seedling connected to the host *Eucalyptus loxophleba* died (Brand et al. 2000). Moreover, there is a general consensus that growth rate of naturally grown sandalwood (without any fertilisation), is significantly higher compared to planted sandalwood. This disparity can be primarily attributed to the existence of multiple host plants, which offer enhanced nutrients for the optimal growth of sandalwood (Sukarna 2002).

Moreover, among the nitrogen fixing and non-nitrogen fixing hosts, generally the nitrogen fixing hosts such as *Sesbania formosa* (Radomiljac et al. 1999a, b), *Acacia acuminata*, *Allocasuarina huegeliana* (Brand et al. 2000, 2003), *Dalbergia sissoo* (Guleria 2013), *Sesbania grandiflora* (Balasubramanian et al. 2021), *Cajanus cajan* (Mohapatra and Anil 2022) lead to significantly higher growth and development (height, diameter, root and shoot dry weight, stem diameter, height, and leaf area) as well as haustorial number and size. Radomiljac et al. (1999a) found that the after 38 weeks the higher growth and lower root: shoot ratio in sandalwood were recorded with N_2-fixing hosts, compared to non N_2-fixing host. Similarly, Annapurna et al. (2006) demonstrated that sandalwood seedlings in general exhibited better growth with N_2-fixing hosts species. Simultaneously, the large number of haustorial formations was observed with a N_2-fixing hosts species (*M. pudica*), thus significantly enhancing the nutrient status and growth of *S. album* seedlings. Moreover, Lu et al. (2014) indicated that the higher number of shoots, number of roots and haustorial biomass production in sandalwood with the N_2-fixing hosts compared to non-N_2-fixing hosts. However, the growth of sandalwood parasitized with two non-N_2-fixing hosts was observed to be significantly greater than the equivalent values of their parasitized treatments. Furthermore, Balasubramanian et al. (2021) also stressed that *S. album* grown with N_2-fixing host performed better than the *S. album* grown with non-leguminous host. Among the N_2-fixing host, the *Sesbania grandiflora* was the good host followed by *Albizia lebbeck* and *Casuarina junghuhniana*. Contrary, Gomes and Adnyana (2017) indicated that the nitrogen fixing host *Sesbania grandiflora* was found un-suitable as a nursery host whereas a greater number of haustorium and better growth parameters (number of leaves, diameter, plant height) of sandalwood with host *Alternanthera* sp. (28 haustorium per plant root) followed by host *Casuarina junghuhniana* (20 haustorium per plant roots).

Moreover, Nakandalage et al. (2021) also found that with the exception of hosts *Mimosa pudica* and *Crotalaria retusa*, non-leguminous hosts produced taller sandalwood seedlings than leguminous hosts. Simultaneously, the impact of hosts on parasite communities not only depends on what is parasitized but also when parasitism occurs (Surendran et al. 1998). As, the early haustorial connection to an efficient host should improve sandalwood's survival and growth under both nursery and field environment (Fox et al. 1996). Moreover, there are instances in which the short-term and long-term preferences for sandalwood host differ. For instance, Singh et al. (2018) revealed that for initial 6–7 years the survival, collar diameter, crown size, height, and clear bole of *S. album* were observed maximum with the host *C. aurantium,* compared to the other two hosts. However, for the long-term *P. Granatum* considered as a most suitable host plant.

29.3 Physiological and Biochemical Parameters of Sandalwood

Sandalwood performance is not only influenced by the host characteristics, such as slow-growing nature, lateral root system, and translocation of water and mineral nutrients, but also depends on the competition for above-ground resources such as light, which is crucial for its growth and various physiological processes (Doddabasawa and Chittapur 2021; Sahu et al. 2021). The physiological and biochemical activities of the plants and haustorium act as a unidirectional physiological bridge between sandalwood and host vascular systems allowing for resource transfer and releasing parasitic plants from many growth restrictions. Water, as an important regulator of various physiological processes, is measured using traits namely relative water content (RWC), water potential (p), and osmotic potential (s), which also explains how plants maintain or regulate the hydration of their cells to an optimal level (Pooja et al. 2019; Dhansu et al. 2021). However, lipid peroxidation has been employed as a marker in the selection of suitable plants because damage to the plasma membrane is frequently associated with electrolyte leakage and cell death under stress conditions (Lata et al. 2019). Moreover, altered physiological responses in terms of the imbalanced source-to-sink ratio, nutritional imbalances, and reduced enzymatic and metabolic activities also affected the rates of ion uptake, transport, and compartmentalization. Furthermore, gas exchange which is a crucial physiological phenomenon, are directly influenced by several factors, including diminished leaf expansion, reduced chlorophyll content, decreased nitrogen content, impaired photosynthetic machinery, altered RuBPase activity, and senescence. Thus, understanding the link between the physiological parameters, such as water potential, osmotic potential, relative water content, chlorophyll content, and gas exchange parameters along with antioxidative enzymatic activities could contribute significantly to deriving suitable plant–host relationships. Therefore, for the successful establishment of sandalwood plantations, it is important to understand the physiological and biochemical interactions between the semi-parasite sandalwood and its host species.

The sandalwood responds differently toward its hosts at the molecular level, as evidenced by variations in the activity physiological and biochemical parameters (Table 29.2). For instance, Mohapatra and Anil (2022) depicted a clear shift in SOD isozyme bands and higher peroxidase activity in sandalwood in association with host *Cajanus cajan* as compared to host *Alternanthera* sp. and without any host. Generally, the water potential of the *S. album* tree was always found lower than that of the hosts in all types of sandalwood–host associations (Varghese 1997). However, Brand et al. (2000) found that the stem and leaf water potentials of sandalwood (*S. spicatum*) trees were either same or significantly lower than their paired hosts, at pre-dawn and midday. Whereas, among the hosts, the stem water potentials of *Acacia acuminata* were significantly lower compared to *Allocasuarina huegeliana*. Particularly, a good host leads to better physiological and biochemical activity in the sandalwood. Moreover, the complex interaction between parasite (*S. album*), and the host during the early growth phases leads to promote the sandalwood growth during the initial stage of plantation (Balasubramanian et al. 2021). Deepa and Yusuf (2015) indicated that sandalwood plants attached roots of host *Mimosa pudica* showed better growth (higher chlorophyll content, protein content, and carbohydrate content) compared to sandal plants attached with other hosts. Similarly, Balasubramanian et al. (2021) found highest photosynthesis rate of 18.42 μmol m^{-2} s^{-1} and transpiration rate of 5.56 mmol m^{-2} s^{-1} in sandalwood with host *Sesbania grandiflora* followed by sandalwood + *Albizia lebbeck* and sandalwood + *Casuarina junghuhniana* with a photosynthetic rate of 15.55 and 15.49 μmol m^{-2} s^{-1}, respectively. Whereas Rocha et al. (2014) showed that the *S. album* trees growing with host *C. equisetifolia* showed 17.66 μmol cm^2 s^{-1} carbon

Table 29.2 Influence of host on the physiological and biochemical attributes of sandalwood

Attributes	Suitable host	Bad/ without host	Source
Water potential (MPa)	−0.9	−1.6	Brand et al. (2000)
Water potential (MPa)	−0.85	−1.27	Rocha et al. (2014)
Carbon assimilation rate (μmol cm^2 s^{-1})	17.66	15.13	
Photosynthesis rate (μmol CO_2 m^{-2} s^{-1})	52.59	20.80	Lu et al. (2014)
Transpiration rate (mmol H_2O m^{-2} s^{-1})	14.83	102.42	
Water-use efficiency (μmol/mmol)	44.34	91.35	
Net photosynthesis rate (μmol CO_2 m^{-2} s^{-1})	10.52	1.87	Ouyang et al. (2016)
Stomatal conductance (mol H_2O m^{-2} s^{-1})	0.21	0.04	
Intercellular CO_2 concentration (μmol CO_2 mol^{-1})	304.10	319.20	
Transpiration rate (mmol H_2O m^{-2} s^{-1})	2.79	0.80	
GS activity (μ mole γ-glutamyl hydroxamate min^{-1} mg^{-1} protein)	0.88	0.24	Deepa and Yusuf (2015)
Chlorophyll "a" (mg g^{-1})	0.79	0.50	Balasubramanian et al. (2021)
Chlorophyll "b" (mg g^{-1})	0.47	0.20	
Total chlorophyll (mg g^{-1})	1.26	0.70	
Photosynthesis rate (μmol m^{-2} s^{-1})	18.42	5.56	
Transpiration rate (mmol m^{-2} s^{-1})	7.56	3.73	
Chlorophyll content (mg g^{-1})	3.29	1.08	Mohapatra and Anil (2022)
Photoassimilation rate (IRGA)	18.35	10.31	
SOD activity (μg protein for 50% inhibition)	20.93	10.16	
POX activity (μg mg^{-1} protein)	11.30	3.09	
Water potential (MPa)	−2.92	−3.81	Verma et al. (2023)
Osmotic potential (MPa)	−3.54	−4.63	
Proline content (mg g^{-1})	40.21	15.92	

assimilation rate in comparison to 15.13 μmol cm^2 s^{-1} as observed in sandalwood grown without host plant. The pre-dawn plant water potential also showed less negative value (−0.85 MPa) in sandalwood planted with host plant compared to more negative WP (−1.27 MPa) without host plant, indicating that sandalwood tree depends on host plant for maintaining its water level and reducing the water stress. Furthermore, Ouyang et al. (2016) also identified that the highest net photosynthetic rate (10.52 μmol CO_2 m^{-2} s^{-1}), transpiration rate (2.79 mmol H_2O m^{-2} s^{-1}), and stomatal conductance (0.21 mol H_2O m^{-2} s^{-1}) were also recorded in leaves of sandalwood planted with host *Dalbergia sissoo* with which sandalwood grew best. Lu et al. (2014) demonstrated that *Santalum album* grown with host *Dalbergia odorifera* had greater photosynthetic rates and ABA (abscisic acid) concentrations. Moreover, irrespective of host species, parasite *S. album* optimizes xylem sap extraction from its hosts by higher transpiration and lower water-use efficiency compared to the host plant. Simultaneously, foliage ABA concentrations were also noted significantly higher in all hosts parasitized by *S. album* than in their unparasitized counterparts.

Moreover, apart from influence of the host plant on the physiological and biochemical parameters, the different abiotic stress such as cold stress (Zhang et al. 2017), and salinity stress (Verma et al. 2023) or application with different organic compounds (Liu et al. 2018; Zhang et al. 2019). Zhang et al. (2017) distinguished that the transcriptomic and physiological changes in sandalwood seedlings exposed to cold stress (4 °C for 0–48 h) and found no obvious morphological alterations. *S. album* seedlings displayed to prior 24 h of cold stress, but wilting in young leaves occurred as cold treatment prolonged from 24 to 48 h. Simultaneously, the activities of peroxidase (POD) and superoxide dismutase (SOD) showed continuously increasing trends, whereas the net photosynthetic rate, stomatal conductance, transpiration rate, and water-use efficiency of sandalwood decrease with cold stress. Moreover, from 0 to 24 h, intercellular CO_2 concentration (Ci) was maintained at a constant level, but at 48 h, it spiked sharply by 60–80%. However, Liu et al. (2018) stressed that the treatment of sandalwood plant with 1 mg L^{-1} BA treatment can enhance the net water-use efficiency and photosynthetic rate in sandalwood and also stimulate the accumulation of photosynthetic pigment (carotenoids, chlorophyll *a*, and chlorophyll *b*). Meanwhile, lipid peroxidation in sandalwood was inhibited greatly, whereas the activities of most antioxidant enzymes (peroxidase, catalase, and superoxide dismutase) were strengthened significantly. Similarly, Zhang et al. (2019) also noted the enhanced activities of antioxidant enzymes, viz. APX (ascorbate peroxidase), SOD (superoxide dismutase), POX (peroxidase), and CAT (catalase) in sandalwood plant tissues after treating with 1 mM salicylic acid (SA), and 1 mM methyl jasmonate (MeJA) or exposure to 4, 38 °C temperature, and high light intensity.

29.4 Nutrient Dynamics in Sandalwood

The root absorbs nutrients primarily as inorganic ions from the soil solution and rate of their absorption is mostly determined by the soil solution concentration immediately surrounding the root. Sandalwood being a semi-parasite is unable to absorb the water and most of nutrients (macro- and micro-nutrient) required for its growth directly from the soil due to lack of root hairs. However, the extent of transfer varied depending on the host plant. The contribution of different nutrients from the host plant to sandalwood tree would be better when planted with more than one type of host plant (Gomes and Adnyana 2017). Lu et al. (2014) found significantly higher nitrogen content and total amino acid in sandalwood grown with the good host compared to bad host. However, there may be a significant decrease in root N of parasitized host *D. odorifera* compared to its unparasitized plants. Similarly, Rocha et al. (2014) in a 6-year-old field grown sandalwood tree demonstrated higher leaf nutrient content was recorded in *Santalum album* with host *Casuarina equisetifolia* (N = 2.65%, P = 0.24%, K = 2.31%) compared to sandalwood tree planted without host plant (N = 2.48%, P = 0.16%, K = 1.68%) and suggested that host plant supplemented the N, P, and K requirement of sandalwood trees through haustorial connections. However, the major impact of host plant was observed in increasing potassium (K) content of sandalwood tree. As sandalwood depends mainly on host plant for maintaining the plant water level and reducing water stress, the high potassium content of sandalwood leaf played a significant role in regulating these processes.

Particularly, the association of *S. album* with N_2 fixing hosts improved the total nitrogen, carbon, phosphorus, potassium, fructose, and malic acid contents in xylem sap of sandalwood (Radomiljac et al. 1998). Specifically, water potential of sandalwood remains lower than its host, which allows it to obtain water and nutrients from the host through well-developed haustorial connections (Srikantaprasad et al. 2022; Ma et al. 2005). Simultaneously, still sandalwood has an ability to directly draw the nutrient from the soil. For instance, Rao (1933) demonstrated that the sandalwood tree depends on its hosts for N, P, and K, while Ca and Fe appear to be directly derived from the soil. Contrary, Iyengar (1965) revealed that sandalwood tree depends on the hosts for N and P, while Ca and K are directly absorbed by roots from the soil. Simultaneously, Ca:N ratio in the sandalwood tree may represent the balance of activity between root ends and haustoria. Moreover, the soil-plant analysis of spike diseased as well as healthy sandalwood plants revealed that Ca is absorbed directly from the soil by sandalwood trees in both scenarios. Similarly, Deval et al. (1975) found that Ca could be directly absorbed by the sandalwood roots, while phosphate, organic substances, amino acid, sugar, and mineral phosphates were drawn from the host plant. Varghese (1997) in a sandalwood-host study demonstrated that there was a translocation of carbon compound between sandalwood tree and hosts, depending on the host plant.

29.5 Salinity Tolerance of Sandalwood

Salt tolerance is a very complex, multi-response of molecular, biochemical, and physiological processes in plants. Although sandalwood leaves possess chlorophyll and is able to photosynthesize on their own, the species is not able to absorb the nutrient and water required for its growth and development, especially under abiotic stress condition (Zagorchev et al. 2021). Further, the salinity-induced ionic impact is continuous and long-term; therefore, the presence of excess sodium and chloride ions in the root zone alters the nutrients availability, absorption, and transport in the plant. Although a special structure, i.e., "Haustorium" in sandalwood establishes the structural and physiological linkage with host to support growth by providing water, minerals, proteins, and genetic material in the sandalwood (Mower et al. 2004; Aly et al. 2011). Mahesh et al. (2018) identified 53 new protein-coding genes which are well known to serve a protective function when subjected to various abiotic stresses such as drought and salinity. Previously, Fox and Millar (2001) suggested that *S. spicatum* can tolerate low to moderate salinity level. Similarly, *S. album* can tolerate slight (2–4 dS m^{-1}) to moderate (4–8 dS m^{-1}) salinity (WAC 2020). However, species are also known to show tolerance to drought, salt (Lake 2019), and cold stress (Zhang et al. 2017) conditions. Verma et al. (2023) suggested that the growth of sandalwood is govern mostly by host species under both control and salinity stress. Under saline stress, the growth and biomass of sandalwood on the hosts *M. dubia*, *D. sissoo*, and *A. indica* decreased minimally. Hence sandalwood has enormous potential for cultivation in degraded saline soils with suitable hosts. Therefore, *S. album* plantations can be a viable way for the farmers to diversify resources in areas subjected to ever-increasing problem of soil salinity, but host selection is the crucial step in achieving success in establishing sandalwood under the saline environment.

29.6 Sandalwood Haustorial Characteristics

Haustoria are bell-shaped structures that penetrate into host roots vasculature to absorb water, mineral, and nutrients. A typical haustorium has vascular tissue, collapsed layers, clasping folds, a central hyaline body, and an endophytic tissue through which transfer of essential mineral nutrient between host and parasite takes place (Zhang et al. 2012). Previous research work demonstrated that sandalwood is a xylem-feeding parasite relying on interfacial parenchyma to transport water and solutes (Tennakoon and Cameron 2006). However, sandalwood haustoria connected to different host's roots may possess different internal structures (Rümer et al. 2007). Therefore, it becomes crucial to understand the functional anatomy of the sandalwood haustoria and the mechanisms through which the parasite is able to develop such close connections with its hosts to fulfill its water and nutritional requirements.

The shape of the mature haustoria of *S. album* is bell-shaped (Tennakoon and Cameron 2006) or inverted conical flask-shaped (Zhang et al. 2012). The *S. album* is a xylem-feeding semi-parasite that depends on the interfacial parenchyma to transfer the water and solutes to *S. album* from the host plant (Tennakoon and Cameron 2006; Zhang et al. 2012). Although haustorial suckers lack phloem (Ouyang et al. 2016). Generally, mature haustoria are composed of the outer hyaline body (high metabolic activity) and a centrally located penetration peg (Tennakoon and Cameron 2006). Although Zhang et al. (2012) suggested that mature haustorium consists of collapsed layer, clasping folds, an inverted flask-shaped form of the vascular tissue, endophyte tissue, and a hyaline body of the central region. The penetration peg pierces to the xylem of the host plant by the cell degrading enzyme activity or direct pressure (mechanical force) (Tennakoon and Cameron 2006; Zhang et al. 2012), which subsequently differentiate into a vessel member (Dobbins and Kuijt 1973). Although Ouyang et al. (2016) argued that black-stained material aids in penetration into the host root by creating a firm connection by liquid (water) tension. Moreover, Rocha et al. (2015), Tennakoon and Cameron (2006) stated that the transfer of the xylem sap could occur via the pits (vessels or tracheids) of host xylem elements.

S. album haustoria went through enormous physiological and structural changes. Generally, the haustoria came within 30 days from the germination (Nagaveni and Srimathi 1985; Barrett and Fox 1997). As the proto-haustoria came into contact with the suitable host roots, it flattens against the surface (Tennakoon and Cameron 2006) or spread out horizontally (Zhang et al. 2012). Afterward, the transition to young haustorium started, and dark staining (purple color) mucilaginous produced (Tennakoon and Cameron 2006). Although researchers had a variable stance on the role of dark-stained material. Baird and Riopel (1983) did not identify the role, according to Tennakoon and Cameron (2006) it helps in adhesion of the parasite tissue to the host, while, according to Ouyang et al. (2016), Heide-Jørgensen (1989) it helps in the host root penetration. A pre-attachment haustorium of 10 days old (about 45 days after germination) sizes about 1 mm in diameter, developed from the cortex of the parent root (Zhang et al. 2012).

Post-attachment, haustoria's intrusive cells (finger-like projections) pierced the epidermis, cortex between the host cells, and extended up to the cambium tissue of the host root. Simultaneously, host root partly encircled by the cortical fold of the haustorium (Tennakoon and Cameron 2006) called the collapsed layer (Rao 1942) originated from the starch-containing cells (Yang et al. 2014). However, there are controversies about the function of these collapsed layers. As, it may be formed due to the internal pressure (Barber 1907) or disparity in the rate of growth of the central meristematic region and its peripheral parts (Rao 1942). Although according to Yang et al. (2014), the collapsed layer helps to efficiently assure that cell inclusion and energy concentrate at the inner meristematic region and are recycled to affect penetration. Moreover, it reinforces the physical connection between the haustorium and host root and for haustorial development provides supply space. Afterward, the penetration peg reached to the xylem of the host root and a thin ellipsoidal disc formed (Tennakoon and Cameron 2006). Generally, the vascular strand gradually appeared after 15 days of attachment (Zhang et al. 2012). Additionally, Zhang et al. (2012) stated that the haustorial gland in *S. album* haustoria was a regular feature when the haustorium came into contact with the host root surface contrary to other studies (Barber 1906; Rao 1942).

Haustoria of *S. album* and host plant root have a close vascular connection (Rocha et al. 2015) made up of almost entirely of parenchyma (Tennakoon and Cameron 2006; Ouyang et al. 2016; Zhang et al. 2012). Although the haustorium in the interface has three to four layers of cells, rich in the cytoplasm, and had a nucleus with evident nucleoli. Also, at the marginal zone of haustoria, many lysosomes of different sizes are present (Zhang et al. 2012). Moreover, the haustoria–host root connection is so intense that both became a single physiological unit (Rocha et al. 2015). Though, apart from the vascular continuity, there may be a finger parenchymal cell or a combination of both vascular continuity and finger parenchymal cells (Pate et al. 1990). Moreover, Rocha et al. (2015), Tennakoon and Cameron (2006) did not find any direct lumen-lumen xylem connection (luminal continuity). It suggests the absence of the mass flow of the salutes and presence of the selective across-membrane uptake from the xylem host plant to the of *S. album* (Tennakoon and Cameron 2006) (Fig. 29.1).

Moreover, Zhang et al. (2012) suggested that *S. album* haustoria have the ability to synthesis the phytohormones necessary for haustorial development. Also, various endogenous hormone levels, viz. GA-like substances, abscisic acid, cytokinins, zeatin, zeatin riboside, and indole-3-acetic acid increase as haustoria attached to

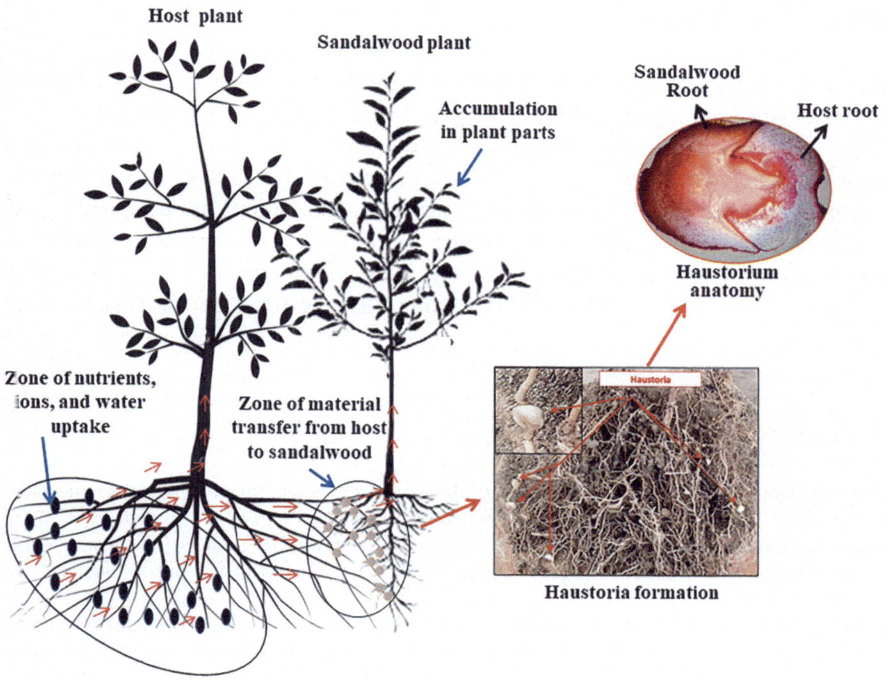

Fig. 29.1 Association between sandalwood and the host species. (Adapted from Verma et al. 2023)

host compared to unattached. Additionally, high auxin to cytokinin ratio leads to haustorial development in *S. album*. Also, auxin signal transmission may be crucial for the haustorial initiation, and GA-like substances play a pivotal role throughout the haustorial development (Zhang et al. 2017). Haustoria connected to the roots of different hosts may have different internal structures (Rümer et al. 2007; Ouyang et al. 2016), Ouyang et al. (2016) observed the differences in the interfacial region and internal structure between finger parenchymal cells and host root cells, vascular tissue, the reaction of host root cells and penetration modes. In *D. sissoo* and *L. japonica*, the finger parenchymal cells of haustoria penetrate the host roots through cell membrane fusion, pressure exertion, and the action of degrading enzymes. Conversely, in *A. sinensis*, the finger parenchymal cells of huastoria surrounded the host root cells, leading to self-digestion of the latter. Although the mature haustoria penetrate the root xylem of hosts *D. sissoo* and *A. sinensis*. Though, the haustoria that connected with *D. sissoo* had mature xylem while *A. sinensis* had not yet formed any vessel elements. Also, the finger parenchymal cells in the haustoria that connected with hosts *D. sissoo* and *L. japonica* roots were rich in organelles, including mitochondria, endoplasmic reticulum, and lysosomes (Ouyang et al. 2016).

29.7 Conclusion

Sandalwood presents a viable option for cultivation in saline soils and has the potential to become a high-value tree species which helps in sustaining the livelihood of farmers in saline environment. However, its success in saline environments depends majorly on host species and salinity tolerance of the host species. Thus, it becomes imperative to select suitable host species that not only aid in providing the essential nutrient to sandalwood but also helpful to lower subsequent stress at molecular level. Overall, the high economic value of sandalwood and its increasing demand in the global market make it a promising investment for farmers in saline condition.

References

Aly AH, Debbab A, Proksch P (2011) Fungal endophytes: unique plant inhabitants with great promises. Appl Microbiol Biotechnol 90(6):1829–1845

Ananthapadmanabha HS, Rangaswamy CR, Sarma CR, Nagaveni HC, Jain SH, Venkatesan KR, Krishnappa HP (1984) Host requirement of sandal (Santalum album L.). Indian Forester 110: 264–268

Annapurna D, Rathore TS, Joshi G (2006) Modern nursery practices in the production of quality seedlings of Indian Sandalwood (Santalum album L.) – stage of host requirement and screening of primary host species. J Sustain For 22(3/4):33–55. https://doi.org/10.1300/J091v22n03_03

Baird WV, Riopel JL (1983) Experimental studies of the attachment of the parasitic angiosperm Agalinis purpurea to a host. Protoplasma 118(3):206–218

Balasubramanian A, Prasath CH, Radhakrishnan S, Sivaprakash M (2021) Host-specific influence on early growth and physiological attributes of sandal (Santalum album) grown in farmlands. J Environ Biol 42:1162–1167

Baldovini N, Delasalle C, Joulain D (2011) Phytochemistry of the heartwood from fragrant Santalum species: a review. Flavour Fragr J 26:7–26

Barber CA (1906) Studies in root-parasitism: the Haustorium of Santalum Album I. Early stages up to penetration. Memoirs of Department of Agriculture, India, Botanical series, vol 1. Agricultural Research Institute, Pusa, pp 1–30

Barber CA (1907) Studies in root-parasitism: the Haustorium of Santalum Album II. The structure of the mature haustorium and inter-relations between host and parasite. Memoirs of Department of Agriculture, India, Botanical series, vol 2. Agricultural Research Institute, Pusa, pp 1–58

Barrett DR, Fox JE (1997) Santalum album: kernel composition, morphological and nutrient characteristics of pre-parasitic seedlings under various nutrient regimes. Ann Bot 79(1):59–66

Brand JE (2013) Effect of different Acacia acuminata variants as hosts on performance of sandalwood (Santalum spicatum) in the northern and eastern Wheatbelt, Western Australia. Aust For 72(4):149–156

Brand JE, Crombie DS, Mitchell MD (2000) Establishment and growth of sandalwood (*Santalum spicatum*) in south-western Australia: the influence of host species. Aust For 63(1):60–65. https://doi.org/10.1080/00049158.2000.10674814

Brand JE, Robinson N, Archibald RD (2003) Establishment and growth of sandalwood (*Santalum spicatum*) in south-Western Australia: acacia host trials. Aust For 66(4):294–299

Burdock GA, Carabin IG (2008) Safety assessment of sandalwood oil (*Santalum album* L.). Food Chem Toxicol 46:421–432

Das SC, Tah J (2016) Effect of host plants on growth and survival of sandalwood (Santalum album L.) in West Bengal. Indian Forester 142(2):193–195

Deepa P, Yusuf A (2015) Histological and biochemical evaluation of *Santalum album* L. seedlings co cultivated with different hosts. Ann Plant Sci 4(3):1016–1021

Deval KS, Bhat JV, Vasantharajan VN (1975) Uptake and translocation of 45 Ca and 32 P by angiospermic root parasite Santalum album and by one of its host plants Dolichos lablab. In: Use of radiations and radioisotopes in studies of plant productivity, proceedings of a symposium held at GB Pant University of Agriculture and Technology, Pantnagar, April 12-14, 1974

Dhansu P, Kulshreshtha N, Kumar R, Raja AK, Pandey SK, Goel V, Ram B (2021) Identification of drought-tolerant co-canes based on physiological traits, yield attributes and drought tolerance indices. Sugar Technol 23:747–761

Dobbins DR, Kuijt J (1973) Studies on the haustorium of Castilleja (Scrophulariaceae). II. The endophyte. Can J Bot 51(5):923–931

Doddabasawa, Chittapur BM (2021) Sandalwood plantations – points to ponder 7. Curr Sci 120(7): 1184–1193

Dutt S, Verma KS (2004) Effect of different host species on the survival per centage and growth increment of sandal (*Santalum album Linn.*) in the nursery condition. Adv For India 27

Fox JED, Millar KL (2001) Sandalwood emergence and growth in relation to salinity, waterlogging and depth of burial. Occasional report no 5. Mulga Research Centre, Curtin University of Technology, School of Environmental Biology, Perth

Fox JED, Doronila AI, Barrett DR, Surata IK (1996) *Desmanthus virgatus* (L.) Willd. An efficient intermediate host for the parasitic species *Santalum album L.* in Timor, Indonesia. J Sustain For 3(4):13–23

Gairola S, Ravi Kumar G, Aggarwal P (2007) Status of production and marketing of Sandalwood (*Santalum album* L.). In: Gairola S, Rathore TS, Joshi G, Arun Kumar AN, Aggarwal PK (eds) Conservation, improvement, cultivation and management of sandal (*Santalum album* L.). IWST, Bangalore, pp 1–8

Gardner FP, Pearce RB, Mitchell RL (1991) Physiology of plant cultivation. Lowa State University Press, Ames. 363 IA

Glatzel G, Geils BW (2009) Mistletoe ecophysiology: host–parasite interactions. Botany 87:10–15

Gomes D, Adnyana (2017) The effect of legume and non-legume to the sandalwood (*Santalum album Linn.*) growth in Timor Leste. Int J Basic Appl Sci 4531:207–237

Guleria V (2013) Analysis of plant, host and management relationships for sandalwood (*Santalum album*) cultivation in new subtropical locality of hill region of Indian Himalayas. Indian Forester 139(1):53–57

Heide-Jørgensen HS (1989) Development and ultrastructure of the haustorium of Viscum minimum. I. The adhesive disk. Can J Bot 67(4):1161–1173

Iyengar AVV (1965) The physiology of root parasitism in sandalwood Santalum album. Indian Forester 91:245–262

Jain SH, Angadi VG, Rajeevalochan AN, Shankaranarayana KH, Theagarajan KS, Rangaswamy CR (1998) Identification of provenances of sandal in India for genetic conservation. In: Aciar proceedings. Australian Centre for International Agricultural Research, Canberra, pp 117–120

Jeeva V, Saravanan S, Devaraj P, Lakshmidevi R (1998) Malady and remedy of sandal cultivation in farmlands and private lands—an overview. Sandal and its products, pp 16–21

Kasim M (2002) Responses to the sandalwood plant hosts Dinokulasi without Versikular mycorrhizae fungi and Asobaktor in dry climate condition Timor Island. Disentrasi Padjadjaran University, Bandung

Lake M (2019) Australian forest woods: characteristics, uses and identification. CSIRO Publishing, Clayton, VIC

Lata C, Soni S, Kumar N, Kumar A, Pooja, Mann A, Rani S (2019) Adaptive mechanism of stress tolerance in Urochondra (grass halophyte) using roots study. Indian J Agric Sci 89:1050–1053. https://epubs.icar.org.in/index.php/IJAgS/article/view/90834

Liu X, Xu D, Yang Z, Zhang N, Pan L (2018) Investigation of exogenous benzyladenine on growth, biochemical composition, photosynthesis and antioxidant activity of Indian sandalwood (Santalum album L.) seedlings. J Plant Growth Regul 37(4):1148–1158

Lu JK, Xu DP, Kang LH, He XH (2014) Host-species-dependent physiological characteristics of hemiparasite Santalum album in association with N2-fixing and non-N2-fixing hosts native to southern China. Tree Physiol 34(9):1006–1017

Ma GH, He YM, Zhang JF, Chen FL (2005) Studies on semi-parasitic sandalwood seedlings. J Trop Subtrop Bot 13:233–238

Mahesh HB, Subba P, Advani J, Shirke MD, Loganathan RM, Chandana SL et al (2018) Multi-omics driven assembly and annotation of the sandalwood (Santalum album) genome. Plant Physiol 176(4):2772–2788

MoEFCC (2012) Report on the Committee headed by Shri AK Bansal Addl. Director General (FC) on the regulatory regime regarding felling and transit regulation for tree species grown on non-forest/private lands

Mohapatra U, Anil VS (2022) Primary host interaction of root parasite sandalwood (Santalum album L): morphological and biochemical responses during interaction with legume host Cajanus and non-legume host Alternanthera. Mysore J Agric Sci 56(1):367–380

Mohapatra SR, Bhol N, Parida AK (2018) Influence of primary host plants on growth and quality of sandalwood (Santalum album L.) seedling. In: Souvenir of the international conference IIESD-2018, p 87

Mower JP, Stefanović S, Young GJ, Palmer JD (2004) Gene transfer from parasitic to host plants. Nature 432(7014):165–166

Nagaveni HC, Srimathi RA (1985) Note on haustoria-less sandal plants. Indian Forester 111:615–618

Nagaveni HC, Vijayalakshmi G (2003) Growth performance of sandal (Santalum album L.) with different host species. Sandalwood Res Newslett 18:1–4

Nakandalage N, Sampath AMU, Anuruddi HIGK, Subasinghe S (2021) Growth performance of sandalwood during nursery stage as affected by different host plants. Trop Agric Res Extens 24(4):360

Nge FJ, Ranathunge K, Kotula L, Cawthray GR, Lambers H (2019) Strong host specificity of a root hemi-parasite (*Santalum acuminatum*) limits its local distribution: beggars can be choosers. Plant Soil 437(1):159–177

Ouyang Y, Zhang X, Chen Y, Da Silva JAT, Ma G (2016) Growth, photosynthesis and haustorial development of semi parasitic Santalum album L. penetrating into roots of three hosts: a comparative study. Trees 30(1):317–328

Padmanabha HA, Nagaveni HC, Rai SN (1988) Influence of host plants on growth of sandal. My Forest 24(2):154–160

Pate JS, Kuo J, Davidson NJ (1990) Morphology and anatomy of the haustorium of the root hemiparasite Olax phyllanthi (Olacaceae), with special reference to the haustorial interface. Ann Bot 65(4):425–436

Pooja, Nandwal AS, Chand M, Singh K, Mishra AK, Kumar A, Kumari A, Rani B (2019) Varietal variation in physiological and biochemical attributes of sugarcane varieties under different soil moisture regimes. Indian J Exp Biol 57:721–732

Radomiljac AM (1994) The influence of pot host species, seedling age and nursery nutrition on Santalum album Linn. Plantation establishment. Ord River irrigation area, Western Australia. In: Sandalwood workshop held at Noumea, New Caledonia, August, vol 1, no 12, p 1994

Radomiljac AM (1998) The influence of pot host species, seedling age and supplementary nursery nutrition on Santalum album Linn. (Indian sandalwood) plantation establishment within the Ord River irrigation area, Western Australia. For Ecol Manag 102(2):193–201

Radomiljac AM, McComb JA, Shea SR (1998) Field establishment of Santalum album L. the effect of the time of introduction of a pot host (Alternanthera nana R. Br.). For Ecol Manag 111(2): 107–118

Radomiljac AM, McComb JA, McGrath JF (1999a) Intermediate host influences on the root hemi-parasite Santalum album L. biomass partitioning. For Ecol Manag 113(2):143–153

Radomiljac AM, McComb JA, Pate JS (1999b) Heterotrophic carbon gain and mineral nutrition of the root hemi-parasite Santalum album L. in pot culture with different hosts. Aust For 62(2): 128–138

Rao YS (1933) Contributions to the physiology of sandal (Santalum album, Linn.) part 1. Nature and extent of parasitism. J Indian Inst Sci 16:167

Rao LN (1942) Parasitism in the Santalaceae. Ann Bot 6(21):131–150

Rashkow ED (2014) Perfumed the axe that laid it low: the endangerment of sandalwood in southern India. Indian Econ Soc Hist Rev 51(1):41–70

Rocha D, Santhoshkumar AV (2022) Host plant influence on haustorial growth and development of Indian sandalwood (Santalum album). In: Indian sandalwood. Springer, Singapore, pp 229–244

Rocha D, Ashokan PK, Santhoshkumar AV, Anoop EV, Sureshkumar P (2014) Influence of host plant on the physiological attributes of field-grown sandal tree (Santalum album). J Trop For Sci 26(2):166–172

Rocha D, Ashokan PK, Santhoshkumar AV, Anoop EV, Sureshkumar P (2015) Anatomy and functional status of haustoria in field grown sandalwood tree (Santalum album L.). For Res 4: 148

Rocha D, Ashokan PK, Santhosh Kumar AV, Anoop EV, Suresh Kumar P (2017) Anatomy and functional status of haustoria in field grown sandalwood tree (Santalum album L.). Curr Sci 113(1):130–133

Rümer S, Cameron DD, Wacker R, Hartung W, Jiang F (2007) An anatomical study of the haustoria of Rhinanthus minor attached to roots of different hosts. Flora 202(3):194–200

Sahu SV, Maheswarappa RH, Kencharaddi RN, Sathish BN (2021) Effect of host plants and potting mixture on growth of sandalwood seedlings at nursery stage. Int J Curr Microbiol Appl Sci 10: 545–559

Singh B, Singh G, Rathore TS (2018) The effects of woody hosts on Santalum album L. tree growth under agroforestry in semi-arid north Gujarat, India. Indian Forester 144(5):424–430

Srikantaprasad D, Gowda AM, Umesha K, Thimmegowda MN, Pusha TN (2021) Influence of nursery hosts on physiology of sandalwood seedlings. Pharm Innov 10(4):1055–1057

Srikantaprasad D, Gowda AM, Pushpa TN, Thimmegowda MN, Umesha K, Ravikumar RL, Prasanna KT (2022) Identification of suitable host for sandalwood cultivation in northern dry zone of Karnataka. Ind Crop Prod 182:114874

Sukarna M (2002) Structure roots, nutrient content of Ca, Mn, N and chlorophyll semai sandalwood (*Santalum album L.*) with and without "host plant". National Research Seminar Papers Mathematics and Pend. Mathematics, Natural Sciences UNY

Surachman (1989) Growth response against fertilizers and hospes. Thesis S-2 FPS UGM

Surendran C, Parthiban KT, Bhuvaneswaran C, Murugesh M (1998) Silvicultural strategies for augmentation of sandal regeneration. In: Aciar proceedings. Australian Centre for International Agricultural Research, Canberra, pp 69–73

Taide YB, Babu LC, Abraham CC (1994) Influence of host species in the initial growth and development of sandal (*Santalum album Linn.*). Indian J For 17(4):288–292

Tefnai J, Mansur I, Wijayanto N (2021) The effects of cajuput and citronella grass as host plants on sandalwood growth. Media Konservasi 26(2):156–163

Tennakoon KU, Cameron DD (2006) The anatomy of Santalum album (sandalwood) haustoria. Botany 84(10):1608–1616

Varghese S (1997) Parasitic interference of sandal (*Santalum album Linn.*). on common agricultural crops from the homestead. M.Sc. thesis. Kerala Agriculture University, Thrissur, 66p

Verma K, Kumar R, Kumar A, Bhardwaj AK, Verma RC (2023) Host plant regulates growth processes, ion homeostasis, and salinity tolerance of sandalwood (Santalum album L.). J Plant Growth Regul 1–13. https://doi.org/10.1007/s00344-023-10906-3

Viswanath S (2014) Sandalwood, an unexplored treasure. Inside fact. Natl Monthly 1:10–13

Viswanath S, Chakraborty S (2022) Indian sandalwood cultivation prospects in India. In: Arunkumar AN, Joshi G, Warrier RR, Karaba NN (eds) Indian sandalwood. Materials horizons: from nature to nanomaterials. Springer, Singapore, pp 281–292

WAC (2020) World Agroforestry Centre. http://www.db.worldagroforestry.org//species/properties/Santalum_spicatum

Yadav AN, Kour D, Sharma S, Sachan SG, Singh B, Chauhan VS, Sayyed RZ, Kaushik R, Saxena AK (2019) Psychrotrophic microbes: biodiversity, mechanisms of adaptation, and biotechnological implications in alleviation of cold stress in plants. In: Sayyed R, Arora N, Reddy M (eds) Plant growth promoting rhizobacteria for sustainable stress management. Microorganisms for sustainability, vol 12. Springer, Singapore, pp 219–253

Yang X, Zhang X, Teixeira da Silva JA, Liang K, Deng R, Ma G (2014) Ontogenesis of the collapsed layer during haustorium development in the root hemi-parasite *Santalum album Linn.* Plant Biol 16(1):282–290

Zagorchev L, Atanasova A, Albanova I, Traianova A, Mladenov P, Kouzmanova M, Goltsev V, Kalaji HM, Teofanova D (2021) Functional characterization of the photosynthetic machinery in Smicronix galls on the parasitic plant Cuscuta campestris by JIP-test. Cell 10:1399

Zhang X, da Silva JAT, Duan J, Deng R, Xu X, Ma G (2012) Endogenous hormone levels and anatomical characters of haustoria in *Santalum album L.* seedlings before and after attachment to the host. J Plant Physiol 169(9):859–866

Zhang X, Da Silva JAT, Niu M, Li M, He C, Zhao J, Ma G (2017) Physiological and transcriptomic analyses reveal a response mechanism to cold stress in *Santalum album L.* leaves. Sci Rep 7: 42165

Zhang X, Niu M, Teixeira da Silva JA, Zhang Y, Yuan Y, Jia Y et al (2019) Identification and functional characterization of three new terpene synthase genes involved in chemical defense and abiotic stresses in Santalum album. BMC Plant Biol 19(1):1–18

The manufacturer's authorised representative in the EU is Springer Nature Customer Service Centre GmbH, Europaplatz 3, 69115 Heidelberg, Germany. If you have any concerns regarding our products, please contact ProductSafety@springernature.com

Printed and bound by CPI Group (UK) Ltd, Croydon, CR0 4YY

01/10/2026

02225892-0019